VERTICAL CITY: A SOLUTION FOR SUSTAINABLE LIVING

垂直城市：可持续生活之道

VERTICAL CITY : A SOLUTION FOR SUSTAINABLE LIVING

垂直城市：可持续生活之道

Kenneth S. H. King, M. Arch., NCARB
and Kellogg H. Wong, FAIA

金世海，建筑学硕士，美国国家注册建筑师委员会员
黄慧生，美国建筑师协会荣誉会员

图书在版编目(CIP)数据

垂直城市：可持续生活之道 /（美）金世海，（美）黄慧生著. -- 北京：中国社会科学
出版社, 2015.3

ISBN 978-7-5161-5107-5

Ⅰ.①垂… Ⅱ.①金… ②黄… Ⅲ.①城市规划－研究 Ⅳ.①TU984

中国版本图书馆 CIP 数据核字(2014)第 269812 号

出 版 人	赵剑英	
特约编辑	董　晖　陈爱儿	
责任编辑	冯春凤	
责任校对	徐彦彬	
责任印制	张雪娇	
顾　　问	张以国	
翻　　译	林建武　陈珊珊　徐彦彬	

出　　版	中国社会科学出版社
社　　址	北京鼓楼西大街甲 158 号 （邮编 100720）
网　　址	http：//www.csspw.cn
	中文域名：中国社科网　010－64070619
发 行 部	010－84083685
门 市 部	010－84029450
经　　销	新华书店及其他书店

印　　装	北京雅昌艺术印刷有限公司
版　　次	2015 年 3 月第 1 版
印　　次	2015 年 3 月第 1 次印刷

开　　本	635×965　1/16
印　　张	38.5
字　　数	1252 千字
定　　价	860.00 元

For Professor Edward Curtis, who inspired me beyond design and planning
to become a socially and ecologically conscious architect.

Kenneth King

致爱德华·柯蒂斯教授，是他启示我超越设计与规划，
成为一名有社会责任和生态意识的建筑师。

金世海

For Donna, our extended families, and in loving memory of Nancy and Jack.

Kellogg Wong

献给多娜，我们的大家庭，以及同南茜、杰克
在一起的美好时光。

黄慧生

Table of Contents

目录

INTRODUCTION

简介

A Public Housing New Town for 160,000 People
新加坡，容纳16万人的公屋新镇

序言

《垂直城市》的出版恰逢其时，对于亚洲尤其如此。在这里，许多国家正经历着快速的城乡流动迁移。据我估计，如果中国、印度和印尼要在接下来的20到40年间变得和美国一样城市化，那么，中国必须建造相当于美国三倍的建筑，印度是五倍，印尼则是两倍多。总计下来，他们必须建造的东西超过美国十倍之多。这意味着，今天在美国有多少栋建筑，这三个国家总共要建造的就是这个数字的十倍。人们不禁要问，我们去哪里找如此多的规划师、建筑师、承包商和建筑材料来设计和建造这些？由此还会得出一个明显的结论，即：这些城市一定会变得紧凑压缩，其人口密度很高，人口数量庞大。尽管国土面积不小，但实际上就城市化和农业需求而言，中国、印度和印尼的土地都是不够的。

然而，土地短缺的问题可能也会带来好处。由于城市被迫变得垂直，拥挤的城市让更多的设施与服务被安置在家的附近，并使公共运输系统变得更便捷。如果与土地使用模式密切结合起来，公共运输系统能够快速运送很多人，节省时间和能源，缓解道路拥堵，提高许多人的生活质量。然后，人们就会发现，垂直城市是一个很好的解决之道。不过，在设计一个美丽且适宜居住的垂直城市环境时，政府、专业人士和普通市民必须努力让城市有效运作。在这里我想补充一点，一个垂直城市还要有公园、康乐设施、物业这些东西，以创造出一幅多种城市图景拼接而成的镶嵌画，让人们有更多的生活选择。

和大国相比，这一挑战对面积很小的新加坡而言更为严峻。我们只有约700平方公里（270平方英里）的土地，上面已经居住了500多万人，而在接下来的数十年间还会有更多的人来到这里。土地总量是有限的，但人口的增长可能是无限的。

我们的政府必须找到办法，在抑制人口增长的同时不妨碍经济发展，总结我们过去50年得出的可靠经验，规划者除了设计出更多的高层建筑，别无其他选择。我们的目标是成为一个有吸引力的贸易中心，同时保证优质的环境和优质的生活；而要达到这些目标，我们必须步步为营。

Preface

The publication of *Vertical City* is very timely, especially in Asia, where many countries are experiencing rapid rural-urban migration. By my estimate, if China, India, and Indonesia were to become as urbanized as the USA in the next 20 to 40 years, China has to build the equivalent of three USAs, India five USAs, and Indonesia more than two USAs. In total, they have to build over 10 USAs. That means that for every building existing in the USA today, these three countries have to collectively build 10 of them.

One may rightly ask, where do we get planners, architects, contractors, and materials to design and construct them? It also leads to the obvious conclusion that these cities must be compact, with high densities and often large populations. Despite their size, China, India, and Indonesia are acutely short of land for both urbanization and agricultural needs.

However, the problem of land shortage may be used to great advantage. As cities are forced to go vertical, compact cities enable more facilities and amenities to be located near homes and help mass transit systems become more viable. If well related to land use pattern, mass transit can move a lot of people quickly, save travel time and energy, alleviate road congestion, and improve the quality of life of many individuals. It would then appear that vertical cities are a good way to go.

But governments, professionals, and citizens have to try hard to make the city work efficiently while designing a beautiful and livable vertical urban environment. I hasten to add that a vertical city must still have parks, amenities, landed properties, etc., to create a rich mosaic of urban milieus and to give people a good spread of environmental choices.

Tiny Singapore faces this challenge even more acutely than large countries. We have just over 700 square kilometers [270 square miles] of land already housing more than 5 million people, with many more to come in the decades ahead. The land mass is limited, but the population growth may be unlimited.

While our government will have to find ways to curb the population increase rate without frustrating economic growth, physical planners have no choice but to plan for high-rises, drawing upon our rather creditable experience accumulated over the last 50 years. We have to walk a tightrope to achieve our multiple goals of being an attractive business center with quality environment and good life.

iv

Let me share with you a few thoughts about how Singapore has turned what was a squatter-ridden city into a modern vertical city over several decades.

In preparing the Singapore Master Plan of 1991, the concept of urban cells (or modules) came in very handy. The city is spatially subdivided into cells at different hierarchical levels. The city of 5.5 million people is first divided into four regions. Each region, with a population of just over 1 million, is about the size of a small city. Each region, in turn, contains about five new towns. Each new town, with a population of around 200,000, accommodates an average of 10 neighborhoods, and each neighborhood is made up of around half a dozen precincts.

With this approach, our planners can assign different densities to different precincts. Gardens, schools, sports fields, lower-density housing, and neighborhood centers are interspersed with high-rise residential precincts so as to moderate the oppressive sensation often associated with extensive areas of tall buildings. This approach seems to have worked. Many visitors to Singapore have commented that our city appears to be less dense than the data suggest.

This leads to a second interesting planning concept used in Singapore: that planners have been able consciously to create an illusion of density, of city size, and of greenness. The juxtaposition of high- and low-rise precincts not only reduces the oppressive sensation of density but gives people an impression that the city is larger than its real size. By locating parks and green buffers strategically, our city appears much greener than the actual amount of green areas used.

But we have to face the fact that high-rise buildings are buildings with handicaps—of heavy dependency on elevators, of being divorced from the ground, of losing one's own identity in a big building, and so on. As early as the 1970s and 1980s, we attempted to identify the handicaps, or shortcomings, of high-rise living, and we did our level best to overcome them through clever site planning and design solutions.

The Housing and Development Board has attained a fair measure of success. Singapore's residents are happy, crime rate is low, and community spirit is strong. An important benefit of our high-rise public housing is that since 1985, Singapore has become a city with no squatters, no homeless people, no poverty ghettos, and no ethnic enclaves. With 82 percent of the people living in public housing flats, Singapore is now a city where 93 percent of the total population owns their own homes, and where the satisfaction level among public housing residents has been 95 percent in the last two decades. Clearly, it is possible to create a vertical city with

我想和你们分享一些关于新加坡如何在数十年间从一个棚屋城市转变为现代垂直城市的感想。

1991年准备新加坡总体规划时，城市单元（或模块）的概念令我们受益匪浅。城市按照空间划分为不同层级的单元。550万人的城市首先分为4个区域，每个区域人口刚过100万，相当于一个小城市的规模。然后，每个区域又包含5个新市镇，每个新市镇的人口约为20万，平均安置在10个社区中，每个社区又由6个小区构成。

通过这种方法，我们的规划师能够为不同的小区配置不同的密度。在高层住宅区之间点缀着花园、学校、运动场地、低密度住宅和邻里中心，用以缓解大片高层建筑带来的压迫感。这种方法看起来是卓有成效的。许多来新加坡的参观者评价，我们的城市看起来没有数据显示的那么拥挤。

这导致新加坡第二个有趣的规划概念的使用：规划者能够有意地创造一种关于密度、城市规模和绿意的图景。将高层与低层的小区混搭在一起，不仅缓解了压迫感，而且给人们营造了一种印象：城市比实际面积要大。通过策略性地布置公园和防护绿地，我们城市比实际拥有的绿化区显现出更多的绿意。

但我们必须面对的事实是，高层建筑也有缺陷——它严重依赖电梯，人们远离地面，在大建筑中迷失个性等等。早在20世纪七八十年代，我们就试图指出高层生活的这些缺陷或弊端，通过明智的场地规划和设计方法，我们也尽力克服这些缺陷。建屋发展局已经有了一个衡量成功与否的标准：幸福的居民区有着较低的犯罪率，强烈的社区精神。我们的高层公屋还有一个重要的收获，那就是，从1985年开始，新加坡已经成为一个没有非法占用公地者、没有无家可归者、没有贫民窟、没有少数民族聚居区的城市。当82%的人居住在公屋中后，现在的新加坡，使93%的人拥有自己的家，同时在过去二十年间，公屋居民的满意率达到95%。很显然，如果人们认同挑战，并以创造性的方法克服这些挑战，那么，创造一个拥有优质环境的垂直城市是完全可能的。

我最后想说的是，规划一座城市就像组装一件工业设计，比如汽车。不管是一座城市还是一辆汽车，其规划和设计者必须理解人的需要，以便让产品变得简单、人性化、舒适、便于使用。对汽车而言，引擎必须完美运转。与此相似，一座城市的规划者必须熟知组成城市机器的所有部件。只有那样，他/她才能将它们组装成一个完美的运行整体。另外，如果两辆汽车同样易于操控，同样有完美引擎，购买者会选择车身设计更漂亮的那辆。同样道理，一座更漂亮的城市对投资和人才更有吸引力。

　　这本书进一步提升了寻找新城市形式以适应垂直城市的需求。由于现代世界中城市面临着前所未有的挑战，这一点确实很有必要。

刘太格

a quality environment if one were to understand the challenges and were committed to overcome them with creative methodologies.

The last point I would like to make is that planning a city is like assembling a piece of industrial design, such as a car. Whether it is a city or a car, the planner-designer must understand the needs of the people in order to make it simple, user-friendly, comfortable, and convenient to use. In the case of a car, the engine must function like a dream.

Similarly, a city planner must know intimately all the various parts required to complete the urban machine. Only then can he or she assemble them into a perfectly functioning whole. Further, given two cars with the same ease of driving and the same perfect engine, a buyer will choose the car with the more beautiful body design. Likewise, a more beautiful city will attract investments and talent.

This book raises the need to seek new urban forms to tackle vertical cities. It is indeed necessary, as cities in the modern world are facing challenges not previously experienced.

Liu Thai Ker

For Liu Thai Ker's biography, please refer to the Appendix.

本文作者刘太格先生的简介，请参阅附录。

Nature in balance, Garibaldi Lake, British Columbia.
加拿大不列颠哥伦比亚省加里波第湖：自然的平衡。

前言

序言 "可持续性" 这个词汇指向人和自然的关系，它给人们的心灵带来共存、平衡、和谐——相比较而言，在东西方哲学之间，这些说法更加密切地与东方哲学联系在一起。一个农业社会，其存在是如此密切地与自然的兴衰起伏相关，人们总是要学习如何顺从自然那不可预测的力量，记住干旱与洪涝，富足与饥荒，阴与阳相互交替的知识，然后才能得到安心与宁静。

相似的是，在很早的西方文明中，人也将自己看作是受自然支配的。自然过于宽广无边，强健有力，神秘莫测，因而难以把握；人们优先考虑的乃是生存。人们耕种庄稼，放牧羊群；而随着季节更替，他们或者是在收获粮食、剪下羊毛，或者就是忍饥挨饿，濒临死亡。然而，随着科学的进步，人们开始相信，装备了科学技术的人类能够征服自然。这导致人们排干沼泽，砍伐森林，围堤筑坝，在开阔的平原上大量耕种作物，修建高速公路和城市。

在人类历史初期，摩西十诫是指导人与其伙伴之间相互关系的指南；戒令预告了文明的终结，乃是对人类轻率和不人道行为的惩罚。但还是难以想象，人有能力毁掉自己的宇宙——要么通过原子战争迅速地毁掉，要么通过数十亿无知个体的浪费习惯缓慢地毁掉，这些人消耗自然有限的资源，污染我们赖以获得食物的土地，以及我们饮用的水和呼吸的空气。他们没有预见到将会导致全球变暖和另一个冰河时代的可怕后果，也没有预见到对地球臭氧层的破坏将会让地球上的生物饱受癌症的困扰，而且，紫外线将会使生物发生基因突变。

如果古人预见到了人类破坏自然的潜力，那么就会有第十一条戒令："不可浪费！"正是这种对浪费的关注推动金世海先生去完成这部著作。

金世海先生是一位美籍华裔建筑师，出生于上海。三十多年来他频繁往返于中美之间，亲眼见证了东方惊人的城市化进程，以及新建筑、新城镇的盲目建设速度和无节制扩张，也看到了耕地无可挽回的损失。他听到很多热心的对可持续性观点的公开支持，也看到了之后通常不太明显的效果。和许多熟悉近代中国的人一样，他感到惊讶的是，曾经指导人类最古老文明之一的那些坚定的价值观正在发生变化。难道说那些价值观原先反映了一个慢节奏的农耕社会，而现在它们

Foreword

The word "sustainability," in reference to man's relationship to Nature, brings to mind powerful notions of coexistence, balance, and harmony—terms more closely associated with Eastern philosophies than those of the West. In an agrarian society whose existence is so closely tied to the ebb and flow of nature, there would be no peace of mind—or tranquility, if you will—until people learn to bend with the unpredictable forces of Nature, taking heart in the knowledge that rain will follow the drought, feast will follow famine, yang will follow yin.

Similarly, in the early days of Western civilization man saw himself at the mercy of Nature. Nature was too immense, too powerful, too mysterious to grasp; the primary focus was on survival. People planted crops and tended flocks and as the seasons passed, they either reaped and sheared or starved and froze to death. However, with the advancement of science came the conviction that human determination enabled by technology could conquer Nature. This led to the draining of wetlands, clearing of forests, damming of rivers, widespread plowing of open plains, and the building of highways and cities.

Early on in human history, the Ten Commandments served as a guide for man's behavior toward his fellow man; warnings foretold the eventual end of the Earth as punishment for people's indiscretion and inhumanity. But it could not possibly have been imagined that it was within man's capacity to destroy his own universe—either quickly by nuclear warfare or slowly by the wasteful habits of billions of mindless individuals consuming Nature's limited resources, polluting the soil that we depend on for food, the water we drink, the air we breathe. They did not envision the dire consequences that would lead to global warming and another ice age, nor anticipate the depletion of the earth's protective ozone layer thereby allowing the bombardment of life on earth by cancer and mutation-causing ultraviolet radiation.

Had the ancients foreseen the potentiality of man's destruction of Nature, there might well have been an Eleventh Commandment: "Thou shall not waste!" It is precisely this concern over waste that motivated Kenneth King to undertake this book.

Kenneth King is a New York–based Chinese-American architect who was born in Shanghai and has for the last thirty-plus years traveled frequently between the United States and China. He has been an eyewitness to the East's astonishing urbanization with its blinding pace of construction of new buildings, towns, and unstemmed sprawl; he has also watched the irreplaceable loss of farmland. He has heard the many ardent public endorsements of sustainability and has observed the often less evident results. Like many familiar with pre-modern China, he was left to wonder what had happened to those steadfast values that once guided one of humanity's oldest civilizations. Could it be that they were reflections of a slower agrarian society and that they must now make way for the demands of a faster, less patient world? What will be the paradigms that will guide modern China forward?

To better understand the complex forces at work, and to examine whether the course currently taken is inevitable or whether there might be more prudent alternatives, it was determined that the most constructive approach would be to write a book. It is hoped that the results of this exercise will generate discussion among interested readers, thereby sparking new ideas and possibilities for the future. It was in this context that I, also a Chinese-American architect with a half-century of experience working in the East, was called in to help. The hope was that our collective views and shared personal and professional insights would add resonance to the exploration.

It is important to understand that we coauthor this book as concerned observers with no pretext of being experts. Instead, for authoritative insights, we interviewed a wide range of academicians, practicing architects and urban planners, building technology specialists, and experts in such related areas as life safety, community development, land use, and, of course, sustainability.

The further we ventured, the more we learned about the enormity and complexity of the field, and the far-reaching significance of our investigation. Each conversation opened new possibilities, new experts to consult. Ultimately we had to draw our research to a close, even though there were many other experts we would like to have engaged. We hope that this book, however incomplete, may nonetheless make a contribution to the raising of awareness, and to stimulating the kind of discussion, argument, and debate that is essential to bring about change.

Kellogg H. Wong

必须为一个更快、更缺乏耐心的世界做出让步？引导现代中国前行的又将是哪些价值模式呢？

为了更好地理解正在起作用的各种复杂力量，为了审视当前的路线方案是否不可避免，或者说，是否有其它更明智的替代方式，可以确定，最富有建设性的方法是写一本书。希望这种行为会在感兴趣的读者之中引发讨论，从而在将来激发一些新的观点和可能性。在这种情形下，金先生请我一起帮忙——我同样也是一名美籍华裔建筑师，有着半个世纪在东方工作的经验。希望我们的一些共同看法和个人观点，能够为研究工作添砖加瓦。

作为这本书的合著者，我们只是充满关切的观察者而不是任何专家，理解这一点很重要。相反，为了获得权威的观点，我们访谈了许多不同的学者、从业建筑师、城市规划师、建筑技术专家，以及相关领域的一些专家——这些领域包括生命安全、社区发展、土地使用，当然还有可持续性问题。

我们探究得越深入，就越发了解这一领域的广阔与复杂，也更明白我们探究的深远意义。每个对话都打开了一些新的可能性，并使更多领域的专家参与进来。最后，我们不得不结束我们的研究，即使还有很多其他的专家是我们想要涉及的。尽管不完整，我们还是希望这本书能有助于增强人们的意识，激发那些对于带来改变而言至关重要的交流、争议和辩论。

黄慧生

For Kellogg Wong's biography, please refer to the Appendix

关于黄慧生的传记，请参阅附录。

x

New York is the largest, safest, and most racially and ethnically diverse city in the United States. It is also the greenest, with some 27,000 acres [10,925 hectares] devoted to public parks and a mobility rate of 82 percent on public transit, bicycles, and on foot. If New York were an independent state, it would rank lowest in per capita energy use.

纽约是美国最大、最安全，种族和族裔最多元化的城市。它也是环保的绿色城市，约有27,000英亩（10925公顷）的公园，82%的城市移动依赖公共交通、自行车和步行。如果纽约是一个独立的州，它的人均资源消耗将排在最低。

自序

A Personal Note

我是一个城市居民，在上海出生成长，之后在香港生活了三年，伦敦生活了五年，1960年定居纽约。上世纪70年代后期，中国向世界开放后，我很多时候就在上海和曼哈顿之间来回奔波。期间，我也在巴黎和苏黎世待了很长的时间；还有太平洋沿岸的城市古晋和马来西亚的打京那峇鲁、吉隆坡、韩国首尔，新加坡和日本东京，更不用说像北京、天津、台北以及许多规模庞大正在发展的中国城市。

我喜欢城市，因为它的效率，生活所需皆近在咫尺。你可以通过公交或步行十分便捷的出行，完全不需要拥有汽车。我喜欢城市生活，它丰富多样、有趣而充满精彩，人们在餐厅里、公园中、音乐会上或只是街头偶遇，就被赋予了发生社会交往的机会。而非每日里孤零零地驾车通勤、再孤零零地驾驶回家，将自己关进那绿草白栅封锁下的独户居所中，浪费着宝贵的时间，同时污染着环境。

我确信有人会不同意我的说法，尤其是在美国。从托马斯·杰斐逊开始，美国人就有将城市当作疾病和罪恶中心的广泛误解。许多环境主义者强化了这种反城市的偏见，他们认为，城市是最大的能源消费者，也是地球上最大的污染源。这话没错，城市现在大约只占整个星球3%的面积，但却消耗了大部分的自然资源，排放出全球80%的二氧化碳和其他数量巨大的温室气体。

但是，从个体的角度看，情景却非常不同。在纽约这样人口稠密的大城市中，人均能源消耗量大大低于美国的其他地区；人们可能会感到惊讶，美国人均能源消耗量最高的地区是怀俄明州的空旷平原。和居住在郊区的人们相比，普通纽约人消耗的电能和汽油，使用的水资源要少得多。他们占用着更少的土地，产生较少的垃圾和有害气体，总之，对环境的破坏程度要更小。很显然，就这一点来说，在濒临危险的生态系统中，我们需要学学这样的城市经验。

对中国和其他发展中国家正在发生的、规模空前的城市化进程而言，这些经验显得尤为迫切。例如，如果中国一直建设不断扩展的新城市以容纳大量外来人口——人们预料这会在接下来

I am an urban dweller. I was born and raised in Shanghai, after which I lived in Hong Kong for three years and then in London for five before settling in New York in 1960. Since the late 1970s when China opened to the world, I've pretty much split my days between Shanghai and Manhattan. Along the way I've spent a good deal of time in Paris, Zurich, and Pacific Rim cities like Kuching and Kota Kinabulu in Malaysia, Kuala Lumpur, Seoul, Singapore, and Tokyo, not to mention Beijing, Tianjin, Taipei, and a host of other large—and growing—Chinese cities.

What I like about cities is their efficiency and adjacencies and the sheer convenience of being able to move around on foot or public transit, quickly and inexpensively, without ever having to own a car. I like the rich variety, interest, and excitement—the fun—of city life, and the opportunities for social interaction in restaurants, concerts, parks, or in spontaneous exchanges on the street rather than being isolated in a car, wasting and polluting precious hours while commuting back and forth to work, pulling into a driveway, and disappearing into a single-family house sealed off behind a tidy green lawn and white picket fence. From an environmental perspective, the American dream is a petrol-dependent nightmare.

I'm sure there are people who would disagree with me, particularly in the United States, where, dating back to Thomas Jefferson, there's been widespread misapprehension about cities as centers of unhealthiness and vice. Many environmentalists reinforce this anti-urban bias. Cities, they say, are the greatest energy consumers and the greatest polluters on Earth. It's true. Cities currently cover only about three percent of the planet yet they devour the lion's share of its natural resources and emit about eighty percent of global carbon dioxide and significant amounts of other greenhouse gases.

But by the yardstick of the individual, the picture is very different. The per capita energy use of dense cities like New York is significantly lower than elsewhere in the country; it might come as a surprise that the highest per capita consumption is in the wide open plains of Wyoming. Typical New Yorkers consume far less electricity and gasoline and use far less water than their suburban counterparts. They occupy less land, produce less waste, produce lower noxious emissions, and in general inflict less damage on the environment. Clearly at this point in our imperiled ecology, we need to learn the lessons that such cities can teach.

Given the unprecedented urbanization taking place in China and other developing countries, the need is urgent. If China, for example, keeps on building sprawling new cities to accommodate the massive migration that is expected to continue for the next thirty years, there won't be enough land to grow food. There won't be enough water or other resources. And if, with their new found wealth, people continue to buy cars in record numbers, the ecoconsequences will be calamitous, not just for China but for the world.

We no longer have the luxury of thinking about such problems, least of all about their solutions, in isolated terms. This is a war—a battle for survival—that will not be won in small uncoordinated skirmishes. It is not just an issue of per capita usages but of global stewardship, reduction, conservation, and commitment.

It is neither reasonable, nor possible, to try to stop human energy consumption. But we can use our resources more intelligently, more sustainably. We have the knowledge and the wherewithal. And, with the right leadership, we have the ability to succeed. We need to revisit familiar patterns and determine whether they are valid or are actually destructive and merely sanctioned by habit. We must understand and embrace the need for change and give full rein to our imagination so that we may benefit from new and emerging technologies and creative ideas.

For years I toyed with the notion of setting forth my ideas in a book. Ironically, my most altruistic and lofty aspirations were unleashed by the grim urban reality of bumper-to-bumper full-stop traffic. What could possibly be the problem, I wondered in the back seat of a taxi. It turned out that service crews were making an underground repair. The roadbed had to be opened and excavated, the utility repaired, and then the street resurfaced with heavy equipment. The entire operation took days, while the repair itself required just over an hour. The inconvenience and cost, the exasperation, the lost time, and the idling pollution of so many cars for such a long period all struck me as so destructive and unnecessary. There just has to be a better way! This incident was the seed that launched my quest to eliminate as much as possible the urban problems that we've inherited, to find a better way to accommodate human needs, and to design a modern sustainable city.

We can no longer live our lives without regard to the impact we have on the environment. Nor can we feel self-satisfied in our new Green awareness and determination to curtail or halt environmentally damaging actions. For the

的30年中持续下去——那将不会有充足的土地用于种植作物，也将没有充足的水源和其他资源。而如果人们凭借新积累的财富，不断无节制地购买汽车，那不仅对中国，而且对世界而言，造成的生态后果都将是灾难性的。

我们不再享有孤立地一个个思考这些问题的奢侈了——更别说去思考解决方案。这是一场战争，一场生死存亡的战争，不可能通过没有谋划协调的小战斗来赢取。这不是一个人均使用量的问题，而是在全球范围内进行管理、减排、保护和承诺的问题。

想制止人类消耗能源，这既不合理也不可能。但我们可以更加明智、更加可持续地使用能源。我们拥有知识和资金，在正确的领导下，有能力获得成功。我们需要重新检视业已熟悉的行为模式，确定它们是否是正当的或者实际上是有害的（仅仅因为习惯使然）。我们必须理解并坚信改变的需要，充分发挥想象力，这样我们就可以从新近的技术和创造性观念中获得帮助。

多年来，我一直想在一本书中阐述我的看法。讽刺的是，我最无私和高尚的志向是在一次严重的交通瘫痪这一城市恶疾中诞生的。我坐在出租车后座思考，问题可能出在哪里？原来维修人员在进行地下维修：道路必须清空挖掘，然后修复管道，再之后街道还要用重型机械重新铺设整个工程要费些时日，而修复工作自身则只需要一个多小时。造成的不便和多出的成本，浪费的时间，以及产生的令人恼火的情绪，如此多汽车长时间空转产生的污染，都让我觉得这些是具有破坏性的、不必要的。一定有更好的办法！这件事是个起源，使我开始努力探索，来尽可能解决我们继承下来的城市问题，同时，找到更好的方法满足人类的需求，并设计一个现代、可持续发展的城市。

我们再也不能不顾对环境造成的影响而生活了，也不能仅仅因为力图减少或停止环境破坏行为的"绿色环保"意识和决心，而感到洋洋自得。为了我们的孩子，仅仅满足于"不那么糟糕"还远远不够；我们必须调整方程式，做出积极的贡献。除了完全扭转形势，我们别无选择。我们为什么不采纳在这世界上任何角落正在发生的优秀经验，并注入想象力地去创造新的可能，

Utility repairs disrupt traffic in midtown Manhattan, creating delays, pollution, and safety risks for pedestrians and motorists.
管道维修扰乱曼哈顿中城的交通，给行人和驾驶者造成时间延误，污染和安全风险。

发展出现代生活的新模式。这就是这本书想要说的全部内容。

金世海

sake of our children, it is not enough to settle for doing less that is bad; we must reformulate the equation to make a positive contribution. We have no choice but to turn things around. Why not take the best aspects of human habitation from wherever in the world they might occur, fuel them with imagination to create new possibilities, and develop new paradigms for modern life. That's what this book is all about.

Kenneth S.H. King

For Kenneth King's biography, please refer to the Appendix.
关于金世海的传记，请参阅附录。

鸣谢

本书是我许多杰出的同事、聪慧的朋友以及亲爱的家人们共同努力的结晶。就这本有关我们星球未来的书籍的付梓，我无法用语言来充分概述这个全球大家庭共同合作所做出的贡献。对那些富有责任感的居民和专家所取得的成就，我感到惊奇和骄傲，同时也让我对人类作为一个整体所能收获的东西充满希望。

　　首先，让我感谢我的团队，他们不懈的协调、组织和编辑是这项工作的基础。首先要提及的非珍妮特·亚当斯·斯特朗博士莫属，从一开始她就参与到本书工作的各个方面——从收集研究材料、采访、写作、编辑到协调本书其它的贡献者——这是一项需要高度耐心与自律的任务，没有她孜孜的奉献，本书是无法面世的。大洋彼岸中国的协调工作由尊敬的张以国博士主持，中文翻译工作由陈珊珊博士领导的优秀团队完成。

　　特别感谢莱因哈德·罗伊在本书写作过程中提出的出色建议，以及他为解释本书中部分概念而绘制的画作。对那些同我们分享知识的建筑师、规划师、学者和其它专业人士，我们也想对他们给出的专业意见表达尊重和谢意。我们的访谈进程先从纽约开始，在那里我们与一些好友进行了讨论，这些人包括杰出的结构工程师，莱斯利罗伯特森建筑事务所的莱斯利·E·罗伯特森；著名的系统工程师马文·马斯以及他博学的同事——克森蒂尼事务所的斯科特·西塞。之后，我们继续与我们在贝聿铭建筑事务所的前同事进行谈论，这些人包括：来自科恩佩特森福克斯建筑事务所的威廉·佩特森、海因吉斯建筑事务所的幕墙专家罗伯特·塞弗里、曹·麦克恩建筑事务所的曹慰祖。

　　研究工作使我们接触并与很多人建立了深厚友谊，这些专家有巴克莫汉德斯公司的垂直运输专家里克·巴克；宋腾添玛沙帝工程事务所的结构工程师潘子强；以及垂直农场的先驱者迪克森·德波米耶，正是他将我们引荐给奇斯+凯斯卡特建筑事务所的格里高利·奇斯。形—式建筑事务所的阿杰马勒·阿克塔什分享了他关于最新建筑技术的渊博知识，而且还让我们会见了他的老师，霍克国际有限公司的卡尔·卡略托。在与斯基德莫尔、奥因斯与梅里尔建筑事务所的T.J.哥特斯蒂埃纳、奥雅纳的阿肖克·瑞吉、以及山姆·施瓦茨工程公司的爱里希·阿西门的会谈中，我们收获颇丰。在成书的最

Acknowledgments

XV

This book is the result of a rich harvest from many brilliant colleagues, as well as wise friends and dear family members. There is no way to adequately express my appreciation to this global community for the myriad contributions that helped bring forth this vision for our planetary future. I am in awe of what we have achieved as a group of committed citizens and professionals, and that gives me hope for what humanity can achieve as a whole.

Let me begin by thanking the team whose untiring efforts to coordinate, organize and edit have been the backbone of this effort. At the top of the list is Janet Adams Strong, Ph.D. who went beyond the call of duty to integrate the many parts of this book`s vision from research, interviewing, writing, editing, to managing all the contributors. A challenging task that required great patience and discipline this book could not have happened without her dedicated effort. Across the world our efforts in China were smoothly coordinated by our esteemed friend Dr. Yiguo Zhang along with the professional translation team spearheaded by Dr. Shanshan Chen.

A special acknowledgment to Reinhard Roy for his excellent advise throughout the writing and his artistic drawings help to illustrate some of the concepts in this book.

A heartfelt thanks goes to the accomplished architects, planners, academics, and professional friends who generously shared their knowledge, expertise and time. We began the interviewing process in discussions with good friends in New York, including the distinguished structural engineer. Leslie E. Robertson, and Marvin Mass, the renowned systems engineer, and his learned associate Scott Ceasar, at Cosentini Associates. We continued with our former colleagues from I. M. Pei & Partners:William Pedersen of Kohn Pedersen Fox Associates, curtain wall specialist Robert Heintges of Heintges & Associates, Maria Sevely of FORM | Proforma, and Calvin Tsao & Mckown Architects.

Our research led us to contact, and forge good friendships with, vertical transportation specialist Rick Barker of Barker Mohandas, structural engineer Dennis Poon of Thornton- Tomasetti, and vertical farming pioneer Dickson Despommier, who favored us with an introduction to Gregory Kiss of Kiss + Cathcart. Ajmal Aqtash of form-ula and core. form-ula shared his extensive knowledge of emerging building technologies and similarly put us in touch with his guru, Carl Galioto at HOK. We were enriched by additional meetings with T.J. Gottesdiener of SOM, Ashok Raiji of Arup, and Erich Arcement of Sam Schwartz Engineering, traffic consultants. We followed up, rather late in the book`s development, with Valentine Lehr of Lehr Consultants International and our good friend William Faschan of LERA, both of whom generously contributed to our understanding of the special engineering challenges posed by supertall buildings.

Tall-building research led us inevitably to Chicago, where we had the good fortune to speak with Antony Wood, executive director of the Council on Tall Buildings and Urban Habitat, as well as to Adrian Smith and Gordon Gill, and also to Philip Enquist of SOM. Continuing west, we benefited from conversations with Professor Rhonda Phillips at the sustainability-focused Arizona State University, San Francisco architects Mark and Peter Anderson, and ecocity trailblazer Richard Register. We made a special trip to the University of Maryland at College park to learn about land management in China Professor Chengri Ding, and to Chapel Hill to meet with urbanist and China specialist Thomas J. Campanella at the University of North Carolina. We are happy to acknowledge Professor Campanella`s mentor, and our old friend colleague Tunny Lee, former head of the Department of Urban Studies and Planning at MIT, who we visited in Cambridge, Massachusetts.

Special interest in China brought us to Beijing, Shanghai, and Tianjin to confirm firsthand our growing knowledge bank and to ensure that our long personal experience in China would analytically encompass current conditions. We are especially grateful to the many people who offered assistance, among them Kenneth P. Wong, real estate COO and Director of International Development of The Related Companies in Shanghai, and the prominent architect and urban planner James Jao of J.A.O. Design international in Beijing.

Viewing urbanization in its global contexts prompted us to investigate other BRIC countries (Brazil, Russia, India, and China) until the constraints of time and budget limited our preliminary research to Brazil. We gratefully acknowledge the generosity of the distinguished urban innovator Jaime Lerner, former mayor of Curitiba, and also Alfredo Trinidade from the Secretariat of the Environment, who could not have been more helpful. We appreciated the opportunity to discuss urban planning with architect-professor Bruno Padovano of Faculdade de Arquitetura e Urbanismo da Universidade de Sao Paulo. For a portion of our trip we were fortunate to be guided by the very knowledgeable Flavia Liz Di Paolo. Through her good efforts we were introduced to Jorge Rubies, the leader of Sao Paulo`s grassroots preservation movement, who struggles valiantly every day against the growing casualties of unbridled urbanization.

The spectacular color visualizations of the Vertical City project were executed by FORM | Proforma under the direction of Maria Sevely, with team members Karen Zhang, Carlos Osorio, and Erwin Sukamto, who worked closely with the authors in developing the tower concepts, scale plans, and other details in realization of the final images. We are also pleased to acknowledge Wonwoo Park and Blake Kurasek, both of whom kindly Permitted us to include the sky farm proposals they envisioned for Chicago.

后过程中，我们又接触了莱尔国际的瓦伦丁·莱尔以及我们的好朋友莱斯利罗伯特森建筑事务所的威廉·法杉，他们慷慨大方的分享，使我们对高超层建筑带来的特殊工程挑战有了更深入的理解。

高层建筑研究最终将我们带到了芝加哥，在那里，我们很幸运可以和安东尼·伍德（他是高层建筑与城市环境委员会的执行主任），以及艾德里安·史密斯和戈登·吉尔，还有斯基德莫尔、奥因斯与梅里尔建筑事务所的菲利普·恩奎斯特进行交流。继续向西，我们受益于亚利桑那州立大学的郎达·菲利普斯教授对可持续性的强调；受益于旧金山的建筑师马克·安德森和皮特·安德森兄弟，以及生态城市的先行者理查德·瑞杰斯特。我们专门去了马里兰大学帕克分校，向丁成日教授讨教了中国的土地管理问题，并到北卡罗莱纳州大学教堂山分校拜见了城市规划及中国问题专家托马斯·J·康帕内拉（吴凯堂）。我们非常想要感谢康帕内拉教授的导师，我们的老朋友和同事巴尼·李，他是麻省理工学院城市研究与规划系前主任，我们在麻省剑桥曾拜访过他。

对中国的特殊兴趣将我们带到了北京、上海和天津，这将使我们能够现场验证自己不断增长的知识储备，而且确保我们自己在中国的长期经验会分析性地囊括当前的情形。在许多提供帮助的人士中，我们特别感谢上海关联公司国际开发部的主管和房地产COO肯尼斯·P.王，以及优秀的建筑师和城市规划师，居住在北京的龙安建筑规划设计顾问有限公司的饶及人先生。

在全球化的背景下观察城市化，这个目标促使我们研究其它金砖国家，不过时间和经费上的限制使我们的初步研究停在了巴西。我们非常感谢杰出的城市改革家，巴西库里提巴市前市长豪梅·勒纳的慷慨大方，以及来自环境秘书处的阿尔弗雷德·特林达德，他为我们提供了巨大的帮助。对于能有机会与圣保罗大学建筑与城市规划学院的建筑学教授布鲁诺·帕多瓦诺就城市规划问题展开讨论，我们也心存感激。在旅途中，我们有幸得到知识渊博的弗拉维亚·利兹·迪·保罗的引导。在她的努力下，我们被引荐给圣保罗乡村保护协会的领导人豪尔赫·爱德华·卢比斯，她每天都在与无度的城市化带来的灾难进行英勇斗争。

垂直城市项目华美的色彩效果，是在玛丽亚·塞弗莉领导下由FORM | Proforma设计公司的团队制作完成的，这个团队成员还包括张莹，卡洛斯·奥索里奥，艾文·苏康托，他们与作者协同合作，

一起完成了最终图像效果中的楼层概念，规模规划和其它一些细节。我们也要感谢朴勇友和布莱克·库拉塞克，他们都很友善地允许我们在书中引述了他们设想的，位于芝加哥的空中农场计划。

像这样一个概要性的成果依赖于许多人的通力合作。第一个要提到的人是苏·费拉罗，我们可爱的执行助理，她在联系协调以及全球范围内资料收集上的不懈努力对于这本书来说弥足珍贵。在许多伸出援手的管理人员中，我们要特别感谢安德森建筑事务所的利克斯·坎贝尔和唐川身；柯森蒂尼咨询公司的格雷琴·本柯、科里·卡尔、凯·劳伦斯；高层建筑与城市环境委员会的纳萨尼尔·霍利斯特；金斯勒建筑事务所的莫妮卡谢弗；海因吉斯建筑事务所的贝琪伊利；霍克国际有限公司的香农·考斯特、克里斯丁·亨利·蒂松、亚历山大·罗伯；科恩佩特森福克斯建筑事务所的大卫·尼尔斯和克里斯·纳克斯；莱斯利罗伯特建筑事务所的卡尔·罗宾逊；杰米·勒纳建筑事务所的安娜·克劳迪娅·弗朗哥·德·卡斯特罗；贝聿铭建筑事务所的纳迪亚·伦纳德；波特曼建筑设计事务所的安迪·华莱士；雅思柏设计事务所的艾伦·周·伯克·安；之前在艾德里安·史密斯—戈登·吉尔建筑事务所的凯文·南斯；芝加哥斯基德莫尔、奥因斯与梅里尔建筑事务所的贝丝·热尔文·穆林；纽约斯基德莫尔、奥因斯与梅里尔建筑事务所的伊利沙白·库巴尼和杰西卡·普莱森特；宋腾添玛沙帝工程事务所的托尼亚·戈奇斯和徐莹莹；我们也很感谢出版助手博格丹利·李斯科夫。

我们也非常感谢极具天赋的图形设计师琳达·津格，她使平淡的文字和图片有机组合起来，化腐朽为神奇。还感谢编辑安·柯斯齐娜，校稿大卫·霍金斯，杰瑞·卡马拉塔，赵海荣和克里斯蒂纳·斯托对最终校正文本也卓有贡献。

最后，在参与此书的专家中，没有谁比我们之前的同事与多年的好友刘太格先生更权威的了。刘先生在新加坡的现代转型过程中扮演着重要角色。我们非常感谢他抽出时间与我们分享他的经验，并为本书作序。

作者特别感谢家人一贯的鼓励、无私的爱和支持，特别是要感谢儿子金瑞琪在张莹和方舟的协助下帮助设计并维护垂直城市网站(www.verticalcity.org)；感谢女儿金瑞玲激发我去探索我的愿景；最重要的是感谢我最亲爱的妻子金王娴歌女士的坚定支持，风雨同舟和相濡以沫。她批判的眼光和无微不至的照料使我能够想得更多，做得更好。

The production of a compendium like this requires a great deal of cooperation and administrative support. At the head of the list is Sue Ferraro, our wonderful executive assistant, whose unflagging services were invaluable in coordinating contacts and gathering materials worldwide. Among the many staff members who helped, we would like to thank Chris Campbell and Chung Shen Tang of Anderson Anderson Architecture; Gretchen Bank, Cori Carl, and Kay Lawrence of Cosentini Associates; Nathaniel Hollister of the Council on Buildings and Urban Habitat; Monica Schaffer of Gensler; Betsy Ely of Heintges & Associates; Shannon Quast, Kristin Henry Tison, and Alexander Robb of HOK; David Niles and Krissie Nuckols of Kohn Pedersen Fox Associates; Carol Robinson of LERA; Ana Claudia Franco de Castro of Jaime Lerner Arquitetos Associados; Nadja Leonard of Pei partnership Architects; Andy Wallace of the Portman Archives; Ellen Choo Bock Eng of RSP Architects Planners & Engineers (Pte) Ltd.; Kevin Nance, formerly of Adrian Smith + Gordon Gill; Beth Gervain Murin of SOM/Chicago; Elizabeth Kubany and Jessica Pleasants of SOM/New York; and Tonia Gotsis and Xu Yingying of Thornton-Tomasetti. We also acknowledge the Production assistance of Bogdan Lysikov.

We are also very grateful to Linda Zingg, the talented graphic designer who tamed a huge amount of imagery and bilingual data into a flowing and elegant design and the dedicated efforts of our copy editor, Ann Kirschner, and proof-reader, David Hawkins. Although Dr. Jerry Cammarata, Stephen Zhao and Cristina Stoll came in late, their skill and insight in helping to complete the final steps of this book is invaluable.

Finally, among our expert participants there is none more expert than our former colleague and friend of many years Liu Thai Ker, who has played such a momentous role in the transformation of modern Singapore. We are very grateful to Mr. Liu for the time and expertise he has shared with us, and for the sage words he generously contributed in the Preface to this book.

Last but not least, the author is especially grateful to family members for their lifelong encouragement, unconditional love, and devoted support. Special thanks to my son Raymond King assisted by Karen Zhang and Summer Fang for the masterful development and maintenance of the Vertical City website (www.verticalcity.org), to my daughter Lynn King for inspiring me to explore my vision, and most of all to my dearest wife, YienKoo, a pillar of strength throughout all the ups and downs, whose critical eye, and steadfast care enabled me to dream big, and do my best.

SECTION ONE / A CALL FOR A VERTICAL CITY

第一篇 / 倡导垂直城市

The world's population, currently 7 billion people, is expected to reach 9 billion by 2050.
当前世界人口为70亿，到2050年预计将达到90亿。

8-8-8宣言

新城市模式的8-8-8宣言

我们主张如下的政策和战略目标，它们都是相互联系的，但为了方便，我们将它们分成三组适合操作的议程，每个议程包括八项内容（8-8-8）。

环境

1. 采用各种可能的手段**抑制全球变暖**。这是最为重要的一项要求。

2. **使用清洁能源**。包括太阳能、风能、地热和潮汐能源，以及植物产生的各种生物能源，所有这些对于缓和气候和能源危机都意义重大。

3. **能源供给本地化**，使之靠近终端消费者，这可以防止长距离传输电能造成的巨大损耗。

4. **保护耕地**。反扩张计划是我们最为关切的内容之一，因为在发展中消失的土地是永远不会再生的。

5. **保护清洁水资源**，包括雨水收集，蒸发技术，以及在非必须使用饮用水的地方重复使用中水。

6. 通过室内和室外植物**净化空气**，这些植物可以将呼出的二氧化碳转化为生命需要的氧气，并净化受污染的空气。

7. 通过再生和再用**减少垃圾数量**，分拣可降解物。把自然当作我们的模型，自然中并不存在垃圾，我们的目标是实现零废弃。通过协调管理，应用新技术，全面的教育项目（旨在说明这项工作的积极效益，以及不这么做导致的消极后果），这是可以实现的。

8. **食物生产本地化**，鱼菜共生和人工鱼塘可以提供更新鲜、口感更佳的食物，不再使用化学防腐剂和制冷剂，同时减少长距离运输产生的污染和腐坏。

8-8-8 Manifesto

3

8-8-8 MANIFESTO OF NEW URBAN FORM

We advocate the following policies and strategic goals, all interrelated but conveniently grouped in three manageable agendas of eight (8-8-8).

ENVIRONMENTAL

1. **Curb global warming** by every means available. This is an imperative of the first order.

2. **Use clean energy** such as solar, wind, geothermal, and tidal sources, or biomass energy from plant-based materials, all of which can help significantly to alleviate both climate and energy crises.

3. **Localize power sources** in close proximity to end-user consumption in order to prevent significant energy loss through transmission over long distances.

4. **Preserve arable lands.** This antisprawl initiative is one of our greatest concerns, as land lost to development will never be returned.

5. **Conserve clean water**, by means of rainwater capture, evaporative technologies, and the reuse of gray water for nonpotable purposes.

6. **Detoxify air** through indoor and outdoor plantings that recycle exhaled carbon dioxide into life-sustaining oxygen, and purify the atmosphere of airborne pollutants.

7. **Reduce waste through recycling and reuse**, separating compostable matter. Adopting Nature as our model, wherein waste does not exist, our goal is to achieve zero waste. Such an undertaking is possible through coordinated management, applied technologies, and full-spectrum education programs that make clear the positive benefits (and also the negative consequences) of noncompliance.

8. **Localize food production**, utilizing aquaponics and fish ponds for fresher, better-tasting food, discontinuing the use of chemical preservatives and refrigerants, and reducing the pollution and spoilage involved in long-distance shipping.

FORMAL

1. **Maximize density and compactness** for maximum efficiency in clustered ultra-tall towers.

2. **Limit the project footprint** to 15-minute walkability from one end to the other.

3. **Dedicate the main surface level to car-free pedestrian-friendly use** with easy access to shops, services and outdoor resources for recreation and enjoyment.

4. **Confine passenger vehicles and deliveries** to lower levels of a raised podium for increased safety, reduced pollution and surface congestion, and improved quality of life.

5. **Locate utilities accessibly in order to facilitate secure access** and minimal disruption during maintenance and repair.

6. **Provide mass transit for clean, efficient access** around the Vertical City, with intermodal links to outlying areas.

7. **Link the towers with sky lobbies for structural stability**, convenience, increased efficiency, and security, and expand opportunities for social interaction in a series of linked town squares in the air.

8. **Use the latest high-performance technologies** to optimize efficiency and sustainability.

建筑形式

1. **密度和紧凑度最大化**，让成群聚集的超高层建筑实现最大功效。

2. **限制项目的占地面积**，从项目一端步行到另一端的步行距离限制在15分钟以内。

3. **项目地面为专属的行人友好区域**，这样人们可以很容易去往商店、服务区和其他户外设施进行锻炼和休闲。

4. **将客车和物流车辆限制**在一个架空裙房的下层区域，增加安全性，减少污染和地面拥堵，提高生活质量。

5. **把公共设施置于较易维护的位置**，使之在维护期间便于到达并最小化因维护造成的服务中断。

6. **在垂直城市周围提供高效清净的公共交通**，并设置到达周边地区的中转枢纽。

7. **利用空中天桥将高楼连起来以增加结构稳定性**，使之更方便、更有效、更安全，这些互通的空中城市广场也为社会交往开拓了空间和机会。

8. **利用最新的高性能技术**优化效率和可持续性。

社会经济的/政治的

1. **复合使用**为居住、就业、教育、娱乐、保健和其他功能提供必要条件，通过取消在家与工作场所之间通勤造成的浪费污染，集中劳动与消费市场，进而获得效率上的优化。我们赞成服务范围和便利设施最大化，以丰富个人的生活体验，最终形成一个投资型和创新型社会——它能完全地利用所提供的每一个机会。

2. **减少运营、管理和维护费用**，让垂直城市的更多资金能用于提升整体的生活水准。

3. **培育多样性和社会包容**，让不同经济水平的人在住房和娱乐上有广泛的选择自由。在垂直城市中没有贫民窟，所有人都有尊严地生活在体面的地方，提高社会福利和生产力，最终使垂直城市成为对所有人而言都是更舒适的地方。

4. **最大化就业机会，避免失业**。失业对于公共利益没有好处，通过整合经济和环境事务，提供绿色产业和岗位，将有益于垂直城市和大环境。

5. **培育民主、有责任心的政府**。在制定政策和立法过程中就整合生态上的需要，认识到人的价值、环境的福祉与良好的商业政策、经济增长并非毫不相关。

6. **推进健康的生活方式和个人发展**，通过积极参与、加强运动以防止肥胖、心脏疾病和其他一些由于久坐不动而产生的苦恼。

7. **承认不同国家和城市在社会、经济、生态、环境上的差异**。允许每一方将我们的基本概念应用到他们的实际需要中去，超越过去破坏性的经济发展和城市化进程。

8. **要求城市和国家政府部门**，提供奖励措施，鼓励新的思考方式，以及新的城市模式进行可持续开发。

SOCIOECONOMIC/POLITICAL

1. **Mix uses to meet essential needs** for employment, housing, education, recreation, health care, and other necessary requirements; optimize the efficiency gains of centralized labor and consumption markets by doing away with long wasteful and polluting commutes between home and work. We endorse the maximum range of services and amenities to enrich individual life experience, ultimately resulting in an invested and creative society that takes full advantage of the unique opportunities afforded.

2. **Reduce operating, management, and maintenance costs** with the Vertical City, allowing more funds to raise standards of living as a whole.

3. **Foster diversity and social inclusiveness** with a wide range of choices in housing and recreation at various economic levels. There are no slums in the Vertical City; every resident will have a decent place to live securely and with dignity, promoting well-being and productivity and ultimately making the Vertical City a better place for everyone.

4. **Maximize the job opportunities and avert misemployment**, which adds little to the common good, by merging economic and environmental concerns and providing green industries and jobs that will benefit the Vertical City and its greater environment.

5. **Nurture democratic and accountable governance** that integrates ecological imperatives in the decision-making and legislative processes, recognizing that human values and environmental well-being are not unrelated to good business policies and economic growth.

6. **Promote healthy lifestyles and individual human development** by active engagement and exercise against obesity, heart disease, and other afflictions related to an overly sedentary existence.

7. **Recognize differences in the social, economic, and ecological conditions of different cities and nations,** allowing each to adapt our basic concepts to their particular needs while moving beyond past—often destructive—processes of economic growth and urbanization.

8. **Challenge leadership in municipal and national governments** to offer incentives that would encourage new ways of thinking and enable the sustainable development of new urban forms.

在发达国家和越来越多的发展中国家（这些地区，越来越多地依赖于快速工业化，从而实现现代化），工业是空气、土壤和水资源污染的主要源头。

Industry is the major source of air, soil, and water pollution in the developed world and increasingly in developing countries where modernization depends on rapid industrialization.

在发达国家和越来越多的发展中国家（这些地区，越来越多地依赖于快速工业化，从而实现现代化），工业是空气、土壤和水资源污染的主要源头。

概述

Overview

据估计，在农业开始发展之前，世界人口从未超过1500万。但10000多年前，随着新石器时代从狩猎采集走向畜牧耕种，稳定生活和可靠食物供应带来了相对安全，这导致生产力的提高和增长率的提升。原先聚集的营地成了永久的住所，最后，出现了拥有成熟治理模式、商业、建筑和交通的复杂城市。不断精细化的技术和工具促进了耕种和灌溉的发展，这就产生了多余的食物，然后进一步推动了劳动分工和社会层级化。还有，城市化的不同机制也改进了环境卫生和公共医疗，结果地球上的人口数量从中世纪开始几乎一直在增长。

在18世纪早期，地球人口第一次达到了10亿之多。100多年之后的1927年，这一数字就变成20亿了。然而，很快地，仅仅用了33年，到1960年，地球人口就飙升到30亿，之后每隔10年人口几乎就要增加10亿：1974年是40亿，1987年是50亿，1999年是60亿，2012年达到了70亿。即使考虑到趋向正在变慢，现在预计，到2030年全球人口将达到80亿，而到2050年则将达到90亿。

这些统计数据一般来说就是一种警报，尽管它们往往被认为很遥远、很抽象。严峻的报告已经清晰地说明了事实：今天全球接近一半的人生活在贫困之中：有30亿人每天的生活费用不足2.5美元。同样令人不安的是，接近8.7亿人正深受慢性营养不良之苦，这意味着全球人口的八分之一并没有足够的食物以获得健康生活所需要的能量。

问题还不仅仅在于人口数量这一点。实际上，有些人会说，人口的增长并不是环境恶化的根源，只是人口太多这一点放大了其他一些关键因素中产生的消极影响，这些因素包括：世界财富、土地、水资源的不均衡分布，不健全或未落实的政府政策，工业化以及人们的生活方式等。

在像印度和中国这样的新兴经济体中，人们渴望那些由欧洲人和美国人确立起来的生活标准，这种生活消费水平可能很高但却是不可持续的。简单说，这归结于更多人想要（要求）更多的物质，使用更多资源，产出更多垃圾。当世界人口超过70亿，而且这些人都要食物、水、住房、能源、衣物、电能和其他消费品时，环境所面临的压力是巨大的。

It is estimated that before the development of agriculture, the world's human population did not exceed 15 million. But some 10,000 years ago, as Neolithic man moved away from hunting and gathering toward planting and husbandry, the relative security of settled life and reliable food supplies led to increased productivity and growth. Clustered encampments evolved into permanent settlements and, in time, complex cities with developed modes of governance, commerce, building, and transport. Farming and irrigation were advanced by increasingly sophisticated techniques and tools, which gave rise to food surplus and, in turn, fostered a division of labor and social stratification. The diverse machinery of civilization similarly promoted advances in sanitation and public health, with the result that the number of people on Earth has grown almost continuously since the Middle Ages.

The population reached one billion around the start of the nineteenth century. More than another century passed before it climbed to two billion, in 1927. But then, very quickly, the population jumped to three billion in just thirty-three years (1960), with another billion added virtually every decade thereafter: 4 billion in 1974, 5 billion in 1987, 6 billion in 1999, and 7 billion in 2012. Current projections, allowing for a slowing trend, envision 8 billion people by 2030, with the global population reaching 9 billion by 2050.

Statistics like these are typically met with alarm, although they can seem just distant and abstract. The reality is brought home by grim reports that nearly half of today's global population lives in poverty: more than 3 billion people survive on less than US$2.50 per day. Equally troubling is the fact that nearly 870 million people suffer chronic undernourishment, meaning that one out of every eight does not eat enough to get the energy needed for an active life.

The problem is not merely a question of numbers. In fact, some would argue that the growing population is not a root cause of environmental decline but that having so many people just magnifies the negative impact of other key factors, such as the uneven distribution of the world's wealth/land/water, unsound or unimplemented government policies, industrialization, and, hardly least, lifestyle.

In emerging economies, such as India and China, people aspire to living standards set by Europe and North America, where consumption patterns are high and unsustainable. In simple terms, it boils down to more people wanting—demanding—more goods, using more resources, and producing more waste. With a world population of more than seven billion and their need for food, water, housing, energy, clothing, electronics, and other consumer products, the strain on the environment is enormous.

Throughout history humans have shaped their physical environment for better or, frequently, worse, unaware of the long-term consequences of their actions. We have destroyed natural diversity by single-crop farming, brought about the extinction of entire species, scarred the land, polluted water supplies, and caused erosion and deforestation. Early on, people would simply move from exhausted sites to richer, more fertile areas. But the past couple of centuries have brought about a behavioral sea change as modern developments have greatly increased our control over the environment and our ability to exploit the planet's resources.

The First Industrial Revolution began in the mid-eighteenth century when hand power gave way to improved water- and steam-powered machines, and coal largely replaced wood for fuel. Tremendous advances took place in metallurgy, while new agricultural techniques and labor-saving tools yielded bigger crops with less manual labor. Rural workers began to migrate to cities and retrained for new jobs in the belching machine-driven factories that propelled the Industrial Revolution to its next, even more momentous phase.

The steady and widespread development of mechanization and the continuous and prolonged growth of the economy that resulted from the First Industrial Revolution were completely without precedent in human history. Almost no aspect of life was untouched.

The Second Industrial Revolution, which began in the 1830s and spread through Western Europe and the United States and then beyond, depended on chemical breakthroughs, the internal combustion engine, steel, railroads, and the single most important engineering achievement of the twentieth century: electrification. Electric lighting in factories greatly improved working conditions and productivity, and, at the same time, made possible electric transit, and such burgeoning communications technologies as the telegraphs, telephone, and radio.

Giant corporations, larger than any commercial entity previously known, transformed the marketplace as new organizational systems and efficiencies in automation and assembly lines drove quality up and prices down in a world of increasing commercialism. Mass production consumed huge amounts of natural resources and left behind an equally vast tonnage of toxic and nontoxic waste.

The next phase of the Industrial Revolution started after World War II. In mid-century America, returning soldiers took

望眼历史，人类会将自然环境变得更好，也常由于其无法意识到自身行为所造成的长期后果，产生糟糕的结果。由于单一农业耕种法，我们已经破坏了自然的多样性，让很多物种整个种群归于灭绝，损伤土地，污染水资源，土壤被侵蚀和乱砍滥伐。从前，人们会简单地从一处荒废之地迁往另一处更富饶肥沃的区域。但是，现代社会的发展已经大大增强了我们控制环境和开采地球资源的能力，因此，过去数个世纪人们的行为已经给自然环境造成了翻天覆地的改变。

第一次工业革命从18世纪中期开始，那时手工劳动让位给改进的水力或蒸汽驱动的机械，而煤在很大程度上替代木材成为燃料。冶金术取得了巨大的进步，新的农业技术和节省劳动力的工具使得较少的人工可以生产出更多的粮食。农民开始移居城市，在热气腾腾的机器工厂中被培训从事新工作，这将工业革命推向下一个更加重要的阶段。

第一次工业革命带来了机械化的稳定、广泛发展和经济的持续增长，这在人类历史上从未有过先例。生活中几乎没有一个领域不受影响。

第二次工业革命开始于19世纪30年代，先在西欧和美国蔓延，然后拓展到其他地方。它依赖于化学上的突破，内燃机、钢铁、铁路以及20世纪最重要的一项工程成就——电气化。工厂里的电气照明大大提升了工作环境和生产效率，与此同时也使得电力运输及一些新兴的通讯技术，比如电报、电话、收音机成为可能。

新型组织系统及自动化和流水线所带来的高效能不断提升产品质量降低产品价格，在一个日渐兴盛的重商主义世界中，比之前所知的任何商业机构都要庞大的大型企业转变了市场。大规模生产耗费了大量自然资源，留下了同样大量的数以吨计的有毒或无毒废弃物。

工业化的下一个阶段开始于二战之后。在20世纪中期的美国，退伍军人利用政府提供的低息贷款购买新房子，安置家庭，追求美国梦。汽车普及化，规模巨大的高速公路网络以及将偏远地区联系起来，在这些基础条件的支持下，战后繁荣引发的发展浪潮深刻地改变了这个国家的社会和经济形态，包括郊区化和无节制的扩张，在家中，电视的发明迅速将美式生活的所有方面都整合在一起。多年的萧条和战争之后，电视为所有有望带来更好生活的产品和业务找到了观众。

In cities globally, motor vehicles are the single greatest air and noise polluters, whether moving or idling in traffic. Cars contribute to 20–25 percent of world's greenhouse gas emissions. The National Urban Mobility Report indicates that Americans collectively spent 5.5 billion hours sitting in traffic in 2011, roughly 38 hours per person, each wasting 19 gallons of fuel in the process. Analyzing driving patterns in hundreds of urban areas across the country, researchers at Texas A&M Transportation Institute estimated the total cost of congestion in the United States was $121 billion.

在全世界的城市中，机动车是最大的空气、噪音污染源，不管它是行驶还是阻滞在车流中时都是如此。在全球温室气体的排放中，汽车的排放量占到了20—25%。《国家城市交通报告》指出，2011年美国人在交通堵塞中浪费的时间共计55亿小时，平均每人浪费38个小时，在这个过程中每人还要浪费19加仑汽油。分析全国数百个城市的驾驶状况，德州农工大学交通研究所的研究者估计，交通拥堵的总损失为1210亿美元。

　　工业革命的伟大发明创新是现代生活不可分割的一部分。随着世界日益走向繁荣，人们，尤其是发展中经济体的人们，纷纷购买冰箱、洗衣机和其他之前难以企及的产品，最为明显的是汽车，那是趾高气昂的中产阶级的骄傲徽章。今天，整个世界每年制造出6000万辆汽车，而且这个数字还在上升。

　　汽车工业的爆炸式发展主要发生在中国和印度。（尽管美国人总被认为热爱汽车，但实际上，

advantage of government-issued low-cost mortgages to buy new homes in which to raise their families and pursue the American Dream. Post-war prosperity fueled interrelated developments that profoundly changed the country's social and economic patterns, including suburbanization and unchecked sprawl, all made possible by widespread car ownership and vast new highway systems that knitted formerly remote areas together. Inside the home, the invention of television rapidly integrated all facets of American life. After years of depression and war, TV found a receptive audience for all the products and practices that promised a better life.

The great inventions and innovations of the Industrial Revolution are an inseparable part of modern living. As world prosperity has grown, people, especially in developing economies, are purchasing refrigerators, washing machines, and other products previously out of reach, most conspicuously cars, the proud badge of the swelling middle class. Today, the world manufactures more than 60 million cars per year, and the number is rising.

Much of the auto industry's explosive growth is taking place in China and India. (Despite the car-loving American stereotype, ownership in the United States actually declined in 2010 for the first time ever. Indeed, a recent study by the Carnegie Endowment for International Peace revealed that American car ownership is actually among the lowest in the developed world.)

Since the 1980s, or perhaps a little earlier, we have been in the midst of the Information Age. Like the Industrial Revolution that preceded it, this new technological/digital revolution has affected every aspect of life: where and how we live, the way we work, who and what we know, and how we create, conduct business, learn, and communicate.

Remote areas, separate regions, and entire countries are now joined in a global community. Computers, semiconductors, fiber optics, the Internet, and cell phones have all presented unimaginable opportunities for invention, no longer requiring physical presence in a central location, corporate management chains, or even substantial financial investment for a start-up business—just a computer and an idea. The new world order permits collaboration among people thousands of miles apart, instantaneously, fluidly, and efficiently, on equal terms without age, gender, or other bias. Unfortunately, it has also wrought a new form of toxic waste, exacerbated by rapid changes in technology, planned obsolescence, and the eagerness of consumers to possess the latest, most advanced gadgets.

Just as the Industrial Revolution brought about the creation and growth of major cities, and in many cases also their decline, the new information superhighway holds the key to new demographics, new industries, and new urban constructs. Readily at hand are vast stores of formerly inaccessible, or even suppressed, knowledge. People around the globe now have a binding interconnectedness that makes new initiatives for a sustainable future possible—and necessary.

美国汽车保有量在2010年第一次出现下降趋势。世界和平组织卡内基基金会的最近一项研究表明，美国的汽车保有量在发达世界中属于最低的行列。）

从20世纪80年代开始，或者可能还要更早一些，我们就已经处于信息时代了。像之前的工业革命一样，这次新的技术/数字革命影响了人们生活的方方面面：从我们生活的地点、方式，到我们工作的方式，我们认知的人和物，我们如何创造、经营生意，如何学习和交流等等。

无论是偏远地区、独立区域还是整个国家，现在都加入到了地球村中。电脑、半导体、光纤、因特网、手机，这些都为发明创造提供了不可思议的机遇，你不必在中心区域，不必需要企业的管理链，甚至为一个新兴行业投入巨资——你只需要一部电脑或一个想法就可以了。新的世界秩序可以让距离数千英里之外的人们展开及时、流畅、高效的合作，不再有因年龄、性别和其他偏见导致的不平等，一切都处在同等条件下。不幸的是，它同样也产生了一种新形式的有毒废弃物，并因为技术上的迅速进步而加重：计划报废，以及消费者希望拥有最新最高级的电器工具。

就像工业革命给大城市带来了兴与衰一样。新的信息高速公路是新的人口特征、新产业和新城市建设的关键所在。你的手边随时都是大量之前无法获得的东西，甚至一些曾被禁止发表的知识。现在，全世界的人都存在互联性，这就使得采取一些新措施实现一个可持续的将来成为可能和必要。

A cell-phone tower disguised as a lonely pine tree in the remote hills of Inner Mongolia. Mobile phones constitute the world's fastest-growing industry, allowing developing countries to leapfrog over former hardware requirements directly to more advanced modern technologies.

在中国内蒙古地区，一座手机信号发射塔被掩饰成偏僻山上唯一的一棵松树。手机已经成为世界发展速度最快的产业，它可以让发展中国家跳过之前的硬件限制，直接掌握更高级、更现代的技术。

Electronic waste constitutes roughly 5 percent of municipal solid waste, or 20–50 million metric tons globally each year.
Only about 13 percent of e-waste is recycled. The rest is incinerated or dumped in landfills, releasing deadly toxins into
the air, soil and water.

电子垃圾约占到城市固体垃圾的5%，总量大约为全球每年2000—5000万公吨。只有大约13%的电子垃圾被
回收利用，剩余的则被焚烧或倾倒在垃圾填埋区中，向空气、土壤和水中释放致命的有毒物质。

Aerial view of suburban homes near Markham, Ontario, in Canada. The widespread development of single-family homes on individual lots depended on automobilization. The pattern, which defined mid- and late-twentieth century subdivisions in the United States, has spread around the world.

从空中鸟瞰加拿大安大略省万锦市附近的郊区住宅。散布在各自用地上的私家住宅都依赖汽车。20世纪中期和末期在美国形成的这种住宅小区模式，如今已经扩展到了全世界。

发展模式

Development Patterns

过去两百年间，在食物生产和产品制造方面，人类很大程度上已经走上了一条康庄大道。但这背后的基本假定（即地球是我们的开采对象）导致了严重的环境、经济和社会政治问题。作为建筑师，本书的作者认为，鉴别并处理可持续发展中的环境问题是一个极具挑战又刻不容缓的任务。

冒险去探究不太熟悉又非常复杂的主题，且包括经济、社会、政治，实际上还有文化各种因素相互作用的各个方面，绝非易事。更困难的是，如何将各个学科领域区分开来，以专注于每个领域中独特的困难挑战及可能的解决之道。因此，请读者们原谅，我们将评论局限于那些自己最有见识的城市化方面，而将其他方面的内容留给更加卓越的专业人士去讨论。

城市化

2005年，前世界银行首席经济学家、诺贝尔经济学奖获得者约瑟夫·E·斯蒂格利茨，将中国的城市化和美国的高新技术发展作为影响21世纪的两个关键因素。不到十年之后的2012年，中国的城市人口在历史上第一次超过了农村人口。

推动中国快速转变的力量可以归于邓小平开创的经济改革。他的战略性目标是实现中国的农业、工业、国防和科学技术现代化，而这些要通过城市化、工业化和经济政策来实现。不幸的是，在推进的过程中，中国正经历着其他发展中国家同样会遇到的诸多问题，包括城市和农村之间不同的诉求，贫富之间巨大的收入差距等等。现代化带来了社会组织、生活方式、财富、消费和生产的重新调整。同时，它使得人们对新住房、基础设施、保障设施的需求，与同样紧迫、同样重要的（也是相对立的）环境保护、自然资源、长期可持续性的要求产生冲突。

我们对城市化问题特别感兴趣。作为一个出发点，或者说，为了正确地看待事物，我们相信首先快速回顾一下无节制城市化带来的后果将是有益的，甚至对生活水平较高的国家，如美国，也是不无裨益的。

For the past two hundred years, mankind largely took the path of expediency in the production of food and the manufacture of goods. But the basic assumption that the world is ours to exploit has led to serious environmental, economic, and sociopolitical problems. As architects, the authors find it a challenging but necessary task to identify and address the environmental aspects of sustainable development.

It is altogether a different matter to venture into the less familiar, very complex, and closely interrelated economic, social, political, and cultural aspects of the subject. More difficult still is the task of separating each discipline from the others in order to focus on its unique challenges and possible solutions. Therefore, with the indulgence of the reader, we restrict our remarks to those areas of urbanization to which we can speak most knowledgeably, leaving other aspects to those with greater expertise.

Urbanization

In 2005, the Nobel Laureate Joseph E. Stiglitz, former chief economist at the World Bank, identified urbanization in China and high-tech development in the United States as the two key factors affecting the twenty-first century. Less than a decade later, in 2012, China's urban population surpassed rural dwellers for the first time in history.

The forces behind China's rapid transformation can be attributed to the economic reforms initiated by Deng Xiaoping. His strategy targeted the modernization of China's agriculture, industry, national defense, and science and technology fields, all carried out through urbanization, industrialization, and economic policies. Unfortunately, in its race forward, China is now experiencing many of the same problems that confront other developing countries, including competing urban and rural demands, and extreme income disparity between the haves and the have-nots. Modernization has brought a rebalancing of social organization, lifestyle, wealth, consumption, and production. It has simultaneously pitted the urgent need for new housing, infrastructure, and support facilities against the equally pressing, important—and rival—need for environmental protection, natural resources, and long-term sustainability.

Our particular interest is urbanization. As a point of departure and to put things into proper perspective, we believe it would be helpful first to conduct a quick review of the effects that unbridled urbanization can have even in

Top: Typical housing development in the eastern United States. Bottom left: residential development near Pearl City, Oahu, Hawaii.
Bottom right: Sun City, Arizona.

上图：美国东部典型的居住区。左下：夏威夷瓦胡岛珍珠市附近的居住区。右下：亚利桑那州太阳城附近的居住区。

美国的城市化和郊区化

通过学术著作、个人专栏、历史小说甚至大众电影，世人都很熟悉美国早期的历史，那时一部分朝圣者生存于一片广阔的荒蛮之地。从欧洲抵达这里后，他们的首要任务是，用广袤森林中的丰富原料修建住所与防御工事。随后不久就是清理土地以展开耕种。数个世纪里，对这个国家的定义主要是一种乡村生活。后来，高楼与城市不断涌现出来，但这只是在20世纪90年代初期才发生的，那时，75%的美国人生活在城市。

在早期的这些城市中，城区中心有步行距离很近的商店和其他商业设施，提供了最便利的生活场所。需要离工作地点很近的劳动者和服务人员，挤在条件不那么令人满意的小巷中。城区中心之外的是中产阶级家庭。而在城区边缘，在没有基础设施或服务的地方生活的则是穷人。

到19世纪和20世纪初，工业革命在制造业、交通运输业和建筑业制造了大量就业机会，这吸引很多人从农村移居到城市（这一模式今天依然在重复上演）。铁路运输方面的技术进步不仅为制造业提供了支持，也使那些能够负担此项费用的人——首先是富人，然后是中产阶级——可以住在离工作地点越来越远的地方。

工业革命促使人们从农村迁移到城市地区，这是均衡发展的组成部分。在发展中，城市壮大了，农业也进步了，铁路为二者之间提供了关键的生命线。但随着越来越多的人涌入城市寻找工作，城市基础设施和配套系统达到极限并开始支撑不住。

郊区扩张

随着铁路系统不断扩建，交通运输线从城市中延伸出去，这使得每天乘车上下班成为可能。而郊区也因此开始固定下来。在二战之后，这一趋向成为主流，那时，从战场上回家的士兵们面临严重的房屋短缺，他们中的许多人参战时还是男孩，而回家的时候已经长成男人，迫切想要在自己的房子里结婚，组建家庭。作为回应，美国政府颁布了士兵法案，并为郊区建设提供便利的低息贷款。大多数新开发的项目都是卧城，在那里，养家糊口的人往往要到别处去上班。

销售员威廉·莱维特和他的兄弟，建筑师阿

a developed country with a high standard of living, like the United States.

Urbanization and Suburbanization in the United States

Through scholarly books, personal journals, historical novels and even popular cinema, the world is familiar with the early days of America, when a handful of Pilgrims braced against a vast untamed land. Their immediate task upon arrival from Europe was to build shelters and fortifications, using the abundant raw materials from the forests that blanketed the land. There followed soon after the clearing of fields for cultivation. A principally rural existence defined the country for centuries. In time, towns and cities sprang up and grew, yet it wasn't until the early 1990s that 75 percent of Americans lived in urban areas.

In these early cities the center of town, within easy walking distance of shops and other places of business, offered the most convenient places to live. Laborers and servants, who had to be close to those same places of employment, were crowded onto the less desirable back streets. Away from the center of town were middle-class families, who needed more room to grow. Beyond, at the very edge of town, without infrastructure or services, were the poor.

By the nineteenth and early twentieth centuries, the Industrial Revolution had created widespread opportunities in manufacturing, transportation, and construction, which attracted migrants from rural areas to the city (a pattern replayed today). Technological advancements in rail transportation not only fueled the manufacturing sector but also made it possible for those who could afford it—initially the wealthy, followed by the middle class—to live farther and farther from their places of business.

The migration from rural to urban areas spurred by the Industrial Revolution was part of a balanced development in which cities grew, farming advanced, and railroads provided the critical lifeline in between. But as more and more people crowded into the city for work, urban infrastructure and support systems were stretched to their limits and started to fail.

Suburban Sprawl

With the construction of ever-expanding railway systems radiating out from cities, daily commuting became possible. With this, the suburbs began to take hold. The trend became dominant only after World War II, when a severe housing shortage was met by returning soldiers, many of whom had joined the war as boys, only to come back as men, eager to marry and raise families in houses of their own. In response, the U.S. government enacted the G.I. Bill and made available low-interest loans that fueled suburban construction. Most of the new developments were bedroom communities where the breadwinner typically commuted elsewhere to work.

William Levitt, a salesman, and his brother, Alfred, an architect, pioneered suburban developments in the United States, selling two-story homes for an average price of $8,000 (two to three years' median salary) with low, or sometimes no, down payment. By comparison with crowded urban conditions, the new houses were clean and spacious, with ample kitchens and living rooms. (Houses in Levittown, N.Y. came equipped with built-in televisions!) Children could play safely in the yard as adults relaxed in a culture of patios and barbecues. People in Levittown, like other suburban residents across the country, enjoyed their homes; suburban expectations of better schools, lower density and crime, and the more stable population that home ownership brought were mostly fulfilled. Far from city factories, tanneries, and slaughterhouses, suburbanites breathed cleaner air and embraced the notion of being "closer to nature." Car outings to distant malls became a mainstay of shopping and entertainment. People enjoyed the "freedom of the road" and its far-flung benefits. Life was good, or so it seemed.

The 1970s oil crisis triggered fleeting alarm and obvious questions about fuel dependency, but more recent concerns over sustainability have focused a steadier, more critical eye on suburbia. What was once considered the paradigm of the Good Life appeared less so under scrutiny. Suburbanization had drained cities of people, business, wealth, and vitality. Overburdened and unmaintained urban infrastructure was allowed to deteriorate, resulting in the breakdown of inner cities and accelerated "white flight," leaving the poor and underprivileged behind. Very often, funds allotted for city repairs ended up being used for improvements further out as development continued to expand.

The unwitting victims of sprawl were the "happy homeowners," who were completely dependent on their cars. Distances were far and low-density public transit was not viable (or wanted), so daily tasks—buying a loaf of bread, dropping the kids off at school, Little League, music lessons, dance classes, play dates, tea parties, and all the many other details of full family lives—inescapably required driving.

The wage earner had to be shuttled to and from the train station, or very frequently just accepted a two- to four-hour round-trip commute in bumper-to-bumper traffic

尔弗雷德是美国郊区开发的先行者，他们以平均8000美金（2-3年的平均工资）的价格出售双层小楼，而购买者只需支付很少的首期款（有时甚至不需要）。与拥挤的城市环境相比，新房屋干净宽敞，有足够大的厨房和起居室。（莱维顿的房子甚至装有电视机！）孩子们可以在院子里安全地玩耍，大人们则可以在露台和烧烤文化中得到放松。莱维顿的人们和全国其他郊区的人一样，喜爱他们的家园，主要是觉得郊区有更好的学校，人口少，犯罪率低，而且房屋所有者接触的人群也更为固定。远离城市工厂、制革厂和屠宰场，郊区居民能呼吸到更新鲜的空气，并接受生活"更接近于自然"这一观念。开车到远处的商场成为购物和娱乐的主要方式，人们享受着"道路的自由"和多方面的福利。生活是美好的，或者至少看起来是美好的。

20世纪70年代的石油危机敲响了短暂的警钟，也提出了关于燃料依赖的显著问题，但是现在对于可持续性的关切乃是用一种更坚定更具批判意味的眼光来审视郊区。曾经被认为是美好生活的样板在仔细的考量中显得不那么美好。郊区化已经抽干了城市的人口、商业、财富和生机。负担过重，没有得到维护的城市基础设施被默认不断恶化下去，这导致城市中心地区的崩溃，加速了"白人逃亡"，却将贫困者和社会底层的人留在了原先的地方。常见的是，随着发展的脚步继续向外迈进，被分配用于维护城市的资金最终却用在进一步扩张上。

扩张之中最毫不知情的受害者是那些"快乐的住户"，他们完全依赖于汽车。因为距离遥远，低密度开发使公交系统不可行或不受欢迎，所以，每天的任务不可避免都要开车实现：买一个面包、送孩子上学、少年棒球赛、音乐课、舞蹈课、玩耍、茶会以及完整家庭生活的一切细枝末节。

工薪阶层同样必须穿梭往返于火车站之间，或者经常要接受拥堵交通中2-4个小时的通勤作为日常生活不可避免的一部分。道路的自由使得司机成为轮子上的囚犯。郊区化的发展给中产阶级带来的不是有益的繁荣，而是污染、交通拥挤和压力。这使得大量时间流失，个人交通和公共服务成本上升，也是石油、土地、水和其他资源过度消耗的原因所在。低效的水平扩张产生的最严重后果是，对开放空间和农田构成的威胁。

渐渐地，这些社区变得更加自足，不那么依赖于它们身后的城市。最后，更多的工作转移到郊区，包括那些被低税率所吸引的大型企业。在那些都有充足的就业机会供应给居民中的工薪阶层的地方，卧室社区变成了更为自足的生活/工作社区。

现在，80%的美国人住在大型都市圈中。实际上只有不到一半的人住在市区，大部分人都在郊区。这就是问题所在，依赖于汽车的扩张不受控制，这导致人们离工作和服务设施的距离越来越远。它也产生了交通堵塞、空气污染、社会隔离等问题；而由于住房和交通需要，它使得对原材料、资源的消费增多；最为严峻的是，原先可能用于食物生产的土地和用于清洁空气的碳汇土地也被消耗了。

由于美国的大部分不动产是私人所有的，所以，农场主可以出售土地，其收益要比农业生产产生的收益多很多。预计每年有1800平方英里的农田在城市化进程中消失。尽管有这样的损失，但研究表明，这个国家在很长一段时间内依然可以养活它的人民。然而，如果这种耕地损失继续下去，很快它就不会再是世界的粮仓了，这可能会导致全球危机。

导致粮食减产情况进一步恶化的原因是，西部平原中用于灌溉的水资源正在减少，那里的自然含水层不久前还被认为是取之不尽用之不竭的。年复一年的水资源浪费，加上不能自然地补齐水源储量（事实上有些地方的井水已经干涸），最终导致庄稼欠收。目前，干旱的环境，加上表层土壤不断被侵蚀，可能导致上世纪30年代和40年代的沙尘暴环境重新出现。

很显然，美国不断扩张的郊区化不能再继续泛滥下去，它也不应当被看作是发展中国家的参照模板。其更大的价值更倾向于作为一个代价高昂同时又不可持续的样本以予警示。

as an unavoidable part of daily life. Freedom of the road made drivers prisoners behind the wheel. Far from the benign flowering of the middle class, suburban growth wrought pollution, traffic, and stress. It accounted for huge losses of time, increased costs for personal transportation and also for public services, and the high consumption of fuel, land, water, and other resources. Among the most serious consequences of inefficient horizontal growth were, and remain, threats to open space and farmland.

Suburban communities gradually became more self-sufficient and less dependent on the cities they had left behind. In time, additional jobs moved out to the suburbs, including large corporations attracted by lower taxes. In many cases where there was ample employment opportunity to support resident wage earners, bedroom communities were transformed into more self-sufficient live/work communities.

Currently 80 percent of the U.S. population resides in greater metropolitan areas. Less than one-half actually lives in the city proper, the majority occupying the suburbs. That is the problem. Unchecked car-dependent sprawl results in increased distances from work and services. It also generates traffic congestion, air pollution, and social isolation, and leads to the greater consumption of raw materials, resources for housing and transportation and, most critically, land that might otherwise be used for food production and carbon sinks to clean the air.

Since most real estate in the United States is privately owned, farm owners can sell their property at a greater profit than agriculture would ever yield. It is estimated that 1,800 square miles [4,660 square kilometers] of farmland is lost every year to urbanization. Despite such loss, studies indicate that this country will still be able to feed its population for some time. However, if such loss of arable land continues, it will soon be unable to serve as the breadbasket of the world, leading to a possible global crisis.

Decreasing crop production here is further exacerbated by shrinking water supplies for irrigation in the Western Plains, where the natural aquifer was until recently thought to be inexhaustible. Years of water waste coupled with the inability to naturally recharge water reserves has virtually dried up the wells in some areas, resulting in failed crops. Today's drought conditions and the steady erosion of topsoil threaten a return to Dust Bowl conditions of the 1930s and 1940s.

Clearly, rampant suburbanization in the United States cannot be allowed to continue undeterred, nor should it be seen as a model for developing countries. The greater value is, as a cautionary tale, both costly and unsustainable.

A billboard in Shunde commemorates the 1992 visit of Deng Xiaoping when the leader praised local officials for "blazing a trail for Chinese industry." A center of manufacturing and trade, the entire Guangdong province has become the most prosperous and populous in China, home to nearly 106 million people in 2012.

顺德的一块广告牌记录着1992年邓小平来访时的情形，当时，地方官员因"为中国工业开辟了一条新的道路"而受到表扬。作为制造业和对外贸易的一个中心，整个广东省已经成为中国最繁荣、最时髦的地方，2012年常住人口达到了1亿6百万。

DEVELOPMENT PATTERNS

中国的城市化

在邓小平的经济改革按照两个精心规划的阶段开始执行后，中国才开始城市化进程。第一个阶段是在上世纪70年代晚期到80年代早期，包括解散农村公社、引入外资、允许私人创业等。第二个阶段是上世纪80年代末90年代，包括将国家所有制企业转变为私人所有制，以及放松贸易保护主义政策和价格控制策略。这些改革根本上改变了国家的基础，使其从一个农民的国家转变为一个城市居民的国家，所有这些带来了经济上的腾飞，使得有13.4亿人口的中国成为仅次于美国的第二大经济体。

下面的统计数据可能有助于更好地理解问题的严峻性：现在，中国的城市人口占了总人口的51.3%，即13亿总人口中的6.91亿，超过美国总人口的两倍还要多。到2035年，中国的城市居民将增加到70%（现在美国的城市人口为75%）。作为三项控制增长措施中的一部分，中国政府计划限制大城市（那些人口超过50万的城市）的人口数量。政府希望大多数不断增加的城市居民能够被安置在中型城市（人口在20万到50万之间），以及小城市（人口在10万到20万之间）。从现在开始到2030年，预计将需要20000到50000座新高层建筑。也就是说在接下来的20年中，每天都将产生3-6座新的摩天大楼。

前所未有的人潮涌入城市，这使得现有的城市开始膨胀，突破界限，进入到周边的开阔地带中，一般来说是进入到临近的耕地中。就中国而言，其陆地面积（3706580平方英里，或960万平方公里）为全球第三，但却由于城市化、植树造林、草原重植项目和自然灾害，以每年0.5%的速度失去了485万公顷（18430平方英里）的耕地。

尽管由于周期性的洪水或干旱，中国在历史上曾饱受饥荒之苦，但这个国家数个世纪以来总体而言成功地以其自身的土地养活了那里的人，这一点已被认为是一个现代奇迹。中国政府准确地意识到，持续丧失耕地将危害到这个国家继续获得上述成功的能力，而且，在中国，土地属于国家和集体，所以他们执行了一项政策，要求任何耕地上的损失都

Urbanization in China

China's urbanization began after Deng Xiaoping's economic reforms were implemented in two carefully planned stages. The first, in the late 1970s and early 1980s, involved the dismantling of the farm collectives, inviting foreign investment and allowing entrepreneurship. The second stage, in the late 1980s and 1990s, involved the conversion of state-owned industries to private ownership and the relaxation of protectionist policies and price controls. Those reforms radically changed the very backbone of the country from a nation of farmers into a nation of city dwellers and, with this, set in motion an economic boom that has catapulted the People's Republic of China, with a population of 1.34 billion, second only to the United States in economic size.

The enormity of the problem is perhaps better understood from the following statistics: Currently, the urban population of China is 51.3 percent, or 691 million, of the nation's total of 1.3 billion people, more than twice the total population of the United States. By 2035, the number of China's urban dwellers will increase to 70 percent (compared to the current 75 percent in the United States). As part of a three-pronged strategy to control growth, the Chinese government plans to limit the number of big cities (those with more than 500,000 people). The government intends that the bulk of the ever-increasing urban population be accommodated in medium-sized cities of 200,000 to 500,000 people, and also in small cities with populations of 100,000 to 200,000. From now until 2030, it is estimated that 20,000 to 50,000 new high-rise buildings will be required. That's roughly three to six new skyscrapers each day for the next twenty years.

The unprecedented influx of humanity into cities results in expansion of the existing city limits into the surrounding open spaces, most often into neighboring farmland. In the case of China, which ranks third in total land mass [3,706,580 square miles, or 9,600,000 square kilometers], at an annual rate of 0.5 percent it has already lost 4.85 million hectares [18,430 square miles] of arable land to urbanization, as well as to reforestation and grassland replanting programs, and also natural disasters.

Although China historically suffered widespread famine due to periodic flooding or drought, the country's overall success, for centuries, in being able to feed its population using its own land has been regarded as a modern miracle. Chinese authorities are well aware that the continued loss of farmland will jeopardize the country's ability to continue to do this and, since in China land belongs to the state and collective, they have implemented a policy requiring that

精耕细作的家庭农业在中国南方稻田和其他作物之间留下了密密交织的小路。

any such loss of land be replenished by comparable acreage elsewhere in the country. Additionally, China has set 120 million hectares [296.4 million acres] as the absolute minimum of arable land to be preserved beyond which there can be no further shrinkage, no impingement. That limit, known as the "red line," is dangerously close.

The authors have witnessed China's phenomenal modernization with both fascination and awe. They marvel at the magnitude and quick pace of the nation's transformation but at the same time wonder about its long-term consequences.

要在国家其他地方获得同等面积的补偿。另外，中国已经将1.2亿公顷（2.96亿英亩）的土地当作需要保护的可耕种土地面积的最低值，这些土地绝不能再减少，不能再受到冲击。这一限制，即所谓的"耕地红线"，已经快被触及了。

作者们带着惊讶和敬畏，见证了中国的现代化进程。他们对这个国家发生转变的范围和速度感到惊奇，但同时也想知道其长期的结果。

Intensive family farming in southern China where narrow paths are woven between rice paddies and other subsistence crops.
精耕细作的家庭农业在中国南方稻田和其他作物之间留下了密密交织的小路。

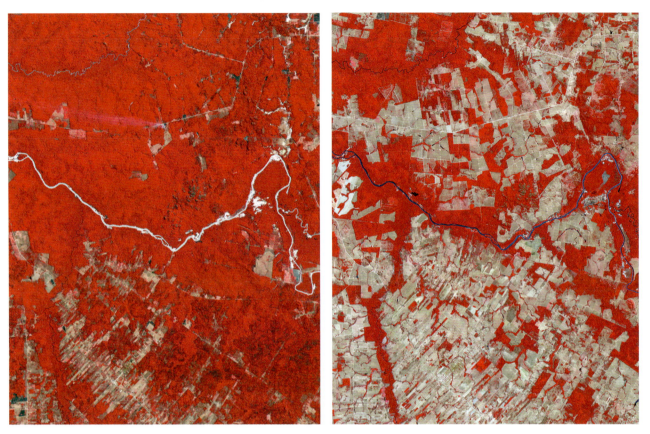

Advance Spaceborne Thermal Emission and Reflection Radiometer (ASTER) images from 1992 (left) and 2006 (right) reveal the widespread deforestation taking place in Mato Grosso, Brazil. The tropical rainforest appears in bright red; gray-green areas indicate forest clearance from logging and agriculture; black and dark gray areas pinpoint recent burning.

高级太空热辐射和反射辐射计（ASTER）拍摄的图像，反映了从1992年（左）到2006年（右）巴西马托格罗索州发生的大规模乱砍滥伐现象。热带雨林呈现出亮红色；灰绿色区域表明森林正由于砍伐和农业而被从地图上抹去；黑色和深红区域代表曾经着火的地区。

城市化与环境

在工业革命之前，世界上只有3%的人口住在城市。今天则总计有50%的城市人口，在接下来的20年中，这一比率有望攀升到60%。据估计，仅仅在中国，接下来的二三十年间，就会有4亿人搬进城市。实际上在2001年，民政部部长多吉才让就宣布，中国的目标是在接下来的20年中建设400座新城。根据高层建筑和城市环境委员会的数据，全球城市化人口每天接近20万，这大约相当于全球每个月增加一个加尔各答的城市人口数。

现在的规划理论和实践并不足以说明如此大量的人口流动。问题是真实的，也是非常严峻和迫切的。迄今为止，解决方案一般就是"建得更高更多"：更多的高楼、更多的公路、更多的地铁等等。然而，问题依旧存在。我们不能再继续按照老办法扩张我们的城市了。必须探究新的解决方案，而且，利用全新的理念，以便使可持续的城市生活成为可能的现实。幸运的是，技术、设备和所需的材料都已具备，或者已是触手可及了。现在所需要的是实现这一宏伟壮举的信念和意志。我们将从相互关联的领域中甄别出最为重大的问题开始。

环境影响

如前所述，不断增长的人口及其向城市中心的迁移，农村地区的工业化，已经导致了无序发展。这样的盲目扩张总是以牺牲周边地区的耕地和未受破坏的土地为代价。在美国，每年流失的农田接近100万英亩（40万公顷）。在中国，这个数字要超过两倍还多，从上世纪80年代末开始，每年被所谓发展所蚕食的耕地面积大约是250万英亩（100万公顷）。持续的土地流失威胁着食物供应，也危害着现代中国最伟大的成就之一：在多年饥荒之后，伴随着1978年一项政策提议公布的土地状况，中国已经能养活自己的人民了。而现在，中国大约12%的粮食依赖于进口。正如中国共产党农村工作领导小组办公室主任陈锡文在2012年年初所言："在城市化进程中，我们必须注意农业的发展……不过，我们当然也不能只是追求自给自足。"不难看出，经济发展和现代化在中国历史的这一阶段还是居于主导地位的。

Urbanization and the Environment

Before the Industrial Revolution, only 3 percent of the world's population lived in cities. Today, the urban total is 50 percent, and it is expected to climb to 60 percent in the next twenty years. Estimates project that 400 million people will be moving into cities in China alone within the next two to three decades. Indeed, in 2001, Doje Cering, State Minister of Civil Affairs, announced China's intention to build 400 new cities in two decades. According to the Council on Tall Buildings and Urban Habitat, global urbanization is approaching 200,000 people every day, roughly equivalent to all the people in Calcutta added to the world's urban population every month.

Existing planning theory and practices are not adequate to address such a huge influx. The problems are real, very serious, and urgent. Thus far, attempts at resolution have typically involved building higher and more: more towers, more roads, more subways, and so forth. Yet the problems remain. We cannot continue to expand our cities following old precedents. New solutions must be explored and radical new thinking brought to bear in order to make sustainable urban living a viable reality. Luckily the technologies, equipment, and required materials are available, or will soon be within reach. What is needed is the conviction and will to achieve the grand feat. We begin by identifying the most significant problems within interrelated domains.

Environmental Impact

As previously stated, the growing human population and its migration into urban centers and newly industrialized rural areas has resulted in unplanned development. Such haphazard expansion invariably occurs at the expense of arable or unspoiled lands in peripheral areas. In the United States, nearly a million acres [approximately 400,000 hectares] of farmland are lost annually. In China, the number is more than double, as every year since the late 1980s some 2.5 million acres [1 million hectares] are gobbled up by development. The continued loss threatens food supply and undermines one of modern China's greatest achievements: feeding its own people following a policy laid out with great public display in 1978, after years of famine and starvation. China now imports roughly 12 percent of its food. As Chen Xiwen, director of the rural affairs policy-making committee of the Chinese Communist Party, explained in early 2012: "During the process of urbanization, we must pay attention to agricultural

乱砍滥伐占了全球森林砍伐量的14%。一些树木作为珍奇的木材，被有选择性地砍伐下来，
但每当砍倒一棵树，总会有其他5—10棵树受到损伤。

Logging accounts for about 14 percent of global deforestation. Trees are selectively harvested for exotic woods,
but for every tree intentionally felled, some 5–10 others are casualties of the process.
乱砍滥伐占了全球森林砍伐量的14%。一些树木作为珍奇的木材，被有选择性地砍伐下来，
但每当砍倒一棵树，总会有其他5—10棵树受到损伤。

发展带来的损失就是，破坏了自然林区和城市周边的牧场；在那里，经过多年的进化演变，动植物已经达到一种平衡的可持续状态，对于都市人来说，这是一个共赢的结果。这种缓冲区对于保护水源、土壤免受污染和侵蚀，分解、吸收污染物，储存和再生珍贵资源大有裨益。它们甚至还有助于气候稳定，缓和全球变暖，为更多野生动物提供栖息地。

通过将二氧化碳转化为氧气，原始森林在净化空气方面扮演着重要的角色。虽然被抱怨却持续破坏雨林的行为，加大了这一问题的严重性。雨林曾经覆盖地球14%的面积，但现在这个数字已经锐减到6%。12亿英亩的森林不管怎么说都是数量庞大的，但它却以每秒1英亩的速度在消失——也就是大约每天85000英亩——它们不停地被砍伐以获得木材，变成建筑材料、家俱和制造纸张的纸浆。可预见的结局是，在不到40年后，这些雨林就有可能消失殆尽。而其带来的结果则必定是灾难性的，因为雨林产出全球20%的氧气，保持着五分之一的淡水，养育着全球一半的动物、昆虫和植物大约1000万种。

所有这些都将被破坏，而且与之相伴的是，一些现在未知的食物，一些可能治愈威胁生命疾病的东西，都将消失。到今天为止，只有不到1%的雨林植物已经被科学地研究过，而这么小的一部分样本却产生出西方世界所用全部药品的四分之一，其中包括一大部分今天用于治疗癌症的药物。

从不起眼的阿司匹林（萃取自柳树的树皮和树叶），到能够重新长出四肢的火蜥蜴，再到可以从泡沫巢中产生大量生物燃料的南美泡蟾（这一发现获得了2011年的地球奖），自然拥有许多问题的答案，其中有一些答案将会澄清生命的问题，这一点正变得越来越清晰。因此，人类干预的悲剧结果是：污染水资源的废弃物和使土地流失养分的杀虫剂导致一些必需的微生物无法存活，作为传粉者的蜜蜂、昆虫捕食者蝙蝠、海洋猎食者鲨鱼都在消失灭绝。这一表单上的物种还在增加。

关键的一点是，在地球这个自我调节、自我修复的有机体内，所有的生物对于维护有机体的生物多样性都有自己独一无二的作用。多样性越是丰富，有机体的强度和韧性就越好。当改变

development... but, of course, we certainly cannot pursue self-sufficiency." Economic growth and modernization take precedence at this stage of China's history.

The loss from development is equally devastating in natural wooded areas and grasslands on the outskirts of cities where fauna and flora, through years of evolution, have achieved a balanced sustainability that is also beneficial to urbanites. Such buffer zones collectively help to protect water and soil from contamination and erosion, break down and absorb pollution, and store and recycle precious nutrients. They further contribute to climate stability, mitigate global warming, and provide habitats for a great many species of wildlife.

Old-growth forests play an especially important role in scrubbing the air by recycling carbon dioxide into oxygen. The often bemoaned but continuing destruction of the rainforest underscores the seriousness of the problem. The rainforest once covered 14 percent of the planet, but it has now been reduced to a mere 6 percent. The 1.2-billion-acre forest is vast by any measure, but it loses more than an acre every second—that's roughly 85,000 acres each day—in the relentless harvesting of timber for construction materials, furniture, and wood-pulp paper. Harrowing estimates project possible depletion in less than 40 years. The consequences are inescapably catastrophic as the rainforest produces 20 percent of the world's oxygen and holds one-fifth of its fresh water and more than half of the world's estimated 10 million species of animals, insects, and plants.

All of this will be destroyed, and with it the loss of still unknown foodstuffs and possible cures for life-threatening diseases. To date, less than 1 percent of the rainforest's growth has been scientifically investigated, yet this tiny sampling has yielded 25 percent of all pharmaceuticals used in the West, including a large portion of today's cancer-fighting drugs.

Recent investigations into the humble aspirin (derived from the bark and leaves of willow trees), the ability to regenerate lost limbs in salamanders, and the production of immense amounts of biofuel from the foam nests of Túngara frogs (a discovery that won the 2011 Earth Award) make it increasingly clear that nature holds the answer to many—some would argue all—of life's questions. Thus, the tragic consequences of man's interventions: the wastes that contaminate water supplies; the insecticides that sterilize the land so that essential microorganisms cannot survive; and the loss of bees as pollinators, bats for insect control, and sharks as cullers of the ocean's weak. The list goes on.

The critical point is that all living creatures play a strategic role in the biodiversity that sustains the planet as a self-regulating, self-healing organism. The richer the diversity, the greater the strength and resilience. When changes are slow there is opportunity for compensating balances to take place. But with the vast amount of development rapidly taking place around the world, in different topographies and climates, no one really knows the full extent of the damage. As Jaan Suurküla reported in 2004 to Physicians and Scientists for Responsible Application of Science and Technology: "The world's ecosystem has been kept in balance through a very complex and multifaceted interaction between a huge number of species. The rapid extinction rate is therefore likely to precipitate collapse of ecosystems at a global scale."

Urbanization at the expense of farmland and natural greenbelts puts pressure on the environment to a point that is not sustainable. If we continue our development in this direction, there will be irreversible changes in our environment that will drastically affect our health and well-being.

Soil Degradation

The Earth's natural cover consists of a complex and dynamic ecosystem of minerals, air, water, and organic materials that collectively sustain all of life. It is among our most precious resources, yet roughly half of the world's soil has been lost over the past 150 years, most of it as a result of human actions.

Soil degradation is the loss of nutrients or organic matter or, alternatively, pollution by toxins so that the soil loses its quality and productivity and can no longer be used for growing. The main cause of soil degradation is erosion, whether by weather, wind, water, or gravity, naturally and gradually wearing away the Earth's surface. Human intervention has greatly accelerated the loss, so fertile soil is removed faster than it can be replenished.

Deforestation, logging, and other land-clearing methods, together with overgrazing and irrigation mismanagement, have brought about salinization and desertification. The soil loses its capacity to retain water and nutrients and basically loses its ability to support life. Construction is another factor, as is compaction from heavy equipment, which destroys essential aeration.

The pressures of the growing population have led farmers to use harmful pesticides and fertilizers to increase crop yield. It is estimated that only 20 percent of the additives are actually absorbed by the plants; the rest runs off to erode and

还很缓慢的时候，平衡还有可能得到调整不致失控。但是世界各地，无论地形与气候如何，都在进行大规模、快速的发展开发时，就没有人真的知道破坏究竟会达到怎样一个规模。正如安·苏尔库拉在2004年向科技应用医生及科学家委员会做的报告所言："世界生态系统乃是通过大量物种之间复杂且多层面的互动才得以保持平衡的。过高的灭绝率有可能意味着全球范围内生态系统的崩溃。"

以牺牲耕地和自然绿化带为代价的城市化，给环境施加了巨大压力，使之处于一种不可持续的状态。如果我们继续以这种方式发展下去，我们的环境将会发生不可逆转的改变，而这会给我们的健康与福祉带来灾难性的后果。

土壤破坏

地球的自然地表是一个由矿物、空气、水和有机物构成的复杂、动态的生态系统，他们供给着地球上所有的生命。土壤是我们最珍贵的资源之一，然而地球上大约一半的土壤在过去150多年间已经彻底消失了，这主要是由人类的行为造成的。

土壤破坏是指土壤中营养物质或有机物质的流失，或者土壤受到毒素的污染以致丧失原有的品质和生产能力，不能够再用于耕种。土壤破坏的主要原因是侵蚀，不管是天气、风、水或者重力，都会自然地、逐渐地破坏地表。人类的干预则大大加快了土壤破坏，使得肥沃土壤的消失速度比补充再生速度要快得多。

乱砍滥伐及其他开荒拓地的方法，加上过度放牧和管理不善的灌溉方式，已经使得土地盐碱化和荒漠化了。土壤失去了保持水分和营养的能力，也就根本上失去了养育生命的能力。建设则是另一个因素，利用大型设备夯实土地会破坏其基本的通气性。

人口持续增长带来的压力使得农民开始用杀虫剂和化肥来提高农作物产量。据估计，实际上只有20%的添加剂能被植物吸收，其他则进入地下，侵蚀和污染土地与水源。精耕农业揭去了土地保护层，而单一耕种式农业和其他不恰当的农业实践则破坏了生物多样性。

Slash-and-burn, among the oldest methods of agriculture, is still widely practiced around the world. It makes previous dense-growth lands available for farming, but contributes to erosion, an increase in greenhouse gas, and the loss of nutrients, natural habitat, and biodiversity.

在最古老的农业方式中，刀耕火种依然在全世界被广泛采用。这种方法使之前长满绿草的土地适于耕种，但同时也导致地表侵蚀，温室气体增加，养分流失，自然栖息地和生物多样性减少。

poison the ground and water. Intensive tilling also strips away protective ground cover, while single-crop farming and other inappropriate agricultural practices destroy biodiversity.

In urban areas, industrial pollutants and household garbage, especially nonbiodegradable products such as plastics and electronic waste, have reached a crisis stage compared by the World Bank to the severity of climate change; they also contribute to it, as landfills emit more and more methane gas into the air every day. Accurate global data is difficult to compile, as practices in different countries and regions vary widely, but since the amount of solid waste increases with economic development and urbanization, the environmental and human health risks will undoubtedly grow. Municipal solid waste is expected to double by 2025.

Air Pollution

Although developed countries have made major advances in the reduction of air pollution, it remains at dangerously high levels in developing countries, where the necessary capital investment to combat it is not available and where regulatory standards, if they exist, are difficult to enforce. As many as two million people die each year from the effects of air pollution.

Major causes of air pollution include the burning of "dirty" fuels like coal and oil in power plants and factories. Coal burning for home cooking and heating contributes to the smoke and soot, whose severity and duration are affected by local climate and topography. Agricultural spraying and dusting, slash-and-burn land clearing, and—critically—airborne construction dusts are also major causes. Air pollution will remain a serious problem as long as feverish urbanization continues, releasing dangerous particulates into the air.

By far, most air pollution in urban areas comes from traffic, both from diesel-fueled heavy transport and from ever-growing numbers of passenger vehicles. In developing countries the problem is exacerbated by the large number of older vehicles on the road. In addition, vehicle maintenance is poor, and leaded fuels are still used in some cases. Auto exhaust is especially dangerous to human health because it is emitted close to the ground where people, children in particular, are most directly exposed. The disastrous effects of lead on childhood health and development spurred the development of catalytic converters and the global phase-out of lead from gasoline (only eleven countries still use leaded gas and just three use it exclusively).

据世界银行估计，在城市地区，工业污染和生活垃圾，尤其是像塑料和电子废物这些非生物降解物，其储量已经达到了一个危险阶段，足以产生严重的气候变化；由于垃圾填埋场每天都向空气中排放大量的沼气，它们实际上已经发生作用了。全球范围内的精确数据不容易收集，因为这种行为随着不同的国家和地区而变化，不过由于固体垃圾的数量随着经济发展和城市化不断攀升，对环境和人类健康的威胁也毫无疑问会随之加剧。到2025年，城市固体垃圾预计将会翻倍。

空气污染

尽管发达国家已经在减少空气污染方面有了长足的进步，但在发展中国家，空气污染依旧处在一个非常危险的层面上，在那里，能够应付污染的必要资金投入往往不到位，即使有的话，管理标准也很难强制落实。每年有多达200万人因空气污染而死去。

空气污染有许多原因，包括：在发电厂和工厂中燃烧像煤和石油这样的"肮脏"燃料。家庭生活中，烹饪和取暖烧煤产生的油烟煤灰（其严重性与持续时间受当地气候与地形的影响）。另外，农业喷洒，刀耕火种的开荒方式，以及——非常关键的——通过空气传播的建设产生的灰尘，都是造成空气污染的重要原因。只要狂热的城市化继续下去，并仍旧向大气排放粉尘颗粒，空气污染就将一直是个严峻的问题。

目前来说，城市地区的大部分空气污染来自交通工具，既有使用柴油的大型交通工具，也有日益增加的小型汽车。在发展中国家，公路上数量众多的老旧汽车，进一步加重了这个问题。另外，汽车维护也跟不上，有些地方甚至还在使用含铅燃料。汽车尾气特别危害人体健康，因为它排放的气体离地面很近，而那里是人群尤其是儿童主要的活动区域。铅对于儿童健康与成长的灾难性危害，刺激了催化转换器的发展，同时也迫使人们在全球范围内开始逐步淘汰含铅燃料（只有11个国家还在使用含铅燃料，其中只有3个国家不使用其他燃料）。

到2035年，公路上小型汽车的数量有望翻倍，达到令人惊愕的17亿辆，这种爆炸式增长主要发生在中国和印度。除了绝对数量之外，产生交

Top: Pollution (gray) is readily distinguished from white clouds in this satellite photo of China's southeast coast. Cars, industry, construction, and coal-burning power plants all contribute to poor air quality, exacerbated by the mountains to the west, which trap air pollution in place. Bottom left: Volcanoes emit sulphur dioxide and other pollutants that cause hazardous vog (volcanic smog) and acid rain. Bottom right: An electric car charging in Amsterdam. The environmental benefit of e-vehicles has been challenged since their manufacturers can emit more toxic waste than factories producing gasoline-fueled cars. The question is even bigger in China, where 75 percent of electricity is generated by coal-burning power plants.

上：在这张中国东南沿海的卫星图中，污染（灰色区）很容易与白云区分开来。汽车、工业、建设、燃煤电厂都对糟糕的空气质量负有责任，而它西部的山区会将污染的空气桎梏其中，进一步加重这种污染。左下：火山爆发喷射出二氧化硫和其他污染物，导致产生有害烟雾（火山烟雾）和酸雨。右下：一辆电动汽车正在阿姆斯特丹充电。电动汽车带来的环境益处已经受到质疑，因为相比生产汽油燃料的汽车，生产电动汽车的制造商排出的有毒废弃物要更多。这一问题在中国尤其严重，中国75%的电能都是由燃煤电厂供应的。

Although recycling programs are growing, it is estimated that some 80 percent of water and other bottles are simply thrown away. Glass bottles take thousands of years to decompose; petroleum-based plastics begin the process after roughly 450 years, but never fully biodegrade.

尽管循环再生项目正在发展，但预计约有80%的水和各种瓶子被简单地扔掉了。玻璃瓶要数千年才能分解；石油制造的塑料大约450年后才会开始分解，要完全生物降解则是不可能的。

通问题的其他原因在于，城市规划上的失败或不作为。由于不适当的管理和法律控制，以及财政投入和处理此类重大问题技术上的限制，公路网的承载能力经常被严重地高估了。集成型"智能城市"中的步行、自行车和公共交通系统，似乎是解决之道。

水

水对生命而言不可或缺——饮用、烹饪、农业灌溉、医疗卫生、制造生产（从衣物到核反应堆），总之，对于地球上一切生物的健康而言，皆是如此。总体而言，地球上水的总量很长一段时间都保持得相对稳定：水通过融化、排放、蒸发到空气中，再通过降水形成自然循环，这一过程总在不断地重复着。

但随着气候变化，污染加剧，问题开始产生了。根据联合国的资料，上个世纪，全球人均用水量的增长速率是人口增长率的两倍。这个增长率在中国和印度要更高，过去50年间，这两个国家的用水量成十倍增加。简单说，越来越多的人正在竞争有限的水资源，他们耗尽水资源的速度超过了它的补充速度。饮用水已经无法维持了，情况正变得越来越糟。

在接下来的20年间，淡水使用量预计将增加40%，这主要来源于城市化和经济发展的快速步伐。随着生活标准的提高，用于改善卫生条件和室内自来水管道的水资源也在不断消耗——包括冲水马桶、水槽、淋浴、洗衣机、洗碗机和其他豪华电器。经济发展也带来饮食上的变化，米饭和豆腐让位于肉食品；一般地，煮好一碗米饭需要大约550毫升水，而一顿有牛肉的饭则需要消耗2200毫升水。

迄今为止，农业消耗的淡水量是最大的，世界的平均值为70%，在亚洲这个数字是84%。由于人口不断增加，可能需要的淡水量还将增加。随着农民越发聪明，他们不再依赖雨水和地表水，而是开始利用可灌溉的河流、小溪和湖泊。他们还挖井从地表下面20-30英尺的含水层获得地下水。工业化带来了新设备，使得打井可以更快更深，以至于对含水层的抽取已经大大超过了它自身的再生能力。这种情况就好像是，不断重复地从银行账号中提取钱款，却从来不往里面增加存款。

By 2035 the total number of passenger vehicles on the road is expected to double to a staggering 1.7 billion, much of the explosive growth taking place in China and India. Beyond sheer numbers, part of the reason for traffic problems is failed or nonexistent urban planning. Network carrying capabilities are often grossly overestimated as a result of inadequate administrative and legislative controls, financial resources, and the technical skills to deal with such enormous problems. Walking, biking, and public transit in compact "Smart Cities" would seem to hold the solution.

Water

Water is indispensable for life—for drinking, cooking, agriculture, sanitation, the manufacture of such diverse products as clothing and nuclear reactors, and, in general, for the health of every living thing on Earth. By and large, the amount of water on the planet has remained fairly constant over time as precipitation has recycled naturally through freeze-thaw, drainage, and evaporation back into the air, endlessly repeating the process.

The problem comes with climate change and the increased population. According to the United Nations, water use per person grew at double the rate of population increase over the last century. The rate is even higher in China and India, where water use has increased tenfold in the last 50 years. Simply put, more people are competing for limited freshwater resources, and they are depleting the supply faster than it can be replenished. Consumption is already unsustainable, and the situation is getting worse.

Freshwater use is expected to grow by 40 percent over the next two decades, largely driven by the blistering pace of urbanization and economic development. As living standards increase, so does water consumption for improved sanitation and indoor plumbing—flush toilets, sinks, showers, washing machines, dishwashers, and other luxury appliances. Economic growth also brings a change in diet as rice and tofu give way to meat; a traditional bowl of rice requires roughly 550 liters of water for its full production cycle compared to 2,200 liters for a meal with beef.

Agriculture consumes by far the greatest amount of freshwater, the world average being 70 percent, and 84 percent in Asia. And given the growing population, more will be needed. As farmers, with increasing sophistication, turned away from dependence on rainfall and surface water, they took to irrigating rivers, streams, and lakes. They also dug wells to access groundwater in natural aquifers some 20 to 30 feet below the

The widespread discharge of often undisclosed and frequently hazardous chemicals into water supplies can lead to irreversible environmental damage. Developing countries are inclined to regard industrial waste as the necessary price of progress, too difficult and costly to combat in the throes of modernization.

大范围将未净化甚至有毒的化学物质排入水中会导致不可逆转的环境破坏。发展中国家倾向于把工业废物当作发展的必要代价，要在痛苦的现代化进程中与之作战，困难太大，成本也太高。

因为蒸发和径流导致的低效灌溉使得地表被侵蚀，养分从土壤中流失，而杀虫剂和化肥又产生了污染，这就使问题进一步复杂化。另外，对动物粪便的处理也产生了严重的环境问题，因为它向水中排放了危险的微生物。在接近城市的农业地区，耕种必须与家庭用水和工业用水争夺淡水资源。尤其是在发展中国家，农民别无选择只能使用地下水，而这些地下水很有可能受到污染，污染源则包括人类排泄物、城市垃圾、泄露的基础设施中排出的重金属和化学物质，甚至还有医疗垃圾中的抗生素、细菌、病毒和其他病原体。受污染的水源使人们在食用受污农作物时面临巨大的健康威胁。有毒的径流，未处理的污水，被固体垃圾污染的湖泊、小溪、河流和沿海水域，对于鱼类、其他水生生物和人类都是灾难性的。

水污染乃是具有历史性意义的一次大危机。毫不意外的是，发展中国家的情况往往最糟，在那里，即使有一些净化设备，也往往是不够的，良好的卫生环境对人们而言总是不太现实，或者说其重要性往往不被理解。按照联合国环境规划署的说法，"在发展中国家，流过大城市的河流不会比下水道干净多少"。这些地区，有80%的病人是因为水传疾病入院治疗的。

尽管地球富含水资源（地表70%为水所覆盖），但其中大部分都是海水。海水淡化的高成本对大多数国家而言都是一种奢谈。自然供应的淡水往往遥不可及，因为多藏于极地冰冠或永久冰层之中。因此，人类所能获得的水资源要低于1%。另外，水资源供应在世界上并不是均衡分布的，而且也不是常年供应的。不同地区往往会形成干旱和洪水、雨季或高海平面的显著对照。洪水泛滥和缺水造成的损害是一样的。气候变化进一步将问题复杂化。

随着淡水需求的增加和供应的减少，水资源利用的问题越来越紧迫。世界银行已经将水资源管理和食物供应当作世界目前亟待解决的问题。保护措施和更好的工农业行为越发重要了，像诸如种植什么、在哪里耕种和如何种植的决定，皆是如此。可持续性措施像雨水收集、滴灌、回收利用、无水马桶和其他一些新技术，都可以提供帮助。降低污染物和废弃物的总量，采用明智的管理措施，这些要

surface. With industrialization came new means for pumping more and deeper, to the degree that aquifers have been drained beyond their renewable capacity. The situation has been likened to making repeated withdrawals from a bank account without keeping up with deposits.

The problem is compounded by the inefficiencies of irrigation through evaporation and runoff, leading to erosion, the leeching of nutrients from the soil, and pollution from pesticides and fertilizers. The disposal of animal waste also poses serious environmental concerns, as it spreads dangerous organisms into the waterways. In agricultural areas near cities, farming has to compete for freshwater with domestic and industrial uses. Especially in developing countries, farmers have no choice but to use groundwater quite possibly contaminated by human waste; heavy metals and chemical discharge from municipal dumps and leaking infrastructure; and antibiotics, bacteria, viruses, and other pathogens from hospital waste. The tainted water presents significant health hazards as people consume the fouled crops. Toxic runoff, untreated sewage, and solid wastes similarly poison lakes, streams, rivers, and coastal waters, with devastating consequences for fish, and other forms of aquatic and human life.

Water pollution is a crisis of epic proportions. Not surprisingly, it is worst in developing countries, where treatment facilities, to the degree they exist, are most often inadequate and where proper sanitation is not always possible or its importance is not understood. According to the United Nations Environmental Program, "[I]n developing countries, rivers downstream from major cities are little cleaner than open sewers." Fully 80 percent of all patients in such areas are hospitalized because of waterborne diseases.

Although Earth is a water-rich planet-some 70 percent is covered by water, most of it is ocean-based. High cost puts desalinization beyond the reach of most countries. Natural supplies of freshwater are remote and inaccessible, trapped in polar caps or permanent ice. Humans therefore have access to less than 1 percent. Moreover, the supply is not evenly distributed around the world, nor is it dependably available throughout the year. Drought in one area contrasts starkly with monsoons, flooding, or higher sea levels in others. Too much water is just as crippling as too little. Climate change compounds the problem.

As freshwater demands increase and supplies decrease, questions about water usage intensify. The World Bank has identified water management and food production as pressing global concerns. Conservation and better agricultural and industrial practices are increasingly important, as are decisions about what to grow, and how and where to grow it. Sustainable measures such as rain capture, drip irrigation, reclamation, waterless toilets, and other new technologies can all help. The overall reduction of pollution and waste, along with enlightened governance, will be required as never before.

Beyond land use, soil, air, and water pollution, and the dire need to conserve precious resources, the physical consequences of human development are legion. In rapid urbanization the challenges are especially pronounced, if only for the great numbers of people involved and the short time frame in which to accommodate them. New cities are springing up maybe not quite overnight but in just a few years, instead of growing slowly by accretion and self-correction over centuries.

There are urgent needs for adequate housing, public transit, and infrastructure. Power generation, water supply, sanitation, and waste disposal must be appropriately, and quickly, addressed. A greater number of urban dwellers demand increased food supplies, and all the production, shipping, and distribution systems required. In all of this, responsible planning is essential to prevent arbitrary or otherwise misguided decision making and to ensure the health and well-being of city residents. Top-notch professionals and creative minds need to collectively address issues of design and environmental control. Special care needs to be taken in new developments and start-up cities, as there will be no mature trees to mitigate the heat island effects caused by impermeable paved surfaces and buildings.

The challenges and sheer physical toll of recent urban migrations are daunting. Yet it is essential that important matters not be shunted to the side by urgency and expedient solutions. We need to be sure that we don't surrender our arable lands and open spaces—once gone, they will never be reclaimed—and we need to protect the purity of our air, soil, and water. We need to make sure that through the lethal combination of climate change and man's hostile actions we don't destroy the physical world, and all hope for the future along with it. Rapid urbanization is proceeding at full throttle with no immediate end in sight. So the question is how best to maximize the opportunities presented for massive change in a way that will ensure a healthier, more sustainable life.

The Brundtland Commission, officially the World Commission on Environment and Development, was created by the United Nations in 1983 to raise public awareness and encourage multinational cooperation. It defined sustainable development as "development that meets the needs of the present without compromising the ability of future generations to meet their

求从未像今天这样迫切过。

除了土地使用和土壤、空气、水污染，保护珍惜资源的紧迫需求外，人类的发展还导致了许多现实的后果。在快速的城市化进程中，人们遇到的挑战尤其显著——仅从众多的人数和解决他们的居住问题所允许的极短时间而言。那么，新城市将如雨后春笋般涌现——尽管不是一夜之间，但也是在短短几年中，就取代了之前数个世纪的缓慢增长，自我调节。

我们对充足住房、公共交通系统和基础设施的需求十分紧迫。电力、水资源供应、环境卫生、废弃物处理，这些都必须得到及时恰当的解决。许多城市居民要求增加食物供应量，这就需要一整套生产、运输和配给系统。在所有因素中，负责任的规划对于避免做出武断或误导性决策，保证城市居民的健康福利是最为根本的。一流的专业人员和创造性思维需要集合起来，共同应对设计与环境控制方面的问题。在新兴发展与刚起步的城市中，人们需要采取特殊的措施，因为，那里不会有成熟高大的树木来缓解因为不透水铺装和建筑形成的热岛效应。

当前城市迁徙带来的挑战和单纯的现实代价是令人震惊的。但从根本上说，重要的是不能受制于紧迫、权宜的解决方案。我们需要确定，不能总是放弃耕地和开放空间，它们是不可再生的，我们要保护空气、土壤和水资源的纯净。通过认识气候变化和人类破坏行为的致命关联，我们应该意识到，我们不能再破坏地球了，所有对未来的期望都与此相关。快速城市化正全面进展且不会立即中断。所以问题是，如何在确保一种更健康、更可持续生活的前提下，最好地利用大变革中的机遇。

世界环境与发展委员会的官方机构布伦特兰委员会，于1983年由联合国创立，旨在提升公共意识，鼓励多边合作。它将可持续发展定义为"既满足当代人需要，又不对后代人满足其需要的能力构成危害的发展"。美国建筑师协会更为严格详细地指出，可持续发展是"社会有能力继续在将来发挥作用，而这种能力不会因为系统所依赖的关键资源被耗费殆尽或超负荷利用，就变得衰弱萎靡"。

尽管证据确凿且每天都在加剧，但还是有人继续否认气候变化，就像有人在面对难以挽回的损失时并不停手，肆意践踏雨林或向大气中排放有毒物质。还有一些人，无论是出于无知、冷漠、政治或贪婪的原因，在消耗地球资源时总是极不明智地麻痹大意，其消耗资源的速度大大超过我们补充资源的能力！这毫无道理。我们正在伤害、毁灭生态系统，而我们自身就是这个系统的一部分，我们的生存就依赖于这个系统。

如果有前进的希望，人类现在就处于一个关键的时刻。正如阿尔·戈尔在他获得普利策奖的著作《濒临失衡的地球》中所说的"……我们现在正面临着全球内战的威胁，这场内战发生在如下两类人之间：一类人拒绝考虑文明无止尽进步带来的后果，另一类人拒绝在面临毁灭崩溃时沉默不言。"正是在这一背景下，本书的作者们决定站出来说点什么。

今天的城市问题可以被比作一场疾病，迄今人们给出的治疗方案都只针对症状，而不考虑长期的系统问题或它们的根本原因。任何想要单独缓解交通、污染、废弃物和其他城市发展问题的意图，总是会引发不良、有害的副作用。甚至更糟，他们可能还会给出错误的希望，声称已经找到问题的解决之道，而这些问题如果仅仅用传统的方法去应对，实际上依然是不可能解决的。数十年来的局部治疗方案已经很清楚地说明，针对这些问题需要一个更全面的处理方法。城市依然深受这些问题带来的困扰，而且问题只会更大、更严重，并持续使之衰弱下去。

在这本书中，我们提出一个可供选择的城市模式，其出发点是整合各种基本的可持续性原则。但我们并未给出实际建设过程中的太多基本详情，尤其是一个技术解决方案的完整清单，不过，我们确实相信，社会已经具有或者很快会具有解决这些问题的技术手段。我们只是为城市生活提供一种方式而已。通过反复实验与分析，我们完善了雏形中的一些关键理念，通过总结教训，我们相信，这将会成为一个崭新的、真正可持续发展时代的开端。

有一个坚定的信念引领着我们的行为，我们称之为新城市模式的8-8-8宣言，它承认城市对于人类未来的绝对重要性。到2050年，预计十个人中有八个人会生活在城市。由于城市越来越成为人们关注的焦点和全球发达及发展中国家的中心，环境、社会和经济上决定性的战役将在城市打响，无

own needs." The American Institute of Architects states the case for sustainability more narrowly, and grimly, as "the ability of society to continue to function into the future without being forced into decline through exhaustion or overloading of the key resources on which the system depends."

Despite convincing evidence that grows daily, some continue to deny climate change, just as others, in the face of irreversible damage, continue to charge ahead, wantonly ravaging the rainforests or discharging toxins into the atmosphere. Still others, whether as a result of ignorance, lethargy, politics, or greed, are criminally negligent in the consumption of Earth's natural resources, depleting supplies at a rate that far outpaces their capacity to be replenished. It makes no sense. We are attacking and killing the living systems of which we ourselves are part and upon which our very survival depends.

Humanity is at a critical juncture if there is any hope of moving forward. According to Al Gore in his Pulitzer Prize–winning book *Earth in the Balance*, "[W]e now face the prospect of a global civil war between those who refuse to consider the consequences of civilization's relentless advance and those who refuse to be silent partners in the destruction." It is in this context that the authors have determined to speak up.

If today's urban problems can be likened to an illness, the therapies prescribed to date have targeted only the symptoms without treating the long-term systemic problems or their root causes. Isolated attempts to alleviate traffic, pollution, waste, and other urban growing pains have sometimes triggered undesirable or harmful side effects. Perhaps even worse, they can provide false hope about having found a cure for conditions that remain incurable if combatted only by traditional means. As decades of attempts at local remedies have made clear, a more holistic approach is required. Cities still suffer the same problems, only bigger, more serious, and increasingly debilitating.

We offer in this book an alternative urban form that incorporates the basic principles of sustainability as its starting point. In presenting this theoretical construct we have not addressed the many essential particulars of actual construction, least of all a complete inventory of technical solutions, although we do believe that society already has, or will have very soon, the technical means for implementation. Ours is but one approach to urban living. By perfecting the key ideas behind our prototype through trial-and-error testing and analysis, and by applying the lessons learned, we believe that

In 1979, the Pearl River Delta was largely undeveloped (top). The region has been radically transformed by paved roads, industry, and urbanization (all of which appear gray in the lower false-color satellite photo from 2003). Development has also caused a decline in rainfall, since the built environment prevents water retention in the soil, leading to less humidity and therefore less precipitation.

1979年珠三角地区还未大规模开发（上）。由于道路铺设、工业和城市化，这一地区的样貌已经发生了翻天覆地的变化（在2003年像素较低的卫星图片中呈现为灰色地带）。开发也导致了这一地区雨量下降，因为建筑环境令水分难以保存在土壤中，所以湿度降低，降水量下降。

论输赢，我们没有任何理由不去关注这些问题。因此，在提出一个针对当前城市问题的解决方案时，我们不得不首先给出一种对可持续性将来的愿景。我们秉承大胆而又适度的崇高期望，通过一种能够激励人们心灵的图景来完成这项工作。

this could well be the beginning of a new era of truly sustainable development.

Our actions are guided by firm convictions, what we call our 8-8-8 Manifesto of New Urban Form, which recognizes the critical importance of cities for the future. By 2050, it is projected that 8 of every 10 people will be urbanites. And since cities are the growing focus of human activity and the heart of developed and developing nations worldwide, it is in cities that crucial environmental, social, and economic battles will be fought and won, or lost. There is no cause that is more deserving of our attention, and certainly none more important in outcome. We thus feel compelled to set forth our vision for a sustainable future as a first step in advancing a solution to contemporary urban problems. We do so with a boldness befitting humanity's highest aspirations and an evocative image capable of stirring its soul.

Locally grown produce at market.
市场上本地种植的蔬菜。

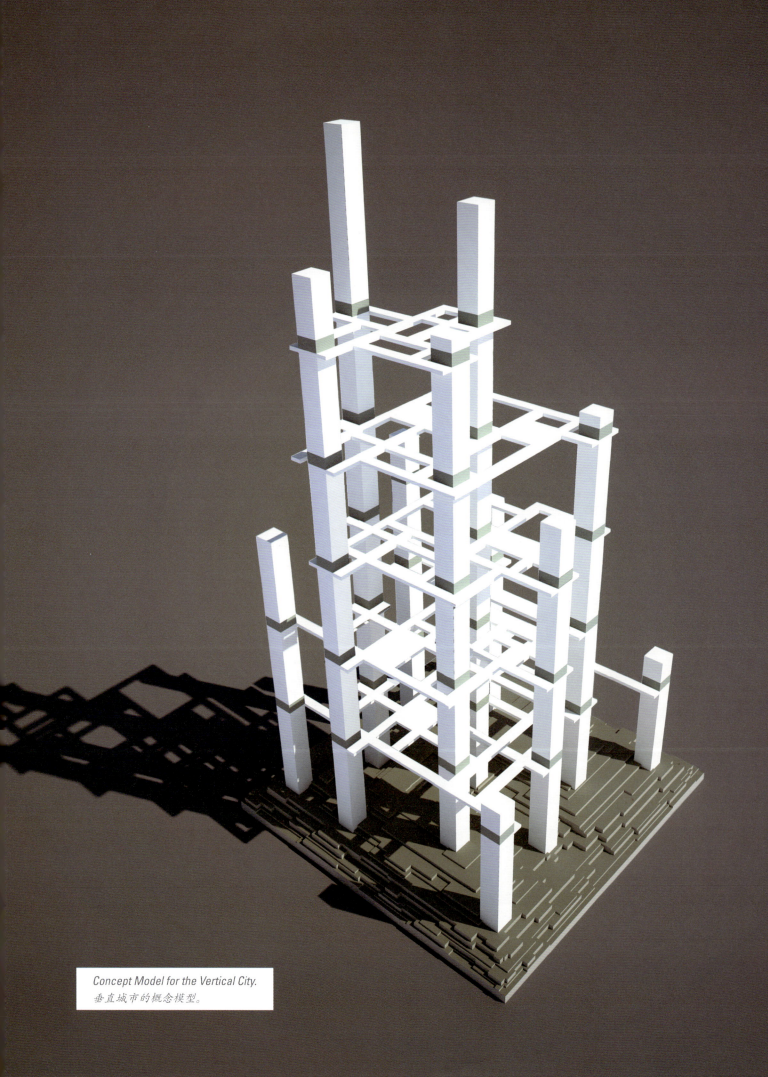

Concept Model for the Vertical City.
垂直城市的概念模型。

垂直城市的愿景

Vision for a Vertical City

上述宣言产生了一个新的垂直城市的愿景，它要求抛弃当前碎片式的城市规划方法——二战之后人们对廉价化工燃料、汽车占有率、郊区独栋住宅的思维定势产生了这种规划方法，这一宣言也激发了一种崭新的关于垂直城市的愿景。这是对于一种新范式的诉求。这种范式能回应空前的人口数和农村向城市迁移潮流带来的紧迫问题。

解决方案在于：通过贯彻垂直城市理论优化人口密度，也就是说，大容量、高效能的超高建筑位于相对较小的没有车辆、方便行人的土地上。在这片区域内，所有自足的基础设施、建筑、设备与服务都是为了提升居民的生活、工作、文化、娱乐、体育、休闲等方面的生活质量。

在我们看来，垂直城市在一个抬高的多层裙楼区上搭建起来。公共服务车辆和私人车辆的流线和停放，位于裙楼以下的地面层，因而裙楼区表层即垂直城市的主平面可以只留给行人和自行车。

在裙楼区中，地上第一层用于基础设施服务，放置水源、下水道和电力线路的设备，还有一些机械、维护、储藏和设备。

位于裙楼区顶部的是通向所有建筑的主入口（街道）。要么是一个稍低的（6-10层）、封闭式、类似购物中心的大型建筑；要么就是多个建筑沿着室外街道走廊排列，进而构成一个更易辨识的街区样式。垂直城市意味着关注整体：各种高度的垂直大楼（1公里高？1英里高？）。大约每100层大楼就相互联结起来，或者通过每个电梯换乘空中大堂连接，或者由结构稳定性的要求来决定连接部位。高高架空的水平行人天桥便于人们在不同高楼之间自由移动。

空中大堂及其支架/桥梁联结（它们扩展了空中大堂）形成了垂直的"村落中心区"。这将是购物、餐饮、社交、闲聊或消磨时光的自然聚集点。当地的艺术家、手艺人、表演家和小摊贩将在这里展示他们的产品。更多精力旺盛的居民将选择在垂直城市众多的游泳池和健身中心锻炼身体，或者在空中大堂延伸的景观网络中，在小公园和小花园中进行散步、慢跑、滑板滑行、骑自行车等活动。

The vision of a new Vertical City inspired by the above Manifesto calls for a shift away from the current piecemeal urban planning approach that is rooted in the environmentally harmful, energy inefficient, and unsustainable post–World War II mentality of cheap fossil fuel, car ownership, and single-family suburban homes. It is a call for a new paradigm to address the urgent problems of unprecedented human population totals and tidal waves of rural migration into cities.

The solution lies in the optimization of population densities by implementing the theory of a Vertical City, that is, high-capacity, high-efficiency ultra-tall buildings occupying a relatively small car-free, pedestrian-friendly parcel of land. Within this footprint are all the self-sustaining features of infrastructure, buildings, facilities, and services necessary for improving the living, working, cultural, entertainment, sports, recreation, and leisure qualities of life for residents.

In our mind's eye, the Vertical City starts with a raised multilevel podium. Service vehicles and privately owned cars are relegated to circulation and parking on the ground level below so that the surface of the podium, the main level of the Vertical City, is reserved exclusively for pedestrians and bicycles.

Within the podium, the first level above the ground is reserved for utilities and services, housing the infrastructure for water, sewer, and electrical lines together with mechanical, treatment, and storage plants, and equipment.

Located on the top of the podium are the main ("Street") entrances to all buildings. Included is a low (6- to 10-story) enclosed mall-like megabuilding or, alternatively, a more recognizable street pattern of multiple buildings arrayed around outdoor street corridors. Focusing the whole is the namesake of the Vertical City: vertical towers of various heights (one kilometer? one mile?). The towers are connected approximately every 100 floors, at each elevator-transfer sky lobby, or as may be dictated by requirements for structural stability. Elevated horizontal pedestrian bridges facilitate movement between towers.

The sky lobbies together with their bracing/bridge connections-extended sky lobbies, as it were, comprise vertical "village centers." These will be natural gathering places for shopping, eating, drinking, or socializing. Local artists, craftspeople, performers, and street vendors will display their skills or wares here. More energetic occupants may opt for a workout in the Vertical City's numerous swimming pools and fitness centers, or perhaps take a casual stroll, jog, skateboard, or cycle through the extensive network of landscaped extended sky lobbies, vest-pocket parks, and outdoor gardens.

Surrounding the Vertical City are open farmlands, which serve as a buffer between existing urban centers and future Vertical Cities. The greenbelts will be of sufficient size for growing vegetables and fruits and will also house fisheries to provide residents with a nutritious local source of food.

　　垂直城市周边是空旷的农业用地，它是现存城市中心与将来垂直城市之间的缓冲地带。绿化带面积要足够大，以便种植蔬菜和水果，同时还可以用作渔场，这样可以为居民提供营养丰富的本地食物。

人类发展与幻想的艺术概念

Artist Concept of Human Development and Vision

Reinhard Roy 关于垂直城市的幻想

这是关于一个未来城市的幻想项目。我以一个艺术家兼设计者的身份作为本书作者的顾问。我与这个项目的结缘得益于我的朋友，纽约的建筑师金世海先生。

如同本书书名所指的，这不是一个与一般城市规划等而视之的概念。它顾及到城市概念所包括的所有必要功能和设施，要让社区能够可持续生存，并兼顾所有的技术与社会功能。

这个问题的答案同时也必须兼顾人类当前不受控制的增长，及其与全世界的协同效应。

A Vision of Vertical City illustrations by Reinhard Roy

This artwork is about a visionary project of future cities. With my experience as an artist and designer, I have created this work as an adviser to the authors. My connection to this project is my friend and New York architect, Kenneth King.

Like the designation already says, it is about a concept which cannot be associated with conventional urban planning. It concerns the concept of a city which must include the necessary functions and complexities which allow for sustainable living within a community, as well as incorporating all of the technical and social components.

This answer must also consider our problems of the uncontrolled growth to this day of mankind and the effects connected with it on the world as a whole.

Evolution 进化

Prehistoric
史前

Agriculture
农耕

Commerce
商业

Industry
工业

Information Age
信息时代

ARTIST CONCEPT OF HUMAN DEVELOPMENT AND VISION

Artist's Vision of Vertical Cities 艺术家想象的垂直城市

Artist's Vision of Vertical City
艺术家想象中的垂直城市

Frank Lloyd Wright's Proposed Mile High Building 1956
弗兰克·劳埃德·莱特建议的一英里高的建筑，1956年

Beijing, with its distinctive ring roads and grid pattern, appears at the bottom of this 2010 satellite photograph. A half-hour high-speed train ride connects the capital to the port city of Tianjin, 122 kilometers [75 miles] to the northwest. In 2012 the population of the twin-city megalopolis totaled approximately 30 million.

北京，以其独特的环形道路和格子图案，出现在这张2010年的卫星照片的底部。从首都乘坐半小时的城际列车即可到达距其122公里（75英里）的港口城市天津。2012年，这两个城市的人口总和约3000万。

沉睡巨龙的觉醒

21世纪初，吸引世界目光的是金砖四国（巴西、俄罗斯、印度和中国）惊人的城镇化进程，在2030年第一季度，这些国家的经济总量有望超过了G7（七国集团：加拿大、法国、意大利、日本、英国、美国和德国）。探究这些新兴国家成功故事背后的细节，这显然会很有挑战性，有益且令人兴奋，不过我们的热情受制于展开探究需要的时间和资源。即使我们很早就开始对印度和巴西展开研究，但还是感到我们必须专注于最熟悉的东西——至少现在是这样。

作为美籍华人，这本书的作者不仅与中国有着亲缘关系，而且还有着特殊的机会亲身观察这个国家的重大转变——这一转变实际上从这条沉睡的巨龙在上世纪70年代末向西方打开大门的时候就已开始，直到现在。另外，我们觉得，有必要把我们的记忆与观察作为一种对历史的记录与大家分享，希望其他人能够理解，过去发生的变化是多么巨大，并且在这一理解的过程中更好地领会我们所面临的挑战。

金世海（Kenneth King）的中国印象

我的祖籍在中国东部安徽省的休宁县，19世纪中叶，我的曾祖父举家迁到了东北方向450公里（280英里）外的上海。我的曾祖父和祖父都不会想到，他们的后代现在散布在世界的各个角落。

20世纪30到40年代，我在上海度过我的童年，生活异常简单。那时没有冰箱，大家每天都到集市去。人们喜欢新鲜的食物，但却根本不会意识到，享用这些新鲜的、不含化学添加剂却富有自然滋味的食物有什么特殊的乐趣。

我们的家在胡同和狭窄到不通汽车的里弄边，下了学，孩子们可以在那里玩耍。特别是在非常炎热的上海夏夜，胡同还有另一个重要的功能，当时没有空调，屋里太过潮湿，无法令人安睡，所以大家都会带着小凳子走出家门，给自己扇扇风，凉快凉快。每个人都会聚集在门口的胡同里，或是闲聊，或是给小孩子讲故事。我还记得那时听着大人说话，看着满天星空的情景。今

The Sleeping Dragon Awakes

Since the turn of the twenty-first century, the world's attention has been drawn to the phenomenal urbanization of BRIC countries (Brazil, Russia, India, and China), whose combined economies is expected to exceed those of the G7 (The Group of Seven: Canada, France, Italy, Japan, United Kingdom, United States, and Germany) as of the first quarter of 2030. While it would surely be challenging, informative, and indeed exciting to explore the details behind the success stories of all these emerging nations, our enthusiasm has been constrained by the practicalities of available time and resources. Even though we undertook early investigative forays into India and Brazil, we felt compelled, at least for the moment, to concentrate on that with which we are most familiar.

As Chinese Americans, the authors of this book have not only ancestral ties to China but the unique opportunity to observe firsthand the country's extraordinary transformation, virtually from the moment the Sleeping Dragon opened its doors to the West in the late 1970s, right up to the present. In addition, we feel obliged to share some of our recollections and observations as a matter of historical record, hoping that others can understand just how momentous the changes have been and, in the process, perhaps better appreciate the challenges that lie ahead.

Kenneth King's impressions of China

My family was originally from Xiuning County, in eastern China's Anhui Province, but in the mid-nineteenth century, my great-grandfather moved to Shanghai, about 450 kilometers [280 miles] further north and east. Neither my great-grandfather nor my grandfather could ever have imagined that their descendants are now settled all over the world.

During my own childhood in Shanghai in the 1930s and '40s, life was simple. Since there were no refrigerators, everyone went to the market every day. People enjoyed fresh food without any conscious awareness of the singular pleasures of eating fresh, chemical-free foods with great natural taste.

Homes were organized around *hutongs*, or narrow car-free lanes, where children played after school. The *hutongs* had another important function, especially on summer nights, which can get very hot in Shanghai. Without air-conditioning it was too steamy to go into the house to sleep, so whole families would bring small stools outside and fan themselves to feel a little cooler. Everyone would gather in the *hutong* in front of their house, and chat and tell stories to their children. I remember, while listening to the adults, looking up and seeing the sky full of stars. That's no longer possible today.

At the end of the World War II, peace finally came after eight years of fighting. The whole country was exuberant. American goods and servicemen flooded into town. This is the first time we had a close look at Westerners. The British and the French, of course, had been in Shanghai for years, but they mostly confined themselves to their enclaves and clubs. In contrast, the Americans were very open and moved about freely all over the city. Although they didn't speak the language, they seemed very accessible.

Postwar euphoria was short lived. As discussed in my biography (see Appendix), the Nationalist Party headed by Chiang Kai-shek and the Communist Party headed by Mao Zedong became pitted against each other in civil war. The communists ultimately took over China in 1949, shortly after which my family moved to Hong Kong. I finished high school there and then studied architecture in London and in the States, and eventually settled in New York.

Meanwhile, Mao Zedong launched the Cultural Revolution in China in 1966. Five years later, when the U.S. Table Tennis team was in Japan for a world tournament, China invited the American team to visit. It was an important breakthrough, as the two countries had been hostile for decades. Thus began Ping Pong diplomacy. It culminated in the historic trip to Beijing in 1972 of President Richard M. Nixon, the first American leader ever to visit the People's Republic of China. At the conclusion of his visit, Nixon and Premiere Zhou Enlai signed the *Shanghai Communiqué*. While there were a number of significant issues about which the two sides did not agree, the *Communiqué* nonetheless established the principles of normalization between the United States and the PRC, and dramatically reversed American policy toward China.

Nixon's visit opened the door that had been closed since 1949. Still, it wasn't until 1979, three years after the death of Chairman Mao, that diplomatic relations were officially put in place. Along the way, President Gerald Ford visited Beijing in 1975, and in 1977, shortly after taking office, President Jimmy Carter again reaffirmed interests in normalization. President Carter then dispatched Zbigniew Brzezinski, his National Security Advisor, to Beijing to lay the actual groundwork.

天，这些都不再可能了。

第二次世界大战末期，在八年抗战之后，和平终于降临了。整个国家充满生机。美国货和美国军事人员涌入城市。这是我们第一次近距离看到西方人。当然，英国人和法国人已经在上海待了很长时间了，但他们大多数时候都将自己关在租界和俱乐部里。相反，美国人很开放，在城市各处游荡闲逛。尽管语言不通，但他们看起来很是友善亲切。

战后的兴奋并未持续很长时间。就像在我的传记中所讨论的（见附录），蒋介石领导的国民党和毛泽东领导的共产党在内战中展开厮杀。1949年，我家搬到香港不久后，共产党最终接管了中国。我在香港完成了高中学业，然后在伦敦和美国学习建筑学。一切自然而然，后来我结婚生子，定居于纽约。

与此同时，1966年，毛泽东在中国发动了文化大革命。五年后，当美国乒乓球代表队在日本参加世界锦标赛时，中国邀请美国代表队前往参观访问。两个国家对峙数十年后，这真是一个重要的突破。由此，乒乓外交拉开序幕。它促成了1972年尼克松总统前往北京的历史之旅，这是美国领导人第一次访问中华人民共和国。访问结束的时候，尼克松总统和周恩来总理签订了《中华人民共和国和美利坚合众国关于建立外交关系的联合公报》（简称《上海公报》）。尽管双方就很多重要议题并未达成共识，《上海公报》还是确立了中美之间外交关系正常化的原则，并且戏剧性地改变了美国对中国的政策。

尼克松的访问打开了中国自1949年以来关闭的大门。不过，在毛泽东主席去世三年之后的1979年，中美外交关系才正式确立。在这个过程中，福特总统于1975年访问了北京，而1977年，在就职后不久，卡特总统就再次重申了对外交关系正常化的兴趣。之后，卡特总统委派他的国务卿布热津斯基到北京完成了实际工作。

1978年，邓小平副总理开始了中国的四个现代化工程：农业、工业、国防和科学技术现代化。由此，他使这个国家进入持续到今天的高速发展阶段。

1979年1月1日，美国正式承认中华人民共和国，双方正式建立外交关系。很快，邓小平就接到访问华盛顿的邀请。1月底，他到达美国，并与卡特政府进行了深入的会谈。

2月1日，白宫在乔治亚州亚特兰大的桃树广场酒店举办午宴欢迎副总理先生。通过一位与民主党全国委员会联系密切的朋友，我们被邀请参加了午宴。伍德科克大使和卡特顾问团的很多人都在那里。现场气氛轻松，而我们的中国客人看起来也很自在。在这次活动中，中美两国数十年的政治恩怨似乎都化解了。

我意识到，中美双方都非常认真对待协议，作为一名华裔，我对中美建交非常高兴。我迫切想要知道中国的发展方向，而它的发展模式将会表现为持续向前。

那个时候，到中国去和在中国国内旅行都很困难。人们可以经由香港，穿过边境到达深圳，从陆路进入中国，或者如我有时做的那样，乘飞机前往布加勒斯特，然后换乘罗马尼亚航空公司的飞机飞往北京。当然，还有其他的办法，但所有办法都很麻烦。

1979年，在亚特兰大的午宴结束后不久，我第一次回到中国。在我离开的30年里，一切几乎没怎么改变。即使在北京和上海，饭店也很少。唯一可以像样吃一顿饭的地方就是在宾馆里，而从西方的标准来说，这些宾馆又太过朴素了。在当地人光顾的几个小餐馆里，没有配给票券是买不到米饭或面条的。

和今天不一样，那时马路上几乎没有私人汽车。少数几辆是美国人留下的，包括道奇、福特和克莱斯勒；新一点的款式一般都是苏联制造的。不管怎么说，汽车是外国的奢侈品。大多数人都通过公共交通系统或自行车去上班。生活脚踏实地，有一种美妙的单纯。在中国，那时几乎没有犯罪。相机或其他值钱的东西落在宾馆不会有什么问题，直到失主重新找到它们。不过，这段天真无邪充满信任的时期并没有持续太久。

我在北京和上海都没有看到什么建设项目。大多数房子和我1949年离开中国大陆时没什么两样，甚至我家在上海以前住的房子也是如此。当然，现在大多数早期的居民区都被拆除，取而代之的是现代高楼。

作为第一批经济特区的深圳和其他港口城市

In 1978, Vice Premier Deng Xiaoping launched the Four Modernizations program - Agriculture, Industry, National Defense, Science and Technology. By this, he set in motion the explosive development of China that continues to this day.

On January 1, 1979, the United States formally recognized the People's Republic of China, and officially established diplomatic relations between the two countries. Immediately afterward, Deng Xiaoping accepted an invitation to visit Washington. He arrived at the end of the month for further discussions with the Carter administration.

On February 1, the White House hosted a luncheon in honor of the Vice Premier at the Peachtree Plaza Hotel in Atlanta, Georgia. We were invited to attend by a friend closely associated with the Democratic National Committee. Ambassador Leonard Woodcock was there, as were several members of Carter's cabinet. The atmosphere was informal and the Chinese guests look relaxed. Decades of political animosity between the two countries seemed to melt away in this one event.

I realized that China and the United States were serious about accord and, being a natural-born Chinese, I was happy for the establishment of Sino-U.S. relations. I was anxious to see the direction China would take and the form its development would assume in moving forward.

In those days, transportation into and within China was difficult. One could enter by land via Hong Kong and walk across the border to Shenzhen or, as I sometimes did, take a flight to Bucharest and change to a Romanian airline flight into Beijing. No doubt there were other ways, all of them difficult.

My first trip back to China took place in 1979, shortly after the luncheon in Atlanta. Little had changed in the thirty years I'd been away. There were very few restaurants, even in Beijing and Shanghai. The only place one could get a decent meal was at the hotel, which itself was pretty modest by Western standards. Of the few eateries for local people, none were permitted to serve rice or noodles without rationing coupons.

Unlike today, there were almost no private cars on the road. What few there were had been left behind by the Americans, including Dodges, Fords, and Chryslers; newer models were usually Soviet-made. Either way, cars were a foreign luxury. Most people went to work by public transportation or bicycle. Life was down-to-earth and had a kind of wonderful simplicity. There was little or no crime in China. Cameras or other valuables left unattended in hotels were safe until retrieved by their owners. However, this period of innocence and trust did

The Sheung Yue River separates Shenzhen from Hong Kong. Prior to 1979 Shenzhen was a small border fishing village with 25,000 residents. Declared a Special Economic Zone (SEZ), one of the first in China, the population of Shenzhen has grown to more than 15 million in 2012.

双鱼河分隔深圳香港。1979年之前，深圳还是一个边陲渔村，拥有25,000名居民。在中国，深圳是最早的特别经济区之一，到2012年，深圳的人口已超过1500万。

的成功，给了中国在更多沿海城市建立经济特区的信心。最终，它们延伸到了东北沈阳这样的内陆地区。经济特区及其带来的经济振兴，成为唤醒沉睡巨龙的第一次强烈震动。

自1979年开始，我每年至少回中国两到三次。我知道这个国家正在发展，但其发展的范围和速度总是令我吃惊。作为一个建筑师和华裔，我非常乐于为中国的不断发展做出力所能及的贡献。

80年代初期，一个相熟的亲戚让我招待前来美国交流访问的一群沈阳设计院的工程师。他们唯一的兴趣就是学习美国的高层建筑设计和建设。我很乐于帮忙，在一些朋友的帮助下，我为他们安排了一个恰当的行程表。

六名工程师从疲劳的旅途中（那时从中国出发，多站停靠的旅途是非常辛苦的）恢复过来后，我们参观了贝聿铭建筑设计事务所的办公室。贝先生和蔼地接待了考察团，并谈到一些正在进行的项目，提出也回答了一些问题。我们的下一站是宾州的利哈伊大学，那里的建筑工程系主任是一位中国教授，他热烈欢迎考察团，并就高层建筑的结构设计问题谈了许多。之后，我们去了芝加哥，在那里参观了斯基德莫尔-奥因斯-梅里尔建筑师事务所。办公室有很多正在进行的项目，而且正在放映一个幻灯片，解释他们设计背后的理念。在拉斯维加斯（我们的客人从未见过这种东西！）短暂停留后，我们去往最终的目的地旧金山。参观了林同炎国际顾问公司，这是一家专长于大跨度结构的工程公司。林同炎先生和考察团谈了大跨度技术，以及抗震设计。20年后，当超高建筑在中国如同雨后春笋般涌现时，我意识到，中国派出这些考察学习团，乃是为国家快速的城市发展在做准备。

许多年之后，我与沈阳的联系依然继续着，有一个朋友将我介绍给当地一位地产开发商。那时，砖还是中国最常见的建筑材料，就像它数千年来所扮演的角色一样，但在传统砖窑中烧制砖的过程，会向空气中排放大量密度很高的黑色碳污染物。而且，这一过程还需要从周边的山上挖出很多粘土。中国的专家意识到这一问题，迫切地寻找一种可以替代砖的建筑材料。

贝瑟公司是密歇根阿尔皮纳的一家混凝土产

not last long.

I saw no construction in either Beijing or Shanghai. Most of the houses remained as they were before I left in 1949, even the house where my family had lived in Shanghai. Now, of course, most of the early residential areas have been demolished and replaced by modern towers.

The success of the first group of Special Economic Zones in Shenzhen and in other port cities gave China the confidence to establish additional SEZs in many coastal cities. Eventually they reached interior regions like Shenyang in northeastern China. The SEZs and the economic stimulus they provided sent the first strong jolt in waking the Sleeping Dragon.

Since 1979, I've been going back to China at least two or three times a year. I know the country is developing, but I am always amazed by the scope and speed at which it's taking place. As an architect and native son, I am interested to contribute whatever I can to help China move forward.

In the early '80s, a well-connected relative asked me to host a group of Chinese engineers from Shenyang Design Institute on a visit to the United States. Their sole interest was to learn American tall-building design and construction. I was happy to oblige and, with the help of several friends, developed a suitable itinerary.

After the six engineers recovered from their trip (in those days, the multistop journey from China was very arduous), we visited the office of I. M. Pei & Partners. Mr. Pei graciously received the group and spoke about a few projects underway, fielding questions and answers. Our next stop was Lehigh University in Pennsylvania, where the head of the structural engineering department, a Chinese professor, welcomed the group with a talk on the structural design of tall buildings. We then moved on to Chicago, where we visited SOM. The office had numerous projects under way and presented a slide show, explaining the ideas behind their designs. After a short stopover in Las Vegas—my guests had never seen anything like it!—we traveled to San Francisco, our final destination. We visited T. Y. Lin International, an engineering firm specializing in long-span structures. Mr. Lin Tung-Yen talked to the group about long-span technology and seismic design. Two decades later, when ultra-tall structures started springing up all over China, I realized the reason China had sent out these study groups was to prepare for the country's rapid urban development.

My connection with Shenyang continued when, a number of years later, a friend introduced me to a local real estate developer. At the time, brick was China's most common building material, just as it had been for thousands of years, but the process of firing in traditional kilns emitted huge quantities of dense black carbon pollution into the air. Also, there were concerns about the vast amounts of clay being excavated from the surrounding hills. Chinese authorities were aware of the situation, and desperately sought a substitute building material.

The Besser Company, a concrete products manufacturer in Alpena, Michigan, had a possible solution. Besser participated in an exhibition in Beijing in the 1980s and introduced a small model Bescopac machine for making concrete block. For whatever reason, whether the timing was too early or the country was just not ready in other ways, the technology had no impact on the Chinese building industry. Yet, to me, its importance was obvious.

In a joint venture with the previously mentioned real estate developer in Shenyang, I supplied a concrete block-making machine for use in developing new local housing. I also introduced polystyrene insulation and thermopane windows, since Shenyang is located in northeastern China and has very cold winters. I worked simultaneously with the Northeast Design Institute on changing building codes in anticipation of the new technologies that would transform the whole building process. Unfortunately, it was difficult to convince people to accept new ideas and methods of construction. Success was thwarted yet again. Of course, today, twenty years later, concrete block, polystyrene insulation, and thermopane windows are used everywhere in China. In fact, China now manufactures concrete block–making machines for domestic use and also for export to many other parts of the world.

While in Shenyang, I had the opportunity to travel by car to nearby towns and cities. We drove mostly on bumpy country roads dotted here and there with small villages. A distance of a hundred miles could easily take an entire day. At one point in the mid-1980s, I needed to travel to Dalian on China's northeast coast and the expressway linking it to Shenyang had just been completed. I remember vividly how few cars there were on the high-speed road and how, on several occasions, we saw oxcarts approaching us against the flow of traffic. Our driver merely slowed down and changed lanes so the carts could pass. I realized that I was, at that very moment, an eyewitness to the merging of nineteenth and twentieth-century China. Today, of course, China has more than 60,000 miles [97,000 kilometers] of expressways, the largest inter-city highway system in the world.

To its great credit, China is not just paving new roads; it is also developing a world-class high-speed rail system. My

品制造商，拥有可能的解决方案。他们参加了20世纪80年代于北京举办的一次展销会，介绍了一种用于制造混凝土砖的小型砖块成型机。要么是因为时机不成熟，要么是因为这个国家还没做好准备，不管是因为什么，这种技术在中国的建筑行业并未产生影响。不过，在我看来，其重要性是显而易见的。

在和一个之前提到的沈阳房地产开发商的合资企业里，我提供了一种混凝土砖制造机，用于开发当地的新住房。由于沈阳位于中国的东北，冬天异常寒冷，我还引进了聚苯乙烯绝缘材料和瑟莫潘双层隔热窗玻璃。与此同时，我与东北设计研究院合作，修改建筑规范，期待着新技术能够改变整个建设的过程。不幸的是，在当时，很难说服人们接受这些新观念和新的建设方法，成功再次离我们而去。当然，20年之后的今天，中国到处都在用水泥砖、聚苯乙烯绝缘材料、瑟莫潘双层隔热窗玻璃。实际上中国现在不仅为自己制造水泥砖制造机，也将它们出口到世界各地。

在沈阳时，我有机会坐车在附近的城镇旅行。大多数时候，我们的车行驶在颠簸不平的乡间小道上。这些道路零零星星地散布在各个小村庄中。一段100英里的路程经常要花上一整天的时间。80年代中期的某个时候，我需要到中国海岸线东北部的大连去，那时连接大连和沈阳的高速公路刚刚完成。我非常清楚地记得，高速公路上的车是那么少，我们经常看到牛马车逆着车流向我们奔来。我们的司机只是减速变向让牛马车可以通过。在那一刻，我意识到，我亲眼见证了19世纪中国和20世纪中国的融合。当然，今天中国已经有超过60,000英里（97,000公里）的高速公路，这是世界上最大的城际高速公路系统。

可以肯定的是，中国现在不仅在铺设新的道路，也在发展一个世界级的高速铁路系统。2011年在北京-天津之间的列车上，我第一次感受到这一点。110公里（70英里）的路程开车一般需要一个多小时，但我们乘坐高铁，30分钟就到达了目的地。与此类似，从北京到上海，超过1000公里（620英里）的路程只需要花5个半小时。如果两个城市之间的距离小于1500公里（930英里），坐火车实际上比坐飞机更节约时间。

In 2012, China realized its connection plans with four north-south and four east-west high-speed rail lines, accommodating 90 percent of the population. The 8,000-mile [13,000-kilometer] network is expected to grow nearly fourfold by 2020.

2012年，中国实现其有四个南北和东西高速铁路线的连接计划，可容纳90％的人口。8,000英里（13,000公里）的网络，到2020年预计将增长近四倍。

first experience with it took place in 2011 between Beijing and Tianjin. The trip of some 110 kilometers [70 miles] would normally take over an hour by car but we arrived in thirty minutes by train. Similarly, the trip from Beijing to Shanghai, a distance of over 1,000 kilometers [620 miles] now takes just five and a half hours. It actually takes less time to travel by rail than by air to cities less than 1,500 kilometers [930 miles] away.

Racing from one city to another at speeds of 380 kilometers [240 miles] per hour, there is evidence everywhere of construction moving further and further into the countryside. China's building boom is unmatched in history. Over the past twenty-five years, thirty-two of the world's 100 tallest buildings were erected in the PRC, and more are planned or under construction. Each year, urbanization consumes roughly 70,000 square kilometers [27,000 square miles] of productive farmland in China alone. The world cannot sustain such losses. Something must be done to abate this terrible trend spiraling out of control .

Another area where I've seen incredible change is in communications infrastructure, which is very important for the development of business and industry and, more generally, for connecting people across China's huge land mass; the Chinese have made tremendous investments in telecommunications in preparation for future long-term development. I remember first noticing business people starting to carry mobile phones in the early 1990s. The devices were heavy and bulky, much too big to fit in a pocket, and looked more like military-issued walkie-talkies from World War II. But it was nonetheless a proud status symbol to own one. Today, there are over a billion mobile phones in China, the greatest number of any country in the world, and also more than 500 million computer users (also the most in the world).

As an architect, I believe in science and technology. It has inarguably improved the quality of life for people, not only in China but across the world. It enables us to travel and communicate with people all over the globe. We can jump in a car and go wherever we want, day or night, just as we enjoy many other conveniences. But all of this comes at a price: a polluted world, with poisoned oceans and waterways, and the extinction of irreplaceable life forms.

Urbanization and its attendant problems are not restricted to China. Many of the same patterns are taking place in countries such as India, Pakistan, Thailand, Indonesia, and Bangladesh, as well as in Japan, Brazil, and also Nigeria. Even Russia, with its huge sprawling land mass, is now

以每小时380公里（240英里）的速度从一个城市飞奔到另一个城市，你自然会看到，到处都在开发建设，而且这种势头越来越向乡村深处蔓延。中国的建设浪潮是历史上任何一个阶段都无可比拟的。在过去25年间，世界100座最高建筑中，有32座在中国屹立起来，而且还有更多正在规划和建设中。仅仅在中国，城市化每年就要消耗大约70,000平方公里（27,000平方英里）的可耕农田。而世界不能承受这种损失。我们必须做点什么来缓解这种可怕的失控倾向。

我所看到的另一个令人难以置信的改变领域就是通信系统，对于工商业的发展而言，更笼统点说，对于在中国广袤土地上进行联络的人而言，这非常重要；为了适应将来的长期发展需求，中国已经在通信行业投入了巨资。我记得在上世纪90年代初，我才第一次注意到商人开始携带手机。那时，这种设备既重又笨，根本不适合放在口袋里，而且看起来像是二战以来发展出的军用对讲机。但毫无疑问，拥有一台手机是身份的象征。今天，中国有超过10亿部手机，它成为全球手机拥有量最大的国家，另外，它还有5亿的计算机用户（同样也是全球最多的）。

作为一名建筑师，我相信科学和技术已经无可置疑地提高了人们的生活质量，不仅在中国，在全世界都是如此。它使我们能够旅行，能够同全世界的人们进行沟通交流。我们可以跳上车，去任何我们想去的地方，不管是白天还是黑夜，就像我们享受其他许多种便利一样。但所有这些都有代价：一个被污染的世界，其海水和河道被玷污，无可取代的生命形态消失灭绝。

城市化及与之相伴的问题并非只有中国才有。在印度、巴基斯坦、泰国、印度尼西亚、孟加拉国，以及日本、巴西和尼日利亚，正在发生着许多相同的事情。甚至是幅员辽阔的俄罗斯，现在也面临着城市化问题，尤其是在莫斯科。

沉睡的巨龙正在觉醒，世界上其他的发展中国家也是如此。爱因斯坦有一个很著名的说法："你不能用一种引发问题的思考方式去解决问题。"急速的发展一直与地球支撑这种发展的能力成反比。很显然，我们需要新策略、新思维、新方法来实现可持续的城市生活。我们关于垂直城市的观点就是这样的一种建议。

黄慧生的切身体验

黄慧生的切身体验开始于贝聿铭建筑设计事务所办公室中一个小型建筑/规划团队，1979年，为了应对将来西方游客的大量涌入，他被邀请到中国给宾馆饭店选址。有一位客户是香港的开发商，他像众多"外来者"一样，意识到中国的商机，正跃跃欲试，焦急地等待着发令枪响起。

那时候，要进入中国必须通过深圳乘坐火车才可以，如今，深圳已经是一个有1500万人口、熙熙攘攘的大城市了，它还是中国最大的集装箱港口之一，但在那时，它只是与香港交界的一个小乡村，也正是在那里，旅客被要求换乘火车并办理入境手续。人们根本想不到这个在边境墙后面几乎看不到的小乡村，很快被确定为中国的实验性经济特区（SEZ）之一——戒备森严的围墙加上封闭的现代工厂建筑、办公区域，以及在中国其他地方不存在的基础设施——直接通向铁路或水路的运输系统。深圳以及其他三个实验性经济特区承诺有廉价劳动力和资源，政府税收、关税上的重大优惠等等。他们想要通过这些，鼓励和促进外国投资和技术进入中国。与深圳一样，其他三个经济特区的分布也是战略性的，以便让居住在紧邻中国的"海外"华人最容易找到它们——这些地方主要是那时依然处于葡萄牙控制下的澳门。

顺道乘坐汽车进入乡下是缓慢、遍遍、吵闹、颠簸的，为了保证汽车的通行，喇叭总是响个不停，路上挤满了背着重担的农民和他们的水牛，自行车则因为负载过重而摇摇欲坠。总是能看到喷着黑烟、砰砰作响、由车把控制的双轮拖拉机，车上堆满了稻草、刚摘下来的蔬菜、家禽或人，还能看到所有东西都挤在一起的推车。这样一趟旅行，将辛劳的农民以及他们的生活都清晰地呈现了出来。汽车中的乘客与街上农民的距离如此接近，他们完全身陷其中，这种生活节奏成为令人难以忍受的一部分。

在早些时候，从火车或汽车上看过去，一望无际的农田让人确信，中国是一个农业国家。但事情就是这样：绿色田野中的疤痕向人们诉说着改变：表层土壤被翻出来运到附近的窑中烧制成红

encountering urban problems, particularly in Moscow.

As the Sleeping Dragon awakes, so does the rest of the developing world. Albert Einstein famously once said that you can't solve a problem with the same thinking that created it. Rapid development continues in direct inverse proportion to Earth's ability to support it. Clearly, we need new strategies, new thinking, new approaches to sustainable urban living. Our idea for a Vertical City is one such proposal.

Kellogg Wong: Eyewitness Experience

Kellogg Wong's eyewitness experience began as part of a small architectural/planning team from the office of I. M. Pei & Partners, invited to China in 1979 to scout hotel sites for an anticipated influx of Western tourists. The client, a Hong Kong–based developer, was among the many "outsiders" who recognized the business opportunities in China, anxiously toeing the line waiting for the starter's gun to go off.

Back then, China was entered by train via Shenzhen, today a bustling metropolis of some 15 million people and one of China's busiest container ports, but at the time it was no more than a small farm village on the border with Hong Kong, where mandatory train transfers were made and immigration procedures conducted. It made little impression that the village, not visible behind the border-crossing walls, had recently been earmarked as one of China's experimental Special Economic Zones (SEZ)—heavily guarded walled compounds containing modern factory buildings, office spaces, and infrastructure not yet in existence elsewhere in China, with direct access to rail or water transportation systems. Shenzhen and three other experimental SEZ came with the promise of inexpensive labor and resources, and major government concessions regarding taxes, custom duties, etc. Together they were intended to encourage and facilitate the introduction of foreign investment and technology into China. Like Shenzhen, the other three SEZ were strategically positioned for high visibility among "overseas" Chinese living in close proximity to China, notably Macau, then still under the control of Portugal.

Side trips by car into the countryside were slow, dusty, noisy, bone-jarring affairs, with constant horn honking to clear a right-of-way on roadways clogged by heavily burdened farmers with their water buffalo, and bicycles teetering from overload. Always there was the smoke-spewing, putt-putting of the two-wheeled mechanical tiller/tractor, steered by handlebars and hitched to a cart overflowing with straw, freshly

58

picked vegetables, farm animals, or people, all thrown together. Such journeys put the hard-working farmers and their lives on clear view and in such close proximity that passengers in the car were completely caught up and made part of the grinding pace of life.

In those early days, the mile after mile of farmland seen from train or car left no doubt that China was an agrarian country. But invariably, raw scars on the green countryside signaled change: shallow pits resulting from topsoil being removed to nearby kilns to make red brick (traditionally the major construction material for bearing walls and interior partitions alike). One can only imagine the sense of loss experienced by the farmer as he watched the eradication, shovel by shovel, of acres of arable land—the irreplaceable legacy of countless generations.

Wong and his team visited many of the major cities that foreign tourists would potentially find interesting, including the Pearl River in the south, Guilin on the Li River, and the West Lake region of Hangzhou, Suzhou, and Wuxi. There was Shanghai with its exciting Bund, and Suzhou, the Venice of China, as well as Nanjing in the east, and finally, in the north, Beijing, the capital of China and home to the historic Forbidden City, Temple of Heaven, Great Wall, Ming Tombs, and, of course, pandas. All of these cities, large and small, basically had remained unchanged for years—centuries—untouched by developments in the West.

Local authorities were directed by their superiors to "attract foreign investment," but the procedure, the very concept, was new; there was neither experience nor example to follow. It was difficult to distinguish opportunities from obstacles, resulting in many false starts.

The architects were shown sites of every conceivable shape and size, from cabbage patches to irregularly shaped brown sites within cities proper. Many were rejected because of difficult or remote locations, but there were others, such as those in Hangzhou, that were deemed in too close proximity to the very attractions the tourists would have traveled so far to see. Not surprisingly, such explanations were met with puzzlement by the Chinese hosts. How could closeness be considered a negative? But the jarring notion of some big tourist hotel lumbering on the open hillock of the Washington Monument, or encroaching on the Statue of Liberty, was all it took to underline how intrusive, how completely and forever wrong, adjacency of a modern building would be. Unfortunately, other developers would not be as sensitive or caring, with predictable consequences.

Wong remembers standing at the Bund in Shanghai and being told by his hosts as he pointed at the open fields

砖（传统意义上，这是承重墙和类似的建筑内部隔墙的主要建筑材料），由此形成了各种浅坑。人们只能想象农民所体验到的那种失落感，他们看着自己的耕地，看着那些世世代代不可取代的遗产，被一铲一铲地蚕食殆尽。

黄先生和他的团队参观了许多外国人可能会觉得有趣的城市，包括南方的珠江地区，漓江边的桂林以及杭州的西湖地区，苏州和无锡。他们还去了上海令人兴奋的码头，中国的威尼斯苏州，以及东部的南京，最后到了北方的北京——中国的首都，那里有历史上著名的故宫、天坛、长城、十三陵。所有这些大大小小的城市，基本上已经很多年——很多世纪——没有改变过，没有受到西方发展的影响。

当地政府接到他上级的指令，要"吸引外资"，但连操作程序这一概念本身，都是新颖的；没有任何经验，也没有任何范例可供仿效。把机遇和障碍区分开来是很难的，这会导致许多错误的开端。

建筑师们看到了各种可能形状、可能大小的位置场所，从农村的菜园到城市中不规则的各种空地。许多方案被否定是由于开发难度和位置偏远，但也有一些例外，比如说杭州的一些地方则被认为太过接近，而让旅客们没有兴致来参观游览。这种解释让中国的接待方大感费解。距离近怎么会被认为是一个消极因素？但是，这些景致如果过于靠近一栋现代建筑，那将非常不合时宜，它会是完全、永远的不合适。比如：对于华盛顿纪念塔空旷山丘上那些笨拙的大型旅游饭店，或者那些渐渐逼近自由女神像的大型旅游饭店，就十分不和谐。不幸的是，其他开发商并不敏感，也不在乎这些可以预见的后果。

黄先生还记得当他站在上海码头，指着黄浦江对岸的空地时，接待方告诉他，一旦政府开始按计划修建通往浦东的桥梁和隧道，就将有很多开发的机遇，而那时候，他和他的团队可以优先选择任何地方。他清楚地记得那个时候自己满是狐疑，"我此生都不会！"这是完美的地点，但却是错误的时机！

在1979年一次相同的选址之旅中，一个官方代表团带领这个建筑师小团队进行了一小时的故宫之游。在参观过程中，有人提到在故宫北面，钟鼓楼之外有一些空闲的开发场所。黄先生的正式回应是为了保护故宫，及其高大城墙的神秘、宏伟、庄严、壮丽气息，不应该修建什么建筑，因为这些无论如何都会把现代城市的标识强加于故宫与天空之间永恒、本质的联系之上。他认为，从故宫中轴线到周边城墙顶端水平视线都需要得到保护。这一特殊的指导方针被北京官方规划机构所采纳，这一点是令人欣慰的。在2012年初，当审查哪些规划上的问题限制了朝阳区（离故宫不远）一个潜在的综合性规划项目时，黄先生被告知建筑被允许的最大高度是由"贝氏故宫视线"规定的。

1982年贝聿铭建筑设计事务所的香山饭店最终竣工，这是建筑师们辛勤工作的成果。它位于一个树木茂密的前皇家狩猎场，在北京的西北郊区，离颐和园不远。这是中国第一座由美国建筑师设计的大型现代建筑，也是解放后中国第一个超越居住、寻求文化表达的建筑。

还有一次是在1989年，黄先生发现自己正在珠江口寻找能够种植在香港中银大厦花园侧翼的树木。它距离黄先生父亲的出生地，也是中华民国建国之父孙中山的出生地广东中山似乎很近又很远。当黄先生询问小村子的具体位置时，他被告知那个地方已经不存在了，为修建新城市，它被铲平了。如果这是真的，它对于黄老先生而言将是一个令人惊愕的坏消息。在通往现代化的客观进程中，在历史上重要的、或者在人们心中重要的一些地方可能会被囊括吞噬。

上世纪90年代，当黄慧生第一次在250英里新建的沈阳—大连高速公路上旅行时，他终于意识到，新的多车道高速公路穿过那些慢悠悠的乡村时，高速旅行终于成为可能，而这势不可挡的转变才刚刚开始。毫无准备的农民们，依然继续着他们在土地里日复一日，卷起裤管从事的劳动，他们没有意识到自给自足的农业世纪已经终结了，随之而来的是难以想象的生活挑战。

改变的迹象四处可见，在乘直升机从澳门飞往中国最南端的海南岛（整个海南岛都是一个巨大的经济特区）的旅途中，这一点十分明显。黄先生和他的同事们一起看到了垃圾填埋区的清晰

across the Wampu River that development opportunities would open up once the government implemented its plans to build bridges and tunnels to Pudong, and that he and his team could have first choice of any site there. He remembers all too well thinking skeptically, "Not in my lifetime!" Perfect location, wrong timing!

During the same site-scouting trip in 1979, the small group of architects was guided by an official contingent on a tour of the Forbidden City. In the course of the visit it was mentioned that there were development sites available north of the palace, beyond the Bell and Drum Towers. Wong's official response, duly recorded by the note-taking hosts, was that in order to preserve the palace and its massive surrounding walls with their defining air of impenetrability, grandeur, solemnity, and pomp, no buildings should be constructed that would in any way impose signs of the modern city on the timeless and essential relationship of the palace with the open sky. Eye-level sightlines from the central axis of the Forbidden City to the tops of the surrounding palace walls needed to be protected. It is gratifying to know that this particular guideline was adopted by the Beijing planning authorities. In early 2012, when examining the planning restrictions governing a potential mixed-use project in the Chaoyang District, not far from the Forbidden City, Wong was informed that the allowable maximum building height was defined by the "Pei Forbidden City Sight Line."

Ultimately, the fruit of the architects' labor was the completion of construction in 1982 of I. M. Pei's Fragrant Hill Hotel on a wooded site, a former imperial hunting lodge, on the outskirts of northwest Beijing, not far from the Summer Palace. It was the first major building in China of the modern era to be designed by an American architect, and the first building in post-Revolutionary China that aspired beyond shelter to cultural expression.

On another occasion, in 1989, Wong found himself at the mouth of the Pearl River looking for trees to plant in the gardens flanking the Bank of China in Hong Kong. It was close and yet so far from his father's birthplace in Zhongshan in the Guangdong Province, or "Central Mountain," also the birthplace of Sun Yat-sen, the founding father of the Republic of China. When Wong inquired about the disposition of the village, he was told that it no longer existed, having been bulldozed to make way for a new city. If true (substantiation was impossible), it would have been devastating news to the elder Wong. Locations important in history or heart were being subsumed in the impersonal march toward modernization.

Among the first to travel in the early 1990s on the newly

opened 250-mile Shengyang/Dalian Expressway, Kellogg Wong could not but be aware that the high-speed travel made possible by the new multilane highway through the once slow-paced countryside was the beginning of unstoppable transformation. Unsuspecting peasants, still going about their daily grinds knee-deep in mud, were unaware that centuries of agrarian isolation had ended, and with this the coming of unimaginable challenges to their very existence.

Signs of change could be seen everywhere, as was evident during a helicopter ride from Macau to Hainan Island in the southernmost reaches of China, the whole of which was a huge economic zone. No sooner were Wong and his colleagues airborne than they saw the unmistakable evidence of landfill, its serpentine ribbon stretching north and south as far as the eye could see. What awareness, if any, was there of the destructive forces being wreaked upon the coastal ecosystem that historically had provided fishermen with their livelihood? As for Hainan Island itself, with its palm trees and long sweeps of clean sandy beaches, a tropical paradise at the same latitude as the Caribbean Islands, there wasn't yet a single beachfront building in sight a mere thirty years ago. Today the island is festooned with resort hotels, and is fast becoming—loud colorful shirts and all—the Hawaii of China.

On a trip to Shanghai in 2011, tall buildings were everywhere, most especially on the skyline of Shanghai's Pudong, where just three short decades earlier Wong and his team had rejected the opportunity to select any building site desired. Now the former open fields are filled with an unprecedented collection of imposing buildings designed by world-renowned architects, each taller than those built before—an internationally recognizable icon of Shanghai and indeed of modern China itself.

With such direct personal experience the authors feel justified, indeed compelled, to focus their attention primarily on China, not to find fault but with the kind of objective examination that can help in moving forward.

踪迹，它像蜿蜒的丝带一样在眼睛所及的地方南北延伸。有一种破坏性的力量正作用在海岸的生态系统上——在历史上，这一生态系统曾经是渔民们的衣食父母。就海南岛自身而言，它有漂亮的棕榈树和干净的绵延沙滩，是和加勒比群岛处于同一纬度的热带天堂，但就在30年前，那里还没有一座海滩建筑。今天，岛上满是度假酒店，满是穿着各种颜色衬衫的人们，它正迅速成为中国的夏威夷。

在2011年的上海之旅中，到处都是高层建筑，尤其是在浦东的天际线上——那里正是短短30年前黄先生和他的团队拒绝的、选择你们想要的任何建筑位置的地方。如今，开阔的空地挤满了由世界知名建筑师设计的、一系列前所未有的壮观建筑，每一座都比之前建设的更高——这是上海，实际上也是现代中国的一个国际化的标志。

正是因为这种直接的个人经验，作者感到有理由（实际上也是必须）将注意力主要放在中国，不是去挑毛病，而是进行一种客观的审视以帮助中国不断前进。

Top: Shanghai in 1939, with the Huangpu River separating the Bund (at left) from the undeveloped Pudong.
1939年的上海，黄浦江把外滩（左）和未发展的浦东分隔开（上图）。

Bottom: In 1993 Pudong was declared a Special Economic Zone (SEZ) and opened to foreign investors. The 1,535-foot [468-meter] Oriental Pearl Tower for radio and television transmissions was built in 1994, signaling plans for Pudong to become the commercial and financial center of China.
浦东新区在1993年被宣布为特别经济区（SEZ），向外国投资者开放。1,535英尺（468米）高的东方明珠塔建于1994年，它是无线电和电视信号传送塔，更是浦东成为中国商业和金融中心的一个信号（下图）。

SECTION TWO / NEW MATERIALS AND TECHNOLOGIES

New Materials and Technologies

第二篇 / 新材料与新技术

新材料与新技术

The 40-hectare (99-acre) solar plant in Thungen, Germany, was the world's largest upon completion in 2010. Its movable panels follow the sun throughout the day to power some 9,000 homes in the Lower Franconia region of Bavaria.

2010年完工的世界上最大的太阳能发电站——德国Thungen，占地40公顷（99英亩），其可移动面板制造的太阳能电源可供巴伐利亚州下弗兰肯地区约9000户居民的需求。

新材料与新技术

世界上新建筑的建设范围与速度是伴随着新技术、新材料一起出现的。我们之前已经说过，为达成目标所需的切实可行的技术解决方案已经具备，因此，在开始讨论垂直城市各种细节之前，让我们简单看一看一些有代表性的能源和建筑工业方面新进展的范例。

能源

化工燃料的燃烧 是世界上首要的空气污染源。这种污染要么来自工业、汽车和日常生活中使用的燃烧煤、石油、天然气，亦或是在发电过程中产生的。由于在地球内部，化工燃料的储存量相对较多，所以它们依然是世界最重要的能源，供应着美国90%的能源需求和世界其他地方80%的能源需求。像"水力压裂法"（液压破碎法）这样的技术可以使得天然气变成液体，煤可以清洁燃烧；这些会持续改进来提高可行性，而化工燃料依然是世界最主要的能源，同时还可能会是全球变暖的最大推手。我们迫切需要更多可持续性的替代方案。

核能 现在提供美国所需能源的大约5%，它在世界范围内应用广泛，尤其是在欧洲；例如，法国接近80%的用电来自核能。而能源紧缺的中国已经有16座正在运行的核电堆，另有26座正在建设中，200座在提议中。就像著名的福岛核电站事故所显示的，核能的问题在于其巨大的灾难隐患。人类的过失、错误的操作或维护、陈旧的设备、自然灾难，都会加大这种威胁。另外，如何安全处理核废料的问题一直存在，它也让核能的未来非常不确定。

核聚变能 通过可与太阳中心相媲美的高温高压，使较轻的原子发生聚变，产生带电气体。从1970年代第一次能源危机以后，人们就充满热情地对这项技术加大研发力度，并且许诺，将会有耗之不尽的清洁能源可用于替代那些对环境造成破坏的能源。但是现在，在探索了60年后，科学家们甚至认为在2050年之前，依然看不到核聚变能商业化的可行性。

清洁可再生能源 像太阳能、风能、水力发

The scope and pace of new construction in the world is paralleled by the emergence of new technologies and construction materials. We have stated previously that the required technical solutions for the actual construction of our proposal are already in hand. We therefore pause briefly to review a small representative sampling of energy sources and construction industry developments before proceeding to the details of the Vertical City.

ENERGY

Fossil fuel combustion is the world's primary source of air pollution, whether through the burning of coal, petroleum, and natural gas as fuel for industry, cars, and residential uses, or in the production of electrical energy. Because of the relative abundance of fossil fuels deep underground, they remain the world's major source of energy, meeting some 90 percent of energy needs in the United States and 80 percent elsewhere. As a result of such technologies as hydraulic fracturing ("fracking"), the conversion of natural gas to liquid, and new methods for the clean burning of coal, all of which are constantly being improved to prolong viability, fossil fuels will remain the world's primary energy source and will likely continue to be the greatest contributor to global warming. Alternatives that are more sustainable are desperately needed.

Nuclear power currently fulfills about 5 percent of energy needs in the United States and is used extensively around the world, especially in Europe. France, for instance, derives nearly 80 percent of its electricity from nuclear sources. And an energy-hungry China, with sixteen nuclear reactors already in operation, has another twenty-six under construction and nearly two hundred more proposed. The problem with nuclear energy, as is well known from Fukushima, is the potential for catastrophe. Human error, faulty operations or maintenance, aging facilities, and natural disasters all add to the threat. Additionally, the ongoing problems of safe nuclear-waste disposal make the future of nuclear power uncertain.

Fusion energy produces electrically charged gas by fusing light atoms at tremendous temperatures and pressures comparable to those at the core of the sun. The technology has been under intense research and development since the first energy crisis in the 1970s and promises inexhaustible supplies of clean power that could well replace other, more environmentally damaging energy sources. But now after some sixty years of work, scientists still do not foresee the commercial availability of fusion energy before 2050.

Clean renewable sources such as solar, wind, hydroelectric, and geothermal collectively generate only about 12 percent of the energy used in the United States, and 19 percent of the world's electricity consumption. But given the state of the global environment, it is incumbent on all of us to support in every way possible the research and development of sustainable alternative power supplies.

Hydropower, or energy harnessed from moving or falling water, has been used for thousands of years to power water wheels. It is renewable and is among the cleanest forms of energy, but it can damage surrounding ecosystems by raising water temperatures, changing oxygen content, and causing siltation, all of which can adversely affect fish and other water life. Large dams, of course, have tremendous consequences, whether because of the human relocation they frequently require (nearly 1.5 million people lost their homes to China's Three Gorges Dam alone), due to erosion, the loss of animal habitats, and significant changes in downstream ecosystems. Hydropower supplies only about 6 percent of energy in the United States but considerably more elsewhere in the world—for example, China (22 percent and rising), Canada (60 percent), and Brazil (82 percent). Some developing countries depend almost exclusively on hydropower, yet increased ecological awareness makes it doubtful that large-scale hydropower projects will find support in the United States.

Wind turbine generators also provide clean energy, emitting no pollution or greenhouse gases. If properly sited, they have very limited environmental impact. There are many locations around the world with abundant wind supplies, although conditions vary with topography and climate. The main problems with this renewable energy source are high initial cost (although this is dropping) and variability: the wind doesn't always blow when it's needed. Also, because of noise, wind turbines are frequently located in remote areas, which leads to loss in transmission on the way to distant consumers. Still, wind power is among the fastest-growing sources of renewable energy. It is projected to supply as much as 20 percent of electricity in the United States by 2030, and a third of the world's electricity by 2050.

Geothermal energy has been used for thousands of years for bathing and heating. Tapping the heat generated and stored beneath the surface of Earth is extremely economical in the right location—for instance, Iceland and the Philippines, where as much as 25 percent of the energy is geothermal. But not every location is suitable. Moreover, drilling is difficult, it emits noxious gases into the air, and, because drill sites are typically far from population centers,

电、地热能加起来也只占美国能源消耗量的12%，占世界电能消耗量的19%。但鉴于全球环境的现状，我们每个人都有责任以各种途径支持科学家研发可持续的、替代性的电能供应方式。

　　水力发电　利用流水产生的能量，已经使用几千年了，现在人们利用的是水轮发电机组。这是一种可再生能源，也是最清洁的能源之一，但由于水温提高，氧气量改变，造成了淤积，它会破坏周边的生态系统，也会对鱼类和其他水中生物造成不利影响。当然，大型水坝会产生严重的后果，因为它经常要求人口迁移重置（仅是中国的三峡大坝就让1500万人失去家园），更不要提动物栖息地、水土侵蚀、下游生态系统的重大改变。在美国，水力发电只提供总能源的6%，而在世界的其他地方这个数字要高很多，比如：中国（22%，而且还在增加）、加拿大（60%）、巴西（82%）。有些发展中国家几乎只依赖水力发电，但是强烈的生态保护意识让大型水力发电项目要在美国获得支持是不太容易的。

　　风力发电机　也提供清洁能源，它不会产生污染或排放温室气体，如果安置得恰当，它们对于环境的影响非常有限。世界上很多地方有丰富的风能，即使情况会随地形和气候而改变。这种可再生能源的主要问题是初始成本太高（尽管它正在下降）以及不稳定性：当我们需要的时候，风不一定总是会刮起来。而且，由于噪音，风力发电机经常位于偏远地区，这就使它在传输到消费者的过程中有所损耗。不过，风能依然是增长最快的可再生能源。预计到2030年，风能将为美国提供20%的电力，而到2050年，风能将占全世界电力供应的三分之一。

　　地热能　用于洗浴和采暖已有数千年的历史了。在一些特定地区，比如冰岛和菲律宾（那里25%的能源来自地热），开采那些产生、储存于地表下的热能是非常经济的。但这并非在每个地方都可以。而且开采是很难的，它会向空气中排放有害气体，况且由于开采的位置一般都远离人口聚集地，能源在传输的过程中也会大量损耗。在全球范围内，地热目前只占世界能源的一小部分，但低成本和新的清洁技术预示着它将进一步的发展。预计地表之下6.25英里（10公里）储藏

Geothermal borehole house in Iceland. Numerous hot springs and 25 active volcanoes are sourced to provide heat and hot water to 85 percent of Iceland's buildings and more than 25 percent of the country's energy requirements.
冰岛的地热井房子。众多的温泉和25座活火山为冰岛85％的建筑和超过25％的地区提供能源。

的热能，是全世界天然气和石油相加总量的五万倍还要多。

太阳能光伏 将太阳辐射转化为电能，可以追溯到1876年，那时人们发现固体硒可以将阳光转化为电能而不需要热能装置或机械部件，但直到1970年它才被应用到商业上。当然，现在太阳能板有着广泛的用途。太阳能光伏上的新发展已经实现了更低的成本和更高效的能源生产，初始的投资费用几年之内就可以获得收益。南加州大学的科学家最近研发了一种新型液体太阳能电池，它由纳米晶体制成，因而非常微小（一个针头上可以有2500亿个晶体），可以画在或打印在光滑的表面上，比如塑料，因而几乎可以随处安装。日常的一些东西，比如汽车车顶、公交车和建筑的整个向阳面会因此具备巨大的电能生产潜力。

there are significant losses in transmission. Globally, geothermal now accounts for just a fraction of the world's energy, but lower costs and new clean technologies promise increased development. It is estimated that heat supplies within 6.25 miles [10 kilometers] of Earth's surface contain 50,000 times more energy than the world's natural gas and oil supplies combined.

Photovoltaics, or converting solar radiation into electricity, had its beginnings in 1876, when it was discovered that the solid element selenium could convert light into electricity without the use of heat or mechanical parts. But it wasn't until 1970 that it had any commercial appeal. Now, of course, solar panels are in widespread use. New developments in photovoltaics promise less expensive and more efficient energy production, allowing the payback of initial investment costs within just a few years. Scientists at the University of Southern California recently developed a new type of liquid solar cell made from nanocrystals that are so small (250 billion of them fit on the head of a pin) that it is possible to paint or print them onto clear flexible surfaces, like plastic, for installation virtually anywhere. The potential is wide open as everyday objects, car roofs, buses, and the complete sun-facing sides of buildings, have the potential to generate electricity. Meanwhile, Japanese researchers are developing solar cells that are spherical so they can harvest light from any direction, including that bounced off other surfaces. Solar energy is the cleanest, most abundant energy source available: sunlight reaching Earth surpasses human energy consumption by a factor of six thousand.

Beyond the means mentioned above, many other techniques are being developed for the production of renewable energies, including tidal power, wave power, biomass power, fuel cells, and biophotovoltaics. Not all are equally promising or commercially viable, but the environmental imperative and diminishing fossil fuel supplies mandate their investigations.

As new renewable-energy technologies emerge, so too do other technologies that open the door to the possibility of improved quality of life. Since transportation accounts for an estimated 25 percent of the world's energy use and is a major source of air pollution, advances are especially important in automobile technologies. Cars are now—and will likely remain—an indispensable mode of rural and suburban transportation. The introduction of hybrid and electric cars will help to reduce emissions, although the benefits are offset by the pollution produced in generating the required electricity, most often from coal-burning power plants. Measures to discourage the use of private automobiles include increasing gas prices, making public rapid transit available and efficient, and encouraging car sharing. All of these practices are heading in the right direction, but none is more effective than the creation of a car-free, pedestrian-friendly society, as is recommended by the authors of *Vertical City*.

Trains have always been more fuel-efficient than cars or trucks and can achieve 480 ton-miles per gallon. By comparison, a 1-ton car would have to get 480 miles to the gallon, and a 2-ton SUV half as much, for comparable efficiency. And that would be just to move the vehicles, without a payload.

China currently has high-speed rail (HSR) service traveling at 240 miles per hour (380 kilometers) from Beijing to Guangzhou, equal to the distance between New York and Key West, Florida. China's train covers 1,200 miles in about six hours, while Amtrak in the United States takes almost three times as long, or roughly sixteen hours for the 1,277-mile trip between New York and Miami. As previously stated, for trips of around 1,000 miles (1,600 kilometers), travel time by high-speed train door to door is comparable to, if not better than, air travel, without the waste or inconvenience of long trips to and from the airport, security check queues, and long walks to the gates. A network of such HSR systems will

同时，日本的研究者已经研发出球形的太阳能电池，能够从各个方向收集阳光，包括向另一面反射的阳光。太阳能是能够获得的最清洁、最丰富的能源：到达地球的阳光具有的能量大大超过了人类的能源消费量（前者大约是后者的6000倍）。

除了上述提到的方法，现在还有许多技术已经被研发出来用于可再生能源的生产，包括潮汐发电、波浪发电、生物发电、燃料电池和生物光伏发电。当然并非所有技术都前景光明或易于商业化，但环境的需要、化工染料的日益减少要求做出这些探索。

伴随着新的可再生能源技术的涌现，其他一些技术也为提高人们的生活质量打开了大门。由于交通占据全世界能源消耗量的25%左右，并且它还是重要的空气污染源，汽车技术方面的进步就显得特别重要了。汽车现在是，而且在可以预见的将来依然会是农村及郊区不可或缺的交通运输方式。引入混合动力汽车和电动汽车将有助于减少尾气排放，但生产所需电能造成的污染抵消了潜在的益处——这些电能经常来自燃煤发电企业。抑制使用私家车的措施包括提高油价，让快速公交系统更方便更有效，以及鼓励汽车共用。所有这些在方向上都是对的，但没有一个措施比创造没有汽车、利于行人的社会更有效，这正是《垂直城市》的作者们大力推荐的。

火车在燃料利用上总是比汽车或货车更有效，它可以达到480吨一英里每加仑。比较一下就知道1吨的汽车如果是同样的利用率，它每加仑要跑480英里，而一台2吨重的SUV跑的距离可能要减半，那还只是空车没有负重情况下的效率。

中国现在从北京开往广州（这个距离相当于纽约到佛罗里达州基韦斯特）的高速铁路运行速度为186英里（300公里）每小时。8个小时内，中国的火车可以到达1200英里内的任何地方，而在美国，美国铁路公司要花上两倍的时间，即从纽约到迈阿密1277英里的旅程得花上大约16个小时。正如前面提到的，1000英里（1600公里）左右的旅程，从头到尾乘坐高速铁路花的时间可以和航空旅行相提并论了——虽然还不能完全赶上——这刨去了往返机场的长途旅行、安全检查、步行一大段路到机舱门等等带来的时间浪费和不方便。

这些奔跑在地上的旅行方式有一个不好的地方，就是消耗了有价值的土地。不过这一点可以

Raised high-speed rail and motorways help to preserve open land while minimizing negative impact on pedestrians and wildlife.
抬高的高速铁路和高速公路，同时最大限度地减少对行人和野生动物的负面影响，有利于维护土地。

通过抬高道路和铁轨予以弥补。必须承认，这样做开始时需要的花费比建在地上要高，但就长期看，保护肥沃的土地，避免车辆/行人交通事故以及误伤日益减少的野生动物，这些会让这种方法物有所值。

随着高速旅行成为今天生活的一部分，减少污染将会成为旅游业的目标。对于商务旅行者来说，超音速客机重新走红；而对冒险家而言，到外太空旅行也并不遥远。高运营成本，安全和高音速是导致协和飞机被不幸替代的原因。但这些因素已经被飞机制造商完全克服了。它们期待着一种新的、更小的商务飞机可以在接下来的10-12年中驶上飞行跑道。不过，拥挤的机场有着很现实的不便之处，而飞机引擎的污染和噪音依然是未解的难题。

新技术也刺激了其他领域的发展，让过去50年成为人类历史上最有创造性的时期。这主要可以归功于通信的发展，最重要的可能是互联网，包括电子邮件、即时通讯和互联网协议，从上世纪90年代中期开始，所有这些都对文化和商贸产生了重大影响。毫无疑问，IT行业有助于促进教育、科学、技术和商贸。我们现在正好在这个新时期的开端处；之前难以想象的发展进步现在每

greatly reduce pollution caused by global transportation.

The disadvantage of these surface modes of transportation is their consumption of valuable land. This could be alleviated, in part, by elevating roads and tracks. Admittedly, the initial cost would be higher than putting them on grade, but the preservation of productive land coupled with the avoidance of vehicular/pedestrian accidents and the killing of diminishing wildlife may, in the long term, weigh in its favor.

Because high-speed travel is part of today's lifestyle, making it pollution free should be the goal of the travel industry. For business travelers, supersonic travel is staging a comeback, while for the adventurer, travel to outer space is close at hand. The high operational costs, as well as the safety and sonic-boom factors behind the demise of the ill-fated Concorde, are being competitively overcome by aircraft manufacturers with the prospects of a new, smaller breed of corporate jet rolling onto the tarmac within the next ten to twelve years. However, the physical inconveniences of airport queues, jet engine pollution, and noise remain unsolved problems.

New technologies have stimulated developments in other areas and have made the last fifty years the most inventive in the history of humanity. This can be attributed in large part to the development of communications, perhaps most significantly the Internet, including electronic mail, instant messaging, and Voice over Internet Protocol (VoIP), all of which, since the mid-1990s, have had a drastic impact on culture and commerce. There is no question that information technology (IT) has helped to further education, science, and commerce. We are just now at the beginning of this new epoch; previously unimaginable developments are occurring daily.

As architects, we have a special interest in new construction materials, technologies, and methodologies. In 2011, China erected a seismically engineered 30-story hotel in structural steel in a record-breaking 15 days from beginning of construction to occupancy. A year earlier, the country announced plans to build Sky City, the world's tallest building, in Changsha, capital of the south-central Hunan Province. Like the above mentioned hotel, the 13-million-square-foot [1.2 million square-meter] mixed-use building, 220 stories and 2,748 feet [838 meters] building is to be executed with largely prefabricated components. Its projected schedule, ranging from 90 to 210 days, constitutes a breakneck achievement, although a good deal of skepticism surrounds the project.

The scientific world is constantly striving for new and improved materials and technologies for use by the construction industry. Following are just a few examples in various stages of experimentation or development.

Graphene is a substance consisting of pure carbon, with a single layer of atoms arranged in a hexagonal, honeycomb mesh to form a thin sheet just one atom thick. It is the thinnest but simultaneously the strongest material known, thanks to its immensely strong carbon bonds. The miracle-like material conducts electricity as efficiently as copper, and it far surpasses all other materials as a conductor of heat. According to the lead researcher Ali Reza Ranjbartoreh: "Not only is it lighter, stronger, harder, and more flexible than steel, it is also a recyclable and sustainably manufacturable product that is ecofriendly and cost-effective in its use." Ranjbartoreh reported that graphene will ultimately allow the development of lighter, stronger cars and planes using less fuel and generating less pollution, and which are ecologically sustainable and cheaper to run. Ultra-small and ultra-fast, graphene may also replace silicon chips in electronics. The problem comes from the fact that graphene is such an excellent conductor that it is difficult to stop the charge, which is a hurdle that needs to be resolved before it can be put to commercial use.

Biocrete, or self-healing concrete, is a technology that commands great interest in the construction industry. Concrete is the world's most widely used building material, but it is prone to cracking and, over time, even microcracks can cause big problems if not repaired quickly. Several promising new approaches are being developed, including a new biological system, pioneered by Dutch researchers, that involves mineral-eating bacteria that become activated when exposed to air and water. The bacteria feed on minerals

天都在发生。

作为建筑师，我对新建筑材料、技术和方法有特殊的兴趣。2011年，中国修建了一座钢结构的30层酒店，其建设速度在工程上令人震惊，从建设开始到完成用了破纪录的15天时间。在一年前，这个国家打算在湖南省的省会城市长沙建造世界最高建筑：天空之城。和前面提到的酒店一样，占地1300万平方英尺（120万平方米）、高2748英尺（838米）的多用途220层高楼，大部分将由预制构建组装而成。整个工程计划预计在90-120天完成，尽管很多人对工程有所质疑，但这还是一项惊人的成就。

科学界一直在努力为建筑行业寻找更新、更好的材料、技术。下面是一些处在不同实验或研发阶段的例子：

石墨烯 是一种碳原子构成的物质，单层的原子在其中呈六角型蜂巢晶格网，最终形成只有单层原子那么厚的平面薄膜。石墨烯具有非常牢固的碳键，这使它成了目前所知最薄也最坚硬的材料。这种神奇材料的导电性能与铜一样好，而作为一个热导体，它也远远超过所有其他材料。根据首席研究员阿里·瑞扎·瑞巴特里的说法："它不仅比钢铁更轻、更坚固、更硬、更柔韧，还是一种可回收、可持续的制造材料，有益于生态环境且在使用过程中有性价比。"瑞巴特里指出，石墨烯最终将会有助于研发更轻更强的汽车和飞机，让它们可以用更少的燃料，产生更少的污染，在生态环境问题上更具有可持续性，而且运行的费用更少。超小而又超快，它甚至可以替代电子行业中的硅片。但石墨烯出色的传导性能也产生了一个问题，即很难控制电荷，这是石墨烯商业化之前首先需要克服的一个障碍。

生物混凝土 或者叫自我修复的混凝土，这是一项会在建筑行业引起巨大兴趣的技术。混凝土是世界上使用最广的建筑材料，但它容易开裂，随着时间推移，如果不及时修理的话，即使很小的裂缝也会造成大麻烦。现在已经有很多前景光明的新技术正在研发中，比如由荷兰研究者倡导的新型生物系统，其中有暴露在空气和水中就会活跃起来的、能够吞噬矿物质的细菌。这些以矿物质为食的细菌在生产过程中被添加到混凝土中，会产生能自动修复裂缝的方解石——石灰岩。韩国则在研发另一项技术，即用一种聚合物涂层来保护混凝土表面。当暴露在阳光下时，涂层释放出一种不溶于水的溶液，自动填充任何的裂缝，进而在阳光照射下加固建筑。这些自我修复技术依然处于试验阶段，但它们在提高混凝土结构使用寿命，节省每年数十亿美元的维护、维修费用方面，有巨大的潜力。

绿色或矿物聚合物混凝土 是一种利用煤粉灰的创新、环保型建筑材料。数量众多的煤粉灰是工业的副产品，如果不转化成普通水泥的话，它们最终也会被填埋。矿物聚合物水泥比普通水泥有更好的防火性能，更强的抗腐蚀性，且不易收缩，还有高强度的压应力和拉应力。它最大的特点是，其整个的生命周期会减少90%的温室气体排放。全世界普通水泥（其中大部分是混凝土）的产量接近每年20亿吨，这是一个重大的环境问题，因为单单水泥生产过程中燃烧燃料最后向空气中排放的二氧化碳量，就占到全球温室总量的8%。重要的是：基于氧化镁的水泥实际上具有碳负性，它能吸收的二氧化碳比制造过程中排放出来的还要多。所有这些以及目前的其他一些进步让我们在减少或回收水泥废料、排放气体方面取得了令人激动的进步。不过，这些新兴的、依然昂贵的技术还没有得到大规模的应用。同时，由于中国、印度和其他发展中国家大规模的建设浪潮，人们对混凝土的需求与日俱增。

高强度混凝土 建筑和建设行业对可持续性的关注不断增加，这激发了其他方面的进步，包括利用新型纳米技术和矿物学生产出来的新型高强度混凝土——它减少了传统混凝土建设过程中所需钢筋总量的75%。另一个令人兴奋的进步是新的隔热方式，它会降低室内外的热传导效应。现在，人们研究关注的重点是革命性、能够自我调节的相变材料（PCMs），它能在高温时融化，在低温时凝固，以一种更低的成本适时地吸收或释放热量，从而保证室内环境的舒适度。如果把这些技术推向市场，那将意味着小型建筑能够调节自身内部环境温度，从而减少甚至是取消人们对空调、暖气的需求。

混凝土打印 或者说累积制造，会将在计算机

added to the concrete during production and produce calcite—limestone —which automatically seals any cracks. Another approach, developed in South Korea, involves protecting concrete surfaces with a polymer coating. When exposed to light, the coating releases a water-resistant solution that automatically fills any cracks and then solidifies in sunlight. These and other self-healing technologies are still in the experimental stage but have tremendous potential for increasing the life span of concrete structures and saving billions of dollars in annual maintenance and repairs.

Geopolymer, or green concrete is an innovative, environmentally friendly building material that uses fly ash, an abundant industrial by-product that would otherwise end up in landfills. Geopolymer concrete offers significantly higher fire resistance than Portland cement, greater corrosion resistance, and lower shrinkage, with high compressive and tensile strengths. But its greatest attribute is its life-cycle reduction of greenhouse gas by as much as 90 percent. The global production of Portland cement, the major component of concrete, is nearly 2 billion tons each year, which translates into a huge environmental problem because the fossil fuels burned in cement production release carbon dioxide into the air, accounting for some 8 percent of the world's greenhouse gases from this one source alone. Significantly, in development are magnesium oxide–based cements that are actually carbon negative, absorbing more carbon dioxide from the air than is released during the manufacturing process. These and other recent advances have made possible encouraging progress in reducing or sequestering concrete's waste and emissions, but the still-emerging and still-costly technologies have yet to find widespread use. Meanwhile, the demand for concrete continues to rise, largely as a result of the massive amounts of construction taking place in China, India, and other developing countries.

High-strength concrete, Growing sustainability cocerns in architecture and in the building industry have led to other encouraging advances, including a new type of high-strength concrete that uses nanotechnology and mineralogy to eliminate 75 percent of the steel reinforcement traditionally required in concrete construction. Also promising are new means of thermal insulation that allow for reduced indoor/outdoor heat exchange. Researchers are now focusing attention on revolutionary self-regulating Phase Change Materials (PCMs) that liquefy in warm temperatures and solidify when temperatures drop, respectively absorbing or releasing heat to keep interior environments comfortable at a fraction of current costs. Market availability of such technology could have huge implications since smart buildings regulate their own interior

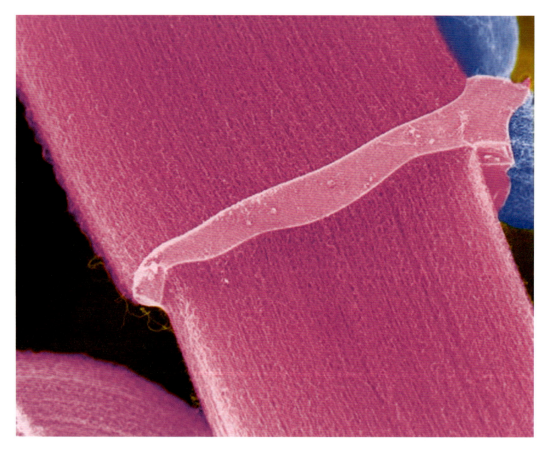

Self-healing concrete is among the many new products that offer extended life for built structures (magnification at top). Another promising innovation is Carbon Nanotube (CNT) fiber (magnified below), spun into a flexible thread that is lighter and 100 times stronger than steel and offers high heat and electricity conductivity.

自愈混凝土是众多为建筑物提供延长寿命的新产品之一（上图是放大图）。另一个充满希望的创新是碳纳米管纤维（下图是放大图），它可以纺成一根灵活的线，这根线很轻，比钢强度大100倍，并提供高的热和电传导性。

上生成的具有精密细节指令的设计用3D打印机输出，它一层一层地"打印"或者说搭建物品。这种打印机使用的不是墨水，而是靠精密控制水泥灰浆的喷洒，所以它可以造出之前无法产生的复杂构件。整个过程极为高效经济，同木材或钢铁等固体材质造成的物体不同，它在切割过程中不会产生浪费或剩余。理论上说，这种技术潜能无限，能够产生出各种独立构件，墙甚至是整栋建筑，可以现场打印然后组装，也可以利用悬吊的移动吊架逐层打印。在打印过程中，所有电缆、管道、导管还有其他配件都被整合在一个高度统一的整体中，不会像传统建筑方法那样产生浪费和污染。

无绳电梯 是自行推进的电梯轿厢，它取消了对沉重起吊电缆的需求。摆脱了这些钢"绳"在长度，或者说尤其是在重量上的限制后，电梯轿厢可以上升到无限高的地方。在一些深矿井中，在美国海军航空母舰上，已经开始使用这种无绳电梯。在高楼中，许多无绳电梯可以共用一个井道，因而会极大增加楼层的可用空间。无绳电梯上升时需要动力，但在下降时会产生电能，综合而言会降低能耗。这种已经研发数十年的垂直交通技术是一项重大突破，它将给建筑设计带来革命性变化，而且会解决楼层高度问题上遇到的最大障碍。不过，其商业可行性则需要大量时间、资本的投入。

自清洁玻璃 在幕墙技术上也有一些重大的工业创新，其中包括：自清洁玻璃——玻璃上很薄的三氧化钛涂层可以与阳光反应，发生化学分解，然后在雨天时无需清洁剂就可以自动洗去玻璃上的污垢。这项技术已经商业化了，不过这一领域前景广阔，需要进一步发展。现在可以面向的是那些希望能够方便有效、无需脚手架和其他设备自清洁幕墙的建筑，因为这会大大节省建筑每年的维护费用。在玻璃上的突破还包括：抗反射、高性能、能减少热损益的涂层，提高玻璃抗碎性能以增强安全性的新型层压技术，能将自然光更充分反射进建筑内部结构的新系统。除此之外，还有许多创新的内部遮阳设备，改进的太阳能技术，将玻璃作为承重材料使用上的重大提升，综合高效的透光性能，一个单一整体中的结构和外表层。

LED灯 与此相似，节能设备和内部系统也

climates, reducing or perhaps even eliminating the need for air-conditioning and heating.

Concrete printing, or additive manufacturing, exports computer-generated designs with precisely detailed instructions to a 3-D printer that then "prints" or builds up physical objects, layer by layer. Instead of ink, the printer uses highly controlled extrusions of cement-based mortar so that it can create complex components that were not possible before. The process offers tremendous efficiency and economy since, unlike objects made from solid materials like wood or steel, there is no waste in cutting away voids or overages. Theoretically, the potential is limitless, capable of producing individual components, walls, or even entire buildings, printed out onsite and assembled or printed in successive layers from a moving gantry overhead. In the printing process, all of the necessary cables, piping, ducts, and other service requirements are incorporated in a fully integrated whole without the cost or pollution of traditional construction methods.

Ropeless elevators are self-powered elevator cabs that eliminate the need for heavy hoistway cables. Unlimited by the length, or more especially the weight, of these steel "ropes," elevator cabs can travel to limitless heights. Such elevators are already in use in some deep mine shafts, and also in U.S. Navy aircraft carriers. In high-rise buildings, several ropeless elevators can share the same shaft, thus dramatically increasing available floor space. Ropeless elevators need power when ascending but because they are power generating on the way down, energy costs are reduced. Such vertical-transport technology, in research and development for decades, is a breakthrough that will revolutionize tall-building design, as it will remove the single greatest limit to increased height. However, commercial viability will require huge investments of time and capital.

Special purpose glass, some of the greatest industry innovations have taken place in curtain wall technology, including self-cleaning glass, where a thin titanium oxide coating reacts with sunlight to chemically break down and wash away dirt in the rain without the need for detergents. The technology is already commercially available but the field is wide open for further development. Within reach are buildings that self-clean their curtain walls easily and efficiently, without scaffolding or other equipment, for a tremendous savings in annual maintenance. Among the other breakthroughs in glass are anti-reflection and high-performance coatings to mitigate heat gain and loss; new laminations to produce shatter-resistant glazing for improved security; and new systems for bouncing natural light deeper into interior spaces. Beyond this, there has been a wealth of innovative internal shading devices, improved solar technologies, and significant advances in the use of glass as a load-bearing material, effectively combining transparency, structure, and exterior skin in a single, unified whole.

LED lighting, Similarly, important advances have taken place in energy-efficient mechanical and interior systems, notably with smart electric metering and light-emitting diode (LED) lighting, which produces illumination that is equivalent to conventional incandescent lighting but requires only 10 percent of the energy.

It is safe to say that architecture and engineering, hand in hand with the construction industry, have been revolutionized by new technologies over the past few decades. CAD (computer-aided design), for example, has redefined building design to previously unimaginable complexities of built form and the efficient integration of the various trades and building systems. New materials and practices have likewise changed onsite operations. These include laser-based survey equipment, automated field welding, modularization of large-scale components, and 24-hour robotic devices to monitor quality control and to orchestrate timely corrections. Management and operations systems have also improved in estimating, procurement, scheduling, staging activities, change orders, and virtually every other aspect of information handling.

Structural and mechanical considerations As buildings grow taller, there is a greater need for structural redundancy, advanced security and tracking systems, and improved methods of fire protection and life safety. Sprinkler systems continue to provide the most effective method of fire protection, perhaps supplemented in tall apartment, hotel, or office buildings with swimming pools as a backup water source, and also by localized fire-fighting stations. Apropos of the whole problem of water management in tall buildings, the distinguished systems engineer Valentine A. Lehr explained in a letter to the authors:

> Engineers rarely think of the transport cost of the water and air that moves throughout a building in lower-rise structures, but these can be quite significant in ultra-tall buildings, especially with needed break-pressure tanks and isolating heat exchangers. Techniques that minimize or reduce transport cost have real application here, and gray water is an area where this finds good application. Instead of collecting all building gray water and conveying it to the building base where it is treated and then pumped again throughout the tower, localized treatment can be used, producing savings in pumping, piping, and energy. An Australian entrepreneur has even developed a self-contained device for use within an individual apartment to

有重要的进步，尤其是智能电表和LED（发光二极管）灯，后者的照明功效与传统的白炽灯不相上下，但所需的能源却只有白炽灯的10%。

可以毫不夸张地说，由于新技术的出现，与建设行业密切相关的建筑和工程在过去数十年间已经有了革命性的变化。例如，CAD已经重新定义了建筑设计，使之达到之前建设形式难以想象的复杂程度，而且可以有效融合多种技术和建筑系统。新材料和新方法也改变了现场施工操作，例如：激光测量设备，现场自动焊接，大规模的模块化组装，进行质量控制、及时协调修正的24小时机器人设备。管理和操作系统的各个方面也都有所提升，从评估、采购、调度到展开行动、改变订单甚至可以说信息处理的各个方面皆是如此。

结构与机械的考虑 随着大楼越建越高，对于简化结构、提高安保和追踪系统，改进消防和安全办法的需求也越发强烈。自动喷水装置依然是消防中最有效的方法，可能在高层公寓、酒店或办公室建筑中，会加上游泳池作为后备水源，当然，还有安置小范围的消防站点。就高层建筑水资源管理的问题，出色的系统工程师瓦伦丁·A·莱尔在一封写给本书作者的信中说：

> 在低层结构建筑中，工程师们很少想到水和空气运输所需的费用问题，但在超高层建筑，尤其是需要压力缓冲装置、独立热交换器的超高层建筑中，这些都会是大问题。减少、降低运输费用的技术在这里有现实的需求，而中水是一个可以得到很好应用的领域。人们不再收集整栋建筑的中水，然后将其运到建筑底层，处理完毕后再输送回大楼；新方法是，可以采用局部处理，这会节约水泵、管道和能源。一家澳大利亚企业已经研发了一种独立设备，可装在每个独立单元内，进而将供水、排水量、管道减到最少。这种创新方法，加上简化运输的分配设备，可以应用在任何地方。另外，功用传输管道的减少可以让地板和天花板夹层之间更紧凑，让整个结构更经济实惠。

怀着对可持续性的关切，作者探索了高层建筑自然冷却的可能性。大体上说，每1000英尺

74

Wind turbines in southeast Washington State. The modern ancestor of ancient windmills (dating back to 200 B.C.), wind turbines are among the fastest-growing sources of renewable energy. As of 2012, they were used by more than 100 countries worldwide, supplying 3 percent of global electricity demand.

华盛顿州东南部的古老风车（可以追溯到公元前200年）的现代版本，风力涡轮机是增长最快的可再生能源之一，截至2012年，它们被应用于全球超过100个国家和地区，满足全球电力需求的3%。

minimize both supply and drainage quantities and piping. This kind of innovative approach can be applied elsewhere as well, with distributive equipment that minimizes transport. Reduced utility distribution pathways additionally make for more compact floor-to-ceiling sandwich height and more economical construction.

Concerns for sustainability prompted the authors to investigate the possibilities of natural cooling in tall buildings. Roughly speaking, temperature decreases 2 degrees Celsius [36 degrees Fahrenheit] with every 1,000 feet [305 meters] of height. If ground temperature on a hot day is 35 degrees Celsius [95 degrees Fahrenheit], then at 5,000 feet [1,524 meters], the air temperature will be a very comfortable 25 degrees Celsius [77 degrees Fahrenheit]. Taking advantage of the laws of nature whereby cold air sinks, chilled air at the top of a building could be brought in to cool lower floors, thus saving the energy and expense of air-conditioning. Val Lehr explained to the authors this theory, and the whole question of stack effect, plus related issues of wind:

Ultra-tall buildings present many significant engineering challenges for building services design. . . . Controlling stack effect (in winter) and reverse stack effect (in summer) goes well beyond energy considerations, as it is essential to occupant life safety. At the same time, tall structures present opportunities for innovative approaches that capitalize on the unique characteristics of such construction, and especially by applying innovative approaches of integration.

Wind is a feature inherent in altitude. Higher elevation above ground means higher wind speeds, with velocities at the 1,600-meter [one-mile] height significant. At these building heights, passing through two or more wind regimes is common. A recent 1,400-meter [4,600-foot] as-yet-unconstructed building was able to utilize the structure's unique geometry so that external wind velocity alone drove the internal air-conditioning without the need for conventional fans, regardless of the wind direction. Cooler outdoor temperatures also lessen the mechanical cooling requirements, allowing for greater quantities of outside air usage when conditions permit.

Wind can also generate problems, especially in terms of building sway in high velocity. Height and building slenderness influence sway, which can be the source of severe discomfort to occupants. Integrating the mechanical systems in innovative ways can help. For example, strategically placed and designed water storage tanks can assist in dampening. More innovatively, gas turbine electrical

（305米）温度会下降2摄氏度（36华氏度）。如果天气炎热，地面温度为35摄氏度（95华氏度），那么在5000英尺（1524米）的地方，空气温度将会是非常舒适的25摄氏度（77华氏度）。利用冷空气下沉的自然规律，建筑顶部的凉爽空气可以用来给较低的楼层降温，这会节省空调的能耗和费用。高层建筑专家瓦尔·莱尔进一步向作者解释了这一运作原理——烟囱效应及与之相关的风的影响：

超高层建筑向建筑配套设施设计提出了许多重要的工程学挑战……由于它对居民生命安全的重要性，在冬天控制烟囱效应，在夏天维持烟囱效应的问题，大大超出了能源考虑的范围。同时，高层结构也为新方法提供了机会，这些方法充分利用了此类建筑结构的特性，尤其是那些创造性的整合方法。

风跟高度有密切关系。离地面越远意味着越快的风速，这在高度达到1600米（1英里）时尤其明显。这一高度的建筑，一般都要穿过两个或更多的风速区。现在一栋高1400米（4600英尺）、尚未建造的建筑，可以利用独特的几何结构，让外部的风速推动内部的空调系统，而不需要常规的扇叶，当然，这是不考虑风向因素的前提下。更低的室外温度也可以减轻机械制冷的要求，条件允许时可以让更多的室外空气进来。

风也会导致一些问题，特别是风速很快时建筑会摇晃。高度和建筑的细长比例会影响摇晃，而这会让居民感到严重的不适。如果以创新的方式整合机械系统可能会有所助益。例如：审慎、策略性地安置、设计储水池将有助于减轻摇晃。更新颖的是，像著名的鹞式垂直起降战斗机那样，用垂直安置的燃气涡轮电动机替代传统水平的排气分流阀，将它安置在建筑顶端，当它产生动力，推动热电设备时，可以大大减少建筑的摇晃。另外一个十分激进（在美学上很有争议）的方法是，将风力发电机放到大楼的顶部，这样既可以持续发电，又可以减少摇晃。

受这种可能性的鼓舞，作者探究了由一组不同高度超高建筑组成的垂直城市的基本前提，这些超高建筑通过内部互通的支撑桥梁或空中大堂连结在一起。我们向莱斯利罗伯特森建筑事务所的结构工程师威廉·法杉求证了理论的可行性。结果再一次令人鼓舞，正如法杉先生写的那样：

通常而言，我相信这个计划在技术上是可行的，虽然花费比较高昂。最高层的大楼其纵横比例将会接近于17：1。这些大楼太过细长，在风力导致的大楼摇晃过程中无法达到大家期待的舒适度。各个大楼依靠桥梁结构互相连结起来可能还不足以让它们变牢固，进而缓解这些摇晃。如果桥梁连结被证明是不充分的，其他的解决方案还包括增加高层大楼占用的空间，增加其他类型的建筑连接物，比如对角支撑，仅在较低楼层增加占地面积或实际使用位置……

建筑形式是作者们最关注的问题之一。我们并不想要大多数高层建筑都会采纳的椎体外形，但在提供必要设施让人从地面上更容易进入建筑的同时，我们还要让建筑牢靠，固定在足够大的占地面积上。我们设想，细长的楼层之间可以有一种多楼层的服务——零售—娱乐平台，可以通过桥梁凝结和连接起来，防止摇晃。这里我们再次从两位顾问工程师宝贵的见解中受益颇多。瓦伦丁·莱尔写道：

由于纵横比上的限制，高层建筑需要宽阔的地基，而这会产生很大的楼面层和建成区。一般来说，这种挑战不是与建设相关的技术问题，而是按照一个合理的时间表

generators set vertically in lieu of the conventional horizontal equipment, with exhaust diverters such as on the famous Harrier "Jump-Jet" Fighter, can be positioned at the building apex and programmed to significantly dampen sway while producing power and driving cogeneration equipment. Another really radical (and aesthetically controversial) approach involves tethering an airborne wind-generation device to the top of such a tower to both continually generate power and dampen sway.

Encouraged by the possibilities, the authors explored the basic premise of a vertical city consisting of a cluster of ultra-tall towers of various heights tied to each other with interconnecting bracing bridges or sky lobbies. We then tested the feasibility of our theory with William Faschan, a highly respected structural and civil engineer and a partner at Leslie E. Robertson Associates. Again, the results were encouraging. Mr. Faschan wrote:

Generally, I believe that the scheme is technically feasible, though expensive. The tallest towers would have an aspect ratio of approximately 17:1. These towers would be too slender to achieve the expected level of human comfort under wind-induced swaying motion. The interconnection of individual towers with the bridge structures may not be sufficient to stiffen and dampen these motions. If the bridge connections proved insufficient, other solutions would include increasing the footprint size of the tallest towers, adding other types of building interconnects, such as diagonal bracing, increasing the footprint or effective stance of these towers only in the lower floors.

The issue of building form was one of great concern to the authors. We did not want the tapered pyramidal profile adopted by most tall buildings, but we needed to stabilize and anchor the building with a sufficient footprint while providing necessary services for easy access from the ground. We envisioned a multistory service/retail/recreation podium giving rise to slender towers, clustered and interconnected by bridges against sway. Here again we benefited from the valuable insights of our two consulting engineers. Val Lehr wrote:

Tall buildings, by virtue of limitations in slenderness ratio, also need wide bases that generate huge floor plates and significant built-up area. Often the challenge is not the technical issues associated with construction but rather the issues of filling the constructed area in a reasonable time frame. That has led to designs that produce the needed base and its corresponding built-up area. These design techniques also afford opportunities for clever mechanical-electrical applications. One recent study of a

2,000-meter [6,560-foot, or 1.24-mile] building resulted in a three-independent-legged structure rigidly interconnected every 30 stories. This configuration produced a natural wind collector at the open juncture point, allowing for vertical-axis wind generators with significant output, especially at altitudes over 800 meters [2,625 feet].

Unique designs with open cores reduce floor area—especially less desirable internal area—and also afford opportunities for wind energy capture and utilization of natural stack effects external to the enclosed area. This is a positive stack-effect application.

Another recent study foresaw a glazed exoskeleton structure with a series of 20- to 50-story buildings "suspended" in the enclosed void. Because the internal volume was climate conditioned, the internal structures achieved their comfort conditioning by merely opening windows (for residential and some office occupancies), or drawing "outside" air through the occupied areas of other occupancies. This approach reduced the mechanical equipment in the buildings themselves and allowed for external terraces (in the exoskeletal void) in a pleasant climate.

It takes great imagination to conceive and build a mile-high building. That same imagination can be applied to the mechanical and engineering services to generate new approaches to traditional problems.

We approached our design with just that kind of commitment, determined to address traditional structural problems (primarily wind) with interconnecting links that simultaneously serve essential community functions. Bill Faschan observed:

The skybridge structures would be major construction projects in their own right and would potentially expand and contract horizontally, pushing and pulling on the towers, unless expansion joints are constructed into the bridges. The bridges may be used to index adjacent towers to one another so as to minimize wind excitations of individual towers. The distance between adjacent towers would have a considerable influence on the amount of wind excitation that individual towers would experience. Where towers are very close to one another, they provide shielding of their neighbors from winds. Where towers are farther apart, their presence can amplify the effects of the wind on neighboring towers. Considerable study would be required to determine favorable combinations of tower-to-tower adjacencies and where to put expansion joints in the bridges, and where to build them without joints.

With such areas of concern thus targeted, we proceeded to develop our Vertical City concept. First we stipulate the

填满建筑区域的问题。这使得在设计时有了所需的地基及其相应的建成区。这些设计技术还为灵巧的电机设备提供了机会。日前，有一项对高度2000米（6560英尺或1.24英里）建筑的研究，其结果是每30层都用一个独立的三角结构牢固地连接起来。这种结构配置在开放的结合点上放置了自然集风装置，特别是当其高度超过800米（2625英尺）时。垂直方向上的集风装置有很大的输出功率。

一些有开放核心区的独特设计减少了楼层面积，特别是吸引人的内部面积变少了，这也为风能收集器提供了可能，而且人们可以利用封闭区域之外自然的烟囱效应。这是一种对烟囱效应的积极利用。

当前的另一项研究预测，一种光滑的框架结构，可以使包括一系列有20-50层的建筑"漂浮"在密闭空隙中。由于内部区域因气候而变化，仅仅通过开窗，内部结构就可以获得舒适的条件（这是对住户和一些办公人员而言的），或者将"外部"空气抽送到其他被占用的区域中。这一方法减少了"建筑"中的机械设备，在好天气时也可以享受外部（在框架的空隙中）空间。

构思和建设一栋1英里高的建筑需要不同寻常的想象力。同样的想象力可以被应用到机械和工程服务上，以便产生解决传统问题的新方法。

我们秉承这种允诺来开始设计，决心探讨传统结构的问题，特别是风，并研究同时服务几个重要社区的内在联系。比尔·法杉认为：

天桥结构将因为自身的优势成为重要的建设规划……而且它有可能横向拓展和收缩，推拉不同的建筑——如果伸缩接头不是被固定在天桥上的话。天桥可能用于连结相互紧邻的大楼，以此降低风对每一栋大楼的影响。邻近大楼之间的距离对于每一栋大楼将承受的风力总量有重要的影响。当大楼相互靠得很近时，它们就为邻近大楼遮挡风。当大楼离得很远时，它们的存在就只会强化风对邻近大楼的影响。确定邻近大楼最恰当的整合方式，以及在天桥哪个位置安放伸缩接头，在哪些地方又不需要伸缩接头，这些需要大量的研究工作。

就所涉及的领域而言，我们发展了垂直城市的概念。首先，我们要求建筑师把新的城市形式看成一个案例或项目，然后有条不紊的讲述了建议的发展模式，参见本书第479页，新的城市形式。

requirements of a new urban form as a brief, or program, to the architect, followed by the step-by-step development of our proposal, which appears on page 479 under section Four New Urban Form.

SECTION THREE / FORUM OF EXPERTS

An Expert Approach
Architecture

Mark and Peter Anderson, Anderson Anderson Architecture
T.J. Gottesdiener, SOM/NY
William Pedersen, KPF
Adrian Smith + Gordon Gill Architecture
Antony Wood, CTBUH

第三篇/专家论坛

一种专家的方法
建筑

马克·安德森和皮特·安德森，安德森·安德森建筑事务所

T. J. 哥特斯迪埃纳，SOM/纽约

威廉·佩特森，科恩·佩特森·福克斯建筑事务所

艾德里安·史密斯和戈登·吉尔建筑事务所

安东尼·伍德，高层建筑与城市环境委员会执行主任

一种专家的方法

An Expert Approach

在美国和中国，快速城市化带来的问题是一种全球增长危机的征兆；解决方案需要建筑学、城市规划、技术、社会学、环境科学和相关领域中最优秀的人才，以一种动态跨学科的方法协同工作。所需的答案要既宏大又细致，既足以处理复杂的整体，又足够灵活到重视个体的要求。

在写这本书的过程中，我们认识到，并非只有一个答案，一个首选的方法。实际上，对于问题的本质，尤其是关于高层建筑的未来或者说它们与环境的关系，也并没有统一的意见。从历史上看，建筑——尤其是大型建筑——总是以进步的名义，通过占用土地，挖掘、运输、转移和消耗自然资源而建立起来。我们在想，是否有可能改变这种现状，使我们可以不再掠夺自然，而是以一种积极保护自然的方式来进行建筑活动。

我们以一生在建筑行业中获得的私人和专业经验来迎接这一挑战。然而我们很快认识到，我们所探究的领域是如此宽阔，其中的专业领域非常多，新的建筑技术如此纷繁复杂，因此负责任地说，我们只是作为见多识广又满腔热情的非专业人士在发挥作用，只是努力去发现问题，而不是给出答案。因此我们决定，通过采访相关领域公认的领头人来设立一个专业论坛，将他们的思想和观点呈现出来以供参考。

今天，史无前例的大批人群正涌入城市，高层建筑是为这些人提供住处的最有效方式，从这一基本前提开始，我们想要探究其中包含的一些更大的问题。对纽约人来说，由于"911事件"在我们的记忆中依然十分清晰，所以我们想要知道将来高层建筑可能发展到什么情况。进一步说，如果增加密度是一个普遍接受的、解决人口增长、城市化扩张问题的方案，我们想要探究，超高层建筑是否会成为一种可能。就超高层建筑而言，技术上有哪些挑战？是否有促进超高层建设的最新技术？住在离地面非常远的高层上有什么社会学层面的意义？

我们对中国有着特殊兴趣，也很好奇地想从正在中国工作的专家那儿了解，他们是如何比较在东方与在美国的工作经验，有哪些特殊的挑战？我们也想从那些有第一手经验的专家那里，学习关于中国可持续性的东西。比如：指导原则是什么？它们是强制执行的吗？在努力解决城市化问题的过程中，中国面临的最大障碍是什么？

The problems resulting from rapid urbanization in the United States and China are symptomatic of a growing crisis worldwide; solutions will require the best minds in architecture, urban planning, technology, sociology, environmental sciences, and related fields, all working together in a dynamic cross-disciplinary manner. The required response needs to be simultaneously very large and small, sufficient in scope to address the complex whole and yet flexible enough to appreciate fine-grained individual needs.

In developing this book, we realized that there is no one answer, no single preferred approach. Indeed, there is not even uniform thought about the nature of the problems, least of all about the future of tall buildings or their relationship to the environment. Historically, buildings—particularly big buildings—were erected at the expense of the land, excavating, channeling, diverting, and consuming natural resources in the name of progress. We wondered whether it might be possible to invert the status quo so that instead of despoiling Nature, we somehow build in a way that would actively preserve it.

We came to this challenge with a lifetime of personal and professional experience in architecture. Yet we quickly realized that the scope of our inquiry is so great, the areas of specialization so many, and emerging building technologies so numerous and complex that, responsibly, we could best serve in the capacity of informed and concerned laymen, asking questions rather than supplying answers. We therefore decided to build a forum of expertise by interviewing acknowledged leaders in related fields and to present their thoughts and insights for consideration.

Beginning with the basic premise that tall buildings are the most efficient means of accommodating the unprecedented numbers of people surging into today's cities, we wanted to explore some of the larger issues involved. As New Yorkers, with 9/11 still vivid in our memories, we wanted to know the prognosis for future high-rise buildings. Going further, if increased density is universally accepted as one solution to the challenges of population growth and expanding urbanization, we wanted to explore whether ultra-tall buildings might be a possibility. What are the technical challenges specific to ultra-tall building? Are there recent technological advances that would facilitate ultra-tall construction? What about the sociological implications of living very high up off the ground?

Given our particular interest in China, we were also curious to know from experts currently practicing there how they compare the Eastern work experience to that in the United States. What are the specific challenges? In addition, we wanted to learn about sustainability in China from those with firsthand experience. What are the guidelines? Are they enforced? Which are the greatest hurdles facing China in its effort to solve its urbanization problems?

Each question generated others—not that we ever ended up with definitive answers—that couldn't somehow be expanded or modified by new or different input. We hope, by adjacency between the covers of this book, to provide just such an opportunity and to encourage cross-pollination among disciplines that traditionally have had little overlap. We also hope to generate discussion, deliberation—even disagreement and heated debate—among our readers, as personal engagement is the first step toward taking action.

The ideas expressed in the following interviews are solely those of the experts. They appear here for information purposes only and do not in any way suggest input or endorsement of our proposal for a Vertical City.

每个问题都会产生其他问题，对这些问题，我们不能得出一个最终的答案（最终的答案是指：不能被新的、不同的意见拓展和修改的答案）。通过将这本书中所涵盖的诸多领域并置在一起，我们希望提供一个机会，鼓励不同学科之间的交流，虽然在传统意义上，这些学科基本上没有交叉重叠之处。我们也希望在读者中引起讨论、思考，甚至异议和激烈的争论，因为亲身参与其中是迈向行动的第一步。

下面访谈中所表述的观点仅仅属于那些专家。他们出现在这里只为了提供信息，并不会以任何方式给出意见或是对我们关于"垂直城市"的看法表示支持。

The crest of Jin Mao Tower above Shanghai's endless skyline. The city's population has grown from roughly 11 million in 1980 to almost 24 million in 2012.

上海连缠天际线中的至高点——金茂大厦。这个城市的人口，已从1980年的1100万，增至2012年的2400万。

Mark and Peter Anderson
ANDERSON ANDERSON ARCHITECTURE

Mark Anderson, FAIA and Peter Anderson, FAIA, brothers, are San Francisco–based architects, builders, and educators at the University of California, Berkeley, and California College of the Arts, respectively. Their award-winning projects include urban planning and a wide range of building types in the United States, Europe, and Asia. Growing from a strong foundation in the practicalities of construction, Anderson Anderson Architecture has worked directly with manufacturers and government agencies in the U.S. and Japan to design and build numerous prefabrication systems that creatively explore new construction technologies and affordable building methods. People-based designs are the focus of Anderson Anderson's interest in sustainable construction. Individually and together, the two principals have directed significant research projects on new communications and construction technologies, materials, and processes, and have served as industry consultants. The firm's extensively published works have appeared in leading international journals, exhibitions, and art installations, and are included in permanent museum collections.

马克·安德森
皮特·安德森

安德森·安德森建筑事务所

马克·安德森和皮特·安德森兄弟是扎根于旧金山的建筑师、建设者，他们还分别是加州大学伯克利分校和加州艺术学院的教师。他们获奖的项目包括城市规划以及遍布美国、欧洲和亚洲的许多建筑形式。基于实际建设的坚实基础，安德森安德森建筑事务所在美国和日本直接与制造商和政府部门合作，设计和建设了许多预制的系统，这些系统创造性地探索新的建设技术和有效的建筑方法。安德森安德森建筑事务所在可持续性建设中关注的是以人为本的设计。两位负责人或单独或协作主导重要的研究项目，这些项目汲及新的交流、建设技术、材料和程序。同时，他们也是行业的咨询顾问。公司出版的大量作品出现在前沿的国际杂志、展览和装置艺术中，其中一些被博物馆永久收藏。

For complete biographies, please refer to the Appendix.
完整的传记，请参阅附录。

Proposal for Lips Tower in San Francisco, a thirsty urban utility sucking water and solar energy from the sky, winner of the 2012 AIA San Francisco Unbuilt Design Honor Award (Anderson Anderson Architecture, 2011)
旧金山的唇塔模型。这是一个形似干渴的城市设施，从天空中吸取水分和太阳能，获得美国建筑师协会2012年旧金山未完成建筑荣誉奖（安德森安德森建筑事务所，2011年）

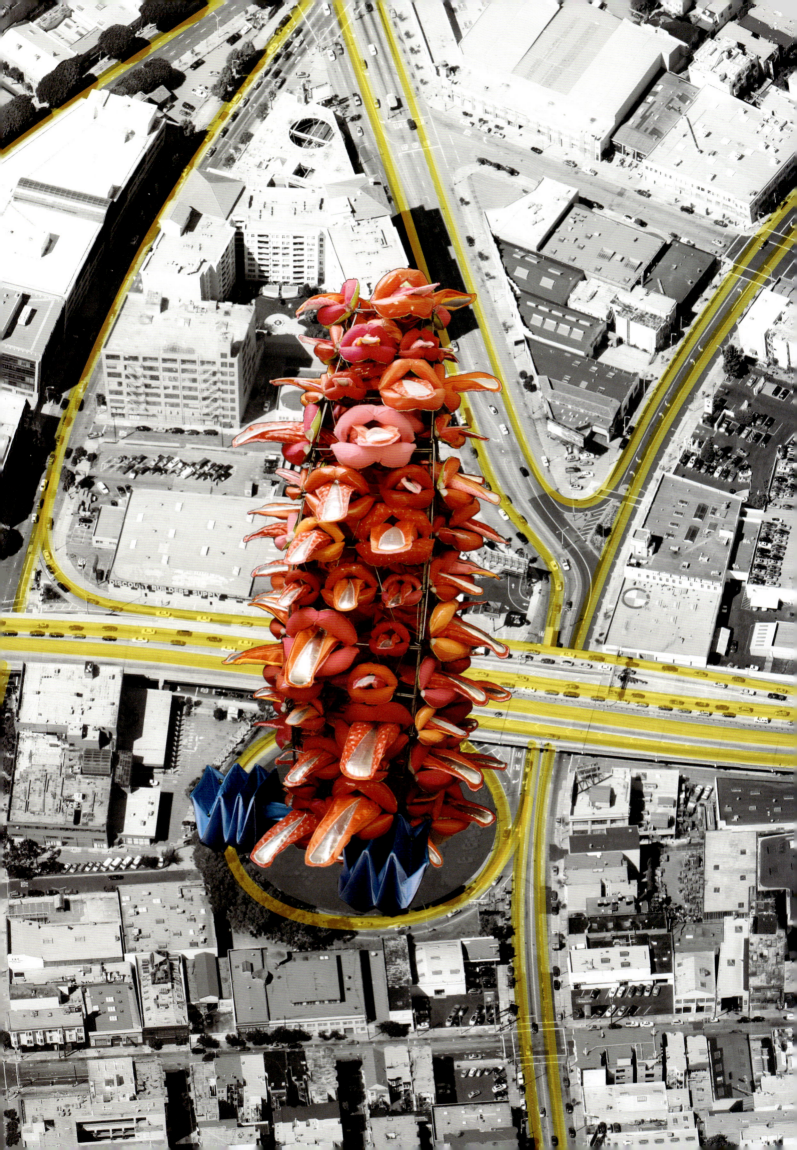

Entry to the Vertical Cities Asia competition by the Department of Architecture, University of California, Berkeley. (2011)
通过加州大学伯克利分校建筑系参加"亚洲垂直城市竞赛"（2011年）。

Please explain the Vertical Cities Asia competition and your role in it.

MA: It's an interesting project established by the World Future Foundation in Singapore, which is dedicated to the future of cities worldwide. They've sponsored the National University of Singapore to do a 5-year series of competitions and symposia about the future of very-high-density high-rise cities in Asia, each with different themes and in a different location. This past year it was in Chengdu, and the theme was clean atmospheres in an urban environment.

What were the criteria for participation?

MA: Ten universities from around the world were invited. Harvard and MIT dropped out at some point so the American universities came down to Berkeley and Penn. Ultimately ETH Zurich won; Delft Technical University and Tongji University in Shanghai placed second and third. [The competition results are posted on www.verticalcitiesasia.com; there will also be a print publication.]

请你谈一下"亚洲垂直城市竞赛"以及你在其中扮演的角色。

MA: 这是一个极为有趣的项目，它由新加坡的世界未来基金会发起，该基金会主要关注全球城市的未来。他们已经赞助新加坡国立大学开展了一项为期五年的竞赛和研讨会，主要讨论亚洲地区高密度、高建筑的城市未来。竞赛和研讨会关注不同的主题、在不同的地区举办。去年，活动选在中国成都，主题是城市环境中的空气清洁问题。

参加这个竞赛有什么条件吗？

MA: 全球共有十所大学受到邀请。其中，哈佛大学和麻省理工学院中途退出，因而美国的大学只剩下加州伯克利大学和宾夕法尼亚大学。在竞赛中，苏黎世联邦理工学院拔得头筹，荷兰的代尔夫特理工大学和中国上海的同济大学分列第二、三名。（竞赛的结果发布在www.verticalcitiesasia.com

上，当然，相关的出版物我们也将在不久后推出。）

我在伯克利任教，同时也是该竞赛的教员指导和工作室指导；彼得（Peter）任教于旧金山加州艺术学院，担任工作室的一名常驻顾问。

教员在多大程度上参与活动？

MA: 在我们的活动中，主要的参与者是学生。不过，每个学校的情况也不尽相同。与伯克利相对宽松的环境相比，欧洲的大学在设计方面有更多的教员参与。在我们伯克利的工作室中，学生们来自景观建筑、城市规划和建筑学这些不同的院系，因而整个工作室的规模相当可观且纷繁多样。我们组建了两个团队，让他们发挥各自的创意。

能看到各所学校的不同想法，是件非常有趣的事情。因为出版和传媒的影响范围已经遍布世界各地，所以人们所期盼的多样性变得极为稀缺。我希望，尤其是在中国，可以看到更差异化、更原创的思想。这些思想来自独特的中国经验，且较少受到西方影响。

PA: 在一定程度上，由于受到我们所教的建筑学方法影响，学生们大多以广泛的调查研究开始，我认为他们做的很出色，这在所有的大学中皆是如此。他们的工作很大一部分就是，在构造塑型之前，努力获取和消化信息。因此我认为，对于其中的一些人，实际整合而成的建筑将颇为不同。马克（Mark）提出：不同方案之间存在惊奇的相似，大家都不仅对社会话题感兴趣，而且关注技术、能源、环境和水循环系统——所有这些都是当下全球设计界的热点。我想我更希望亚洲的竞赛者在这些问题上提出独特的见解，但似乎他们也深受当下国际建筑界语境的影响。

是否存在一种中式建筑学？

MA: 我认为现在并没有。当然，中国有着悠久的建筑历史，特别是伟大的园林传统，我认为这是中国建筑最独特、最有价值的历史贡献。中式园林的演绎方式如此纷繁多样，若能在中国新时期的城市规划中更多地看到它们的影子，那一定是非常棒的。

我一直将中式园林理解为一种人造的自然。与近来西方看待自然的观念或以更质朴的方式存在于自然之中的观念不同，中式园林重建了一种自然观念，一种远景。在对自然与人造世界截然不同的理解中，它塑造了一种与建筑世界的特定关联。

I teach at Berkeley, and served as faculty adviser and studio instructor for the competition. Peter, who teaches at the California College of the Arts in San Francisco, served as a frequent consultant to the studio.

What was the degree of faculty involvement?

MA: In our case it was mainly student work, but that varied from school to school. Berkeley tends to be a less dictatorial place than some of the European schools, where there was greater faculty input regarding design approach. At Berkeley, we had students from the departments of landscape architecture, city planning, and architecture, so it was a fairly big and diverse studio. We set up two teams and let them develop the ideas.

It was interesting to see what different schools are doing. Because the range of influences through publishing and media is so pervasive around the world, there wasn't quite as much variety as one might hope. I would rather have seen more diversity, especially from China, more original thinking that is unique to the experience of China and a little less influence from the West.

PA: Partly as a result of how we teach architecture, the students started with broad research investigations, which I thought were very well done. That was pretty consistent across all the schools. A substantial portion of their work was finding and digesting information before coming up with forms, so I think the actual integration into buildings came fairly far down the line for some of them. To Mark's point about there being surprising similarities, everyone was interested in social issues, but also technology, energy and environment, water recirculation—all the things that are now hot buttons throughout the design world. I guess I expected that the Asian entries would have had a uniquely different take on some of those things, but they seemed very influenced by current architectural discussions globally.

Is there a Chinese architecture?

MA: I don't think there is now. Certainly there was historically, especially the great garden tradition in China. To me, that's the most unique and valuable historical contribution of Chinese architecture. There's so much richness in the way the Chinese garden developed; it would be wonderful to see more of that in China's new urban design projects.

I've always understood Chinese gardens to be a kind of human-constructed nature. Unlike the more recent Western idea of looking at nature, or being in nature in a more pristine way, it reconstructs an idea of nature, a kind of distant view, and creates a certain relationship with the built world in a very different layered understanding of the natural and man-made worlds.

PA: One interesting thing about the competition was that all the students did a very good job of bringing the landscape in vertically, which is somewhat of a departure from many ideas of tall buildings, especially in the West, where tall buildings take you farther from the ground. As Mark suggested, the opportunity to integrate the history of Chinese landscape design has great potential. I thought the Tongji proposal was particularly striking, with its ground plane raised up into the air.

One wonders what the results would have been if the test project had been located outside China.

MA: Even though it's sponsored by Chinese donors, it won't always be focused on Chinese cities. It's always a problem to design something for a foreign place. There are all sorts of political and cultural debates involved, but again, it's a universal issue. We're not Americans or Chinese anymore, we're all designing everywhere; we all think we can be culturally appropriate to wherever we work. One questions that. On the other hand, so much of what people in China are doing is culturally inappropriate. There's so much amnesia about its great traditions. It's almost the outsiders who are more sensitive.

Is it amnesia or rejection?

MA: That's a very good question. Personally I don't think it is rejection, just a lot of amnesia and fascination with foreign influences.

Please explain a bit about the modular architecture you've been developing.

MA: We've worked on some standardized modules that can be high-rise or have a broad application in many building forms, and also some that are very specific to high-rise or low-rise production. A couple of years ago we worked on a project sponsored by Living Steel, a European-based steel manufacturer's association working with Bao Steel in China. There was a competition in which five firms from around the world were invited to produce an 11- to 12-story south-facing block, kind of a standard Chinese high-density dwelling.

It was mid-income housing, but the idea was for it to be very affordable (which hasn't in recent years been the real

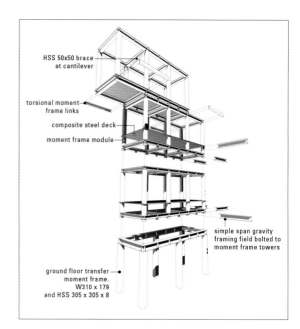

Wuhan Blue Sky Prototype, exploded section of modular moment frame box assembly.
武汉蓝天模型，装配成整体的模块分解成部分。

PA: 竞赛中有件非常有意思的事情：所有的学生都很好地将景观垂直处理，这实际上有点偏离了许多关于高层建筑的观点——这些观点在西方尤其突出，因为在那里，高层建筑让你远离了地面。像马克指出的，融入中国景观设计的历史将会产生巨大潜力。我认为同济的方案很有吸引力，它将地平线提高至空间。

假如实验项目不在中国进行，会产生什么后果。

MA: 虽然实验项目得到了中国资助者的支持，但它并不一定只聚焦于中国的城市。为一个陌生的地方设计，总是有些困难的。这其中包含着各种政治与文化的争议，但这是一个普遍的问题。我们不再仅仅是美国人或者中国人，我们在各处进行设计；无论在哪里工作，我们都会与之产生文化上的契合。另一方面，今天人们在中国所做的很多东西都在文化上不合时宜。人们对于自己曾经的伟大传统，已经遗忘了太多，反而是那些外来者显得更为敏感。

这究竟是遗忘还是拒绝?

MA: 这是个很好的问题。我自己不把它看作拒绝，只是遗忘太多，而且沉迷于外来文化的影响。

Solar-oriented Wuhan Blue Sky Prototype, shortlisted in Living Steel's second International Competition for Sustainable Housing.
(Anderson Anderson Architecture, 2007)

以阳光为导向的武汉蓝天模型，入选以可持续居住为宗旨的第二届住宅钢结构项目国际竞赛。（安德森安德森建筑事
务所，2007年）

请简要解释一下你所开发的模块化建筑概念。

MA: 我们已经着手开发了一些标准化模块，它们
可以用来建筑高楼，或广泛应用于众多建筑样式
中；我们也开发了另外一些模块，但它们只适用于
特定的高层或低层建筑。多年前，我们开发了一个
由生活钢铁（Living Steel）资助的项目。生活钢
铁是一家欧洲和中国宝钢合作的钢铁建筑商。当时
还有个竞赛，全球有五家公司被邀请，生产一种
11-12层南向的模块，这样的建筑是一种标准的中
国高密度住宅楼。

这种住宅适合中等收入者，但该项目的初衷是
让人们能够负担得起（最近几年，它并未成为中国
建设过程中真正的焦点），工厂生产且用于建筑的
是钢铁而不是水泥。至于中央政府为什么要鼓励
钢铁行业，有许多的原因。对我们而言，这个将要
坐落在武汉的项目，更多地成为了与附近汉江和长
江联系在一起的水系统，而且使得分水岭被抬升，
以适应垂直建筑的生活方式。在一系列的结构分析
之后，我们提供了一些模块，它们以一种悬挑的方
式发生偏离。这些模块堆叠成塔，中间则注入钢结
构。在各层中，模块略微偏移，使得整座塔悬浮着
越出地平线上的基底。工厂定制的楼梯和机械管道

focus of construction in China), factory produced, and created in steel instead of concrete. There are lots of reasons why the central government wants to encourage the steel industry. For us, the project, which was to be located in Wuhan, became very much about the water systems related to the nearby Han and Yangtze rivers and drawing this watershed up into the vertical life of the building. After a lot of structural analysis we came up with modules that are offset in a cantilevered way. The modules are stacked to form towers, with an infill steel structure in between. The modules are slightly offset on each level, so that the overall tower cantilevers substantially beyond the base at ground level. Factory-produced stairs and mechanical and plumbing systems are located within these tower cores. Once you have the towers in place, these serve as working platforms from which to build the rest of the floor area with steel beams spanning between towers. The complex parts of the construction are built in the tower core modules delivered to site and hoisted into place, so it's very fast to build.

Is there any limit to how high you can go?

MA: Not if you're going vertically, but we were stacking laterally in a cantilevered form, which limits total height but offers a number of self-shading and sunlight-access advantages By leaning our building to the south in the cantilevered tower form we were creating a self-shading wall for summer but which would get full light in winter. This self-shading form plus a porous and through-ventilating unit organization results in less energy use, and more comfortable dwellings with natural ventilation and optimal sunlight.

So there is an inherent depth limitation?

MA: Yes, but also, because they're leaning, the buildings in a residential neighborhood can actually slide closer together, so it takes about 20 percent less land area than a more typical 11-story residential neighborhood, given the same sunlight access to all units. The building code in China requires a certain amount of sunlight every day in every dwelling unit, so major living spaces effectively have to be on the south. To avoid buildings shading other buildings' living spaces, there's a certain land spacing required between buildings in typical 11-story residential developments. The cantilevered form allows the buildings to have closer spacing while still not shading each other.

Do the steel units comply with fire rating?

MA: Yes. They have concrete decks and are fully fireproofed to meet international building codes.

PA: Most of the units are 2 stories. On every other floor there is an open-air horizontal link, which concentrates activity to create more of an urban feeling of a street than a high-rise setting. All of the apartments have doors onto this street, with 2-story units entered on the north side of the street, leading upstairs, and with living spaces crossing over the street, allowing all units to have both south and north exposures; every unit has four sides of air and thus good cross-ventilation.

MA: It's like Le Corbusier's l'Unité d'Habitation, with the main corridors on every other floor. Because of Chinese building code and the requirement to face south, it means there's usually an elevator core for every 2 to 4 apartments per floor, so there is very little horizontal interaction between people. Here, with a kind of skip-stop approach, people are encouraged to walk, use stairs, and meet each other; it's more like a front porch onto the street.

One of the great things about traditional Chinese cities is the richness of people walking and talking on the street,

工程系统位于塔芯之内。一旦塔正处其位，这些设备便可以作为工作平台，以便修建钢柱连接塔楼的剩余楼层部分。建造过程中，较为复杂的环节在于：将塔芯之中的构建输送到场，并吊入其指定的模块位置，这样使建造过程变得非常快捷。

这种建筑对高度是否有所限制？

MA: 假如垂直向上，是没有限制的。但是我们以一种悬挑的方式从侧面堆叠，这就会限制总体的高度。不过这种方式带有很多自遮阳和获取阳光的益处。由于建筑以悬挑的塔式结构向南倾斜，我们创造出一堵在夏天时的遮阳墙，而它在冬天时又能获得充足的阳光。这种自遮阳形式，加上一套渗透性强、通风好的装置系统，既节省了能源，又提供了具有自然通风和最佳光照的宜居环境。

那么，这样的建筑在内部深度上还是有所限制的吗？

MA: 是的，不过，由于它们是倾斜的，住宅区的建筑实际上会越靠越近，因而，若考虑到所有单元中同样的阳光照射量，相比于普通的11层住宅楼，它们大概节约了20%的土地使用面积。中国的建筑规格要求每个住宅单元，在每天都获得一定量的阳光，因而大部分住宅空间实际上必须是面南的。为了防止某栋建筑的生活空间被其他建筑遮挡，在普通的11层住宅楼之间，往往需要一定数量的土地空间以拉开间隔。而悬挑形式允许建筑物在互不遮挡的前提下有更近的距离。

钢结构单元是否也遵守防火等级要求？

MA: 是的。它们具有混凝土层面，并且符合国际建筑规格，是完全防火的。

PA: 这些单元大多数是双层的。每隔一层，就会有露天的横向连结，用于举办活动来创造出街道的都市感而不是高层环境。所有的公寓都有门通向街道，双层单元占据着街道的北侧，引向楼梯，并且有生活空间跨越街区，因此，所有单元皆南北通透；每一单元都四面空敞，通风良好。

MA: 就像勒·柯布西耶的巴赛公寓，每隔一层都有主廊。由于中国建筑的规格，以及朝南的要求，每层都有一部电梯为2-4户公寓服务，因而人与人之间就少有横向交流的空间。在这里，通过一种跳站停车的方式，鼓励人们步行，使用楼梯以便遇到其他人，这更像是一条通向街区的前廊。

传统中国城市的一个伟大之处就在于它的繁华热闹，人们可以在街道上、阳台上、屋子里行走和交谈。无论天气冷暖，这都是一种室内外适宜的生活，虽然在一定程度上，它源于人们的生活空间比较狭窄。这一点是现代都市中最大的一个缺憾，因而，我们非常乐于尝试去发现一种更街区式的生活方式——即使在高空，且只有一些简单的设施，例如人在行走时发出声响的木板平台，或者能够从一条阶梯式街区望向另一街区的阳台。尽管这些都是很小型的建筑，但它们依然有着非常鲜明的外在潜力，也可以通过安置商店或办公室不断加以丰富和充实。

这些理论如何应用于一栋占地面积更小的高层建筑上？

MA： 促进高层居民之间的交流需要更多的思考，因而，这绝不是靠一部电梯通向一条很短的走廊，然后入户到独立单元中就能解决的，而是要找到能为高层建筑提供更多横向交流空间的方法。况且，在垂直的布置中拥有多样性是必要的，它为传统亚洲城市的高层建筑带来了另一伟大的影响：每个人都工作和生活于同一区域；它并不是一块隔离开的使用空间。

人们会担心在同一栋建筑中，生活、工作、睡眠、娱乐的功能是否可以长期有效使用，这些建筑是否能够提供人们所需的灵活性。

PA： 关于将生活和工作结合起来这个问题，我们经常假设，同样的人总是在一个地方生活、工作，如果她/他想要换个别的地方工作怎么办？城市真正运转的方式是：你可以在一条街上而不是在你住的地方工作，而且其他人也可以来到你的街区工作。丰富性来自于融合，因而，有了良好的交通运输，人们的来往就成为可能，也就是说，你并不需要刻板地生活和工作在同一栋楼中。

密度是如何判定的？

MA： 关于恰当的密度以及它究竟意味着什么，存在着许多不同的观点。算上旅游者和日班工人，有多少人是从外部来的？这肯定没有明确的定义。垂直城市竞赛的一个评委，来自巴塞罗那的华金·萨巴特（Joaquín Sabaté）教授，基于街区大小，人口数量和区域划分，针对欧洲城市做了一项研究，因此他对于这个问题有了一个确定的统计学答案。

on balconies, in windows; it's a very indoor-outdoor life even in the cold or hot weather, in part because of the small size of living areas. That's one of the biggest losses in modern cities, so we have been very interested in trying to encourage a more streetlike way of living, even up in the sky, with just simple things, like wood-plank decking that gives sound to people walking or balconies that look down from one stepped street to the next. Although these are fairly small buildings, they have quite a lively exterior potential and could be enriched by programming with shops or offices.

How do these theories apply in a high-rise building, with a smaller footprint?

MA: Promoting interaction among people higher up requires a lot of thought, so it isn't just an elevator ride to a short corridor and into an individual unit, but rather finding ways for more horizontal interaction at higher elevations. Also, having lots of diversity in vertical programming is essential, bringing to the high-rise is another great thing about traditional Asian cities: everybody works and lives in the same area; it's not a segregated zoning of uses.

One wonders about the long-term validity of living, working, sleeping, and playing in the same building, and whether it offers the flexibility people want.

PA: By combining living and working, we often assume that the same person is living and working in that place. What if he or she wants a job elsewhere? The way cities really function, you might work on a street other than where you live, and another person might come to your street to work. The richness comes from the combination, so with good transit making it possible to go back and forth, it doesn't mean you have to literally live and work in the same building.

How is density determined?

MA: There are a lot of different opinions about appropriate density and what that means. How many people come from the outside, including tourists and day workers? There is no clear definition. One of the Vertical City competition jurors, Joaquín Sabaté from Barcelona, did a study on European cities based on block size, population, and zoning, so he had a kind of absolute statistical answer to this. People talk about a kind of tipping point; we're already there in many Asian cities, so it's clear that we are pushing beyond the statistical balance.

In the U.S. there are fewer people for more space; everything is much more concentrated in China, so it's a great laboratory for all these issues. But one thing that struck me throughout the competition discussions was a very narrow understanding of political ramifications. There was a lot of awareness that a vertical city needs diversity of income groups, so there was a lot of discussion of class groups, which really surprised me. Talking about class has been important in China for the last 50 years, but from a different perspective, and now there's talk about needing places for low-income people. It's a sort of an old-fashioned capitalist idea of accommodating different classes, instead of recognizing that there are different social strata that need to be integrated and that the relationship between them has to change, that there needs to be a more fairly distributed economy and society rather than an emphasis on having separate places for the rich and the poor. These class discussions take place without irony, without any political understanding of what this means in relation to the history of the last 100 years in China.

Thinking more creatively about political and social structure could make the current approach to architecture much more interesting. Social structure is embedded in our assumptions of design, yet raising these issues is a little dangerous because there was so much emphasis on that in the 1960s, '70s, and early '80s with largely empty or discredited results. This kind of social and political positioning at the expense of architecture created a certain antagonism between being a progressive architect and being a progressive social activist. It seems like we've totally forgotten what was learned and thought about at mid-century in terms of the political and social ambitions for a modern society. We're looking at a modern style of building but without more comprehensive thought about sustainability, the overall social, political, economic structure that's reinforced with certain assumptions about the way we design and build.

PA: One of the primary focuses of the competition was this: as urbanization takes place, what happens to farmland? Everyone is moving into the cities, and the cities are spreading out, so how is farmland integrated into the planning? Most proposals had some aspect of urban farming.

It should be one of the advantages of a strong, central government in China to set policy that will accomplish greater goals and, in terms of development, regulate the mix of different kinds of units. That's very common in development planning. A city sets goals to achieve larger goals beyond those of individuals, and developers learn to work within that structure; it creates a kind of level playing field for the market. Anyone who wants to build a new building in San Francisco has to provide X number of affordable housing units.

人们常谈论一种临界点；其实许多亚洲城市已经在临界点上，很显然，我们正在突破统计学上的平衡。

美国地多人少，而中国的一切都更为集中，因此它也成了所有问题的一个大实验室。不过，在整个竞赛讨论中，令我触动的一个事情是对于政治分歧极为狭隘的理解。垂直城市需要各种收入层级的人，这一点已经被很多人意识到，因此，就有了许多对阶层的讨论，这些讨论着实令我惊讶。过去的50年中，在中国谈论阶层是非常重要的，不过，那时人们是从不同的视角出发；现在，人们谈论的是为低收入阶层寻找空间。适应不同阶级是一种旧式的资本主义观点，它没有意识到不同的社会阶层需要被整合，他们的关系需要改变，需要一个更为公平分配的经济和社会环境，而非过于强调将富人和穷人分置于不同的空间。这些阶层的讨论不带任何讽刺，也没有任何关于中国过去一百年历史的政治性理解。

如果更创造性地思考政治与社会结构，将会令当前的建筑方法更加有趣。社会结构体现在我们对设计的假想中，不过，提出这些问题略有风险，因为在上世纪六七十年代和80年代早期，人们往往过于强调它，结果造成空洞和不可信。这种以牺牲建筑为代价的社会政治观点，促使一名进步的建筑师与进步的社会活动家之间产生对立。似乎我们完全遗忘了，自己曾在世纪中期为构建一个现代社会政治和社会抱负而所作的思考与努力。我们在寻找一种现代的建筑风格，却没有对可持续性、整个社会、政治、经济结构有更广泛的理解，这种结构也强化了我们对设计和建筑方式的设想。

PA: 在这次竞赛中，有一个重要的关注点：随着城镇化的推进，农田发生了什么改变？因为所有人都向城市迁移，城市正延伸拓展，那么，农田如何被整合到规划中？大多数的方案都提到了都市农业化问题。

中国的一个优势在于：它有一个强有力的中央政府制定政策，以完成宏大的目标，并从发展的角度，整合不同的部门。在规划中，这些是非常普遍的。一个城市设立目标，可以超越那些个人，完成更大的目标，发展者在这种结构内部开展工作，为市场营造了一个平稳有序的运作环境。任何想要在旧金山盖新房的人，必须支付得起一定数量的房屋单元费用。

在中国以及其他一些地方，当前追求可持续性具有多少真实性？又有多少是公共关系？

MA：我的感觉是，在全世界的建筑师、城市规划师、学生和市民中，兴趣是最真实的。持续性是一个重要的问题，而城市被设计的方式乃是这个问题的重头戏。我是说，这也许并不仅仅是几个政治家和公司的环保理念，但即便只是谈论，它依然有一个积极的方向。

PA：假如企业的兴趣支持环境方面的议题，即使一开始只是从一种公关宣传的角度去考虑，也会将这些议题锁定，并很难被避开。因此，对我而言，无论一开始的理由是什么，它都产生了判断行为恰当与否的标准。其中，一项最大的成就是让大公司提升绿色环保的能动性，而不是为清洁空气立法去斗争。例如：他们可能依然在幕后做一些小动作，但实际上都支持一项公司绿色环保政策。我觉得无论缘由是什么，结果都是一种成功。

从中国的观点看，中央政府多少受实用主义的影响，试图管理有限的资源，这一举措是令人兴奋的。能源政策和可持续性设计，是中央集权制定政策的优势！

你经常在建筑和饮食之间进行类比。

MA：这是一个由来已久的兴趣。过去我们在日本做了很多工作，亚洲一直令我深切痴迷的是当地的烹饪。我认为世界各地的人们都十分重视饮食传统，而非建筑传统或城市生活。我们以这样的方式思考城市潜能：如果人们能够对于他们的食物如此关心，并且真的保持与传承饮食传统，那么，相同的情况也能发生在建筑中，发生在城市形式中。所以我们倾向于依照烹饪的类比进行思考。

那么，这就超出了建筑的范畴，延伸到城市之精细的核心，延伸到城市气息，延伸到生活的质量问题。

MA：是的。人们尚未完全丢弃自己的文化传统或发展历史，他们生活在一起，分享食物。既然关注吃什么不是一个巨大的跨越，那么我们就可以真的在乎我们生活的空间、园林、建筑，以至于整个的文化生活，这些非常重要。

PA：拓展一下这种类比，我们关注的焦点已不完全在于厨房、工具器物，而在于烹饪、分享食物的味道和整个人类经验。同理，在建筑中，我们并不

In the recent quest for sustainability in China, but also elsewhere, how much is genuine and how much is public relations?

MA: My impression is that among architects, urban designers, students, and citizens around the world, the interest is genuine. Sustainability is a very major issue, and the way cities are designed is a big part of that. I don't have any sense that it's just a green-washed view other than maybe for a few politicians or corporations. Even if it is just talk, it has a positive direction.

PA: If corporate interests embrace environmental issues, even if initially from a PR perspective, it locks them in and makes it harder for them to diverge. So regardless of the reason in the first place, to me, it raises the standard by which we judge appropriate behaviors. That's one of the greatest accomplishments of having the big corporations promoting green initiatives instead of fighting against clean air legislation, for example—which, of course, they still do somewhat behind the scenes—but to actually embrace a corporate green policy. I think that's a success no matter what it is.

From the point of view of China, it's exciting that the central government, partly for pragmatic reasons, is trying to manage limited resources. Policies for energy and sustainable design are real advantages of a centralized policy-making government.

You often draw analogies between architecture and food.

MA: This is a long-standing interest. We used to do a lot of work in Japan, and one of the things that always interested us about Asia in general is the deeply ingrained regard for local cuisine. I think that people worldwide really value the tradition of food in a way we haven't valued the tradition of architecture, or of urban life. The way we think about the potential of cities is that if people can care this much about their food, and really preserve its traditions and have a reawakened interest in these things, the same thing can happen about architecture, urban form. So we tend to think in terms of gastronomic analogies.

That goes beyond just buildings to the fine-grained core of what a city is, to the smells and lullabyes, the quality of life.

MA: Right. People haven't thrown away all their cultural traditions or the history of how they developed, living with each other, sharing food. It's not such a big step to think that if we really care about what we eat, we could really care about the space we're in, the garden we're in, the building we're in, and to savor the whole culture of life. It should be really important.

PA: By extension of that analogy, the focus isn't so much on the kitchen, the physical artifacts, but on the cooking, sharing, and taste of food, the whole human experience. That's what we can hope for in buildings: that we don't treat them as objects separate from human experience. It's not just a trophy, the tallest, or some other physical attribute like that.

MA: The other hopeful thing about that is the change in the way food has been thought of in the United States over the past 40 years. It hasn't just been a sudden urge to get back to the old ways. It's much more inventive than that, a new, very vibrant entrepreneurial approach. Resurrect what was lost but in a creative way. That's what's interesting about the possibility for architecture. We're not trying to recreate cities, but to say there are ways of living that are really rich in old traditions and that we can be really original and creative about.

You talk of architects as humanists, sociologists—chefs— at the very same time that architects are more inclined than ever toward engineering.

MA: As one of the oldest professions, architecture has had a more intact tradition than any other field of human endeavor. That's really been shaken in recent history, so that in the last 150 years urban design took its separate path, and engineering really separated, so architecture became increasingly narrow as a profession and in terms of what society expects of it. But the real tradition of architecture is quite comprehensive. It's responsible for thinking about how human beings live in the constructed world, and that's very broad in its responsibilities. If you believe in that tradition it means we can't become as specialized as the world might like. We shouldn't be apologetic about that. Technology has become so complicated that to be halfway competent in any area you really have to specialize. But if we split up into specialties for curtain wall, structure, urban landscape, and so on, then nobody's responsible for how all these things fit together. What we should realize is that architecture, our specialty, traditionally has been the big picture. Rather than lament our limits, we should celebrate our breadth; we orchestrate everything in the city.

PA: As educators of architects, we teach the importance of making connections between disparate things, maybe politics, or social or environmental agendas. The important thing

将它们当作从人类经验中分离出来的客体对象。它不只是一座纪念碑，具有"最高"之类的物理性。

MA: 还有一件颇有希望的事情，过去40年的美国，人们思考食物的方式已然发生了变化。回到古老的方式，并非仅是一时冲动。它更多的是一种别出心裁、崭新而又活力十足的创意方法。以一种全新的方式复活失去的东西，这就是建筑之可能性中很有意思的地方。我们并不试图重建城市，而是想，在传统中有着丰富多彩的生活方式，我们完全可以从中获得原创性和创造性。

你以人文学家、社会学家、厨师的身份谈论了建筑师，而同时，建筑师较之从前更倾向于工程技术人员。

MA: 作为最古老的行业之一，建筑学相比人类其他的领域有着更完整的传统。然而，在最近的历史中，它却发生了动摇，在最近的150年中，城市设计开拓出自己独特的道路，和工程学分离开来，正如社会对它的期待一样，建筑学作为一个行业变得越来越窄小了。但建筑学真正的传统非常广泛，思考人类如何生活在被建构的世界中，这种责任是相当广义的。假如相信那种传统的观点，这将意味着我们不会像世界所期待的那样专业，所以不必为不专业而道歉。技术已然如此复杂，以至于在任何领域要想具有中等的竞争力，你就必须变得专业。但是，如果我们区别开幕墙、结构、城市景观等不同专业，那么，就没有人为这些东西结合在一起而负责。我们应当认识到，建筑学这一专业，从传统上讲就是一幅很大的图景。与其对我们的局限深感痛惜，不如庆幸我们的广博；在城市中，我们将一切协调地编排在一起。

PA: 作为建筑师的培育者，我们教授连接各种不同事物的重要性，比如政治、社会或环境等相关事项。关键是，我们成为了专业的多面手，具有别人没有的专业性。如果说我们精于哪一样东西的话，我认为就是在各种事物之间描绘出其他人看不到的联系。就像成为一名新闻工作者，一位优秀的新闻工作者能够写出超越她/他专业领域之外的文章，通过研究，她/他会知道如何提出正确的问题，并找出恰当的细节层次。同样的，作为建筑师，我们必须平衡来自不同人的不同观点。

当下对超级建筑的热衷有没有重新让人们对勒·柯布西耶产生兴趣?

MA: 是的,在新加坡的讨论会上,一些年长的发言者依旧在谈论现代都市规划之恐怖以及勒·柯布西耶的一些东西:如公园中的塔状建筑、线性城市以及诸如此类的观念。然而,几乎所有做超级建筑的学生,特别是中国学生,总是期待大方案、大图表,就像是在上世纪60年代或者勒·柯布西耶的时期。有一种兴趣正在复兴,即:回归解决超级问题的那些超级方案上。这些问题和方案在今天看来显得更加宏大。但如果认为勒·柯布西耶的一些方案或者新陈代谢派的观念可以被简单地重复使用,那简直就是令人吃惊的天真。我确信,对于这些问题一定会有大规模的结构方案予以解决。我们不应该排斥现代派的方案,但不能仅仅重复前人,设想用同样简单的方案解决问题。必须有细微差异的理解,而且考虑得更周全。

某种程度上说,这与我们的唇塔(Lips Tower)有关,它涉及到持续性,以及利用新技术、新方法整合自然环境。我们做的所有项目,几乎都有一种特定的自我矛盾、自我怀疑,这对于我们而言相当重要。解决方案总是不确定的,没有唯一的方式,也不存在普遍的设计方案。但必须有一个出于善意、全局性的、关于需要做什么的构想,围绕这个构想或许在一些内部细节上可以略微不同,也正是这些不同,让我们对建筑有了真正深刻的理解。因此,必须对某些怀疑、幽默和讽刺有所认识。

PA: 我想我们为旧金山高层超密度楼房结构提出的模块化方案,是对这个问题最好的说明。我们不是想着设计出一个方案,能够照顾到所有的细枝末节,而是着眼于使大型方案运行起来:如与环境、能源系统等相关的结构问题,但是要为社区的个体化保留空间,这个空间并不一定是横向的,可以是一种内部结构垂直的居住区。我们对于这种混合搭配很感兴趣。

MA: 勒·柯布西耶最令我们感兴趣的,是他对于世界细腻而又复杂的理解。他于20世纪40年代后期创作的一幅杰纳斯面具素描,对我们很有启发意义。那是一个太阳,它的一半正在微笑,另一半却是美杜莎的头颅。这幅素描象征着在二战之后,他的创作方式发生了改变,并回到了生活的复杂性中。他认识到,世界一方面是由希望、阳光、现代

is that we're specialized as generalists. Nobody else has that area of specialization. If we are good at one thing, I think it is drawing connections between things that other people aren't seeing. It's like being a journalist. A good journalist can write an article outside his or her own technical field, through good research, knowing how to ask the right questions, and finding the right level of detail. In the same way, we as architects have to balance different viewpoints from multiple people.

Have current interests in megastructures led to renewed interests in Le Corbusier?

MA: Yes. At the symposium in Singapore, some of the older presenters were still talking about the horrors of modern urban planning and some of Le Corbusier's ideas for the tower in the park, the linear city, and things like that. And yet almost all of the students were working on megastructures, especially the Chinese students, always looking for the huge solution, the big diagram, like 1960s or Le Corbusier's ideas. There's a renewed interest in going back to some of these megasolutions to megaproblems, which are so much more mega now, but it is surprisingly naïve to think that some of Le Corbusier's proposals or the Metabolist ideas can simply be reused. I'm sure there's going to have to be a very big-picture structural solution to the problem. We shouldn't reject the modernists' schemes, but we shouldn't just go back to them and imagine that the same singular vision is going to solve the problem. There has to be a nuanced understanding, something more.

In a way, that's related to our Lips Tower, which is all about sustainability and harnessing new technology to integrate with the natural environment in new ways. Almost all of the projects we do have a certain self-contradiction, a seed of self-doubt, that's fairly important to us. There always has to be a recognition that the solution is somewhat tentative, that it can't be the only way, that there's no universal design solution. There has to be a well-intentioned, big-picture idea about what needs to be done, but also some nuanced ways around it, within it, that allow for a much more complex understanding of what this architecture is. It has to recognize certain doubts, humors, ironies.

PA: I think a good illustration of this includes our modular proposals for a high-rise super-dense structure for San Francisco. Rather than thinking we were going to design the scheme down to every last detail, we focused on putting into motion the big-picture: structural things related to environmental issues, energy systems, etc., but leaving room for the individuation of the neighborhood, which isn't just horizontal but could be a vertical neighborhood within the structure. We're interested in that mix.

MA: What's always interested us about Le Corbusier is his nuanced and complex understanding of the world. He did a pen sketch of a Janus mask in the late 1940s that's been very important for us. It's a sun, and half of it is smiling and the other half is a head of Medusa. That sketch is emblematic of the way his work changed, for one thing, after World War II, and gets back to the complexity of life. The recognition is that the world is composed of the hopeful, sunny, modernist view on the one hand, but there's also this recognition that it's much more complicated, much more dangerous in certain ways. It's the traditional understanding of destructive forces and creative forces in balance; you can't have one without the other.

派观点构成的，另一方面，世界也是复杂的，在某些方面要危险得多。这就是对破坏力和创造力处于平衡之中的传统理解，你不能拥有一个而失去另一个。

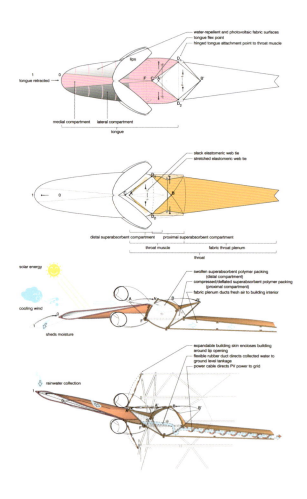

Above: Lips Tower, functional details.
上：唇塔，功能细节。

Right page: Lips open/close and tongues extend/retract in response to weather conditions.
右面：根据天气条件来控制唇的开关和舌头的伸缩。

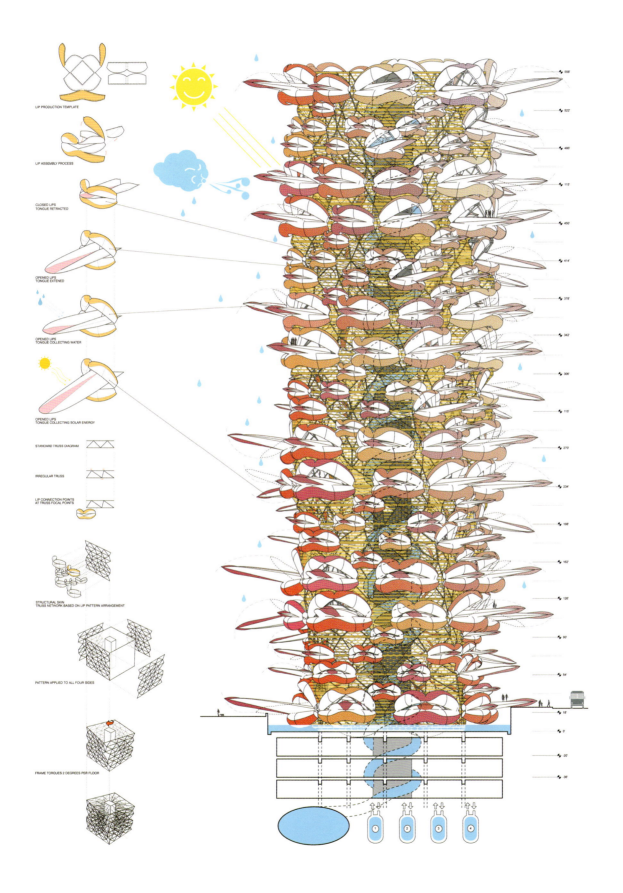

LIP PRODUCTION TEMPLATE

LIP ASSEMBLY PROCESS

CLOSED LIPS
TONGUE RETRACTED

OPENED LIPS
TONGUE EXTENDED

OPENED LIPS
TONGUE COLLECTING WATER

OPENED LIPS
TONGUE COLLECTING SOLAR ENERGY

STANDARD TRUSS DIAGRAM

IRREGULAR TRUSS

LIP CONNECTION POINTS
AT TRUSS FOCAL POINTS

STRUCTURAL SKIN
TRUSS NETWORK BASED ON LIP PATTERN ARRANGEMENT

PATTERN APPLIED TO ALL FOUR SIDES

FRAME TORQUES 2 DEGREES PER FLOOR

ARCHITECTURE Mark and Peter Anderson

T. J. Gottesdiener

SOM/NY

T. J. Gottesdiener, FAIA, is the managing partner of the New York office of Skidmore, Owings & Merrill. A graduate of Cooper Union's Chanin School of Architecture, he joined SOM in 1980 and became a partner in 1994. Mr. Gottesdiener specializes in high-rise and super-tall towers, but his diverse practice also includes some of SOM's most visible master plans, renovation, corporate, and transportation projects, among them the mixed-use Tokyo Midtown project, one of the largest private developments in Japan's history; the landside terminal at Ben Gurion International Airport in Tel Aviv; and extensive work in Brazil and Asia. He has also been responsible for such complex and challenging projects in New York as the revitalization of Lower Manhattan, and has overseen the design and construction of both 7 World Trade Center (2006) and 1 World Trade Center, which will be North America's tallest building upon completion in 2015. Mr. Gottesdiener's significant contributions to architecture were recognized by the Cooper Union's President's Citation in 2007.

T. J. 哥特斯迪埃纳

SOM/纽约

T.J.哥特斯迪埃纳是美国建筑师协会会员，斯基德莫尔、奥因斯与梅里尔建筑师事务所纽约办公室的执行合伙人。他毕业于库伯高等科学艺术联合大学建筑学院，在1980年加入斯基德莫尔、奥因斯与梅里尔建筑师事务所，1994年成为一名合伙人。哥特斯迪埃纳先生的专业领域是高层和超高层建筑，但他的丰富实践还包括一些SOM最著名的总体规划，一些具有革新性、整体性的交通运输方面的项目，其中包括多用途的东京中城项目（这是日本历史上最大的私人开发项目之一）;特拉维夫本古里安国际机场的候机楼;以及巴西和亚洲的众多作品。他也在纽约主持过像重振曼哈顿下城区这样复杂而极具挑战的项目，监管指导过世贸中心七号大楼（2006年）和一号大楼的设计与施工，后者将于2015年完成并将是北美最高的建筑。在库伯高等科学艺术联合大学2007年的校长致辞中，哥特斯迪埃纳先生对于建筑业的重要贡献得到了认可。

For complete biography, please refer to the Appendix.

完整的传记，请参阅附录。

Chhatrapati Shivaji International Airport, Mumbai, India (SOM, est. completion 2014)
印度孟买的贾特拉帕蒂·希瓦吉国际机场。（斯基德莫尔、奥因斯与梅里尔建筑师事务所，2014年完工）

Unprecedented urbanization has given rise to almost spontaneous satellite cities in China but also numerous planned communities and ecocities, like the one in Tianjin with a projected population of 350,000. Is there an ideal population for new developments?

I don't think 350,000 is enough. A new town might start with 100,000 people but it has to be able to be flexible enough to expand. I was just in Brasilia, which is a relatively new city, and it is bursting at the seams because it is the capital of the country and the government is growing. The population has doubled, maybe tripled, since it was built. Lucio Costa made the master plan for Brasilia back in 1957, but it was so rigid that Brasilia couldn't grow, so people started moving out into makeshift bedroom communities—suburbs by accident—which just grew by themselves, many with no zoning. Some people moved 15 to 20 miles outside the city.

We actually have a client there who has a very interesting story. He bought an old ranch with hundreds and hundreds of acres that is currently filled with squatters—people who occupy, but do not own, the land. They have built homes, and they've made these homes into gated communities. These are middle-class administrators who commute to work in Brasilia. Our client is doing something really phenomenal; he went back to the original land owner, bought the title, and then sold it to the homesteaders. The people are willing to pay because the title makes possession legal, and land value, which might have been worth $30,000, all of a sudden increases to $50,000 by virtue of having a legal title. By happenstance they've got power, and little by little the local government, what there is of it, brought in water. So now our client is going to transform this low-scale impromptu community of about 20,000 people into a legitimate city. He is going to use the proceeds from his land sales to put in a town hall, a fire department, parks, and a hub for transportation back into Brasilia, and slowly change the structure into a live-work community instead of just a residential community.

Now when you drive through, you see little shops and storefronts on dirt roads. The potential is there. When you think about how much of India and China are like this, this could be a real model. China believes that by 2050 there will be 400 million people migrating to urban areas; India anticipates 200 million. There is no question that the only viable future is an urban one.

中国前所未有的城市化几乎同时催生了许多卫星城市、规划区和生态城，例如天津就有一个规划中人口35万的社区。对于新的开发项目，是否有理想的人口数量呢？

我认为35万是不够的。一个新的城镇开始时可能只有10万人，但它必须具有足够的灵活性，以便可以进一步扩展。我在巴西利亚的时候，它还是一个相对较新的城市，但它现在已经非常拥挤，因为它是一个国家的首都，并且政府也在增长。自从这座城市建立以来，人口数量已经翻倍，也可能已经增到了三倍。卢西奥·考斯塔(Lucio Costa)早在1957年就为巴西利亚进行了总体规划，但它过于刻板，限制了巴西利亚的拓展，因此人们开始迁到临时的近郊居住区，有时会迁到郊区。这些地区一直自我扩张，但很多并没有进行分区规划。还有一些人搬到了距离城市15到20英里以外的地方。

实际上，我们在那里的确有一个客户，并且他的经历非常有趣。他以前购买了一个面积达数以百英亩计的旧农场，那里充满了土地占用者，他们占用土地，但并不拥有土地。他们建造了房屋，并形成了封闭式小区。这些中产阶级管理人员在巴西利亚上班，每天往返两地。我们的客户正在做一些真正特别的事情，他又找到了原来的土地所有者，购买了土地所有权，然后将它卖给这些土地占用者。这些土地占用者都愿意付款，因为土地所有权可以让土地占有合法化，所占有的土地最初可能仅值3万美元，但凭借合法的所有权，会突然增至5万美元。出于偶然性，他们得到了权力，地方政府也渐渐开始为他们供水。所以，我们的客户现在准备将这个拥有2万人口的小规模临时社区改造成一个合法的城市。他将使用他的卖地收入筹建市政厅、消防部门、公园和通往巴西利亚的交通枢纽，并慢慢地将该社区转变为生活工作综合社区，而不仅仅是住宅社区。

现在，当你开车经过时，你会看到泥土路边的小商店和店面。但那里是有发展潜力的。当你想到中国和印度与此多么的相似，这里就变成了一个真实的模型。中国相信，到2050年将会有4亿人迁移到城市，印度预计有2亿人迁移到城市。毫无疑问，未来唯一可行的道路就是城市化道路。

除非有非常好的交通运输系统，否则城市只能依靠汽车进行通勤和本地运输。

这就是为什么每个城市都需要良好规划的原因：

尽可能地整合铁路、轻轨和其他公共交通方式。这是未来的挑战：交通运输、水（水非常重要）和电。零能耗甚至是积极能耗，虽然现在未完全实现，但未来是可以做到的。在一定程度上，它已经开始发生。纽约市前不久委托我们设计了位于史坦顿岛的一所学校，该学校将会是一所零能耗学校。我们已经开始设计，并且要做到持续改进能耗水平。

许多小型和水平的建筑物都覆盖有太阳能集热器，但这仅仅占用了建筑物较少的净面积。是否有什么技术可以优化利用高层塔楼的较大表面，来产生电力？

是的，但我认为这并不会来自太阳能。我可以向你展示我们与伦斯勒理工学院(RPI)一起开发的太阳能集热器。它的尺寸与手表一样，但能量非常强大。但我认为紧凑型能量系统，例如燃料电池，将会成为解决方案。它以一种热电联产的形式，将氧气转变为氢气和电力。世界贸易中心一号塔楼使用了我们所开发的太阳能集热器，可为该建筑物提供3兆瓦的电力，大约占该建筑物所需总电力的13%。诚然，13%不是个大数字，但毕竟它是一个非常庞大的建筑。整个问题本身没有尽头。这种技术将会继续得到完善和发展，从而变得更加高效，提供更多电力。

SOM承担了世界上一些最高建筑物的设计工作。自从弗兰克·劳埃德·赖特 (Frank Lloyd Wright) 以来，就一直有关于1英里高的建筑物的谈论，现在的确有一些项目正在上马。这些建筑物还能更高吗，比如说2英里高？

嗯，我不知道是否有人真正专注于此。在一定层面上，高层建筑的建造主要是关乎经济性，在另一个层面上，它关乎自我表现和个体表现，不管是个人、企业或国家。城市建筑的最佳高度为50至70层，这就是为什么曼哈顿的商业楼宇几乎都处于这个高度。SOM设计了许多这个高度的建筑物，同时，我们也设计过110—150层的建筑物。对于后一种情况，并不是出于实用性或经济性，或者合适性（我不太确定这个词是否合适），而是出于其他原因。问题是你在什么时候会停下来？

Cities tend to rely on automobiles for commuting and local transportation unless there is a good enough transit system.

Well, that's why every city needs good planning: to integrate as much rail, light rail, and other modes of public transport as possible. That is the challenge of the future. Transportation, water—water is very important—and power. Zero energy consumption or even positive energy consumption. It's not fully achievable today, but in the future it will be. At some level, it is already happening. The City of New York just awarded us a school in Staten Island which is going to be zero energy. We are figuring this out, and we just have to keep ratcheting it up and up to the next levels.

Many small and horizontal buildings are covered with solar collectors, but this is just a tiny amount of net square footage. Are there technologies that optimize the large surfaces of high-rise towers to generate power?

Yes, but I don't think it is going to be from solar. I can show you a solar collector that we developed with Rensselaer Polytechnic Institute (RPI). It's the size of a wristwatch yet very powerful. But I think compact energy systems, such as fuel cells, are going to be the solution. It's a form of co-generation that converts oxygen into hydrogen and power. Tower One has 3 megawatts in the array we are implementing at the World Trade Center. They will generate about 13 percent of the building's power. Admittedly, 13 percent is not a big number, but it's a huge building. The whole issue is generational. The technology is going to have to get better and continue to grow and become more efficient and put out more power.

SOM is responsible for some of the world's tallest buildings. Ever since Frank Lloyd Wright, there's been talk of a mile-high building, and now there are actually some on the boards. Could they go even higher, say two miles?

Well, I suppose if somebody really wanted to put their mind to it. On a certain level, doing a high-rise building is about economics, and on another level it's about ego and individual expression, whether personal, corporate, or national. The optimum urban height is 50 to 70 stories, and that's why commercial buildings in Manhattan are virtually all at that height. We've done our share of those buildings at SOM, but we've also done the 110- and 150-story buildings. In the latter case it is not because it's practical or economical, or—I hesitate to use the word—even appropriate. It's done for other reasons. The question is at what point do you stop?

What is the greatest technological constraint on building tall?

Well, elevatoring has always been an issue, but it's not insurmountable because you can take an elevator that climbs 50 stories, cross over at a sky lobby to another elevator bank, and then continue up. But there is a limit to the number of changes people will make, typically not more than one. Now in development are ropeless elevators that run on electromagnetic fields. They're not out in the marketplace yet but I think they will be in the near future. But elevators are not really the main problem. Today, I think our greatest challenge in designing high-rise buildings is life safety and security.

Fortunately, with computer systems we can now start modeling the fuzzy logic of people-moving. If there is an event, large or small, where should people go? How do they get there? How much time does it take to get out? Think about 2-hour fire-rated stairs: why are they two hours if evacuation takes four hours? So these are the kinds of things that are going to be our challenge.

Before 9/11, we all opened up a code book and it prescribed everything we had to do, and we did it. After 9/11 we took the code book and threw it away, and looked beyond pre-dictated requirements, about the other issues that we need to consider as responsible architects and engineers. Now the New York City building code has been rewritten—largely as a result of SOM's thinking on 7 World Trade Center—and it is completely different from what it was when first released in 1968.

Refuge, or "safe," floors are used throughout Asia, but their real value has never been tested, and never used in the U.S.

Safe floors were considered as one of the many code revisions in New York but they were rejected. I'm not sure I know all the reasons. I am not a fire engineer and I would never propose to be one, but personally I'm not sure how many people would actually stay on a refuge floor. It's against human nature.

After 9/11, everybody said, "We're not going to build anymore high-rise buildings." Do you remember that? That was the immediate, emotional response to the events of that day. From 2001 on, we designed high-rise buildings in Tokyo, China, Hong Kong, the UAE, all around the world, but not here. Then it slowly started to change; it's a cycle. Owners and developers want their buildings to be safe, but they're no longer so worried about building tall. Still, with the exception of buildings like Tower One at the World Trade Center, I don't think there will be many who want to build above the tabletop. In

建造高层建筑的最大技术限制是什么？

嗯，电梯一直是一个焦点问题，但它并不是无法克服的问题，你可以乘电梯到达50层，然后穿过空中大堂，再转乘电梯到达更高的楼层。但对人们换乘的次数有限制，一般不超过一次。目前，人们正在开发一种在电磁场上运行的无绳电梯。这种电梯目前尚未上市，但我认为在不久的将来一定会出现。电梯并不是真正的主要问题。现今，我认为高层建筑设计的最大挑战是生命安全和安保。

幸运的是，通过利用计算机系统，我们现在可以开始模拟人员移动的模糊逻辑。如果事件发生了，无论大事件或小事件，人们应该去哪里？他们如何到达那里？需要花多少时间才能出去？对于2小时耐火楼梯来说，如果疏散需要4个小时，那为什么耐火时间仅为2个小时？因此，这类事情将会成为我们的挑战。

在911之前，我们都使用一本法规手册，它规定了我们必须做的一切，并且我们做到了。911以后，我们不再使用该法规手册，作为负责任的建筑师和工程师，我们开始考虑预先规定之外的问题。现在，纽约市建筑法规已经改写，这主要归因于SOM对于世界贸易中心的思考，现在的建筑法规与最初1968年发布的法规是完全不同的。

整个亚洲地区都使用避难层或者叫"安全"楼层，但其实际价值从未得到测试，并且美国从未采用过这样的楼层。

安全楼层是纽约建筑法规修订时的提议之一，但它被否决了。我并不确信自己知道所有的原因。我不是消防工程师，也没有成为消防工程师的打算，从个人的角度讲，我不确信有多少人将会真正停留在避难层，这有悖于人性。

911之后，大家都说："我们不打算再建造高层建筑。"你还记得吗？那是对当天所发生的事件的直接和情绪化反应。从2001年起，我们在东京、中国大陆、中国香港地区、阿联酋和世界其他国家和地区设计了多栋高层建筑，但在这里却没有。然后事情慢慢开始改变，这是一个周期。业主和开发商们希望他们的建筑物能够安全，不过他们已经不再对高层建筑表示忧虑。除了像世界贸易中心一号大楼这样的建筑物之外，我认为不会有太多人要求建造更高的建筑物。在美国，他们希望建造60—70层的建筑物，而不是110层。在世界其他地方，对建筑高度并没有进行限制的原因有两个：

Top: PS62 Corridor, 10 A.M., Staten Island, NY (SOM, est. completion 2015)
上：PS62 走廊，上午10点，斯塔顿岛，纽约。（斯基德莫尔、奥因斯与梅里尔建筑师事务所，2015年完工）

Left: PS62 Section, 10:30 A.M.
左：PS62横截面，上午10点半。

首先，他们没有受到911事件的影响；第二，正如我前面所说，很多客户建造建筑物的原因超越了经济性的考虑，是出于自我表现或国家、区域甚至地方自豪感。例如，重庆市并没有超高层建筑。他们可能会觉得需要什么东西来标识出这座城市，市长也希望这样。一位商人或开发商可能也准备这样做，超高层建筑物是现代化的一个象征。它可以起到彰显作用，以区别于其他城市。

摩天大楼首先在美国诞生，并且美国建筑师仍然处于高层建筑设计的前沿，但这个接力棒是否正传给其他人？

我不得不说，的确如此。我们都在中国工作，并希望在那里做项目，但我们知道，总有一天会出现大量接受过高等教育的杰出中国建筑师，他们自己就可以独立完成这些工作。他们会越来越好。现在，

the U.S., I think they'll want to do 60- to 70-story buildings not 110. For the rest of the world, no holds are barred for two reasons. First, they are very far removed from the events of 9/11. And second, as I said before, a lot of clients are doing buildings for reasons that go beyond economics. They're taking that extra step because of ego, or national, regional, or even local pride. The city of Chonqing, for example, does not have a super-tall building. They might feel the need for something to mark the city. The mayor wants it. A businessman or developer might be ready to do it. A super-tall building is a badge of modernization. It distinguishes one city from all others.

The skyscraper was born in the United States, and American architects are still in the forefront of high-rise design, but is the baton passing into other hands?

I would have to say yes. We're all working in China, and we want to do projects there, but we know that someday there will be lots of bright, smart, educated Chinese architects who will be able to do this on their own. They're becoming very, very good. Right now, foreign architects are required to work with a local design institute in China, but we believe at SOM that we should always partner with a local architect. We think this is critically important, and, frankly, we have better projects as a result of collaboration.

Is there a parallel growth in client knowledge base and sophistication?

Absolutely. Our clients not only learn from us but from other architects, the contractors, etc. Before you know it, they're the smartest ones in the room!

Aside from height, what other factors distinguish the design of tall buildings in China from those in the U.S.?

Today, there is more freedom—and I use that word loosely—with our Asian, Middle Eastern, and Indian clients than there is with American clients. I think that's largely because American clients have so fine-tuned what they believe their constraints are, their profit margin, their schedules. They come to the table really prepared.

To the point of eclipsing creative opportunity?

Eclipsing is not the word I'd use, but I would say there is a great deal more opportunity to be innovative and inventive in other places. If you look at New York, Los Angeles, and Chicago, and then hop over to Hong Kong, Shanghai, and the Middle East, you will see the difference in the architecture, the kind of exuberance or freedom the architect is permitted. It's very exciting.

Are height limits imposed in China?

In some cities, absolutely, but always to be negotiated. There are prescribed limits, but in a lot of instances you can present an argument and the restrictions will be modified. It's more difficult in America to get a variance like that, but it can be done, although it's probably not as frequent and it's probably a longer process. In China it's a much more a negotiated position, maybe politically negotiated as opposed to being an architectural zoning negotiation. In New York, for instance, you can increase the size of a building and sometimes the height if you provide certain public benefits, like a plaza, open space, transportation improvement, things of that nature. In China it is different, and I will give you an example. We

外国建筑师被要求必须与中国本地设计院合作，而我们SOM工作人员认为，我们确实应该一直与当地的建筑师合作，这是极为重要的。坦率地说，通过合作，我们可以有更好的项目。

客户的知识和经验是否同时并行增长？

当然。我们的客户不仅向我们学习，而且向其他建筑师、承包商学习。在您意识到这点之前，他们已经是屋子里最聪明的人了！

除了高度之外，还有哪些要素将中国和美国设计的高层建筑区分开来？

当前，与美国客户相比，我们的亚洲、中东和印度客户具有更高的自由度（我很宽泛地用这个词）。我认为这主要是因为美国客户会细致地调节他们所认为的制约因素、利润率和日程安排。他们的这些准备工作更为完善。

这是否达到了阻碍创新机会产生的程度？

我并不会使用"阻碍"这个词，但我会说在其他地方的确有更多的创新和创造机会。如果你先环顾纽约、洛杉矶和芝加哥，然后再环顾上海、香港和中东，你便会发现建筑之间的差异以及建筑师在设计上获取的自由度。这是非常令人兴奋的。

中国对建筑高度是否有所限制？

当然，一些城市对此是有限制的，但总是可以协商的。虽然有一些规定的限制，但在很多情况下，你可以提出论据，去修改限制。在美国，很难做到这样的变化，就算可以做到，也不会如此频繁，并且过程可能更长。在中国，可以对此进行协商，可能是政治协商，而非建筑分区规划上的协商。例如，在纽约，如果你提供某些公共设施，比如广场、空地、交通改善等性质的东西，你可以增加建筑物的大小，有时可增加高度。但在中国，情况并不是这样，我可以举个例子。我们在北京负责一个项目，该项目位于市区三环，并有建筑高度限制。该项目的客户与城市主管部门就建筑容积进行协商，并出于观景目的，希望增加建筑高度。相邻的土地所有者也参与进来，并希望重新开发自己的土地。我们的客户很高兴，因为他现在可以再次与有关主管部门协商，并提出建立更高、更紧凑的建筑以展示城市的全貌。现在，我们有很多合理的发展城市的理由，可以让该城市有更高的密度、高度和容积。这个城市愿意听取我们的意见，因为我们知道如何使路面整洁、流通、停车以及规划发展的所有事情。

一些人支持高层建筑所带来的文化展示，而其他人坚持认为这无关宏旨，因为高层建筑在定义上都是一样的。

这是一个私人问题，答案关涉到有关建筑物的目的。如果它旨在体现象征意义，那么你可以得出明确结论。我认为建筑远不止于此，坦白说，我认为高层建筑并不是相同的，但是，它们也不一定必须具有浓厚的文化意涵。这也是SOM的项目设计理念，建筑物必须对应它所处的特别地点、位置和方案。世界贸易中心一号塔楼似乎可以位于任何一个地方。但我认为，它的形状非常简单，并具有标志性，这使得该建筑物永远不会位于其他地方，因为它只对应特定的方案和特定的位置，并且对应特定位置的生命安全和安保的要求，我想这才是关键。如果"建筑的真理"这个说法中存在什么启迪，那么就是：一个建筑需要对应其所在的场所，而不是文化意涵或象征意义，应考虑到它的方案、用途，以及面对的特殊挑战和在具体的位置如何解决这些问题。

客户是否会专门要求能源效率，或者SOM是否会提出类似的建议？

我们的客户经常要求我们为其提供可持续性的设计。多年来，这一直是SOM的核心价值理念，并且我们一直在努力推广这个理念。即使在客户对绿色设计不感兴趣的情况下，我们仍然努力鼓励他们考虑节能效率。有时，我们有幸遇到开明的客户，他们会说，"我希望我们的建筑成为未来的榜样"。现在，这种观点在中国并不普遍，但我认为很快就会开始普及。现在有一些地区，例如中东，特别是阿布扎比和迪拜，非常认可并推崇这一理念。你可能并不期望石油国家会关注节能，但它们对此表现的很开明。当然，由于高能源成本，欧洲历来对可持续发展非常关注。

一般情况下，企业客户会更真诚地去实践这一点，它们比一般的开发商客户更能理解公民责任。

你如何展望城市的未来？

良好的规划，这个是个很简单的答案。通常，建筑师们仅局限于考虑单个建筑物，我们在对话之初，

were working on a project in Beijing. It's in the third ring where there is a set height limit, and the client was negotiating with the city about bulk and additional height, which he wanted because of views. The adjacent landowner came in and said maybe he would like to redevelop his site, too. Our client was delighted because now he could go back to the authorities with a taller, much more cohesive proposal for a full city block. Now, we had lots of urban reasons why it could support greater height, density, and bulk. And the city was disposed to listen on the basis of what we were able to do in terms of clarity, circulation, and parking, all the things involved in a planned development.

Some people argue for cultural expression in tall buildings, while others insist it doesn't matter because high-rise buildings by definition are all the same.

That's a very personal question and begs a response about a building's objective. If it is meant to be symbolic, well, then you have your answer. I think architecture is much more than that. Frankly, I would say that tall buildings are not all the same, but neither do they have to have heavy cultural references. I believe—and this is instilled in the philosophy of the projects we do at SOM—that buildings have to be about their particular place, their site, their program. I would argue that maybe Tower One looks like it could be anywhere. Its shape is very simple and iconic, but that building would never be anywhere else because it is about a specific program and a specific site, about the life safety and security requirements that went into the building in that particular location. I think this is the key. If there is any merit in the old saying "truth in architecture," it is that a building needs to be about its place. Not its cultural references, not its symbolism, but about its program, its use, what its particular challenges are, and how they are addressed in that specific location.

Do clients ask specifically for energy efficiencies, or is that something SOM brings to the table?

We often have clients who ask for sustainable design. But it has been a core value of SOM's for years and we have always tried to push that agenda wherever we can. Even in cases where the client is not interested in green design, we still try to encourage them to think about efficiencies that will save energy. Sometimes we are fortunate enough to get enlightened clients who say, "I want my building to be a model for the future." Right now, this kind of thinking is not prevalent in China, but I think it's going to start to be. And there are places like the Middle East, especially Abu Dhabi and Dubai, where they believe in it very strongly. You might not expect oil countries to be concerned about energy conservation, but they're very enlightened. Europe, of course, has historically been very concerned about sustainability because of the high cost of energy.

In general, it is corporate clients who are more genuinely committed, who understand civic responsibility much more than the typical developer client.

How do you envision the future of cities?

Good planning. That is the simple answer. Too often, architects have limited themselves to thinking about individual buildings. We started this conversation discussing the migration of 400 million people into urban areas. This is going to make urban planning much more important on an entirely different scale. It's not about the block, it's not about the district, it's about planning the entire city: how to plan cities that are economically viable, have real forethought about infrastructure, transportation, water, and density.

We're doing another very cool project, in India. You must have seen the movie *Slumdog Millionaire*. Remember, in the opening scene, the kids playing on the tarmac at the airport? Well, we're working with a private developer who is privatizing the airport. We're moving that slum. First, we are building a new community where we will move those inhabitants, and then they'll increase the size of the airport. The new development is basically a living community not far from where the people live now. That's very important. We're providing improved living conditions with water, sewers, and facilities of that nature. But the people are reluctant, of course, to leave their patch of dirt. So this starts out being more about social interaction and a cultural understanding than about architecture.

Density is an issue. People live so close together that you and I would find it impossible. Do you know the Tata Nano car? It's the world's most inexpensive car; it's tiny, with a little lawnmower engine. We've been talking similarly about doing "nano" housing. It is the same basic idea—that someone living in a slum could now afford to own a more structured home instead of living in a shanty. The bottom line is that we could draw all the pretty bedrooms we want, but if we're not hitting the mark in terms of understanding what the people want and what we can do to make them feel comfortable, we're not going to be successful.

讨论到将会有4亿人迁入城市，这将会使城市规划更为重要，并且规模完全不同。这并不是街区规划，也不是分区规划，而是整个城市的规划：如何规划出在经济上可行的城市，并对基础设施、交通、供水和密度有真正的深谋远虑。

我们正在设计另一个非常棒的项目，它位于印度。你一定看过电影《贫民窟的百万富翁》。还记得在影片开始时，孩子们在机场停机坪上玩耍的情景吗？我们正在与一位私人开发商合作，这位开发商正在收购该机场。我们会迁走那个贫民窟。首先，我们建设一个新的社区，以便让这些居民搬进来，然后他们将会扩大机场。新的开发项目基本上是一个居住社区，距离人们现在居住的地方并不远，这一点非常重要。我们为这些居民改善了生活条件，例如供水、下水道以及诸如此类性质的设施。但这些居民并不愿意离开他们的贫民窟，而这就涉及到建筑之外的社会互动和文化理解等问题。

密度是一个问题。你和我都会觉得，人们不可能居住的太近。你听说过塔塔纳米车吗？它是世界上最便宜的汽车，车型很小，采用剪草机引擎。我们一直在谈论类似的"纳米"住房。这是基本相同的想法，可以让居住在贫民窟的一些人们买得起结构更好的房屋而不是住在棚屋里。结果是我们确实可以设计出我们想要的那种精美卧室，但如果我们不了解人们想要什么，以及我们需要做些什么才能让他们感到舒服，那么我们就不会成功。

One World Trade Center, view to north, New York. (SOM, est. completion 2015)
世贸中心一号楼，朝北方向，纽约。(斯基德莫尔、奥因斯与梅里尔建筑师事务所，
2015年完工)

William Pedersen
KPF

William Pedersen, FAIA, is a founding design partner of Kohn Pedersen Fox Associates (KPF), which he established in 1976 with A. Eugene Kohn and Sheldon Fox. He was honored in 2008 when the Council on Tall Buildings and Urban Habitat elected the Shanghai World Financial Center the Best Tall Building Worldwide, and again in 2010, when CTBUH conferred on him the Lynn S. Beedle Lifetime Achievement Award. Mr. Pedersen has also received seven AIA National Honor Awards, the Gold Medal for lifetime achievement in architecture from Tau Sigma Delta, the Arnold W. Brunner Memorial Prize in Architecture, the Rome Prize, the Prix d'Excellence, and the U.S. General Services Administration's Building Design Excellence and Honor awards.

Mr. Pedersen lectures and serves on academic and professional juries and symposia internationally. He is a visiting professor at the Rhode Island School of Design; Columbia, Harvard, and Yale universities; Distinguished Professor in Architecture at the University of Illinois at Chicago; and a board member of the University of Minnesota and recipient of its Alumni Achievement Award.

威廉·佩特森
科恩·佩特森·福克斯建筑事务所

威廉·佩特森是美国建筑师协会会员，科恩·佩特森·福克斯建筑事务所（KPF）的创始设计合伙人——他和A·尤金·科恩、谢尔顿·福克斯在1976年创立了这家事务所。当高层建筑与城市住区理事会在2008年将上海环球金融中心选为全球最佳高层建筑时，他获得了荣誉·2010年，高层建筑与城市住区理事会授予他林恩·S·比多终身成就奖。佩特森先生还获得过七项美国建筑协会的荣誉奖项、荣誉协会建筑业终身成就金奖、建筑业的阿诺德·W·布鲁纳纪念奖、罗马奖、卓越成就奖，以及美国总务管理局的建筑设计卓越荣誉奖。

佩特森先生在全球许多学术、专业评审会议和研讨会上做过演讲，担任过职务。他是罗德岛设计学院、哥伦比亚大学、哈佛大学和耶鲁大学的访问教授；是伊利诺伊大学芝加哥分校的建筑学杰出教授；明尼苏达大学的董事会成员，并获得了学校的杰出校友奖。

For complete biography, please refer to the Appendix.
完整的传记，请参阅附录。

ARCHITECTURE William Pedersen

Shanghai World Financial Center, the tallest building in Pudong upon completion by KPF in 2008.
上海环球金融中心，浦东地区最高建筑，KPF建筑事务所于2008年完成。

Developing countries have taken a leading role in sustainable building initiatives. Do your clients typically ask for LEED certification?

Yes, of course, and city governments are mandating it. Certainly in Japan and Korea there's a tremendous focus on energy conservation, much more than in the United States, but even here it's pretty standard. There are various technologies and different agendas at work, not all of them equally productive, but I think this is going to work itself out over time. The important thing is that people are finding sustainability a highly desirable direction to go.

At KPF we achieve or pursue LEED (and its equivalents, such as BREEAM in the UK and Estidama in the UAE) on many of our projects, but we also try to promote sustainability in other ways outside the certification system. By designing projects of utmost quality, contextual sensitivity, flexibility, and performance, we produce architecture that is not just long-lasting but that builds value over time. Over the past 30+ years, as our practice has grown and become more global, we have reinforced this approach by incorporating the latest methods and technologies in environmental design into our work. But ultimately, regardless of the systems we incorporate or the LEED points achieved, it has been the goal of creating buildings and places of enduring significance that has driven our practice and guided our success.

We also try to achieve sustainability through urban design, addressing, for example, the massive migration of people, especially in China, from the countryside into urban centers. As population shifts have placed greater demands on the environment, as well as on infrastructure and transportation systems, public and private entities have responded with transit-oriented, mixed-use developments. We've been in the forefront of this movement at KPF, designing some of the world's most progressive and complex projects, from entirely new sustainable cities (not just designed, but actually being built and populated, such as New Songdo City in Incheon, South Korea) to super-tall towers serving as vertical cities. With the density of such designs, public transport is encouraged, land is conserved, and energy and water usage reduced.

What is the relevance of building tall?

The tall building has been the fundamental, even dominant, component of the modern city for almost 100 years. By its nature, it tends to be insular and autonomous. Ever since my student days, I've been trying to find ways for the tall

发展中国家在可持续性建筑的倡导方面已经处于领先地位了。你们的客户会不会经常要求有LEED（能源与环境设计指导）认证？

是的，城市政府会委托我们这么做。在日本和韩国，人们对节能问题的关注当然要大大超过美国，甚至在这里，节能是非常正常、非常普遍的。他们有各种技术，各种不同的机构发挥作用，虽然效率并不完全一样，但随着时间的推移，它们将会有所作为。重要的是，人们发现可持续性是一个十分可取的前进方向。

在KPF公司，我们在许多项目上都获得或者正在致力获得LEED［以及与之相当的，比如英国的BREEAM(建筑研究所环境评估办法)和阿拉伯联合酋长国的Estidama（绿色建筑条例）等认证］，不过我们也试图在认证体系之外，以其它的一些方式推动可持续性发展。通过设计具有质量最好、环境感受最佳、灵活性和高性能的项目，我们建造的建筑不仅能够持久，而且还树立了某种长期的价值。在过去30年间，随着工作日益增多且变得更加全球化，我们通过将环境设计方面最新的方法"核技术"融入到工作中，强化了这种方式。不过从根本上说，不管是我们融合的系统还是获得的LEED分数，建造具有持久意义的建筑和场所的目标，已经成为推动我们工作实践、引导我们收获成功的动力。

我们试图通过城市设计，为大量外来人员选址（例如，特别是在中国，这些人从乡村迁移到城市中心）来获得可持续性。由于人口流动对环境、基础设施、交通系统都产生了巨大的需求，公共和私营机构对此做出的回应是：开发以交通为导向、多重用途的工作。在KPF，我们就处于这一潮流的最前沿，我们设计一些世界上最前卫、最复杂的工程，从全新的可持续城市（不只是停留在设计阶段，实际上已经在建设，且有人口的迁入，比如韩国仁川的松岛新城），到构成垂直城市的超高层建筑。在这样的设计密度下，公共交通被大力提倡，土地得以保存，能量和水的使用也减少了。

这和高层建筑有什么关联呢？

高层建筑成为现代城市中基础的，甚至是主导性的要素已经有近100年的实践了。究其本质，它

Tall building designs by William Pedersen, prepared for the Lynn S. Beedle Lifetime Achievement Award of the Council on Tall Buildings and Urban Habitat (awarded 2010).
由威廉·佩特森设计的高层建筑，获得高层建筑与城市住区理事会授予的"林恩·S·比多"终身成就奖（2010年获奖）。

倾向于孤立自主。甚至在我还是学生的时候，我就尝试着去寻找一些方法，将高层建筑与街墙联系起来，使之成为都市环境中一个积极的社会参与者。我们鼓励建筑之间的连接性，而不是让它们孤立的像一些物体一样站在那里。我们的目标是，承认特殊环境造成的压力，并通过某种形式归纳总结这些特征，然后改变一栋建筑的环境。

对于高层建筑要在其中实现自身的环境而言，每个形象都必须是明确的。每个地方都产生了一种不同的可能性，一种不同的出发点。这与

building to relate to the street wall and to be an active social participant in the urban context. The goal is to encourage connection between buildings rather than their just standing as isolated objects. We aim to contribute to and transform a building's environment by acknowledging the pressures of its particular context and, in a way, to summarize those specific characteristics.

Each visualization has to be specific to the context in which will be realized. Every place generates a different possibility, a different point of departure. That's significantly different from the International Style, for example, where a

International Commerce Centre (ICC), Kowloon, Hong Kong. (Kohn Pedersen Fox, 2011)
香港九龙环球贸易广场（ICC）。（KPF建筑事务所，2011年）

building is based on its own requirements and where any connection with place is actually the exception. Admittedly, as buildings grow taller, contextual conversation becomes more difficult, given that buildings of great height have their own structural and mechanical biology. They're inherently more isolated. Still, one hopes to find a way of connecting even very large buildings to their place and context, but the manner has to be carefully considered.

We've been fortunate to have had many opportunities to create a number of buildings of great height, such as the Shanghai World Financial Center (SWFC) and, more recently, the International Commerce Centre (ICC) in Hong Kong. In each case, we had to think in different terms.

Can you offer an example of such contextual linkage?

For SWFC, we referred to the Chinese culture and drew on the relationship of two ancient symbols to shape the building. A square prism—the symbol used by the ancient Chinese to represent the Earth—is intersected by two cosmic arcs, representing the heavens; the tower ascends in a gesture to sky. The interaction between these two realms gives rise to the building's form, carving a square sky portal at the top of

国际建筑风格有很大的差别，例如，在国际建筑风格中，一种建筑是基于自身的需要，任何与场所之间的联系实际上都是独特的例外。诚然，随着建筑越来越高，环境、氛围之间的对话变得越发困难，因为高层建筑有自己的结构和机械生态学，它们在天性、本质上就是独立的。但是，人们还是希望找到一种方式，将很大的建筑与它们的场所、环境联系在一起，不过，采取什么方法还要细致考虑。

我们很幸运，有很多机会建造一系列的高层建筑，比如上海的环球金融中心(SWFC)，以及最近在香港建的环球贸易广场(ICC)。在每个项目中，我们都必须以不同的方式进行思考。

关于这种环境关联，你能举个例子吗？

比如上海环球金融中心，我们参考了中国文化，并且利用两个古老符号之间的关系来塑成建筑的外形。一个正方棱柱形——在古代中国这个符号用来代表大地——被两个巨大的圆弧割开，圆弧在古代中国代表了天，大楼冲天而上。两个区

ICC base, and canopy. (right)
香港环球贸易广场的地基和顶盖。(右)

域的相交互动产生出了建筑的形式，在大楼顶部雕刻出一个方形的空中之门，它一方面给予结构某种平衡性，另一方面将两种对立的因素结合起来：天空和大地。

在设计环球贸易广场的时候，我们也考虑到了建筑的场所。九龙(Kowloon)这个词来自于"九条龙"之意，它也是这一结构的建筑灵感。客户要求一栋在结构上非常高效的建筑，所以我们给出的方案是，建造一个向着太阳的锥体正方棱柱。建筑的正面是四个平面要素，它们不对称地延展，超出了凹角，向着楼顶上升攀沿，而大楼的顶冠由整片的玻璃构成。每个平面都可以独立运作，都与大地、天空有独特的关系。这些凹凸有致的表面以稍微倾斜的曲线作为基础。曲线从层叠的屋顶板（或者说龙的鳞片）形成的结构上升起，为在其它三面办公室和酒店入口处眺望码头的人营造出若隐若现的华盖苍穹之感。在北面，表层以一种迅猛的姿势向香港联合广场的中心突进，合上了"龙尾"的中厅。这个中厅乃是整座楼的公共面孔，是与地铁站的主要连结点。

the tower that lends balance to the structure and links the two opposing elements—the heavens and the Earth.

In designing ICC, we also looked at place. The word Kowloon stems from nine dragons, which was a design inspiration for the structure. The client requested a building of great structural efficiency, so we responded by creating a tapering square prism reaching for the sun. The main facades are articulated as four planar elements extending partially beyond the re-entrant corners and rising above the tower roof as sheets of glass to form the tower crown. Each separate plane has the capacity to act independently, and each has a distinct relationship to the ground and sky. These chiseled facades give way to gently sloped curves at its base. The curves lift off from the structure as a cascade of overlapping shingles (or scales of a dragon) to create sheltering canopies for the office and hotel entrances on the three sides overlooking the harbor. At the north, the facade sweeps down in a dramatic gesture toward the center of Hong Kong's Union Square development, enclosing the "dragon tail" atrium. This atrium serves as the public face of the tower and the primary connection to the rail station.

Presently, we're designing a tower for Hyundai in Seoul, 1,770 feet [540 meters] high. It's the first opportunity to create a building's structure and biology in a way that entirely determines its character and form and how it specifically relates to its context. An important aspiration for me is to create the potential for social interaction in a building, a sense of community and common purpose. With this in mind, we created a vertical campus for Hyundai—our first experience in building a series of socially interactive spaces within a tall building.

How do you approach the design of a super-tall building?

We always start with program: what is the building's function and what are its social potentialities? Given that, we begin from the perspective of structure: how to achieve a certain height with maximum efficiency. That's what we did in competition for the International Commerce Centre in Hong Kong, which was completed in 2010. We started by asking Les Robertson [the structural engineer], "What is the most efficient form we can develop?" And then we made architecture out of it.

We started with a whole series of structural diagrams and then we developed alternatives based on those imperatives. There are certain issues that the tall building generates in terms of its basic shape. Wind forces become the dominant issue, not the gravity forces. Wind generally comes from 360 degrees—in some parts of the world it comes predictably from one direction—so buildings tend to be relatively symmetrical. Slab buildings don't work well at great heights, so the ideal diagram for a tall building really becomes quite organic. It roots itself in the ground and tapers away at the top. But invariably the whole is based on structural imperative. We don't start with the architecture but develop it from a structural point of departure.

Where in its evolution is the tall building?

After 9/11, there were predictions about the demise of the tall building. Ironically, more tall buildings were developed in the last 10 years than ever before. In my opinion, we are entering the golden age of the tall building, due to its inherent potential for social interaction. The individual tall building, with communal facilities and public spaces—like their own town squares—is like a city within a city. It has the potential to connect with its specific environment as well as to the larger cultural context.

I believe that as a result of increasing land values and the desirability of conserving energy, water, and other resources,

现在，我们正在为首尔的现代公司设计一座大楼，1770英尺（540米）高。这是第一次有机会建造一个建筑，其结构和生态以某种方式完全决定了它的特点和形式，以及它与环境的特殊关系。对我而言，一个重要的灵感是，要在一座建筑中营造出社会交往的可能，营造出一种社区和共同目标感受。以此为指导，我们为现代公司建造了一座垂直校园——这是我们第一次在一个高层建筑内部建造一系列的社会交往空间。

你如何设计一座高层建筑？

我们总是以项目开始：建筑的功能是什么以及其社会潜能是什么？考虑到这些，我们就从建筑的视角思考：如何达到一个特定的高度，同时达到最高的功效。这就是我们在香港环球贸易广场竞标时所作的，这次竞标发生在2010年。一开始，我们就问莱斯·罗伯特森（建筑工程师）："我们可以设计的最有效率的形式是什么？"然后由此做出建筑。

我们从一系列的结构图表开始，然后基于那些要求设计出不同的方案。就基本的形状而言，高层建筑产生了一些特殊的问题。风力而非地心引力，成了主要的问题。风一般会从360度各个方向吹来（在世界的某些地方，它只固定从某个方向吹来），因此建筑倾向于相对对称。平板式建筑在高层不太可靠，所以对一栋高层建筑而言，理想的图表需要在结构上变得很有序。它将自身扎根在大地之上，在楼顶处才逐渐变尖，但整体上却一定都符合结构的规则。我们不是从建筑出发，而是从一个结构的出发点出发进行开发设计。

高层建筑演进发展到什么程度了呢？

911事件之后，有很多预言说高层建筑将会终结。但讽刺的是，在过去的十年间，高层建筑建设的数量要远远超过以往。在我看来，我们正在进入一个高层建筑的黄金时期，因为它在社会交往方面有自身固有的潜力。单个的高层建筑，有公共设施和公共空间（像他们自己的城市广场）就像一个大城市中的小城市一样。它有潜能将其特殊的环境与更大范围的文化环境联系起来。

我相信，作为土地价值增加，以及能源、水和其他资源保护的一个结果，高层建筑将继续证明它们作为城市中高效、可持续性要素的重要性，特别是当我们的城市依赖大众交通的时候。

许多年来，一些建筑师依然坚持认为高层建筑并不是特别有趣，你设计了建筑的底部和顶部，然后把其他的东西放在中间。在开发高层建筑的艺术形式方面有没有一些新的兴趣呢？

绝对有。当我们在上世纪70年代末开办公司时，建筑的样式——特别是投机性的高密度办公楼——在世界上许多建筑师看来，并不被看成是一种有价值的艺术方面的努力。我并不这么认为，特别是考虑到它在城市和我们的生活中所扮演的关键角色。

过去十年，世界上许多顶尖的建筑师都对高层建筑有了新的兴趣，这使得建筑的样式正在提升。这一行业已经取得了长足进步，但还有很长的一段路要走。我相信高层建筑在接下来的20到30年间将会有巨大的转变。

在高层建筑，尤其是多用途城市中心的设计这一领域，KPF建筑设计公司已经成为世界的领导者。面对世界上不断增长的城市，这种范例被当作典型的、文化的、可持续的解决方案，我们已经在功能和技术上不断挑战极限。算起来，我们已经在东京、首尔、香港、重庆、沈阳、广州，现在也在上海设计了最高的一些建筑（有一些已经完成，有一些还在建设中）。深圳的平安金融中心高达2126英尺（648米），当它完成的时候，将会是中国最高的建筑，同时也是全世界最高的办公楼。

除了非凡的高度，近年来，高层建筑还有哪些重要的变化呢？

在我看来，最重要的改进是规划性。实际上，我曾说过，高层建筑的未来在于其规划性的发展演变。它不能被认为是一个独立的事物，每个建筑都必须被看作是一座城市。

我们很荣幸与森稔在六本木新城工作，森稔是当时最重要和最有影响的开发商之一。当丹下健三引领的新陈代谢派将建筑探究城市作为一个有机整体的观念时，森稔先生作为一个建设者开始了他的事业。理论概念被融入到每个建筑之中，构成了更大的城市结构，而这引起了森稔先生对城市化的兴趣，我想，可能还影响了他对工作、

tall buildings will continue to demonstrate their importance as efficient and sustainable components of our cities, particularly when served by mass transit.

For many years, some architects maintained that tall buildings were not particularly interesting, that you design the bottom and the top and then put everything else in the middle. Is there new interest in developing the art form of tall buildings?

Absolutely. When we began our firm in the late 1970s, the building type—particularly the speculative high-rise office building—was not considered a worthy artistic endeavor by many of the world's architects. I could never understand this, especially given the dominant role it plays within our cities and in our lives.

Over the last 10 years, many of the world's finest architects have taken a new interest in the tall building, with the result that the building type is improving. There have been giant strides forward in the industry, but there is still a long way to go. I believe the tall building will transform itself tremendously in the next 20 to 30 years.

The design of tall buildings and of mixed-use urban centers in particular is an area in which KPF has been a world leader. As this paradigm continues to be embraced as a programmatic, cultural, and sustainable solution for the world's growing cities, we have continued to push the envelope in terms of functionality and technology. All told, we have designed the tallest buildings (some completed, some under construction) in Tokyo, Seoul, Hong Kong, Chongqing, Shenyang, Guangzhou, and currently also in Shanghai. Our 2,126-foot-high [648 m] Ping An Finance Center in Shenzhen, when completed, will be the tallest building in China and the tallest office building in the world.

Beyond great height, what significant changes have taken place in tall buildings in recent years?

To me, the most important advances are programmatic. In fact, I'd say that the future of the tall building lies within its programmatic evolution. It can't be thought of as a singular thing. Each building must be thought of as a city.

We have had the great fortune to work on Roppongi Hills with Minoru Mori, one of the most important and influential developers of our time. Mr. Mori began his career as a builder when Metabolist architects, led by Kenzo Tange, explored the notion of the city as an organic whole. Theoretical concepts weaving individual buildings into the larger urban fabric shaped Mr. Mori's interest in

Rappongi Hills, Mori Building, Tokyo. (Kohn Pedersen Fox, 2003)
东京六本木新城，森大厦株式会社。（KPF建筑事务所，2003年）

休闲、文化、生活各个区块之间关系的态度。

森埝先生首先将建筑看成是一种社会艺术，对于人们在社会之中的交往方式会施以深刻的影响。每天长时间地在家和工作地点之间往返，会耗费人的精神，令人萎靡不振，所以他决心把人们带回到城市中心。他预想，六本木新城就是未来的城市模型，这是一个融合、并置、鼓励人类在一个很小的范围内展开多层次交往的地方。

六本木新城不只高（接近70层），而且它有一个很大面积的楼层，面积超过了1英亩（0.4公顷）。670万平方英尺（622450平方米）的多用途开发项目包括了办公室、住宅、酒店、机构、零售单元，可容纳2000户居民和2万工人；它每周的平均人流量大约为10万人。人们出于各种理由来到这里，不管是因为它的生活方式、文化、学习和休闲用途，还是因为其休闲娱乐、商场饭店，当然还有办公室。重要的是，它是将整个复合体融合在一起的方式。生活在一个这么繁华热闹的城市环境中，而且感到很快乐，这是一件多么令人兴奋的事情啊！森埝先生为如下一种哲学信念所触动：如果各种规划要素结合在一起能获得更多的东西，那么，现代城市有能力成为一个更加充满生机和活力的地方。

六本木新城是森埝先生热情推动的，它是在一个极具生机的整体中并置多种要素的例子。当被问到他将森埝艺术博物馆置于六本木新城森大厦顶部这一勇敢举措时，他回应说，这个博物馆占据了5%的地产面积，但却占据了他内心的25%。他的城市眼界和想象，作为城市潜能的一个范例在全世界有着巨大的影响。对我来说，作为一种学习经验，它已经成为了建筑上的灵性和顿悟。

对世贸中心的攻击有没有对六本木新城产生什么影响？

六本木新城2001年才开始建设，所以你说的问题当然会被列入考虑范围。但我们不能仅仅因为担心有人可能企图摧毁它，就制止建设高层建筑。相反，我们必须在不牺牲建筑重要性、美观性或者在保证建筑内部、周边人们生活质量的情况下，设计出安全的建筑。911是个可怕的悲剧，它将一直成为全球心态的一部分，但重要的是要记住，双子塔的高度使它成为了一个容易受攻击

urbanism and, I suspect, his attitude toward the interrelationship of places for work, recreation, culture, and living.

Mr. Mori saw architecture primarily as a social art exerting a profound influence on the way people interact in society. Long daily commutes from home to work and back again are wasteful and greatly enervating to the human spirit, so he was committed to bringing people back into the city center. He envisioned Roppongi Hills as the city of the future, a place that combined, juxtaposed, and encouraged different facets of human interaction within close proximity.

Roppongi Hills is not all that tall (approximately 70 stories), but it has a massive floor plate of more than an acre [0.4 hectare]. The 6.7-million-square-foot [622,450-square-meter] mixed-use development contains office, residential, hotel, institutional, and retail components, and houses 2,000 residents and 20,000 workers; it has an average weekday population of about 100,000. People visit for various reasons, whether lifestyle, cultural, learning, and leisure uses, or for entertainment and recreation, shopping and restaurants, and, of course, offices. What's important is the way that the whole complex is woven together. It's so exciting to feel the joy of living in an urban environment with so much going on! Mr. Minoru Mori was motivated by a philosophical belief that the modern city has the capacity to be a far more vital and dynamic place if various programmatic elements are combined to achieve more.

Roppongi Hills is an example of Mr. Mori's passionate drive to juxtapose diverse components in a wonderfully vibrant and dynamic whole. When asked about his bold move to place the Mori Art Museum at the top of the Roppongi Hills tower, Mr. Mori responded that the museum occupies only 5 percent of the real estate but 25 percent of his heart. His urban vision and imagination have had tremendous influence throughout the world as a textbook example of urban potential. For me, as a learning experience, it has been an architectural epiphany.

Did the attack on the World Trade Center have any impact on Roppongi Hills?

As Roppongi had just started construction in 2001, it was of course a consideration. But we cannot stop building tall buildings because we're afraid someone may attempt to bring them down. Instead, we must design safe buildings without compromising their significance, their beauty, or the quality of life that goes on inside and around them. 9/11 was a terrible tragedy that will always be a part of the global psyche, but it's important to remember that while the Twin Towers' height made them an easy target, the World Trade Center was primarily attacked because of what it represented.

If all current constraints were removed, how would you visualize the future high-rise building?

I frequently refer to Raphael's great fresco "The School of Athens" to help convey the manner in which tall buildings should interact with each other—similar to the philosophers gathered in intimate conversation groups within the painting. Like a human, each building has a specific biology, and, being a social being, it should make a physical gesture to its surrounding context, creating a larger conversation. All of our buildings seek to accomplish this: 333 West Wacker Drive in Chicago, IBM's Canadian headquarters, First Hawaiian Center in Honolulu, and DZ Bank headquarters in Frankfurt are all examples of individual buildings that make very specific gestures to their particular contexts.

It is my aspiration to further develop this focus on gestural action while still working with the building's structural metabolism. Ideally, as in "The School of Athens," each building, each anthropomorphic component, can be fully integrated in its biology and can still gesture. However, from my perspective, the social responsibilities of tall buildings' urban architecture are more important in the evolution of the building type than the biological necessities of the individual buildings.

的目标，但是世贸中心受到攻击首先是因为它所象征的东西。

如果当前所有的限制都消除了，你会如何设想将来的高层建筑？

我经常会提到拉斐尔的伟大壁画"雅典学派"，来帮助传达这样一种态度，即所有建筑都要互相联系——类似于壁画中哲学家们以亲密的对话小组聚集在一起。和一个人一样，每栋建筑都有一种独特的生态学，同时，要成为一种社会存在，它必须对周边环境做出一种物质上的回应姿态，营造一种更大范围的对话交流。我们所有的建筑都试图实现这一点：芝加哥西威客大道333号大厦、IBM加拿大总部、檀香山的第一夏威夷中心以及法兰克福的DZ银行总部都是我们上面所描述特征的范例，它们都是对自身特殊环境做出独特回应姿态的建筑。

我希望保持对结构上的新陈代谢发挥作用的同时，还能进一步拓展这种对姿态反应的关注。理想上，就像在"雅典学派"中一样，每栋建筑、每个拟人的要素，都能够完全融入到其生态中去，而且依然能够有所作为。从我的角度看，在建筑形态的演化进程中，高层建筑对城市建筑的社会责任，要比单个建筑的生态需要更加重要。

Rappongi Hills, Mori Building: Practicing tai chi (upper left); Escalators in atrium lobby; Exterior plaza (upper right).
六本木新城，森大厦株式会社：人们在练习太极（左上）；中庭大堂中的自动扶梯（右上）；外部广场。

Adrian Smith + Gordon Gill
ADRIAN SMITH + GORDON GILL ARCHITECTURE

122

Adrian Smith, FAIA, RIBA, engages the particulars of the places where he designs, interpreting and honoring the societies his buildings serve. His contextual approach has produced buildings acclaimed for their beauty, elegance, and subtle cultural references, winning well over 100 major awards. He has designed four of the world's tallest buildings, including the tallest, Burj Khalifa in Dubai. Upon its completion in 2016, Kingdom Tower in Jeddah, Saudi Arabia, will become the world's tallest, followed in fourth place by China's Wuhan Greenland Center. Prior to founding his firm in 2006, Mr. Smith was a design partner at Skidmore, Owings & Merrill and its chairman from 1993–1995. He has written two books on his work and co-authored (with Gordon Gill) *Toward Zero Carbon: The Chicago Central Area DeCarbonization Plan.*

Gordon Gill, AIA, is a preeminent exponent of performance-based design, driven by the purposeful relationship between design and performance criteria. He is responsible for the world's first net zero–energy skyscraper, Pearl River Tower (designed at SOM Chicago), and the first large-scale positive-energy building, Masdar Headquarters. Gordon's award-winning work also includes performing arts centers, museums, schools, civic spaces, and urban master plans across the globe. In 2009, the Chicago Reader named him Chicago's Best Emerging Architect. Before co-founding AG+GG in 2006, he was an associate partner at Skidmore, Owings & Merrill and a director of design for VOA Associates. He recently co-founded PositivEnergy Practice, a consulting firm that designs and implements energy- and carbon-reduction strategies for clients internationally.

艾德里安·史密斯和戈登·吉尔
艾德里安·史密斯和戈登·吉尔建筑设计事务所

艾德里安·史密斯，美国建筑师协会会员，英国皇家建筑师协会，艾德里安·史密斯十分注重他的设计所处位置的特殊性，他完美诠释了他的建筑服务社区，并以此为荣。这种关注背景的方法，已经在全球范围内建造了很多因美丽、优雅、细腻的关乎文化且著名的建筑，并获得了超过100项大奖。史密斯已经设计了四栋世界最高的建筑，包括迪拜的哈利法塔、上海的金茂大厦、芝加哥的特朗普国际大厦，以及中国南京的紫峰大厦。在他的作品中，沙特阿拉伯吉达的王国大厦将于2016年完工，它将会成为世界最高的建筑，在这之前是现在排名第四的中国武汉的绿地中心。在2006年创立自己的建筑事务所以前，史密斯先生是斯基德莫尔、奥因斯与梅里尔建筑师事务所的设计合伙人，并在1993年至1995年间担任事务所主席。他写过两本关于自己工作的书，并与戈登·吉尔一起完成了《向零碳进军:芝加哥中心区域的脱碳计划》一书。

戈登·吉尔是基于能效进行建筑设计的优秀典范，他追求设计与能效标准之间的特定关系。戈登负责世界上第一座净零能耗的摩天大楼珠江塔(由芝加哥的斯基德莫尔、奥因斯与梅里尔建筑师事务所设计完成)，以及第一座大型的正能耗建筑:马斯达尔总部。戈登先生备受赞誉的作品包括遍布全球的一些表演艺术中心、博物馆、学校、城市公共空间和城市总体规划。2009年，《芝加哥读者》称他是芝加哥最好的新兴建筑师。在2006年成为艾德里安·史密斯-戈登·吉尔建筑设计事务所共同创始人之前，他是斯基德莫尔、奥因斯与梅里尔建筑师事务所的副合伙人，同时也是VOA建筑事务所的设计总监。现在，他与人共同创立了"正能源实业公司"(PositivEnergy Practice)，一家为全球客户设计、实施能源和碳减排策略的咨询公司。

完整的传记，请参阅附录。

For complete biographies, please refer to the Appendix.

Burj Khalifa at night, Dubai, United Arab Emirates
(Adrian Smith + Gordon Gill, 2010)
夜晚中的哈利法塔，阿拉伯联合酋长国迪拜（艾德
AS+GG建筑设计事务所，2010年）

What is the importance of density for the future of sustainable environments?

GG: I'm sure you have seen the charts showing the world's most sustainable cities and their relationship to density, the performance of their transit systems, the infrastructure; the more dense, the more compact. Singapore is a great example of that. In fact, their metro system and postal service are actually profitable because of the density and the compactness of those systems. You begin to see all the positive effects of urban support systems when a city is truly dense. I tell people all the time that we are like fish and we have built ourselves a coral reef. The reef begins to define itself and finds its own equilibrium and balance, whereas if it just bleeds out and goes on forever, it becomes very inefficient and hard to manage and control. Infrastructure demands are limitless; grid systems become duplicated and replicated. And when one section decays and another spreads out, the value assessment of renewal becomes a question, whereas if you're constantly renewing and replenishing the core, you have a vibrant and sustainable mind-set about your city.

AS: Take the Loop in downtown Chicago. It's roughly a square mile [2.6 square kilometers], so you can walk from one side to the other in a few minutes, and there's a good public transport system as well. But it is not a live/work environment; it's all office space. You have to go beyond the Loop to find residential, but then it starts to become a 2-mile [3.22-kilometer] walk or even longer. We just published a book about how to de-carbonize the Loop and what would be required to get the 550 buildings in the central area to meet the 2030 challenge ["Toward Zero Carbon: The Chicago Central Area DeCarbonization Plan"]. The book has a lot to do with existing buildings, since new buildings in Chicago account for only about 1 percent annually; you're not going to de-carbonize for 100 years or more if you ignore the existing fabric of the city!

GG: By looking at every aspect of a building and its environment, we began to develop a language about buildings specific to their sites and performance. We found that the kind of prescriptive approach to solving a problem is not necessarily something that you can apply over and over. There's a basic DNA in terms of the process, how we approach it, and the things we're looking for, but the solution is never the same. If you take a building and move it to a different site, the characteristics that influence the building are completely different. It's one thing to build in isolation, but the behavior between buildings is like one person versus a room full of

---

对于未来可持续环境而言，密度具有怎样的重要性？

GG: 我相信你已经看过有关图表，这些图表说明了世界上最具可持续性的城市以及这些城市与密度、运输系统的性能、基础设施的关系：密度越高，人口就越密集。新加坡就是一个很好的例子。实际上，由于其高密度和人口密集性，该国的地铁和邮政服务是可以盈利的。当一个城市的密度真的很高时，你可以看到城市保障系统的所有积极作用。我总是告诉人们，我们就像鱼一样，为自己建立了珊瑚礁。珊瑚礁先是固定下来，然后保持自身的平衡性，但如果它在损耗并一直这样，那么它将会变得非常低效，并且难以管理和控制。基础设施的需求是无限的；网格状道路系统持续发展并且纵横延伸。当其中一部分出现问题时，另一部分会取而代之，此时进行更新的价值评估会成为一个问题，但如果你持续对核心区进行更新和补充，那么对于你的城市，你就拥有了一个充满活力和可持续发展的思维定式。

AS: 以芝加哥市中心卢普区为例。它大约为1平方英里（2.6平方公里），所以你可以在数分钟内从一端步行到另一端，并且它也具有良好的公共交通系统。但它并不是一种商住两用的环境，而完全是办公空间。你必须走出卢普区，才能到达住宅区，这样你就需要步行2英里（3.22公里）或更远。我们刚刚出版了一本书《向"零碳"进军:芝加哥中心区域的脱碳计划》，主要讲关于如何实现卢普区的脱碳化，以及为了让中心区的550幢建筑物实现2030年的挑战目标，需要做到什么。该书主要涉及现有的建筑物，因为芝加哥每年的新建筑物大约仅占1%，如果你忽略了该城市的现有建筑物，那么你在100年或更长时间内都无法实现脱碳目标!

GG: 通过着眼于一栋建筑及其环境的各个方面，我们开始开发一种针对建筑物位置和性能的建筑语言。我们发现，用于解决问题的规范性方法未必是可以反复使用的。对于具体的过程、我们如何进行处理以及我们正在寻找的东西，有一个基本的DNA，但解决方案不会相同。如果你负责建造一栋建筑，并将其变换到不同的位置，那么影响建筑的特征是完全不同的。单独修建一座建筑物是一回事，但建筑物之间的互动是另一回事，它就像是一个人对着一个充满了人的房间。风向、人们流动的动态方式、如何创造空间、太阳方向、景观的相互作用，确定了建筑物之间的关系。当你进行城市设计时，你会考虑到你认为美丽或具有一定品质和比例的东西，例如城市农业、交通系统、空地等特征。这就是复杂性真正有趣并令人兴奋之处，但它同时也令人非常沮丧，因为总是存在着限制。

该书提供了一种逐步递进法，以解决所发现的问题，并实现2030年挑战中所设定的特定目标。它为八种不同的策略提供了原理和指标，因此就有了特定的标准。但对于不同人来说，目标得以实现的方式会有所不同。这就是它的动人之处。

AS: 我们关注的重点并不仅仅是现有的建筑物。我们有一个图表，涉及到了建筑围护结构，并且它概括了各种项目的能耗，包括单层窗格玻璃和70%玻璃、双层窗格玻璃和60%以下的玻璃。能源使用量基本与该曲线一致。我们对建筑物的照明、HVAC、垂直运输负荷和其他部分也进行了类似说明。因此，该书像是一种路线图，它向你展示了各个系统的相关性。

除了这一具有里程碑意义的出版物之外，相同的原理如何应用于你的新工作？

AS: 在一些项目中，我们试图实现零碳排放量，特别是位于阿布扎比的马斯达尔总部，我们将其设计为一栋零碳、零废物、零能耗的建筑物，实际上也就是实现正能耗。通过使用太阳能和光伏能源，这是可能的，尤其是在中东。

GG: 在能源方面是可能的。在碳计算方面，具体取决于你对净零能源消耗所设定的界线，尤其当包含碳源时，则更为复杂。即使达到净零能源消耗水平，由于有施工和材料，在建筑的生命周期内，你也不可能在达到要求。换句话说，如果建筑物未能起到积极作用，那么它基本上是中性的，可以默认为起着消极作用。它将不会是一个碳平衡建筑。

你们新建的珠江大厦作为一个净零能耗建筑，引起了极大的关注，但最近的报告表明　它并不是这样的。

GG: 我们听说它的效率约为57%至60%，这是相

people. The combined behavior in terms of wind, the dynamics of how people move, how spaces are created, solar orientation, views, and so on all define the relationships between buildings. And when you're designing cities, you factor in things like urban agriculture, transit systems, and open-space characteristics that you think are either beautiful or have a certain quality and proportion. This is where the complexity truly becomes interesting and very exciting, but at the same time highly frustrating because certain restrictions are imposed.

The book provides a step-by-step approach to solving a found condition and to achieving a specific goal defined by the 2030 challenge. It offers principles and metrics for eight different strategies, so there are specific criteria. But the manner by which the goals are achieved will be different from one person to another. That's the beauty of it.

AS: Our focus is not just existing buildings. We have a chart that speaks to building envelopes, for example, and it outlines the energy consumption for everything from single panes and 70 percent glass, down to double-skin, and below 60 percent glass. The amount of energy usage is roughly equivalent to that curve. We do the same thing with lighting, HVAC, vertical transportation loads, and other components of buildings. So the book is kind of a road map. It shows you the relativity of various systems.

Beyond this landmark publication, how do the same principles apply in your new work?

AS: We have projects where we've attempted to achieve a zero-carbon footprint, notably Masdar Headquarters in Abu Dhabi, which was designed as a zero-carbon, zero-waste, zero-energy building—in fact, one that achieves positive energy. It's possible, especially in the Middle East, because of solar and photovoltaics.

GG: The energy side is possible. The carbon calculation side of it and, depending where you draw the line for net zero, embedded carbon sourcing, is more complicated. Even at net zero, because of construction and materials, you're not at all where you need to be over the life cycle of the building. In other words, if the building is not positively contributing, then it's basically neutral and by default negative. It's not going to be a carbon-neutral building.

Your recent Pearl River Tower generated a great deal of interest as a net-zero building, but recent reports indicate it's not.

GG: We hear that it's about 57 to 60 percent efficient, which is still pretty good. It's better than anything else we know of right now. Part of the reason that Pearl is not net zero is because the community in which it is built, Guangzhou, isn't ready to receive contributions to the grid, so the embedded systems in the basement are now sitting there empty, waiting for the moment when the city can accept the building's power.

AS: Pearl River was a competition and they asked for a highly sustainable project. We said this is our opportunity to try to push to zero energy. We built in space for microturbines, and other things they said they'd incorporate in the future but couldn't pay for now. Gordon's right, they don't have the technology to receive the positive energy in the grid, so it's hard to get to a negative.

GG: We've been told they're suspicious about the quality of the energy that would come from the building to the infrastructure.

What are the building's main sources of energy?

GG: It's a combination of things. We have vertical-axis wind turbines, which is only about 3 percent, but we also have photovoltaics. We also have the double wall that helps to insulate the building, radiant panels on the ceilings, and raised-floor systems. We dropped the floor-to-floor height and got five extra floors in the same amount of height because we were so efficient in the ceilings; microturbines were a part of that, in the basement.

When did your concerns with energy and sustainability begin?

AS: I've been talking about zero-energy buildings since the early '90s, and have been trying to promote that ever since. About a year ago we established PositivEnergy Practice as a separate company to really focus exclusively on these issues, primarily from a mechanical viewpoint.

Every project is unique and every client is unique in terms of what they want. We invariably find ourselves pushing the boundaries of sustainability, low energy, low carbon. A lot of this has to do with a culture of design and an attitude about the way you live. Since we first started, there has been a slight culture shift with respect to what it is that we're trying to do

当不错的。这比我们现在所知道的其他建筑物都要好。珠江大厦并非是净零能耗建筑物的部分原因是，它所位于的城市，即广州并未准备好接受它对电网的贡献，因此位于地下室的嵌入式系统目前还没投入使用，它正等待着城市可以接受该建筑所提供的电能的时刻。

AS: 珠江大厦是一项挑战，他们要求一个具有高度可持续性的项目。我们说，这是我们尝试实现零能耗的机会。我们为微型涡轮机和其他设备设计了空间，以便他们将来在建筑物中纳入这些设备，尽管他们目前没有能力购买这些设备。戈登是正确的，他们还没有接收电网中正能源的技术，因此也不能说它是消极的。

GG: 有人告诉我们，他们担心从建筑物输送到基础设施的能源的质量。

该建筑物的主要能源是什么？

GG: 它利用多种能源的组合。我们有垂直轴风力发电机，其提供的能源仅占3%左右，我们还有光伏电池。此外，我们采用有助于建设物绝缘的双层墙，在天花板上使用辐射板，并使用活动地板系统。我们降低了楼层高度，并在建筑物高度不变的情况下，增加了五个楼层，因为我们对天花板的能效设计具有丰富的经验；位于地下室的微型涡轮机也是能源的一部分。

你们对能源和可持续发展的关注是从什么时候开始的？

AS: 自从20世纪90年代初以来，我一直在谈论零能耗建筑，并努力推广该类建筑。大约在一年前，我们成立了"正能源实业公司"，该公司主要从机械学角度，专注于这些问题。

每一个项目都是独一无二的，就客户所想得到的东西而言，每一位客户也是独一无二的。我们一直在推广可持续性、低能耗和低碳。这与设计文化以及你们对生活方式的态度有很大关系。从我们最初开始以来，人们对可持续性的认识逐渐发生变化。最初，人们难以相信它具有价值或认为这样做很重要。但现在情况有所改变，人们逐渐开始接受并理解它。

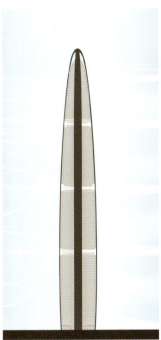

Left: Pearl River Tower, Guangzhou, China (Adrian Smith + Gordon Gill, 2011)
Center: Wuhan Greenland Center, wind diagram
Right: Wuhan Greenland Center, Wuhan, China (Adrian Smith + Gordon Gill, est. completion 2016)
左：珠江塔，中国广州。（AS+GG建筑设计事务所，2011年）
右：武汉绿地中心，中国武汉。（AS+GG建筑设计事务所，2016年完工）
中：武汉绿地中心，风向频率图。

GG: 当然，这取决于你努力争取的东西。人们会为获取LEED绿色建筑金级认证而欢呼，但LEED本身并不能解决能源和碳问题。有一些衡量指标或运营成本，以及类似的东西。例如，自行车停车架是一种很好的选择，让生活更加低碳，但是……当你看到我们的文化如何定义可持续性时，你就会发现还有其他文化，这些文化以非常不同的方式，在历史上已经解决了与自然的关系问题。这主要是关于你的生活方式与自然之间的平衡，以及建筑环境和自然环境之间的平衡。

客户是否接受新技术，或者由于成本原因，他们还在犹豫不决？

AS: 成本在某种程度上是一个问题，但并不是大问题。我们参与设计了波士顿601国会街项目，它是美国第一个双层气候墙、自通风结构建筑。当然，德国也有这样的项目，但并未采用自通风。它利用通风腔，实现了自然通风，而没有借

with buildings. Initially, it was difficult to convince anybody that this had value or was important to do. Now that is less so, and it's becoming more and more accepted and understood.

GG: Of course, it depends on what you are striving for. People will stand up and wave flags about getting to LEED Gold, but LEED on its own doesn't address energy and carbon per se. There are metrics or operational costs and things like that. I mean, bicycle racks are good, but. . . When you look at how our culture defines sustainability, you find that there are other cultures which historically have addressed their relationship with Nature in very different ways. It is about the balance between the way you live and nature, between the built environment and the natural environment.

Are clients receptive to the new technologies or do they balk because of cost?

AS: Cost is somewhat of an issue, but not a huge issue. It's more a perception of high cost. We got as far as 601 Congress

Zifeng Tower, curtain wall detail.
紫峰大厦，幕墙细节。

Street in Boston, which was the first double-climate-wall, self-ventilated structure in America. Of course, this has already been done in Germany, but not self-ventilated. It's naturally ventilated through cavities, not assisted by fans like the Foster projects are. It costs a little more to do this, but the client gets it back in terms of comfort and energy savings. In this scenario we were able to shortcut the approval process by one year—in Boston, it can take a couple of years to get planning approval—because they really wanted an example of a good double-wall system for commercial use. Sometimes justification is based not purely on dollars and cents.

GG: Part of the answer also has to do with the nature of how buildings get built, by whom, and who uses them. When a building is owner-occupied, you have a much better hearing for why the upfront costs are necessary and how they are going to pay off. If a developer is going to build and then immediately sell the building to someone else, the altruistic motive goes away.

AS: If you own a building and hold on to it for 25 years you can reap the benefits, but if you just plan to build it and flip it, you'd want the lowest denominator. Unfortunately the second scenario, flipping, is much more common. Many things the architect wants to do are value-engineered out because the developer doesn't want to spend extra money when he's not going to get the payback himself. Sometimes they just don't want to be the first to do something.

助于风机，例如福斯特项目。这样增加了一点成本，但客户在舒适性和能源节约方面，得到了补偿，因此更加划算。在这种情况下，我们能够将审批程序缩短一年（在波士顿，有可能需要几年时间才能获得规划批准），因为他们的确希望拥有一个用于商业用途的具有双墙壁系统的建筑物作为先例。有时候，理由并不是纯粹基于成本。

GG: 该问题的答案还与建筑物的建造性质、建造者以及使用人员相关。当一栋建筑物由业主自用时，业主就会更多听取建议，例如预付费用为什么必要，以及他们将如何支付等。如果建筑物由开发商建造，然后立即卖给其他人，那么利他主义的动机就会消失。

AS: 如果你拥有一栋建筑物并持有25年，你就可以从中获益，但如果你只是打算建造并出售，那么你希望成本降到最低。不幸的是，第二种情况，即出售，往往更为常见。建筑师希望做的很多事情基于价值工程，因为当开发商自己无法得到回报时，他们并不想花额外的钱。有时候，他们仅仅不想成为第一个吃螃蟹的人。

ARCHITECTURE Adrian Smith + Gordon Gill

Zifeng Tower, Greenland Financial Centre, Nanjing, China.
(Adrian Smith + Gordon Gill, 2009)
紫峰大厦，绿地金融中心，中国南京。(AS+GG建筑设计事务所，2009年)

In designing a building, do you take into account its full life cycle and plan for disassembly?

GG: Sometimes, not all the time. We talk about the kind of metrics for modularity, or material selection, disassembly, reuse. We've done that on some projects, like the Vancouver project that we did for 50 units on a 1-acre [4,047 square meters] site. We were trying to see how we could use the Mithun carbon calculator to get a low carbon footprint. We got a carbon footprint of 1 by virtue of the calculator. But that took into consideration disassembly and reuse of materials. We find that it's a combination of things that allows you to maintain a quality of life, an ecology of the city, more than it is an engineering or technological fix, because the latter tends to address just one stratum of an approach, not the full-bodied life qualities of cities.

Is there a height at which a building cannot be sustainable?

AS: Well, the taller you go, the more complex everything becomes. We are going to study that issue: how tall is impractical?

GG: On my word of honor, we were just discussing that question.

AS: That's our next book.

Do you see any limitations on the height of future ultra-tall buildings?

AS: A lot has to do with the elevatoring systems. How many times are people willing to transfer to get to their floor? The practical height of today's elevators is 500 to 575 meters [1,640 to 1,886 feet], and then you transfer at the sky lobby and go the next 575 meters [1,886 feet]. Most people don't mind a single transfer, but I don't know how a double transfer would work. We haven't really hit that stage. I don't think anyone has done one where you go up to a sky lobby, transfer, go up to another sky lobby, then transfer again and continue up.

GG: But there are many ways in which a super-tall building would have an advantage over a much lower building in terms of sustainability, in terms of harvesting renewable energy.

AS: In terms of carbon.

GG: In terms of the wind being cooler and faster at higher altitudes; you could make use of that and potentially induct some of that air for cooling mechanical systems.

当设计一栋建筑时，你是否考虑到它的整个生命周期，并为拆卸进行规划？

GG: 有时候会。我们会谈论模块化指标，选材、拆装和重复使用。在这方面，我们设计了一些项目，例如温哥华的项目，当时我们负责设计一栋占地1英亩共计50个单元的建筑物。我们试图看看如何可以使用米森公司的碳排放计算器，以获得一个低碳足迹。通过使用计算器，我们的碳足迹最终是1。该建筑考虑到了拆卸以及材料的重复利用。我们认为，它是多种事物的组合，可以让你保持高品质的生活以及城市的生态环境，它并不仅仅是一种工程或技术修复，因为后者往往仅解决部分问题，而非城市的生活质量。

是否有一个特定高度，当建筑物达到该高度之后就不再具有可持续性？

AS: 嗯，建筑物越高，一切就变得越复杂。我们将会研究这个问题：建筑达到什么高度是不实用的？

GG: 说实话，我们正在讨论这个问题。

AS: 我们的下一本书将会介绍这个问题。

对于未来超高层建筑的高度，你认为有什么限制吗？

AS: 在很大程度上，这与电梯系统相关。为了到达目标楼层，人们愿意换乘多少次电梯？当今电梯的实用高度为500至575米（1640至1886英尺），通过在空中大堂换乘，最终到达顶层，即575米的高度（1886英尺）。大多数人并不介意一次换乘，但我不知道两次换乘是否可行。我们还没有达到这个阶段。我认为还没有人建造了需要两次换乘的建筑物，对于这种情况，你首先要到达空中大堂，然后换乘，再到达另一个空中大堂，然后再换乘，以到达目标楼层。

GG: 在可持续发展以及获取可再生能源方面，超高层建筑比低层建筑更具优势。

AS: 还有在碳排放方面。

ARCHITECTURE Adrian Smith + Gordon Gill

GG: 海拔越高，风速就越快，并且空气更冷。你可以利用这个原理，将冷空气引入建筑物，用于冷却机械系统。

AS: 现在，电梯下降时所产生的能量几乎可以抵得上电梯上升时所消耗的能量，从能源角度来看，电梯效率大幅度提高，而以前却达不到。电梯操作完全实现了电脑化，与30年前甚至20年前相比，切换机制的能源消耗更低。

然后是土地使用问题。

AS: 这是一个大问题。以迪拜塔为例，它占地约5英亩（约20230平方米）。如果将该建筑物全部改为2层或3层建筑物，那么需要大约70到125英亩的土地（约283000到506000平方米）。因此，在可持续发展方面，高层建筑腾出的土地使用是一个大问题：不只是耕种，还有森林和其他生产性的使用。如果你能够创造一个高密度的商住两用环境，这样就可以步行上班，并取代开车上班，从而大幅度减少停车场和道路系统所占用的土地面积，并避免开车上班时所带来的石油消耗。世界上很多更高效的城市都很古老的，它们当初被开发为商住两用城市时，并没有很发达的交通。

GG: 还有一些地方，例如香港岛，它们的边界由地形特征限定。

通过打造可持续发展的卫星城市，这样人们就不需要前往其他地方工作，这是否现实？

AS: 当然，尤其是在中国，这一点可以强制执行。很多中国人仍然买不起，但如果他们能够在适当的区域生活和工作，并可以利用公共交通，能够在周末租车，以及诸如此类的东西，这就是可行的。

我记得当我在20世纪90年代初首次去北京时，长安街充满了自行车。现在，我估计只剩下15%到20%左右。每个人都有一辆小汽车，或至少是某种机动车。从文化角度来看，汽车是一种社会地位的象征，暗示了你在社会上的地位以及你的购买能力。所以这是一个更大的问题。为了成为可持续发展的城市，必须将汽车排除在外吗？应该不会。我们在阿布扎比时就遇到了这样的情

AS: Elevator systems now have gotten to the point where they're generating almost as much energy when they're coming down as when they're going up—almost, not quite—so they are now quite efficient from an energy perspective, whereas before they weren't. The operations are all computerized, so the heat generated by the switching mechanisms is much less energy-intensive than it was 30 or even 20 years ago.

Then there's the land use issue.

AS: That's a big one. Take Burj Khalifa, which occupies about 5 acres [about 20,230 square meters]. If you were to put that into a 2- or 3-story context, it would take roughly 70 to 125 acres [about 283,000 to 506,000 square meters]. So the land use that tall buildings allow—not necessarily farming, but forests and other productive uses—is a huge issue in terms of sustainability. Also, if you can create a high-density live/work environment where instead of driving you walk to work, you'd save untold square footage in car parks, road systems, and all the petroleum you'd burn to get there. Many of the world's more efficient cities are ancient, so they developed as live/work cities without viable transportation.

GG: Or there are places like Hong Kong, an island, where the boundaries are topographically defined.

Is it realistic to imagine a sustainable satellite city, where people don't have to travel elsewhere to work?

AS: Absolutely, especially in China, where it can be mandated. A lot of Chinese still cannot afford cars, but if they could live and work in a tight-knit area and have access to public transit, or be able to rent a car on the weekend, or something like that, it could work.

I remember when I started going to Beijing in the early 1990s, Chang'an Avenue was a river of bicycles. Now I'd say there's just 15 to 20 percent left. Everyone has a car, or at least some kind of motorized vehicle. From a cultural standpoint, the car is a status symbol, and says something about your place in society and your ability to purchase goods. So it's a bigger question. Must cars be excluded to have a sustainable city? Probably not. We've been in situations in Abu Dhabi where the no-car concept didn't really work because there are some basic services that a city needs. The idea in that particular case kind of led to flawed concepts about vehicular access in general, because if a city were truly pedestrian, there's no need to design for cars—but what about fire trucks, ambulances, and the like?

Burj Khalifa and the Dubai skyline, 2010
哈利法塔和迪拜天际线，2010年

134

GG: It's one thing to mandate no car, but another way is by disincentivizing. In Toronto, for example, parking can be 30 to 50 dollars a day. The alternative is to take public transit, which is good, so you can actually get around and make the connections you need. If there is an odd-and-even license plate rationing, if the city is vigilant about how parking is organized and its access to transit, it actually does work. In Toronto, people take the train to go downtown. It's just faster.

AS: We are now designing a master plan for a satellite city which we can't discuss with great specificity other than to say that it is in China, the population is about 100,000, and the issues we're discussing are highly germane.

GG: There's a mix of building heights, but some of the same principles that we talked about in Chicago's Loop are applied here, such as walkability, access to green space, live/work, economic accessibility, and education. It also has health care, shopping, waste-disposal systems, and transit. So it gets really rich really fast.

What is the impact of the aging population?

GG: We're talking to the client about this from a cultural standpoint. It's also very important from an economic standpoint. It is not so practical for a married couple to care for four elderly parents, so the design of these cities needs to take that into account. It's different here than in China. There, the whole issue of sons and daughters taking care of their parents and where they live and how they live as a family unit is a unique cultural phenomenon.

AS: Another aspect is that these cities have to be built and developed for middle-income Chinese, not exclusively for the wealthy. So that brings the bar of quality down considerably.

GG: We have a lower-income population calculated into our demographics. We've tried to understand how the people live and what their basis of existence was. Was it farming? Making things? A part of what we talk about here is having that group or that generation and their children begin to engage the city through what makes them most comfortable. It's a quality of life that allows them not to be segregated, and affords them the opportunity to educate their offspring in a way that they then can grow in the city and move to other areas.

AS: The project is not all high-rise office buildings; there's a mixture of start-up cottage industry areas to develop. And there's low-cost office space as well as Class A offices.

况，在那里，"无车"理念并不可行，因为城市需要一些基本服务。在特定情况下，关于汽车使用的错误观点会导致错误的理念，如果一个城市成为了真正的步行城市，那么就没有必要为汽车做专门的设计——但消防车、救护车之类的车辆怎么办？

GG: 可以颁发法令，禁止车辆使用，但也可以采用限制方法。例如在多伦多，每天的停车费可以为30至50美元。另一种是利用公共交通，这是很好的办法，你同样可以到达各个地方。如果采用单双号限行措施，并且城市对停车点以及公共交通的使用进行严格管理，这的确可行。在多伦多，人们坐火车去市区，并且更快。

AS: 我们现在正在为一个卫星城市进行总体规划设计，该城市位于中国，人口约10万，它与我们正在讨论的问题密切相关。

GG: 该城市的建筑物高度并不相同，但我们之前谈到的有关芝加哥卢普区的一些原则仍旧适用，例如步行便利性、绿色空间的使用、商住两用性和经济便利性等，教育也是一个关键因素。同时，它还提供有医疗保健、购物、废物处理系统和交通服务。该城市发展得非常快。

人口老龄化有什么影响？

GG: 我们从文化的角度与客户谈论这个问题。从经济的角度来考虑，也是非常重要的。让一对夫妇照顾四个年迈的父母，并不是那么实际，因此这些城市的设计需要考虑到这一点。这里与中国的情况是不同的。在那里，儿子和女儿照顾父母、他们居住在什么地方以及如何以家庭为单位生活，这些都是独特的文化现象。

AS: 另一个方面是，这些城市的建设和开发必须考虑到中等收入人群，而不是专门针对富人。因此，这样可以相当程度上放宽限制要求。

GG: 我们将较低收入人口计算到我们的人口统计内。我们试图了解人们如何生活以及他们的生存基础是什么。是农业耕作？还是手工制作？我们所讨论的一部分内容，就是通过使用让他们感到最舒适的方法，让这一人群或一代人以及他们的子女开始参与城市生活。这是一种生活质量的提升，可以让他们不会被隔离，并为他们提供机会，更好地教育他们的后代，从而可以让他们在城市成长，并迁移到其他地区。

AS: 该项目并非全部是高层写字楼，还包括新兴产业区，不仅有低成本的办公空间，还有甲级写字楼。

GG: 城市已建成部分与专供农业使用的相关空置土地之间有关系。

AS: 这并不是说将一个城市放在特定的位置，然后观看它在里面成长。这个城市与我们已经谈到、设计并且规划的外围环境有着密切关系。"城墙"内外有着不同的技能组合。我认为某些群体在城市外围发挥出色，然后他们会进入中央商务区。以一种有意义的方式为城市作出贡献的能力，则会为人们赋予价值感。与此相反，如果人们被安置在一个岗位上，而其掌握的技能无法让他们发挥自己的能力并作出贡献，那么他们会觉得自己被落下，并成为一种负担。这也是一个文化问题，因为如果没有可用的汽车，你就无法去工作并作出有意义的贡献，那么你就成为一种负担。在城市的文化影响方面，你会看到许多在火车站闲逛的人。你可以把他们装进火车，但他们稍后还会回来，因此你并没有解决任何问题，只是在延误或拖延他们。

GG: 我觉得这整个想法非常有趣，其核心是发展卫星城市，而不是指责城市。这是一个我们都需要自己来回答的问题，而中国现有的许多城市其内部基础设施全都压力巨大。

你认为究竟是什么构成了一座卫星城市？

GG: 好问题。当我们谈论卫星城市时，我觉得会有这样的成见，认为卫星城市是限定在中心城区之外的一个独立实体。如果它是一个真正的城市并且自身具有可持续性，那么它应该有自己的经济、身份和区域，并有附属的生活工作环境，即一个你可以真的成长、结婚生子、取得成功、学习和养老的地方。

GG: There is a relationship between the built portion of the city and the associated open lands dedicated to agriculture.

AS: It's not about putting a city on a site and watching it grow inside. There's a peripheral relationship of the city to its environment that we've talked about, designed, and planned. There are different skill sets inside and outside the "city wall." I think there are certain groups that function at a much higher level on the periphery than they would inside a CBD [central business district], for example. So we have both; there's all of it. It is part of the dignity and the ability to contribute to the city in a meaningful way that gives people a sense of value, as opposed to being put in a position where their skills don't allow them to contribute so they feel left behind or a burden. That again is a cultural issue, because if there is no vehicle for you to work and to contribute meaningfully, then you do become a burden. And in terms of a cultural impact on a city, those are the people that you see hanging out at the train station. You can put them on a train but they come back later, so you're not solving any problems, you're just delaying or deferring them.

GG: I find this whole idea very interesting, the notion of a satellite city as opposed to attacking the city at the core itself. It's a question that we all need to answer for ourselves, since the internal infrastructure of many existing cities in China is fully stressed.

To your mind, what actually constitutes a satellite city?

GG: Good question. When we talk about satellite cities, I think there's a preconception of a stand-alone entity that's defined outside the body of the central core. If it is truly a city and is sustainable on its own, it kind of has an economy about it, has an identity, a place about it, has a live/work environment associated with it—a place where you can actually grow, have children, be successful, learn, and age.

Can a bedroom community with shopping be considered sustainable?

GG: I think so. It depends on the population and its relationship to live/work. We've spoken about context and the ideals of a culture. In the United States, it seemed that the pastoral ideal was a wonderful achievement. There was a time when, because of the automobile and the desire to own your own piece of ground, people moved out of the core. Planting your own tree, digging a hole, building a fence—that's what it was all about. You had a connection to your pastoral

Kingdom Tower, Jeddah, Saudi Arabia (Adrian Smith + Gordon Gill, est. completion 2017): View from the clouds; from the water; from the air

王国大厦，沙特阿拉伯吉达（AS+GG建筑设计事务所，2017年完工），从云端看的效果图，从水中看的效果图，从空中看的效果图。

带有购物设施的近郊居住区具有可持续性吗?

GG: 我认为是的。这取决于人口以及与生活工作的关系。我们谈到了文化背景和理想。在美国,田园理想似乎是一个很奇妙的成就。曾经有一段时间,由于汽车的普及以及人们渴望拥有自己的一片土地,人们搬离核心城区。他们种树、挖坑并修建围栏,营造自己的家园。逐渐地,我们看到了生活方式的转变,看到了旅行时间对个人时间的影响,我们工作更多,并且工作时间更长,这种生活方式变得越发不切实际。人们居住在郊区并不一定是因为费用低,实际上,住在郊区的成本更高。所以,你现在可以看到人们正迁回城市。

在输出这种文化方面,我们做了许多工作,汽车是其中一部分。它是一种社会地位的象征。"你有汽车吗?如果你没有,那么我不会嫁给你。"所以现在的问题是,到底是田园理想还是对汽车的需求将人们推向了城市边缘,如果你住在城市,并且就在街对面工作,那么你为什么会需要汽车呢?因此,只有结果才能证明方法是否正确。有关汽车的效率,我们已经进行了一些非常有趣的讨论。有人说:"让我们设计每加仑(3.8升)汽油可行使50英里(80公里)的汽车。"这种提议好吗?我不知道它究竟好不好。在25mpg(10.5公里/每升)的情况下,你可以居住在距离工作地点一个小时车程的地方,如果在50mpg(21公里/每升)的情况下,你会住在什么地方呢?距离城市越远,越能感到放松,是吗?那么释放到空气中的额外的碳、增加的流量和拥塞问题等,该如何处理呢?城市扩张的历史可以追溯到古代,那时人们会乘坐马车迁往农村。这永远都是一个问题。

expectations. More and more, we saw the transition of that lifestyle and the impact of travel time and personal time, and working more and working longer, and that lifestyle became much more impractical. It wasn't necessarily because of lower costs; it actually might have been more expensive to live out in the suburbs because the nicer homes were out there. So now you see people migrating back into the cities.

We've done a great job of exporting this culture, and automobiles are a part of it. They're a status symbol. "Do you own a car? I'm not marrying you if you don't have a car." So the question is whether the pastoral ideal or the demand for a car is going to push people out to the fringes of the city, because if you live in the city and work across the street, why would you need a car? So it's kind of the end justifying the means. We've had some very interesting discussions about the efficiencies of cars. Somebody says, "Oh, let's design a car that gets 50 miles [80 kilometers] to the gallon [3.8 liters]." Is that good? I wonder if it's good. If at 25 mpg [10.5 kilometers per liter] you can live an hour away from work, at 50 mpg [21 kilometers per liter] where would you live? The more distant from the city, the greater the relief, right? What about the additional metric tons of carbon released into the air, the increased traffic and congestion? Urban sprawl dates back to antiquity when people took to their chariots to the countryside. It will always be an issue, always.

Antony Wood
CTBUH

Antony Wood, B.Arch., Ph.D. As the executive director of the Council on Tall Buildings and Urban Habitat since 2006, Antony Wood has significantly revitalized the CTBUH. He was previously CTBUH vice chairman for Europe, and head of research. He is also an associate professor in the College of Architecture at the Illinois Institute of Technology, with a specialization in sustainable design of tall buildings. He previously taught architecture at the University of Nottingham in England (2001–2006). Trained in Britain, Dr. Wood practiced architecture in Hong Kong, Bangkok, Kuala Lumpur, Jakarta, and London (1991–2001), during which time he developed his passion for, and experience in, tall buildings. His doctoral dissertation explored the multidisciplinary aspects of skybridge connections between tall buildings. Dr. Wood is associate editor of several leading professional journals and author of numerous publications, including *Tall & Green: Typology for a Sustainable Urban Future* (2008). In progress are *The CTBUH Guide to Sustainability for Tall Buildings in Urban Environments* and *New Paradigms in High Rise Design*, as well as *The Tall Buildings Reference Book.*

安东尼·伍德
高层建筑与城市住宅协会执行理事

自从2006年以来，安东尼·伍德就担任高层建筑与城市住宅协会执行理事，他使这一协会恢复了生机。他原先是欧洲高层建筑与城市协会的副主席，而且兼任研究部主管。他也是伊利诺伊理工学院的助理教授，专门研究高层建筑的可持续性设计。安东尼之前在英国的诺丁汉大学教授建筑（2001年至2006年）。伍德博士在英国接受教育和训练，在中国香港、曼谷、吉隆坡、雅加达和伦敦从事实际的建筑工作（1991年至2001年），这一时期，他产生了对于高层建筑的热情并在其中进行历练。他的博士论文从多学科不同角度讨论连接不同高层建筑的空中天桥。伍德博士是许多前沿专业杂志的助理编辑，也是多种出版作品的作者，如《高层与绿意：关于一个可持续城市未来的类型学研究》（2008年）。即将出版的有《高层建筑与城市住宅协会关于城市环境中高层建筑可持续性指南》，《高层设计的新范式》以及《高层建筑参考书》。

For complete biography, please refer to the Appendix.
完整的传记，请参阅附录。

Skybox for student housing in Chicago, submitted to Professor Antony Wood, IIT, by Praima Gupta (2008).
芝加哥学生住的空中房屋，由帕莱玛·古普塔设计，提交给安东尼·伍德教授。

What is the current status of tall buildings in the world?

Throughout history there have been pockets of intense tall-building activity, starting in Chicago in the late 19th century, then Art Deco in New York in the 1920s, Chicago in the 1960s and '70s, and Southeast Asia in the 1980s. This activity was usually concentrated in a specific city or region. However, in the last 10 years there has been a massive and unprecedented building boom affecting almost every corner of the world simultaneously, in Asia and the Middle East predominantly, but also in the U.S., Europe, Australia, even parts of South America and Africa.

What accounts for this?

The answer is different in each place. One of the drivers is investment and return on land development. That's always been a driver for tall buildings. An equal driver is the quest for "iconicness." Historically the push for an icon was mostly at the corporate level, one dynamic corporation against another, whereas now it's about a dynamic city or a dynamic country over others in a very competitive international market. The world's tallest buildings used to bear names like Chrysler, Sears, Petronas; now they're more likely to be called Chicago Spire, Burj Dubai, Moscow Tower. Reflecting city ambitions has been one of the primary drivers in the Middle East, for instance, whereas in the more mature American and European markets the driver is still largely financial.

The third driver is sustainability. The American model of urban development, i.e., a dense downtown core and a horizontal, ever-expanding suburb, is just not a sustainable way forward. UN statistics show that there are almost 200,000 people urbanizing every day. That means as a global species we need to build a new city of a million people every week. It's happening predominantly in China, India, Brazil, and Indonesia, but also in Vietnam, Turkey, and lots of other countries.

The horizontal city is not a sustainable model because of the energy requirements needed to support it; as much embodied energy goes into creating a city as into operating it. As much energy goes into lighting, roads, infrastructure, and so on as there is energy being choked into the atmosphere from two-hour daily commutes.

Cities need to become denser to accommodate urban expansion, but it is not universally accepted that tall buildings are the solution. Lots of European cities are very dense without skyscrapers. Paris, for example, is a relatively uniform 12 to 14 stories throughout. Although tall buildings concentrate more people on a single plot of land, it's not necessarily a denser

世界高层建筑目前的状况如何?

纵观历史,曾经有一些密集的高层建筑运动:从19世纪晚期的芝加哥开始,到20世纪20年代纽约的装饰派艺术,再到20世纪60年代和70年代的芝加哥,然后是20世纪80年代的东南亚。这一运动一般都以一个特定的城市或地区为中心。然而,在过去的十年里,出现了一次规模巨大、史无前例的建筑发展热潮,它几乎同时影响到世界的每一个角落——主要是在亚洲和中东,但也发生在美国、欧洲、澳大利亚,甚至南美洲和非洲的部分地区。

如何解释这一现象?

在每个地方答案不尽相同。其中一个推动因素是土地开发上的投资与回报,这经常是高层建筑的一个推动力量。还有一个相同的推动力就是对于"标志性"的追求。从历史上看,努力争取一个标志大多发生在公司层面上:一个活力充沛的公司和另一家公司竞争;然而,现今它发生在竞争激烈的国际市场上,是一个活力充沛的城市或国家与另一个城市或国家的较量。过去,世界的最高建筑常被冠以克莱斯勒、希尔斯、马来西亚国家石油公司等名号,现在它们更有可能被称之为芝加哥尖顶、迪拜塔、莫斯科塔等等。例如:在中东,反映城市的雄心就是一个非常主要的推动力,而在更加成熟的美国和欧洲市场,推动力依然来自经济。

第三种推动力是可持续性。美国模式的城市发展,是一个密集的市中心加上一个横向扩展的郊区,这并不是一种可持续发展的模式。联合国的统计数据表明,每天大约有20万人被城市化。这意味着对人类来说,我们需要每周建立一个可容纳100万人的大城市。这一现象现在主要发生在中国、印度、巴西和印度尼西亚,但在越南、土耳其和许多其他国家也有出现。

横向发展的城市并不是一种可持续的模式,因为它需要能源来支持。创造一座城市所需要的具体能源与维持这座城市所需要的一样多。许多能源转化为照明、道路、基础设施等等,也有一些能源由于每天两小时的交通换乘而被排放到空气中。

城市需要变得更加密集以适应其扩张的需要,但并非所有人都愿意将高层建筑作为解决方案。许多欧洲城市非常拥挤,但却没有摩天大楼。比如,巴黎普遍是相对统一的12到14层建筑。尽管高层建筑将更多的人集中在一个单独的地界之上,但它不一定是一个更加密集的建筑物。例如,芝加哥没有哪个地方的密度能赶得上曼哈顿或香港。那些坚持高层建筑并不一定是可持续性建筑类型的人,有

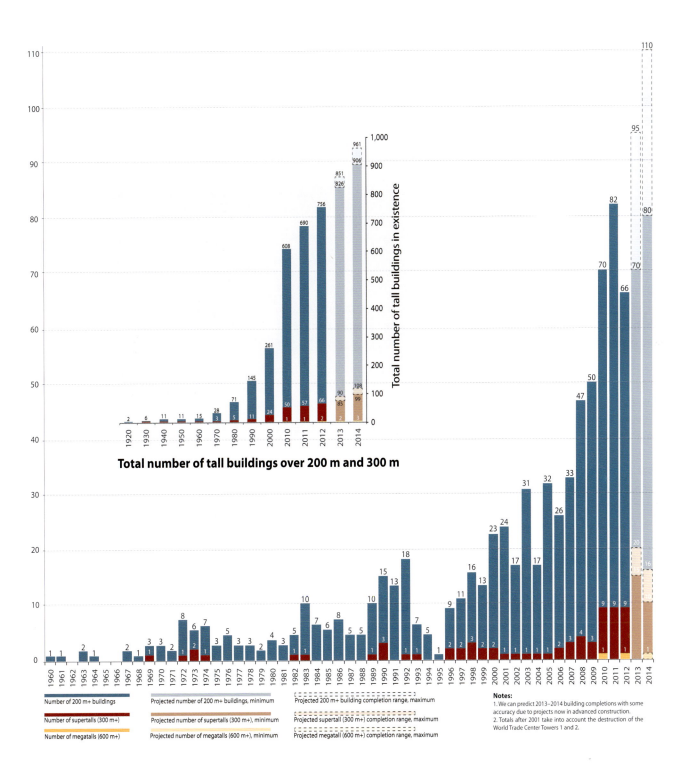

Total number of tall buildings over 200 m and 300 m

Total number of tall buildings in existence

Number of 200 m+ buildings
Number of supertalls (300 m+)
Number of megatalls (600 m+)

Projected number of 200 m+ buildings, minimum
Projected number of supertalls (300 m+), minimum
Projected number of megatalls (600 m+), minimum

Projected 200 m+ building completion range, maximum
Projected supertall (300 m+) completion range, maximum
Projected megatall (600 m+) completion range, maximum

Notes:
1. We can predict 2013–2014 building completions with some accuracy due to projects now in advanced construction.
2. Totals after 2001 take into account the destruction of the World Trade Center Towers 1 and 2.

Tall buildings completed each year over 200 meters [656 feet].
每年完成的的高层建筑超过200米（656英尺）。

ARCHITECTURE Antony Wood

fabric. Chicago, for instance, has nowhere near the density of Manhattan or Hong Kong. Those who maintain that tall buildings are not necessarily sustainable building types have a very good case. It is only in the bigger urban picture beyond the single building—through increased urban density—that the tall typology really comes into its sustainable own.

In presentations, I normally surprise the hell out of people by saying that 95 percent of tall buildings are poorly designed and that they are not at all a sustainable proposition; they only become sustainable in the larger dense urban context. If anyone tells you that a tall building in itself is sustainable, they are not talking correctly. We are very far from net-zero-energy tall buildings, despite all the hype. If we ever do get there it will likely only be net-zero-energy for operating carbon, not for all the embodied energy of construction.

What is the threshold beyond which a building is not sustainable?

Let me say as a precursor that the Burj Khalifas of the world get the most attention. Burj Khalifa is 828 meters [2,717 feet], virtually as tall as the Hancock Tower on top of the Willis Tower. Buildings are definitely getting higher, but right now there are only 50 or 60 in the entire world that are over 300 meters [984 feet] in height, which is the definition of a supertall. Debates focus on whether iconic supertalls are sustainable, but the reality is that 99 percent of tall buildings are not even playing that game. The vast majority of tall buildings around the world are usually utilitarian housing or office buildings of 40 to 60 stories, so it is very dangerous to base the whole argument about the sustainability of tall buildings on 50, 60, or even 1,000 supertalls; it's a small proportion of the market.

As to the threshold for a sustainable tall building, that depends very much on where and what it is. Instinct would say that there is a height threshold, and above that, it doesn't make financial sense, let alone sustainability sense. But the bigger issue is that we are not developing our cities as intensely as we need to. My Ph.D. dissertation was called "Pavements in the Sky: The Use of the Skybridge Between Tall Buildings." For many years I've been fascinated with the idea of horizontal connections between tall buildings. It started as a kid with futuristic films showing linked towers in multilevel cities, but is it just science fiction cinematography?

Actually, it's being done now, but in a very small piecemeal way. The archetypal skybridge-connected building is Petronas Towers: two 88-story towers with transfer levels at 41 and 42. But there are others, like "The Pinnacle@Duxton"

一个很好的例子：只有在超越单一建筑的更大城市图景中——通过城市密度的增长——高层建筑类型才形成自己的可持续性。

在演讲中，当我说出这样的话时，听众往往会被吓一大跳：95%的高层建筑设计很糟糕，它们并不都是一个可持续性的提案，它们只有在更大密度的城市环境中才具有可持续性。如果有人告诉你，一个高层建筑自身就是可持续性的，那他说的话并不对。尽管各种广告宣传天花乱坠，但我们离零能耗的高层建筑还是非常遥远的。如果我们曾经达到过可持续性的标准，那有可能是碳的使用达到了零能耗；但它却没有将建筑中所有的具体能源都算在内。

一栋建筑物超出了某个界限就不再是可持续性的了，那么，这个界限是什么？

让我以一个先行者的身份来说一下，哈利法塔（Burji Khalifas）已经获得了全世界最多的关注。哈利法塔高828米（2716.54英尺），实际上相当于汉考克大楼加在西尔斯大厦（现在称为威利斯大厦）楼顶那么高。建筑确实变得越来越高，但是现在，全世界只有50或60座高度超过300米（984英尺）的建筑——这个高度的建筑可以被定义为超高建筑。人们争论的焦点是，是否标志性的超高建筑是可持续性的？事实上，99%的高层建筑甚至和可持续性毫无关系。世界上绝大多数的高层建筑一般都是40到60层的实用住房或办公楼，因此将关于高层建筑的可持续性论证都奠基在五、六十座甚至1000座超高建筑之上，是极其危险的；超高建筑只占市场份额的一小部分。

一栋可持续性高层建筑的门槛，很大程度上取决于它在哪里以及它是什么。直觉会告诉我们存在着一个高度的门槛，超过这个标准，就没有什么经济上的意义，也没有什么可持续性的意义。但更大的问题是，我们并非像实际所需要的那样，热情高涨地发展城市。我的博士论文题目是"空中道路：对高层建筑之间的空中天桥的利用"。很多年来，我都十分醉心于高层建筑之间的横向连结这一想法。这种感觉开始于我还是一个小孩时看过的一些未来电影，其中展示了多种城市中连结的塔楼，那些仅仅是科幻小说的电影表现形式吗？

实际上，这种想法现在已经以一种很渐进的方式被实现了。由空中天桥连接起来的典型建筑是马来西亚吉隆坡的双子塔：两栋88层高的建筑，有两层（41和42层）用于换乘。还有其他一些例子，比如在新加坡，由政府建设的高层楼房"达士岭尖塔",就有三个重要的天桥层。

就可持续性而言,当楼层进一步升高时,已不具有能量、碳和经济上的意义。我们不停地垂直上升,而对于水平层面上的可能性熟视无睹,这是绝对疯狂的表现。设想如下情况,我们接了一座中国城市,城市的整个区域有100万人而且他们都住在低层的公寓中。在十年内,同样的土地上将会聚集大约500万到1000万人口,这并非是一个幼稚愚蠢的讨论。我们该如何处理这种情况呢?

你要么向上要么向外。如果我们赞同可持续性/密度的观点,并想在城市周边保留农业的空间,那么答案就是向上。但问题是,生活在这块土地上的人需要地面来维持他们的生活,比如公园、人行道、社交区、学校、医院、商店、孩子们玩耍的地方或者遛狗的地方。

这将是城市垂直化所带来的问题。办公室、居民楼和宾馆的功能(我只是描述了高层建筑99.9%的功能),这三种功能并不能完全表现出纷繁多样的城市,这是个大问题。建筑物之间的空中天桥,并不是一些奇异花哨的连结,它是将土地看作基本支撑层,并通过在空中引入有目的的水平层面来复制这种支撑层。我指的不是建造整个一层地面,挡住太阳和光线。我在谈论扩大有目的地修建建筑间的桥梁和平面的机会。如果考虑到建筑与建筑之间,以及建筑内部所拥有的特设的水平连结,那么一栋160层的建筑仅仅是直上直下的结构,是不是有点疯狂呢?不管你是否考虑能源、操作性和社会交往方面的问题,空中天桥都是一个了不起的想法。它们为什么现在没有被普遍采用呢?因为它需要许多协同合作的思考,而我们作为一个物种还没有进化到足以接受它的程度——特别是以城市管理和规划的方式来接受它。

一座城市通常所作的,就是创造出一个城市总体规划,它通常是两个维度上的分区规划,偶尔还会对最大高度进行限定。我们所需要的是一个全面连接的三维规划,它将让水平层面与垂直层面联系起来,并在策略层面上允许一种横向的汇合,以及形成一些公共区域。因此,可持续性是否有垂直高度上的门槛?当然有。但实际上,如果你开始将可持续性看作是更有意义的,那么作为一个垂直堆叠的城市,依然还是会有这些门槛,但它将成为一个完全不同的规划方案。

谁来为基础设施负责呢?

这是关键,它需要多方面综合的考虑。现在,谁在为公园、人行道、下水道、照明设备等负责呢?实际上在全世界范围内,尽管过去的10到20年间,

government-built high-rise housing in Singapore, which has three significant levels of skybridges.

Apropos of sustainability, there comes a point where there's no energy, carbon, or financial sense in just going further upward. But what's absolutely crazy is that we keep going vertical without bringing in the horizontal plane. Let's say we've got an area of a Chinese city with a million people living in low-rise apartments. In 10 years there will likely be 5 to 10 million people concentrated on that same land area. This is not a silly discussion. How are we going to do it?

You go either far up or far out. If we agree with the sustainability/density argument, and also want to preserve agricultural space surrounding the city, then the answer is up. But here's the thing: the people who live on this land require the ground plane to support their existence, for parks, sidewalks, social interaction, schools, hospitals, shops, places for their kids to play, or to walk their dog.

That's the problem with bringing the city vertical. Office, residential, and hotel functions: I just described 99.9 percent of tall buildings. Those three functions do not represent the diverse city. This is a major problem. The idea of skybridges between buildings is not about throwing in a few fanciful connections, it's about seeing the ground as an essential support layer and replicating it by introducing strategic horizontal planes in the sky. I'm not talking about creating a whole ground floor and blocking out the sun and the light. I'm talking about maximizing the opportunity for bridges and planes between buildings strategically. Considering the strategic horizontal connections between and within buildings, isn't it crazy to have a 160-story building and just go up and down? Whether you consider aspects of energy, operation, social interaction, whatever, the skybridge is a fantastic idea. Why aren't they being done? Because it requires a hell of a lot of joined-up thinking, and we're not evolved enough as a species, specifically through forms of city government and planning, to allow it.

The most a city typically does is create an urban master plan, which is usually a two-dimensional zoning plan, occasionally stipulating maximum heights. What we need is a fully connected three-dimensional plan that allows the horizontal plane to come up into the vertical plane and permit a horizontal influx, public zones, at strategic levels. So is there a vertical threshold for sustainability? Absolutely. But actually, if we started to consider this as something much more significant, as a stacked vertical city, there would still be those thresholds, but it would become a completely different scenario.

144

Who would be responsible for infrastructure?

That's the key. It needs joined-up thinking. Whose responsibility is it now for parks, sidewalks, sewers, lighting, etc.? It's city governments, virtually all over the world, albeit with more public/private financing in the last 10 to 20 years. But once you hit the threshold of a building, the responsibility becomes the developer's. Are they going to pay for this necessary urban infrastructure within the building? Likely not. That's why government intervention needs to dictate what happens inside a building.

Don't get me wrong. There are multiple challenges to overcome to incorporate these new elevated networks—challenges of ownership, structural dynamics, safety, and terrorism, linking into a network that might come 40 years later, etc. But I believe there are many benefits and that's what we need to do in the future.

Didn't urban density used to have negative connotations?

If we continue to develop this planet on the same ideals that founded the American dream, we will not survive as a species. We need to recognize that. Something has to give. It will not happen through choice, but through necessity. We're getting closer to it every day. Every day there's some major disaster. I don't know whether it's 5 years or 50 years away, but once it becomes a matter of survival rather than just choice, cities and countries around the world will start to act.

Everyone cannot live in a single-family home. We need to find a balance, and be more conscious of the resources we consume. It's possible to occupy the land and still alleviate density-related problems.

Many people choose to live in the city. Take Singapore, where 80 percent of the people live in government-built high-rise housing, and most of the other 20 percent live in privately built high-rise housing. Put "government" and "housing" together in any other country in the West and it spells disaster. People are not stark raving mad as a result of high-rise living in Singapore, Hong Kong, and many other Asian cities. Somehow they reconcile the vertical realm with the other functions that humanity needs to function. Understandably, in the West, we don't want to give up what we have.

In the interests of complete disclosure, I must say that although I'm the executive director of the Council of Tall Buildings and Urban Habitat, I live in a single-family home in Oak Park [a suburb of Chicago], one of the world's most beautiful architectural neighborhoods. It has a certain urbanity, especially in the context of the American suburb, but it's certainly

社团/私人团体开始更多地在经济上进行投入，但负责任的还应该是城市的政府机构。而一旦碰到建筑的范畴，责任就变成开发商的了。他们会在建筑的内部为这些必要的城市基础设施买单吗？似乎不会。这就是为什么政府需要介入那些建筑内部的规定和要求的原因。

不要误解我。将这些不断提升改进的新网络结合起来，面临着诸多需要克服的挑战——这些挑战来自于所有权、结构动力学、安全性和恐怖主义，以及与一个可能是40年之后的网络进行对接等等。但我认为，它有许多好处，而且也是将来我们需要做的事情。

城市的密度过去不是常常具有否定的含义吗？

如果我们继续基于同样的理想来开发地球，并且这一理想形成了美国梦，将来我们作为一个物种就会灭绝，必须认识到，有些东西一定要放弃。这不会通过选择而发生，却会通过需要发生。我们每天都越来越靠近它，每天都有一些大灾难。我不知道它离我们还有5年还是50年，但一旦它成为一件生死攸关的事而非仅仅是一种选择，世界上的城市和国家将会开始行动起来。

每个人都不能生活在独栋住宅中。我们需要找到一种平衡，也需要对于消耗的资源有更多的认识。我们还是有可能在占有土地的同时减轻与人口密度相关的诸多问题的。

许多人选择生活在城市。以新加坡为例，有80%的人住在政府建的高层住宅楼中，而另外20%的人则大多住在私人建的高层住宅楼中。在西方任何国家，将"政府"和"住宅楼"放在一起，都意味着灾难。在新加坡、香港和其他许多亚洲城市，人们并没有因为高层生活而完全疯狂。他们以某种方式将垂直领域的生活与其他一些人类所需的功能结合起来。在西方，我们并不想放弃我们所拥有的，这一点是可以理解的。

为了开诚布公，我必须说，尽管我是高层建筑和城市住宅理事会的执行理事，但我也住在芝加哥郊区橡树公园的一座独栋住宅中——橡树公园是世界上最美丽的建筑社区之一。这个社区有一种特殊的都市风格，尤其是在美国式郊区这一大背景中，但它的确不是垂直的都市。即使你付钱给我，我也不会让我的孩子住在一栋高层楼房中，因为那里没有空间和设施满足我所要求的高品质生活。

有趣的是，我最近读到一份澳大利亚人的报告，上面说住在高层建筑中的孩子要比独栋住宅中的孩子享受到更多的自由和流动性。这怎么可

Left: Locally inspired Swadeshi Tower in Mumbai was designed in an advanced studio at IIT to link center city functions.
Right: Pinnacle@Duxton, Singapore. (ARC Studio Architecture + Urbanism, in collaboration with
RSP Architects Planners & Engineers (Pte) Ltd, 2009)
左：印度风格的大楼，受孟买用于洗衣和晾干的巨型露天洗衣场的启发，伊利诺伊理工学院高级工作室设计。
右：新加坡达士岭组屋。（由ARC建筑、城市化事务所与雅思柏设计事务所合作完成，2009年）

能呢？因为你更可能让你的孩子去隔壁商店或者往下走60层，而不是打开门来，让他们沿着街道自由地散步走动。"在高层建筑中，行走的路程多，而且年轻人拥有更多的自由。"在读到这些之前，我总是不由自主地设想那些灌输给我，被我认为是显而易见的理论：高层建筑中的孩子能享受到的流动性要少一些，但这并不总是一种显而易见的解答。我们必须做更多的研究，以便更好地理解此类问题。

从传统上说，高层建筑已经与有权势的公司以及有影响力的城市居民联系在一起。低收入家庭的因素

no vertical city. I wouldn't put my kids in a high-rise building if you paid me. There is not the space and facilities for what I deem a good quality of life.

Interestingly, I recently read an Australian report that kids living in high-rise buildings enjoyed more freedom and mobility than those in single-family homes. How can this be? Because you're much more inclined to let your child go to the shop when it's next door, or 60 stories down, as opposed to opening the gate and letting him walk freely along the roads. The distances traveled, the freedom at an early age, is much greater in high-rise buildings. Prior to reading this, I automatically presumed what has been drummed into me as obvious:

kids in tall buildings enjoy low mobility. But it's not always the obvious solution. We need to do more research, and better understand issues such as this.

Traditionally, high-rises have been associated with powerful corporations and affluent urbanites. How do lower-income families factor into the equation?

There are two kinds of people living in tall buildings: those who choose to and those who have no choice, i.e., the comfortably rich and the uncomfortably poor. It would be well to look at places like Singapore that manage to achieve what we in the West have not. The typology has been seen at fault. In the UK, even now, it's difficult to build a high-rise residential building because people cannot disassociate it from the post-war blocks thrown up in the 1950s. Was the typology at fault, or the particular circumstances? Was it overcrowding? Lack of maintenance? Europe is now cutting back because the social infrastructure is not affordable. So where does that leave us and our ability to provide for people who can't provide for themselves? That's a very difficult question, and I don't know the answer. But there are cities in the world that have dealt with this successfully.

Actually, in this scenario of going vertical and government investing, I think there is a path to create a vertical city that is not only sustainable but socially inclusive. In London now, you have to provide 20 percent social housing in any major development over a certain size; in parts of America as well. A private apartment might go for a million dollars and a half-size apartment lower down in the building might go for $100,000. Admittedly, it's not the same quality, but it is in the same block. Is this a way forward?

Is there a certain level of inherent segregation?

There often can be, not just in high-rise buildings, but low-rise as well, with horizontal segregation rather than vertical. We are in an age of experimentation in response to this crisis. The problem is that everybody deems themselves an expert, rather than acknowledging that we don't have all the answers and need to work together better to find them.

To digress for a moment, if you imagined a tall building 20 years ago, you would have said three things with certainty: it would be located in North America, it would be an office building, and it would be steel. Now the opposite is true: It will be in Asia, be residential or mixed-use, and be concrete or composite construction. So the world has completely turned. What we're seeing now is sustainable technologies used as trophies on buildings, not just high-rises. They're

如何被考虑进去呢？

居住在高层建筑中的有两类人：选择住在那里的人和那些没得选择只能住在那里的人，比如，安乐的富人和难过的穷人。最好还是看看像新加坡这样的地方，它获得了我们在西方得不到的东西。在一些地方，建筑的类型已经被误解了，比如英国，即使是今天，也很难建造一座高层住宅楼，因为人们很难将其与上世纪50年代所摈弃的战后大楼区别开来。到底是某种建筑类型出了问题，还是由于特殊的环境？是过度拥挤？还是缺少维护？欧洲现在正削减预算，因为它们再也支付不起社会的基础设施了。所以，我们处于什么位置？我们需要怎样提帮助那些尚不能独立自主的人？这是一个非常困难的问题，我也不知道答案。但是世界上有一些城市已经成功地解决了这个问题。

实际上，在这种垂直向上和政府投资的方案中，我认为存在一种方式，可以创造出一个垂直的城市，它不仅是可持续性的，而且在社会层面上还具有包容性。现在在伦敦，任何超过一定面积的大型开发都必须提供20%的福利房屋；在美国的部分地区也是如此。一栋私人公寓可能要达到100万美元，而一栋面积为前者一半，并且楼层更低的公寓可能只要10万美元。当然，两者之间在质量上有一定的差异，但它们都是在同一栋大厦中。这是一种进步的方式吗？

有没有专门的一层用于建筑内部的隔离？

经常是会有的，不只是在高层建筑中，在低层建筑中也是一样，是水平隔离而不是垂直的。我们处在一个回应危机的实验年代。问题是，每个人都将自己视为专家，而不是意识到我们没有全部的答案，我们也必须更好地协同工作找到这些答案。

暂时换个话题，想象一栋20年前的高楼，你将可以肯定地说出三点：它一定坐落在北美，它一定是一座办公楼，它一定是钢结构的。现在情况恰好相反：它肯定在亚洲，是居民楼或者是多用途混合楼，它是水泥结构或者复合材料结构。所以，世界已经全然改变。我们现在看到，可持续性技术像是荣耀标志般地被用在建筑上，而不仅仅是用在高层建筑上。它们乐于展示自己的可持续性证书，但大多数情况下，这些东西都没什么意义。以风力涡旋机为例，在高层建筑上，这是好东西吗？没人真的知道。只有十年之后，我们得知了事实，掌握了重要的输出数据，才能做出判断。

现在有这样一种观点：从一台风力漩涡机获得的能量，与风速的立方成一定的比例。所以，

ARCHITECTURE Antony Wood

TATA Tower Project, Mumbai, provides housing and electric vehicle parking for employees of Tata, India's largest auto manufacturer. (Seth Ellsworth & Jayoung Kim, CTBUH, 2010)
孟买的塔塔大楼。孟买为塔塔公司的员工提供住宿和电动车辆停放，塔塔公司是印度最大的汽车制造商。（赛斯·埃尔斯沃思和乔恩·金，世界高层都市建筑协会，2010年）

如果你让风速加倍，那么产生的能量将是2×2×2倍，也就是说，将是8倍之多。所以，实际上所要做的就是增加风速，之后你就会获得能量输出上的巨大提升。风速会随着高度的上升而自然提升，所以，在这种理论中，将一台风力漩涡机放在高楼的楼顶是一个非常好的主意。

不过，也有许多反驳的观点，像噪音、震动共鸣等理论。人们发现，能量的产生并不能首先解释制造风力漩涡机所用的碳。相反，英国已经在离海岸很远的海上风力漩涡机上进行了大量投资。为了在接下来的20年里，提供总需求20%-30%的能源产量，他们在海上建设了巨大的风力农场，这些风力农场不会打扰到任何人，但也正因如此，很大一部分电能在转化运输回城市的过程中就损失掉了。因此，在城市中创造满足需求的能量——例如，通过高层建筑上的风力漩涡机——可能是一个更好的解决方案。但事实是，没有人可以确认这一点。

我们都可以合理地解释，直到我们了解了艰难的事实之前，盲目跟风是很危险的。问题是，当我们认为自己掌握了所有答案的时候，就不打算通过实验找出答案了。从现在开始50年后，人们会将这个时期（在欧洲从大约30年前开始，在美国则是从五六年前开始）理解为一种完全实验性的回应，并认识到环境变化的严重性。有一些道路将会

meant to display sustainable credentials, but mostly it's utter nonsense. Take wind turbines. Are they a good thing on tall buildings? Nobody really knows. We won't know for 10 years after the fact, once we've had significant data output.

Here's the argument: energy from a wind turbine is proportional to the cube of speed in, so if you double the wind speed, the energy coming out is 2x2x2, i.e., to the power of 8, so it's actually in your interest to increase wind speed so you massively increase the energy output. Wind speed naturally increases the higher up you go, so in theory, putting a wind turbine on top of a tall building is a good idea.

But there are multiple counterarguments, like noise, vibration, etc. It was also found that the energy produced wasn't even justifying the carbon for making the turbines in the first place. Conversely, the UK has invested heavily in off-shore wind turbines. To provide 20 to 30 percent of its total energy needs within the next 20 years, it's building huge wind farms out at sea where they don't bother anyone, but then a high percentage of power is lost in transmission back to the city. So creating energy at point of need in the city—for example, through wind turbines on tall buildings—might be a better solution. The reality is nobody knows.

We can all rationalize, but until we see the hard facts, it is very dangerous to presume. The problem is that when we think we have all the answers, we're not going to experiment to find out. Fifty years from now people will see this period, which started in Europe about 30 years ago and in America about five or six years ago, as a completely experimental response in realization of the seriousness of climate change. Some of the paths will prove productive, others will not. The danger is that everybody thinks every path is productive and they are not willing to share data and experiences. We need to accept that this is bigger than any single person's reputation and work together on it.

What is sustainability? The word has been so overused that it's no longer clear what it actually means.

Absolutely right! It is probably the most abused term ever. You can pick up a magazine and see green advertisements for products that have nothing to do with sustainability.

A couple of decades ago a teacher friend told me he ate only organic produce. It's twice as expensive, so why? A tomato is a tomato, right? He explained that wasn't necessarily so, although it certainly seemed so to me. Then I read *Cradle to Cradle: Remaking the Way We Make Things*, by William McDonough and Michael Braungart, and realized that tomatoes sometimes grow in overused soil treated with chemicals so many times that you'd have to eat eight tomatoes now to get the same goodness you would have gotten from one tomato 30 years ago. Sometimes we don't do a good enough job of educating people about choices. It's no use just saying this is better for you, eat this. I'm a relatively intelligent guy, I know what a tomato looks like, but when you understand the reasons it doesn't have the same nutritional value, then things start to change.

Are you familiar with Dickson Despommier and vertical farming?

Yes, I know Dickson quite well. Vertical farming is really quite an exciting proposition. It addresses a lot of the issues we've been talking about, how cities need to be intensified. Importing pears from New Zealand doesn't make sense on any grounds. South Korea actually rents something like 50 percent of Madagascar to grow its vegetables because it only has 3 percent arable land and it's cheaper—not necessarily more energy efficient—to rent half of Madagascar than it is to grow locally. Creating agriculture inside a building, free from pestilence, relatively free from climate cycles, producing vegetables at a point of need all makes sense.

I actually did a vertical farm with my students 10 years ago; where Dickson and I disagree is that he thinks tall buildings should be built to house vertical farms. I think it's too big

被证明是富有成效的，其他的一些则不是。危险就在于，每个人都认为自己的道路是富有成效的，而且他们并不打算分享自己的数据和经验。我们需要接受这样一个事实，这个问题比任何个人的名声都要重大，因此必须协同合作来解决。

什么是可持续性？这个词已经被过分滥用，以至于人们都不清楚它究竟是什么意思了。

绝对正确！这可能是有史以来被滥用最多的一个词。你可以随便拿起一本杂志，看看产品所作的绿色广告，这些广告与可持续性并无半点关联。

数十年前，一位教师朋友告诉我，他只吃有机产品。这种产品的价格是普通产品的两倍，所以为什么他这么做？西红柿就是西红柿，对吧？他解释说，并不一定如此——尽管在我看来，事实就是如此。然后我读了威廉·麦克唐纳（William McDonough）和迈克尔·布朗嘉特（Michael Braungart）的书《从摇篮到摇篮：重塑我们的生产方式》，我意识到，西红柿有时候生长在过度使用的土壤中，这些土壤被多次使用化学肥料，所以，你现在必须吃八个西红柿，才能获得30年前你吃一个西红柿所能得到的营养。有时候，我们没有在选择方面对人们进行足够好的教育。仅仅是说"这对你更好，吃了它"，这是于事无补的。我算是一个还比较聪明的人，我知道西红柿长什么样，但是当你明白为什么它在今天已经没有那么高的营养价值时，事情就开始发生改变了。

你对迪克森·德波米耶(Dickson Despommier)和他的垂直农场熟悉吗？

是的，我和迪克森很熟。垂直农场确实是一个非常令人振奋的主张。它涉及到许多我们已经谈论到的问题，城市应当如何得到强化。从新西兰进口梨并非对任何地方都有意义。韩国实际上租用了马达加斯加的一半土地来栽培植物，因为它自己只有3%的可耕种土地，而且租用马达加斯加的一半土地也比在韩国本土的种植要来的便宜——但未必会更有节能效果。在一栋建筑之内创造农业，摆脱了病虫害，相对而言也摆脱了气候循环的影响，生产出满足需要的植物，这些都非常有意义。

实际上，十年前我就和我的学生开始兴建了一个垂直农场；我和迪克森观点不一致的地方在于：他认为高层建筑应当建设得能够把垂直农场置于层内。我认为这样的话投资太大。高层建筑在能源和金钱上是消耗大户，但它们提供了所需的功能空间。如果完全不顾这些，而只想着给温室大棚留下一个很大的空间，这在经济上是不现实的。我相信，垂直农场的未来在于表层农场，它会将建筑外壳的价值最大化。

我们最近刚在首尔设计了一栋相当大的办公楼，农业存在于建筑的表层，所以你并没有将建筑的所有部分都占用，而只是使用了建筑的外层。这一点很合理，不过，它还有另外一些重大的益处。

下面的这个行为是很荒谬可笑的：建一座玻璃塔，然后讨论能量和结构动力学，以便引入一种复杂的遮阳装置来解决使用玻璃所产生的问题——在中东尤其如此。

双层的建筑外壳极具潜力。如果你想象在双层之间夹着一个垂直表层农场，植物吸收了阳光的同时也以极低的代价保护了内在空间。植物需要成为城市中一种必要的物质因素，我们需要让城市变得温和，让城市中有生机勃勃的部分。这对于社会心理学以及能源问题都是有益的，会减少热岛效应等等。它也引入了一种与我们时代相关的新型城市美学，我在考虑未来的高层建筑时会将植物作为一个非常重要的方面。

将植物放在高层建筑的外层可能是有问题的，但已经有人这么做了。例如，在新加坡的牛顿轩项目中，就有一堵40层高的绿墙。它只是整栋建筑的一小部分，但确实是其中的一部分。

这是在法兰克福德国商业银行大厦上的一种升级吗？

德国商业银行大厦依然是我们所拥有的最具可持续性的高层建筑之一。实际上，它是革命性的。在建筑的每一个三角平面转角处，都有一个垂直、循环流通的芯板，在不同的芯板之间是一窄条形的办公室，因为德国法律规定，你离一扇窗户的距离不能超过7米（23英尺）。自然通风并非是一种无私的利他行为，而是取决于法律规定。然后，有一系列的空中花园在建筑中逐节上升。这样，建筑中不仅有了绿色之肺，进行自然的循环流通，而且它们也可以变成会议室，社交空间。我为什么不将我的小孩放到一栋高层建筑中？因为那里几乎没有可用于发展社区感受的空间，尽管这是在德国商业银行大厦！

许多年前，我有一个学生，重新将德国商业银行大厦设计成住宅用大楼，我认为那是一个很好的设计图样。我可以想象我的孩子和他们的三轮脚踏车，他们是在空中花园中好些，还是在有汽车的街上更好呢？现在，这个方案让人们很难理解！

an investment. Tall buildings are energy and finance guzzlers, but they provide much-needed functional space. To take that all away and build a big space for a greenhouse denies financial reality. I believe the future of vertical farming lies in facade farms that maximize the building envelope.

We just designed a sizable office building in Seoul where the agriculture exists within a facade so you're not using the whole building, just the skin. That makes sense, but there are other major benefits as well.

It's ridiculous to build a glass tower, especially in the Middle East, and then go through the energy and structural gymnastics to employ complicated shading devices to solve the problem caused by using glass in the first place.

Double-skin facades have a lot of potential. If you imagine one with a vertical facade farm sandwiched between the skins, the vegetation takes the sunlight while protecting the internal space at a much lower cost. Vegetation needs to become an essential physical part of cities. We need to soften our cities, bring back the organic realm. This has benefits for social psychology as well as energy, reducing heat island effect, etc. It also introduces a new urban aesthetic relevant to our age. My vision of tall buildings in future cities sees vegetation as an important aspect.

Putting vegetation on the outside of tall buildings can be problematic, but it has been done. For instance, there's a 40-story green wall in the Newton Suites project in Singapore. It's only a small part of the building, but it is part.

Is that a variation on Commerzbank in Frankfurt?

Commerzbank is still one of the most sustainable tall buildings we have. Actually, it's revolutionary. It has a vertical circulation core in each corner of a triangular plan, and a narrow strip of offices between the cores because German regulations dictate you can't be more than 7 meters [23 feet] from an operable window. Natural ventilation is not altruism, it's due to regulation. Then there's a series of sky gardens stepping up the building. These are not only green lungs in the building, which enable the natural ventilation, they're also turned into meeting spaces, social space. Why won't I put my kids in a tall building? Because there is little space to allow a sense of community to develop. Yet here it is in Commerzbank!

A couple of years ago I had one of my students redesign Commerzbank for residential use; actually, I think it's a great model. I can imagine my kids with their tricycles. Would they be better off in the skygardens or on the street with cars? Now we've turned things on their head! It's 14 years since Commerzbank was completed, and I still think it's one of the

Left: Moksha Tower vertical cemetery frees land for the living in densely populated Mumbai (Yalin Fu & Ihsuan Lin, designed in an advanced architecture studio at IIT with Professor Antony Wood, 2010). Right: Proposal for the vertical remaking of Mumbai, designed in an advanced architecture studio at IIT with Professor Antony Wood, 2009.

左：在人口稠密的孟买，垂直公墓的解脱塔为活着的人省出了土地（付亚林和林怡萱，与安东尼·伍德教授合作在伊利诺伊理工学院高级建筑工作室中设计，2010年）。

右：对孟买的垂直重塑方案，与安东尼·伍德教授合作在伊利诺伊理工学院高级建筑工作室中设计（2009年）。

most sustainable tall buildings, in the widest sense of sustainability. It's as much about social sustainability as energy use. My only problem with Commerzbank is that it doesn't advance the aesthetics of tall-building architecture beyond the steel-and-glass expression that has predominated for the past half century or more.

When I was 21, I bought a one-way ticket to Hong Kong and spent the next 10 years practicing as an architect in Southeast Asia. Before I was 30, I'd traveled across 60 countries on all seven continents. I was interested not in the similarity but in the differences of culture, climate, food, language, etc. What has happened in the last 100 years—and it's been accelerating recently—is that cities and countries and cultures are becoming homogenized. Architecture is one of the worst forces of global homogenization, and tall buildings are the worst typology within architecture.

Look at Shanghai, Sydney, Moscow, wherever. The skyline may become synonymous with the place, but it's not born or inspired by it, physically, culturally, or environmentally. If you think about traditional forms of architecture, for example, they are usually a product of thousands of years of vernacular tradition. What happened with the modern movement in the late 1950s onward is that we were able to free ourselves from traditional architecture, and we created the

德国商业银行大厦用了14年才修建完成，而且我依然认为，它是高层建筑中最具有可持续性的建筑之一——从最广义的可持续性而言。它的社会可持续性和在能源使用上一样多。关于德国商业银行大厦，我唯一的问题是，它并未超越钢铁和玻璃的建筑表现方式（这些东西占据主导地位已经有半个多世纪了），因而对高层建筑的美学没有提升。

我21岁的时候，买了一张去香港的单程车票，而且在接下来的十年内一直作为一名建筑师在东南亚展开实践工作。在30岁之前，我已经走过了7大洲的60个国家，我对于相似性并不感兴趣，而对文化、气候、食物、语言等方面的差异却颇有兴致。过去一百年中所发生的——而且现在得到促进的——乃是：城市、国家和文化变得越发同一化。建筑是全球同一化中最坏的力量之一，而高层建筑则是建筑中最坏的类型。

看看上海、悉尼、莫斯科，任何地方。天际轮廓线就是这些地方的代名词，但这一轮廓线并非自然、文化或者环境所产生和激发的。如果你想想建筑的传统形式，它们通常是当地数千年传统的产物。上世纪50年代后期，不断向前发展的现代运动，意味着能够将自身从传统建筑中解放出来，而且创造了直线型的空调箱，并将它们安置在全球无数个地方，这就是所谓的国际风格。

如果在亚洲有一栋摩天大楼项目，那么，失败的记录将会突然出现在迪拜或其他地方，这是错的，完全错的。但确实有一些相反的例子——高层建筑看起来是从特定地方产生出来。台北101大楼是一个我称为"字面文化象征"的例子，他们想要做的就是建造一座本土的摩天轮。这是在正确方向上迈出的一步，因为它至少想要赋予城市独立性，想要在其文化之中寻找灵感。不过，他们以一种非常字面的方式来做这些事情。他们把"一座中国的宝塔"当作是对一栋101层高楼的恰当比喻。

我认为我们需要做的，是回到对于形成本土建筑有真正影响的因素上，比如说气候和文化，这些应当是高层建筑的灵感。我在很多方面都非常喜爱马来西亚的吉隆坡双子塔。不过，西萨·佩里（César Pelli）会告诉你这是一座马来西亚式的建筑，因为它的规划受到伊斯兰图案的启发，这就是我所谓的"抽象文化象征"。由于感受到要对当地文化脱帽致敬，佩里将这些文化因素突出表现在88层的高楼中，并在这一过程中创造出许多美妙的东西。毫无疑问，它成了马来西亚的代名词，但这并不一定就意味着它必须来源于这个地方，如果双子塔建在迪拜，它也会成为迪拜的代名词。在几英里外，就是另一座本地建筑师建造的宏图大厦，其灵感来自相同的规划形式。但在这栋建筑中，就不只是二维的抽象游戏了，伊斯兰的表达体现在三个维度中。建筑的外层实际上是一个敞开的雕花百叶窗，玻璃表层大概在外层3英尺以外的地方，灵感来自历史上著名的杰利屏风，而这是对伊斯兰文化的一种真挚表达，我认为这是一座了不起的建筑。

我们开始这次对话时说到：过去10到15年发生的事情是空前的。去年是迄今为止摩天大楼完成竣工最为成功的一年，但这会被今年的情形超过。现在我们建造比过去多出十倍的高层建筑。也就是说，有10%已经完成了，但这个数字依然意味着有大量的高层建筑。

哈利法塔从构思到完成实践总共只用了六年。对于这种规模的建筑而言，真的非常特殊。建筑顶端的外部温度要比建筑底部低了华氏4度，但之前没有人在2700英尺（823米）高的地方使用过进气管，所以它们被安置在低一点的地方。其实，通过将进气管安置得高一点，他们本来可以得到4度免费的降温。我们总是在学习之中。有许多更好的项目正在建设之中，秉承着更加可持续性的原则。

rectilinear air-conditioned box and plunked it in thousands of places around the world; the International Style.

If there's a skyscraper project in Asia, the unsuccessful entries will pop up in Dubai or somewhere else. This is wrong, completely wrong. But there are examples in the opposite direction—tall buildings that do seem to be born of place. Taipei 101 is an example of what I call "literal cultural symbolism." What they tried to do is make a local skyscraper. It's a step in the right direction because at least they tried to distinguish the city and tried to find inspiration in its culture. But they did it in a very literal way. They took a Chinese pagoda as a suitable metaphor for a 101-story building.

What I believe we need to do is go back to the genuine influences that shaped vernacular architecture to begin with, like climate and culture. These should be the inspiration for tall buildings. Look at Petronas Towers, which I love on many levels. However, César Pelli will tell you this is a Malaysian building because its plan was inspired by Islamic patterning. This is what I call "abstract cultural symbolism." Having felt he tipped his hat to local culture, Pelli then extruded it to 88 stories and in the process created something quite beautiful. It's become synonymous with Malaysia, no question, but this doesn't necessarily mean it grew from there. If Petronas had been built in Dubai, it would have become synonymous with Dubai. Just a couple of miles away is another building by a local architect inspired by the same plan form—the Menara Dayabumi Complex. But in this building it's not just a two-dimensional abstract game; the Islamic expression is extruded into three dimensions. The building skin is actually an open fretwork brise-soleil. The glass facade is about three feet behind this screen inspired by historic jali screens, and it is an honest expression of Islamic culture. I think this is a fantastic building.

We started this conversation by saying that what's happened in the last 10 to 15 years is unprecedented. Last year was the most successful year of skyscraper completion ever, only to be surpassed by this year. Now we're building 10 times more tall buildings than ever before. Let's say 10 percent get built. That will still be a hell of a lot more tall buildings.

Burj Khalifa was conceived and realized within six years. That's pretty special for a building of that size. The external temperature is 4 degrees cooler at the top than at the bottom, but nobody ever had air intakes 2,700 feet [823 meters] high before, so they're positioned lower down. They could have gotten 4 degrees of free cooling by putting the air intakes higher up. We're learning all the time. A lot better projects are being built with more sustainable principles.

Are there technical limitations of steel and concrete that would preclude a 2-mile [3.2-kilometer]-high building?

There is only one real limit on height, and that's financial. With a wide enough base or connective fabric we could build above a mile high, absolutely. Whether we want to is another question. You would need a hell of a lot of thought and consideration, but I don't believe materials would be a barrier. Elevators, absolutely not. With sky lobbies, you can put one on top of another. The most convincing argument I've heard is from Bill Barker, the structural engineer, who maintains that the biggest barriers are about the human condition, not technical building aspects. For example, the inner ear drum at low pressures can pop. If you flew into La Paz, the world's highest capital city, in the Andes, you'd probably get altitude sickness, but if you worked up from Lima, Peru, you acclimatize. So even that can be overcome in tall buildings by staging vertical movement.

So there might be a process of acclimatization, not taking weeks, but maybe hours, to get up there if need and desire were strong. So I don't think there are barriers; material, structural, physical, or even human. The issue is whether there's a need to do it, and if so, who's going to pay for it? We just abandoned the space program; does anyone really want to build a 2- to 3-mile-high [3.2 to 4.85 kilometers] tower?

How did 9/11 affect tall buildings?

Everyone thought 9/11 spelled the death of the tall building, but I think it's actually the opposite. That event like no other was projected into the world's consciousness. There was a lot of research and debate, lots of press coverage, in the months following the event, which ended up kind of promoting the typology. Everybody questioned it for a while, but in the course of questioning they decided tall buildings weren't necessarily at special risk and came to the conclusion that impacts all aspects of life, i.e., you can't live your life in fear of what might be.

It's interesting that New York is now probably the most active tall-building market in the U.S. In this year's Best Tall Building awards, there were more entries from New York than any other American city, whereas in the past few years that honor fell to Chicago.

After 9/11 the world drew a deep breath; the closer to New York, the deeper the draw, and the bigger the delay in progressing. But that's why we're now seeing towers being completed 10 years after in New York. It's the lag effect. There's only one tall building being built in downtown Chicago currently. Why? The global financial meltdown, that's what killed tall-building development in the West for a few years. It wasn't 9/11.

钢铁和水泥是否在技术上有局限性，以至于不能承受2英里（3.2千米）高的建筑？

在高度上只有一个真正的限制，那就是经济因素。有足够宽的地基或者连结结构，我们绝对可以建成超过1英里高的建筑，但是否想做是另一个问题。你需要许许多多的想法和思考，但我不相信材料会成为障碍，电梯绝对也不是障碍。就空中大堂而言，可以将其中一个大堂放在另一个之上。我所听过的最有说服力的观点来自比尔·巴克尔（Bill Barker），一位结构工程师，他坚持认为最大的障碍是人的境况，而不是技术建设层面的问题。例如，内耳鼓膜在低压之下会破裂，如果你进入安第斯山脉中的拉巴斯，世界上海拔最高的首都城市，你很可能患上高原病，但如果你是从秘鲁的利马来到这里，你便会适应这里的环境。所以，通过逐渐进行垂直运动，这些情况都能够在高层建筑中被克服。

所以可能会有一个适应的过程，假如需要或欲望非常强烈的话，不需要几周，或许几个小时就可以做好准备了。所以我不认为存在障碍，材料上的、结构上的、身体上的或者说是人自身的。问题在于，是否有必要去做，如果有，谁将为此付钱？我们刚刚放弃了太空计划；是否真的有人需要建立一栋2到3英里（3.2至4.85千米）的高楼？

911如何影响高层建筑？

每个人都认为911宣判了高层建筑的死刑，但我认为实际上恰好相反。这一独特的事件是世界性的问题。在事件之后的几个月，有大量的研究和争论，大量的新闻报道，最终的结果却是以继续推进这一类型的建筑告终。每个人都对它进行了一阵很长时间的质疑，但在质疑的过程中，他们断定高层建筑并不一定处于特别的危险之中，并最终得出影响生命各个方面的结论，比如，你不能够生活在未知事件的恐惧中。

非常有意思的一点是，纽约现在可能是美国最活跃的高层建筑市场。在今年的最佳高层建筑奖竞赛上，来自纽约的参赛项目要比来自美国其他城市的都多，尽管过去数年，这一荣誉总是落到芝加哥头上。

911之后，世界深吸了一口气，离纽约越近，这口气就越是深沉，同时也越是耽搁了前进的脚步。但这也是我们在十年之后的纽约看到大楼纷纷落成的原因。这是一种滞后效应，目前在芝加哥市中心，只有一栋高层建筑正在施工建设中。为什么？全球经济灾难，是经济问题在数年时间内阻挠了西方高层建筑的发展步伐，而不是911事件。

Above: Petronas Towers, Kuala Lumpur, Malaysia (César Pelli, 1994)
上：双子塔，吉隆坡，马来西亚。（西萨·佩里建筑事务所，1994年）

Right: Dayabumi Tower, Kuala Lumpur, Malaysia
(Urban Development Authority of Malaysia, 1984)
右：宏图大厦，吉隆坡，马来西亚。（马来西亚城市发展局，1984年）

SECTION THREE / FORUM OF EXPERTS

Technology

Rick Barker, Barker Mohandas
Marvin Mass and Scott Ceasar, Cosentini Associates
Carl Galioto, HOK
Robert Heintges, Heintges & Associates
Dennis Poon, Thornton Tomasetti
Leslie E. Robertson, Structural Engineer, LLC

第三篇/专家论坛

技术

里克·巴克，巴克·莫汉德斯公司

马文·马斯和斯科特·西塞，克森蒂尼事务所

卡尔·卡略托，霍克国际有限公司

罗伯特·海因吉斯，海因吉斯建筑事务所

潘子强，宋腾添玛沙帝工程师事务所

莱斯利·E·罗伯特森，莱斯利·罗伯特森建筑事务所

Rick Barker
BARKER MOHANDAS

Rick Barker is a principal of Barker Mohandas Vertical Transportation Consultants and is involved in most of the firm's work. Prior to co-founding Barker Mohandas, he was the director of technical services at Otis World Headquarters. He also chaired the Otis product strategy group for dispatching control products and simulation tools, and co-led strategy for the company's tall building elevator Skyway™. Mr. Barker led a major study to improve elevator energy efficiency; co-led the Otis Odyssey™ system development, which integrated horizontal and vertical transportation; and was a key liaison between Otis world headquarters and United Technologies Research Center. He previously led vertical transportation at Jaros Baum & Bolles Consulting Engineers. He was involved at Westinghouse Elevator (now Schindler) in the company's largest projects in western New York State, and at Delta Elevator in Boston (now Otis) in the company's largest modernization contracts. The author of multiple professional papers and lecturer at conferences and at two leading universities, Mr. Barker is named, singly or jointly, in 24 patents held by Otis.

里克·巴克
巴克·莫汉达垂直运输咨询公司

里克·巴克是巴克·莫汉达垂直运输咨询公司的负责人，并参与了公司的大部分工作。在协同创立巴克·莫汉达咨询公司之前，他曾担任奥的斯世界总部技术服务部门的主管，同时也主持奥的斯产品策略小组，负责协调控制产品和模拟工具，同时协同主导公司高层电梯品牌Skyway™的运营策略。巴克先生对提高电梯能耗做了大量研究，共同领导了奥的斯Odyssey™系列产品的开发——这一产品将水平和垂直运输整合在一起，他还是奥的斯世界总部和美国联合技术研究中心之间关键性的联络员。他之前在JB&B工程师事务所负责垂直运输工作，参与了西屋电梯（现在的迅达集团）在纽约州西部最大的项目，以及波士顿的三角洲电梯（现在的奥的斯），公司最大型的现代化合同工程。他是多篇专业论文的作者，并在多个会议以及多所著名大学发表过演讲。巴克先生拥有24项专利，有些是单独持有的，有些是与他人共同持有的。

For complete biography, please refer to the Appendix.

完整的传记，请参阅附录。

Dubai, view from the Burj Khalifa during construction (2008)
迪拜，哈利法塔建设期间从上面看的景致。（2008年）

In recent years buildings have soared to unprecedented heights. Is there a limit to how high they can go?

From the perspective of vertical transportation we have found that, even with the components available today, a kilometer [3,280 feet] is not that difficult. We have also done studies for mile-high buildings [1.61 kilometers]. The trick using conventional components is the system strategy, until improvements in elevator or lift technology are developed and available.

Is it safe to say that vertical transportation experts should be consulted early in the design process, even at the conceptual stage?

Exactly right. The sooner the better—basically once an architect has a form and a general occupancy massing in mind with the owner. For very tall buildings, we're consulted in parallel with the structural engineer. The taller a building and/or the more mixed use, the first thing you want to know is how to stack up the occupancy and what's the impact on the building core or cores. You want the most dense occupancy at the bottom, so it's usually most efficient to start with commercial and offices; then hotel, sometimes with connected serviced apartments above that; then residential apartments; and then, for perhaps the best revenues, a luxury hotel and, of course, an observation deck on top. There have been times when a client has actually switched the locations of the uses after we've suggested the most efficient stacking of the types of occupancy.

How do sky lobbies come into play?

Sky lobbies naturally separate the types of occupancy in a tall mixed-use tower, and allow some lift hoistways to be located above other lift hoistways in most tall towers. Most of the very tall buildings we see these days are about 100 stories. If that building has mostly offices, it will generally be divided into thirds by two sky lobbies served by shuttle lifts, where passengers will transfer to local lifts. By the way, in any system design, we try to avoid more than two primary lift rides. For example, the tower may have offices in the lower two-thirds and a hotel on top. For the hotel, a guest would take a lift to the sky lobby, check in, and then take a local lift to his or her room. Passengers for the offices may need to ride incidental lifts in the car park, or escalators, but for the luxury hotel, arriving guests will usually be dropped off curbside and can board the sky lobby shuttle lifts directly. For a 100-story apartment tower, a sky lobby is seldom needed.

近年来，建筑在高度上已然飙升到前所未有的水平。对于最大高度，是否有个限制呢？

从垂直运输的角度来看，我们发现，就算只采用现有的零部件，建设一座1000米高（3280英尺）的建筑也并没有那么困难。我们已对1英里高的建筑进行了研究（1.61千米）。在升降机技术得到改进或升降机进一步发展且付诸实际应用以前，使用传统零部件的方法是最系统性的策略。

能否肯定地说，在设计过程中甚至是在概念构思阶段，应该早点咨询垂直运输专家？

这绝对是正确的，而且越快越好。基本上，只要建筑师有了雏形，在头脑中有一个大概的业主入住率概念图就可以了。就非常高的建筑而言，向我们咨询和向结构工程师的咨询都是同步进行的。建筑越高，它的用途越多样，你首先要知道如何计算入住率，以及它会对建筑核心筒有什么影响。人们希望在建筑的底部，让入住率达到最高，所以一般来说在底部以商用、办公室建筑开始是最有效的；然后是宾馆，有时在上面会有连在一起的酒店式公寓；再往上是住宅公寓；然后可能是收益最大的豪华酒店，当然，顶部会有一个观景台。有时，在我们给出最有效的使用楼层排列布置类型的建议后，客户实际上已经变更了不同使用类型的位置。

空中大堂如何在其中发挥作用？

空中大堂很自然地区分开一栋多重用途高楼中的不同使用类型，它能让一些电梯井道置于另外一些电梯井道之上。今天我们看到的大多数超高层建筑基本上都在100层左右。如果该建筑的用途主要是办公，一般来说它会被两个空中大堂分为三部分，穿梭式电梯将会在其中发挥作用，乘客们可以通过它换乘到区域电梯中去。顺便说一下，在任何系统设计中，我们都试图避免两次以上的电梯换乘。例如，高楼在较低的三分之二可能是办公室，而在顶部则是一家宾馆。对于宾馆而言，客人将乘坐一部电梯到达空中大堂，登记入住，然后乘坐一部区域电梯到达自己的房间。要去办公室的客人可能需要使用停车场的附属电梯或自动扶梯；但对于豪华酒店来说，客人到达时一般会停靠在路边，因此可以直接乘坐到达空中大堂的穿梭式电梯。而一栋100层的公寓楼，则几乎不需要空中大堂。

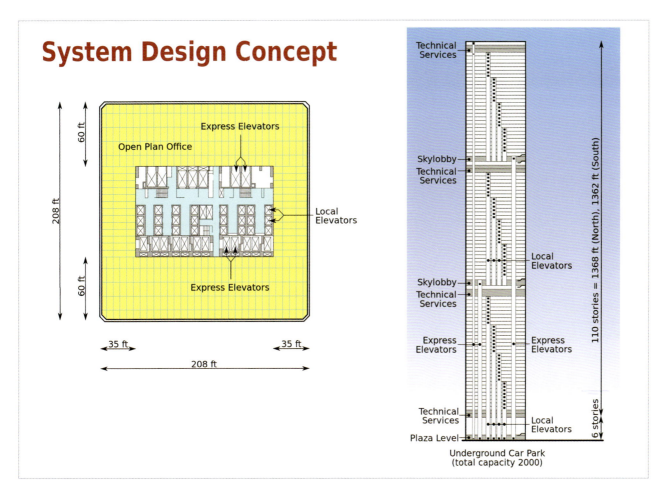

World Trade Center, New York (1966–73); Elevator System Design Concept
纽约世贸中心（1966年—1973年），电梯系统设计理念

在规划中东一座前所未有的1英里高（1.61千米）大楼时，贝建中（Didi Pei）第一次提出了市镇中心的概念来解决这些问题，这种市镇中心不是一个空中大堂，但它根本上是将大堂主体提升到一个使大楼合理的位置，考虑到实际街道层面上的人流量，这个合理的位置是60层。由此，我们当然可以规划出带有空中大堂，又是"传统的"，1000米高（3280英尺）的大楼，并用一座实际在楼顶之下的观景台对大楼顶部进行修饰装点。再说一遍，利用现有的电梯技术，我们可以做很多事情。

在公寓中，空中大堂可以进一步分开，我们可以用更少的电梯为更多的楼层服务。但是对于办公室来说，根据楼层的区域以及可用于区域电梯的空间大小，会对楼层的多少有一个限制，大约是每

In a plan for an unexecuted mile-high [1.61-kilometer-high] tower in the Middle East, Didi Pei approached the issues by first proposing a town center, not quite a sky lobby, but basically raising the main lobby up to a place that starts to make sense for such a tower, about 60 stories high, considering the mass circulation at the actual street level. From there we could essentially help plan what would be considered a "conventional" 1-kilometer-tall [3,280-foot-tall] building with sky lobbies, and trimming off the top of the tower with an observation deck located well below the actual top of the tower. Again, much can be done with technique, using the lift technologies available at the time.

In apartments, the sky lobbies can be farther apart; we can serve many more floors with fewer elevators. But for offices there tends to be a limit of about every 32 or so floors, depending on the floor area, and space available for local lifts. The population here is much greater than in a hotel and much more than in apartments. After planning a lift system to handle the populations quantitatively, we then want to look at queue lengths in the main lobby, and how long people will wait for a lift at any floor; unfortunately, their patience is much shorter than waiting for a subway or bus.

What is the maximum wait time?

In an office building 30 seconds should be the maximum average wait, while some people will wait as long as 90 seconds, and very few people should wait longer, maybe 1 to 2 percent of the calls. For apartments and hotels the average wait can be longer. We do traffic studies for the peak demand rates. An office building, of course, has incoming peaks, lunch peaks, and an evening peak. And even at the start and end of the workday there will be counterflow traffic, and when one company occupies many different floors we will see significant interfloor traffic. With a hotel or apartment it's kind of a two-way early-evening peak. Of course, in international hotels traffic can occur anytime. Regardless, we try to match the lift system to the peak demand rates.

The variables to study are many. Let's consider just a 15-story office building served by one group of passenger lifts. How many people are on each floor? How many, how big, and how fast are the lifts, and how many lifts are there in the group? Then, for high-rise office buildings, the questions become a combinatorial problem: How many groups of lifts are there, which ones address all those questions, and how many floors does each lift group serve? To be productive, we found it necessary to develop advanced optimization software to help answer the questions and identify the potential choices of lift systems. Basic multigroup programs were developed at Otis and Westinghouse over 30 years ago, for example, by Dr. Bruce Powell. Earlier still, it appears that the first widely accepted method to calculate lift traffic performance was developed by Bassett Jones, a consulting engineer involved with planning elevators for the Empire State Building. He developed a calculation method to study one group of lifts at a time. For example, he calculated that a low-rise group of lifts should serve the first zone of office floors. Another group of lifts, traveling faster, should run express past those office floors to serve another zone of office floors, and so on up the building, with more groups of lifts. Well, if you kept doing that in a tall building, where all lifts start at the main lobby, you would run out of room. So the sky lobby is a brilliant concept. It allows a building to be divided into packages that are stacked up, one on top of the other.

32层左右。这里的人口要远远超过宾馆，而且比公寓中要多的多。在规划完一个电梯系统解决人口数量的问题后，我们开始想主厅的排队长度问题，以及每一层中的人需要花多长时间来等一部电梯；不幸的是，他们的耐心往往比等地铁或公交车要差很多。

最长的等待时间是多少?

在一座办公楼里，平均来说，30秒钟是最大的等待时间，当然，有些人可能会等上90秒，更少的一部分人会等得更久——这可能占到要乘坐电梯人数的1%-2%。公寓和宾馆的平均等待时间要长一些。我们做过关于高峰需求率的人流量调查，一栋办公楼当然会有上班高峰、午饭高峰以及傍晚高峰。甚至在工作日的开始和结束时会有人流的逆流，当一个公司占据不同的楼层时会有人流的合流。而宾馆或公寓只有双向的早晚高峰。当然，在一家国际型宾馆，人流会出现在任何时候。无论如何，我们想让电梯系统满足高峰需求率。

要研究的变量太多。让我们先来考虑一栋由一组客梯服务的15层办公楼。每一层有多少人？有多少电梯，多大载客量，速度有多快，以及一组里面究竟有多少部电梯？而对于高层办公楼来说，这些问题就成了一个综合性的难题：有多少组电梯，每组电梯为多少层楼服务？为了更有效率，我们发现有必要开发一套高级优化软件，来辅助回答这些问题，并标示出各种潜在可选择的电梯系统。基本的电梯群控程序30多年前就在奥的斯公司和西屋电气被开发出来了，例如布鲁斯·鲍尔(Bruce Powell)博士的发明。再往前，巴塞特·琼斯(Bassett Jones)提出了最早被广泛接受的计算电梯流通运输性能的方法，他是一位顾问工程师，参与了帝国大厦电梯的规划。琼斯开发了一种计算方法，可以一次对一组电梯进行研究。例如，他通过计算得出结论认为，一组低层的电梯应该为办公楼层的第一个区域工作。另一组运行更快的电梯，应该迅速通过这些楼层，为办公楼层中的其他区域工作，以此类推，在大楼的上面部分，将会有更多组的电梯。但如果你在高层建筑一直这么做，那么，你将用光所有的房间，因为那里所有电梯都从主厅出发。因此，空中大堂是一个很聪明的概念。它允许一栋建筑被分为多个部分，它们相互之间层层堆叠起来。

什么时候才引入了空中大堂的概念？

许多人都谈到了这个问题，据我了解，它的第一次重要应用归功于赫伯特·泰斯勒的纽约港口管理局（现在是纽约和新泽西港口管理局），我参考了一篇1968年的老文章，上面说的很清楚，他们将它用在纽约的世贸中心上。

电梯的最快速度是多少？

多年前当我还在奥的斯的时候，我被鼓励去写一篇论文，主题恰好就是这个，或者说得更准确些：是所需要的速度。我给那篇文章的标题是："每分钟2000英尺（610米）够吗？"三菱已经生产出超高速的单层电梯，速度可以达到12.5米（41英尺）每秒。奥的斯感到自己被迫加入到速度竞赛中，设计出了一种速度为15米（49英尺）每秒的电梯。在台北国际金融中心，东芝/通力的产品达到了16.8米（55英尺）每秒。在上海，三菱刚刚卖了些单层电梯给一栋高楼，其速度达到了18米（59英尺）每秒，也就是3600英尺（1097米）每分钟。它就像是一对法拉利或布加迪呼啸着冲向观景台。这些是非常非常特殊的电梯。对于大多数高层办公楼来说，我们最需要的是速度在10米（33英尺）每秒或2000英尺（600米）每分钟的电梯。为了运送如此多的人——（数千人）——我们在空中大堂的穿梭式电梯和区域电梯中，都采用了双层设计。让这么重的电梯轿厢以更快的速度运行显然是不现实的，因为钢缆的重量、机械设备、电力系统、轿厢结构等等都会非常巨大。当我们有很多人要运输的时候，小而快速的单层电梯没有太大的帮助。这就是我过去在奥的斯那篇文章的主旨，当我们欣赏有着类似赛车的工程产品，和有吸引力的观景台时，上述观点仍然是对的。现在，更好的方案可能是提供一种向上过程中的景致，那样的话，你就不希望电梯走得太快了。

是否出现过这样的情况，同一个井道中的两部电梯运行的方向恰好相反，一个向上，一个向下？

有一家电梯公司已经重新采用了这样的想法：一个井道中有两部缆绳牵引电梯，它们沿着同样的路线运动，甚至有时还向着相反的方向运动。和许多人一样，我们相信，即使人们的控制能力已经得到提

When was the sky lobby introduced?

It's been claimed by different people, but from my information, including an old paper dating from early 1968, it seems very clear that credit for its first significant application goes to Herbert Tessler of the Port of New York Authority [now the Port Authority of New York and New Jersey], for the World Trade Center in New York.

What is the maximum elevator speed?

When I was at Otis years back I was encouraged to write a paper on just that issue, or rather, the need for speed. I called it, "Is 2,000 feet [610 meters] per minute enough?" Mitsubishi had produced super-speed single-deck lifts 12.5 meters [41 feet] per second. Feeling compelled to join the speed race, Otis designed such a lift for 15 meters [49 feet] per second. In Taipei Financial Center, Toshiba/Kone reached 16.8 meters [55 feet] per second. Mitsubishi just sold some single-deck elevators for a tall building in Shanghai that will reach 18 meters [59 feet] per second, which is almost 3,600 feet [1,080 meters] per minute. It's like having a pair of Ferraris or Bugattis racing up to an observation deck. These are very, very special lifts. For most tall office buildings, the most we need to go is 10 meters [33 feet] per second or 2,000 feet [61 meters] per minute. To move so many people—thousands—we use double decks for both the sky lobby shuttle and local lifts. Moving such heavy lift cars at faster speeds is just not practical; the machinery, power electronics, car structure, etc., would be huge, considering the weight of the wire ropes. And little, fast single-deck elevators don't help us when we have so many people to carry. That was essentially the subject of my old Otis paper, and we still find that true, while we do appreciate the product engineering with the race cars, and some of the attraction for an observation deck. Now, it might be even better to provide a view on the way up, but then you would not want to go so fast.

Has there ever been a case where two elevators go up and down in the same shaft?

One elevator company has resurrected the idea of two roped elevators in one hoistway, which are traveling along the same path and are even traveling toward each other. We believe, as many others do, that this is being done with increased risk, even though controls have improved. The risk is obviously that the cars could crash into each other, which is what happened when Westinghouse Electric tried this in the 1930s. After that it was written out of the elevator code in the U.S. For very tall

Representative double elevator
(at Midland Square, Nagoya)
双层电梯的代表
（米德兰广场，名古屋）

buildings, controlling movements of elevator ropes is more difficult, which adds control complications to this old idea.

When I was at Otis, I was very involved in a different idea. The company sought to provide elevator products in the short term suited to the "mile-high" [1.61-kilometer-high] building. Ropeless elevators were not viewed as a short-term product development, so the Otis answer was to "climb the mountain," so to speak, traveling in steps, and using mostly predeveloped parts from their roped elevator product for tall buildings.

For such buildings, the basic idea was that the cabs could be moved off the car frame of an elevator onto the car frame

高，但要做到这些也要冒着不断增加的风险。最明显的风险是：不同的轿厢有可能相互碰撞——上世纪30年代西屋电器公司在做试验时曾经发生过此类事故。在那之后，美国的电梯规则就禁止了这种方案。对于更高的建筑来说，控制电梯缆绳的运动要困难得多，这就给这一古老的想法增加了控制上的复杂性。

在奥的斯的时候，我被另外一种观点深深吸引。公司想要在短期之内提供适合"1英里高"（1.61千米）建筑的电梯产品。而无绳电梯并不被看作是一项短期的产品研发，所以奥的斯的回答是，要"爬山"，也就是说，一步一步前进，最大限度地利用他们为高层建筑生产缆绳电梯时那些前期开发的零件。

对于这些建筑，轿厢可以从一部电梯的结构中分离出来，然后放到一部相连电梯的轿厢结构中去，这将会把轿厢运输到更高的位置上去。当一个轿厢被从电梯的结构中分离出来，另一个轿厢会移动上来，它或者来自一个未连接的装载区，或者来自另一个电梯的轿厢结构，因此考虑到平衡力位于缆绳的另一端，一个缆绳电梯中的机械牵引机制并没有失去其牵引力。用简单的话说，这意味着当人们想要去更高的楼层时，并不需要在空中大堂停下来，只需从电梯组的这边换到另一边；相反轿厢会让电梯内的乘客一直处于变换的电梯之中，这被称为奥的斯的奥德修斯系统。它不同于当前复兴的"一个井道两个缆绳电梯"方案，因为电梯绝不会在同一条垂直轨道上运动。它们只在井道中连接的分支管道上相遇。人们可能会说，这一系统采用的是"一个井道多部轿厢"方案，但实际情况与此非常不同。

坦率地说，我并不认为这是为高层建筑给出的最佳方案，但我也确实看到，如果轿厢被从低层电梯的轿厢结构中移出，放到水平的自动载人系统（APMs）中去，其中的一些关键原理对于大型机场以及类似的"宽大直立型"项目非常适用。一位麻省理工学院的教授要求我向他建筑学专业的学生阐述一下奥的斯的奥德维修系统，然后帮他管理一个工作室，以便将这一系统应用在水平载人装置上，这一装置广泛应用于麻省理工学院中斜长的隆起地带。上述所有优点都被预见到了，当然，挑战也同样被预见到了，但这是一个很容易满足的挑战，它就是水平系统的灵活性问题，因为这个系统往往非常固定。所有东西上面都有一条缆绳，即使是在水平载人系统上也有。在地上，对灵活性来说，没有固定住的轿厢是合理的。而这又回到了不能将你自己固定在空中的问题。

无缆绳的悬浮电梯前景如何？

我希望我年轻一些而且有更多的钱。大概在13或14年前，我就投入到无缆绳电梯的研究工作中，尽管不像参加奥德修斯系统那样投入许多的精力。在看到最后的研究成果时，我认为无缆绳电梯是没有希望的。不过当时的研究是电动机和其他相关技术的产物，而且建立在那时的系统技术研究之上。

of a connecting elevator that would transport the cab even higher. When one cab was moved off a car frame, another would be moved on, from either an offline loading area or from another car frame, so traction was not lost at machine sheave with a roped elevator, considering that the counterweight is attached to the other end of the ropes. What this meant in simple terms was that people wouldn't have to stop at a sky lobby to change from one bank of elevators to another to reach higher floors; rather, the cab would keep changing elevators with the passengers inside. It was called the Otis Odyssey™ system. It was different from the current resurrection of two roped elevators in one hoistway, because the elevators never traveled in the same vertical path. They only met at connecting offsets in the hoistways. While one might say the system used a single hoistway with multiple cabs, there were significant differences.

Frankly, I didn't think it was the best answer for a tall building, but I did see that if cabs were moved from car frames of low-rise elevators to horizontal automated people movers (APMs), some key elements were very applicable for large airports and similar "wide-rise" projects. A professor at MIT asked me to present Otis Odyssey™ to his architectural students and then help him conduct a workshop applying the system, with a horizontal people mover, to the long campus spine of MIT. All kinds of merits were foreseen. One challenge was also foreseen, but one that could be easily met. This was the flexibility of the horizontal system, which was quite tethered. Everything had a rope on it, even the horizontal people movers. Untethering cabs on the ground only makes sense for flexibility. This comes back to untethering yourself in the air, as well.

What are the prospects for ropeless maglev (magnetic leviation) elevators?

Well, I wish I were younger and had more money. About 13 or 14 years ago I was involved in a research study on ropeless elevators, although not as involved as with the Odyssey™ system. After seeing the final study, I thought ropeless is hopeless. However, the study was a product of the motor and other technologies at the time, and was based on the system techniques studied at that time.

For larger applications people back then thought of linear induction motors, which are all wound electrically and have certain inefficiencies as a result. With the introduction of permanent magnets to all kinds of motors, efficiencies have improved, as have flexibilities in arranging and locating motor parts. The introduction of synchronous permanent magnet motors for elevators has been a significant change. For ropeless elevators, they would be called linear synchronous motors. Included in a second set of changes, which our industry has been slow to pick up on, are lightweight materials. The biggest problem with today's roped elevators is the weight of all its steel ropes—that's our limiting factor.

We rely on a counterweight attached to the other end of the ropes so the motor does not have to lift the entire weight of the car and the people inside. And we also have a reverse loop of compensation ropes attached below the counterweight and elevator car to counterbalance the weight of the ropes themselves. Considering the weight of all those steel ropes, the weight of the car frame must increase, along with the capacity of safety devices on the car to stop the car. Control complications also increase significantly with travel with roped elevators. So, if you untether the elevator, car weight naturally decreases and then you can focus on lightweight materials used in the structure of the car frame. While you will not have the benefit of a counterweight to defy gravity going up, the elevator linear motor and its power electronics drives can be used going down to regenerate power with untethered elevator cars.

Second in significance are lightweight materials, improvements in noncontact and wireless means to bring power and signals to elevator cars, and some other things. Third, and not necessarily last, is just being smart about the technique or system. If, for example, you had built a ropeless elevator that starts at the ground and then stops at every floor, you would have to wait a long time for it to come back down. You would need more and more elevators, eating up space. You have to employ a smarter technique. For example, if you want to put out a fire at the top of a hill and you run up with one bucket of water at a time, you would have a tough time putting the fire out. But if you keep passing buckets of water to your friends next to you, there is a much higher through-put capacity of water. That was one basic technique in the Otis Odyssey™ system, to pass cabs on to connecting elevators, to be able to return in less time in the important first leg of the elevator journey.

I believe we have and know the right people to plan, design, test, and build an ideal system. Back when the Dubai market was very hot, we thought we would have somebody to fund this who was also a self-interested investor in tall buildings. Since the major elevator companies find more than 90 percent of their business in low-rise, mid-rise, and high-rise buildings, not tall buildings per se, the major elevator companies don't tend to focus much of their R&D on this, primarily because almost everything has to be developed

为了有更大的应用，人们回过头反思直线感应电动机，它在电力应用上完全是个缺陷，最终的结果就是导致某种低效。随着永久磁铁被引入到各种电动机中，效率得到了提高，而且在排列和安置电动机零件时也具有了灵活性。为电梯引入永磁同步电动机，产生了重大的改变。对于无缆绳电梯，它们将被称为直线同步电动机。第二组改变中所包括的是更轻的材料——我们的工业已经放慢脚步来了解这种改变了。今天缆绳电梯的最大问题，就是所有这些钢缆的重量——这是限制我们的一个因素。我们现在靠的是系于缆绳另一端的平衡力，因而电动机就不需要举起全部的重量——包括轿厢本身和里面的乘客。在平衡力和电梯轿厢的下方，我们还有补偿性缆绳构成的反向循环系统，以便抵消缆绳自身的重量。考虑到所有这些钢缆的重量，轿厢结构的重量必须要增加，而且轿厢上安全装置的性能也要提高，以便能够在需要的时候让轿厢停下。沿着缆绳电梯运动，控制的复杂性显著提升。所以，如果你切断缆绳和电梯的连结，轿厢的重量自然就降低了，然后你就可以把重点放在轿厢框架结构中的轻质材料上。当你不能够利用一种平衡力来抵消上升所产生的重力时，电梯的直线电机及其电力驱动装置，可以与松开的电梯轿厢一起向下运动以重新产生电能。第二个重要意义在于轻质材料，无接触、无缆绳方法上的进步，它为电梯轿厢带来了电力和电讯号，以及其他一些东西。第三点是——这可能不是最后的一点——技术或系统正变得越来越灵巧高效。例如，如果你已经完成了一部无缆绳电梯，它从地面开始，在每一层都停一下，那么，你要花上很长一段时间等它再次下来。你将需要越来越多的电梯，它们会把空间都占用消耗掉，所以必须采用一种更灵活的技术。例如，如果你要在山顶扑灭大火，而你一次只能带着一桶水跑上去，那么想把火彻底扑灭可能时间上就很成问题。但是如果你不停地将一桶桶水传递到你旁边的朋友手里，那么水的流通输送效率将会高得多。这就是奥的斯的奥德修斯系统中一项最根本的技术，将轿厢传递到邻近连结的电梯中去，这能够让电梯在重要的第一次往返中以最短的时间回来。

我相信我们已经拥有了合适的人选，他们了解如何去规划、设计、测试和建立一个理想系统。在迪拜市场非常火爆的时候，我们认为应该让一些人在这个项目上进行投资——他们同时也是在高层建筑上顾及自身利益的投资者。由于主要的电梯企业

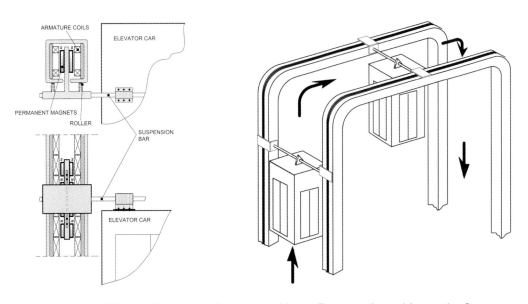

Ropeless elevator, courtesy of Zbigniew Piech, Linear Synchronous Motors: Transportation and Automation Systems, 2nd ed. (2011), Fig. 7.7, Ch. 7, p. 230.

无缆绳电梯，兹比格纽·皮耶希，《直线同步电动机——传输和自动化系统》，第二版，2011年，第七章，第230页，图7.7。

意识到，它们超过90%的生意都在中低层以及高层楼房上，而并非主要集中于超高层建筑，所以，大型的电梯企业并不倾向于过多地将研发精力放在高层建筑上，因为几乎所有东西都需要重新开发。除了规则和安全性，以及你从缆绳电梯的问题中吸取的教训，你几乎要抹去你知道的关于传统电梯产品的一切。就安全性而言，你想要让无缆绳电梯至少与缆绳电梯相当，甚至从理论上说应当更好。对于许多人来说，如果没有关于安全性的行业知识，这就完全是一种全新的局面。就我而言，我希望在我们变老之前，能够对此项工作贡献绵力。

目前来看，使用细缆绳和现成零部件，电梯可以走到多高的地方？

我们实际上不需要细缆绳，当谈到钢索时，我们需要的是更大的缆绳。这是我们的一个问题。最大的向上距离大约是600米（1968英尺）左右；如果是双层电梯的话，还要短一点，接近于500米（1640英尺）。不过，为了真正回答你的问题，为了克服缆绳电梯中的这一独特障碍，即缆绳重量问题，几乎所有的电梯制造商都研究且试验了采用芳纶纤维的轻质缆绳，他们获得了不同程度的成效，但也发现了一些重大的问题。似乎到目前为止，我们尚未

new. You have to almost erase everything you know about traditional elevator products themselves, except the codes and safety, and also your lessons learned from problems with roped elevators. For safety, you then want the ropeless elevator to be at least equal, and ideally better. Except for that industry knowledge on safety, it is to many people a whole new ball game. For us, I just hope we can contribute to this effort before we all get much older.

At the moment, using thin rope and available components, how high can elevators go?

We actually don't need thin ropes; we need larger ropes when we are talking about steel wire ropes. That's one of our problems. The upward maximum is around 600 or so meters [1,968 feet]; less for a double-deck elevator, where we may see close to 500 meters [1,640 feet] soon. However, to answer your real question, to overcome this particular problem with roped elevators, which is the rope weight, almost all lift manufacturers have researched and tried lightweight ropes using aramid fibers, with varying success and some significant problems. It appears that we have not heard the last word on this yet. However, even if the latest attempt proves successful, all we will have done is to extend travel for a single lift in a hoistway, and again, the higher you go with a single elevator, the longer you need wait for it to come back.

Is there a vertical transportation system for the automatic delivery of goods in a high-rise building?

For our 1-kilometer [3,280-foot] and mile-high [5,280-foot or 1.61-kilometer] towers planned for Dubai, we proposed a unique system of "sky docks," Copyrighted© by Barker Mohandas, LLC, USA, located at or near mechanical levels or sky lobbies, served by express service shuttle elevators. At the "sky docks," the service shuttles connect to local service elevators. If Mrs. Smith wants to move in, she can't just send up her lamp or chair. It just wouldn't work; things need to be in containers and on wheels, prepackaged for productive transport. So we had a system where all the containers were checked through security, tagged, and then automated so that the containers would be loaded on automated bogies, which then carried the containers to the service shuttle elevators using RFID-based (Radio Frequency Identification) guide paths. The service shuttles then made just one or two stops to the "sky dock" goods lobbies, where the containers were unloaded and goods transferred to the local service lifts. So within your residential zone of maybe 25 floors you would come up or down to get your mail, and so forth. The system was semi-automated to allow staff also to be transported. It's just a matter of cost to go to the next step for fully automated deliveries. We were just providing the main vein for an automated container transport system that we also suggested based on products readily available now. From the sky docks, goods could in fact be transferred onto local lifts and delivered to your door on automated wheeled carts, and with enough sensors the system could also be designed to be staff-friendly.

Tall buildings involve long elevator rides. Are there limits to the amount of time people willingly spend in transit? How is that time best spent?

Our observation is that people do have limits on transit time, particularly in an elevator stopping at every floor. But if you're traveling express or nonstop, it's more tolerable. You know where you are going, you keep moving, it's comfortable. The provision of some key information would enhance the experience. I'm not sure about random audio/visual shows and that kind of thing, which could be more of a distraction, but information about where you are and what is going on, with graphics that can be quickly understood. I think way-finding in these buildings is a very important subject.

Getting around in big buildings isn't always easy. Things don't necessarily align easily when you transfer between elevators, or from elevator to stairs, or move from one area

听到关于这个问题的最终结论。不过，即使最终的目标被证明已经实现，我们必须要去做的还是延伸一个井道中单部电梯的运行长度，不过，问题又回来了：一部电梯上升的越高，你就需要花费越多的时间等待它回来。

在高层建筑中，有没有一个垂直运输系统，用于货物的自动运送？

为了我们在迪拜规划的，高达1千米（3280英尺）和1英里（5280英尺或1.61千米）的大楼，我们提出了一套独特的"空中码头"系统（美国巴克莫罕达斯有限责任公司专利所有），它位于或接近于机械层或空中大堂，配备快速的服务梭型电梯。在"空中码头"，服务梭型电梯与局部服务电梯连接在一起。如果史密斯夫人想要搬进来，她就不能仅仅是把她的灯或者椅子运上去，那是没用的。为了有效的运输，必须早就有些东西被完整地安置在轿厢中或在轮子上。所以我们有了这么一个系统，其中所有的轿厢都通过安全检测，被打上标记且自动运行，因此，轿厢将被装载在自动转向架上，然后利用基于无限射频技术的引导路径，被带到服务梭型电梯中去。然后，服务梭型电梯在到达充满货厅的"空中码头"前，只会停一两次；而在"空中码头"中，轿厢中的货物被卸下来，然后换乘到局部服务电梯中去。所以，在你的住宅区内（这个住宅区可能有25层之多），你只需要向上或向下来收取你的邮件等东西。这一系统过去是半自动化的，因而也可运输人员进入到下一步的全自动化运输，只是一个费用问题。我们只是为一个自动轿厢运输系统提供主干结构，我们也建议说，这一系统应当尽量采用当前可以获得的一些产品。从空中码头出发，货物实际上可以被运输到区域电梯中，然后通过滚动的小车投递上门，如果有了足够的传感器，这个系统也可以被设计成供人使用的。

高层建筑包括很长的电梯运程。就人们能接受的，花在运输上的时间总量而言，有没有一些限制？这个时间应该如何安排利用？

就我们的观察，在运输时间的花费上确实有一定的限度，特别是在一部每层都停的电梯上。但如果你乘坐的是快速或不停顿的电梯，那反倒好得多。你知道要去哪里，你一直在移动，这是很舒畅的。提供一些关键的信息会提升这种经验。我对于随机视听展示，以及相关类型的东西（它们更多的可能只是用来分散人的注意力）有何作用并不确定，但例如"你在哪里"以及"将会发生什么"这类信息配上图像以后，将会很容易被人们理解。我认为在此类建筑中，路径查找是一个非常重要的课题。

在一栋大楼里四处走动并不总是很容易的事。当你在电梯之间换乘，或者从电梯走到楼梯，又或者从一个区域到另一个区域时，有些东西并非一定就是井井有条的。也许你对于主厅楼层中的那些双层电梯尚未熟悉。准确清晰的信息需要在第一时间，并且持续地向人们提供。我认为工业设计师和建筑师需要在路径查找方面扮演更为重要的角色。而且，当发生紧急情况时，人们知道该做什么，这也是非常关键的。他们该去哪里？他们如何出去？有一些标示会在你进入大楼进行路径查找时，一直引导着你，同样地，必须有一些标识能够帮助你在紧急情况下逃离大楼。

目标调度有何重要性？

它已经在办公大楼的主厅中找到了自己的位置。它会将那些想去普通楼层的乘客安排到特定的电梯轿厢中，因此在上升过程中，电梯停靠的次数就会减少。往返主厅次数上的改进，有助于减少早高峰时段等候的队列。从另一个角度看，电梯作为一套轿厢调度系统，当它与一套安全栅门系统融合在一起时，这项技术也是非常有用的。每次是一个人，而不是一群人冲进来，当然，电子身份卡也会使管理和控制一栋大楼中的人员更加容易。在因火灾需要疏散人员时，或许其中有些人是残疾人、幼童、老者时，至少应当有一些信息能够说明谁还在大楼里，以及他们可能在哪个位置。这是一套真的需要改进和应用的系统，当然，它同时还是要尊重个人的隐私。

关于这一点，美国的规则在消防员自身的消防安全方面已经落在了后面。数十年来，英国都有消防队专用的安全电梯，它也保护了通向楼梯的后备方案。不幸的是，它非常小，只能容纳一辆轮椅或一位装备齐整的消防员。相反，美国只是将每部电梯的钥匙给消防员，而这些电梯并未受到保护。这有点像在一个烟囱里上下移动，特别是规则规定，烟的出口只能在烟囱的顶部。1995年，我曾写过一篇论文，推动为消防员设立安全电梯的主张，

to another. Or maybe you're not already familiar with double-deck elevators at the main lobby levels. Good, clear information needs to be provided initially, and needs to be ongoing. I think that industrial designers and architects need to play a much, much stronger role in way-finding. Also, it's critical in an emergency that people know what to do. Where should they go? How do they get out? The same displays for way-finding to get in, and that guide you along your way, should be the same displays to help you exit in an emergency.

What is the importance of destination dispatching?

It's already found its place at the main lobby in office buildings, assigning arriving passengers with common floor destinations to certain elevator cabs, so that the number of elevator stops is reduced going up. The improvement in round-trip times from the main lobby helps reduce queues there during the morning peak time. The technique also has benefit when integrated with a security turnstile system that is also in view of the elevators as car assignments are given. It's one person at a time, not a crowd rushing in. Electronic ID cards, of course, also make it easy to manage and control who is in a building. During evacuation for a fire emergency, and/or when disabled, very young, or elderly people are involved, at least there would be some information about who is still in the building and where they might be. It's a system that really needs to evolve and be adopted, while still respecting personal privacy.

On that note, American codes had been lagging behind in fire safety for firefighters. For decades the British have had a protected lift for fire brigade use that also has protected backup access to stairs. Unfortunately, it was very small, capable of accommodating just one wheelchair or a fully-outfitted firefighter. Americans, by contrast, just handed the firefighter keys to every elevator, which were not protected. It's kind of like riding in a chimney, especially since a smoke vent was specified at the top under the code. Back in 1995 I wrote a paper to push the idea of protected elevators for firefighters, recommending the use of service elevators in office buildings since these elevators are already large enough for an ambulance stretcher. New U.S. model building codes have finally caught up with that and require a protected "first responder's elevator" or "fire service access elevator." And very recently, thanks to encouragement by NIST (National Institute of Standards and Technology, USA), at least for high-rise office buildings, we now need either space for 3 stairways, which is hard to accommodate, or 2 stairways, with the passenger lifts protected for occupant evacuation. So the U.S. has finally leaped ahead in the codes via the 2009 and later editions of the IBC (International Building Code, USA). When the U.S. makes a statement like that, particularly when we're doing so much international work, it establishes an important model. In our designs we now need to look for protected backup access to stairs from every lift lobby, in case either the elevators or their power fails, or considering that human behavior during a fire may be unpredictable. I would say that as China is thinking about super-tall buildings this needs to be addressed, and the precedent has just been set in the United States.

Is the firefighter's lift lobby smoke free?

It should be, with doors that close during a fire and with lobbies that are smoke pressurized. Protected backup access to stairs should also be provided in case the lift shuts down or loses power, or simply to allow careful firefighters to ascend the remaining handful of floors to the actual fire floor, using stairs.

But the shaft itself could never be pressurized?

We defer the wisdom and requirements for that to qualified MEP and fire protection engineers, in conjunction with code authorities, considering also complications with stack effect and reverse stack effect, which becomes a challenge in very tall buildings. While British-based codes may suggest you do this, I think they were initially written for low- to medium-rise buildings in London, and the practicality and effectiveness should be revisited as we build much higher. Under U.S. codes, for example, I think if you protect and pressurize the lift lobbies, there isn't really a need to go further; it would be too late. The pressurized lobby is your first and most important line of defense. Again, we defer further comment to the disciplines mentioned.

Are lift systems sustainable?

I think today, for elevators with wire ropes, the lifts are about as energy efficient as they could be. The motors have become as efficient as they can be for the latest technology available, using permanent magnets that have been introduced to the rotating element in the motor; which has added 3 percent to 5 percent efficiency to the motor component itself. Today's fully regenerative powered electronics drives, which have a power factor close to unity, are another contributor. As an interesting example, during a fire, for lifts operating in evacuation mode, when a lift with a full load goes down, the car is heavier than the counterweight; when an empty car goes

我建议使用办公楼中的服务电梯，因为这些电梯很大，足够容纳救护车担架。新的美国标准建筑规则最终也迎头赶上，要求有一部安全的"最先可以做出反应的电梯"或者"可用于消防服务的电梯"。最近，多亏了美国国家标准与技术协会（NIST）的推动，至少就高层办公楼而言，我们现在要求，要么有三个楼梯的空间（这一点很难得到满足），要么有两个楼梯的空间，同时客梯必须得到保护以保证疏散时能够容纳人员。因此，通过2009年以及美国国际建筑规则后来的一些版本，美国终于跃升到了前排位置。当美国做出类似的声明，特别是当我们做如此多的国际性工作时，它就确立了一个重要的标准。为了防止电梯或电梯的电源系统无法工作，或者考虑到人在火灾时的行为是难以预测的，在我们的设计中，现在需要寻找从每个大堂的电梯安全向楼梯撤退的方法。我要说，随着中国正在考虑超高建筑方面的问题，这些需要被提出来，而美国正好现在就开创了这方面的先例。

是否消防员的电梯间是无烟的？

应当是，门在火灾期间是关闭的，电梯间则是密闭隔烟的。但也应当提供安全撤向楼梯的方法，以防止电梯关闭及电源中断；或者只是简单地允许消防员谨慎地利用楼梯爬向其他各层，以到达实际着火层。

但是转动轴自身永远不会是密封的吧？

我们遵从合格设备层及消防安全工程师的智慧和要求，同时也与规则的权威性结合起来，并考虑到了烟囱效应和反烟囱效应的复杂性——这点在超高层建筑中正成为一个挑战，英国的规则可能会建议你这么做。我认为它们一开始是为伦敦的中低层建筑设定的，而当我们建设更高的高楼时，这一规则的实用性和有效性应当得到重新评估。例如，在美国的规则下，我认为如果你保护且密封了电梯间，那么真的就不需要再进一步了；也已经太迟了。封闭的电梯间乃是你防线上第一和最重要的东西。再说一次，我们把评述留给上面提到过的规范。

电梯系统是否是可持续的？

我认为在今天，由于升降机都是采用钢缆的，电梯应当尽可能有效地利用能量。由于最新可以利用的技术采用了永久磁铁，将它应用到电动机的转动元件中，可为电动机自身的零件增加3%-5%的效率，所以，电动机已经尽可能变得高效了。今天，完全再生式的电力驱动元件，有着接近于整个机械装置的功率因素，它是另外一个贡献者。有一个有趣的例子，在火灾期间，由于电梯运行于疏散模式之下，当一部满载的电梯向下运行时，轿厢要重于平衡力，当一部空电梯向上运行去承载其他人或货物时，平衡力要重于轿厢重量。因此，在这种重要的模式下，实际上有一个电力的净再生过程，此时，电梯的电力设备允许电力回到大楼的电网中，以供应其他的接电负载装置。在你最需要它的时候，电力实际上每次都再生了。对于高层建筑，我们也要求规划能够显示，电梯的传动系统可以在将来的某一天被拆除和替换。所以我认为就今天电梯所处的位置看，鉴于当前可以获得的电梯产品，电梯还是非常具有可持续性的。

世贸中心对于载客电梯有何影响？

高层建筑被不停地建设，甚至更多，所以其实没什么影响。我认为它主要是对消防规则发生了影响，旧有的规则已经过时了。那起事件发生之前，在我所参加的各类规则委员会中（针对紧急情况进行操控的美国机械工程师协会电梯规则委员会；各种旨在帮助纽约市消防局改进电梯的委员会），我有很长一段时间都在推动这种规则的改进。不幸的是，美国电梯行业中特殊的商业利益看起来让这些改进推迟了很多年，实际上，即使在保护电梯还不是他们职责的时候，同样的这些公司就已经在世界其他地方，在其他的规则要求下为消防员提供这类安全电梯了。但是，正如前面所提到的，美国的规则最终为消防员进行了改进，并且实际上已经跃升到了其他规则的前面，涵盖了人员疏散电梯，它本质上还是建立在与英国消防队使用了数十年的电梯相同的设计上。对于超高层建筑来说，这是重要的一步。当你需要疏散的时候，谁想要向下爬200层的楼梯？另外，就七星帆船酒店以及后来的纳赫勒港湾大楼（它们都是为迪拜规划的）而言，我们认为，将高层建筑与其附属大楼连接起来时，位于空中大堂的空中天桥，在整个的消防出口规划中将起到非常重要的作用。

up to get another full load, the counterweight is heavier than the car. So during this important mode, there is actually a net regeneration of power, where the lift power electronics allow that power to go back to the building grid to help power other connected loads. Power is actually regenerated at a time when you may need it most. For tall buildings we also ask for plans to show that lift powertrains can be dismantled, and replaced, at a later date. So I think that for where the lifts are today, they are very sustainable for the current lift products available.

What impact did the World Trade Center have on tall buildings and specifically on elevatoring?

Well, tall buildings are certainly being built, even more of them, so there's been no effect on that. I think its main impact was on fire codes, which was way overdue. I had long been pushing for certain code improvements before that event on various code committees I had joined—the ASME [American Society of Mechanical Engineers] elevator code committee for emergency operations and various committees to improve elevators for the FDNY [New York City Fire Department]. Unfortunately, certain commercial interests in the U.S. elevator industry seemed to delay such improvements for many years, even when the work to protect the elevator was not their work, and the same companies provided such protected elevators for firefighters elsewhere in the world under other codes. But, as mentioned earlier, U.S. codes have finally improved for firefighters, and have actually leaped ahead of other codes to cover occupant evacuation elevators, essentially based on the same designs used for decades for lifts for the British fire brigade. This is an important step for super-tall buildings. When you need to evacuate, who wants to take the stairs down 200 stories? By the way, for Al Burj and later for the Nakheel Harbour Tower, both planned for Dubai, we suggested that skybridge at sky lobbies connecting to other sub-towers would be very useful in the entire fire egress plan.

Marvin Mass, Scott Ceasar
COSENTINI ASSOCIATES

Marvin A. Mass, P.E., chairman of Cosentini Associates, is one of America's foremost mechanical engineers, having collaborated on many of the most important buildings of the past half century. Since beginning his distinguished career in 1952, he has overseen the ongoing development and enhancement of his firm's standards of design, construction, and supervision. Mr. Mass is a long-standing faculty member at Yale University and the Harvard Graduate School of Design, and has lectured at the University of Pennsylvania, Pratt Institute, and Cooper Union. He was honored with the Franklin Institute's prestigious Brown Medal, AIA Institute Honors, and a Lifetime Achievement Award from the American Society of Heating, Refrigerating, and Air-Conditioning Engineers (ASHRAE). Mr. Mass earned his professional degree from New York University and is licensed in 23 states. He is a member of the National Society of Professional Engineers, ASHRAE, the Building and Architecture Technology Institute, and the Institute for the Study of the High-Rise Habitat. Mr. Mass is an Honorary Member of the American Institute of Architects.

Scott Ceasar, P.E., joined Cosentini Associates in 1988 and is now a senior vice president with over 20 years of project experience that includes major hotels and convention centers, courthouses, corporate headquarters, and high-rise office buildings. Mr. Ceasar was one of the first engineers in the country to receive LEED accreditation from the U.S. Green Building Council and is responsible for assisting Cosentini's engineering teams with the development of sustainable engineering solutions for their projects. Mr. Ceasar holds a Bachelor of Science in mechanical engineering from the Johns Hopkins University in Baltimore and is a licensed professional engineer in New York. He is an active member of the U.S. Green Building Council and ASHRAE.

马文·马斯，斯科特·西塞
克森蒂尼事务所

马文·A·马斯，克森蒂尼事务所主席，是美国最重要的机械工程师之一，在过去的半个世纪中他参与了许多最重要建筑的建设工作。自从1952年他开始自己精彩卓越的职业生涯以来，他见证了他的公司在设计、建设、管理标准上的持续发展和提高。马斯先生是耶鲁大学和哈佛设计研究院的长期教员，他也在宾夕法尼亚大学、普拉特学院、库伯高等科学艺术联合大学做过讲座。他获得过富兰克林学会著名的棕色奖章、美国建筑师协会荣誉奖章，以及美国采暖、制冷与空调工程师学会颁发的终身成就奖。马斯先生在纽约大学获得自己的专业学位，并在23个州得到认证许可。他是国家专业工程师学会，美国采暖、制冷与空调工程师学会，建筑与建筑技术协会，高层住宅研究协会的会员。马斯先生还是美国建筑师学会的荣誉会员。

斯科特·西塞于1988年加入克森蒂尼事务所，现在是一名具有20多年项目经验的高级副总裁，这些项目包括有大型酒店、会议中心、法院、公司总部和高层办公楼。西塞先生是最早从美国绿色建筑委员会获得LEED认证工程师之一，他负责协助克森蒂尼的工程团队为他们的项目提供可持续性的工程解决方案。西塞先生从巴尔的摩市的约翰霍普金斯大学获得机械工程方面的科学学士学位，并且是纽约认证的职业工程师。他是美国绿色建筑委员会和美国采暖、制冷与空调工程师学会的活跃会员。

Cosentini Associates served as MEP for the projects illustrated
克森蒂尼事务所为所示项目提供工程实践手册

For complete biographies, please refer to the Appendix.
完整的传记，请参阅附录。

The Bow, Calgary, Alberta, Canada (Foster + Partners with Zeidler Partnership Architects, 2012)
加拿大阿尔伯塔省卡尔加里弓形大楼。（福斯特建筑事务所和蔡德勒建筑师事务所，2012年）

How has mechanical engineering changed over the past half century?

MM: I started in this business in 1948–49. Because of war and the Depression, very little had been built for decades, but then, all of a sudden, there was a tremendous building boom.

Almost all of my early buildings were corporate headquarters, which meant each had one major piece of mechanical equipment. If someone needed to work overtime, you had to run the whole building, not just half or a quarter; it was all or nothing. This became a real issue with increasing international business. Why run the entire building for some guy in New York working after hours with Paris or London? George Kline came up with the idea of putting in a floor-by-floor system so that only specific areas would be conditioned while the rest of the building was shut down. We first used the new system for 499 Park Avenue (1981), designed by James Ingo Freed of I. M. Pei & Partners. Other engineers criticized me for designing "drugstore units," but the floor-by-floor system we designed became the industry standard.

You have to understand that engineers had no experience designing air-conditioning for big buildings. It was all done by the manufacturers—Carrier, York, Westinghouse—and each specified its own chillers and fans. That's why these buildings all had big central plants. Because of 499 Park, our smaller floor-by-floor systems became very popular with developers who rented to multiple tenants with different needs. It wasn't too long before single-tenant headquarters buildings wanted the same flexibility.

What are some important new MEP developments?

MM: Douglas Mass, who now heads up Cosentini, developed an underfloor air system in lieu of the customary practice of running ductwork in the ceiling. Now, instead of the 4- to 5-inch [10- to 13-centimeter] raised floors for all the telephone, computer, and other electronics wiring, Doug's alternative was to raise the floor 12 inches [30 centimeters], pump the air in underneath, and then just put supply diffusers wherever they're needed. To change a partition, you just switch the diffuser from here to there rather than tearing down the ceiling. This is what we've been doing recently. AT&T, Goldman Sachs, Bank of America: all of these buildings have underfloor air systems.

The big thing these days is sustainability. It's very interesting, but at this point, I'm not sure if it is economically sound. It is important to think green, and we go out of our way to use the most efficient equipment available. There

在过去的半个世纪，机械工程发生了怎样的改变？

MM: 在1948年至1949年间，我开始进入这个行业。由于战争和经济萧条，已经有几十年基本没有兴建什么东西了，但是在那之后，突然出现了一个巨大的建设繁荣期。

在早期，我完成的所有的建筑几乎都是公司总部，这意味着每个项目都有很大一部分是关于机械工程的。如果有人需要加班工作，你必须让整栋楼都运行，而不是一半或者四分之一，也就是说，要么全部要么什么都没有。随着全球贸易的增加，这成了一个现实的问题。如果纽约的一些人要和巴黎或伦敦的人一起工作，但时差让他们的工作时间晚了几个小时，那么，为什么要让整座建筑都为这些人运行呢？乔治·克莱恩提出一种分层系统的观点，这样的话，只有一些特殊的区域处于正常状态，而建筑的其他部分会被关闭。我们第一次将这一新系统应用在公园大街499号上（1981年），这栋建筑是由贝聿铭工作室的吉姆·弗里德设计的。当时，其他工程师批评我设计的是一些"杂货店"，但我们设计的分层系统后来成为了行业标准。

你必须理解，工程师们并没为大型建筑设计空调系统的经验。它完全是由制造商做好的——开利公司、约克公司、西屋公司——每家公司的冷却器和风扇都是指定的。这就是为什么这些建筑都有着很大的中央设施。由于公园大街499号的缘故，我们更小的分层系统在开发商那里变得越发受欢迎，他们按照租户不同的需要将楼房租出去。此后不久就出现了需要同样灵活性的单身租户总部建筑。

工程实践手册有哪些重要的新发展？

MM: 道格拉斯·马斯现在领导克森提尼公司，他开发出一套地板下的空气系统，用来替代一般在屋顶的管道运行方式。目前，用于电话、电脑和其他电缆线的，4到5英寸的活动层已经被替代了，道格拉斯的方案是，将楼层提升12英寸，将空气抽入地下，然后在仅需要的地方放上供风分布口。如果要变换一个区域，你只需要将分布口从这里换到那里，而不是把天花板都撕下来。这是我们最近已经做的。AT&T大楼、高盛集团、美国银行，所有这些大楼都有地下空气系统。

最近流行的话题是可持续性。这很有趣，但我并不能确定它是否足够经济。有绿色意识是很重要的，而且我们也按照自己的方式采用现在可以获得的最有效的设备。有一些建筑的拥有者，实际上在头五年中削减了花在节能系统上的总金额。不幸的是，一些操作人员和维修人员只是依赖于过去的经验，他们不知道如何升级、维护新系统。由于这些问题，本来可以在五年内收回成本的系统可能需要十年的时间才能实现目标。

当我刚开始我的事业时，所有的施工工程师都是为轮船工作的服务人员，在今天看来，这些已经不太常见了。但现在关于怎样降低能源费用，一些操作人员并没有相关的能力或知识。除非管理建筑的人知道该做什么，否则我们的所有研究都没什么意义；而多半时候，他们也确实不知道做什么。

SC: 我不会做出如此强硬的评价。有许多业主知道如何管理建筑，也雇佣了一些合格的人员。例如德斯集团、铁狮门、汉斯公司，就都知道如何管理他们的建筑，他们找对了人。有一些非常优秀的管理者，不过也有一些人，他们的观念还停留在30年前处理问题的方法上。

当你的工作变得国际性时，有什么转变？

MM: 我们被认为是美国最有见识的一批工程师，所以被雇佣。我们详细说明了一组规划示意图以及一个理论体系，然后将工作转交给海外的某个人去具体实施。问题是，他们并不能制造出同样的产品。当地的代理人是更有效率还是更没有效率，这点我们并不知道。而且，许多发展中国家没有足够的经验，也不知道如何调节系统以减少能源消耗。

我们现在正与金斯勒事务所一起为上海中心大厦工作。我们尽量设计得节能环保，但这栋建筑高达102层，由于烟囱效应和系统压力，不可能只有一个系统来应付这么高的大楼。所以我们将建筑分为20层一组，每一组都完全是独立的，有自己的机械系统——包括电力和冷却塔，实际上每20层就排放热量。唯一贯穿整座建筑的管道是用于消防喷洒的，上海的工程师有节能方面的必备知识，但是我不确定操纵员是否具备。

对于一栋高层建筑来说，什么高度是最有效率的？

MM: 如果我们可以将建筑保持在大约40层以下，系统的压力是正常的（250磅）。40层以上，你就需要300到350磅，因此管道系统变得更大，厚度也增加了，同时电机马力和能量消耗也随之上升，所以最有效率的建筑大概是40层。

have been some buildings where the owner actually saved the amount of money spent for energy conservation systems in the first five years. Unfortunately, some of the operations and maintenance people just rely on past experience; they don't know how to upgrade and maintain the new systems. Because of these problems, systems that are supposed to pay for themselves in five years may take twice as long for payback.

When I started in business, all the operating engineers were servicemen who had worked on ships. You don't always get that these days. Now, some of the operations personnel do not have the ability or the knowledge about what needs to be done to lower energy costs. All the studies we've done don't mean a damn thing unless the guys operating the buildings know what to do. And half the time, they don't.

SC: I would not make so strong a statement. There are many property owners who know how to operate buildings and hire qualified personnel. The Durst Organization, Tishman Speyer, and Hines, for example, all know how to operate their buildings; they get the right people. There are some very good operators and then there are others who are still thinking the way things were done 30 years ago.

What happens when you work internationally?

MM: We're hired because we're recognized as some of the most knowledgeable engineers in the United States. We specify a schematic set of plans and a theoretical system, and then we turn the job over to somebody overseas to produce. The problem is that they don't manufacture the same products. Whether their local substitutes are more efficient or less efficient, we don't know. Also, a lot of developing countries don't have sufficient experience to know how to adjust the systems for reduced energy consumption.

We're currently doing Shanghai Tower with Gensler. We designed it to be energy efficient, but the building is 102 stories high and due to stack effect and system pressures, you can't run a system 102 stories. So we divided the building into 20-story packages, each totally self-contained with individual mechanical systems, including electricity and cooling towers; we actually discharge heat every 20 floors. The only pipe that continues the whole height of the building is for sprinklers. The engineers in Shanghai have the knowledge necessary for energy saving, but whether the operators do, I can't say.

What is the most efficient height for a high-rise building?

MM: If we could keep buildings down to about 40 floors, the pressure on the systems is normal (250 pounds [113 kilograms]). Above 40 floors, you need 300 to 350 pounds [136–159 kilograms], so the piping gets bigger, the thickness increases, and the motor horsepower and energy costs go up considerably. So the most efficient buildings are in the neighborhood of 40 floors.

Is sustainability achievable?

SC: I think so. But it's a question of what sustainability is. There are lots of definitions, but basically, it's looking to a functional building that will meet future needs. This is a good trick because we don't really know what future needs will be. But we do know that energy prices have risen historically, and that global warming is an issue. From what I see of the hard science, it's real. A lot of people still debate this, but I tell them it certainly can't hurt to reduce emissions.

So a sustainable building should have low energy usage, low materials resource usage, create as little waste as possible, have a small carbon footprint, and create a comfortable usable environment for occupants. By having a design team that works together to create complementary building systems, sustainability can be achieved. Sustainable design is really just an integrated, well-designed building.

Because LEED [Leadership in Energy and Environmental Design] has become such a big factor in the market, more and more clients say, "I want a LEED building." But it soon becomes clear that some don't really know what LEED is. It's always a push between total project capital expenditure and the available budget for what you put into the building, but that doesn't mean it can't be sustainable.

We start by asking the client about his needs and about the building's function. We look at about 20 different factors, like energy efficiency, capital costs, maintainability, operability, redundancy, disaster recovery, and so on. We determine what's important to the client, and then we start planning the building. It's different from 20 or 30 years ago, when an architect would do a design and say "put your systems in." We now want to be there in the beginning, to look at the local climate, how the building will be sited, facade design for maximum efficiency. What we are really looking at is how energy consumption can be reduced, and how the building can be optimized before putting our systems in. It's a significant change from the way things are typically done.

可持续性是否可以获得?

SC: 我认为可以。但问题在于,究竟什么是可持续性? 关于这点有许多定义,但根本上说,它面向的是适应未来需要的一种功能性建筑。这是一个很好的说辞,因为我们并不真的知道未来需要的是什么。但我们知道能源的价格又创新高,而全球变暖正成为一个大问题。就我在自然科学中看到的而言,这个问题是真实存在的。有许多人还在为此争论,但我告诉他们减少排放量并不会有什么损害。

因此一栋可持续性建筑要有较低的能源消耗量,较低的原材料资源使用量,尽可能少些浪费,要有较小的碳排放量,为住户创造出一个舒适方便的环境。通过协同工作,来创造互补性建筑系统的设计团队,可持续性是可以达到的。可持续设计其实就是一个综合的、设计良好的建筑。

因为LEED(能源与环境设计指导)在市场中已经成为了一个重要的因素,越来越多的客户说:"我想要一栋LEED建筑。"但是有一点需要很快澄清:有些人并不知道什么是LEED。在总的项目资本支出以及你在一栋建筑上的可用预算之间总是有一种张力,但这并不意味着它就不能是可持续性的。

通过询问客人的需求,以及建筑所需的功能开始,我们考虑了12种因素,像能效、资金成本、可维护性、可操作性、冗余、灾难恢复等等。我们确定什么对于客户而言是重要的,然后才开始规划建筑。这与20或30年前已经不一样了,那个时候,一个建筑师会做出一个设计,然后说"把你的系统放进来吧"。我们想在一开始就达到那一步,考虑当地的气候,建筑将如何选址,表层设计如何达到最大效率。我们真正考虑的是,能源消耗如何可以降低,在你的系统置入之前,建筑如何可以优化。相对于常规的做事方式而言,这是一个重大的转变。

现在,我们为卡尔加里市的一栋建筑工作,那里的建筑师邀请我们加入。究竟哪里才是选择建筑工地的最佳方向? 他们有六个不同的模型,哪个是最有效率的? 我们在表层进行了一些快速的能量流动测试,以确定有多少能量流入和流出。主景最后是一个面朝南的门廊,它像一个日照热量的缓冲区,冬天我需要很少热量,因为可以从太阳收集热量。卡尔加里市非常寒冷,但很多天都有高强度的阳光。在夏天,门廊如同一个缓冲区般发挥作用。让热量稍稍参与到建设中,如果你要把办公室温度保持在72华氏度(22摄氏度),你得让门廊升到

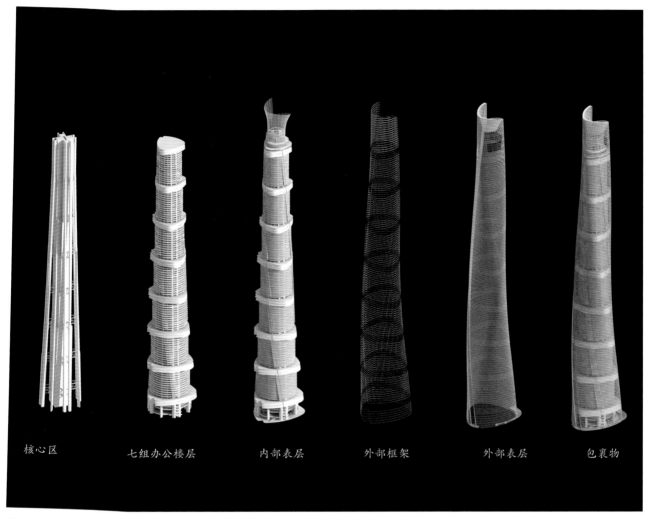

核心区　　　　七组办公楼层　　　　内部表层　　　　外部框架　　　　外部表层　　　　包裹物

Shanghai Tower: raw core, office floors in 7 packages, inner skin, outer frame, outer skin, and wrap. (Gensler, est. completion 2014)
上海中心大厦：核心区，七组办公楼层，内部表层，外部框架，外部表层，包裹物表层和包裹物。（金斯勒建筑事务所，
预计2014年完成）

TECHNOLOGY Marvin Mass, Scott Ceasar

Recently, for a building in Calgary, the architect invited our input. What was the best direction for siting? He had six different shapes; which was the most efficient? We did some quick energy runs on the facade to determine how much energy flows in and out. The main feature became a south-facing atrium that acts as a buffer for solar heating. In winter we use less heat because we're collecting heat from the sun; Calgary gets very cold but has a lot of high-intensity solar days. In summer, the atrium acts as a buffer zone. You let the heat build a bit, so if you're maintaining the offices at 72 degrees [22 degrees Celsius], you let the atrium go to 78 degrees [26 degrees Celsius]. Because it's a tall space, the stack effect will carry heat to the top. You don't condition the whole atrium, just a small part of it—at the bottom, where the people are. Passive strategies like that can really reduce energy consumption before we even look at what systems to put in.

Has LEED certification been incorporated into building codes?

SC: LEED has not yet become integrated into local building codes, but LEED elements have been incorporated into some building codes. In addition, the federal government, some states, and some municipalities have adopted laws requiring government buildings to be LEED certified. For federal work, buildings have to be at least LEED Silver; the same goes for anything built with city or state funding in New York City. The state also has Executive Order 111, which requires energy consumption to be better than ASHRAE 90.1. It's not an energy code but a national baseline established by ASHRAE [American Society of Heating, Refrigerating, and Air-Conditioning Engineers] for a building's minimum energy efficiency. It has become the point of comparison and is also the basis for most of the energy codes in this country.

Are net-zero buildings achievable?

SC: Yes, but again, it depends on the definition. It can be based upon carbon footprint, fuel input—some people even count the purchase of renewable energy credits, or what you're spending in energy compared to what you're putting in. And then there are some people who want to be totally off the grid, zero energy, where they have nothing taken in from outside. To me, it's if you're generating all the energy by renewable means to get net zero; you generate enough energy to make up for what you use even if you're pulling it off the grid at night and injecting it in during the day: you've got an energy balance of zero.

78华氏度（26摄氏度）。因为这是一个很高的地方，烟囱效应会把热量带到大楼顶部。你不需要限制整个门廊，而只需要控制其中的一小部分——在底部，在有人的地方。在我们考虑什么系统将被置入之前，这样的一些被动策略真的可以减少能源消耗。

LEED认证是否已经被纳入建筑准则中了？

SC： LEED尚未被纳入当地的建筑准则中，但LEED因素已经成为一些建筑准则的一部分。另外，联邦政府，一些州，以及一些自治市已经接受了政府建筑需要获得LEED认证的法律。对于联邦政府办公地而言，建筑必须至少是LEED银质认证；在纽约市，任何市政府或者州政府投资建设的建筑都要符合相同的标准。州政府还有一个111号行政命令，要求能源消耗优于ASHRAE（美国采暖、制冷与空调工程师学会）90.1号标准——这不是一个能源准则，而是一个由ASHRAE为建筑最低能效设定的国家标准线。它已经成为对照点，同时也是这个国家大多数能源准则的基础。

是否可以获得零能耗建筑？

SC： 是的，不过再说一遍，这有赖于对它的定义。它可以基于碳排放量，燃油输入——有些人甚至计算购买可再生能源的分数，或者是与你输入的能源相比你所花费的能源量。然后有一些人想要完全摆脱电网，零能源输入，他们不从外界引入任何东西。在我看来，如果所有的能源都是来自可再生的方式，你就达到了零能耗，你生产了足够的能源来填补自己的需要，即使你晚上从电网之中引入能源，而在白天的时候又将其注回电网也是可以的，你依然得到了能量上的均衡。

这在高层建筑上是有可能的，尽管非常困难——因为并没有太多生产可再生能源的方式。一般来说是太阳能光电板、风力涡轮机，以及通过地热能源（这在很多地方都难以获得），换句话说，实际上就是让热从地上流出以产生蒸汽。除了地热的方式，大多数可再生能源都意味着占据很大的空间。而在高层建筑中，你并没有这些空间，因此这个问题变得更加困难。

我们在巴黎有一栋八层的建筑，接近于100万平方英尺，有很大的楼层面。建筑的屋顶覆盖着太阳能光电板，也有一些风力涡轮机，但是坦率地说，风力涡轮机没有发挥太多的作用，主要还是太阳能光电板在起作用。要达到零能耗，需要做一些根本性的调节：

The Solaire in Battery Park City, New York, said to be the first green residential high-rise in the United States (Pelli Clarke Pelli, 2002)

纽约炮台公园市的阳光华夏，美国第一座绿色高层住宅。（佩里·克拉克·佩里建筑事务所，2002年）

It is possible, although very difficult, in a high-rise because there are not a lot of ways to produce energy renewably. Basically it's photovoltaics, wind turbines, and through geothermal energy (which isn't available in a lot of places), in other words, actually tapping heat from the earth to produce steam. Except for the geothermal mode, most renewable energy means take up a lot of space. And in a high-rise building you don't have that, so it becomes a lot more difficult.

We had an 8-story building in Paris just shy of 1 million square feet, with huge floor plates. The roof was covered with photovoltaics. We also had some wind turbines, but frankly, they didn't do much, mostly it was the PVs. To achieve net zero, certain fundamental adjustments were required:

1) Forget the old office building standard where tenants got 0.5 watts per square foot [5.38 watts per square meter] for light and power; now it's 0.7 or 0.8 watts of lighting per square foot [7.5 to 8.6 watts per square meter].

2) For auxiliary power in offices, it's more like 1.5 or 2 watts per square foot [0.14 or 0.18 watts per square meter], although people still think more is better; and the temperature will not be maintained at 72 degrees [22 degrees Celsius] year round. In summer, we let it go up to about 78 degrees [26 degrees Celsius]. In winter, maybe down to 69–70 degrees [20.5–21 degrees Celsius]; you'll wear a sweater. It's easier to do this in Europe, which has a different attitude and much longer tradition of energy conservation. For this to work in the States, we're talking about a fundamental change in mentality.

3) Creating systems that are very low energy. If you have a limited area to generate your energy, the idea is to use as little as possible. Options include low-usage lighting systems, low-usage HVAC systems like radiant ceilings with displacement ventilation, using a desiccant instead of compressor energy to dehumidify the air. All these things taken in conjunction with renewables, and with the change in the way you operate and occupy a building, make it possible. The hardest part is getting people to change the way they think about the built environment.

Is it true that if you were to cover an entire building in solar collectors and put in the best wind turbines, you're still not going to collect more than 25 percent of the energy needed?

SC: In a high-rise building that's true because you don't have the surface area. Wind turbines are great, but typically it is difficult to install them in a high-rise, even on a rooftop, because the vibrations are so great that it increases structural costs to the point of not being financially feasible. If

1）忘掉旧的办公建筑标准：以前每个租户有0.5瓦特每平方英尺（5.38瓦特每平方米）用于照明和供电；现在的标准是0.7或0.8瓦特每平方英尺（0.14或0.18瓦特每平方米）；

2）办公室的辅助电源，一般是1.5或2瓦特每平方英尺，尽管人们依然认为越多越好，温度不会常年保持在72华氏度（22摄氏度），在夏天我们可以让它升到华氏78度（26摄氏度），冬天，可能降到69—70华氏度（20.5-21摄氏度），你将需要穿件毛衣。在欧洲做到这些很容易，他们有不同的态度以及更长的节能传统。如果这些要在美国发生作用，我们就要谈到人们心态上的根本改变。

3）创造非常节能的系统。如果你有一块有限的地方生产自己的能源，这个系统会鼓励你尽可能少用。我已经谈到了低能耗照明系统，像带有置换通风光照顶棚的低能耗空调系统，这种顶棚采用的是一种干燥剂而不是消耗能源的压缩机来去除空气中的湿气。所有这些东西都与可再生资源有关，与你管理和居住一栋楼房方式的改变有关，正是这些改变使之成为可能。最难的部分是让人们改变思考建筑环境的方式。

纵使你把整栋建筑都覆盖上太阳能集热器，放入最好的风力涡轮机，你还是收集不到所需能量的25％以上，这是否是真的？

SC： 在一栋高层建筑中，这的确是真实的，因为没有表面区域。风力涡轮机很好，但一般来说在高层，它们很难被安装，即便在屋顶也是如此，因为震动得太厉害，反倒增加了结构上的费用，以至于在经济上有点负担过重了。如果你有很多钱，那么你可以完成。如果说到可持续性，却做一些不实惠的事——无论是经济上还是因为消耗了太多资源，那么，可持续性都没有什么意义。

现在，太阳能运作良好，即使它还不是一种全效技术。有两种方式可以利用太阳能：太阳能光电板（它会产生电能）或者太阳能集热器（它产生的热能可以直接用于建筑取暖系统也可以产生蒸汽用于发电）。

对零能耗还有另外一个定义，它包括从其他地方购买可再生能源，因此不是局部的零能耗，而是全部零能耗。在一个城市的环境中，你让你的能源消耗尽可能少些，然后从可再生的方式中购买电力。从个人角度说，我认为这有点像欺骗，但这确实是获得零能耗的有效方法，而且实际上，这么做也是在支持更多可再生能源的生产。

在欧洲，从外墙到楼芯的强制性距离是10米，大约合30英尺，而在纽约则几乎达到50英尺。

你真正想要的是19米的总租赁跨度，或者大约90英尺，楼芯两侧各30英尺。这就会引入100%的阳光。现在我们正在圣胡安的一栋四层建筑上做一项零能耗研究。我们想让日光照射房屋，所以设计了宽大的庭院以及反射器将光线反射到屋内。有一种多孔玻璃屏只允许20%的阳光直射进来，当阳光打到玻璃上时，产生热量光波分散开来，进入到孔中，所以它们从来没有射进建筑中。换句话说，我们得到了阳光，却没有获得太阳热量。这是很有效的。

在那栋建筑上，即便我们采用的一套标准变风量空调系统，没有达到零能耗，也依然要比ASHRAE标准好上30%。我们正在寻找一种冷却顶棚以及通风系统，它将会带来低于ASHRAE标准60%的效果，但这会超出项目规划预算。

阳光辐射顶棚是不是要贵多了？

SC: 目前是这样的。

阳光辐射顶棚的制冷效果是否比制热效果好？

MM: 为了建设这些带有当代系统的成功建筑，你需要一个技术熟练且知识渊博的操作员。保持建筑内部的正压，以及将从外面带到建筑内部的空气祛湿，这些是很重要的。为了保持适当的条件，特别是如果你打算让温度升到78度（26摄氏度）或者甚至80度（27摄氏度），适度的控制在这些建筑中就显得越发重要了。只要空间环境是干燥的，就会很舒服，但是一旦环境变得潮湿，人们就会觉得不舒服。

SC: 用阳光辐射系统将热量从建筑中驱除，其功效实际上要比向空间中输送空气好上四倍多。阳光辐射顶棚对于加热和制冷都有效，但是在冷却循环中，如果不盯着湿度，它就会凝结滴落下来。有些建筑是有冷却横梁的。这是一种非常有效的技术，但操作者必须知道它们是做什么的。

有两种不同的冷却横梁类型，消极的和积极的。消极的只是一个散热片。如果你有一个冷却顶棚——比如说水流过的冷却面温度为65华氏度——热传递是基于温度差异的。这是一个真实的辐射效应。

一整根冷却横梁与此不同。冷却水管贯穿其中，它像个散热片一样发挥作用，但同时又创造出一种对流效果；随着热空气被冷空气引入其中，它也形成了一种下冲气流。所以你获得了一点点气流——不太多，但它依然是阳光辐射型的。这就是

you have a lot of money you can accomplish a lot. If you talk about sustainability, to do something that is not feasible, whether financially or because of using too many resources, it just doesn't make sense.

Right now, solar works well even though it's still not a totally efficient technology. There are two ways to do solar: PVs, which generate electricity, or solar collectors, which generate heat either to use directly in the building heating system or to generate steam to make electricity.

There's another definition of net zero that includes the purchase of renewable energy from elsewhere: not site net zero but total net zero. In an urban environment, you keep your energy usage as low as possible and then buy power from renewable means. Personally, I think it's a little like cheating, but it is a valid means of achieving net zero and actually, by doing that, you're supporting the production of more renewable power.

In Europe the mandated distance from exterior wall to core is 10 meters, roughly 33 feet, whereas in New York it's almost 50 feet [15 meters].

What you really want is a 19-meter [62-foot] total lease span, or about 90 feet [27 meters], with 30 feet [9 meters] on both sides of the core. That allows 100 percent daylighting. We're doing a net zero study right now for a 4-story building in San Juan. We wanted to daylight it so we have large courtyards and reflectors that bounce light inside. A fritted-glass screen admits only about 20 percent of the solar directly. When it hits the glass, a lot of the light waves that produce heat get dissipated and get caught in the cavity, so they never enter the building. In other words, we get the daylight without the solar heat gain. It's effective.

On that building, if we don't go net zero, with just a standard VAV [Variable Air Volume] system, we're still 30 percent better than ASHRAE. We're looking at a chilled ceiling and displacement ventilation system that would bring it down to 60 percent better than ASHRAE, but this would exceed the planned project budget.

Radiant ceilings are that much more expensive?

SC: At this point in time, yes.

Do radiant ceilings work better for cooling than heating?

MM: In order to have a successful building with these advanced systems, you need a skilled and knowledgeable operator. It's important that you maintain positive pressurization in the building and that you dehumidify all the outside air you bring inside. Humidity control becomes very important in these buildings for maintaining proper conditions, especially if you're going to let the temperature go up to 78 degrees [26 degrees Celsius] or even 80 [27 Celsius]. It could be comfortable as long as the space is dry, but once you get humidity in the space, it becomes very uncomfortable.

SC: Removing heat from a building with a radiant system is actually about four times more efficient than supplying air into the space. Radiant ceilings work well for both heating and cooling, but in the cooling cycle, if you don't watch your humidity, it can condense and drip. Same thing with chilled beams. It's a very efficient technology, but the operators have to know what they're doing.

There are two different types of chilled beams, passive and active. Passive is really just a radiator. If you have a chilled ceiling—let's say a 65-degree [18 degrees Celsius] surface for the chilled water running through—heat transfer is based on temperature differential. It's a true radiant effect.

A chilled beam is a little different. It's got chilled water pipes through it. It acts as a radiator but it additionally creates a convective effect where, as the warm air gets drawn in by the cold, it creates a down draft. So you get a little bit of flow—not much, but it's still radiant. That's a passive chilled beam.

And then there's the active chilled beam, which sits in the ceiling and takes air from the system. But it's only about 20 percent of what's normally needed for cooling. Supplied air takes care of ventilation requirements, fresh air, but it also removes humidity from the space. Operators ask why they need to cool the air, reheat it, and then put it back into the space. It's to take out the moisture. They say, "If I've only got 65-degree [18 degrees Celsius] chilled water for the beams, let me run my chilled-water temperature higher, and only cool the air down to 65 degrees [18 degrees Celsius] and supply it right to the space." But this doesn't remove enough humidity. You get moisture in the space and it drips off the chilled beams.

MM: The one concern with a chilled-beam system is that when you start to move office partitions around, all the piping in the ceiling has to be changed, as opposed to the old-fashioned system under a window. Sometimes it's just a diffuser that you have to move rather than draining the system down and repiping it. Chilled beams work very well, or better, with open offices; we try to set up a logical grid so that when you start moving walls, 90 percent of the time you'll get a beam.

一种被动型的冷却横梁。

然后还有一种主动型的冷却横梁，它位于顶棚处，从系统中吸收获取空气。但它只有正常所需冷却量的20%。补给的空气主要满足通风设备以及新鲜空气的需要，但它也将湿度从空中祛除。操作者会问，"为什么他们需要冷却空气，然后对它进行加热，再放回到空间中"。这是为了把水分蒸发掉。他们说："我也可以从横梁上获得65华氏度的冷却水，再把冷却水的温度变得高一些，并且仅仅把空气降到65华氏度，将其注入到空间中。"但这并没有充分祛除湿度。你在空间中得到的湿气，它会从冷却横梁上滴下来。

MM： 与冷却横梁系统相关的一点是，当你开始移动办公区的时候，房顶的所有管道都要改变。与一个窗户丁面的旧式系统完全不同，有时候它只是一个你需要移去的扩散器，而非排系统、重布管道。冷却横梁与开放型办公室一起工作得很好，或者说更好。我们试图建立一个逻辑网络，所以当你开始移动墙的时候，90%的时间你将得到一个横梁。

SC： 如果一个业主关注灵活性，以及每两年要移动一下所有东西，我会推荐这些系统中的一个，这就是为什么我们总是会回到业主的要求上。除非移动建筑符合业主的要求，否则它就算不上是可持续性的。比如：能源效率、室内空气质量以及其他所有方面的要求。建筑应当达到某种恰当的功能才能成为可持续性的。它应当在较长一段时期内做到它必须做的。对我来说，这才是对于可持续性的定义。

可持续性可以接受的费用是多少？

SC： 如果你建设一栋确实很好，考虑周详、设计精良的建筑，还有不错耐用的系统，费用上的差异是非常小的。

如果从一开始就正确地规划，你所需的花费会少很多。一栋新建筑要获得一个基本的LEED认证，只需要增加1%或2%的费用，有些人认为，如果你规划正确，基本上没有什么费用上的增加。相反，费用上规格不高的建筑要达到这一要求可能需要增加10%到15%的费用。而对于零损耗建筑，增加的费用可能要达到60%到70%　　这取决于你建设的是什么。

太阳能光伏板非常昂贵。没有政府的补贴，投资回报大概是15到16年，如果你是政府或者机构，那么非常好，但如果你是一个私人开发商，事情就不一样了。

是否太阳能光伏板有20年的使用寿命？

SC: 在第一个20年内，大多数制造商会保证80%的产品没有问题。在那之后，随着时间的推移，品质就会开始下降。

太阳能光伏板是否曾被嵌入到幕墙中？

SC: 是的，但问题是，你是否在花最少的钱买最多的东西，这就难说了。这是位置问题。如果面南或面西，太阳能光伏板幕墙的效果最好，但即使是那时，它们也只有大概65%的功效。最优的太阳能光伏板位置是安装在你的太阳方位角处。

一个普通的光伏板，安置在最佳的位置，可以在每平方英尺获得11瓦特的能量（118瓦特每平方米），也有一些可以产生16到17瓦特（172-183瓦特每平方米），但费用要高出很多。

你能够用新的替代品换掉旧的光伏板吗？

SC: 是的，那就像换一个拱肩板一样。许多制造商可以定制尺寸，但要花不少钱。而且，太阳能光伏板在垂直平面上只有65%的功效，所以，问题是你想要如何花钱。

我们在迈阿密有一栋1000英尺高的可持续性建筑，称为迈阿密摩天楼。它基本上是一个观光胜地——观景台、酒店/酒吧、活力展览、礼品商店——但是在底部和顶部之间，有一个发电机，如何对它进行优化呢？在南边，我们向外弯曲一点点，以便给予太阳能光伏板一些最好的阳光，然后我们沿着垂直方向，开始慢慢让水溢下来。从东向西，楼房装满了风力涡轮机，你可以看到它们。没有办公室，噪音就不是一个问题。所以在能源生产方面，风力供应30%而太阳能供应70%。

MM: 你不能白天时将额外能量储存起来，到晚上再使用吗？

SC: 可以，但是储存在电池中，而这会变得……

MM: 变得很贵。

SC: 它损害了性能。很昂贵，而且占据的空间非常重。实际上我们从纽约经济发展公司得到了一个项目，与广场建设公司、密切尔/朱尔戈拉建筑师事务所一起在布鲁克林陆军码头的屋顶上安装太阳

SC: If an owner is concerned about flexibility and moves everything around every two years, I wouldn't recommend one of these systems. That's why we always go back to the owner's requirements. A building is not sustainable unless it does what the owner wants it to do, energy efficiency, indoor air quality, and everything else aside. A building has to meet its proper functionality to be sustainable. It's got to do what it has to do for the long term. To me, that is the definition of sustainability.

What is acceptable cost for sustainability?

SC: If you're building a really good, well-thought-out, well-designed building, with good robust systems, the cost differential is very low.

You spend less by planning properly from the beginning. To get a basic LEED certification on a new building, it's only a 1 or 2 percent increase; some maintain that if you plan it right, there's basically no increase. By contrast, for an inexpensive spec building there might be a 10 to 15 percent cost increase. For net zero, it could be a 60 to 70 percent cost increase, depending on what you're building.

Photovoltaics are very expensive. Without government subsidies, the payback is 15 or 16 years, which is good if you're government or institutional, but not if you're a private developer.

Do photovoltaics have a 20-year life span?

SC: Most manufacturers will guarantee 80 percent of production within the first 20 years. After that, the capacity starts to degrade over time.

Are PVs ever integral to the curtain wall?

SC: Yes, but the question is whether you're getting the biggest bang for your buck. It's positioning. PV curtain walls work best if they face south or west, and even then they're only about 65 percent efficient. Optimal photovoltaic installation is at your solar azimuth angle.

An average panel, optimally installed, can get 11 watts per square foot [118 watts per square meter], and there are some that can produce 16 to 17 watts per square foot [172–183 watts per square meter], but they cost a lot more.

Can you change old panels with new replacements?

SC: Yes. It would be like changing a spandrel panel. A lot of the manufacturers will custom-make sizes, but it costs a lot. And again, PVs are only 65 percent efficient on a vertical surface, so it's a question of how you want to spend your money.

We have a sustainable 1,000-foot-tall building [305 meters] in Miami called Sky High Miami. It's basically a tourist attraction—observation deck, restaurant/lounge, energy exhibition, gift shop—but between the bottom and top it's an electricity generator. How do we optimize this? On the south side we angle out a bit to give the PVs some optimal solar, and then we go vertical and start slowly sloping down. From the east-west direction, the tower is filled with wind turbines; you can look right through. Without offices, noise isn't an issue. So wind supplies about 30 percent and solar about 70 percent of the energy generated.

MM: You can't store excess energy during the day to use at night?

SC: You can, but in a battery, and that becomes—

MM: Expensive.

SC: It blows the performance. It's expensive and it also takes space and has extremely high weights. We were actually awarded a project from the New York Economic Development Corporation, working in conjunction with Plaza Construction and Mitchell/Giurgola Architects to install PVs on the roof of the Brooklyn Army Terminal and tie in with a battery system to balance production and consumption. When does the energy go back to the building? When does it go to the battery? When is it released to the grid? This is a pilot project that will test the software. If it works out, there could be micro generating stations all over the city that would feed power into the grid and reduce strain on the infrastructure.

MM: The question is who is going to operate it?

SC: Well, the nice thing about photovoltaics is that there's nothing to operate.

Is water-heated energy passé?

SC: No. Heating water is very good, You just have to find a way to use it. There's the old black-box water heater that captures the sun. The other technology now is evacuated tube solar collectors: glass tubes about 1.5 inches [3.8 centimeters] in diameter with a copper pipe and black-painted fins inside; these are extremely efficient. The technology can be made even more efficient with parabolic reflector mirrors, by which you can produce steam and use it to power a turbine.

Another feature is that they're not as directionally critical as photovoltaics. For PVs you always want direct sunlight,

能光伏板，同时配备一套电池系统以便在生产和消费之间做出平衡。能量何时回到建筑？它何时储存到电池中？又何时释放到网络中？这是一个试点项目，将会用软件进行测试。如果它正常工作，城市将会遍布一些小型的发电站，它们会向电网输送电力，减少基础设施的负担。

MM: 问题是谁来操控它？

SC: 太阳能光伏板的一个好处是，没有什么可操控的。

是否水暖式能源已经过时了？

SC: 不，水暖是非常好的。你只需要找出一种方法来利用它，正是古老的黑箱热水器捕获了阳光。当下其他的技术是真空管太阳能收集器；直径大约1.5英寸的玻璃管，内部是铜管以及漆成黑色的散热片，这些是极为有效的。如果加上抛物面反射镜，这种技术的效率会更高，你可以用它产生蒸汽，给涡轮机供电。

另一个特点是，和太阳能光伏板不同，方向对于它们并非至关重要。太阳能光伏板，总是需要直接的阳光照射，而这些真空管收集器可以在间接照射的情况下工作，可以反射阳光。它实际上比光伏板要更有效率，只是需要占据更大的空间，再说一次，当你有热水的时候，在夏季生产高峰期，如何处理它们就是一个问题。

我在之前提到的巴黎项目的整个南面表层上，使用了这种太阳能集热器。夏天，我们用它驱动一台吸收式冷冻机，以满足大楼制冷的需要。冬天，我们将水直接引入大楼的加热系统。一台6000加仑容量的储油罐将帮助我们度过没有阳光的日子。

这种系统的三维结构有没有提出特殊的建筑上的挑战？

SC: 我们做了一个实物模型来研究大楼南侧，看看它将如何工作以及是否可以识破它，这看起来相当有趣，建筑师和业主都很喜欢。我们将管向外弯曲一点，所以可以获得一些空间作为观景廊，同时你也有很宽的地方来安置它们，以便可以看穿它们。

有些公司正努力开发生物燃料，比如基于藻类的燃油——这是其中很大一部分。他们在水池中养殖藻类，然后压缩藻类以获取燃油，同时利用这些燃油来制作生物燃料。这真是非常棒的东西。有许多能源将会变成矿物燃料的替代物，但我们还没走到那一步。它潜能无限，但还是那句话，尚未实现。也有许多工作致力于更大地提升太阳能光伏板的效能。但最终还是回到排放问题上，例如，在美国并没有一股主流的推动力来发展混合动力汽车或电力汽车。大多数研究来自于亚洲和欧洲。

地热怎么样？

MM: 地热是我们已经用了很多年的东西。

SC: 有两种不同类型的地热。地热交换是你把地球当作一个巨大的热能资源或热能保险箱。还有一种地热，是你给地球的热量装上水龙头，用它来产生蒸汽。泰特拉技术公司实际上有几个这样的项目。随着我们不断前行，将会有更多这一领域的研究。

对于一个门外汉来说，似乎你们需要做的就是挖一个很深的洞。

MM: 如果你有足够的资金……

SC: 有许多地方可能你只需要挖到地下800英尺的地方。但为了满足大规模生产所需的数量，你需要一大片这样的地方。这项技术看起来只在像内华达州、犹他州和冰岛这样的地方才有效，而在空间紧迫的大都会地区是行不通的。

你是否曾参与生物废料转换工程？

SC: 我们实际上在一些工程项目中已经关注这一问题了，但是就生物废料产生能源以替代化石燃料而言，你需要足够的供应量。

超高层建筑生产的废料足够吗？

MM: 最大的问题是找到一个心甘情愿的客户。因为这是一种新技术，可能会出错，而且在这样一个充满诉讼的时代，我们必须非常小心。

SC: 而且，你总是需要一个很好的操作者。

可持续性对于玻璃表层有什么影响？

SC: 每个人都想要更多的玻璃，但是从能源的角度看，它并不总是好东西。即使对日光采集来说，也是玻璃上面的三分之一部分效率最高，而非下面的三分之二部分。玻璃的很多问题是视觉和美学问题。有一些新的建筑准则试图劝阻玻璃的使用。

对于炮台公园（BPC）的河流大厦来说，他们想要在西侧表层上安装落地玻璃窗。炮台公园的要

whereas these evacuator tube collectors can work with indirect, bounced sunlight. It's actually more efficient than photovoltaics but takes up more space, and again, when you have hot water it's a question of what to do with it in peak summer production.

We used these solar collectors on the whole south facade of the Paris project I mentioned earlier. In the summer, we use this to operate an absorption chiller to satisfy building cooling requirements. In the winter we feed this water directly into the building heating system. A 6,000-gallon [22,700-liter] storage tank will assist in riding through periods without sun.

Does the system's three-dimensionality pose particular architectural challenges?

SC: We did a mockup to study the building's south side to see how it would work and if you could see through. It was pretty interesting-looking. The architect and owner really liked it. We angled the tubes out so you could still get some view corridor and also spaced them widely enough to see through them.

Several companies are working hard to develop bio-diesel, such as algae-based fuel oil. That's the big one. They grow algae in tanks and then press the algae to take out the oils and use that to make bio-fuel. It's pretty cool stuff. There is a lot of energy going into fossil fuel alternatives but we're not there yet. It's got a lot of potential, but again, it's not there yet. There's a lot of work going into greater efficiencies for photovoltaics. But getting back to emissions, for instance, there isn't a mainstream push in the U.S. to develop hybrid or electric cars. Most of that research is coming out of Asia and Europe.

What about geothermal?

MM: Geothermal is something we've been using for years.

SC: There are two different types of geothermal. Geothermal exchange is where you use the Earth as a big heat source or heat safe. There is also geothermal where you tap the heat of the Earth and use it to generate steam. TetraTech actually has a couple of these projects. There will be more research in this area as we move ahead.

To a layperson, it seems that all you need to do is dig a deep hole.

MM: If you've got enough property…

SC: There are lots of places where you have to go down maybe only 800 feet [244 meters]. But to get the quantities you need for major production, you'd need fields of these. It's a viable technology that seems to work in places like Nevada, Utah, Iceland, but not in a space-starved metropolitan area.

Have you been involved in human waste conversion projects?

SC: We've actually looked at it for a few projects, but for bio-waste to generate power, instead of burning fossil fuel, you need adequate supply.

Would a super-tall suffice?

MM: The biggest problem is finding a willing client. And since it's a new technology, mistakes can be made, and in this litigious era, well, we have to be very careful.

SC: And, as always, you'd need a good operator.

What effect has sustainability had on glass facades?

SC: Everybody wants more glass, but from an energy perspective it's not always a good thing. Even for daylighting, it's the upper third of the glass that's most effective for daylighting, not the lower two-thirds. A lot of it is visual and aesthetic. Some of the new building codes try to discourage it.

For River House in Battery Park City [BPC], they wanted floor-to-ceiling glass on the west facade. BPC requirements specified punched windows not to exceed 40 percent of the facade. So they installed a very energy-efficient double facade that yields the same thermal criteria. They felt it was worth the money. It didn't have a payback in energy savings but in marketing floor-to-ceiling glass facing the river, they made the money back right away. Sometimes other issues come into play.

We got stuck on energy efficiency but sustainability is about producing a better built environment than you have. It might not be net zero, but it will be better than the standard office building. If you can do 30 to 40 percent better than what's out there today, that's a big jump in energy efficiency and if at the same time you can make a better space to be in, a better environment, you're that much further ahead.

求特别指出打孔窗口不能超过表层面积的40%。所以他们安装了能够达到同样热量标准的节能双面表层。他们认为这物有所值，虽然在节能方面并无回报，但通过销售面对河流、带落地窗的楼层，他们立刻就把钱赚回来了。有时候其他因素也会掺入其中，发挥作用。

我们卡在了能源效率问题上，但可持续性可以产生一种比你现在所拥有的建筑环境更好的建筑环境。它可能不是零损耗的，但却比标准的办公建筑要好。如果今天可以做的比外面的建筑好30%到40%，这就是在能源效率上的一个大跃进，如果同时你可以给出一个更好的空间、更好的环境，你就已经领先很多了。

Riverhouse, Rockefeller Park in Battery Park City, New York. (Ennead Architects, 2009)
纽约炮台公园市的沿河建筑和洛克菲勒公园。（恩尼德建筑事务所，2009年）

Carl Galioto

HOK

Carl Galioto, FAIA, is a senior principal at HOK, managing principal of the New York office, and a member of HOK's board of directors. He has over 30 years of industry-leading expertise in the design and implementation of large, complex projects, including such high-rise and super-high-rises as 1 World Trade Center, 7 World Trade Center, and 7 Times Square in New York; Lotte Tower in Seoul; and in China, Evergrande Tower in Zhengzhou and Greenland Tower in Dalian. His experience includes numerous aviation and health sciences projects. Using building science to better design aesthetically, functionally, and from the standpoints of sustainability and life safety, Mr. Galioto was influential in changing the New York City Building Code after 9/11. He helped analyze the NIST investigation of the WTC collapse and translated those findings into model building codes. To innovate sustainable solutions, he co-founded the Center for Architecture, Science and Ecology at Rensselaer Polytechnic Institute. At HOK he is responsible for firm-wide project delivery, including HOK's virtual design and construction initiative: building SMART.

卡尔·卡略托

霍克国际有限公司

卡尔·卡略托是霍克国际有限公司的高级负责人，纽约办公室的主管人员，也是霍克董事会的成员。他有着超过30年且在行业领先的设计经验，并有执行大型复杂项目的专业知识，这其中包括高层和超高层建筑，比如世贸中心一号楼、世贸中心七号楼、时代广场7号楼、首尔的乐天大厦，以及中国郑州的恒大大厦、中国大连的绿地大厦。他的经验还包括各种航空和健康科学项目。卡略托先生利用建筑科学，基于可持续性和生命安全的立场，从建筑设计美学上、功能上改善设计，在911事件之后，他在修改纽约市建筑规范方面很有影响力。他帮助分析国家标准暨技术学会对世贸中心倒塌的调查报告，并将这些发现转化为标准的建筑规范。为了创造出可持续性的解决方案，他在伦斯勒理工学院与人协同创立了建筑、科学和生态学研究中心。在霍克国际，他为全公司的项目交付负责，这其中包含了霍克国际在设计和建设方面的创造性精神——聪明地建设。

For complete biography, please refer to the Appendix.

完整的传记，请参阅附录。

One World Trade Center, New York. (SOM/Carl Galioto, est. completion 2015)
纽约世贸中心一号楼。（斯基德莫尔、奥因斯与梅里尔建筑师事务所/卡尔·卡略托，2015年完工）

Are tall buildings inherently unsafe?

No, they are not. Developments over the last 10 years have been substantially improving the safety of tall, and particularly super-tall, buildings. After 9/11, I was involved in the design of 7 World Trade Center (WTC) and also in developing changes to the building code, looking at the issues objectively rather than emotionally. Lots of good ideas emerged, some less so. One night after I'd been on a cable TV show about safety, someone from Dayton, Ohio, telephoned to suggest suspending chains in front of buildings so that aircraft would get tangled up in them. He was calling on my home number at 10:00 p.m. I was polite and thanked him for his thoughts—it was certainly something I had never thought of!—and said we'd look into it. There were lots of harebrained schemes for escape systems and things like that, but fundamentally, safety can be substantially improved by making existing building systems more redundant and robust.

Take fire suppression systems, for example. There has never been a serious high-rise fire in a commercial office building with a functional sprinkler system. Looking back on significant high-rise fires in South America, Asia, or the United States, in Philadelphia, Los Angeles, or at the Bank of America in downtown Manhattan, none of the buildings had functional sprinkler systems. And at Meridian Place in Philadelphia, the fire spread vertically until it reached a floor where the sprinkler system had been installed, and it put the fire out. That always indicated to me that these systems actually do work, and if they can be made more reliable or more robust, then that's a sensible approach to improving safety.

Also, one needs to look at other elements with a more performance-based approach, such as egress systems. Instead of just relying on some generic informational code, actually looking at how people walk, how they exit stairs, examining the human factors already researched and how they could be applied, and then looking at actual emergency response. I don't know many architects who did this, but I spent a fair amount of time reading fire service manuals and interviewing fire chiefs about how firefighters approach and set up a site, how equipment is brought in, how it's moved vertically, how forces are deployed, and also how the fire chiefs would like to do it. Those lessons were applied to 7 WTC and elsewhere. In some ways it's rather simple.

The manuals discuss bringing in a team to investigate a fire, the number of people involved, how equipment is brought up, how to maintain a safe environment for the firefighters, how to attack a fire, how to retreat. That helped me think about how to lay out a floor, and how to position

高层建筑会不会不安全?

不, 不会。经过过去十年的发展, 高层建筑, 尤其是那些超高层建筑的安全性已经大幅提升。在911之后, 我参与设计了世贸中心 (WTC) 七号楼, 并修正建筑规则, 看待问题更加客观而不是感性。许多好创意接踵而至。一天晚上, 我参加完有线电视一期关于安全的节目后, 有个俄亥俄州代顿市的人给我打电话, 建议在高楼前采用悬挂链条, 将飞行器缠绕在链条中。他晚上十点给我家里打的电话, 我很客气地对他表示感谢, 并说我们将认真考虑他的创意——这确实是我从来没想到过的主意! 就逃生系统以及类似的东西而言, 确实有太多无效的方案, 但从根本上看, 现有的建筑系统变得更周全更牢靠, 可以让安全性得到实质性的提升。

以灭火系统为例, 那些配有喷洒系统、功能健全的商业办公大楼, 从未发生过严重的高层火灾。回头看看发生在南美、亚洲或美国 (费城、洛杉矶或者曼哈顿闹市区的美国银行) 的几起重大高层火灾, 起火的建筑物都没有这种喷洒系统。在费城梅第安酒店, 火势蔓延直上, 直到装有喷洒系统的楼层中灭火装置开启才将大火扑灭。这些案例表明, 喷洒系统确实有效, 如果它们能够被改造得更可靠、更实用, 那么, 它就是提高安全性最直接的一个方法。

另外, 人们应该从更注重行为的角度, 去审视其他的一些要素, 例如逃生系统。它不是仅仅依赖一些普通的资讯规则, 而且要实际地去了解人们如何行走, 如何走出楼梯, 查看已得到研究的人体参数以及如何利用参数, 之后还要查看实际发生紧急情况时的反应效果。我不太认识会这样做的建筑师, 但我花了很多时间阅读消防手册, 并采访消防队长, 问消防员是如何着手工作、设立工作站、带入设备、垂直运送, 以及消防力量如何部署、消防队长怎样展开活动。这些课程被应用于世贸中心的七号楼和其他一些地方。某种程度上说, 它其实很简单。

消防手册详细说明: 要带一个队伍探查火情, 了解被困人数, 要知道如何提供装备, 如何为消防员准备安全的工作环境, 如何与火情搏斗, 如何撤退。这些有助于我思考怎样设计一个楼层, 怎样布置电梯和楼梯, 更符合楼层安全, 而不是令情况更加糟糕, 迫使消防员必须竭尽全力。

是高度本身使问题复杂化? 还是在高层建筑中问题会有所差别?

危险因素随着高度而增加。一栋75英尺（23米）高的楼房，由于其梯状延伸而被归为高层建筑，但它与2000英尺（610米）的高层建筑有着巨大差异。汽车的安全设计引人关注，但飞行器的安全设计引起的关注就更大了。为什么呢? 因为内在的风险性很大，我对超高层建筑有着同样的感觉。因为高度与其潜在的风险，超高建筑对系统的检测以及设计必须达到的性能，显然要比一般建筑的要求更高。

特别是哪些系统呢?

所有系统，每个系统都是如此。例如，与结构系统联系在一起的风险是什么? 由火灾引起的连续坍塌、结构损毁，这些是首要的问题，而且，根据建筑的类型和位置，还有恐怖袭击可能带来的破坏。你看着这个系统，然后说，不错，这就是我们需要用来设计的标准。例如，我们设计了不会失效的防火系统，但在实际情形中，一旦它真的失效了，我们如何设计出可以抵制最终焚毁烧尽的结构系统? 比如，一场大火烧毁了空间中的所有可燃物，最终由于没有东西可烧而燃尽了自己吗? 如何设计出一套不完全依赖于防火系统的结构体系? 通过增加体系的周密性和牢靠性，建筑可以获得独立，成为不完全依赖于其他系统的体系。

这些想法是只适于标志性建筑? 还是它们有着广泛的应用范围?

这些概念中有许多已经进入建筑规范的主流思想，它对整体的安全性确实是有积极意义的。

新规范中有没有包含消防员电梯这一问题?

没有，尽管消防员电梯在许多地方使用，但美国并没有。世贸中心一号楼的消防服务设施，是以欧洲研究的消防员电梯为基础而设计的，这意味着有一台服务电梯（实际上是一部摇摆车）可以正常操作，而当紧急情况发生时，也可被用来救火。现在，如果纽约市发生火灾，所有的电梯都会落到地面上。然后，消防部门就会接管电梯，并依据他们所观察的情况做出关键举措。即使只是一部摇摆车，服务电梯也有着标准入口，有自己的升降机

elevators and stairs in a way that would facilitate safety as opposed to creating compromised situations that firefighters have been forced to deal with as best they could.

Does height simply compound the problem or is the problem different in tall buildings?

The element of risk is increased by height. A 75-foot-high [23-meter] building is classified as high-rise because of ladder extension, but there is a vast difference between this and a 2,000-foot-tall [610-meter] building. Great care goes into automotive safety design; even greater care goes into the safety design of aircraft. Why? Because the inherent risk is greater. I feel the same way about super-tall buildings. Because of their great height and potential risk, the examination of the systems and the performance for which they must be designed is much greater in super-talls than in other buildings.

Which systems in particular?

All of them, every single system. For example, what are the risks associated with the structural system? Progressive collapse, structural failure, damage due to fire, those are the primary issues and, depending on which building and where it is, potential damage from terrorist attack. You look at the system and say OK, here are the criteria we need to design for. For instance, we design the fire suppression system not to fail, but in the event that it did fail, how could we design the structural system to withstand burnout, i.e., a fire that consumes all of the combustible content of a space and then burns itself out because there's nothing left to burn? How to design the structural system so that it does not rely completely on the fire suppression system for safety? By increasing the redundancy and the robustness of the systems, one can achieve independence and not overly rely on other systems.

Does this thinking apply just to landmarks or is there broader application?

Many of these concepts have found their way into mainstream thinking on building codes, and that's a really positive thing for overall safety.

Have firefighter lifts been incorporated into the new code?

No, not in the United States although we're using them in various other locations. Fire service access at One World Trade

190

Center is based on research applied in Europe for firefighter lifts, which means that one has a service elevator that is a swing car that can be used for normal operation and, alternately, for emergency operations by fire service. Now, in the event of a fire in New York City, all elevators go back down to grade. And then the fire department takes over the elevators on a key operation depending upon their observation of the situation, but they're standard elevators with standard hoistways. The objective of a fire service lift, even in this kind of swing condition, is that it has its own dedicated shaft. Its equipment is protected because it's water resistant and the hoistway is pressurized against smoke migration from other shafts. It has a number of features above and beyond.

Is it possible to pressurize that hoistway in an ultra-high-rise building?

That's an excellent question. It requires a very substantial effort but it is possible. Additional intake has to be introduced, you just can't take it from the top; it would be like trying to blow down a 6-foot-long straw [1. 83 meters].

Even if all these new ideas are implemented, there's no guarantee against collapse from a fully loaded aircraft.

I participated in an analysis of exactly that scenario. This kind of impact is not normal; it's an astounding force. But short of that, there are many criteria a building can be designed for.

Does the new building code require a third set of stairs?

That was added to the international code in 2009, although I questioned it. Two is a certain level of redundancy, and three is even more redundancy, so I said facetiously, why stop there? Make it five! The point is to have redundancy in the egress system, not in the number of stairs. Yes, the number of stairs helps, but the international code really doesn't properly address the size of the floor plate, for example. If the building has a spire or small floor plates at the top, do you really need three stairways up there? It depends on the occupancy, but generally, no. So my approach is to look at the stairs themselves, in terms of real widths of people, not just theoretical numbers. They should be at least 66 inches [168 centimeters] wide in tall buildings. Why? Because that is the width of two people walking side by side, allowing for sideways movement back and forth, and also for the fact that people don't walk pressed against the handrail; they maintain a distance on both sides. Also, there may be a situation where

井，因为是防水的，它的设备受到保护。而且电梯入口密封，这样其他机井的烟雾不会浸入其中。除此之外，它还有许多其他的特点。

超高层建筑中的电梯入口是否可以密封？

这是一个很棒的问题。密封需要很多努力，但确实是可能的。必须引入额外的通风口，不可能在顶部通风；这就像是要吹倒一根6英尺长的稻草。

即使所有这些新观念付诸实施，当一架满负荷的飞机撞击大楼时，还是无法保证不发生坍塌。

我参与过对这种特定场景的分析。这种冲击并不正常，其力度令人震惊。但除了此种特殊情况之外，建筑的设计是可以有许多标准的。

新的建筑规范是否要求第三组楼梯？

2009年，这一点就被加入国际规范之中，尽管我对其提出了质疑。两组楼梯在某种程度上已是多余，三组楼梯就更是如此，因此，我开玩笑地说，为什么止步于此？做五组好了！问题不在于楼梯的数目，而在于它让整个疏散系统变得冗余。楼梯数量是能起到一定作用，但国际规范却并未恰当地给出诸如楼板尺寸这样的标准。倘若建筑有一个尖顶，或在顶部有一个很小的楼板，那么，你真的需要三个楼梯通到那里吗？当然这视建筑的容纳量而定，但一般来说并不需要。所以，我的方法是观察楼梯自身，依据现实中人们所需的宽度，而不是理论上的数字。在高层建筑中，楼梯至少需要66英寸（168厘米）宽。为什么？因为这是两个人并排行走时，横向移动的空间——而且实际上人们在行走时并不握住扶手；他们往往与两边都保持着一定的距离。另外，可能还有一种情形，即：两个人正在下楼，而另外一个人，可能是消防员，正爬上楼梯。所以需要更多的宽度，更大的落脚面积，并将门安置妥当，使人们融入交通流，就像高速公路系统一般。我们为世贸中心一号楼进行此项研究，发现只要移动门，就可以减少几分钟的疏散时间。你所要做的只是移动门，这就是那些非常简单却可以提升原设计系统的可靠方法。

当发生高层火灾时，标准的步骤是否依然是从火源处逃离，往上一到两层，然后呆在那里不动？

根本上说是这样的，就地保命乃是最明智的方法。

难道大多数人不会倾向于从建筑中疏散？

是的，在后911时代确实如此，人们的倾向是疏散。

拓宽楼梯是否有助于疏散？

拓宽楼梯对疏散或任何一种撤离方式都是有益的。基于一般的规范，一个普通的办公楼层可能需要五分钟来疏散人员。这其中有诸多原由决定你为什么想要加速疏散进程，因为会发生一些情况，需要采纳整体的疏散方案。

反对用楼梯变宽来容纳更多人，理由是什么？

人们总是要进到楼梯中去，所以，拓宽楼梯底部并不能提高楼梯的流量，它只是快速地使人进入到楼梯系统中去。我在世贸中心楼层底部采用的是一种重复结构，因而两个楼梯形成一个交叉区域，变成了四个楼梯，由此增加了流通量。假如在地面上发生了一起事故，建筑的一侧或另一侧的地面入口发生危机，那么总有备用的可以应急。

那么避难层呢？

这是另一个非常好的问题。避难层主要在中国香港和中东地区采用，除了每隔15层这种情况，它是非常有效的就地保命策略。不过我认为，以其所需的大量建筑容积来看，避难层的价值是有待商榷的。

我在科威特参加了一项超高塔的工程建设，介绍了电梯作为疏散构件的观点：它将空中大堂与紧接其下的避难层或"过渡避难区"结合在一起。目的在于，将人们沿着楼梯带到一个区域中，在那里，他们可以继续沿着楼梯往下走或者进入防火区，也可以乘电梯往下走。从传统的观点看，将人群疏散出大楼并不用电梯，但是人们需要做出新的考虑。在这一特殊情形中，电梯成了高速运行的列车，它在混凝土结构中，完全是

two people are going down and another person, maybe a firefighter, is coming up the stairs. So greater width, greater landing size, and positioning the door to look at and thereby facilitate merging with traffic flow, like a highway system. We did studies on this for One WTC and found that just shifting the door would reduce evacuation time by several minutes. All you have to do is move the door. Those are the kinds of very simple steps that can improve the reliability of previously designed systems.

In a high-rise fire, is it still standard procedure to go up one or two floors from the fire source and then stay put?

Primarily, yes, defend-in-place is the most sensible approach.

Wouldn't most people be inclined to evacuate the building?

Yes, certainly for a period of time after 9/11, the human tendency will be to evacuate.

Would widening the stairs facilitate evacuation?

Widening the stairs would facilitate evacuation or any exiting. Based on general codes, a typical office floor might take about five minutes to evacuate. There's every reason why you may want to speed that up, and also because there are going to be circumstances when a total evacuation is advisable.

What is the argument against stairs becoming progressively wider to accommodate more and more people?

People are entering the stair all the way up, so widening the stair at the bottom doesn't improve the flow because it is only so quickly that they can enter the stair system. What I used at the WTC is a redundancy at the base of the building so that two stairways come to a crossover area and then become four stairways for increased exiting. If there is an incident on the ground and the exit to grade is compromised in one face of the building or another, there is always an alternative to use.

What about refuge floors?

That's another excellent question. Refuge floors, which are employed in Hong Kong, China, and also in the Middle East, are the alternative to defend-in-place. Effectively, it's kind of a defend-in-place strategy, except every 15 floors. I think the value of refuge floors is questionable relative to the great deal of building volume they require.

I worked on a super-tall tower in Kuwait where we introduced the idea of elevators as evacuation components, combining a sky lobby with a refuge floor, or "transitional refuge area," immediately below. The objective was to bring people down the stairs to a point where they could either continue down the stairs or enter a fire-rated holding area, and from there take those elevators down. Traditionally, one doesn't use elevators to evacuate a building, but one needs to consider new developments. In this particular case, the elevators are high-speed shuttles so they don't open to the office floors. It's a completely sealed tube in a concrete structure. So, if there were ever a fire below, people could get on these elevators and safely bypass it because there would be no direct contact with the fire floor.

This system was accepted by both the code officials and the fire service in Kuwait. The building is now complete and has trained personnel, which is really important in these kinds of buildings.

How have recent technological advances changed the design process?

I believe, at every level, that integration and transdisciplinary—not interdisciplinary—design is essential. Engineers need to think like architects and architects need to think like engineers and be knowledgeable about structural engineering decisions for super-tall buildings. We need to look at how buildings perform, not just egress and safety, but from energy performance, from fundamental performance of the buildings in terms of building movement, from all aspects. Everyone needs to be thinking with their antennae up in a completely transdisciplinary way.

It was much more siloed in the past. Architects would say, "I have an idea for a building, let me talk to the structural engineer and see how we're going to hold it up." Or, "We need a mechanical system, where am I going to put it?" rather than thinking about really integrating the systems in the building.

Have new technologies made this possible?

Absolutely. Something that I'm very involved in is building-information modeling (BIM). In addition to building information as a product, BIM is a process of design and a platform for collaborative transdisciplinary thinking. So here at HOK, architects engage in design analysis using various kinds of applications around the models they build to study wind flows or exposures. One fellow just came to me in the lobby

一个密封管，不向办公楼层敞开。由于不与着火楼层直接接触，如果下面发生大火，人们可以进到这些电梯，安全地逃过大火。

这一系统被科威特的规范制定者和消防部门所接受。上述大楼已经完工，并配备了训练有素的操作人员，这一点对于此类建筑来说非常重要。

当下的技术进步如何改变设计流程？

我相信，从各个层面上看，整合和跨学科的设计是根本，而不是学科间的互动。工程师需要像建筑师一样思考，建筑师同样需要像工程师一样思考，要拥有对高层建筑结构工程的相关知识。我们需要观察建筑物是如何发挥功能的：不仅要考虑出口和安全性，而且要考虑能耗情况，依照建筑的运行情况，考虑建筑物的基本性能，以及相关的方方面面。每个人都需要以完全跨学科的方式，从直觉出发进行思考。

过去要比今天封闭得多。建筑师们会说："我有个关于建筑的想法，让我和结构工程师谈一下，看看我们如何去实现这个想法。"或者"我们需要一个机械系统，我要将它放在哪里呢？"而不会真正想着将系统融合到建筑中去。

新技术是否可能实现这一点？

绝对可以。最近我在研究建筑信息模型。它除了可以提供建筑信息，还是一个设计流程，一个跨学科思想协同工作的平台。因而在HOK公司，建筑师们通过各种程序参与设计分析，这些程序围绕他们所建构的风流、风向模型。有个同事从大厅径直向我走来，想要将一个垂直的散热片放在建筑上以遮挡阳光。我说从建筑的结构来看，垂直散热片不会起到任何作用，不过我也说："别全信我的，利用这些程序，看看你能得到什么，一会儿试试相同的横向散热片。"他回到座位，运行应用程序，然后发现垂直的散热片无论如何对限制温度都不起什么作用。因此很快，人们就能提出设计的真正基础。我们的目标不是取代工程师，而是和他们进行更有效的对话。

从什么时候开始，你参照了工程实践手册（MEP）？

开始工作后不久。有许多专门领域，也有跨学科领域，比如，我们可以看着一些透明的东西问道，怎样去处理玻璃？然后去实验各种类型的玻璃。当我们认为有一个很好的解决方案时，就会引入工程实践手册，并不是非要等到完成了对太阳能收集、分配、采光效果的全部研究后才能开展工作，我们一直尽全力地开展对上述步骤的优化工作。你可以在我们的办公室看到，即使在这些上世纪60年代的古老建筑，都引入了充分的日光以减少40%的光照耗能。它还可遮蔽阳光，宛如调光器和灯光控制器。在这一事例中，我们并没有对建筑的外层进行操作，但这些确实是我们在设计研究时分析出来的。

有了合适的知识和工具，设计团队可以变得多小？

可以非常小。有了所有这些专门知识和跨学科方法，我们可以造出更好的建筑。

小团队是否意味着成本下降？

实际上并非如此。我们可能人数不多，但要花费更多的时间来做这些研究。而且，在基础设施、计算机和前沿软件上的投资成本是巨大的。在这些人和物的资源中，我们需要找到最有前景的部分，它不仅仅是富有创造力的个人，而是在设计建筑时，能够运用参数化和性能为向导方法的个体。

你现在在做什么工作？

我们最近刚刚开始在中国设计一栋高达560米（1837英尺）的大楼。这是一栋可持续的、绿色的巨大建筑，可持续、绿色正是我们核心价值的一部分。

建筑在某一时间之后是否不再具有可持续性？

不是，但人们需要看看对可持续性的定义。纽约有许多750英尺（229米）高的建筑，这是有很多原因的，房地产是其中的一个。人们喜欢乘坐电梯直通自己的楼层，而且，从容量的角度看，高于大厅之上750英尺（229米），电梯就耗尽燃料了。另一个重要的原因是空气阻力。一栋高楼，其钢结构的重量大概是每平方英尺20磅（9千克），但是当楼高达到800英尺左右（244米），重量就迅速上升到35磅（16千克）每平方英尺甚至更多。因为高度增加风阻也增加了，因而需要整个建筑始终应对这个问题以控制移动，使人难以察觉。

and was looking to put vertical fins on a building for sun shading. I said the vertical fins wouldn't do anything based on the building configuration but said, "Don't take my word for it, use such and such program, see what you get, and then try the same fins horizontally." He went back to his desk, ran the application, and found the vertical fins pretty much useless for limiting heat gain in any meaningful way. So very quickly, people can come up with a real basis for design. Our objective is not to replace engineers but to engage them in a much more informed dialogue.

At what point do you bring in MEP?

Soon after. There is a great deal of specialization, again transdisciplinary, so for example, we might look at some transparent form and ask, What can we do with glass? Then we'll try out different types of glass. And then, when we think we have a pretty good solution, we'll bring in the MEP, but not until we've done our studies on solar heat gain, glare, and daylighting effects, and we've tried to optimize those as much as we can. You can see in our office, even in this old 1960s building, we've introduced enough daylight to reduce the energy dedicated to lighting by 40 percent. It's shades up, dimmable ballasts, and lighting controls. In this case, we didn't have control over the exterior of the building, but that's exactly what we analyze in doing our design studies.

With the appropriate knowledge and tools, how small can a design team be?

Pretty damn small. And with all the expertise and the transdisciplinary approach, we're producing much better buildings.

Do smaller teams translate into savings?

The truth is no. We may have fewer people, but it takes more time to do all these studies. And it takes tremendous capital investment in infrastructure, computers, and cutting-edge software. Also, we have to find and keep the best and the brightest out there, not just creative individuals but those oriented toward a parametric, performance-driven approach to designing buildings.

What are you working on now?

We started work very recently on a building in China that will be 560 meters [1,837 feet] tall. That's a good-size building—sustainable, naturally; that's part of our core values.

Lotte Super Tower, Seoul, South Korea. (SOM/Carl Galioto; unexecuted design, 2007)
韩国首尔乐天大厦。（斯基德莫尔、奥因斯与梅里尔建筑师事务所/卡尔·卡略托；未被采用的设计，2007年）

捆绑连结可以解决这个问题吗?

完全可以。其实许多种系统都可以提升性能,但需要一定的花费。例如,一种斜肋构架在建筑实践和理论上都很吸引人,但却并不便宜,因为当钢铁的重量每平方公尺(米)呈下降趋势的时候,用于连结的成本则是大幅提高。

承租者能够接受多少个空中大堂?

一个。你可以有空中大堂,但是那些出门上班的人不会忍受超过一个中转站点。这不是文化,是人类的本性就不希望走太多的台阶。我真的认为一个中转站点就足够了,但关于这些中转点,有各种各样可以谈论的有趣特征,像空中大堂和双层电梯。双层电梯常在亚洲和其他一些地区使用。纽约比较急躁,我们不喜欢空中大堂,也不喜欢双层电梯。

是否可能有一种零能耗的建筑?

零能耗建筑是没有问题的。但一座零能耗的高楼则是另一回事。在零消耗方面,水平建筑更有优势,这些优势体现在屋顶面积和楼层面积之比,以及可以采用太阳能集热器之类的东西。

将建筑物的外表面转化为一种能量源泉/太阳能集热器,我们就能实现吗?

为时尚早。我们可以通过建筑的集成光伏获得一定数目的能量,建筑的风能系统也可以产生一些收益。在高楼中,我们总是会遇到此类关于风的问题,如何利用它呢?广州的珠江塔尝试将风转化为能量,虽然所获回报并不丰厚,但这是一个很有益的尝试,他们在正确方向上迈出了重要一步。没有这种尝试,我们将驻足不前,因此,你得为迈出这些脚步的人鼓掌。

客户一般是接受可持续性,还是说费用依旧是一个重要因素?

就建筑物而言,费用当然是一个因素。但随着各种产品和系统的发展,要获得能源与环境设计(LEED)银质认证是很容易的。按照美国绿色建筑委员会官方的说法,升级到LEED金质认证并不需要增加额外费用。但和我交谈的大部分评估者反映,从银质认证升级到金质认证大概增加了3%的

Is there a point above which a building is no longer sustainable?

Well, I say no, but one has to look at the definition of sustainable. New York has a lot of 750-foot-tall [229-meter] buildings for various reasons. One is real estate. Also, people like to take an elevator directly to their floor and, from a capacity standpoint, the elevator system runs out of gas about 750 feet [229 meters] above the lobby. But the other big reason has to do with wind resistance. The weight of structural steel is about 20 pounds [9 kilograms] per square foot for a high-rise building, but once you reach about 800 feet [244 meters], it very quickly goes up to 35 pounds [16 kilograms] per square foot, or even more. With that height comes increased wind pressure, which then needs to be accommodated all the way down the building in order to limit movement below perceptive levels.

Can bundling resolve that?

Absolutely. There are many, many kinds of systems that will improve performance, but at a cost. For example, a diagrid, which is particularly attractive architecturally as well as intellectually, isn't cheap, because while the weight of steel declines per square foot [meter], the cost of the connections greatly increases.

How many sky lobbies will a tenant typically tolerate?

One. You can have sky lobbies, but someone traveling to work is not going to tolerate more than one transition. That's not cultural, it's human nature. It just becomes too many steps. I really think one transition is the maximum, but there are all kinds of interesting features to talk about in those transitions, like sky lobbies and double-deck elevators. Double-deck elevators are frequently used in Asia and elsewhere. New York is grouchy; we don't like sky lobbies, and we don't like double-deck elevators.

Is a net-zero building realizable?

A net-zero building, yes. A net-zero tall building is another matter. Horizontal buildings at this point stand a much better chance because of the ratio of roof area to floor area and introducing solar collectors, that sort of thing.

How close are we to converting the skin of a building into an energy source/solar collector?

Far away. We can recover a limited amount of energy with building-integrated photovoltaics. Building-integrated wind also yields modest gains. We have all these wind issues in tall buildings, so how do we take advantage of it? There was an attempt to direct wind toward energy at Pearl River Tower in Guangzhou. The return was pretty modest, but it was a good attempt and an important step in the right direction. Without such attempts we wouldn't be getting anywhere, so you applaud people for taking those steps.

Do clients typically embrace sustainability or is cost still a significant factor?

Depending on the building, cost is certainly a factor. But with the development of various products and systems, it's very easy to attain LEED Silver certification. The USGBC officially says there is no increase in cost up to LEED Gold. But most estimators I've spoken to say that to get from Silver to Gold might be an extra 3 percent; going to Platinum is considerably more. If a client has a particular aspiration, it's certainly achievable. One stands a better chance of getting to Platinum in a horizontal building, but it's certainly achievable in a tall building. We achieved Platinum in our office, even though this is a non-LEED-certified building. It took some work, but we were able to do it.

What is the most complex curtain wall you have worked on?

There are so many options with curtain walls right now! We could look at double skins, shallow cavity, deep cavity, internally ventilated, externally ventilated, shaded, or unshaded, with external shading devices in metal or terra cotta. I also researched an active curtain wall in which the outer layer actually folded up, like a garage door, and the upper layer served as a sun-shading device with integrated photovoltaic panels. On sunny days, it would be up in its open garage position; in the winter it would be down. We even did a mockup for it. I would say that is about as complicated as you could get. The more mechanical you make a curtain wall, the more expensive it is, and the less likely it is to be implemented on a large scale. And, of course, curtain walls have to be maintained, especially operable ones. This particular system was intended for a tall building, but the cost was prohibitive. I could see something like that implemented on a relatively

费用，而升到铂金认证则要花费更多。如果客户有特殊的期许，这当然是能够做到的。在水平建筑中，要达到铂金认证机会更多一些，不过对高层建筑而言，这也并非难以企及。我们的办公室就已经获得了，虽然这是一栋未经过LEED认证的建筑，虽然实现铂金认证花了一些功夫，但我们还是可以做到。

你设计过的最复杂的幕墙是什么？

现在的幕墙有太多的选择。我们可以找到双层的、浅空腔的、深空腔的、内部通风的、外部通风的、遮蔽的、未遮蔽的、由金属或陶瓷制成的外在遮蔽装置等。我也研究了一种活动幕墙，其外层如同车库门，可以折叠，上层加入集成光伏电池板，可以作为遮阳装置。晴天，它可以收起，像敞开的库门；而冬天，它将被放下。我们甚至做了个实体模型。我想说，这是你能想象的最复杂的构造了。幕墙的制造机械越高，就越贵，越得不到大范围的应用。当然，幕墙应该是被保留的，尤其是那些可操作的幕墙。这种独特的系统被用在高层建筑里，但费用很高。类似的东西被应用在相对低层的建筑上。我很乐于做这种尝试，还有个尝试，是我过去一直研究的一种三层玻璃系统。它已经应用在加拿大北部的气候带，因而算不上革命性产品，但我很想研究一下在纽约使用这套系统的功效。它只是多出一层玻璃，一层空气层，因而比标准的双层玻璃要厚些：半英寸（大致为1.25厘米）的空气层，加上3/8英寸（大致为1厘米）的外层，因而整体厚度为1 5/8英寸（2.25厘米）。

从幕墙到核心区域最深能达到多少？

这些都是受市场左右的。我为纽约证券交易所设计的一座塔，从核心到表面有50英尺（15.24米）厚，但老实说，我不认为这有什么意义。一般来讲，42-46英尺（12.8-14米）就是恰当的距离了。这会让阳光进入工作区、流动空间，以及环绕建筑核心的玻璃墙办公室。办公室最关心的是外部感知，同时依赖于天花板的高度和隔间距离，不要太多自然光。光线发生反射，并非直接照进办公室，但距离还是不能超过50英尺（15.24米）。50英尺只是对一般情况的延伸。

7 World Trade Center, New York (2006) (SOM/Carl Galioto): Curtain wall detail and general view from the southwest
纽约世贸中心七号楼（2006年），幕墙细节和东南部全景图（斯基德莫尔、奥因斯与梅里尔建筑师事务所/卡尔·卡略托）

办公室的位置围绕着建筑核心，这难道不是对过去实践的一种颠覆？

确实如此。在工作区，确实发生了一种文化转变。开放的办公室反映了一个事实，即：高级工作人员并不总是待在办公室，因此那些整天坐在桌子旁的人们获得了阳光。正是人们对健康的认识和生产率的驱使，这种革命性变化才会发生。这可以追溯到上世纪90年代早期和之前在总务管理局 (GSA) 支持下，卡内基梅隆大学所进行的一些非常棒的研究。GSA致力于此项研究，以确立自身的价值。当然，GSA在华盛顿的办公室就依照这种指导方针建立，它确实产生了影响。

low-rise building. I'd love to try it sometime. Something else I'd like to try, which I've studied in the past, is a triple-glazed system. It's used in Canada and in northern climates, so it's nothing revolutionary, but I'd like to study the efficiency for use in New York. It's just an extra layer of glass and extra air space, so it's just a little thicker than standard double glazing, an extra half inch [roughly 1.25 centimeters] of air space plus 3/8 of an inch [roughly 1 centimeter] for the outer layer, so about 1 5/8 inches [4.15 centimeters].

What is the deepest curtain wall–to–core space that can be tolerated?

Well, it's all driven by the marketplace. I worked on a tower for the New York Stock Exchange where the depth was 50 feet [15.24 meters] from core to face, but frankly, I don't think it made any sense. Generally the range of 42–46 feet [12.8–14 meters] is about the right distance. This allows daylight into the work areas, circulation space, and glass-walled offices around the core. The main concern in the offices is outside awareness, not so much natural light, but it depends on ceiling and partition heights as well. Reflectors and such allow indirect light further in, but you still can't go more than 50 feet [15.24 meters], and that's a stretch.

Isn't the location of offices around the core a reversal of past practice?

It is a reversal. There really has been a cultural shift in the workplace. The open office plan recognizes the fact that senior people aren't always in the office, and so it gives daylight to the people who are sitting at the desks all the time. It was quite revolutionary, driven by perceptions in health and also productivity. It goes back to some very fine research done at Carnegie Mellon, and supported by the GSA, in the early '90s and before. GSA really worked with the research to establish its value. And, of course, the GSA offices in Washington were built along those guidelines; it has an impact.

When you work in China, how does the design process differ?

We are required to stop at the design development phase. In past work we were able to do some consulting on construction documents, and actually, we're able to do that now because we have in-house curtain wall engineers who can assist the local design institutes.

It's a very robust design development package, and we always maintain a construction document review process to make sure the design intent is upheld. But yes, the process is quite different. You really have to finish and say everything you're going to say at the DD phase because that's when the design institute takes over.

In going forward, what advice can you offer architects?

Be careful. Architects need to abstract what architecture is, and what service architects provide. We need to conceive ideas, manage information, and be the client's trusted advisers. These components are really important. If we can fulfill that role, the practice of architecture will be strong. But many parties are eager to step in and encroach on the territory of design professionals. So architects need to think about this particular role.

当你在中国工作时，设计流程有何不同？

在设计发展阶段我们被要求停止工作。过去，我们的工作可以在建筑结构的文件上做些咨询，实际上，现在也可以这么做，因为我们有室内幕墙工程师，他们能够帮助当地的设计机构。

这是一个很可靠的设计进展方案，我们总是在对建筑结构文件进行审核，以保证设计意图得到贯彻。不过，流程确实非常不同。你需要在方案深化阶段就完成工作，并说出你想要说的一切，与此同时设计机构接管方案。

至于将来的发展，你对建筑师有什么建议？

小心谨慎。建筑师需要明白建筑是什么？我能够提供什么服务？我们需要构思创意，管理信息，成为客户可以信赖的顾问。这些要素非常重要。如果我们可以扮演这种角色，建筑业将会健康兴旺。但是，许多人急于冒进，总是试图涉足设计专业的领域。因此，建筑师需要思考自己的这一特殊身份。

Greenland Dalian East Harbor Tower, Dalian, China. (HOK/NY; est. completion 2016)
中国大连东港绿地大厦。（霍克国际有限公司/纽约，2016年完工）

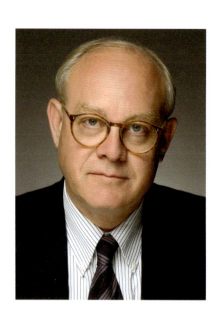

Robert Heintges
HEINTGES & ASSOCIATES

Robert Heintges, FAIA, is the founding principal of Heintges & Associates and, with over 30 years of experience, is an internationally recognized authority on the design and implementation of the curtain wall. Since 1989, Heintges & Associates has designed and consulted on well over 150 million square feet of curtain wall and exterior cladding for projects throughout the United States and in 17 other countries. Mr. Heintges is an adjunct professor of architecture at Columbia University's Graduate School of Architecture, Planning and Preservation. He has contributed his extensive knowledge on curtain wall theory and practice as well as his expertise in environmentally responsible design to numerous publications and organizations. Mr. Heintges is personally involved in virtually every project the firm undertakes and has developed long-standing relationships with many of the world's greatest architects.

罗伯特·海因吉斯
海因吉斯建筑事务所

罗伯特·海因吉斯是美国建筑师协会会员，海因吉斯建筑事务所的主要创始人，他有着超过30年的幕墙工作经验，因此是幕墙设计和施工方面国际公认的权威。自从1989年以来，在美国和17个其他国家，海因吉斯建筑事务所已经为各类项目中超过1.5亿平方英尺的幕墙及外墙包裹层提供了设计和咨询服务。海因吉斯先生是哥伦比亚大学建筑、规划和保护研究所的兼职教授。他将自己丰富的幕墙理论、实践方面的知识，以及他在适应环境设计方面的知识贡献给了许多出版物和组织。海因吉斯先生实际上亲自参与到公司的每个项目中去，并与世界上许多最伟大的建筑师建立了长期的联系。

For complete biography, please refer to the Appendix.

完整的传记，请参阅附录。

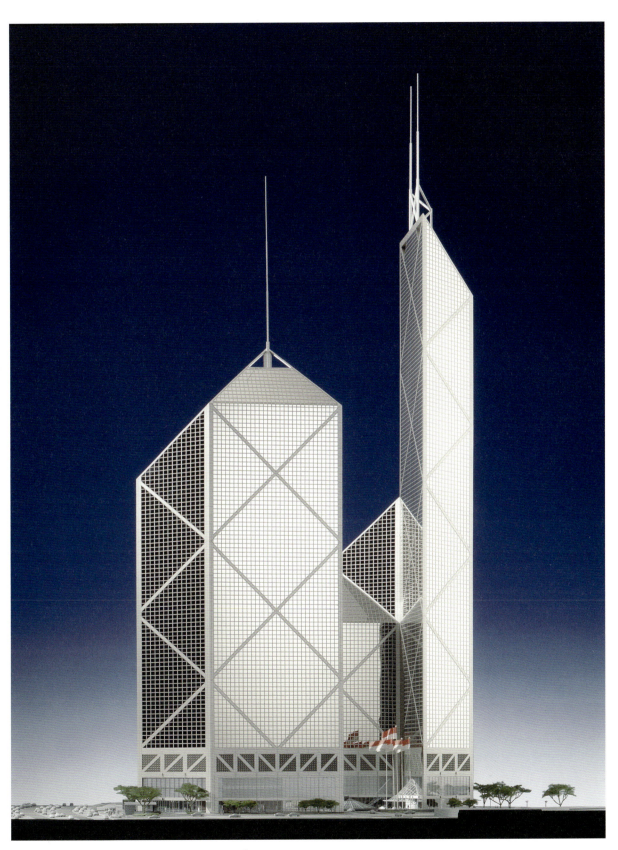

Sentra BDNI, Jakarta, Indonesia (Pei Partnership)
印度尼西亚雅加达森德拉中心印尼国民商业银行。（贝氏建筑事务所）

What is the significance of ultra-high-rise buildings in China, and how do they relate to sustainability concerns?

China has indeed enthusiastically embraced the high-rise curtain wall building as a symbol of economic success. This commodity of Western, or perhaps American, architecture has been ubiquitous throughout Asia for some time and didn't originally have much to do with local context or conditions, much less sustainability. I believe that that is changing now, with the influence of architects like Ken Yeang. Regarding the ultra-high-rise (arguably the ultimate paradigm, for better or worse), it is hard to understand how this typology could be sustainable, especially in locations where there is no corresponding infrastructure to support it. It is ironic perhaps that the ultra-high-rise doesn't even exist yet here in the U.S., but there is still a lot of interest in China and the Middle East. It is much more interesting and rewarding for us as local architects to begin to address local design and performance criteria and to adapt and improve the American typologies, not import them.

So is the sustainable high-rise a contradiction in terms?

I think it has been, especially with respect to the ultra-high-rise. Admittedly, however, things are changing and we're seeing a lot more creativity and interest in responsible design in the U.S., as well as in Asia. Not too long ago, one might see the identical high-rise building design going up in very different climates—Chicago or Singapore, Beijing or Miami. That's changing, just as values concerning sustainability and the built environment are changing.

The Western, or perhaps more accurately American, facade typologies have been facilitated by an amazing evolution in technology that has paralleled the development of the modern high-rise. Sophisticated high-performance glass products, for example, were developed and pioneered in the States. This has led to a certain dependency on technology to solve what might at one time have been considered architectural design problems. But it has also been very enabling in terms of the aesthetic and formal aspects of curtain wall design.

Where does curtain wall design figure in the spectrum of architectural art and technology?

I think the curtain wall is the one building system where art and technology most clearly converge. And as curtain wall consultants, managing that convergence is a large part of what we do in collaboration with design architects. Each

超高层建筑在中国有什么意义，它们怎样与当前关注的可持续性联系在一起？

中国实际上已经热切地接受了高层幕墙建筑，并将其作为经济成功的一个标志。这种西方建筑的产品或者说是美国的建筑产品，在一段时间内风行亚洲，但却与本土的语境、条件无关，更不用说与可持续性有什么关联。我想这一点现在正在改变，特别是有像杨经文（Ken Yeang）这类建筑师的影响，就超高层建筑（无论好坏，这都是最终的样式）而言，很难理解这种类型学要怎么可持续，特别是在那些并无相应基础设施来支持高层建筑的地区。颇为讽刺的是，可能超高层建筑甚至在美国都不存在了，但在中国和中东地区，人们依然对它兴趣浓厚。作为本土的建筑师，我们提出当地的设计和性能标准，开始调整和提升美国的类型学，这是更有意思、更有回报的事情，而不是引入它们。

那么，可持续的高层建筑是否是一种自相矛盾的表述？

我认为曾经是，特别是超高层建筑。不过，应当承认，情形正在改变，在美国和亚洲，我们看到，人们正对所负责的设计投入更多的创造力与兴趣。不久前，就可以看到，相同的高层建筑在不同气候环境中设计建造起来——从芝加哥到新加坡，从北京到迈阿密。事情正在改变，可持续性以及建筑环境相关的价值观都在改变。

西方的类型学，更确切地说是美国建筑外观的类型学，得到了技术上巨大进步的帮助，这种进步伴随着现代高层建筑的发展。例如，尖端的高性能玻璃产品在美国得到开发和发展。这导致了某种对技术的依赖，人们通过技术来解决那些曾经被认为是建筑设计上的问题，关于幕墙设计中美学和形式的问题，技术更具有解决的能力。

幕墙设计在建筑艺术和技术方面，处于哪一个重要的部分？

我认为幕墙是一个建筑系统，它将艺术和技术最明晰地融合在一起。作为幕墙顾问，我们在与建筑设计师合作时要做的很大一部分工作，就是处理这种融合。每个合作都有不同的目标，而且用完全不同的方法。技术作为一种手段，总是其中的一部分，不过，我们当然更有浓厚的兴趣将技术升华为艺术。

响应人们所关注的能量和环境问题，是当今的一个重要话题，但它的费用昂贵。因此，在一个所有东西都被最初成本所驱使的环境中，这种响应就成了一种障碍。开发商和评价者所考虑的幕墙费用模式（在整个建设费用中占据百分之多少的份额），是上世纪70年代到90年代建筑思想的延续，它意味着，我们甚至在有"工程"之前就开始谈论"工程价格"的项目了。

在美国或亚洲，建筑外观花费在全部建筑费用中所占的比例，与欧洲的情况非常不同。注意到这点是非常有趣的，可能欧洲社会将比我们在建筑环境上投入更多的价值，因为人们期望建筑可以长存下去。这不仅体现在设计上，也关乎费用管理，以及用长远的眼光看待创造性的生活费用带来的益处，比如说超过20到30年，这现在成为美国建筑界讨论的一个部分。

在美国，费用的情况如何？

对于一座高层建筑来说，幕墙的费用一般占整座地面建筑物费用的12%左右，而对一些更高性能或"更有创造性"的幕墙而言，费用可能会更高，大约在15%到20%之间。如果这种创造性并不完全是美学或形式上的，而且还融合了节能的目标，那它更容易被接受。与超高层建筑相反的超高性能时代即将来临。

在美国，能源效益原则变得越来越严格，因为美国就是这一原则的推动者。在纽约市，布隆伯格市长增加了绿色规范执行小组，小组由我们这样的专业人士组成，执行提高纽约市建筑规范的可持续性和能源目标，另外也考虑一些荣誉和利益来鼓励绿色设计。许多有趣的东西得到讨论，并且作为激励将很快被执行，以超出能源规范的最低要求。这是一项非常有趣的事业，但与欧洲相比，我们还有很长的一段路要走。

太阳能光电板是一个例子。在德国如果你在建筑中使用太阳能光电板，同时将电能输送回电网，政府和电力部门将以四倍的价格买回你所支付的电能费用。这对于替代技术的研究和发展来说，是一项非常强大的激励手段。不幸的是，在美国，建筑外层的太阳能光电板，更多地是出于营销策略的考虑而非其他。我将不再讨论这

collaboration has a different goal and a completely different methodology. Technology is always a part of this, as a means, but we are definitely seeing more and more interest in celebrating technology as art.

Responsiveness to energy and environmental concerns is a large part of that today but has tended to be more expensive up front. So, in an environment where everything is being driven by initial cost, that mind-set is an obstacle. The developer's and estimator's cost model for curtain wall, as a percentage of total construction, is still a holdover from the 1970s through the 1990s, which means we have begun some projects where "value engineering" is being discussed even before there is any "engineering."

It's interesting to note that facade costs as a percentage of total building costs is quite different in the States or in Asia than in Europe. Perhaps European society places more value on the built environment than we do, since it is expected to last longer. Being creative not only with design but with managing cost and thinking long term about life-cycle cost benefits, say over 20 to 30 years, is now becoming part of the discussion here in the U.S.

What is the cost in the U.S.?

For a high-rise building, the curtain wall is usually estimated at around 12 percent of total construction above grade, but something more high-performance or "creative" might be more like 15 to 20 percent. If that creativity is not strictly aesthetic or formal but also integrated with the goal of energy efficiency, it is more likely to be accepted. Ultra-high-performance, as opposed to ultra-high-rise building, is just around the corner.

Energy codes are also becoming stricter here in the U.S., which is an incentive. Here in New York City, Mayor Bloomberg implemented the Green Code Task Force, made up of professionals like ourselves, with a mandate to enhance sustainability and energy goals of New York City's Building Code, but also in consideration of credits or benefits that will encourage green design. Lots of interesting things are being discussed and will soon be implemented as incentives to actually exceed energy code minimum requirements. This is a very interesting undertaking, but we are still a long way off compared to Europe.

One example is photovoltaics. If you use PVs on your building in Germany, and you put power back into the grid, the government and power utility will buy it back at 4 times what you paid for it. This is a considerable incentive for research and development of alternative technologies. Here in the U.S., PVs on a facade are often more about marketing than anything else, unfortunately. I'm getting off the subject here, but the point, I suppose, is that facade design, not to mention building design and costs, is tied to larger economic issues. Our limitations are not the fault of architects, engineers, or developers, or attributable to any single concern. Everyone has to be a party to meaningful change, and this has to start with society as a whole.

In an office building, from the standpoint of net to gross, what is the maximum distance from curtain wall to core?

In the narrow rectangular European typology it's about 25 feet [7.6 meters]. Who in the States is going to do that? In the American and Asian typologies, it's 45 feet [13.7 meters]. And of course, this is driven by the fact that it's cheaper to have larger floor plates in terms of the ratio of curtain wall area to enclosed floor area, which can be about 30 percent for super efficiency versus 45 to 50 percent with the European typology. If daylight harvesting or natural ventilation is a criterion, priorities for efficiency are different.

What is the time frame for curtain wall design and development?

Usually it's a bit more accelerated than for other components of the building. One might go out to bid for the curtain wall in the middle of what would normally be Design Development because procurement for curtain wall and its materials generally has a longer lead time. That's one of the reasons that the procurement strategy for the wall is important to understand from the outset. This is even more critical if the wall design is to involve the research and development or testing of any new materials or methodologies. For this reason, innovation is very difficult on projects where the curtain wall is part of a lump-sum main contract bid for the whole building. Especially for a complex or ambitious curtain wall design, it is much more successful if the curtain wall contractor can be designated early on so that this process can begin ahead of the rest of the project.

Bidding and award of a curtain wall subcontract might take from two to four months. Once there's an award, administration of the curtain wall subcontract begins with mockup drawings, shop drawings, and materials submittals and review prior to fabrication. The time from award to first curtain wall on site is usually about one year. The whole shop drawing review process is very important and very collaborative, essentially the final step in the architectural design process.

一主题，但我想指出的是，建筑外层的设计不应该只关注建筑自身的设计和费用，而应该与更大的经济问题联系在一起。我们的局限性并不是建筑师、工程师或开发商，不是某个单方面出了问题。每个人都是富有变化意义的一部分，而这一变化必须在整体的社会中开始。

在一座办公楼中，从总量到净含量的角度考虑，幕墙到楼芯的最大距离是多少？

在矩形狭窄的欧洲类型学中，大概是25英尺（7.6米）。可在美国，有谁愿意这么干？美国和亚洲的类型学中，是45英尺（13.7米）。当然，原因在于如下事实：根据幕墙区域与封闭楼层区之间的比率，楼面板越大，费用越是便宜，这个比率在大约30%的时候效益最好，而欧洲类型学则是45%到50%。如果日光量或者自然通风是一个标准的话，产生功效的比率就不同了。

幕墙设计和开发在时间上怎么安排？

经常比建筑中其他部分要稍微快一些。人们可以在正常设计开发的中间时段，去为幕墙招标，因为幕墙及其材料的采购需要很长的研制周期，所以从一开始就了解幕墙的获取策略非常重要。如果幕墙设计包含研究、开发或测试任何新材料、新方法，这一点就更为关键了。项目革新非常困难，因为幕墙是整座大楼项目招标合同中的一个重要构成部分。尤其是对于复杂、要求过高的幕墙设计而言，如果幕墙承包方可以提前规划，那么，整个工程进度可以在其他项目之前开始，这样设计会更加成功。

一项幕墙业务转包的投标和中标大概要2—4个月。一旦有人中标，就开始对幕墙转包进行管理，包括模型图纸、施工图、材料提交、建造之前的审核。从中标到第一个幕墙就位，一般需要一年的时间。整个施工图的审核过程非常重要，且非常具有协同性，一般来说是建筑设计流程中的最后一步。

这是不是意味着，你并不设计结构图纸，而是直接跳到施工图那个步骤？

不是这样的。人们不需要为幕墙做结构图纸，因为幕墙的转包是一个以实际功效为基础的合同。建筑师的外层图纸可能非常具体详细，然而，就算比建

Tomorrow Square, Shanghai, China. (John Portman & Associates, 2004)
中国上海明天广场。(约翰·波特曼建筑设计事务所，2004年）

筑其它部分的结构图都更详细，从根本上，它也只是合同文本中的图纸。它的作用就是：当建筑师其它部分的图纸还在设计过程中时，可能需要这些图纸为招标做准备。设计的最后实现是在施工图的审核过程中逐渐形成的。如我前面提到的，幕墙作为建筑的一个方面，使人们参与到整个建筑的建造。从建筑产业中学习、分享、合作，所有这些对于开拓和创新都至关重要。建筑师与产业之间的合作关系，开始得越早，就有越多的新系统和新材料的发展开拓成为可能。

这个行业是否避开采用具有强烈反射效果的幕墙？

行业基本上追随着建筑的时尚。反射玻璃的幕墙，尽管在数十年前曾经流行，而现在则备受轻视，其

Does this mean you don't do working drawings and just jump right into shop drawings?

Not exactly. One does not do working drawings or construction drawings as such for a curtain wall, because the wall subcontract is almost always a performance-based contract. The architect's facade drawings might be very detailed, however, even more detailed than working drawings for other parts of the building, but they are essentially contract document drawings. These drawings might need to be ready for bidding when the rest of the architect's drawings are perhaps still in Design Development. The final realization of the design evolves during the shop drawing review process. As I mentioned earlier, this is why the curtain wall is the one aspect of architecture where one can be most intimately involved in the making of the building. Learning from the industry, sharing, collaborating: all of that is really important in terms of creativity and innovation.

When there is a more collaborative relationship between the architect and the industry from very early on, a lot more is possible in developing new systems and materials.

Is the industry shying away from highly reflective curtain walls?

The industry basically follows architectural fashion. Reflective glass curtain walls, though once popular decades ago, became vilified because of aesthetics and, to a lesser degree, problems with glare due to reflectivity, but mostly because of image, and perceived sameness. A greater interest in transparency has led to the development of more sophisticated, spectrally selective, coatings and a much wider range of aesthetic possibilities.

There is, however, some renewed interest in reflectivity to address different performance requirements in winter and summer and also to reduce glare. Now there are hybrid coatings that do more than just one thing, and products with coatings on multiple surfaces. It's harder, however, to design a building that is responsive in traditional ways to orientation, or natural ventilation, or daylight harvesting, etc., because those are concerns that may not necessarily be well integrated with a preconceived architectural aesthetic. Increasingly, as I mentioned earlier, we have found that architects today are far more interested in formally responding to and even celebrating performance architecturally rather than simply relying on product performance.

How efficient are multiskinned walls?

In certain climates they can be very efficient, but I think their benefits in terms of performance are somewhat exaggerated. I think the main benefit is actually aesthetic: it is perhaps the only way possible to achieve a totally transparent facade in the early modern ideal of Mies van der Rohe's all-glass skyscraper designs, with thermal and solar performance that can't be matched with a single-skinned facade. But the payback in terms of performance is not immediate, and there are maintenance and other issues (such as loss of usable floor area) that have to be factored in. It's not as though you double your performance with a double-skinned facade. It's more a way of achieving energy efficiency, or at least meeting or possibly exceeding code requirements, and still having the aesthetic impact of a lot of glass.

中的一部分原因是美学上的，另外一点是由反射而引起的炫目问题，但主要是因为外形看起来都一样。对透明玻璃的更大兴趣，已经使得一种更复杂、在光谱上更富有选择性的涂层不断发展，也使得一种更宽泛的美学可能性得以呈现。

不过，现在有人重新对反射玻璃感兴趣，这种兴趣一方面表明人们在冬天和夏天对性能有不同的要求，另一方面也是为了减少炫目。现在，有许多不只适用于一种材料的混合涂层，也有许多产品，在不同的表面上有不同的涂层。不过，很难再设计出一种建筑，以传统的方式回应方位朝向、自然通风、阳光采集等等方面的问题，因为这些关注点也许不会被融合进一种预想的建筑美学中。正如我提到的，我们不断地发现：今天的建筑师更爱从形式上进行回应，甚至从建筑学的角度称颂建筑的性能，而不仅仅是依赖产品的性能。

多重表层的墙壁功效如何？

在特定的气候环境中会非常有效，但我认为某种程度上，它们在性能方面的益处被夸大了。我认为主要的益处实际上是美学的：在密斯·凡·德罗 (Mies van der Rohe) 关于全玻璃摩天大楼设计的早期现代理想中，这可能是一种获得完全透明的建筑外层的唯一方式，它拥有一层外表所不能比拟的热量和太阳能功效。但是有关性能的回报并不是立竿见影的，需要考虑维护费用和其他一些问题（比如失去了本来可用的建筑面积）。好像并非有了双层的建筑表面，你就有了双倍的性能。它更多是一种获得能量效益的方式，或达到规范要求，或可能超过规范要求，同时保留了许多玻璃产生的美学效果。

那些非常复杂的，比如各种反射镜、内部百叶帘和其他遮挡过滤系统的墙，其功效如何？

我们已经对所谓智能、主动型的双层幕墙做了很多研究，我想，可以公平地说，并没有一个简单的答案。计算性能是非常困难的，即使有计算机的流体动态分析也不准确，因为不能将所有无形的东西，比如维修都计算在内，但这些东西也在"功效"之中。而且，可能实际的效率并不符合预期，所以没有人会在这些墙建成之后再去发表效率的数据。我们只能获得很少一部分数据，因此很难从过去吸取教训。但随着更多确认的要求、长久性能的论证，特别是能源与环境设计（LEED）认证，税务信用以及能源信用参与进来，这一切正在发生改变。

一些新技术对于幕墙有何影响？

嗯，总是有新的玻璃材料、新的玻璃涂层，可以用来控制阳光热量的获取。早期的涂层只是限制所有光的进入：自然光，红外线，紫外线，所有一切。后来的低辐射涂层在光谱上具有选择性：它能阻挡红外线，让紫外线和光谱中其他可见的部分通过。但这些产品的问题是，有太多的可见光照射进来，令人刺眼。颇有讽刺意味的是，现在这些非常好的产品在大多数情境下都没起什么作用，特别是当每个人使用电脑时，他们在玻璃被强光穿透的办公室里看不清楚屏幕。所以，拉下遮光物之后，你在哪里呢？回到了上个世纪。有一些很有意思的新涂料技术已经开发出来，它接收所有可见光谱中的光，但是采用全息分离技术将其中的大部分引向屋顶，因而在收集阳光的同时，滤掉了刺眼的光。玻璃自身像放满光的货架一样工作，它很复杂，又很简单。现在也有许多新的相变或可切换的玻璃材质，其中一些已经可以商业化了，不过目前要么费用昂贵，要么还有其他一些局限性。

有没有开发出一些新材料可以替代铝？

据我们所知，现在看来还没有什么神奇的材料可以替代铝，尽管对于制成铝所需的能量而言，大家都承认铝并不是节能的。除此之外，铝使用起来则是非常高效和万能的，并具有巨大的全球产量和制造网络，这使得它不可避免地与幕墙联系在一起。同时，铝还很容易循环利用。我相信铝大概是现代幕墙设计以及系统设计不断更新过程中主要的促进者。

在一块典型的铝、玻璃合成板中，是否铝要比玻璃来的贵？

并不一定如此。复杂的玻璃产品有很多层，有涂层和釉层等等东西，很容易就超出铝的费用。相反的情形也是可能的，没有什么比全玻璃外层需要的花费更合理了，无论是与之相对的铝、玻璃合成物还是石料玻璃合成物。即使在今天，一座

How efficient are those extremely complicated walls, those that variously include reflectors, internal blinds, and other screening systems, for instance?

We've done a lot of research on so-called intelligent or active double walls, and I think it is fair to say that there is no simple answer. Calculating performance is very difficult, and even with computational fluid dynamic analysis, it is not precise and cannot factor in all the intangibles, such as maintenance, that should also figure into "efficiency." Also, nobody publishes performance data after these walls are built, perhaps because the actual performance doesn't meet the projections. There is very little data available anywhere, so it is very difficult to learn from the past. This, by the way, is starting to change, with more requirements to confirm and demonstrate performance long term, especially where LEED, tax credits, or energy credits are involved.

What are some new technologies affecting curtain wall?

Well, there are always new types of glass and new types of glass coatings for managing solar heat gain. Early coatings just limited all light that came through: physical light, infrared, ultraviolet, everything. Subsequent low-E coatings were spectrally selective, blocking the infrared and letting in the ultraviolet and visible part of the spectrum. The problem with a lot of these products is that there's just too much visible light coming in, resulting in glare. It's ironic that these really great products don't work in most contemporary situations, especially in offices where everyone is using a computer and can't see their screens with high visible-light-transmittance glass. So you pull down the shades, and where are you? Back in the last century. There are some interesting new coating technologies being developed that admit all the visible spectrum but use holographic separation to direct most of it up to the ceiling, thus harvesting all the daylight while eliminating glare. The glass itself acts like a light shelf. It's very sophisticated yet simple. There are also a number of new phase-change or switchable glass materials. Some of these are becoming commercially available but are very expensive at this point, or have other limitations.

Have new materials been developed to replace aluminum?

There is no magic material that will replace aluminum anytime soon that we know of, although aluminum is admittedly not that efficient in terms of the energy required to make it. Other than that, it's very efficient to use and very versatile, with a ubiquitous global production and fabrication network

Reflections at Keppel Bay, Singapore. (Daniel Libeskind, 2012)
新加坡吉宝湾映水苑。（丹尼尔·里伯斯金建筑事务所，2012年）

that has made it inextricably associated with curtain wall. It is also very easy to recycle. I believe aluminum is arguably the main enabler of modern curtain wall design and ongoing innovations in system design.

In a typical panel of aluminum and glass, is the aluminum more expensive than the glass?

Not necessarily. A sophisticated glass product, with multiple layers and coatings and frits and so forth, could easily exceed the cost of the aluminum. The opposite can also be true. There is nothing more cost-effective than an all-glass facade, as opposed to one of aluminum and glass, or stone and glass. An all-glass building, even today, is still probably the least expensive, which is one reason why they are ubiquitous. It's important to remember that an all-glass building does not mean all-vision glass; most of an all-glass building is spandrel glass, backed with insulation, vapor barriers, and other materials.

How do you envision tall buildings in the future?

I really don't know. I suppose they'll just get taller and taller. It's amusing, but it would not surprise me if at some point

全玻璃建筑，依然可能是最便宜的，这也是它们为什么无处不在的原因。记住这一点很重要：全玻璃建筑并不等同于全视域的玻璃，大多数全玻璃建筑都是三角拱腹玻璃，由隔离层、蒸汽层和其他材料来支撑。

你如何设想将来的高层建筑?

我真的不知道，我只是想它们将变得越来越高。这很神奇，但如果在某个时候，规则开始限制这种过剩，我也不会感到惊讶。我们对于所有超高层建筑都有些冷嘲热讽，坦白说，我还是必须承认，我们对能为一些极高建筑的幕墙工作而感到自豪。尽管在设计上差别不大，但由于数量和被允许的规模经济，还是存在着一些特殊的机遇的。但我对支持高层建筑的基础设施颇感忧虑，这些高层建筑有很多建立在超低密度的城市基础设施之上。例如，在马来西亚的吉隆坡市中心，就没有可持续性的基础设施以应付密集的发展。同样的事情也发生在莫斯科，一些新的商业区在城市核心之外，类似于巴黎的拉德芳斯，但由于那里没有公路，不能支持机动车交通，当然也就不能支持一个地区有这么多与高层建筑对应的人口。城市的下水道系统经常支撑不住，供水系统也是如此，巨大的用电需要使电能供应紧张，但是，高层建筑依然是城市的基本原则。所以，为什么在这种环境中，有人会将高层建筑当作是可持续性的，这对我来说是一个复杂的谜。

不同的幕墙会不会限制一座建筑的高度？

并不一定，因为一旦你达到了某个特定高度，就真的没有关系了。当然，利用自然通风是很困难的，高层建筑会导致烟囱效应和压力不均衡。从建筑活动方面看，超高层建筑上的幕墙，就像中高层建筑中的一样会正常工作。就风力荷载、抗震性以及其他的所有方面而言，在设计时就有严格的限制，而幕墙，在一座80-100层，甚至更高的高楼或超高楼上，并不显得更巨大。不过，正如我已经提到的，如果一座高楼有许多幕墙，它会给人们创造出设计的机会。大体积能够容纳新颖设计所带来的设计工程费用，因为这些费用被大体积所涵盖，因而也更能支付得起。可能对于超高建筑幕墙最大的限制是定期地对它进行清洗。

在很高的地方，持续的风将对特殊的结构产生挑战，但这对幕墙并没有不利影响，这种看法是否正确？

这对玻璃只会造成很小的影响，因为玻璃不太可能很长时间持续负载它们，从幕墙的角度看，并不是真的影响重大，它只是另一个设计上的制约。

从机械工程师的观点看，压力在40层以上就不再正常。

虽然空气通过墙壁的渗透对于高层建筑来说变得来越重要，但相比于幕墙，高度对于机械工程师和空调系统而言，是一个更大的问题。高层的逃生区，需要向室外敞开，这也同样是一个挑战，不过，就像普通的消防安全和出口一样，这些并不是幕墙的问题。

幕墙建筑是否发展了玻璃清洁系统？

在幕墙设计中，很早就考虑了外部维护的制约，这是一个很好的行为。在具有复杂几何形状的建筑中，这点尤其重要，因为在这些建筑中，正常

codes begin to limit the excesses. We're a bit cynical about the whole ultra-tall building thing, to be honest, although I have to admit we are proud to have worked on the curtain walls of some very tall buildings. Although there is very little difference in design, there are some unique opportunities just because of the quantities and enabling economies of scale. But I do wonder about the infrastructure to support tall buildings. Many are built on an urban infrastructure meant for much lower density. In Malaysia's Kuala Lumpur City Center, for example, there was no sustainable infrastructure to handle the density of the development. It's the same thing in Moscow City. These are new commercial zones outside the city core, similar to La Défense in Paris, but they can't support automobile traffic because there are no roads, certainly not enough for the population of so many tall buildings in one place. The sewer systems often can't handle it, the water supply can't handle it; there's huge electrical demand that strains capacity, but tall buildings are still the norm. So it's a complete mystery to me why anybody would think of the tall building in this context as being sustainable.

Are there different curtain wall constraints the higher a building goes?

Not really, because once you get to a certain height it really doesn't matter. Of course, it is very difficult to use natural ventilation because of stack effects and the pressure imbalances that result in tall buildings. In terms of building movements, curtain walls on ultra-tall buildings function just as they would normally on, say, a medium-tall building. As far as wind loads, seismic, and everything else, those are all design constraints that, for curtain wall, aren't really greater on a tall or super-tall building of 80 to 100 or more stories. As I mentioned, however, a very tall building has a lot of curtain wall, which creates design opportunities. High volume can absorb the design engineering costs associated with innovative design because these costs are pro-rated over a larger area, and thus can be more affordable. Perhaps the biggest constraint on a super-tall building curtain wall is washing it regularly.

Steady wind at a great height poses particular structural challenges, but is it true that this has no adverse effect on the curtain wall?

It has a very small effect on glass because glass doesn't like to be steadily loaded over long periods of time, but this is not really significant in terms of curtain wall, and is simply another design constraint.

From the mechanical engineer's point of view, pressures are no longer normal beyond 40 stories.

Great height is a far bigger problem for mechanical engineering and HVAC systems than it is for curtain wall, although air infiltration and exfiltration through the wall do become much more critical for very tall buildings. Areas of refuge at great heights, requiring openings to the exterior, are also a challenge, but these, as well as fire safety generally and egress, are not really curtain wall problems.

Does the curtain wall architect develop the window-washing systems?

It's good practice to consider the constraints of exterior maintenance early on in curtain wall design. This is particularly important on buildings with complex geometries where a normal window-washing or building maintenance system may not be practical. Ideally the maintenance system on a complex building is conceived integrally with the facade design, as it was, for instance, on I. M. Pei's Bank of China in Hong Kong. That whole system was designed in concert with the facade, to be totally concealed within the building and deployed through operable components of the curtain wall.

Window washing is really an unfortunate misnomer, however, because one has to wash everything, not just the windows, and be able to service the wall as well. Especially in an urban, or corrosive, environment, you have to wash the metal and stone and spandrel glass, not just the vision glass or windows. This is why so many early curtain wall buildings today are falling apart. On the very first high-rise curtain wall building in the U.S., the UN Secretariat, for example, the windows were washed as they are traditionally on a masonry building: the window washer opened the window, attached himself, stepped out onto the exterior sill, and washed the vision glass. The spandrel glass on the Secretariat was never washed, nor was the anodized aluminum, which soon corroded and ultimately became severely pitted and in some areas completely eaten away.

In considering the requirements for a window-washing or building maintenance system, it is also a good idea to consider how the facade will be maintained over time, how glass and other components are going to be replaced, and so forth. In this context, it is essential to the life-cycle planning and sustainability of the building.

的玻璃清洁或者建筑维护系统可能并不适用。就复杂建筑的维护系统而言,其理想形式应是与外观设计联系在一起的,这点体现在贝聿铭设计的香港中银大厦上。整个系统的设计与建筑外观相呼应,被完全包在建筑内部,并通过幕墙的可操作部分得以运作。

然而,窗户清洁真是一种不幸的误读,因为人们必须清洁所有东西而不仅仅是窗户,同时还包括幕墙。特别是在都市或者腐蚀性的环境中,你必须清洗金属、石头和三角拱腹玻璃,而不仅仅是玻璃面或窗户。这就是许多早期的幕墙建筑今天已经损毁的原因。例如,在美国的第一座高层幕墙建筑联合国秘书处大楼上,窗户的清洗就像在传统砌体建筑上所作的一样:清洁工打开窗户,系上自己,慢慢走到外面的窗台上,然后清洗视窗。秘书处大楼上的三角拱腹玻璃从没有清洗过,被氧化的铝板同样如此,它们很快受到腐蚀,并最终变得严重坑洼不平,其中某些地方已经完全被侵蚀了。

在考虑窗户清洁或建筑维护系统的要求时,考虑一下建筑外观如何常年保持,是否玻璃和其他材质将被替代等等,这也是个不错的主意。在这种语境中,它对于建筑的生命周期规划和可持续性至关重要。

玻璃建筑是否有固定的使用寿命?

这是一个很少被提及,却会引起恐慌的话题。所有东西都有使用寿命,包括玻璃,尤其是绝缘玻璃。绝缘玻璃的密封边缘最终无法抵制水分的入侵,冷凝效应会在光线里发生。防止密封边缘失效的时间大概是十年左右,这意味着玻璃有可能保存至少25年,也可能是这一数值的两倍,但之后呢?玻璃最终必须被替换。从现在算起,50年后,有许多新的玻璃建筑需要替换掉它们的玻璃,而其中有些玻璃的四面都被硅脂上釉粘合了,这将使替换过程更加痛苦。有趣的是,这是外层带有整块玻璃的双层建筑外观论证自己合理性的一条理由。因为内层的绝缘玻璃更加容易上釉重装。

针对每50年就要更换外衣的高层建筑，有没有一种新的策略？

绝对有。这点变得越来越重要了，在所有的项目中，我们确定最适合的维护项目，其中包含正常的年度维护、间歇性整修和间歇性部件替换等等，以契合整个项目生命周期的要求。当你建造一栋百年建筑时，你需要确定那究竟意味着什么。百年建筑并不意味着所有东西都要维持一百年，而是说，要获得百年建筑，你必须预测该采取哪些维护措施。

你最具有挑战性的项目是什么？

可能是雅加达的Sentra BDNI中心（贝氏建筑事务所）。不幸的是，这个项目并没完成。这是一个特殊的挑战，因为印度尼西亚几乎没有相关的幕墙经验，更不用说有资源采取一种面积达数百万平方公尺的复杂幕墙，而我们却需要用当地的印尼公司来做这些。我们花了几周时间来约见这个国家所有的幕墙公司，但最终我们必须说服客户做出让步。他们同意可以使用印尼国内与国外的合资企业做此项工程。然后我们开始着手确定那些符合资格、同时愿意与当地合作的外国幕墙公司。最终我们成功地将四个幕墙项目团队配对结合起来，去招标这项工程。我们知道需要非常全面的幕墙图纸来完成此项工作，因此，招标图纸的设置包括300个大AO尺寸的绘制表。所有这些努力换回的是出色的招标意见书，幕墙工程最终给了一家日本公司和一家本地公司组成的合资企业。文化、语言以及在那种环境中做一项复杂的幕墙工程所需的展示和物流过程，都非常具有挑战性。我们对四种不同类型的幕墙模型进行性能测试，并且幕墙也进入施工；但是，当印尼经济暴跌时，工程很遗憾地终止了。

你在中国的经历是什么？

我们在亚洲所有其他国家具有的经验都比中国来的多。在中国的经验被这样的事实所影响：我们与美国建筑师一起工作，设计完成之后，实际上必定会交给上海设计机构或北京设计机构或其他地方准政府性质的设计机构，他们会说："非常感谢，我们再也不需要你了。"这很有趣，但也很令人沮丧，因为工程的管理机构受

Is there a set life span for glass buildings?

It's a scary subject that is rarely addressed. Everything has a life span, including glass, and insulating glass in particular. The edge seal of insulating glass will eventually fail to exclude moisture, and condensation will form inside. The warranties against edge-seal failure are generally 10 years, which means the glass is probably going to last at least 25 years, and possibly twice that, but then what? The glass will ultimately have to be replaced. There are many new glass buildings that will need to have glass replaced 50 years from now, and a lot of those are four-side structurally silicone-glazed, which will make this even more painful. Interestingly, this is one argument for the double-skin facade with an outer skin of monolithic glass. An inner skin of insulating glass is much easier to reglaze.

Is there a new strategy for tall buildings in which one anticipates reskinning every 50 years or so?

Absolutely. That is becoming very important, and on all our projects we identify the most appropriate maintenance program, which includes normal annual upkeep, refurbishment intervals, component replacement intervals, and so on, to meet project life span requirements. When you do a 100-year building, you need to identify what that means. It doesn't mean everything is going to last 100 years, but to get your 100-year building you are going to have to anticipate taking certain actions.

What was your most challenging project?

Probably Sentra BDNI in Jakarta (Pei Partnership Architects). Unfortunately the project was never completed. This one was a particular challenge because Indonesia had little relevant curtain wall experience to speak of, much less the resources to undertake a sophisticated curtain wall of millions of square feet, yet we were required to use a local Indonesian company. We spent weeks interviewing all of the curtain wall companies in the country, but ultimately we had to convince the client to compromise. They agreed that we could make local and foreign joint ventures. We then went about identifying qualified foreign curtain wall companies who would be willing to pair with local firms. We were able to successfully pair up four curtain wall teams to tender the project. We knew that we needed very comprehensive curtain wall drawings to pull this off, so our tender drawing set included 300 large AO size sheets of drawings. All of this effort was rewarded with excellent tender submissions, and the curtain wall was

Grand Lisboa, Macau, China. (Dennis Lau, Ng Chun Man, 2007)
中国澳门新葡京酒店。（丹尼斯·刘，吴俊文，2007年）

到了限制。我们对于中国大陆之外的中国幕墙承包商、加工方以及项目提供者，倒是有更多的经验。我们最大的挑战是澳门的新葡京赌场和酒店，花费数年才得以完成。这是一个非常复杂的工程，有着雄心勃勃的几何结构以及不太可能完成的时间表。幕墙和覆盖层超过500,000平方英尺（46451平方米），工程交给了一家奥地利公司，尽管我们对此表示反对。在施工图提交到一半的时候，这家公司破产了。我们与主要的承包方协商之后，找到了一家由工程师和施工队组成的中国机构来揽下这一工程。虽然这个团队从来没用建筑有机硅玻璃建造过一个幕墙系统，我们还是重新开始了一项新的设计，并使得整个装配、测试、启用过程赶上原来的日程安排。这真是一项紧张繁杂的工作，但却完成得很好；我们从这种经验中学到，中国人可以做任何事。正如我喜欢说的，这是因为他们还不知道那是不可能完成的。这真是美妙极了！参与到一项工程，我们的经验已经被欣赏，但也可以做出一些改变；这真是非常值得去做的。

finally awarded to a Japanese firm and a local consortium. The culture, the language, all the staging and logistics of doing a very complex wall in that environment was really challenging. We went all the way through performance mockup testing of four different wall types, and the curtain wall was well into fabrication, when the Indonesian economy nose-dived and the project, sadly, was terminated.

What has been your experience in China?

We have much more experience in all of the other Asian countries than we do in China. Our experience in China has been tempered by the fact that we've worked with U.S. design architects whose designs are inevitably handed over to the local architectural Shanghai Institute or Beijing Institute or other local quasi-governmental architectural institute, after which it was, effectively, "Thank you very much; we don't need you anymore." It was interesting but very frustrating, because construction administration was limited. We have a lot more experience with Chinese curtain wall contractors, fabricators, and suppliers for projects outside of China. Our most challenging was the new Grand Lisboa Casino and Hotel in Macau, completed several years ago. This was a very complex project with an ambitious geometry and an impossible schedule. The curtain wall and cladding, over 500,000 square feet [46,451 square meters] of it, was awarded to an Austrian firm despite our objections. Halfway through shop drawing submittals and workshops, this company went bankrupt. Working with the main contractor, we helped put together a local Chinese consortium of engineers and fabricators to take over the project. Although this group had never before done a unit system curtain wall with structural silicone glazing, we started over with a new design that was fabricated, tested, and installed in time to meet the original schedule. It was intense and a lot of work, but fantastic; we learned from this experience that the Chinese can do anything. As I like to say, it's because they don't know yet that it can't be done. It was a beautiful thing. It's very rewarding to participate in a project where our expertise has been appreciated and where we were able to make a difference.

Dennis Poon
THORNTON TOMASETTI, INC.

Dennis Poon, B.C.E., M.Eng., is a vice chairman at Thornton Tomasetti structural engineers. His more than 30 years of experience includes high-rise offices, hotels, airports, arenas, and residential buildings throughout the world, applying state-of-the-art engineering technologies for building analysis, design, construction, and project management. Beyond Taipei 101 in Taiwan, he has engineered the Signature Tower in Jakarta, Shanghai Tower, Ping An Tower in Shenzhen, and Greenland Tower in Wuhan, China, all more than 600 meters tall. Mr. Poon's responsibilities at Thornton Tomasetti include the firm's widespread international operations. He has lectured at leading professional conferences and in universities worldwide, including MIT, Columbia, Delft University of Technology in The Netherlands, the New Jersey Institute of Technology, and Hong Kong City University. Mr. Poon's degrees in civil engineering include a bachelor's from the University of Texas at El Paso and a master's from Columbia University. *Engineering News Record* included him among the "Top 25 Newsmakers of 2010," celebrating his work on more efficient high-rise structural design, part of his effort to align China's building code with international standards.

潘子强

宋腾添玛沙帝公司

潘子强是土木工程学士，工程学硕士，宋腾添玛沙帝结构工程部门的副总裁。他在建筑行业超过30年的工作包括遍布全球的高层办公室、酒店、机场、室内运动场和住宅楼。在这些作品中，他将最先进的工程技术应用于建筑分析、设计、建设和项目管理。除了位于台湾的台北101大楼，他也参与了雅加达签名塔、上海中心、深圳平安大厦、中国武汉绿地中心的工程工作，这些建筑的高度都超过了600米。潘先生在宋腾添玛沙帝的工作包括公司众多的国际业务。他也在前沿专业会议以及全球许多大学进行演讲，这些大学包括：麻省理工学院、哥伦比亚大学、荷兰代夫特工业大学、新泽西理工学院和香港城市大学。潘先生在土木工程上的学位有德克萨斯大学埃尔帕索分校的学士学位，以及哥伦比亚大学的硕士学位。《工程新闻记录》将其列为"2010年25大新闻人物"，以表彰他在更高效高层结构设计方面的成绩。他的一部分工作还包括使中国的建筑规范符合国际标准。

Thornton Tomasetti is structural engineer for the projects illustrated.
宋腾添玛沙帝工程师事务所负责图示项目的结构工程项目。

For complete biography, please refer to the Appendix.
完整的传记，请参阅附录。

Plaza 66, Shanghai, China (Kohn Pedersen Fox, 2001)
中国上海恒隆广场。（KPF，2001年）

What role do Western architects play in China?

China has made huge progress, especially in technical matters, but they're still lagging in software. That's why they still need Western architects. You cannot learn art and creativity just from a textbook. It has to grow from generation to generation. It's a talent; you can't study a talent. Chinese culture is quite conservative, so independent thinking is not encouraged too much. Look at Chinese painting, which is very constant from dynasty to dynasty. While growing up, I myself was groomed in a very rigid way. There's a cap on creativity. The Chinese invented fireworks but they didn't take the next step to invent gunpowder and weapons, right? The Chinese are technically very strong, but how do you apply that? They need software, management and organizational skills, and experience. From an architectural perspective they still have to tap into a culture of creativity in order to push themselves to another level of achievement.

Right now, the market is quite open to both local Chinese and overseas architects. Opportunity depends on the client's selection process. At this point, all the developers in China are locally grown. They travel around the world and then they want a Ferrari, a Lamborghini, or modern design. They are highly influenced by overseas architecture and have high expectations. At the same time, some developers are still very conservative and don't want anything ultramodern or too advanced. Either way, they're spending almost the same amount for construction, so why not pay some additional fees to hire a more creative Western architect for a Ferrari-look building?

One part of architecture is form and function, or the process that dictates and constrains form. You can have a Frank Gehry building, but inside, function is limited; it cannot be used as typical office space. The second part is curtain wall design, i.e., how you dress up the form. Then you have landscape design and harmony with the city so that everything blends together. Some spec developers don't care about this; they just want to make money.

Even without the best-designed form, a nice curtain wall, like a nice dress, can really make a difference. So I think the overseas architect still has an edge in China. But Korean architects, who are already quite westernized, are catching up. I cannot comment on their interiors, but their exterior forms are quite attractive.

西方建筑师在中国扮演什么角色?

中国已经取得了巨大进步，尤其是在技术上，但他们在软件上依然落后，这就是为什么中国依然需要西方建筑师的原因。艺术和创造性不能仅仅从课本上学习，这些东西必须经过数代人的发展。这是一种天赋；天赋不能学习！中国文化非常保守，因此不鼓励独立思考。看看中国绘画，其传统世代延续不断。在我成长的过程中，曾被用一种非常刻板的方式进行培养，这是一种对创造性的限制。中国人发明了烟火，但是他们没有往下去发明火药和武器，对吧？中国人在工艺上做的很出色，但如何应用这些呢？他们需要软件，需要管理、组织技术和经验。从建筑的角度看，他们依然要深入了解一种富有创造力的文化，以便可以推动自身，获得另一层次的成就。

现在，市场对于中国本土及海外建筑师都非常开放，机会取决于客户的选择。目前，中国的所有开发商都是在本地发展起来的。他们游历世界，然后想要法拉利（Ferrari）、兰博基尼（Lamborghini）或现代设计。他们深受海外建筑的影响，而且有着很高的期望。同时，一些开发商依然十分保守，不想要任何超现代或太前卫的东西。不管怎样，既然他们花费了几乎同样多的钱去搞建筑，那为什么不支付一些附加费用，去雇佣一名更有创意的西方建筑师去建造像法拉利那样的建筑呢？

建筑的一部分是形式和功能，或者说是指定和限制形式的过程。你可以拥有一栋弗兰克·盖里（Frank Gehry）的建筑，但是，其内部功能的限制，不能被当作典型的办公空间。第二部分是幕墙设计，例如你如何对形式进行装饰。然后你有了景观设计，以及与城市整体的融洽感，以至于所有东西都结合在一起。有些投机的开发商不在乎这点，他们只想着赚钱。

即使没有设计最佳的形式，一面漂亮的幕墙宛如盛装一般，真的可以让建筑与众不同。所以我认为，海外建筑师依然在中国有自身的优势。但是已经很西化的韩国建筑师正在迎头赶上，我不能对他们的内在结构做出评论，但他们的外在形式确实非常吸引人。

在纽约，一般的超高层建筑大概有40-50层，其他地方也是如此吗？

现在在中国，限定高度是180米（590英尺），这大概是40-45层高的建筑。所以任何高过这一标准的建筑，必须由一个特殊的高层建筑委员会审核通过。对于我们来说，30层就被认为是超高层建筑了；而在中国，这个数字一般是40层和50层。

在中国现代建筑规范的发展中，您非常有影响力。这一规范是如何演进的？

中国的建筑规范由来已久。在过去的二十多年里，中国的建设市场欣欣向荣，但在那之前，它已沉寂多时。建筑规范依然是以苏联的规范为基础，大部分只适用于低层建筑。因此，随着高层建筑的引进，必须配套相应的国家规范；还有城市的建筑规范，因为每个城市都有自己的规范。所有东西要匹配完毕是需要时间的。

制定规范的基本原因，是要将质量和安全的最低要求标准化。但是不同的省份有不同的要求，有一些较其他的要严格得多。和乡村相比，在城市里，你可以指望有更好的质量控制和更出色的劳动技能。因为规范试图涵盖所有的情形，它倾向于更加保守以便在普遍应用时依然有效。但是，当你知道如何驾驶一辆法拉利的时候，你并不需要为此担心。如果你设计一栋60层、70层或100层的大楼，或者一些特殊的样式、结构系统，这些保守的规范就不完全严格了。这就是为什么会有一个审查委员会会议来讨论这些问题。因此，我们从一条用于讨论的底线开始，然后试图证明，某一特殊的建筑或结构可以不受现存规范的制约。我们主要是提出事实和分析结果，同时说明我们的方法是正确的，能够避免引起危险或不安全的情形，并达到世界级的工程水准。

在中国有一个规范委员会，其成员大部分是总工程师、各大学的教授和不同省市的领先设计机构。在北京的建筑设计和建设部门也有官方审查小组，来评估超高建筑的设计。作为外国工程师，我们在宋腾添玛沙帝结构师事务(Thornton Tomasetti)已经设计出许多的著名结构、高层建筑和大跨度机构。因此，有机会去会见这些专家，并与这些中国的顶尖人物发展关系，这些人评价我们的工作，并在我们的图纸上签字盖章。所以如果你是我们建造100层建筑的合作伙伴，我们必须一起会见委员会，并以礼貌、逻辑的方式解决问题。这并不

In New York an average high-rise is 40 to 50 stories. Is that true elsewhere?

In China right now the limit is 180 meters [590 feet]. That's about a 40- to 45-story building, so anything higher would have to go through a special high-rise building review committee for approval. To us, 30-story buildings are considered high-rise; in China, it's 40 stories, and 50 is common.

You've been very influential in the development of China's modern building code. How has it evolved?

China's building code has come a long way. The construction market has been booming in China for almost 20 years, but before that it had been static for a long time. The building code was still based on Russian codes, mainly for low-rise buildings. So with the introduction of high-rise buildings, the national code had to deal with that; city codes, too, because each city has its own. It takes time for everything to align.

The basic reason for a code is to standardize minimum requirements for quality and safety. But different provinces have different requirements, some more stringent than others. In cities you can expect better quality control and labor skill than in the countryside. Because the code tries to cover all situations, it tends to be more conservative so it can be universally applied and still function. But when you know how to drive a Ferrari, you don't need to worry about that. If you design a 60-, 70-, or 100-story building or some special form or structural system, the conservative codes aren't entirely rigid. That's why there is a board review meeting that allows you to discuss the issues. So we start with a baseline for discussion and then try to prove that a particular building or structure can go beyond existing code requirements. We basically present the facts and our analysis, and we demonstrate how our approach can be justified without causing danger or unsafe conditions and still achieve world-class engineering.

There is a code committee in China that consists of most of the chief engineers and professors from major universities and leading design institutes in different cities and provinces. There is also an official review panel from the Building Design and Construction Department in Beijing to assess super-tall building design. As foreign engineers, we at Thornton Tomasetti have worked on a lot of prominent structures, tall buildings, and long-span structures. We've had the opportunity to meet and develop a relationship with these experts, all top guys in China, who review our jobs, and sign and seal our drawings. So if you were my partner on a 100-story building, we'd meet the committee together and resolve any problems

中国武汉绿地中心。（AS+GG建筑事务所，预计2016年完成）

Wuhan Greenland Center, Wuhan, China. (Adrian Smith + Gordon Gill, est. completion 2016)
中国武汉绿地中心。（AS+GG建筑事务所，预计2016年完成）

Taipei 101, Taipei, Taiwan, China. (C.Y. Lee & Partners, 2004)
中国台湾，台北101大楼（李祖原联合建筑师事务所，2004年）

是走后门，在许多项目之后，他们意识到他们必须改变规范，使用更中性的语词，以便为更深入的讨论留下空间。中国的建筑规范并非错误；只是它必须在整个国家的范围内说明不同的情形、劳动技能和环境因素。

如何解释过去十年间建筑高度上的突飞猛进？

我认为这很简单：经济原因！因为在每个城市或镇区的人口数量——甚至一个小城镇也有几百万人口——中国的土地如此珍贵，价格如此高昂，所以你总是想要完全利用它们。香港即是如此。在纽约，大楼的地下室常被用作档案室，而在香港，地下室将是三层的餐馆和购物中心，机械设备的空间将会出现在别的地方，每平方英寸（厘米）都被用于商业。

in a respectful and logical way. We are not trying to cut corners. After more and more projects, they realize they have to change the code and use more moderate wording so there's room for further discussion. The Chinese code is not wrong; it just has to address different situations, labor skills, and environments across the entire country.

What accounts for the tremendous escalation in building height over the past decade?

I think it is quite simple: economic reasons. Because of the population in each city or township—even a small town has a couple of million people—land in China is so expensive, so precious, that you want to make full use of it. In Hong Kong it's the same. We house archives in the basement of our building in New York, whereas in Hong Kong there would be three levels of restaurants and shopping. Mechanical space would be somewhere else. Every square inch [centimeter] is used for business.

Who owns the land in China?

Modern China came from a communist background, so everything was owned by the government, except in the villages, where families own land by right of birth, deeded from one generation to the next. But the majority of farmland is owned by the government and leased to the farmers. You basically cannot claim family heritage in the city because the original owners left China and the government assumed control, so the land is government-owned. Now, when the government wants to build on village land it has to resettle the tenants who have been there since the Revolution, and pay them hundreds of thousands of dollars in compensation. After the land is reclaimed, the government can auction it off to different developers for office buildings, condominiums, or commercial use. I think you can only use that land for 40 to 70 years. So even a successful developer only has the land for a limited time before having to renegotiate. After that, I think the government can ask for additional money. The same thing happened in Hong Kong, where the land was leased to the British for 99 years and then reclaimed by China.

Can foreign developers and investors lease land in China?

Everyone is leasing now, even local developers. You bid the land, you do not own it. You lease it for 40 to 70 years. Basically you need a license; even major developers cannot invest in China. They first have to establish themselves with some kind of real estate consulting business, and then, after so many years, their work is evaluated before a business license can be granted. For most projects, foreign developers or investors usually team up with local developers. That is a national strategy to protect the market from extreme volatility. So it is good to have a Chinese partner to safeguard the interests of China. Usually the Chinese partner has a government background, which explains why he has the land in the first place. It's not 100 percent capitalism; there is a lot of government influence, a strong system of checks and balances. However, Hong Kong developers have a lot of advantages because they are not regarded as foreigners. They can easily start up their own real estate company or get their real estate license and don't need local partners.

Is sustainability a serious commitment in China or just lip service?

It is politically correct to emphasize sustainable design. I think the Chinese government does have the will and the

在中国，谁占有土地？

现代中国来自共产党背景，所以，所有东西都为政府所有。除了在农村，在那里家庭根据出生权而占有土地，并从一代人传到下一代人手里。但农田的大部分为政府所有，并租赁给农民。在城市中，你基本上不太能够要求家庭继承，因为原来的拥有者离开了中国，而政府采用控制的方式，因而土地归政府所有。现在，当政府想要在农村土地上进行开发时，它必须重新安置那些革命之后就生活在这片土地上的定居者，并且支付他们成百上千的美金作为补偿。在土地回收之后，政府可以将其拍卖给不同的开发商，用于兴建办公楼，公寓，或用于其他商业用途。我认为：你只可以使用土地40到70年。即使是成功中标的开发商，在重新开始谈判协商之前，也只是在限定的时间内拥有土地。之后，我想政府可以要求更多的金钱。同样的事情也发生在香港：土地租给英国99年后，重归中国。

外国开发商和投资者是否可以在中国租赁土地？

每个人都在租，即使是本地的开发商。你投标土地，但并不占有它，只是租赁40到70年。一般来说你需要一份执照，甚至大的国外开发商也不能在中国投资，他们得先通过某种房地产咨询生意来立身，所以，许多年后可能会有一种商业执照颁布给这些国外的公司，而在颁证之前，他们的工作就通过过去的事务得到了评估。对于大多数项目来说，外国开发商或者投资者往往与当地开发商组成团队。这是一项国家策略，以保护市场，避免过度反复无常。有一个中国伙伴来保护中国的利益，这是一件好事。一般来说，中国伙伴都有政府背景，这样就解释了为什么他们一开始就拥有土地。这不是百分之百的资本主义，其中有许多政府影响，以及一套强有力的相互制衡体系。然而，香港的开发商有许多便利，因为他们不被当作外国人。可以轻松地成立自己的房地产公司，或者在没有本地伙伴的情况下获得房地产营业执照。

在中国，可持续性是一项严肃的承诺还是口头说说而已？

强调可持续性设计在政治上总是正确的。我认为中国政府确实有决心和必要，来推进可持续性设计和控制污染。中央政府非常想要这些，他们知道中国的水和其他资源已经被污染，所以想建立更多的污水处理厂，但地方政府往往没有资金来做这些。地方政府想要吸引工商业，提供更多的就业机会，但如果在环境控制上着力太多，工业将会移到其他的省份去。中国的工厂试图把一切东西都造得便宜些，以便将产品卖给凯马特（Kmarts）和塔吉特（Targets）超市。但是怎么能够在引入污染控制设备的同时，依然保持产品的低价呢？这是一个恶性循环。他们确实想这么做，但这需要时间。现在，中国试着不去推动高污染产业，而是努力推动高附加值的制造业。劳动力在中国也不如以前那样廉价了，这就是为什么有很多工业产业现在转移到越南、孟加拉国、泰国、斯里兰卡这样的国家的原因。

美国存在如此高的能源消耗，是因为每个家庭都有如此多的汽车，以及如此多的家用电器。中国人没有这么多的电器，但是大家赚更多的钱，开始购买汽车。过去，街道上都是自行车，但现在，除了在一些居民区，它们基本上都不见踪影了。因此，政府必须跟上汽车污染的步伐，现在在北京，许多出租车开始使用天然气，就像香港的出租车和公共汽车一样。

除了工厂和汽车尾气排放，发电厂也是一个重要的污染源，甚至影响更糟，因为它们所使用的化石燃料必须依赖进口。中国并没有石油，但它有许多煤。即使现在煤已经不容易找到了，但由于许多工厂需要更多的电能，于是他们燃烧越来越多的煤并污染了空气。在找到充足的可替代能源之前，他们不会停止使用煤。所以，有一种强烈的意愿，想要推动风能和太阳能的使用。即使在乡下，当地的农民和定居者也有自己的小型屋顶风车、热水箱和太阳能板，这些设备很原始但却功效明显。中国政府也推动核能，虽然最近有日本的核能事故，但他们还是想利用核能满足10%-15%的能量需求。

在中国自上而下的结构中，地方政府是否会收到中央的拨款？

不，情况并不一样。一些大城市直接由中央政府管辖，但是小一点的城市则由各个省管辖，而这些省则由中央政府管辖。它们只为自身的财政情况及福利负责，而且由于国家政策，往往需要为税收基础做出贡献。

在某种程度上，中国存在一个省际公路系统。许多公路是在私人资金的支持下兴建的，因而建成之后要征收通行费来偿还建设费。宣传最广的公共项目是高速铁路，政府已经在上面花费了数十亿美元。这些高速铁路中，有三条南北代走向，还有三条东西走向。整个工程已经完成了50%-60%，剩

necessity to promote sustainable design and control pollution. They want it desperately. They know their waters and other resources are polluted and they want to build more sewage treatment plants, but municipalities just don't have the capital to do it. Local governments want to attract industry and business for employment opportunities and so on, but if they impose too many environmental controls, industry will go to a different province. Chinese factories try to do everything cheaply so they can sell to the Kmarts and the Targets. But how do you keep prices low by importing all the pollution-control equipment? It's a vicious cycle. They do want to do it but it takes time. Right now China is trying not to promote heavy-pollution industry and is pushing for high-value manufacturing. Labor in China is not as cheap as it used to be, so that's why a lot of industry is moving to countries like Vietnam, Bangladesh, Thailand, Sri Lanka, etc.

America's high energy use exists because there are so many cars for each family, and so many appliances. The Chinese do not have the same number of appliances yet, but with the population making more money, people are buying cars. The streets used to be filled with bicycles, but now they're disappearing except in local neighborhoods. So the government has to catch up on car pollution. Right now in Beijing, many taxis use natural gas, just like taxis and buses in Hong Kong.

Beyond factory and auto emissions, electric power plants are a main source of pollution, which is doubly bad because they use fossil-based power that has to be imported. China does not have oil, but it has a lot of coal. And even coal is hard to find right now because there are so many factories that need more electricity, so they're burning more and more coal and polluting the air. They cannot stop using coal until they find sufficient alternative energy sources. There's a big push toward wind and solar. Even in the villages, local farmers and residents have their own small rooftop windmills and also hot water tanks and solar panels, primitive but effective. The Chinese government is also pushing nuclear energy, and even with the recent Japanese accident, they are trying to use it for 10 to 15 percent of energy needs.

In China's top-down structure do local governments receive federal monies?

No. They are different. Some major cities are directly governed by the central government, but smaller cities are governed by the provinces, which are governed by the central government. They're responsible for their own accounting and well-being and are required to contribute toward the tax base for national policy.

To a certain extent, there is an interstate highway system. A lot of roads were built with private money, and tolls are collected to pay for them. The most publicized public work is the high-speed railroads, which the national government has spent billions of dollars on. Three lines run north-south and three run east-west. About 50 to 60 percent is finished, with the remainder due in the next 5 to 10 years. In a very short time, China has built the world's largest network of high-speed railroads, with speeds up to 380 kilometers [236 miles] per hour. It's amazing. China must have mass transit. Imagine Hong Kong, Beijing, or Shanghai without it!

There are good things about democracy, but there are also good things about central decision-making in China, which allows them to do something right and do it fast, and get it done. But if it's done wrong, it will have to be fixed! So far, I think they've been doing it right with the mass transit system and other major infrastructure projects throughout China.

So this is a very good success story. The problem with high-speed trains is that the fare is too expensive for the factory workers. So they're reducing speeds by 10 to 20 percent and lowering costs to accommodate the general public. If a high-speed railroad passes through a town, the real estate will boom to pay for it. But before the town can boom it needs people and business. Right now, a lot of trains are not so full, so about six months ago they slowed the construction to wait for economic growth to catch up a bit.

In the same way that China is taking a lead in mass transit, it has become a leading force in building tall. What are some of the challenges confronted in China?

Land is expensive, so if contractors can increase the floor/area ratio, they can build more apartments or office space and make more money. Also, it's a question of supply and demand. Newlyweds used to live with their parents but now they want their own space. Also, the family house is leaking and needs repair, so who wants to stay there, with no bathroom? People want to move into a new condominium, so there is a real need for new apartments for the middle-aged, definitely for the young, and even for the elderly because their very old apartment is not suitable for modern dwelling.

So the demand is there for residential buildings in major cities, less so in smaller cities where old farmhouses remain. But if people rebuild, they do not want a one-story brick farmhouse anymore but a modern building or townhouse in the village area. The townhouse consists of the ground floor that has a living room and bedrooms for the elders, the second floor for the first brother, and the third floor for the second

下的将在今后的5-10年内完成。在非常短的时间内，中国已经建立了世界上最大的高速铁路网络，其速度高达380公里（236英里）每小时。这真是令人惊叹，中国必须有公共交通，想象一下如果香港、北京或上海没有公共交通将会怎样！

民主是有好处的，但中国的中央决策机制同样也有好处，这能让他们去做一些正确的事情，并且做的非常快，迅速完成。但是如果做错了，就必须修正了！到目前为止，我认为在整个中国的公共交通和其他基础设施项目方面，他们做的不错。

这是一个好的成功故事。高速铁路的问题在于，车票对于普通工人而言太贵了。因而，他们降低了10–20%的速度，调低了价格以满足普通公众的需要。如果一条高速铁路穿过一个城镇，当地的房地产将迅速发展来支付建设花费。但是在城镇能迅速发展前，它需要人口和商业。现在，许多火车并不满员，所以大约六个月前，中国放缓了建设进度，以等待经济发展的速度赶上来一点。

与现在中国在公共交通领域的领先位置相同，在高楼建设方面，中国也已经是一股主导力量了。在中国，面临的问题有哪些？

土地非常昂贵，因此如果能够提高楼层与建筑面积比，承包方可以建立更多的公寓或者办公区，从而挣到更多的钱。而且，这也是一个供求的问题。新婚夫妇过去与他们的父母住在一起，但是现在，他们想要自己的独立空间。而且，父母的房子往往有渗漏现象，需要修理，那么谁想住在这种没有卫生间的房子里呢？人们想要搬进崭新的私人住宅，因此，对于中年人，尤其是对于年轻人来说，对新公寓的需求是真实的——甚至于对老年人来说也是如此，因为他们的旧公寓已经不再适合现代的居住条件了。

因此，在主要的城市，对于住宅楼的需求变得显而易见，而在小一点的城市，由于旧式农舍的存在，这种需求会稍微小一些。但是，如果人们重建房子，他们不会再想要一栋一层楼的砖砌农舍，而是想要一座现代建筑或者乡村地区的"联排别墅"。这种"联排别墅"的第一层有客厅和老人的卧室，第二层则是兄弟中老大一家的，第三层则是老二一家的。如果有了孩子，那就是三代同堂，而你最好把房子盖得再高一点。在中国，有许多这种"联排别墅"和高层公寓大楼正在兴建。

Ping'An International Finance Center, Shenzhen, China. (Kohn Pedersen Fox, est. completion 2015)
中国深圳国际金融中心。（KPF建筑事务所，预计2015年完成）

brother. If children come along, it's three generations, so you'd better build high. There are lots of townhouses and high-rise condo buildings being built all over China.

224

Back to high-rise buildings: do clients make specific reference to the World Trade Center, asking for a building that will not fail if hit by a plane?

Surprisingly not, but definitely indirectly. The crime rate in China is relatively low compared to other developed countries. Also, internal security is strong. Major developers have government ties so they believe in their safety system. They never ask about a plane flying into a building because they do not believe that will happen. Actually, their concern is seismic. They worry about earthquakes more than terrorists, and about fires in high-rises more than anything.

Earthquakes can come any time; you never know. Basically they're categorized as frequent (about 50 years' occurrence), moderate (500 years' occurrence), and severe (2,500 years' occurrence). Based on the current building code, all the high-rise buildings in China are built to withstand severe earthquakes. That is why it is much safer to stay in a high-rise building in China than in a village house. Village houses are not as well engineered.

But whereas earthquakes are natural occurrences and can be structurally mitigated, anything can cause a fire. Fire-fighting 100 floors above the ground can be difficult, so the code in China is very stringent. There are minimum numbers of stair egress and the distance traveled to reach a staircase is strictly regulated. Right now for a commercial building, hotel, service apartment, condo, and office, the new code requires that every 10 to 15 stories you put in a refuge floor where people can wait for help, protected by sprinkler and fire safety equipment. There is a fire separation between each 10- to 15-floor zone, and they are supposed to have their own fire-fighting facility within each zone. So the most you sacrifice is one 10- to 15-floor zone.

What challenges are posed when a building exceeds 100 stories?

Well, the taller you build, the heavier the loading on the foundation, so you have to make sure the soil is good enough, and what kind of pilings you need and the kind of foundation for support. And then you have to worry about wind load from typhoons or hurricanes and seismic loading. You have to design for all these requirements.

We do a lot of wind tunnel tests to ensure that the building is designed for the proper wind loads. We also want to shape the building form so the wind can pass around it

回到高层建筑的话题：客户们有没有专门提到世贸中心，想要一栋如果受到飞机撞击而不至于倒塌的大楼？

很奇怪的是，并没有。不过，间接的肯定有：与其他发达国家相比，中国的犯罪率较低，而且，其内部是非常安全的。主要的开发商同政府有联系，所以他们信赖其安全体系。他们从未询问，如果一架飞机飞进一栋大楼会怎样，因为他们不相信这会发生。实际上，他们关心的是地震，他们担心地震多过担心恐怖分子，最担心的是高层火灾。

地震随时会来，你从不知道什么时候即将发生。地震大致分为多遇地震（大概50年一遇），中强地震（500年一遇）和特大地震（2500年一遇）。当前的建筑规范要求：中国所有的高层建筑都要能够抵御特大地震！所以在中国，待在高层建筑中要比在乡村别墅中还要安全，乡村别墅的设计并没有这么好。

不过，尽管地震是自然事件，而且可以从结构上进行缓解，但所有的东西都可以引发火灾，距离地面100层以上的消防工作会很困难，在这一问题上，中国的规范非常严格：楼梯出口的最低数量，前往楼梯间需要的距离，都有着严格的规定。现在，对于商业建筑、宾馆、酒店式公寓、公寓大楼和办公室来说，新的规范要求：必须在每10—15层之间安置一个避难层，人们可以在那里等待救援，并受到喷洒器以及其他消防设备的保护。每10—15楼的区域内有一道防火隔墙，每个区域内都要有相应的消防设备。因此，成本最高的就是这种10到15楼的区域。

当建筑超过100层时带来的挑战是什么？

楼建的越高，地基所承载的就越重，所以你必须确定有足够好的土壤，确定需要什么类型的桩基和地基来支撑建筑。然后你要当心台风、飓风、地震带来的巨大风力载荷，设计必须考虑所有这些要求。

我们已经做了很多风洞测试，以保证建筑设计合乎正常风力载荷的要求。我们也想要好好设计建筑的形状，让风能够环绕着建筑，而不至于产生太多的旋流或涡旋脱落效应。我们有足够的智慧够避开风，使它不会引起太多负面的风效应——比如，利用方形建筑的步进角来分散湍流。因此，我们与建筑师一起工作，为超高建筑设计合适的空气动力形式。

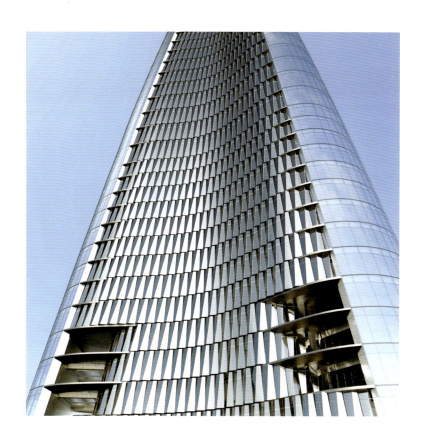

Wuhan Greenland Center, exterior detail
武汉绿地中心外部细节

这主要是针对表层的处理，而不是针对结构自身，对吗？

是表层。当一栋建筑的结构弯曲得宛如椒盐脆饼时，建造起来就很困难，它不是支撑楼层负重的最有效方法。不过，我们弯曲一栋建筑的幕墙，使之类似于手表的表盘。在令建筑构造简单、建设起来经济实惠的同时，建筑中所有的柱梁都可以保持垂直，而且只让平板随着扭曲的幕墙而旋转。

一栋建筑在高度上有没有限制？

是的，当然有限制。我们最近刚刚在宋腾添玛沙帝写了一篇关于这个问题的文章。高度受到地基承载力和材料强度的限制，需要非常高强度的水泥和钢铁，但这些材料都有自己的强度极限。另外，建筑的形状也是个问题：一栋中空的建筑允许风穿过其间，因而你可以将它建的更高，而重量却不增加，风力载荷反而更小。高度自身只是一个虚拟数字。

我们受到材料物理学和现实的限制。限制确实会有，但大楼还是能建的非常高，绝对可以超过1000米。但是没关系，你所有的时间都在云中。

without causing too much swirl or vortex shedding effect. We are smart enough to cheat the wind so it does not cause too many adverse effects, such as using stepping corners in square buildings to break up turbulence. So we work with the architects to design the proper aerodynamic form for super-tall buildings.

Is that mostly the treatment of the skin, not the structure itself?

The skin. When a building structure is twisted like a pretzel it is very difficult to build; it is not the most efficient way to support the floor loading. However, we can twist a building curtain wall similar to the dial of a watch. All the columns can be kept vertical in the building and rotate just the slab with the twisting curtain wall while still keeping the building framing simple and economical to build.

Is there a limit to how tall a building can go?

Yes, of course there is a limit. We just wrote an article about this at Thornton Tomasetti. Height is limited by foundation capacity and material strength. You need very high-strength concrete and steel, and these materials have their strength limit. Also, there's the shape of the building. A hollow building allows the wind to pass through so you can build taller without much weight and less wind load. Height itself is a fictitious number.

We are restricted by the physics of materials and reality. There will be a limit, but it will be astonishingly high. We can definitely go beyond 1,000 meters [3,281 feet]. But it doesn't matter, you'd be in cloud nine all the time; you'll have to use a lot of elevators because you'd have to make so many transfers. You can't have a 1,000-meter [3,281-foot] cable to lift the elevator straight upward because the weight of the cable would eat up the elevator motor's capacity to lift it up. So you'd travel up the super-tall building zone by zone. Then how do you bring the water up and keep it in storage, how do you do it? The reality is, how tall do you really need the building to be?

We actually worked on a mile-high [1.6-kilometer] tower with well-known architects before, but the project stopped after the schematic design phase because of the world economic situation. We are now working with Adrian Smith + Gordon Gill Architects on the Kingdom Tower in Saudi Arabia, which is over 1,000 meters [3,281 feet] high. A mile-high [1.6-kilometer] building is definitely achievable if people have the right reasons and the money to do it. Technically it is feasible. It's just the economics of making it work.

What kind of structural system is required for a mile-high [1.6-kilometer] building?

Basically, you want to activate the entire width of the tower to resist lateral wind and seismic loadings. The principle is to make the building's full width work for you rather than just the tower core, which is much smaller. We have different ways of doing it, depending on how efficient you want to make the building. The exterior super-truss is an efficient system for lateral stability, but I think a mile-high [1.6-kilometer] building would require a combination of structural systems. You could use super-diagonals on the outside in addition to a tubular system, compared to a super-column and outrigger truss system.

It also depends on the shape of the building. One thing I am sure about is that the tower base will be bigger and the building form will taper inward as it goes up. It will not be a box going up because you have to cheat the wind. Only with a tapered shape will you lower the center of gravity and achieve more stability and less wind load at the top. Any smart engineer or architect would plan a mile-high [1.6-kilometer] building in such a stable, wind-friendly way.

你必然要使用许多电梯，因为不得不转乘很多次。不可能有一条1000米的索绳，能够将电梯垂直抬起，绳索的重量会耗光电梯电机的承载量，使之不能将电梯向上抬起。因此，你要分区在超高建筑中穿行。那么，你如何将水带上去，并储存在蓄水池中，你怎么做到这些？现实是，你真的需要多高的建筑呢？

我们实际上曾经与建筑师一起建造一个1英里高的塔楼，但由于世界经济形势，项目在图纸设计阶段就被叫停了。现在我们与AS+GG建筑师事务所合作，为沙特阿拉伯的"王国塔"工作，它的高度超过1000米（3280英尺）。如果人们有合适的理由，足够的金钱，1英里高的建筑肯定可以实现，技术上自然是没有问题的，问题只是在于实现它的经济条件。

一英里高的建筑需要什么样的结构系统？

大致上，你需要充分激活整座大厦的宽度，以抵抗侧风和地震产生的冲力。原则就是，利用建筑的所有宽度，而不是仅仅利用楼芯，因为后者毕竟小得多。根据你想要的建筑功效，可以有各种办法来做到这些。外部的巨型桁架乃是保持横向稳定性的一种有效系统，但我认为，1英里高的建筑可能需要一种多结构系统的组合。与1种巨型柱梁和悬吊桁架系统相对应，你可以为一套管状系统外层加上巨型对角线结构。

它也依赖于建筑的形状。我可以确定的一点是，大楼的基底较大，而建筑形式将随着不断增高而呈锥形逐步减小。它不会像一个盒子那样向上延伸，因为你必须避开风。只有呈现为锥形，你才能降低重心，获得更大的稳定性，以及使顶部的风力载荷变小。任何聪明的工程师或建筑师，都会以这样一种既稳固又与风融洽的方式来设计一栋1英里高的建筑。

将大楼置于带有良好地基的地方，这一点对于控制成本非常重要，但我们仍然还是可以在结构上动一些心思。每隔30-50层之间，就应该为工作人员提供一些站区，否则大楼中就有数百人吃午饭而上上下下，你还必须每次都完成一个阶段的建设。关于如何处理超高建筑中建设的分阶段问题，人们已经做了许多有趣的研究。

在200层的大楼上如何清洁窗户？你花了时间来选择玻璃墙等东西的颜色，但是现在的污染情况严重，你每隔多久能看到大楼的顶部呢？它们脏的太快了。我更希望看到污染得到控制，否则你就不会去欣赏建筑本身的设计。

151 Incheon Tower, Songdo International City, Incheon, South Korea.
(John Portman, est. completion 2015)
韩国仁川松岛国际城151大楼。（约翰·波特曼建筑设计事务所，预计
2015年完成）

新材料如何帮助超高建筑?

好，让我们回到基础。集料，也就是糙石的强度，它对于混凝土的抗压强度有着重要作用，所以我们试图找到更强的集料。另外，类型正确的胶合剂和添加剂能够显著增加水泥的强度。我们也需要高强度的配筋，但尚未看到新材料可以让我们绕过水泥和钢铁。

玻璃钢要比钢铁更坚韧。那么我们为什么不用玻璃钢来做配筋呢? 你的设计必须考虑到防火等级。当温度很高的时候，玻璃钢会在火中融化，所以使用玻璃钢受到限制，除非你处于一个特殊的环境中，在那里火灾不是一个问题。实际上，我们正

It's important to place the tower in a location with a good foundation to control cost, but construction-wise it can be done. They just have to put in station areas for the workers every 30 to 50 floors, otherwise you'd have a few hundred people trying to get up and down for lunch. You'd also have to build one stage at a time. A lot of interesting studies have been done about how to deal with construction staging for super-high-rise buildings.

How do you wash the windows on the 200th floor? You spend time choosing the color range of your glass wall, etc., but we have so much pollution now, how often do you even see the tops of the buildings? They get so dirty so fast. I would really like to see pollution controlled, otherwise you cannot appreciate the architectural design.

Have new materials facilitated super-tall buildings?

Well, we go back to fundamentals. The strength of the aggregates, the coarse stone, contributes to the compressive strength of concrete, so we try to find strong aggregates. Also, the right kind of cement and the admixtures can significantly increase the strength of concrete. We also need high-strength reinforcing bars, but we haven't yet seen new materials that will steer us away from concrete and steel.

Fiberglass is stronger than steel. So why not use fiberglass reinforcing bars? You have to design for fire rating. When the temperature is very high, fiberglass will melt in a fire, so use of fiberglass is limited unless you are in an environment where fire is not a problem. Actually, we are studying a wind farm project where we explored fiberglass reinforcement since there's effectively nothing combustible in the area. But in truth we didn't need fiberglass for strength and we have enough steel and concrete to take care of it, so why use a more expensive material? It's a balancing act.

Has the recent building boom rendered Asia an architectural laboratory for new design?

A lot is happening in developing countries like China, India, Malaysia, Vietnam, and even Jakarta, Indonesia, where we are doing a 600-meter [1,968-foot] tower in a severe earthquake zone. The economies of these countries are growing, and since land is always expensive, they want to build tall. Every iconic high-rise becomes a symbol of a certain city and also of a particular country and specific owner. It is like a trophy, which increases the value of all your adjacent properties. Super-high-rises cost more to build and cost more to maintain, so the economic benefits are more secondary than direct.

Is it possible to chart a building's proportional cost-to-height increase in increments of 25 or 50?

The cost of building one floor below ground is usually 50 percent or 100 percent more than building one floor aboveground. Beyond that, the cost of a 100-story building is easily $300 to $500 per square foot [$3,230 to $5,380 per square meter]. But a mid-rise 30-story building can also cost as much. Many factors affect building construction cost. But if you ask if there's a premium for a high-rise, the answer is definitely yes. It's how you manage that premium: if you have a very rational structural system, a straightforward architectural system and shape, then the premium will be less for high-rise buildings.

If you have a good foundation, soil, or bedrock, then basically the premium for the foundation is negligible, whether it's

在研究一个风力农场项目，我们在那里开发玻璃钢筋，因为那个区域没有什么是易燃的。实际上，我们并不因为强度而需要玻璃钢，我们已经有足够的钢铁和水泥来解决强度的问题，所以，为什么要使用一种更昂贵的材料呢？这是一种平衡的行为。

当前建筑业的蓬勃发展是不是让亚洲成为一个新建筑设计的实验室？

发展中国家，比如中国、印度、吉隆坡、越南甚至印度尼西亚的雅加达（在那里，我们正在为一个严重地震带上的地区，建设一栋600米<1968英尺>高的大楼）正在发生很多事情。这些国家的经济正在发展，而土地十分昂贵，所以他们想要盖高楼。每个标志性的高层建筑都成为一个特定城市、一个国家和某个特殊拥有者的符号象征。这就像一个纪念奖杯，增加了所有周边房产的价值，建造和维护超高建筑要花费更多，因此经济收益并非是直接的，而是次要的。

有没有可能用图表列出在25层或50层增量的定额下，费用随高度增加的比率？

地下一层的建筑费用，一般来说要比地上一层的建筑高出50%到100%。除此之外，一栋100层建筑的花费很容易就达到300-500美元每平方英尺（3230到5380美元每平方米）。但一栋30层中等高度的建筑也会花费这么多，有许多因素影响建筑建设的费用。但如果你问，是否高层建筑有额外增加的费用，答案是肯定的，问题是如何管理这种额外增加的费用：如果你有一个非常理性的结构系统，一个简单明确的建筑系统和形状，那么对于高层建筑来说，额外增加的费用就要少一些。

一般来说，如果有好的地基、沙土、基岩，地基上所要花费的额外费用就可以忽略不计，无论建筑是30层高还是100层高。相反，如果你想要将一栋100层的高楼放在贫瘠的土壤上，额外增加的费用定然会翻倍。通常，楼层横梁和大楼板坯的费用是相同的。不过，你确实要为一座高楼的侧面支撑系统以及支撑90层楼而非20层楼的柱子支付额外增加的费用，同样，你也要为机械系统支付额外费用。但到底要支付多少？可能是30%，不会是所有费用的100%。所以，如果你拥有一块土地，政府允许你盖更多的楼层，你就可以卖出更多的楼层单元。那么，假如可以收回成本并赚到钱，为什么不盖一栋100层的高楼呢？

Petronas Towers in Kuala Lumpur City Centre, Kuala Lumpur, Malaysia. (César Pelli, 1994)
马来西亚吉隆坡城市中心的双子塔。（西萨·佩里建筑事务所，1994年）

按照定义来说，超高建筑是单一的、具有标志性的，但如果有一群互相毗邻的超高建筑，相互之间一栋比一栋高，那会发生什么？

如果一座城市有许多100层高楼，你能将它们租出去吗？一些超高建筑现在是完全租出去了，但其实有很多年的时间，事情并非如此。所以，在开始的时候，不能有太多100层高楼，因为商业市场消化不了它们。

上海是个例外，因为它是中国的金融中心。每个在中国做生意的公司都希望在那里拥有一间办公室，正是这推动了需求，人们需要好的住房和好的办公室。在香港也是如此，它是东南亚的中心，也是进入中国的通道。深圳也是这样，我们正在那里建设一栋126层的保险公司总部大楼，因为他们需要它。需要是否会一直持续下去呢？当然不会，它会到达某个特定的点，然后停一会，消化一下，之后再转到另一个方向去，那是危险的。在西方国家，在家工作变得越来越普遍了。但在中国，老板还是想要看到所有人，但是将来，整个的趋向还是会跟美国一样，进行机构精简。

a 30-story or a 100-story building. By contrast, if you want to put a 100-story building on poor soil, the premium will definitely double. Usually the cost of the floor beams and slabs of a building are the same. However, you do pay a premium for a tall tower's lateral bracing system and for its columns that support the 90 stories instead of just 20 stories. You also pay a premium for the mechanical systems. But how much? Maybe 30 percent; it would not be 100 percent of the total cost. So if you own a piece of land and the government allows you to build more floor area and you can sell more floor units, then why not do a 100-story building if you can recover the cost and make money?

Super-talls by definition are singular and iconic, but what happens if there is a cluster one right next to the other, taller, taller, taller?

If there are multiple 100-story buildings in one city, can you rent them out? Some super-tall buildings are now fully leased, but for many years they were not. So at the beginning you cannot have too many 100-story buildings because the commercial market may not be able to absorb them.

Shanghai is the exception, as it is the financial center of all of China. Every company doing business in China would like to have an office there. So that drives up demand. They need good housing and good office buildings. The same is true of Hong Kong, which is the heart of Southeast Asia and the entryway into China. Shenzhen as well. We are building a 126-story insurance company headquarters building there because they need it. Will the demand continue forever? Of course not. It will get to a certain point and then stop for a while, be digested, and then go in another direction. That's the danger. Working from home is more and more common in Western countries. In China the boss still wants to see everybody, but that trend could follow the U.S. in downsizing in the future.

How do you envision urbanization in 20 to 50 years?

Urbanization cannot sustain itself without a parallel infrastructure. Urbanization in Hong Kong is successful because they put in the infrastructure first. Shanghai, too. They've put so much money into highways, flyovers, tunnels, etc., and are building subway lines to sustain new growth. Beijing and all the major Chinese cities are doing the same, starting with the infrastructure.

India is planning many new developments, but developers are concerned about what kind of highway and road systems, sewer system, electric supply there will be to bring people to their fancy new high-rise buildings. In China, it is easier because the central government officials will sit down and plan together and make a quick decision on whether to build a subway system and other infrastructure projects. The money comes from the government. Land values next to the subway escalate immediately, so they can sell the land and generate revenue to recover their investment faster.

Debate, discussion, and voting usually slow down the process in democratic countries. China has the advantage here and, of course, they want to create more employment. They want to increase internal consumption instead of only exporting goods, and they encourage people to vacation in China and spend money at home. They are trying to beef up their internal economy. When the country is at peace and living in harmony they have better lives. People care first about their lives, their next apartment, their next trip. I am happy to see China growing this way, because with billions of people in a stable situation it's good for the world.

你如何想象未来20-50年间的城市化进程？

如果没有配套的基础设施，城市化是不能持续的。香港的城市化很成功，因为他们最先建设了基础设施。上海也是如此。他们已经在高速公路、立交桥、隧道等项目上投入重金，并且正在建设地铁路线以支持新的发展。北京和中国的所有大城市都是这么做的，首先从基础设施开始。

印度正在规划许多新的发展项目，但是开发商关心的是，什么样的高速公路、道路系统、下水道系统和电力供应系统，能将人们带到那些漂亮而又崭新的高层建筑中去。在中国这很容易，因为中央政府官员会坐下来一起规划，然后快速做出决策，决定是否建设一个地铁系统和其他的配套基础项目。需要的资金来自政府，地铁附近土地的价格会迅速升高，因此他们可以出售土地，产生税收收入，从而迅速收回投资。

在民主国家，辩论、商谈和投票常常会使整个进程变得缓慢，中国在这一点上有自己的优势。当然，中国想要创造出更多的就业岗位，想要增加国内消费，而不再仅仅出口商品；而且他们想要鼓励人们在中国度假，在国内花钱，加强自己的内部经济。当国家处于和平时期，人们和睦相处的时候，就会有更好的生活。人们在乎的首先是他们的生活，他们的下一座公寓，他们的下一次旅行。我很高兴看到中国以这种方式发展，因为十多亿的人生活安定，这对整个世界来说是件好事。

Shanghai Tower crests over Jim Mao Tower (center) and Shanghai World Financial Center. (Gensler, est. completion 2014)
从上海中心大厦金茂大厦（中心）以及上海环球金融中心。（金斯勒建筑事务所，预计2014年完成）

Leslie E. Robertson

STRUCTURAL ENGINEER, LLC.

Leslie E. Robertson, P.E., C.E., S.E., D.Sc., D.Eng., Dist.M.ASCE, NAE, AIJ, JSCA, AGIR, Chartered Structural Engineer (UK and Ireland), First Class Architect and Engineer, Japan. Dr. Robertson (Les) established his own firm, Leslie Earl Robertson, Structural Engineer, LLC, in January 2013 after more than a half century leading the New York practice that still bears his name (Leslie E. Robertson Associates/LERA, 1958–2012). During his distinguished career Les developed many notable structures, including the structural designs for the World Trade Center (New York), Bank of China Tower (Hong Kong), Miho Museum Bridge (Shigaraki, Japan), and Shanghai World Financial Center. Les currently directs the structural design for the Jamsil Lotte Super Tower (Seoul) and the KL 100 Tower (Kuala Lumpur). A member of the National Academy of Engineering and Distinguished Member of the American Society of Civil Engineers, Les has served in leadership positions on these and many other professional organizations. The IABSE International Award of Merit in Structural Engineering, IStructE Gold Medal of the UK, and Fazlur Rahman Khan Medal are but a few of the honors Les has received for his professional contributions. A Distinguished Alumnus of Berkeley (B.S., 1952), Les has been academically acknowledged as well through honorary doctoral degrees from four universities, including Notre Dame, Lehigh, Western Ontario, and Rensselaer Polytechnic Institute.

莱斯利·罗伯森

莱斯利·罗伯森建筑事务所

莱斯利·罗伯森，职业工程师、土木工程师、系统工程师、科学博士、工程博士、美国土木工程师协会杰出专家，美国国家工程院、日本建筑学会、智能社区联盟、电离辐射顾问小组，特许结构工程师（英国和爱尔兰），第一流的建筑师和工程师（日本）。在以他的名字引领纽约实践半个多世纪后，莱斯利·罗伯森在2013年1月建立了自己的公司莱斯利·罗伯森建筑事务所。在他杰出的事业中，设计了很多建筑结构，包括世贸中心（纽约），中银大厦（香港），日本美秀博物馆桥（信乐）以及上海环球金融中心（上海）进行了结构设计。莱斯利现在正负责乐天蚕室超级大楼（首尔）以及吉隆坡100大厦（吉隆坡）的结构设计工作。作为美国国家工程学院的会员，以及美国土木工程师学会的杰出会员，莱斯利在这些机构以及其他一些专业机构中担任过领导职务。曾经荣获国际桥梁与结构工程协会颁布的国际杰出贡献奖，英国工程师协会的金质奖章以及法兹勒·瑞海姆·坎恩，这些只是莱斯利因出色的专业贡献而获得的一小部分荣誉。作为伯克利的杰出校友（科学学士，1952年），莱斯利也在学术上得到了行业的认可，他获得了四所大学的荣誉博士学位，包括圣母大学、利哈伊大学、西安大略大学和伦斯勒理工学院。

For complete biography, please refer to the Appendix.

完整的传记，请参阅附录。

Puerta de Europa, Madrid, Spain. (Philip Johnson, John Burgee, 1996)
西班牙马德里欧洲门。（利普·约翰逊和约翰·伯吉，1996年）

What has been your experience in China?

We're now working on five or six buildings over 500 meters [1,640 feet] high, all in the Far East. We have done over $1.5 billion worth of buildings in China, including Hong Kong, where we've worked for many years.

Today, there are few Western firms doing Construction Documents in China, because Chinese architects and mechanical engineers must be involved in order to successfully pass through the approval process. I believe that the Bank of China Tower, Hong Kong, was the last Chinese project where we did the CDs.

How have new concerns for sustainability impacted architecture?

I would say that 99 percent of people have no clear idea of the meaning of "sustainability." And that's true also for many architects. Sustainability impacts us as structural engineers because we are a part of the design process. Basically, we have two construction materials: steel and concrete (including reinforcing steel). Imagine a structure lasting 75 or 100 years; at the end of that period the steel frame is able to be reformed into new structural shapes so as to produce a new building. In a concrete building, perhaps the reinforcing bars can be salvaged, but the basic concrete material is more difficult to reuse. Cement production has got to be close to, or perhaps is, our largest source of greenhouse gases.

Many years ago, we would specify a minimum cement content in our concrete. Now, we specify a maximum cement content, directing contractors to use alternative products such as fly ash, additives, and the like. In other words, we use the same amount of cementaceous material, just less Portland cement. Strong arguments can be made for concrete construction, and most of our designs are totally dependent on the use of concrete. At the same time, it is important to look at every aspect of your building to see if there is some way you can reduce the energy required to make, transport, and install the structural system. The later demolition and recycling of the structure should be considered.

Which materials or technologies have allowed tall buildings to soar beyond the World Trade Center (WTC) and the Willis (formerly Sears) Tower?

Well, it's not the structural materials; steel and concrete have been improved, but they fundamentally remain the same as in the past. We have used very high-strength steels, but in ways that would not cross your mind immediately. For example, we

你在中国有什么经历？

我们现在正在做的五六栋高度在500米（1640英尺）以上的建筑，都是在远东。在中国，包括香港（我们在那里工作了很多年了），我们已经完成了总价值超过15亿美金的建筑。

今天，很少有西方公司在中国做施工图，因为中国的建筑师和机械工程师必须参与进来以便成功地通过审核阶段。我相信，香港的中银大厦是最后一个由我们做施工图的中国项目。

对于可持续性的新关注如何影响建筑？

我想说有99%的人对"可持续性"的含义并没有清晰的认识，对于许多建筑师而言也是如此。但可持续性对于我们这些结构工程师产生影响，因为我们是设计过程的一部分。根本上说，我们有两种建设材料：钢铁和水泥（包括钢筋）。想象一个屹立75或100年的建筑物，到其最后阶段，钢铁框架能够被重塑为一些新的结构形状，因此可以产生出一个新的建筑。在一栋水泥建筑中，可能钢筋会被回收利用，但是基本的水泥材料很难被重新使用了。水泥生产可以确定几乎是，或者就是，我们温室气体的最大来源。

许多年前，我们会在混凝土建筑中标示出一个最低水泥含量。现在，我们则会标示出一个最高水泥含量，指导承包商使用替代产品，比如粉煤灰、添加剂之类的东西。换句话说，我们使用同样数量的凝结材料，只不过普通水泥的数量下降了。支持混凝土建筑的人可能会给出有力的论据，而我们大多数设计都完全依赖于混凝土的使用。同时，注意到你自己建筑的每一个方面，看看你是否能以某种方式减少建造、运输和安置结构系统时需要的能源，这一点是很重要的。后期结构的拆除和回收利用也应当被考虑在内。

哪种材料或技术可以让高层建筑超越之前的世贸大厦（WTC）和威利斯（之前的西尔斯）大厦？

这不是结构材料的问题；钢铁和水泥都已经改进过了，但它们本质上还是和过去一样的东西。我们已经使用了高强度的钢铁，但是这并不会以某种方式很快地超出预期。比如说，我们用非常高强度的钢铁来避免负重过大，通过把它们做的更小来让结构组件不那么僵硬呆板。而对于新材料来说，从铆钉到高抗拉螺栓的改变已经是跨越了一大步了。韦拉扎诺海峡大桥可能是最后一个大型的采用铆钉的项

TECHNOLOGY Leslie E. Robertson

U.S. Steel Tower, Pittsburgh, Pennsylvania (Harrison, Abramovitz & Abbe, 1996)
宾夕法尼亚州匹兹堡美国钢铁大厦。（哈里森，阿布拉莫维茨和阿贝事务所，1996年）

目。这可能是因为工程师奥斯马·安曼想让这一结构维持更长的时间，他不太信任使用高抗拉螺栓。

让我们看看玻璃，过去十年间技术上的进步已经是非常明显的了。玻璃现在能够应付的内部/外部环境令人难以置信。就能源系统来说，新玻璃产品已经成为主要的贡献者。

随着建筑越来越高，电梯越发受到关注。除了美国，双层电梯现在几乎已经在其他地方成为标准配置。在世贸中心，我们有先升到中间第三点，然后再升到楼顶的短程电梯。所以我们也算有通向顶层的直达电梯。现在，电梯可以升到更高的地方。随着建筑越来越高，电梯轿厢不会变得越来越重，但是电梯的缆索会变重。缆索不仅需要牵引轿厢，它还需要牵引自身。

use very high-strength steels to avoid load, making structural components less stiff by making them smaller. As for new materials, well, the change from rivets to high-tensile bolts was a big step. The Verrazano-Narrows Bridge is, perhaps, the last major riveted structure; this may be because Othmar Amman, the engineer, intending the structure to last for many years, distrusted the use of high-tensile bolts.

Turning to glass, the technological improvements over the last 10 years have been extraordinary. It's just incredible how glass can now respond to the inside/outside environment. In terms of energy systems, the new glass products have been major contributors.

In terms of growing taller, it is much about elevators. The double-deck elevator is now almost standard everywhere except in the U.S. In the WTC, we had shuttle elevators that

*Shanghai World Financial Center, Shanghai, China
(Kohn Pedersen Fox, 2008)*
中国上海环球金融中心。(KPF建筑事务所，2008年)

went up to third-points and up to the roof. So we had nonstop lifts right to the top. And now, elevators can go even higher. As buildings get taller, elevator cabs don't get heavier, but the ropes do. Not only does the rope have to carry the cab, it has to carry itself.

The next big step, in my view, will likely be the cableless elevator. That would change height limitations in an instant. The equipment is already there—and used for various, mostly military-type elevators—but it's a huge financial investment on the part of the elevator manufacturer. Someday a Mitsubishi, a Schindler, an Otis, or others may do it. But for now, we're at the 500-meter [1,640-foot] range because that's a two-ride system: a shuttle to the sky lobby and then a local elevator. More than that is a hard sell because the executives, demanding fast service, occupy those top floors. One of the first ideas that comes to mind is the use of a linear induction motor to propel the cab.

在我看来，下一个重大进展有可能是无绳电梯。这将立刻改变高度上的限制。这种设备已经有了（主要被用于各种军用类型的电梯上），但是对于电梯制造商而言，这是一项重大的财政投资。也许有一天，三菱、迅达、奥的斯或其他的制造商会去着手做这些东西。但是现在，我们还是在500米（1640英尺）的距离范围内，因为这是一套双程系统：一部通向空中大堂的短程电梯，然后再换上一部区间电梯。由于主管们一般都在顶层，而且要求最快捷的服务，所以这项措施不仅仅是难以推广那么简单。人们立刻想到的是利用直线感应电动机来驱动轿厢。

我们为高达1英里（1.6千米）的建筑提供了一套概念设计方案。要明智地达到这样一个高度，就需要一个横向支撑结构。我们开发了一套系统，在其中，建筑内的使用空间是在对角线支架之内的。也就是说，建筑和支架是一体的。

高层建筑的设计是不是从一个结构概念开始的？

我相信，结构概念在建筑形状的发展中会是一个重要的因素。我们提供了一套概念设计方案，允许空气在建筑的不同层面上进行流通。通过打乱漩涡，在降低风载荷问题上，我们的效率将提升四倍。建筑师采纳了这个主意作为建筑形状的主导因素。将大型建筑放在一起需要不同领域富有想象力的人之间协同合作。

假定在高度问题上没有技术限制，当有人来找你，要求你建一栋2英里（3.2千米）高的建筑时，你会对他说什么呢？

首先，我将需要很大一片土地。要达到2英里（3.2千米）的高度，一个合适的地基是根本的。但是这个尺寸的楼层实在是太大了。因此使用面积应当被打碎，分为各种区块。不管这些要素是垂直修建起来的且中间有一些支撑还是倾斜的，都需要建筑师和工程师和谐共处，通力合作。从根本上说，它将会像是多元的建筑，可能并非直通顶层，而是各个部分以各种方式结合在一起形成一个单一的结构。你可以寻求简易型以节省材料，也可以实验新方法来进行设计和建设。

这种建筑会不会有核心区？

不一定。我们为电梯设计了一套系统，实际上，它走的是一条随机、倾斜的线路。奥的斯开发了一个方案，在其中，轿厢上升到了一个转换楼层，然后在轿厢再次纵向移动之前，它先进行了横向移动。这有可能会很不舒服，因为横向加速必须非常缓慢地开始。我们持有一些关于电梯斜向上升的专利。轿厢是倾斜的，所以不会有横向加速度的感觉——你甚至都不知道你是倾斜的。如果世界继续建设自由样式的建筑，我们最终可能会需要这种电梯。我们从未在这个专利上收取一分钱，而且坦白说我们也从未有这样的打算。申请专利的唯一原因是，你可以自己利用这种想法，当其他人不采用你的想法时，既是有问题的，也是代价昂贵的。

关于高层建筑的材料问题……

由于水泥在长时期内会收缩，所以这个问题是很严重的。水泥总是在收缩。让我们把建筑弄成一圈钢柱围着一个水泥核心。在某个程度上，这并没有问题，但长时期内的收缩，会使得楼层的倾

We produced a concept design for a mile-high [1.6-kilometer] building. To reach that height sensibly, it should be a braced structure. We developed a system where the occupied space within the building was within the diagonal braces. That is, the brace and the building ended up being one and the same.

Does tall building design begin with a structural concept?

I believe that the structural concept should be an important element in the development of the building shape. We produced a concept design that allowed air to flow through the building at different levels. By breaking up the vortices, we were able to cut the wind load by a factor of 4. The architects adopted the idea as a dominant element in the building shape. Putting great buildings together requires cooperation among imaginative people in different disciplines.

Assuming there are no technological limitations on height, what would you say to someone who came to you for a 2-mile-high [3.2-kilometer] building?

First of all, I would demand a significant amount of land. In order to rise two miles [3.2 kilometers], an adequate base is essential. But floors of that size are just too large. So the occupied space should be broken up into pods of some kind. Whether these elements go up vertically, with some kind of bracing between, or are inclined, needs to be worked out with the architect and the engineer in complete harmony. Fundamentally, it would be like multiple buildings, maybe not all getting to the top, but linked together in some way to form a single structure. You look for simplicity to save materials, and you try new ways of approaching the design and construction.

Would it have a central core?

Not necessarily. We worked out a system for elevators that, in fact, could follow a random, nonvertical path. Otis developed a scheme in which the car ascends to a transfer floor; then the car moves horizontally before again moving vertically. It would likely be uncomfortable because horizontal acceleration has to begin very slowly. We hold patents on an elevator that will ascend on a slope. The car tips so that there is no sensation of lateral acceleration —you don't even know you're tipping. Should the world continue to build randomly shaped buildings, we might end up needing this kind of elevator. We've never collected a penny on a patent and frankly never expect to. The only reason to patent is that you can use the idea yourself; it is both problematic and expensive to keep others from using your ideas.

Bank of China Tower, Hong Kong, China. (I. M. Pei & Partners, 1989)
中国香港中银大厦。（贝聿铭建筑事务所，1989年）

With respect to materials for very tall buildings . . .

The issues are huge for the long-term shortening of concrete. Concrete never stops shortening. Let's take a building with perimeter steel columns and a concrete core. There's nothing wrong with this up to a point, but with long-term shortening, the slope of the floors becomes excessive. Additionally, the steel columns go through temperature change, shrinking in the summer and growing in winter. So the floors slope up and down. We ran levels in an existing building to determine acceptable levels of slope. Where we change all of the columns to concrete, are they going to shorten with time at the same rate? Nonsense; shortening has to do with the surface area of the material compared to its volume, etc. There are many different reasons that concrete vertical elements shorten differently.

In the Bank of China in Hong Kong, at each refuge floor, we transfer loads from small steel columns out to the big concrete corner columns, so they effectively restart every 15 floors or so. (I'm making it sound much more simplistic than it is, but that's the concept.) Still, it's not the same, because the big concrete columns and the big concrete walls shorten

斜变得越发严重。另外，钢柱也会经历温度的变化，在夏天收缩，在冬天膨胀。因此楼层上下都是倾斜的。我们在一栋现有的建筑中水平跑动，以确定倾斜可接受的程度。在我们将所有钢柱改为水泥的地方，它们会不会以同样的速度和时间收缩变短呢？收缩变短与材料表面区域同其体积的比值有关。水泥垂直线会有不同的收缩变化，这是有许多不同原因的。

在香港中银大厦每个隔火层，我们都将负重从小的钢柱转移到大的水泥角柱上去，所以它们大约每隔15层就重新开始了（我把它说的听起来比实际情况要简单得多，但我所说的只是概念而已）。情况还是发生了改变，因为大型的水泥柱和大型的水泥墙会以不同的程度收缩。我们采用伸臂桁架以及类似标准的东西来解决这个问题。在一些建筑中，伸臂桁架和对角线结构在一定张力下必须分开。然后在你上紧螺丝之后，负重会变为零。从设备层入手是非常容易的，伸臂桁架对角线出现在空中大堂的楼层中，顺带说一句，非常美观大方。

结构都是精心设计的，因此它能够承载支柱

和桁架的缺失，重量实际上已经转移到结构的其他部分上去了。那些显而易见容易损毁的区块，有着多种复合的负载途径。有一种重复但很稳固的方式，我们从上世纪50年代开始就已经应用在设计中了。在一栋大型建筑中向世界展示你所有的结构系统并不是很明智的。我们并不主张要说出所有的秘密，一个人需要对这些东西比较敏感。

如果真的有重要的想法，那就是重复性和稳固性，以及多种负载途径。我知道的，在高层建筑上的第一个例子是伦敦的罗兰波英特公寓大楼。那里的一个火炉发生了天然气爆炸，摧毁了建筑的一整个角。这是因为在那里没有第二种负载途径。世贸中心则是重复性的一个完美例子，它满目疮痍地挺立着却没有倒塌。是的，火灾烧毁了建筑，尽管许多支柱都被摧毁了，但结构依然能够挺立。这样的一种能力应当被应用在所有大型和不是那么大型的建筑中。令人期待的是，工程师们已经理解了这样的观念。看看俄克拉荷马城的建筑，在那里一次小型的汽车炸弹爆炸就摧毁了一半建筑。

你是纽约世贸中心的结构工程师，自从911事件以来，有没有客户单独提出要求，想要让他们的建筑能够承受一架满负荷的波音747的撞击？

没有。实际上，我会说我们可以根据这种情况进行设计，而你将要为此进行补偿。

让我们回顾一下：在1945年，一架B-25米切尔轰炸机撞到了帝国大厦。确实有一些结构性的损毁，导致了一个区域内数人丧生，但建筑的结构整体并没有受连累。撞到世贸大厦的飞机与帝国大厦的相比，在能量上有着巨大的提升，因为飞机当时正在超速飞行。然而，尽管结构损毁非常严重，建筑依然挺立着。

想象一下，你正在世贸中心中组装飞机，它占据了几乎整个楼层！如果看世贸中心北侧，你会看到又大又长的翼板，而机翼已经从结构支柱中伸出去了。飞机变得越来越大。问题是，为它们而进行设计是否是明智的。

世贸中心的结构是为低速低空飞行的波音707设计的，而非实际撞击建筑的高速飞机。小型飞机撞击建筑总是有一定规律性的，但是如果飞机的体积和新空客车一样甚至更大，那么……我想，至少在数年之内，如果有人想要劫持飞机，乘客的反应会是站起来并进行战斗，而不是坐在那里。乘客可能是最大的安全因素，也可能是唯一的安全因素，

at different rates. We use outrigger trusses and the like that are calibrated to deal with that problem. In some buildings, the outrigger trusses, the diagonals, have to be disconnected after a certain amount of strain. The load then goes to zero, after which you retighten the bolts. Access is very easy from the mechanical floors. The outrigger truss diagonals appear in the sky lobby floors—very handsome, by the way.

The structure is designed so that it is able to tolerate the loss of columns or trusses, with the loads transferred to other portions of the structure. There are multiple load paths for sensibly destructible pieces. There is a kind of redundancy, robustness, built into the design that we have used since the mid-1950s. It's not necessarily smart to tell the world all about your structural system in a major building. We're not advocates of telling all the secrets. One needs to be sensible about these things.

If there is a really important thought, it is the issue of redundancy and robustness, and multiple load paths. The first example in a high-rise that I know of was Ronan Point, an apartment tower in London. There was a gas explosion in an oven and it took down a whole corner of the building. That was because there was no secondary load path. The WTC was a wonderful example of redundancy. It stood with gaping holes in its sides and did not collapse. Fire brought the towers down, sure, but the structure was able to stand despite the destruction of many columns. That kind of capability should be built into all major and not-so-major buildings. Hopefully, engineers have picked up on that concept. Look at the Oklahoma City building, where a little truck bomb brought down half the building.

You were the structural engineer for the World Trade Center in New York. Since 9/11, have clients specified that their buildings withstand the impact of a fully loaded 747?

No. Actually, I would say we can design for that circumstance and you would have to compensate us for doing so.

Let's go back: in 1945, a B-25 Mitchell bomber hit the Empire State Building. The impact did some structural damage, caused the death of persons in the area, but the structural integrity of the building was not compromised. The planes that hit the WTC represented a huge increase in energy over that experienced by the Empire State Building because the planes were flying above the rated speed. Still, while the structural damage was extraordinary, the buildings stood up.

World Trade Center, New York, NY (Minoru Yamasaki, 1973)
美国纽约世贸中心。（山崎实，1973年）

因为我不知道我们是要根据波音747还是空中客车来进行建筑设计。这是可以做到的，但这是不是一个好主意却是存疑的。

911事件之后，我认为我作为结构工程师的职业结束了。但香港一家最大的开发商要求与我会谈。他们有许多大型建筑正在建造之中，所以他们想要知道该做些什么。他们的结构工程师提出建议说，要将结构系统改变为水泥的。我则试图向他们说明其中存在的风险和他们能够进行的选择。最终做出的决定是继续水泥方案。但是使用水泥的建筑结构达到一定程度后就被取消了，因为建设花费了太多的时间。

当一个建筑师向你提出某个建筑概念作为建议，你首先会怎么做？

位置与你对建筑的思考方式有很大的关系。在新加坡，风速很慢，地动也不太剧烈；在洛杉矶，地动和风速都很轻微，大约是香港的四分之一。无论你在哪里，楼层负载都是有一定标准的。办公室就是办公室，公寓就是公寓，酒店就是酒店，只要我们在地球上进行建设，在哪里并不会有太大的差别。但是在一座非常高的建筑上，最大的侧面负载一半来自于风。香港的风力负载可以比东京的地震负载要大上四五倍。所以你一定是在你想要建设的环境中开始工作。

你对风力涡轮机有什么体会？

现在我们打算建造一栋高层建筑，其屋顶有三个垂直方向的风力涡轮机。这和我们确定气流问题，以及如何处理这些问题密切相关。这些东西很有意思，但顶层的风力涡轮机并不能解决我们的能源需求问题。边远区域的风力涡轮机，比如在海上或是山顶，才是最好的。假如在一个无风日，你却需要充分的电力服务。所以，其维护费用非常之高。

一种扭曲的形状是否会为大楼增加强度？

这取决于它是如何建造的。一座传统的高楼，包括传统结构，如果支柱是朝着一个方向倾斜的话，当被扭曲时强度会降低。第二个结构需要让水平拉力抵制住倾斜的支柱。当然，实际扭曲的是楼层而不是支柱。

Imagine you were to assemble the plane inside the WTC; it occupies almost the whole building floor! If you looked at the north side of the WTC, you'd see the big long sweep where the wing took out structural columns. Planes are getting bigger and bigger. The issue is whether it's sensible to design for them.

The structure for WTC was designed for a slow-flying, low-flying Boeing 707, not the high-speed planes that actually struck the buildings. Light airplanes hit buildings with some regularity, but if you go up in size to the new Airbus and ever-larger planes, well . . . I think that, at least for some years, if someone attempts to hijack an airplane, the passengers' reaction will be to stand up and fight, not to just sit there. The passengers, then, are the biggest safety factor they've got, and perhaps the only one, because I don't know that we want to design buildings for 747s or the Airbus. It can be done, but whether it's a good idea is questionable.

Following 9/11, I thought my career as a structural engineer was over. But I was called to talk with the board of one of the biggest developers in Hong Kong. They had several major buildings under construction and wanted to know what to do. Their structural engineer had recommended changing the structural system to concrete. I tried to reason with them about their risk and their options. The decision was made to go ahead with concrete. But the building's construction, using concrete, reached a certain point and then they trashed the idea because it took so long to construct.

When an architect approaches you with a building concept, what is the first thing you do?

Location is going to have a lot to do with the way you think about the building. In Singapore, the wind speeds are low and ground shaking is modest; in Los Angeles the ground shakes and the wind is pretty mild, about one-quarter of that in Hong Kong. Floor loads are pretty much standard no matter where you are. An office is an office, an apartment is an apartment, and a hotel is a hotel; there's not much difference where we construct it on the planet. But the largest lateral loads on a very tall building are most commonly from the wind. The wind loads in Hong Kong can be 4 to 5 times greater than the earthquake loads in Tokyo. So you start out with the environment in which you're going to build.

What are your experiences with wind turbines?

We're doing a tall building right now with three vertical-axis wind turbines on the roof. It has to do with the clarity of airflow and how you deal with it. These things are interesting, but rooftop wind turbines are not going to solve our energy needs. A remote location of the turbines, in the sea or on mountaintops, is best. On a windless day you need full electric service. Also, maintenance costs are very high.

Does a twisted form add strength to a tower?

It depends on how it's done. A conventional tower, including the structure, when twisted will lose strength where the columns are sloped in just one direction; a second structure is needed to resist the horizontal thrust from the inclined columns. Of course, it is practical to twist the floors but not the columns.

What is the effect of smooth air?

Prior to the WTC, because wind tunnels were constructed to test airplanes, not buildings, all wind tunnel testing of buildings was done with smooth air. I attended the first world Wind Engineering Conference in England; there, I met Alan Davenport. He was clearly so far ahead of everybody else! So I told him that we were working on the WTC project and that it presented an opportunity of a lifetime. I encouraged him to take a sabbatical from the University of Western Ontario to work with us, and that we'd turn his ideas into reality. We changed the whole concept of wind engineering.

To obtain a comparison, the WTC was tested with smooth airflow; it was the last time we ever did that. The tests were conducted in England in a facility run by Kit Scruton. The name probably means nothing to you, but you see his work all the time on car antennas, and elsewhere. That clear-air wind environment could demolish some tall buildings. You see, turbulence is very beneficial to buildings. It isn't a uniform thing but changes with wind speed and building dimension, and it quiets the wind-induced response of buildings.

We do not use smooth airflow. The speed of the wind is close to zero at the ground surface; the speed increases with height. This increase is due to the distance away from the roughness of the ground; were there no roughness, the wind speed would be constant with height.

What is the premium on today's distinctively shaped buildings?

Building shape affects the cost of everything, in part through the amount of required material. The most economical building is probably a straightforward Eiffel Tower–shaped form. A

平稳的气流有什么作用？

在世贸中心之前，由于风洞被建设起来是为了测试飞机而不是建筑，因此所有对建筑的风洞测试都是与平稳的气流相关的。我参加了在英国举行的第一界世界风力工程大会，在那里，我遇到了阿兰·达文波特。他很显然远远领先于其他所有人！因此我告诉他，我们正在为世贸中心项目工作，而这提供了一个千载难逢的机会。我鼓励他暂时停止在西安大略大学的工作，和我们一起共事，然后我们把他的想法变成现实。我们完全改变了风力工程的观念。

为了获得一个比较对照，世贸中心在平稳的气流下进行测试，那是我们做过的最后一次测试。我们在英国做了这个测试，设备由斯克鲁顿装备提供。你可能对这个名字没有感觉但你总是会在汽车天线和其他一些地方看到他的产品。晴朗无风的环境会毁掉一些高层建筑。你看，涡轮机对于建筑来说是大有裨益的。它并不是一个整齐划一的东西，而是随着风速和建筑的体积而发生改变，而且它也会平息由风引起的建筑回应。

我们并不利用平稳的气流。在大地表面，风的速度接近于零，速度会随着高度而增加。这种增加取决于离开高低不平的地表的距离，如果地表没有高低不平，那么风速随着高度按一定的比率发生变化。

今天一些独特形状的建筑有什么额外的费用吗？

建筑的形状会影响一切东西的费用，部分是受到所需材料的数量影响。最经济的建筑可能是直截了当的埃菲尔铁塔形状。有一个很好的论据可以支持这种形状的建筑。如果你在每一个层面上都考虑风力负载，同时你通过一个重力负载来平衡它，那么你是将风力负载引入到结构中而没有在结构系统中创造出任何弯曲或剪切。所以每一次你把它建得更高，也就更重，你都要扩展地基，这有点类似于埃菲尔铁塔所使用的办法。

所以一栋建筑的顶部越来越尖是有结构上的理由的？

绝对有。如果目标是把许多高价的房产放在空中，那么，你就要在埃菲尔铁塔模式下进行建设。这取决于你想要建得有多高。在500米（1640英尺）的范围内，就是我们现在所处的情况，建一个锥形的结构花费要少一些，但是另一方面，我们不是在建造结构，而是在建造建筑物，这意味着真正的地产。有许多建筑并不是锥形的。在那样的一种高度范围内，有直上直下的圆柱形，而不一定都是方形。我们现在正在做的都是同样的横截面，除了在最顶部的地方。

你是否把所有的高层建筑都看作是文化表达？

我不认为高层建筑有什么文化特征，也不存在所谓的美国高楼或中国高楼。从根本上说它们都是一样的，自身就是一个整体。一栋高层建筑就是它自己，它需要对将要在它上面建构起来的周边环境做出回应。可能阳光、阴影之类的问题，比所谓的文化差异要更重要一点。

当设计一座在地球上某个地方建造起来的高层建筑时，这并不意味着同样的建筑能够在其他地方建造起来。这是因为环境的负载（风力和地震）、地基的考虑、建筑规范，特别是为高层建筑设定的建筑规范是各不相同的。实际上，建筑规范很少说明与非常高的建筑联系在一起的问题。为什么会这样？因为这些高层结构乃是例外。比如说，如果你去深圳，那里有一条关于地震的规定；而如果你去澳门，那里也有一条关于地震的规定。在两个城市之间是香港，却没有这方面的规定！尤其是在香港，当然也在许多其他地方，我们会做两套计算评估。比如中银大厦，我们为自己做了一套计算评估，同时为建筑部门做了另一套计算评估，以说明这一建筑达到了当地的规定要求。

在近几年，工程师和建筑师的行业发生了怎样的变化？

电脑为开发建筑形状，既带来了祸害也带来了自由，在电脑之前，许多东西可能并不会被开发设计出来。年轻的建筑师今天在办公室工作，他们不太需要手绘草图。他/她一整天坐在电脑前面（可能从早到晚！），因为这就是今天建筑师制作建筑图纸的方式。比如说，在我们的办公室中，我们的工程师就不画图纸。我们雇佣的CADD（电脑辅助设计与制图）操作员大部分就是建筑师。但我们的工程师还能够绘图，并且实际上有一些比我们的建筑客户画的还要好。

生活总是快速变化的。曾经有一段时间，大学就训练学生的建筑和工程技术。比如，在英国，他们要进入技术行业，成为技术人员就要取得一定的成绩。他们要进行考试，获得认证，然后拥有一种经过确认的技术。我们应该在这里也这么做，但我

good argument can be made for a building with such a shape. If you take the wind load at any level and you balance against a gravity load at that level, you're getting the wind load into the structure without creating any bending or shear in the structural system. So each time you make it taller, and therefore heavier, you need to spread the base out further so it's kind of an Eiffel Tower approach.

So there is a structural justification for tapering in a building?

Absolutely. Where the goal is to put a lot of high-price real estate in the air, you're going to construct on the Eiffel Tower model. It depends on how high you go. In the 500-meter [1,640-foot] range, which is where we are now, it is less expensive to construct a structure that tapers, but on the other side of the coin, we are not constructing structure, we are constructing a building, which means real estate. There are many buildings that don't taper. In that height range, there are cylinders, not necessarily square, that go straight up. We're actually doing one right now that is the same cross section all the way up, except at the very top.

Do you see tall buildings as cultural expressions?

I don't think there is anything ethnic about high-rise buildings; there isn't a U.S. high-rise or a Chinese high-rise. Fundamentally, they are all the same, an entity unto themselves. A high-rise is a high-rise, needing to be responsive to the environment where it is to be constructed. Perhaps issues of sunshine, light and shadow, and so forth are more important than ethnic differences.

When designing a very tall building that's to be constructed at one place on the planet, it doesn't mean that the same building can be constructed at other locations. This follows because of environmental loading (wind and earthquake), foundation considerations, and building codes, particularly for very tall buildings. Realistically, building codes seldom address issues associated with very tall buildings. Why should they? These tall structures are the exception. If you go to Shenzhen, for example, there's a seismic regulation. And if you go to Macau, there's also a seismic regulation. Between the two is Hong Kong, with no seismic regulation! Particularly in Hong Kong, but also in a lot of other places, we do two sets of calculations. For the Bank of China Tower, we did one set of calculations for ourselves and another set for the building department, showing that the building meets local codes.

244

How have the engineering and architecture professions changed in recent years?

The computer has brought both a curse and a freedom to develop building shapes, many of which probably shouldn't be developed. The young architect working in an office today has little need for sketching by hand. He or she sits in front of a computer all day (and probably well into the night!) because that's how today's architect makes architectural drawings. In our office, for example, our engineers don't make drawings. We hire CADD operators who are largely architects. But our engineers are able to sketch, and, in fact, some are now better sketchers than many of our architectural clients.

Life is changing very rapidly. There was a time when colleges trained architectural and engineering technicians. In the UK, for example, they have a grade for entering technicians. They take an exam, they're licensed, and have a proven skill. We should have that here but we don't. Why?

But something should happen, particularly in architecture, because there is a gap between the Harvard graduate at one end and architectural technicians at the other. The lack of technicians is hurting the professions. We could stand fewer graduates of architecture and engineering and more technicians.

Where are the engineering and architecture professions headed?

Right now, for us, we sailed right through the last set of economic problems because, for many years, fully half of our work has been overseas, or at least outside New York. Seeing into the future is tough, but it's not too far away that the engineers and the architects of the rest of the world will catch up to the West, thus reducing our ability to obtain contracts for work overseas.

Architects and engineers from China, Japan, Korea, Vietnam, and the like are hardworking and talented. They ask, "Why can't we do better ourselves since we're located here and speak the language?" U.S. firms are trying to compensate for that by setting up offices overseas, some of which are quite large. We have found that much engineering work on buildings, even those in the United States, is outsourced to overseas offices. Everyone is looking for the lowest production price. That's a popular technique for obtaining design commissions, particularly among the large offices.

们没有。为什么？

因为有些事情将会发生，特别是在建筑业，因为在哈佛的毕业生与建筑技术人员之间存在着鸿沟。技术人员的缺乏正在损害这个行业，我们可以接收比较少的建筑和工程专业毕业生，但需要更多的技术人员。

工程和建筑行业要走向何方？

现在，我们正一帆风顺地度过最近一系列的经济困境，因为许多年来，我们几乎一半的业务都是在海外，或者至少是在纽约之外开展的。预见未来是很难的，但世界其他地方的工程师和建筑师将赶上西方世界，这一点距离我们并不遥远，因此减少了我们获取海外工作合同的能力。

来自中国、日本、韩国、越南和其他一些地方的建筑师和工程师都非常勤奋并且极具天赋。他们问："我们扎根于此，说着当地的语言，那么为什么我们不能自己做得更好呢？"美国公司现在试图通过设立海外办公室来弥补这一点，其中有些办公室还是非常大的。我们已经发现，建筑中的很多工程工作已经外包给了海外部门——甚至包括那些美国本土的工程工作。每个人都在寻找最低的生产成本。这是一种流行的获得设计协议的技巧，尤其是在那些大型的部门之间发生竞争时。

Lotte Super Tower, Seoul, South Korea. (Kohn Pedersen Fox, est. completion 2014)
韩国首尔乐天超级大楼。（KPF建筑事务所，预计2014年完成）

SECTION THREE / FORUM OF EXPERTS

Planning

Erich Arcement, Sam Schwartz Engineering
Thomas J. Campanella, University of North Carolina
Chengri Ding, University of Maryland
Philip Enquist, SOM/Chicago
James C.Jao, J.A.O. Design International
Jaime Lerner, Instituto Jaime Lerner
Liu Thai Ker, RSP Architects Planners & Engineers
Bruno Padovano, Universidade de São Paulo

第三篇 / 专家论坛

规划

埃里希·阿斯门，山姆·斯瓦茨工程公司

托马斯·J·康帕内拉，北卡罗来纳州大学教堂山分校

丁成日，中国土地政策与城市管理项目

菲利普·恩奎斯特，斯基德莫尔、奥因斯与梅里尔建筑师事务所

饶及人，龙安建筑规划设计顾问有限公司

杰米·勒纳，杰米·勒纳研究所

刘太格，雅思柏设计事务所

布鲁诺·帕多瓦诺，圣保罗大学

Erich Arcement
SAM SCHWARTZ ENGINEERING

Erich Arcement, P. E., PTOE, is senior vice president and general manager of the New York office of Sam Schwartz Engineering (SSE). He is directly responsible for all projects in regional New York. Previously, Mr. Arcement was the director of SSE's Traffic and Transportation Engineering group. He has extensive expertise in engineering, ranging from geometric design and analysis of roadway, bicycle, and pedestrian facilities to traffic impact studies, environmental assessments, pavement marking, and signage design. He has been with the firm since it opened in July 1995 and has managed hundreds of projects ranging from small impact studies to multimillion-dollar on-call contracts. Mr. Arcement's responsibilities include SSE staffing, quality control, and oversight of project finance. Mr. Arcement is a professional engineer and professional traffic operations engineer and earned a Bachelor of Engineering degree in civil engineering (Cooper Union for the Advancement of Science and Art) and a Master of Science degree in transportation planning and engineering (Polytechnic University).

埃里希·阿斯门
山姆·斯瓦茨工程公司

埃里希·阿斯门是职业工程师，职业交通运营工程师，山姆·斯瓦茨工程公司（SSE）纽约办公室的高级副总裁和主管。他直接负责纽约地区的所有项目。阿斯门先生之前是SSE交通和运输工程小组的负责人。他在工程方面有着丰富的经验，这些经验包括从道路、自行车和行人设施的几何规划、分析到交通影响研究、环境评估、路面标线、标识设计的诸多方面。自从这家公司1995年7月成立以来，他就已经在那里负责了数百项工程——包括从小型的后果研究到数百万美元的合同。阿斯门先生的职责包括SSE人事、质量控制和项目资金监督。阿斯门先生是一位专业工程师和专业交通运营工程师，他获得了城市工程方面的工程学士学位（库伯高等科学艺术联合大学），并获得交通规划与工程方面的科学硕士学位（科技大学）。

For complete biography, please refer to the Appendix.

完整的传记，请参阅附录。

*Helmsley Building uptown portal onto Park Avenue, New York (during Summer Streets,
when seven miles of roadways are closed to vehicular traffic, 2012)*
纽约赫尔姆斯利大楼通往公园大道的门廊（"夏日街道"活动期间，有7英里的
道路被封闭，无法通行，2012年）

What is the focus of Sam Schwartz Engineering?

Transportation-based design, to some degree overseas, in Pakistan, Russia, Canada, but our major focus is in the U.S., with our headquarters in New York. Traffic engineering used to be about moving cars, but cities have now started to focus on alternative modes and are doing transit-oriented developments specifically to reduce the number of cars.

Over the last four or five years, there has been a major change across the country, focusing on "complete streets" for all users, not just motorists. They've taken out capacity for cars to provide capacity for bikes, which is a huge shift in the way the city operates. There is a whole segment of midtown Manhattan called Broadway Boulevard, where certain streets are closed to cars. Studies show that traffic now actually flows better, in addition to providing much more space for pedestrians and bicycles. It's a question of capacity but also priority: how much time is given for pedestrians to cross versus how much time is given to cars to move through the intersection.

In developing a new city, how might different traffic flows be accommodated?

I wouldn't begin with roads but with transit, so hopefully the entire city would be connected by alternative modes instead of being connected by cars. Put in a system of pedestrian, bicycle, and transit networks to serve the majority of the population. Obviously, subways, buses, and rail can service many more people simultaneously with less impact. Biking and walking are low-impact and healthy. All of these modes are much more efficient than vehicles in general.

A major issue is getting goods and services off the road. If your major draws are in the city center, make it pedestrian-, bike-, and transit-oriented, and have the roads on the outside as necessary. Bringing vehicles into a densely populated area is what causes traffic.

Los Angeles is a great example. Since there wasn't a transit base and since everyone is very car-oriented, even with five- or six-lane highways it can take an hour to drive 10 miles [16 kilometers]. One of the amazing things we talk about all the time is the number of meetings I can attend in a single day in NYC. I can get to midtown in minutes. So certainly if you're starting from scratch, it's important to make sure that everything is connected by transit, pedestrian, and bicycle. Cars are almost secondary.

What is the optimum distance between transit stops?

It depends on what exists. A quarter-mile [.4-kilometer] radius is typically considered a comfortable walk. However, as

山姆·施瓦茨工程公司的业务重点是什么？

基于交通之上的设计，我们有一些海外业务，比如在巴基斯坦、俄罗斯和加拿大，不过我们的总部在纽约，业务重点还是在美国。交通工程过去主要是指汽车的流通，但现在城市开始关注一些其他不同的替代模式，特别是进行公交导向式的发展，以减少汽车的数量。

过去的四到五年，在美国有一个重大的转变：人们开始关注所有道路使用者的"完整街道"权，而不仅仅是机动车驾驶员的街道权。他们减少汽车的占用空间，以便为自行车腾出更多的空间，这是城市运作方式上的一次重大改变。在曼哈顿中城区有一整块称为百老汇林荫大道的地方，那里的街道拒绝向汽车开放。研究表明，这项措施不仅为行人和自行车提供了更多的空间，而且事实上，现在的交通也更通畅了。从设计的角度看，这是一个容量的问题，但也是一个优先权的问题：通过十字路口时，给汽车多少时间，与给行人多少时间，这二者是相对的。

在建设一座新城市时，怎样提供不同的交通流量？

从为一座城市画草稿起，我就不会从道路开始，而是从运输线开始，所以希望整座城市不是通过汽车，而是通过可替换模式连结起来。你可以加入一个行人、自行车系统，以及各种交通网络，来为城市中大多数的人服务。很显然，地铁、公交车和铁路可以在不产生太多不利影响的情况下为许多人服务。骑自行车和步行，都是影响较小且更为健康的选择。一般来说，所有这些方式都比机动车要更有效率。

一个很重要的问题是，让商品和服务远离道路。如果你主要设计的是城市中心，那么就让它以行人、自行车和公交为导向，让外围的道路成为必须。要知道，在一个人口稠密的地区引入汽车，乃是引起交通拥堵的重要原因。

洛杉矶是一个很好的例子。由于那里没有公共交通作为基础，而且每个人都以汽车为工具，甚至有很多五车道或六车道的高速公路，在上面行驶10英里（16千米）要花上一个小时。我们一直在说的一件很神奇的事情就是：在纽约市一天我能够参加的会议数量。我可以在几分钟之内到达中城区。所以可以确定，如果你从草稿出发开始设计，非常重要的一点是确保所有东西都被公交系统、行人和自行车系统联系在一起，并且使汽车总是排在第二位。

Traffic in New York City (2012)
纽约市的交通（2012年）

公交站点之间最佳的距离是多少？

这取决于实际情况。0.25公里辐射半径之内一般被认为是很适合步行的。然而，随着人们对公交系统越来越习惯，情况已经开始发生变化，他们现在期望步行的再远一些。自行车骑行的合适距离有时在2.5到5英里（4—8公里）之间，耗时接近15到30分钟。这很大程度上依赖于社区、便利设施的类型以及居住在社区中的人。就公交站点来说，一般是间隔半英里，但我还是想说，这也取决于土地的使用、人口密度和其他因素。

那么，该如何区分不同的交通流量，比如在香港，市中心如此拥挤，以至于他们发明了一整套提升式的街道网络？

将行人分隔开来，既有好处也有坏处。它消除了所有行人、机动车和自行车之间的冲突，所以现在过马路

people become more accustomed to transit, this has begun to change, and they are willing to walk longer distances. A comfortable biking distance is somewhere between 2.5 to 5 miles [4 to 8 kilometers], approximately 15 to 30 minutes. This is heavily dependent on the community, the type of amenities, and the people who live in the communities. With regard to transit stops, usually a half mile between them, but again, this is dependent on land uses, density, and other factors.

What about stratifying the different flows as in Hong Kong, for instance, where downtown is so concentrated that they created a whole network of raised streets?

There are pros and cons to separating pedestrians. It removes all the conflicts between pedestrians, vehicles, and bicycles, so it is much safer to cross the street. However, when a community is built around pedestrian bridges and interconnected tall buildings, the basic fabric of a city can be lost. Instead

252

of having a bustling street life with retail-lined streets and activities, you simply have cars on the streets with people inside. It detracts in some ways and can make parts of the city feel dead.

We worked on a project in one of the Canadian border cities with the United States where trucks were traveling through town separating certain communities. Our design was basically to bury highway traffic and put a park on top as a linkage. Local vehicular traffic would still be able to drive through areas, but the highway-speed vehicles were buried below park space.

One of the things that's happening throughout the U.S. now is studies on decommissioning large highway segments that have low demand but high capacity, and to actually bring traffic in on a grade-level boulevard, because highways tend to be barriers that break up cities. If you do it safely, it activates rather than separates the community.

What are the criteria for pedestrian safety?

There are many criteria and factors for pedestrian safety. I can outline some here but it would take a while to fully describe the criteria for pedestrian safety.

You don't want to have too long a distance to cross. Often you'll see wide boulevards with a large central median for people and pedestrian refuges. There are simple design techniques such as neck-downs or bulb-outs where the sidewalk expands at the intersections to provide a shorter crossing distance. It's also safer because by moving pedestrians out into the visible area of traffic, they can be seen by approaching vehicles instead of being hidden behind parked cars. These types of designs also calm traffic, which is an umbrella term for the many methods we use to reduce traffic speed and improve the pedestrian experience.

Speed bumps, speed humps, and speed tables all fall in the same category. Instead of speed bumps (which are rounded), or humps (which are wider and rounded), there's a speed table, where cars basically drive up an inclined plane and continue at the same level as the pedestrian and then go right back down, typically at crossing locations.

One of the design techniques we use is changing materials on roadways to those used in the pedestrian space. We recently used this type of design for a driveway at a hospital. Vehicles entering the hospital campus drive up a short incline to bring them up to sidewalk grade. The sidewalk texture is continued onto the driving path. The benefit is motorists feel they have entered into a space meant more for pedestrians than vehicles and therefore drive at a lower speed and are more careful. Pedestrians are then protected with decorative

要安全得多了。不过，当一个社区是围绕着过街天桥和高层建筑之间的连接物而被建立起来时，那么一座城市就可能失去了它的基本构架。靠着街边排列的零售小店，过一种熙熙攘攘的街区生活，这一场景已经没有了，你所见到的只是街上的汽车以及汽车中的人。它在某种程度上损害了城市，使得城市的一部分生活失去了生机。

在紧挨着美国的加拿大边境有一座城市，我们正在为该城的一个项目工作，在那里，卡车穿过城镇，将特定的社区分开。我们的设计大致是隐藏高速公路，在上面修建一座公园作为连结点。当地的来往车辆依然可以驾驶通过这个地区，但高速公路上的交通工具则隐藏于公园区域的下方。

当前遍布全美都在发生的一件事，就是研究是否该关闭大型的高速公路路段。它们往往需求较少但却有很大的吞吐量，并且实际上为水平维度上的林荫大道提供了交通，因为高速公路正渐渐成为分割城市的分界线。如果你能妥善地处理这些，结果会是激活而不是分裂社区。

针对行人的安全，有哪些准则要求呢？

有很多准则和因素是针对行人安全的。我可以在这里概述一下其中的一些，但要完全把这些描述清楚还需要一点时间。

你肯定不想在过马路的时候距离太长，经常我们会看到宽阔的林荫道上有一个供行人使用的大型中央隔离带以及一些行人安全岛。有一些很简单的比如收缩或凸出之类的设计技术，能够在十字路口扩展人行道，从而提供更短的穿越距离。这对于行人来说也更为安全，因为通过将行人放到车流可视的区域中，他们就可以被渐渐靠近的车辆看到，而不是被隐藏在停留的车辆后面。这类设计减轻了交通拥堵，对于我们所采用的多种降低交通速度、提升行人体验的方法而言，可以说这是一个概括性的表达。

减速带、减速丘和减速台都属于同一类。与减速带（它是圆形的）或减速丘（它更宽大，同时也是圆形）不同，就减速台而言，汽车基本上可以驶上一个斜面，并和行人一起继续在同样的水平面上运行，然后在交叉的位置上向下转向。

我们所采用的一种设计技术，是将路面上的材料改为人行道上所使用的材料。我们当前正在一所医院的车道上使用这项技术。进入医院的机动车被开到一个很短的斜坡上，使它们上升到了人行道的层面上。这种人行道的质地结构继续在车道上铺展，它的好处是，开车的人感觉他们已经进入了一个空间，这个空间更多的是属于行人的而非属于汽车，所以他们往往会开的更慢一些，更小心一些。行人也因此受到了装饰花盆、矮柱或其他一些实际分隔的街道装饰品的保护。

如果车流量被放在地下，与行人分开，那么，驾驶员会不会因为缺乏让他们保持警惕的视觉刺激而导致事故增加？引入自然光会在这方面有所帮助吗？

答案在于设计。如果你设计的地下系统有宽敞的道路，却没有流量控制，汽车可能会高速通过。我们遵循的许多设计标准，以及所使用的像路面标记、标示这样的工具，都是为了去控制可能引起交通事故的一些因素。所以，正如我所说的，这真的是一个设计问题。

把汽车放到地下去有许多原因。隧道很明显可以钻过河流，穿过山脉，而且没有行人。将汽车和行人分开的一个例子是在大学里，车都停在地下，人群则在街道上流动。在一些例子中，我们甚至把所有的交通流量都设计到地下去，而在地上的所有东西都是与行人和自行车相关的，同时，如果校园够大的话，可能还会提供往返大巴服务，而不是让人们各自开车到不同的目的地。如果能有一些斜坡，直接让汽车开到地下去，那么，当你走进一所大学校园时，将几乎看不到汽车。

自然光已经成为交通工程中的一个主题。例如，当驾驶员行驶在阳光下的道路上，接着进入一个隧道，他们的眼睛必然要调整来适应不同的光线。对于驾驶员来说，这可能会很危险。解决这个问题的关键之一就是，要保持一条持续的水平线，这也就是为什么在新修隧道及对旧隧道进行整修时，在隧道的入口和出口处会有大量的光线射入隧道并从隧道中射出。这有助于缓解驾驶员眼睛的极度调整，同时提供一种更加安全的过渡。对于行人而言，在地下，自然光当然也是有益的。但是从一个驾驶者的立场出发，它就有可能是消极的。让人们开车连续地穿过黑暗和光明的地方，这在设计上讲是一个挑战。

与地下铁路系统相比，轻轨如何？

与一个地铁系统相比，轻轨系统的承载量更小。地铁往往有比轻轨更多、更大的车厢。轻轨线路将可能是地铁线路的补充，为略少的乘客量服务。

planters, bollards, or other street furniture that provides a physical separation.

If traffic were put underground, separated from pedestrians, would increased accidents result from motorists' lacking the visual stimulus to keep them alert? Would the introduction of natural light help?

The answer is in design. If you design the buried system with wide roadways and no traffic controls, cars might speed through it. Many of the design standards we follow and tools we use such as pavement markings, signage, and the like are meant to help control factors that may cause accidents. So as I stated, it is really a design concern.

There are various reasons to put cars underground. Tunnels are obviously used to go below rivers or through mountains and do not contain pedestrians. An example where you may separate cars and pedestrians is in a campus, where parking happens below grade and pedestrians circulate at street level. In some cases we've designed all the circulation below grade, where everything on grade is linked to pedestrians and bikes, and if it's a large enough campus, maybe a shuttle service rather than letting people drive to different destinations. If ramps are provided for cars to go below grade right when they enter a campus, you would barely see the cars.

Natural light has been an issue in traffic engineering. For instance, when motorists drive on a roadway exposed to the sun and then enter a tunnel, their eyes must adjust to the difference in light level. This can become dangerous for drivers. One of the keys to this is to maintain consistent light levels, which is why in many more recently built tunnels, or often in retrofitted older tunnels, there will be extensive lighting at the entrance and exit of the tunnel that tapers into and out of the tunnel. This helps ease the adjustment of a motorist's eyes and provides a safer transition. For pedestrians underground, natural light is certainly a benefit. But from a driver's standpoint, it's probably a negative. It would be a design challenge to allow people to drive through dark and light spots consistently.

How does light rail compare to a subway system?

A light-rail system has less capacity than a subway system. Subways tend to have more and larger cars than light rail. A light-rail line would likely be implemented to serve lower ridership than a subway line.

Are trolleys and light rail staging a comeback?

These types of systems are gaining in popularity. They are an effective way to move people. Just as popular, and usually less expensive, is Bus Rapid Transit (BRT).

We are involved in New York's recent introduction of Select Bus Service (SBS), which doesn't have all the elements of bus rapid transit (BRT)—it's kind of a hybrid. One of the major differences is the interaction with traffic and the fact that the buses don't have their own right-of-way. In BRT, as an example, there may be a roadway serving motorists, but in the middle of or adjacent to the roadway there would be a separated and dedicated right-of-way for buses. In New York City, it's very difficult to find space, so here the buses in many instances mingle with traffic. They have their painted bus lanes but sometimes they have to come out of those lanes, or cars make right turns from the bus lanes. It also has off-board payment, so you purchase a ticket at a kiosk before getting on the bus.

Another difference is that passengers don't board from a same-height platform so they still have to climb the stairs, but they have lowered the buses, and they've also changed the system for handicap entry. Previously, the back stairwell would even out into a platform and descend to street level; now they use ramps, which are much quicker. The smaller the grade difference, the quicker it is to get on the bus. SBS has more points of entry and, because you're not paying on board, you can get on in the back or the front. Some systems have more entries than others. I've seen buses with four different entry points.

With every entry you take away seats but gain capacity; seats take up a lot of space. Japan recently removed seats from subways during peak times because they take up so much space.

In a pedestrian scenario, can bicycles be intermingled without too many problems?

Yes, but it depends on how much space you have. In Hudson River Park, which runs along the west side of Manhattan, they have separate bike lanes and pedestrian paths. That's an easy way. When bikes and pedestrians are expected to share the same space, if it is narrow and not delineated, that can be an issue.

In New York there are a number of ways that pedestrians and bikes are being prioritized. In addition to the Walk/Don't Walk signals for crossing intersections, bike signals have been added throughout the city, which allows for vehicles, bikes, and pedestrians to all be accounted for within a roadway. New York has been working to balance the needs of all people who use the street system. The ultimate benefit is that you're actually moving more people.

电车和轻轨是否正在回归?

这些系统类型正越来越受欢迎,它们都是运送人员的有效方式。非常流行而且不贵的一种方式,是快速公交系统(BRT)。

我们正在参与当前纽约引入选择性公交服务系统(SBS)的工作,它并不具有快速公交系统的全部特点,而更像是一种混合物。其中一个很大的差异是它与交通流量的互动,以及公交车并不具有自己的路权这一事实。例如,在快速公交系统中,可能有一条路是为机动车驾驶员服务的,但在道路中间或者在这条路的旁边,一定会有一条独立和专属的公交车道。在纽约市,很难找到这样的空间,因此很多时候这里的公交车是与车流混合在一起的。他们也有划定的公交车道,但是有时公交车必须从这些车道中出来,或者汽车会从公交车道向右转。它有车外的支付方式,所以在你上公交车之前,先要在一个售票亭里购票。

另一个差别是,乘客并不从一个相同高度的平台上上车,所以还要爬上阶梯,不过,他们已经降低了公交车的高度,并为了让残疾人方便上车而改变了系统。之前,辅助阶梯的梯井甚至会向外伸展为一个平台,并且降到与街道的水平高度一样;现在,他们使用更快的斜坡道,斜坡高度差别越小,就可以越快地登上公交车。选择性公交服务系统有更多的进入口,而且由于你不是在上车的时候交费,你可以自由地选择从前面或后面上车。有的系统入口更多,我曾经看过有些公交车有四个不同的入口。

每个入口移掉的座位都可以得到更多的容量,座位总是占据很多的空间。日本最近在高峰时间将地铁上的座位全部撤销,就是因为它们占据了太多的空间。

在一个严格意义上的行人方案设计中,自行车能不能被引入其中且不引起太多的问题?

可以,不过这取决于你有多少空间。在紧靠曼哈顿西边的哈德逊河公园,他们就将自行车道和人行道分开,这是一个很简单的方法。但当自行车和行人要共用相同的空间时,如果空间很小而且没有区分清楚,那确实可能会产生问题。

在纽约,有许多路段为行人和自行车划出一个优先序列。除了在路口设置"步行/禁止步行"的标示,自行车的标示也被添加在整座城市中,这就使得机动车、自行车和行人各自在道路上占一席之地。纽约已经在努力平衡所有街道系统上的需要,最终的好处是,实际上正在运送更多的人。

Select Bus Service with pre-boarding ticket dispensers (2012)
在上车前使用售票机的选择公交车服务（2012年）

在城市中对于自行车有没有速度上的限制？

除非另有警示，否则我相信对于自行车速度的要求
与对机动车的要求是一样的。

什么是混合交通环境中的最优布局？

那种能够依据一个街区中人口、密度、土地使用及
其他许多要素的差异而做出改变的布局是最优的。
你可能经常会让人行道沿着建筑，然后在人行道旁
修建公路，但如何满足行人、自行车和机动车的需
求却不是一成不变的。有时自行车和机动车共用道
路，而现在更通常的做法是让自行车在人行道和汽
车车道之间拥有自己专属的空间。

**为了引入一套地铁系统，要不要对人口进行粗略估
算？**

这依赖于你要连结的东西和人们的位置。人口密度对

Is there a speed limit for bikes in the city?

Unless otherwise posted, I believe bikes are required to
follow the same speed limit requirements as vehicles.

What is the optimum configuration for mixed traffic?

The configuration can change depending on the population,
density, land uses, and numerous other factors in a neighborhood. While you'll usually have sidewalks up against the
buildings and the roadway adjacent to the sidewalk, how the
pedestrians, bikes, and vehicles are accommodated varies.
Sometimes bikes and vehicles share the road, while now it
is much more common for bikes to have dedicated space
between a sidewalk and vehicular travel path.

**Is there a population rule of thumb for the introduction of
a subway system?**

It depends on what you're connecting and where the people
are. The subway system is important around population

A "complete street" in midtown Manhattan
曼哈顿中城区的一条"完整街道"

Midtown street signage, Manhattan (2012)
曼哈顿中城区的街道招牌。（2012年）

density. In New York, it has brought more people to the city because everything is so connected, going back to what I said about being able to attend multiple meetings in one day. One of the difficulties in Los Angeles, for example, is lower population density. It's hard to imagine where to put access points for a subway because people are so spread out; you'd need bus connections to get to the subway. However, Los Angeles is currently focusing on this and making strides.

It really depends on existing conditions. Having a subway from the start obviously helps; then you can design around it. This goes back to the question of how to start a city. You decide where the different centers should be and then you'd have transit stops and population density around those locations. You wouldn't need a car to get from point A to point B; you'd use the subway. New York City is a prime example of how the subway system helped to build the city itself.

What is your experience with bicycle rentals?

We have been following New York City's effort to implement a bike-share program, which started in May, 2013. It has bike

于地铁系统来说很重要。在纽约，它已经将更多的人带入城市，因为所有东西的联系都如此紧密，回想一下我所说的能够在一天之内参加多场会议的例子。再比如，洛杉矶的难题就是人口密度较低。你很难设想要在哪里设置地铁的出入口，因为附近的人太过分散了，你需要乘公共汽车到地铁，不过，洛杉矶现在正在关注这个问题，并且取得了很大的进步。

这真的取决于现存的条件。从一开始就有地铁显然会很有利，然后你可以围绕着它展开设计。这又回到了那个问题：如何开启一座城市。首先，你需要决定不同的中心所处的位置，然后设计交通站点，以及在这些站点周围的人口密集区。要从A点到B点，不需要一辆汽车而是使用地铁。就地铁系统如何帮助建设城市自身来说，纽约市就是一个非常好的例子。

关于自行车租赁你有什么经验？

我们依据纽约市的努力方向，提供了一套自行车分享系统，它已经进入议事日程，随时可以付诸实践。自行车点将遍布全市，这有点类似于自行车的租赁，但也有很多不同。它需要有一笔成为会员的费用，并且只有在你保存自行车超过一定的时间时，才会增收额外的费用。你可以在遍布城市各个区域的自行车点挑选一辆自行车，然后在其他地点把车还掉。这个项目的资金通过广告以及会员和用户的交费来支付。我们公司是这个计划的大力支持者，因为它减少了驾驶车辆的需要。

一旦这个项目被执行，我们就需要观察这个系统，看看真正起作用的是什么，然后根据情况进行调整改进。谁在哪儿会使用？以及它会变成休闲方式？

沿着河边骑行是一回事，但600个自行车点会让数千人在街上骑自行车，并加入到严重的交通里。

会有这样的问题，但是如果做的恰当的话并不会带来危险。这也是为什么城市正增加数百英里的自行车线路，同时也将自行车道分隔开来。所以我们又回到了设计问题。自行车将会靠近自行车道，或者被穿插到系统之中去，使得自行车骑行者不至于骑着骑着就转到街道上去，这是一个进步的过程。现在有更多人骑自行车，特别是在夏天的时候。还有一点是，在那些自行车不断增加的地方，喧闹的车流也会安静下来，因为当自行车足够多足够密集的时候，机动车驾驶员也开始更多地了解到了路况信息。

季节因素有多重要呢？

当然，存在季节的因素，但是也有很多勇士，他们在冬天或者在雨天骑车。这就是他们上下班的方式。不过要是天气更温暖或更干燥，就会有更多骑自行车的人。

环形公路的效率如何？

再说一次，这还是要看你如何设计它们以及人们想要去哪里。你想要尽可能地消除交通拥堵，所以如果一个城市有许多的集中点，那么环形公路就合理，因为它们会按照目的地的不同而分离车流。但是如果只有一个中心点，而每个人都在使用那条环形公路，拥堵就会随之而来。

交通环岛有没有减轻车流压力？

在整个国家，人们对交通环岛的关注，已经超过了交通信号灯。不可思议的是，对行人和车辆而言，它们似乎更安全，因而事故也更少了。

人们变得更加小心翼翼，不仅避免了一些冲

stations located throughout the city. While it is similar to bike rental, it differs in some ways. There is a cost for membership and an additional cost only if bikes are kept for longer than a specified amount of time. You can pick up a bike from any bike station located throughout parts of the city and return it to any other station. It is paid for through advertising as well as the membership and user fees. Our company is a huge proponent of this because it reduces the need for driving.

Now that it's been implemented, the system needs to be observed to see what really works and then be modified accordingly: who uses it and where, and whether it is recreational or commuter based.

What are the risks of putting thousands of cyclists on the street in heavy traffic?

It's not dangerous if done properly. That's why the city is implementing hundreds of miles of bike lanes and separated bike paths. It all goes back again to design. The bike stations are near bike paths, or worked into the system so cyclists aren't just riding right out into the street. It's an evolving process. There are many more bikers now, especially in the summer. There's a point where having more bikes will start to calm traffic because when they become dense enough, motorists become much more aware.

How important is the seasonal aspect?

Certainly there is a seasonal factor, but there are also the hardcore people who ride bikes even in the winter, or in the rain. That's just how they get to and from work. But the warmer or drier the weather is, the more bikers you will have.

How effective are ring roads?

Once again, it goes back to how you design them and where people are going. You want to dissipate traffic as much as you can, so if a city has many focal points, ring roads make sense because they separate traffic to where it needs to go. But if there is one central point and everyone is using those ring roads to get to there, congestion will become an issue.

Do traffic rotaries ease traffic?

Throughout the country, there has been more of a focus on rotaries than traffic signals. They are found to be safer, surprisingly, for pedestrians as well as vehicles; there are fewer accidents.

People become more cautious, and you remove some of the conflicts. You're also more observant. When you drive up

258

to a traffic signal, you just follow what the light tells you; in a rotary you have to pay attention to what is happening because there usually is no signal telling you to stop.

Rotaries seem to favor drivers who know where they are going. What provisions are made for motorists unfamiliar with the area?

You try to satisfy everyone in the simplest way. The key is to provide clear signage, not too much, but signs that explain where you need to go. You wouldn't want someone to drive around the circle continuously, because that adds to traffic volume, and if someone is lost, they are less focused on driving. It's dangerous to have people looking at a map, trying to figure out where they are going, and not paying attention to traffic.

I'm a traffic engineer, and sometimes I even find myself lost, driving in areas that just aren't signed well. Signage is much more important than we all think.

There is talk of powering electric cars by battery pads in parking areas or even embedding them in the roadway.

I've heard about that being used for light rail, so that every time a train pulls into a station a pad charges it so it can continue on to the next station without needing a third rail or catenary lines overhead. It makes sense. I'm sure the technology could exist for cars, but I just don't know enough to say if it's viable. I don't know who's going to pay for all the electricity, but you could get billed for how much you drive. The city is looking at GPS and other technology to possibly track and charge motorists for how much they are using the system. It's a way to discourage driving or to more fairly charge people based on how much they actually use the roads and bridges.

How do you envision traffic in New York's future?

I actually think traffic volume is going to decline. It's happened already; a lot of it was the economy. But there's such a different focus now in terms of what cities are looking at. It's really moved from cars to focus on transit, walking, and biking, and as a city focuses more and more on that, people's expectations shift so they're more willing to walk longer distances and bike to work if the amenities are there. We just can't sustain so many cars, so as the focus changes and the money goes into different places, you're going to see fewer vehicles. People will take mass transit, whether subways, buses, or ferries.

It really depends on what is being served and where development is happening. A number of ferry services failed over the years because they weren't serving the right population. Ferries provide a great service, but they service a more

突, 你也会更加敏锐机警。当你开车走向一个交通信号灯的时候, 你只是服从信号灯告诉你的东西; 在交通环岛上, 你要注意正在发生的事情, 因为往往没有信号灯告诉你应该停下来。

交通环岛看起来对那些知道自己想去哪里的人是有帮助的。有没有一种办法, 可以让一个人在找到自己的前进路线之前, 在这个环圈内绕个三四次呢?

你想要以最简单的方式满足所有人的要求。关键是要提供清晰的引导标示, 不要太多, 但标示要能说清楚你应该去哪儿。你不会希望有人一直开车绕着环圈运动, 因为这增加了车流量, 如果有人迷路了, 他们就不会将注意力集中在驾驶上。让人们看地图, 然后认真找出目的地, 而没有注意到交通状况, 这是很危险的。

我是一个交通工程师, 有时我甚至发现自己都会开在一个引导标示不清楚的地方而迷路。引导标示比我们所有人想的都更重要。

人们现在正谈论停车场中使用电池板或嵌入马路来发电的电力汽车。

我听说这个技术已经被用于轻轨了, 每当一部火车被拉进站台, 电池就会给它充电, 所以它可以继续行驶到下一站而不需要第三条线或前面的悬吊线, 这是合理的。我相信这一技术可以为汽车而存在, 只是不确定是否可行。我不知道谁会支付所有的电能费用, 但你可能需要为你开了多少路程而支付账单。城市正着眼于GPS和其他技术, 以便可以根据驾驶员对系统的使用量, 对他们进行收费。这是一种抑制驾驶的方式, 或者说是一种基于人们实际对道路和桥梁的使用而进行公平收费的方式。

展望未来, 你对于纽约的交通有什么想法吗?

实际上, 我认为交通流量将会下降, 这个现象已经在发生了, 其中很大一部分是由于经济原因。不过, 根据现在各个城市所着眼的问题看, 存在着许多不同的关注点。纽约已经从关注汽车转移到关注公共交通、步行和自行车上。随着一个城市越来关注这些东西, 人们的期望也在发生改变, 如果设施健全的话, 他们现在更加愿意走较长的路以及骑自行车去上班。我们真的无法供养这么多的汽车, 所以随着关注点的转变, 资金会被投放到不同的地方去, 机动车将会变少, 人们将会乘坐公共交通工具, 不管是地铁、公交车还是渡轮。

Times Square shared by pedestrians and vehicles, USA (2012)
美国行人和汽车共同使用时代广场。（2012年）

　　这确实取决于提供什么服务，以及在哪个地方发展。有许多渡轮服务很多年都不景气，因为并没有为正确的人群服务。渡轮提供了很好的服务，但它们服务的对象只是一些特殊的人群，而且还变得越来越有季节性。它们应该根据你要去的地方，在营运时间上长一点或者短一点。现在正在提供的东河渡轮服务就非常成功，而且已经超出了人们的预期。

随着人口老龄化，有没有一些针对老弱人士的特殊考虑呢？

绝对有。关注长者的设计，换句话说，为他们提供更多过马路的时间，更方便、更好的行人空间——这当然是正在发生的事。交通工程学的各个领域都在关注一些特殊的设计，以便可以将老年社区纳入考虑范围。

specific population, and tend to be more seasonal. They might take a little bit longer or shorter depending on where you're going. The recently implemented East River Ferry service has been very successful and has been exceeding expectations.

As the population ages, are special considerations given to the aged and infirm?

Absolutely. Senior-focused design, in other words, providing more time to cross the street, more accessibility, better pedestrian space—that's certainly happening. There are areas of traffic engineering looking at special design to take older communities into consideration.

Have smart phones had an impact on transit?

Yes. You can pull up your Google Maps and see traffic conditions, and some systems post real-time transit schedules or send alerts so you know when to wrap up a meeting and go to the bus stop. It's definitely the wave of the future. Personal phones have changed everything, and they are already being used as a way to make transit more viable.

There is a wide variety of transit-based apps meant to facilitate travel on mass transit.

Is there a trend toward automated driving?

That technology is being studied but it's still far off, because it also requires people's acceptance of that sort of thing. Driving yourself is about being in control. It's difficult for people to accept the fact that they are going to be in a car that they can't drive, just trusting to technology. Trust is a big deal.

We'll have to see how technology pans out. How you transfer over to that system is difficult, too, as there may be a lot of investment in the actual roadway system to allow safe implementation. Some of the cars need to follow certain tracks and can actually read the markings on the road, but if the markings aren't good, you have problems. It's a neat idea but it needs further development.

One thing that has become a big deal is car sharing. When Zipcar was started a dozen years ago, there was discussion that for every car share, you could take 15 to 20 cars from a community; at a recent conference I heard someone say 30 cars. Car shares are still relatively new, although they are quite prevalent now. It represents a huge change, as they reduce car ownership but without taking away the independence of driving. If everyone shares a car, it's cheaper for the user. It's a great compromise.

You buy the service, not the equipment.

That's right, so you don't pay for gas or insurance, just the rental fee. Add up over the year the cost of car sharing versus owning your own car and paying for insurance, gas, maintenance, oil changes, etc. The difference in personal cost is huge; the environmental difference is also huge because by not owning a car, you might be less likely to rely on autos in general, although you could always get one if you really need it. People will just naturally rely on transit, walking, or biking if that is the primary choice provided to them and it is most convenient.

In terms of returning the car for someone else to use, for every hour you're late there's a stiff penalty, which is what makes people bring cars back on time. Of course, it's possible that the car you reserved won't actually be available, in which case you're in trouble. But I don't think that happens often because it's a community that respects the fact that they are sharing a car—and they have to pay hefty penalties.

手机对交通有影响吗?

是的,有影响。你可以打开你的谷歌地图,查看交通情况,有一些系统还会发布实时交通行程或者给你发送警示,所以你可以知道什么时候可以结束会议去公交车站。这绝对是将来的潮流。个人电话改变了一切,它们已经被看作一种使公交系统更加有效的方式。

有一系列基于公交系统的程序,非常有助于人们乘坐大众交通工具。

有没有一种朝向自动驾驶的趋势呢?

人们已经研究了这一技术,但这似乎有些遥远,因为接受起来需要一个过程。自己驾驶意味着自己在控制,而人们坐在他们不能驾驶的汽车里,只能去相信技术,这个事实对于人们来说接受起来还是有困难的,信任是个大问题。

我们必须看到技术是如何发展的,如何转换到那个自动系统上是很难的,就像为了确保安全,实际的道路系统需要大量的投资。有一些车需要跟随特定的路径,因此可以完全解读路上的标记,但如果这些标记不够好,就有问题了。自动驾驶是一个好主意,但它还是需要进一步的发展。

还有一个重要的议题是汽车共享。当热布卡公司在十几年前成立的时候,有一个讨论认为,如果每辆车都分享共用的话,那么一个社区可以减少15-20辆车;在最近的会议上,我听到有人说可以减少30辆之多。汽车共享依然是相对新颖的,尽管它们现在已经很流行了。这代表着一个重大的转变:在减少汽车占有量的同时并没有取消驾驶的独立性。如果每个人都分享一辆车,那它对于用户来说是更便宜的,这是一个很棒的让步。

你购买的是服务,而不是设备。

没错,所以你不需要支付油钱和保险,只需付租金。一年下来,汽车分享的花费与拥有自己的汽车,然后支付保险、汽油、维护费、机油更换费等等加起来相比,个人花的钱是有着很大差异的。环境上的差异同样巨大,因为如果不拥有一辆汽车,你可能就不太会在日常生活中依赖汽车,尽管如果你真的需要的话你确实可以买一辆。人们将会很自然地依赖公交系统、步行或骑自行车——假如其中有一些是提供给他们的首要且最为便利的选择。

就归还汽车让其他人使用而言,超过期限的每个小时都有严厉的罚款,也正是这项措施使人们及时归还汽车。当然，有可能你预订的车实际上得不到，那样的话你就有麻烦了。但我不认为这种事会经常发生，因为这是一个群体，群体成员都尊重这样一个事实：他们是在分享汽车——并且，如果归还不及时要支付很重的罚款。

另一个进步是自动停车。在中国城的巴克斯特街123号有一个自动车库。它占据着更小的空间，而且对环境是无害的。如果进入一个传统的车库，那么，你会驶上斜坡以获得位置，你需要地方来进出车库，还需要安全区域来步行——楼梯，自动扶梯等等。有了自动停车，你只需要将车开到井口建筑中的电梯上，锁上你的车门，你的车就会自动停好。这真是太神奇了，中间并无人为干预。

自动停车的一个好处就是，同样的停车数量，它所需要的空间要少得多，而这些空间，对于实际的停车需要来说就足够了。因此，有更多规划内的空间可以被用于办公室、零售商店等等。另外一个好处则是环境上的：当汽车到达车库的时候，它们就熄火了，没有了寻找停车空间的徘徊，也就减少了排气量。建设这些自动停车场在费用上要更贵一些，但日后的运营费用则要便宜不少，因为你没有专门负责管理停车的人员等等。每个车库不再有全套的维修人员，只需要一个人在房间的某个地方盯着图像系统上10到20个点即可，如果他认为有人需要注意一下，只要发个警示就行。自动停车正在欧洲使用，但在这里还刚刚起步。

它是完全电脑化的。基本上，你会得到一张票，这张票确认了你的汽车，然后电脑会把它放在正确的位置上。为了提高效率，或者为了知道什么时候不同的汽车需要离开，这套可变的系统可以被继续编程修改。例如，如果你要在7：00的时候取车，而我则要在10：00的时候取车，我来的更晚，那么，电脑可以调整位置以满足不同人的需要。

移动它们的机制是什么？

这基本上是一种小型电动机驱动的托盘跟踪系统。有一些是有转盘的，所以你的汽车会被送到临街的位置。也有一些车库，你可以来的时候在某个地方，离开的时候却是在另一个地方。它们也包括一些视频信息，因此实际上你可以看到车的停放状况，人们为他们的车辆安全而感到高兴。这项技术是很神奇的。谁知道五年之后会发生什么呢？

Another advance is automated parking. There's an automated garage at 123 Baxter Street in Chinatown. It takes up much less space and is environmentally friendly. If you go to a conventional garage you drive up the ramp to get to your space and you need room to get in and out, and safe areas to walk—stairs, escalators, etc. With automated parking, you just drive onto a lift in the head house, lock your door, and the car is automatically parked for you. It's amazing. There's no human intervention.

One of the great things about automated parking is that a much smaller space is required for the same amount of parking, just enough for the actual parking of cars, so there's a lot more programmatic space for offices, retail, whatever. The other great benefit is environmental: vehicles are turned off when they get to the garage, and the lack of circulating looking for a space reduces emissions. It is more expensive to build, but afterward the operational costs are much cheaper because you don't have valets, etc. Instead of a full maintenance staff for each garage, there is a guy in a room somewhere watching 10 to 20 lots on video systems, and he just sends an alert if something needs attention. Automated parking is used in Europe, but it's just starting here.

It's all computerized. Basically, you get a ticket that identifies your car and the computer puts it in the right place. The variable system can be programmed for efficiency, or for knowing when different cars need to exit. For example, if you pick up your car at 7:00 and I pick up mine at 10:00, but I arrived later, the computer can shuffle the cars to accommodate individual needs.

What's the mechanism for moving them?

It's basically a pallet track system driven by small electric motors. Some have turntables so your car is delivered facing the street. There are also garages where you come in at one location and leave through another. They also include videos so you can actually watch your car being parked. People feel happy that their car is safe. The technology is just amazing. Who knows what will happen five years from now?

261

PLANNING Erich Arcement

Thomas J. Campanella
UNIVERSITY OF NORTH CAROLINA AT CHAPEL HILL

Thomas J. Campanella, Ph.D., M.L.A., is an associate professor at the University of North Carolina at Chapel Hill. He previously held teaching and visiting appointments at MIT, Columbia University, the Harvard Graduate School of Design, Nanjing University, and the Chinese University of Hong Kong. Dr. Campanella is a recipient of Guggenheim and Fulbright fellowships and is a Fellow of the American Academy in Rome. Among his award-winning publications are *The Concrete Dragon: China's Urban Revolution and What It Means For the World* (2008); *Republic of Shade: New England and the American Elm* (2003), winner of the Spiro Kostof Award from the Society of Architectural Historians; and *Cities From the Sky: An Aerial Portrait of America* (2001). He is also co-editor, with Lawrence Vale, of *The Resilient City: How Modern Cities Recover From Disaster* (2005). He is a regular contributor to the *Wall Street Journal*, and his essays and criticism are widely published elsewhere.

托马斯・J・康帕内拉（吴凯堂）

北卡罗来纳州大学教堂山分校

吴凯堂，博士，园景建筑硕士，是北卡罗来纳州大学教堂山分校城市规划与设计专业的助理教授。他之前在麻省理工学院、哥伦比亚大学、哈佛设计研究院、南京大学和香港中文大学任教、访问。康帕内拉博士获得过古根海姆奖以及富布莱特资助计划，也是罗马美国学院的成员。他有许多享誉盛名的作品，如《钢筋水泥巨龙：中国城市革命及其对世界的意义》（2008年）；《阴影共和国：新英格兰和美国榆树》（2003年）——这本书获得建筑史学家学会颁布的斯皮罗・科斯塔夫奖；《天空城市：一幅美国的空中肖像》（2001年）。他和劳伦斯・威尔是《弹性城市：现代城市如何从灾难中复原过来》（2005年）一书的主编，也是《华尔街日报》的固定撰稿人，他的文章和评论也在其他很多地方发表过。

For complete biography, please refer to the Appendix.

完整的传记，请参阅附录。

Shanghai Urban Planning Museum, model of central Shanghai in 2020
上海城市规划博物馆，2020年上海中心的模型

Suburbanization in China is very different than in the United States, with dense mid- and high-rise apartment towers spreading out from the core. What is the defining line where the city stops and the suburbs begin?

It depends on how you define "city." By land area, some of the largest cities in the world are in China. But you have to look at it closely, because what city officials often do is basically draw a line on a map and call it all city; so this is why there is plenty of farmland remaining in the city of Chongqing or Shanghai or Beijing. Within the city proper, you have not only city and suburb but lots of rural countryside too. Now, if you define city not from an administrative or legal standpoint but according to the actual built fabric, then you've got a more traditional gradient of dense urban center to lower-density suburban development mostly oriented around the automobile, eventually petering out to rural. Chinese cities traditionally have been very dense; the suburbs are much less dense, but compared to American suburbs, they are much denser—denser than many of our center cities, even.

There are also very low-density single-family subdivisions, like our McMansion developments here. They're just not as common and they tend to be farther out from the center. The standard unit of suburbanization in the U.S. is the large-lot single-family home. In China, it is more the mid-rise, even high-rise, gated housing estate. If you planted this kind of development in an American city, it would be considered fairly dense infill. It's all relative.

And all this Chinese sprawl is a new thing. The division between city and country was very sharp and clear until only recently. The whole era of American suburbanization, when cities in the U.S. spread out thinly as a result of the automobile—that didn't happen in China. During the Mao years, cities were artificially contained and growth was very severely limited. Mao was very skeptical about big cities.

Where is the residence of choice?

That's a great question, and a hard one to answer. It depends on a lot of things—on relative affluence, station in life. Certainly China does not have the same tortured history that we have about cities and urban culture. Americans have always been ambivalent urbanists; we've always had a certain skepticism about the value of cities, going all the way back to Thomas Jefferson. The Chinese have never had that, so their move to the suburbs is not underwritten at a philosophical level as it is here by a fascination—an obsession, even—with the "middle landscape," as Leo Marx called it. The romance

中国的郊区化与美国非常不同，密集的中高层公寓塔楼从中心向外延展。城市与郊区的分界线到底是什么呢？

这取决于如何定义"城市"。就土地面积而言，世界上几个最大的城市都在中国。但你需要仔细考虑这个问题，因为城市官员经常只是在地图上划出一条线，然后称之为城市；所以在重庆、上海、北京这样的地方，城市范围不仅包括城镇、郊区，还有大量的农田。如果不从行政或法律的立场来定义城市，而是根据实际的建筑结构，以汽车数量为参照，那么你就会得到一个更传统的密度梯度，从城市中心到低密度郊区，最后逐渐减小到乡村。传统意义上，中国城市就是高密度的，而郊区的密度则要小很多，但是它与美国的郊区相比，显然要密集得多——甚至比美国许多中心城市的密度都要高。

也有密度很低、由单个家庭构成的独立小区，就像美国的巨型豪华住宅区。但这并不普遍，而且倾向于远离市中心。美国郊区化的标准单元，是大量单个家庭的住宅。在中国则更多是中密度甚至高密度的封闭式小区。假如在一座美国城市中规划这种发展方案，人们会认为它实在太过拥挤，这都是相对的。

所有这些中国式的城市扩张都是崭新的。过去，城市和乡村之间的划分非常明确，而今天则并非如此。在美国郊区化时代，由于汽车的时兴，城市稀稀落落地向外扩展，所有这些并没有发生在中国。毛泽东对于大城市充满疑虑，在他的时代，城市被人为地抑制，其增长被严格限定。

住宅应选在哪里？

这个问题很大，也很难回答。它依赖于许多因素——需要相对的富足，需要社会生活地位。当然，在城市和城市文化方面，中国并没有我们所经历的曲折历史。美国人都是自相矛盾的城市规划专家：我们总是对城市的价值存在一定的疑惑，并千方百计想回到托马斯·杰斐逊的时代。中国人从不那样，他们搬到郊区去，并不是要在哲学层面上得到保证，而美国人对此非常迷恋、痴迷，就像被利奥·马克斯（Leo Mark）称为的"中间景观"一样。将田园风景中的小屋置于一抹绿意之中，这种浪漫在中国并不容易看到，中国人热爱城市，那是大家都想去的地方。

由于太多人想要住在城市里，致使那里的房地产价格令人咋舌。因此，人们想到郊区去，找到更便宜的住房。这和我们郊区化经验中所发生的一切迥然不同。美国的白人中产阶级搬离城市，不是因为他们不能支付城市房价，而是他们不想再待在那里，便宜的燃料、汽车以及高速公路让他们有条件撤出城市。在美国，价格暴跌和社会结构分离，是另外的一套机制在发生作用。在中国，另一个使问题变得更复杂的因素是，城市中的贫民被旧城区的发展所淘汰，并被分流到郊区，那里的土地更加便宜，而开发商可以为落伍的城市居民提供替代性住宅。因此，与典型的美国式郊区化不同，在中国，是穷人搬到城郊而富人住在市区。

专为那些分流的人所建的区域，有没有特别的发展？

当然，在高速公路上看，这些区域没什么特别，但高端项目都是封闭的。这些封闭的房地产项目都有一个非常明显的共同之处：楼房与某种形式的社区中心之间有宽敞的花园式空间，其中有些非常漂亮。大多数的美国住宅社区完全没有公共空间，那些为贫民而建的分流点并不舒适便利，建造也很粗糙。建筑质量则完全是另外一个问题。

那些建设工程为什么都封闭着？

这是一个大问题。中国社会正在发生改变，变得越来越两极分化：一边是拥有着巨额财产的富人，另一边则是数量不断增加的贫民。因此，许多地方的犯罪显著上升，而无论对错，受到指责的往往是农民工。所以，保证安全是为何封闭的一个最简单的回答，它也是一种排外的观念。更复杂一些的说法是：门、墙和围栏乃是深深扎根于中国空间文化的主题——比如：紫禁城、四合院，类似于西安或北京这样满是围墙的皇城，甚至用墙把整个国家挡在长城以内的观念。但是同样，在毛泽东时代，单位或工作所在地也是用墙围绕并封闭的。工作单位的观念来自苏联，但那些封闭的结构，则完全是中国式的。

有没有一些中国特性被整合到建筑原型中去？

从建筑的角度说，中国人非常关注跨代、延续的核心家庭。楼层的规划要使姻亲能够生活在一起，这些是我们在美国不曾看到的。姻亲经常就在同一个住宅区中购买另一个单元，或者就住在

for that little house in the pastoral landscape, set in its little patch of green, doesn't really translate in China. The Chinese love cities; that's where everyone wants to be.

So many people want to live in the cities that the price of real estate there is just obscene. So they are going out to the suburbs to find less expensive housing. That's very different from what happened in our experience of suburbanization. The American white middle class moved out of the city not because they couldn't afford it but because they didn't want to live there anymore, and with cheap gas and cars and highways they now had the means to get out. Prices plummeted and the social fabric came apart, so a very different process was at work here. Also—and this is where it gets even more complicated—the urban poor in China are being displaced by inner-city development and decanted to the suburbs because that's where land is cheap and where the developers have been allowed to put up replacement housing for the displaced city dwellers. So again, unlike classic American suburbanization, it's the poor who move to the outskirts and the rich who live in town.

Are there specific developments for those decanted?

Definitely. From the highway they all look similar, but the higher-end projects are all gated. But these gated estates tend to have a fairly robust commons, with open and park-like spaces between the towers and a community center of some sort. Actually, some of them are quite nice. Most U.S. subdivisions have no public shared areas at all. Those used as decanting sites for the poor have far fewer amenities and are often very poorly constructed. Construction quality is a whole other issue.

Why are the developments gated?

That's a great question. Chinese society is changing and becoming more bifurcated, with extreme wealth at one end and increasing poverty at the other. There's been a sharp increase in crime in many places. It's the migrant workers who, rightly or wrongly, are always blamed. So security is the easy answer, and a sense of exclusivity. The more complicated answer is that gates and walls and enclosures are themes deeply rooted in Chinese spatial culture—the Forbidden City, the courtyard house, walled imperial cities like Xi'an or Beijing; even the whole notion of walling off the nation itself behind the Great Wall. But also, during the Mao era, the physical campus that contained the *danwei*, or work, was itself walled and gated. The work-unit idea was adopted from the Soviets, but its physical expression as an enclosed compound was purely Chinese.

Are there specifically Chinese qualities incorporated into the prototype?

At the architectural scale, there is greater attention to the multigenerational, extended nuclear family. Floor plans almost always accommodate in-laws living with the family, which is something you almost never see here. Or often in-laws simply buy another unit in the estate or live across the hall. The nuclear family is still very strong in China; and, ironically, it's the presence of grandma and grandpa that allows both spouses to pursue careers and still raise a child.

How does the size of rooms compare?

You would think they would be smaller, but the market-rate units are pretty generous and would certainly measure up well to a Manhattan apartment, and are even more generous perhaps.

There would seem to be a disconnect between rapid urbanization and good urban planning, which takes time.

There's a huge disconnect, and in general, development has been leapfrogging way ahead of planning. In my book *The Concrete Dragon*, I liken it to somebody trying to rake leaves in a tornado—the raker of leaves being the planner. One of the reasons development on the urban fringe peripheries has been problematic is because it has leapfrogged ahead of public transportation systems. No one is building more public transit anywhere in the world than in China. All the rest wouldn't total even half of what China is building, but it's still not enough to keep pace with development. If you look at satellite images, the growth of cities appears like exploding nebulae—just unbelievable. There is no way China can build enough subways and light-rail lines to keep pace with that. It would bankrupt the nation.

So most of the suburban areas are oriented around the automobile. Massive highways are being extended and looped around cities. There's a joke that eventually one of Beijing's outer ring roads will pass through Inner Mongolia. People are buying cars at a clip that we haven't seen in the U.S. since the 1950s and '60s. Highway building is also roughly equal in pace to the peak of our interstate building campaign here, back when we were able to do that kind of thing.

走廊对面。核心家庭在中国依然十分强大，但很讽刺，正因为有了爷爷奶奶，夫妻双方可以在追求事业的同时抚养子女。

与之相应的房间有多大？

你也许认为它们应该比较小，但市场化的住宅单元都很宽敞，达到了曼哈顿公寓的标准，甚至有可能还要更宽敞。

似乎快速城市化与合理的城市规划之间是分开的，后者需要更多的时间。

差异是巨大的，而且一般情况下，发展总是以跨越的方式超出规划。在我的《混凝土做的龙》一书中，我将之比作想要在飓风之中收集树叶的人——收集者就是那些城市规划者。城市边缘地带的发展出现问题，一个很重要的原因是：它超越了城市的公共交通系统。没有哪个地方比中国正在发展的公共运输还多，全世界其他地方公共运输加起来的总和，都不及中国正在建设的一半，但这依然不能跟上中国高速发展的速度。如果看卫星图片，城市的发展就如同爆炸的星云一样让人难以置信。如果中国没有办法去建设足够的地铁和轻轨列车，以跟上其发展的脚步，这会令整个国家破产。

大多数城郊地区都依靠汽车，因此大量的高速公路环绕着城市不断延伸。有个笑话讲：最终北京的外环线将会穿越内蒙古。自从20世纪五六十年代以后，我们就没有见过人们买这么多的汽车了。中国高速公路发展的速度也几乎赶上美国州际公路建设的顶峰时期，那时我们还能做这样的事情。

有没有这样的情况，城市或政府为建一座新城而留出一片空地？

有很多地方这么做，为了巨大的人口，往往都有六位数之多。在我的书中，关于上海的那一章里我谈到，沿着一条环线，有八个或九个新开发的新城——每一个新城都以一个特别的族群作为主题。建设新城的做法显然谈不上新鲜。从英国的伯恩威利（Bournville）阳光港城和埃比尼泽·霍华德（Ebenezer Howard）的花园城市开始，西方就已经这么做了。不过，很少有人建设真正整全齐备的提供就业、发展工业的地方，这一挑战促使投资者去发掘工作和就业来源。但在中国，由于很多人从农村来到城市，他们除了建设卫星城市之外没有其他选择。不过许多中国的新城镇已经步履蹒跚，甚至成了空城。

有没有可能建设一个理想的生活、工作社区，有工作机会，也有常驻人口？

我认为有。如果有人能做到这些，那一定是中国人。过去，户口体制将人一生都固定在某一个地方。但它现在不再被强制执行，这也是为什么那些外来人口能够来到城市的原因。从理论上说，他们是"非法"的外来人口，但没有人强制执行法律，因为中国官方知道，如果将大家都遣送返乡，经济将在十分钟之内崩溃。这些人有工厂的工人，建筑工人，有大楼的清洁工和各行各业的服务人员。

在规划的新城市中，是否要将注意力放在居民心理和社会性上？

这让我想起简·雅各布斯(Jane Jacobs)和她的著作《美国大城市的生与死》。在中国，人们悲痛地抱怨失去了像"老街坊"这样产生在胡同里的联系纽带，他们被迫迁移到城市的边缘，在那里安家落户。过去许多人只要骑十分钟的自行车，就能从单位回家，但现在，他们住的很远，有时需要花两到三个小时乘车。因此，他们往往也在自己的岗位上干不了太久。

这个领域很值得研究，我希望看到中西方更多年轻的社会科学家、城市规划专家开展此项工作，因为我们所谈论的是正在影响着数以百万人的巨大变化，它使美国的城市重建经验显得颇为苍白。另一方面，在中国文化中，至少在城市，人们熟悉高密度的生活并感到舒适。我们都知道现代主义者关于城市化命运的观点，尤其是美国的公共住房；那真是一场十足的灾难，而建筑也常常饱受批评，就像圣路易斯的普鲁蒂-艾戈(Pruitt-Igoe)，虽然多次获奖，但后来却被妖魔化并最终被摧毁。

究竟是建筑上失败，还是社会结构上失败？这是一个老问题。

非常正确。我过去也常这么说，直到李灿辉(Tunney Lee)带我参观了香港沙田区禾輋的公共屋村。那里有非常棒的户外用餐处，很随性、很短暂，每天晚上都在广场上兴旺起来，广场位于这

Are there cases where the city or the government sets aside undeveloped land for a new town?

There's been a lot of that, for generally huge populations, six digits. In my book, in one of the Shanghai chapters, I talk about the series of eight or nine scratch-built cities planned around one of the ring roads—each theme on a different nation. The whole tradition of building new towns is obviously not new. The West has been doing it since Bournville and Port Sunlight in England, since Ebenezer Howard's influential work on garden cities. Very few have worked in the sense of building truly rounded, full-service places with on-site employment and industry. The challenge was always getting investors to build sources of work and employment. But in China, with so many people coming in from the countryside, they've got no choice but to build satellite cities. Even so, many Chinese new towns have floundered and some are largely empty.

Is it possible to build an ideal live/work community, with jobs and a permanent population?

I think so. If anyone can do it, it's the Chinese. In the past, the *hukou* system nailed you in place for life. It's not really enforced anymore, which is why all these migrants are coming into the cities. Technically they're illegal immigrants, but nobody's enforcing the law because the Chinese authorities know that if they sent everyone home, the economy would collapse in ten minutes. These are the factory workers, the construction workers, the people cleaning buildings and doing service-sector jobs.

In planning new cities, is attention paid to the psychological or sociological needs of residents?

I'm reminded of Jane Jacobs and her *The Death and Life of Great American Cities*. People have complained bitterly in China about losing the communal bonds that they had in the "old neighborhood," in the *hutong* that were bulldozed in Beijing, for example, and that they are forced out to housing way out on the urban periphery. Many of these people used to be a 10-minute bike ride from their jobs, but now they're out so far that they sometimes have a two- or three-hour commute. So of course they don't last very long in those jobs.

It's a very rich area for research and I hope there are many young Chinese and Western social scientists and urbanists who are going to take it up because we're talking about very dramatic changes affecting millions of people. It makes the American experience of urban renewal pale in comparison. On the other hand, in Chinese culture there is a greater famil-

Market at Wo Che, Sha Tin New Town of Hong Kong, 2007
香港沙田新城，禾輋市场，2007年

iarity and ease with hyperdensity, at least in the cities. We all know the fate of modernist urbanism, and especially public housing in the U.S.; it was an unmitigated disaster, and the architecture was often blamed, like at the Pruitt-Igoe houses in St. Louis, which won multiple awards but were later demonized and eventually destroyed.

It's the old question of whether the failure is in the architecture or in the social structure.

Exactly. I used to sing that song, too, until Tunney Lee took me to the Wo Che public housing estate in Sha Tin, Hong Kong. There's a wonderful outdoor eating area, very informal, very ephemeral, that nightly comes to life in a plaza at the very center of these otherwise incredibly banal 1970s British-designed mid-rise apartment blocks. The whole thing works magically. It works because the estates are well-maintained by the housing authority, because there is a greater mix of incomes, and because Hong Kong people have a certain cultural familiarity with and even an affinity for hyperdensity, which we just don't have here. Here in the U.S., the same

座上世纪70年代英国人设计的、非常陈旧的中层公寓大楼的中心。一切都进行的非常美妙，这是因为房屋得到了管理部门很好的维护，因为这是一个不同收入阶层混合的地方，香港人有一种文化上的亲密感，甚至感觉高密度富有吸引力——这点美国是没有的。在美国，同样的建筑被诬蔑成是为穷人修建的，充斥着罪犯仓库的"那些项目"。

在禾輋，街道的地面很大程度上是机动车的地盘，但居民并没有减少。街道的生活实际发生在上一层，包括大部分的商店和餐馆。而在这之上有半公开的住宅走廊，就像新加坡的住宅楼——外部走廊是空中的公共/私人街道。除了没有汽车来打扰你，它们真的就像正常街道一样。史密森(Smithsons)和雅各布斯(Jacobs)的观点基本上是一致的，只不过雅各布斯接受有车辆的街道，而史密森认为汽车已经损害了为人服务的街道，因此为什么不创造一种全新的空中街道，远离这些车辆？当然，在大多数情况下，像在伦敦的罗宾汉住宅楼，这些规划并不可行。而在香

Wo Che Estate overview, Sha Tin New Town of Hong Kong, 2007
香港沙田新城，禾輋，2007年

港，它确实神奇般地发挥了作用。

　　中国汽车的流行，转变了人们对自行车的观点，自行车现在被视为贫困潦倒的象征，而在许多西方城市，自行车却在不断增加。

　　在与自行车关系的演化上，中西有着巨大差异。人口统计资料显示，在美国，有很大一个精英群体已经接受了自行车——他们是进步的，受过良好教育的人才，有着强烈的环境道德感。而在中国，这类人更倾向于别克和宝马的代理商。我有个朋友在南京大学，他一直住在学校附近的一座职工公寓，且经常步行上班。他喜欢车，也追求车的名誉价值，因此就买了一辆。结果，他每天早上要花半个小时开车绕着学校转，去找一个停车位，然后停在离他的老公寓有四分之三路程的地方。最后，他在郊区购置了一处房产，买车才显得更合理一些。现在，他在郊区的超级广场买东西，汽车引发了一种全新的生活方式。我不是在开玩笑，我的妻子也在南京大学教书，她是她们系里唯一骑自行车上班的人，她所有的同事都为此而笑话她——要知道，这是在一所建筑

kind of architecture became stigmatized as "the projects"—as crime-riddled warehouses for the very poor.

At Wo Che, the street-level ground plane is not depopulated, but it's largely the domain of motor vehicles. Most of the street life is actually up one level, including the majority of shops and restaurants. And above this you have the semi-public residential corridors, like in Singapore housing estates, where exterior corridors are public/private streets in the sky. They really are like normal streets, except there are no cars to bother you. The Smithsons and Jacobs all sang essentially the same song, except Jacobs was willing to accept the street with cars in it, whereas Alison and Peter Smithson argued that the automobile had ruined the street for people, so why not create a whole new street structure up in the air away from those machines. Of course, in most cases, like the Robin Hood Housing Estate in London, it didn't work. In Hong Kong it works magnificently.

The popularity of cars in China is the inverse of bicycles, which are now equated with poverty and deprivation, even as cycling is on the rise in many Western cities.

China and the West are about as far apart as possible in their evolving relationship to the bicycle. Here, if you look at the demographics, it's quite an elite group that has embraced the bicycle—progressive, well-educated professionals with a strong environmental ethic. In China those are the very people making a mad dash to the Buick and BMW dealerships. I have a friend at Nanjing University who lived until recently in a faculty housing complex right near campus, and usually walked to work. He wanted a car for the prestige value and because he just loves cars; so he bought one. And soon he was circling the campus for a half hour every morning looking for a place to park, and parks three-quarters of the way back to his old flat. Eventually he bought a place in the suburbs so that his car made more sense. Now he shops at big-box suburban stores. The car dictated a whole new lifestyle. I'm not kidding. My wife, who also teaches at Nanjing University, was the last person in her department to ride a bicycle to work, and all her colleagues ribbed her for it—this in a department of architecture, where half the courses are about green design and sustainability.

People talk about electric cars and say "I don't use gas anymore." Well, where is the electricity coming from? It's still energy, folks! Anyway, if even a fraction of the people in China adopt the automobile as we have here, the planet will be fried. Hybrid or electric cars are not going to save us. If you look at everyone wanting automobiles in China and India, I don't know how we're going to work our way out of that. What's hopeful in this picture is that China, even as it's building its vast highway system, is also building even more miles of high-speed rail, light rail, subway. That's the hopeful part. Energy-wise, China is ahead of everyone in terms of developing alternative energy industries.

It's largely driven by economic imperative.

Chinese officials know that for the Communist Party to remain in power and to keep people happy, they need to keep delivering. Truth be told, they've delivered a lot in recent decades. Chinese people are not happy with everything the Communist Party is doing, but they tolerate the corruption and nepotism because it's a small price for one of the greatest economic revolutions of all time. More people have been lifted out of poverty in China than ever before in history, and the Communist Party has to keep delivering on that.

The Chinese authorities are savvy enough to know that the great Chinese economic engine, which has created so

系里，其中一半的课程是关于绿色设计和可持续性的。

人们谈论电动汽车的时候说："我再也不用汽油了。"可电是从哪里来的呢？各位，它依然是能源啊！不管怎样，即使只是一部分中国人像我们在美国这样使用汽车，整个地球就会被烤熟。混合动力或电动汽车不会拯救我们。如果在中国和印度所有人都要车，真不知道我们该如何走出困境。在这幅景象中，还能让人看到希望的是，虽然中国正在建造大规模的高速公路系统，但同时也在修建更长的高速铁路、轻轨和地铁线路。就发展替代性的能源工业而言，中国在能源智慧上领先于所有国家。

这主要是经济需要所推动的。

中国官方知道为了共产党能够继续掌握政权，为了让人民幸福，他们必须付出。说实话，在最近的数十年间，执政者已经付出了很多。中国人对于共产党的举措并不都很满意，但他们忍受了腐败和裙带关系，因为在所有时期，这是伟大的经济革命所要付出的小代价。和任何一阶段的历史相比，现在的中国都有更多的人摆脱了贫困，而共产党将在这一点上继续努力。

中国官方机智地意识到，已经创造如此多财富的中国经济引擎需要一些东西才能运行，而石油已经快耗尽了，这一点我们都知道。人类正进入烦躁喧嚣的全球政治时代，为剩下的石油储备而斗争。当然，中国在发展替代性能源上非常明智，他们有资金和资源，坦白说，我们曾经也有机会那样做。还记得上世纪70年代吉米·卡特（Jimmy Carter）说"不要消耗过多能源"，"汽油限量供给"等诸如此类的话。我们本可以那么做，但那时还是没学会。记得2008年汽油涨到4美元一加仑的时候吗？所有人都开始坐火车了，那时美国铁路客运公司的上座率是二战之后最高的，但是当油价格下降后，运动型多用途汽车（SUV）的销量开始上升，火车上座率下滑，人们重新开车出行。替代能源是我们在很久以前就花巨资投入的一个项目，但是现在却落后了，我们将被中国彻底击败。中国人生产的太阳能板价格只有我们的一小部分，因为实际上是政府在为结果进行投资。

Beijing, Fortune Garden, 2002
北京丰台，富锦嘉园，2002年

中国是否在其现代化的一个相对早期阶段，并未关注整体的环境问题？

我认为中国官方一开始并没想要为环境保护分心——他们只是将其作为对经济增长的刹车手段。但最近几年，他们已经开始认识到，自己是在环境污染中缓慢自杀。中国正在污染维持社会和增长所需要的空气和水源，一旦这种情况发生，北京就会认真对待环境保护问题，但这肯定不是出于任何利他主义的考虑。全世界的目光都在盯着中国，西方早就处过这个位置，做过这些事情。上世纪80年代和90年代早期，中国才真正开始关注环境，而此时美国已经污染地球几十年了，并开始与环境友好相处。这是中西方之间另一个重大的对照点。

much wealth, needs to run on something, and the oil is running out. We all know that. We're just getting to the era of fractious global politics and fighting over the remains of the oil reserves. The Chinese are very smart about developing alternatives now while they have the money and resources, as we should have done, frankly, when we had the chance. Remember the 1970s with Jimmy Carter, "Don't be fuelish," gas rationing, and all that? We could have done it then. We still haven't learned. Remember in 2008 when gas went up to $4 a gallon? Everybody started riding the train. Amtrak ridership was the highest since World War II, and then as soon as gas prices went down, SUV sales started climbing, rail ridership dropped, and people started driving again. Alternative energy is something we should have invested heavily in a long time ago, and now we're lagging and we're going to get whipped by China on this. The Chinese can produce solar panels for literally a fraction of what it costs here, because the government is literally invested in the outcome.

Isn't China addressing the whole environmental issue at a relatively early stage of its modernization?

I don't think the Chinese authorities wanted to bother with environmental protection initially—they saw it as a brake on economic growth. But in recent years they've come around to recognize that they were slowly killing themselves with pollution. China is poisoning the very air and water needed to sustain the society and sustain growth. Once that kicked in, Beijing got serious about environmental protection. But certainly it wasn't coming out of any deep sense of altruism. The eyes of the world are on China now. The West had already been there, done that. By the time China really got going in the 1980s and early '90s, the U.S. was making nice with the environment after decades of polluting the planet. It became another great point of contrast between China and the West.

Could your curriculum be taught in China and be equally applicable?

The least applicable would be the vast apparatus that we've created in this country and in the West in general, largely in the wake of the urban renewal disaster fiasco, to assure greater community input to govern planning development.

That's the least applicable, although it is arguably the most desperately needed. Things like transportation planning would be very transferable. We have folks here at Carolina doing more and more consulting work in China, and we've got an initiative under way with Peking University, part of which might actually involve training courses for working Chinese planners. So a lot is very applicable—with the exception, as I said, of participatory planning, which is a big exception, because that's been a mainstay of post-1970s American planning.

How does the production of built environments differ in China and the U.S.?

There are many differences—in scale, speed, and sheer ambition. There is also the phenomenon of replication—bootleg DVD culture applied to architecture and building. In China, things are copied and become almost cookie-cutter very quickly. We generally don't do that here, nor do we want to build on the scale that China has become renowned for. Standard building types and floor plates are literally copied. There's a whole black market in replicated designs.

你的课程是否可以在中国讲授，是否同样适用于中国？

我们在重建城市的失败教训中领悟到，那些在美国和在西方制造的庞大机构，是最没用的。我们应保证更大型的群体参与进来，控制计划中的发展。

公共参与是最没用的，虽然它在中国非常急需。类似于运输规划之类的东西是可以被转移的。加利福尼亚有很多人在中国做咨询工作，我们与北京大学有一个正在进行的初步合作计划，其中可能包括，为工作中的中国规划者提供培训课程。所以，有很多东西还是可以被应用的，除了我刚才所说的，分享参与——这是很大一个例外，因为它主要是20世纪70年代后期美国人的设想。

中美之间建筑环境的差异是什么？

差异有很多——规模、速度、绝对的雄心！也存在复制的现象——盗版DVD的文化被应用到建筑学和建筑物中。在中国，东西很快会被复制，然后几乎快速形成模板式。我们一般不这么干，也不想像中国那样建造如此大规模的复制。在中国，标准建筑样式和平面被完全复制，并在复制设计方面，存在一个非法交易的黑市。

PLANNING Thomas J. Campanella

黑市？它带有贬义吗？

这是个好问题。它是灰色的，没有什么贬义，除非一名西方建筑师卷入其中，然后中国人会很震惊地发现，他们的西方同事居然因为自己的方案被抄袭而暴跳如雷。在南京那所建筑学院，我见过这种事情发生，和大多数中国建筑学院一样，他们有一个咨询部门，是积极参与实践活动的设计机构。几年前，这个设计机构参与了一个项目，在南京市郊建立一所新的农村小学。这是一个大工程，是美国人梦想中要花数月甚至数年时间才能完成的项目。但这个机构在数周之内就解决了问题。为什么？其中很大一部分是因为拷贝复制，一些想法来自另一家中国学校项目，或者可能来自荷兰的一些学校。谁知道呢？工程从梦想到完成只花了一年半的时间，真是不可思议。

我最近在北卡罗来纳的杜罕做了一次名为"七年之桥"的演讲。建造杜罕高速公路上的这座小小人行桥，花费了七年的时间。我们都知道原因：构思过程极端重视公众和社会的参与，以及所有的环境要求——这些会对我们的陈述产生影响。我在最近几年一直提倡：在两种极端之间获得一个更好的平衡，在自上而下集权的中国式发展和缓慢的美国式发展之间获得平衡，后者总是喜好咨询每一位利益相关者的意见。

中国需要更多的民主进程来参与设想，美国也需要执行得更快一些。假如要建造一条从里士满(Richmond)到罗利(Raleigh)的高速铁路，我们是否真的需要一个堆起来有4英尺高的环境后果报告？这是一条长廊，从19世纪60年代之后就开始通火车了，这并不是一片处女地。铁路确实有些过剩，但当你考虑铁路可以带来的好处时，它又很值得去建设。很多公共参与进程，民主的审核与平衡，并不能够考虑到长期的益处，比如：可持续的基础设施和社会、地区甚至最终整个地球的能源规划。公众往往目光太过狭隘，他们只关注那些小蜗牛是否濒临危险（我并非冒犯濒危物种），或者居民会不会因为不喜欢在自己的后院中出现哪些东西而吵闹？我们需要更好地去平衡长期利益与本地、当下的利益。

Black market? Is a stigma attached?

It's gray. That's a good question. There is little stigma unless a Western architect is in the picture, then the Chinese are shocked to learn that their Occidental colleague is outraged that his or her scheme has been copied. I've seen this happen in Nanjing in the architecture school where, like most Chinese architecture schools, they have a consulting wing known as a design institute that's actively engaged in practice. Several years ago the design institute got involved with a project to build a new rural elementary school on the outskirts of Nanjing. It was a great project, sort of a dream commission that an architectural office in the U.S. would spend months or years on. And they turned it around in a matter of weeks. Why? It was largely replicated, probably from some other Chinese school project, or maybe some school in the Netherlands. Who knows? The project went from dream to complete in about a year and a half. It's amazing.

I recently gave a talk about what I've named "The Seven-Year Bridge" here in Durham, North Carolina. It took seven years to build this little footbridge over the Durham Freeway. We all know the reasons: a planning process that places a high value on public participation and input from the community, plus all the environmental requirements—impact statements and the like. What I've been advocating in recent years is that we need a better balance between the two extremes of authoritarian, top-down China-style development and the measured slowness of U.S.-style development, with its penchant for consulting every last stakeholder.

China needs more democratic process and more participatory planning, just as we need to push things through more quickly. Do we really need a stack of environmental impact statements four feet high to build a run of high-speed rail from Richmond to Raleigh? This is a corridor that we've been running trains through since the 1860s; it's not exactly virgin terrain. It's excessive, and when you consider the benefits high-speed rail offers, it's well worth doing. Much of the public-participation process, the democratic checks and balances, fails to take into account the long-range benefit of good, sustainable infrastructure and energy projects to society, to the region, and eventually to the planet. It's too locally focused, whether on the little snail that's endangered—no offense to endangered species—or the neighborhood that's crying because they don't want this or that in their backyard. We need to better balance larger long-term benefits against the local and the immediate.

Chengri Ding
UNIVERSITY OF MARYLAND AT COLLEGE PARK

274

Chengri Ding, Ph.D., is associate professor at the Urban Studies and Planning Program of the University of Maryland. He specializes in urban economics, urban and land policies, urban planning, and China studies, and has written extensively in learned publications, in both English and Chinese. He has edited three books on China regarding land and housing policies, urbanization, and smart growth. Dr. Ding has led many international policy projects on China, including urban master plans, farmland protection, property taxation reform, and local public financing. His contributions are highly valued by senior government officials and scholars alike; he is a sought-after keynote speaker. Dr. Ding has served as consultant to the World Bank, Global Business Network, FAO, and many leading Chinese agencies, such as NDRC. He also serves on the advisory board for the International Institute of Property Taxation. Dr. Ding is the founding director of the Lincoln Institute of Land Policy's China Program.

丁成日，博士
马里兰大学

丁成日是马里兰大学城市研究与规划学院的副教授。他研究的领域是城市经济、城市和土地政策、城市规划以及中国研究，他写了许多学术作品，这些作品都是以英文和中文双语发表的。他主编了三本关于中国土地和房屋政策、城市化以及智能增长方面的书籍。丁博士将许多国际政策项目引入中国，这些项目包括从城市管理规划、耕地保护到财产税、地方公共财政。他的贡献受到高级政府官员和学者的高度赞赏，他是一位深受大家喜爱的演讲者。丁博士曾经为世界银行、全球商业网络、联合国粮食与农业组织，以及许多重要的中国机构，比如国家发展和改革委员会当过顾问。他也为财产税国际研究院的咨询委员会工作。丁博士是林肯土地政策研究院中国部主任。

For complete biography, please refer to the Appendix.
完整的传记，请参阅附录。

New development on former farmland in the Yangtze Delta, China
中国长三角地区耕地上的新开发项目。

Where are the main centers of migration?

Migrants are going to big and small cities alike. Basically, people follow jobs. Of course, you can see many migrants in Beijing, Shanghai, Guangzhou, or Shenzhen. In Shenzhen, 90 percent of residents are migrants. They have lived there for 20 years but still do not have official *hukou*, so technically they are still considered foreign. In almost all major cities, growth is dominated by migrants. Kunshan county, for example, is something like 960 square kilometers [371 square miles]. In 1997, its population was about 500,000, half urban, half rural. By 2007–8, the population quadrupled to over 2 million, which means that in ten years over 1.5 million people came from outside. This is a small city, but it is representative of the whole county.

What accounts for the surge?

Jiangsu Province wanted to take advantage of Kunshan's location near Shanghai, so they specifically targeted it as its key growth port. Kunshan received a lot of special policies, incentives, and investments from the provincial government.

Is Jiangsu's government independent of the central government?

Politically not, but local governments have a lot of autonomy in economic affairs as China is decentralized. There are 55 national industrial development zones, but each province, municipality, and individual city can create its own development zones. Kunshan, for example, has twelve industrial zones. One is national, three to four are provincial, the rest are local. Some are very successful. In one zone there are 30,000 to 50,000 workers. Kunshan is a typical case. The government, or sometimes an individual factory, builds apartments and rents them cheaply to the migrants so people come there to work.

What kind of apartments?

High-rise buildings. The quality and conditions are better than the dormitory I lived in as a college student almost 30 years ago, and much better than when these people were back on the farm. It all depends on where they're from. But generally, dormitory conditions now are better even than in Beijing's *hutongs* of 20 to 30 years ago. However nice, those compact houses had no indoor plumbing. So in the winter, in the middle of the night, you'd have to walk maybe 100 meters [328 feet] for a toilet. Dormitory housing offers a big improvement. That's why people choose to come to the city. They

移民的主要中心在哪里？

移民们去大城市和小城市并没有什么不同，从根本上说，人们追随的是工作。当然，你可以在北京、上海、广州或深圳看到许多移民。深圳90%的居民都是移民，他们已经在那里住了20年，却依然没有获得正式的户口，所以从技术上说，他们依然被认为是外来人口。在绝大多数大城市里，移民主导了城市的发展。例如，昆山县的面积大约有960平方公里（370平方英里），1997年，昆山县的人口大概是50万，一半是城镇人口，一半是乡村人口。到2007年、2008年的时候，人口已经翻了四倍，超过了200万，这也就意味着在十年期间，有超过150万人从外部进入到昆山。这是一个小城市，但它却是整个国家的一个缩影。

怎么解释这种数量上的激增？

浙江省想要充分利用昆山临近上海的地理位置，所以他们特别将昆山当作关键性的增长点。昆山获得了许多来自省政府的特殊政策、奖励机制和投资。

江苏省政府是否独立于中央政府？

在政治上不是，但从经济上说，地方政府在经济事务上有很大的自主权，就像中国是分权式的。总共有55个国家工业开发区，但是每个省、直辖市以及每个城市都能设立自己的开发区。例如，昆山就有12个工业区，其中一个是国家级的，还有三四个是省级的，剩下的都是地方的。有一些开发区非常成功，一个区内就有3－5万工人。昆山是个典型的例子。政府有时是私人工厂，修建公寓，以低廉的价格租给移民，因此人们可以来到这里工作。

什么样的公寓？

高层楼房。质量和条件要比我30年前当大学生时住的宿舍好，也比这些人回到农村的住宅条件好，当然这要看他们来自哪里。但一般来说，现在宿舍的条件甚至比二三十年前的北京胡同要好。不管多漂亮，这些拥挤的胡同住房在那时也没有自来水。所以在冬天的深夜里，你必须走大概100米（328英尺）去上厕所。宿舍住房是一个很大的进步。这就是为什么人们选择来到城市的原因。他们可以在屋里使用洗手间和洗澡，另外一个差别是可以在屋里烹饪。有一些公寓配有厨房，像美国的套房，其他的则是共享的设施。

如果居住条件如此好，到底是什么阻止每个人都跑到城市里来呢？

这是一个好问题。宿舍只是为那些在工厂里工作的人准备的。要得到一间公寓，你首先必须得到一份工作。

政府修建和租赁公寓的例子有很多，但公寓并不用来处理成千上万的移民，而是给工厂一些单元，让它们提供给工人们，所以是工厂收集租金，再向政府支付这些租金。这使得事情更容易些，在一些楼房中，政府将不同的公寓区分给不同的工厂。另外一个模式则是，一个公司建立自己的公司宿舍。

有没有一些公寓楼是由独立的开发商修建的？

有，特别是在靠近香港的广州地区。农村居民和农户，有机会出租他们的房子，并获利颇丰。有时候，那些楼房现代得你都不敢相信，而且都更大更好。

这就像一块宅基地。每个农民都有自己的土地，因而他可以修建一座房子，住在第一层，然后把第二层、第三层或第四层租出去。有时候，房子甚至有十层，但大多数是四到五层。在中国有许多人以此发家致富。

像苏州这样靠近上海的地区，公寓由政府修建，所以价格相对便宜，因为它们是有补贴的。但是在广州，由于政府没有在那里修建公寓，而且工厂也太小，不能修建自己的公寓，所以在背后起推动作用的是市场。这不像30年前了，那时政府决定一切。

如果所有人都搬到城市里，农田怎么办？

这是个问题，食品安全关系重大。最低数量的农田应当得到保护，但事实上在中国，食品安全不依赖于农田保护，它应当通过农业政策予以解决。如果农业没有了农田，食品安全将会通过农田保护这样的土地政策来解决。但是如果农业没有了农民，食品安全就应当通过农业政策进行解决。无人耕种这一事实主要是源于经济上的决定：农民们都搬到城市去，赚取更高的非农业

can use the toilet and take a shower indoors. Another difference is indoor cooking. Some apartments have their own kitchens, like efficiencies in the United States; others have shared facilities.

If living conditions are so much better, what is to prevent everybody from coming to the city?

That's a good question. Dormitories are only for the people who work in the factories. To get an apartment, you first have to get a job.

There are cases in which the government builds and rents apartments, but since it doesn't want to deal with thousands of migrants, it gives the units to a factory for its workers, so the factory collects rent and then pays the government. It makes things easy. In some buildings, the government assigns blocks of apartments to different factories. Another model is a company that builds its own dormitory.

Are any apartment buildings built by independent developers?

Yes, typically in the Guangzhou region, near Hong Kong. Rural residents, peasants, take the opportunity to rent out their houses and make big money. Sometimes the buildings are so modern you cannot believe it, much bigger and better than many houses.

It's like a homestead. Each farmer has his own land so he builds a house, lives on one floor, and rents out the second, third or fourth. Sometimes they're ten stories, but most often four to five. Many people in China are rich because of this.

In areas like Suzhou, near Shanghai, it's the government that builds. Apartments are relatively cheap because they're subsidized. But in Guangzhou, because the government hasn't built apartments there and because the factories are too small to build their own, it's market driven. It's not like 30 years ago, when the government determined everything.

If everybody moves to the city what happens to the farmland?

That's a problem. One concern is food security. A minimum amount of farmland has to be maintained but, in truth, food security in China is not dependent on farmland protection. Food security should be addressed through agricultural policy. If there is no farmland to farm, food security should be addressed through land policy such as farmland protection. But if there are no farmers to farm, food security should be addressed through agricultural policy. The fact that no one

Rice paddies in Yangshuo, Guangxi Province, southeast China.
中国东南部广西阳朔的稻田。

farms is due to economic decisions as farmers move to cities to earn much higher nonagricultural wages. In addition, data shows that China is an importer of agricultural goods, which may give a false impression that China's food security is a legitimate concern. There are two things to keep in mind. One is that good agricultural policies have helped to raise grain production in China over the past 10 years. The other consideration is that imported grains largely reflect changing tastes and quality concerns. For instance, now you can hardly find the kind of rice that was consumed 30 years ago in Shanghai. Everyone now eats steamed rice mainly produced in the Northeast. Rice produced along the Yangtze River is mainly used for industry and herd feeding.

Hasn't China developed a new strain of 3-harvest rice?

Yes, but no one eats it because it doesn't taste good. Also, China produces a lot of wheat. It is good for noodles and such but not for making bread, so they have to import from American and Canada.

工资。另外，数据显示，中国是一个农产品进口国，这可能给人一种错误的印象：关注中国的食品安全是完全合理的。要记住两点，一是好的农业政策已经在过去的十年内帮助中国提高了粮食生产能力；另一点是，进口粮食很大程度上反映了人们不断变化的口味和对质量的关注。例如，现在你很难找到那种30年前在上海食用的大米了。如今，每个人吃的都是产自东北的白米饭，长江流域的稻米产地现在主要用于工业和畜牧喂养。

中国不是已经研发了一种新的、一年三熟的稻米吗？

是的，但没人吃这些稻米，因为不太可口。而且，中国也生产很多的小麦。小麦主要用于面条及相关的食品，但却不适于做面包，所以他们必须从美国和加拿大进口适合做面包的小麦。

政府是否保护好的稻田？

是的，他们有政策。这一般来说是正确的，和美国一样，没有政府补贴，农民不可能在现代社会生存下去。农民的收入不能与非农业行业相提并论。我做过的一项研究显示，马里兰的每个农民必须耕种至少25英亩（大概10公顷，或者151亩），才能有足够的家庭收入，满足最低的生活标准，这是在美国。而在中国，大概是每人1亩或0.3英亩，这还仅仅是满足食物的需求。在如此少的面积上，你生产不出足够的东西去获得收入购买其他东西。中国有如此多的土地，但也有如此多的人口，而且每个人要占据大约0.5英亩的土地。

在农村地区，过去土地为公社成员集体拥有。从那之后，中国出现了家庭联产承包责任制的改革。政府意识到共产主义的生产效率太低了，他们决定以家庭为单位划分农田，并签订一年的协议。所以，基本上说，人们上交收成的一部分给公社，公社再向政府交税，但人们可以保留剩下的所有东西，这导致了生产效率的提高。

这项政策开始于上世纪80年代早期。这也是为什么在早些时候，中国最富的人是农民。一位教授可能一个月赚50块人民币，但通过家庭联产承包责任制，农民可以每个月赚到1000块人民币。到上世纪80年代后期和90年代初期，企业进行了改革。改革在农村开始并扩展出去。90年代，当国有企业开始进行改革时，城市居民变得富有多了，变化太快了！

回到上世纪80年代，那时，只有不到10%的GDP来自私营机构，现在，这个数字接近65%。大体上说，有两种类型的企业主办者：国家所有的企业和集体所有的企业。集体所有企业大致为一些城市、省份或政府部门（像交通部或农业部）所拥有。大约有60%的企业是集体所有的，这些企业的数量有上万之多——在中国，它们是最大的雇主。相反，只有大约100家或更少的国有企业，但规模都非常庞大，主要分布在像石油、天然气、钢铁等行业中。

过去，要在中国成立外企是很困难的。

没错，但在2001年中国加入WTO（世界贸易组织）之后，情况就开始发生改变。按照WTO协议，不同工业有不同的保护方式。例如，银行业和金融业最近才开始进入中国市场，而这在十年

Does the government protect the good-rice fields?

Yes, they have policies. It's generally true, even in the United States, that without government subsidies, farmers cannot survive in modern society. Their income just can't compare with nonagricultural jobs. I did a study showing that each farmer in Maryland had to farm at least 25 acres [roughly 10 hectares, or 151 mu] to make enough family income for minimum standards of life. That's in the U.S. In China it's about 1 mu, or .17 acre, per person! That's just for food. With that area you cannot produce enough to generate income for other things. China has so much land, but with so many people, each ends up with roughly half an acre [.2 hectare].

In rural areas, land is collectively owned by the people in a commune. Then China was reformed with the Household Responsibility System. Realizing that communism's productivity was really low, the government divided the farmland by household and issued one-year contracts. So basically, the people paid a percentage to the commune and the commune paid tax to the government, but the people got to keep whatever was left over, resulting in higher productivity.

This started in the early 1980s. That is why, early on, the richest people in China were the farmers. A professor might make 50 RMB [$8] a month, but through Household Responsibility farmers could make 1,000 RMB [$158] a month. In the late 1980s to early 1990s, entrepreneurs were reformed. So it started in rural areas and spread from there. In the 1990s, city people became much richer when state-owned enterprises started to reform. Fast!

Back in the 1980s, less than 10 percent of the GDP was from the private sector. Now it's more like 65 percent. Basically, there are two types of entrepreneurs: state-owned firms and collective-owned firms. Collective-owned are basically owned by cities, provinces, or some government branch, like the department of transportation or department of agriculture. Roughly 60 percent of enterprises are collective-owned. There were thousands of them, maybe hundreds of thousands, and they were the biggest employers in China. By contrast, there are only about 100 state-owned enterprises or fewer, but they're giants, like petroleum, natural gas, steel, etc.

In the past it was difficult for foreign entrepreneurs to get established in China.

True, but that has been changing since China joined the WTO [World Trade Organization] in 2001. By WTO agreement, different industries have different terms of protection. For instance, the banking and financial sector recently started to penetrate the Chinese market; that was impossible 10 years

Longshen, Guangxi Zhuang Autonomous Region, China. The region's terraced rice paddies and densely clustered housing constitute the two main uses of rural areas: farmland and residential.

中国广西壮族自治区，龙胜。该地区的阡陌和密集的聚集住房构成农村地区的两个主要用途：耕地和居民。

ago. The auto industry is still protected, I believe. Also, many professional services sectors have been open to foreigners, like design, legal, etc.

Who oversees national and local economic zones to ensure against damaging competition or duplication?

Nobody. One of the reasons China has been so successful in the past 30 years is because of economic and fiscal decentralization. That's good. The downside is that decentralization created fierce local competition. These incentives and programs have been the main driving forces for the development of economic and industrial development zones. By the summer of 2004, there were 6,866 zones across the country, covering more than 38,600 square kilometers [14,900 square miles]. In a nationwide inspection in the summer

前是不可能的。我相信，汽车行业依然受到保护。而且，许多专业服务行业已经向外国人开放了，比如设计、法律等领域。

谁来监督国家和地方的经济特区，确保不会出现恶性竞争或山寨盗版？

没有人。中国在过去30年取得如此巨大成功的一个原因就是：经济和财政的权力分散，这是好的。不过它所带来的负面后果是：权力分散造成了激烈的地区竞争。那些激励机制和项目已经成为经济和工业开发区发展的主要驱动力。到2004年夏天，整个国家有6866个开发区，占地面积超过38600平方公里（14900平方英里）。2004年夏的一项全国性调查表明，计划内有面积达24900

平方公里（9600平方英里）的开发区由于缺少投资被撤销，这个数字占到总数的64.5%。其中，有超过1300平方公里（500平方英里）的土地已经被重新用作农业生产。

财政和经济权力分散，导致在许多制造行业出现了产能过剩的现象。有一些产业已经表现出产能过剩的征兆，包括钢铁业和汽车业：钢铁当前的产能大约在4.7亿吨，还计划增加1.3亿吨，但实际的需要则只有3.7亿吨；而汽车的产能与市场预计的消化量相比，多出了50%。其他产能过剩的行业包括：铝、铁合金、焦炭、电石、氯化钙、铜、水泥、电能、煤、纺织、石化、纸盒、化肥、家用电器、微型计算机和造船。

在中国，2005年汽车产能与市场需求的比率是1.47:1；算上当前正在建设的企业，以及那些打算进入到这一市场的企业，这个数字定然会升高。如果所有在建的企业于2020年之前完工，中国将有能力为全世界生产汽车。这真的非常疯狂！

从现在起到2020年，会发生什么？

我们不知道。有些企业会被强制关闭。这是对资金的极大浪费。中央政府没有权力对此进行控制，在美国，奥巴马没有时间去监管所有制造业的情况，这不可能做到。所以在问题变得严重之前，没有人会去处理。

让我们回到土地使用的情况。

在农村地区，对土地的使用主要有两种方式：宅基地和农业用地。从历史上说，人们有自己的宅基地，但从根本上却没有财产权。如果你的儿子们结婚了，你就在公社的边界内，为他们各自建一栋房子。土地非常便宜，没人会对此留心。但是现在，它成为了一个问题，人们想要抓住它，在许多地方，比如广州，发生了这样的事情：人们开始在所有可以建房子的地方大兴土木，然后将其租出去。房子变得越来越大，而一栋建筑物的寿命却可能被减少到只有数年。这主要是市场推动，寻租行为所导致的。在那些地方，建筑物如此拥挤，以致被说成"从窗户上就可互相亲吻"。拥挤产生了严重的问题，像安全、卫生系统、空气质量和消防等。

中央政府开始制定政策来防止此类情况的发生。比如，其中有一个政策就是，每户占用120

of 2004, 24,900 square kilometers [9,615 square miles] of planned development zones were eliminated due to lack of investments, representing 64.5 percent of the total. More than 1,300 square kilometers [500 square miles] have been returned to agricultural use.

Fiscal and economic decentralization have resulted in redundant and excess capacity in many manufacturing sectors. Sectors already exhibiting excess capacity include steel, with current production capacity estimated at 470 million tons and planned capacity of an additional 130 million tons compared to demand of only 370 million tons, and also automobiles, with almost 50 percent more capacity than estimated market absorption. Other industries with excess capacity include aluminum, ferro-alloys, coke, calcium carbide, copper, cement, electric power, coal, textiles, petrochemicals, paper boxes, chemical fertilizers, domestic electric appliances, micro-computers, and shipbuilding.

In China in 2005, the ratio of automobile production capacity to market absorption was 1.47:1; when firms that are currently under construction and those that are planned begin to enter the market, this number will surely rise. If all plants under construction were completed by 2020, China would have the capacity to produce cars for the entire world. That's really crazy!

What will happen between now and 2020?

We don't know. Some plants will be forced to close. It's a big waste of money. The central government does not have the power to control it. In the United States, Obama does not have the time to monitor what's going on with every manufacturer. It can't be done. So until issues become severe, nobody deals with it.

Let's go back to land usage.

In rural areas, there are two main types of land use: residential and farmland. Historically, there was homesteading and basically no property rights. If your sons were getting married, you just built a house for each within the commune's boundaries. Land was cheap so nobody cared. But now it becomes an issue. Now people want to grab it up. That is what happened in many places, as in Guangzhou. People began to build wherever they could and rent out. The houses became bigger and bigger and a building's life could be reduced to just a few years. It's mainly market driven, rent seeking. In those places, buildings are so crowded that they are called "kissing over the window." Crowding causes serious concerns like safety, sanitation, air quality, and fire protection.

The central government began to introduce policies to prevent this from happening. One of the policies is 120 square meters [1,292 square feet] of land per household, and maximum 3-story height, for instance. Nonimplementation becomes an issue.

It should be pointed out that these standards are quite significant. Because of urbanization many young people become educated and never return to farming. So maybe it's just two couples in the house, or the parents and just one child; 120 square meters [1,290 square feet] is pretty significant so people build 3-story buildings, easily 150 to 200 square meters [1,615 to 2,153 square feet], and rent them out. If your land is close to the city, then you become rich. But if you are in a remote area, there's no demand.

Land in cities is state-owned; land in rural areas is collective-owned. Land development can only happen on state-owned land; therefore farmers cannot sell land to developers directly. For commercial development, land has to change to state ownership and then local government, as a representative of the state does the land acquisition. It's a two-step process. In the U.S., developers can negotiate directly with individual owners, but in China, development is prohibited unless the land is converted first. So land conversion is difficult, but possible. The farmers are becoming increasingly smart and are asking for huge compensation.

Roughly 80 to 90 percent of urban expansion has taken place on farmland. When a city's population grows from 5 to 10 million, there's no other choice.

Beyond China's east coast, is development happening in the west, too?

Not much. Most development is in the east, with some isolated construction in Inner Mongolia and other places because of certain resources, like mining, but few people live there. It's all farmland.

What about increasing density within the existing city?

They do that in a minimal way, but redevelopment costs are huge, and there are resettlement problems. Redeveloping existing buildings with something higher is increasingly difficult because of protection. Current tenants don't want to go through demolition. They demand huge compensation and many become unbelievably rich. People ask for maybe 500,000 RMB [$82,000]; that's high but reasonable. But occasionally someone demands much more than the others, maybe 5 to 10 million RBM [$820,000 to $1.6 million], for the same house in the same location, right next to the others.

平方米（1292平方英尺）的土地，而且楼层最多只能有三层。但随后，不按政策执行就成了另一个问题。

应该指出，这些标准非常重要。因为城市化的缘故，许多年轻人开始接受教育，而且再也不会回去务农了。因此，呆在房子里的可能只有两对夫妇，或者是父母与他们唯一的孩子，这样120平方米就已经足够用了，而人们修建的三层楼房很容易就到达150-200平方米（1614平方英尺到2153平方英尺），所以他们就将多余的租出去。如果你的土地靠近城市，那么你就发财了。但是如果你的土地在偏远地区，是不会有人来租的。

在城市中，土地是国家所有的；而在乡村地区，土地则是集体所有。土地开发只能发生在国家所有的土地上；因此农民并不能直接将自己的土地卖给开发商。为了商业开发，土地必须变成国家所有，然后地方政府作为国家的代表，就会进行土地收购。这是分为两个步骤的过程。在美国，开发商可以直接与个人所有者进行协商，但是在中国，开发商被禁止这么做，除非土地所有权首先被转换。因此，土地的转换是很困难的，但却不是不可能的。农民们变得越来越聪明，总是要求一大笔赔偿金。

大约有80-90%的城市扩张已经发生在农田上。当一座城市的人口从500万上升到1000万时，并不存在其它的选择。

除了中国东部沿海，西部是否同样也在开发？

不太多。大多数还是发生在东部；内蒙古和其他一些地方由于特殊的资源（比如矿产业），有一些零星的开发建设，但很少有人住在那里。西部地区都是农田。

在现存的城市内，人口密度也在不断增加吧？

他们想要以最小的代价解决问题，但重新开发的费用巨大，还存在再安置等问题。由于保护政策，重新开发现有的建筑使之变得更高，这一点越来越困难了。现在的居住者不愿等着被拆迁，他们要求巨额赔偿，许多人变得极为富有。人们可能要求50万人民币，这个数字很高，但却是合理的。但偶尔也有些人，他们的房子就在其他人房子的旁边，是同一地段的相同房型，但他们要求的赔偿则比其他人要多得多，可能是500万到1000万人民币。现在不断出现一些极端的例子：

人们要求巨额赔偿款，甚至威胁达不到要求就自杀或烧掉他们自己的房子。如果你是政府官员，你会怎么做？如果你让这些人失去生命或财产，你会被指责为管理不善。那么，是不是你就满足他们不合理的要求呢？我确实认为这是一个人权的议题，但我也相信，中国尚未发展一个合理恰当的机制来解决这些极端案例。过高的赔偿同样也是个公平正义的问题：要对纳税人公平。

为了更大的公众利益，有没有土地征用权？

他们确实有土地征用权，但有些人会采取极端行为。在美国，如果你夺走我的土地去修高速公路，我不会自杀。但在中国，却有数百起这样的案例。要解决这些问题，需要怎样正确的制度安排呢？什么能构成公平正义呢？如果现在有人要求500万的赔偿，那么，下一个人可能要求1000万。什么时候是个尽头？公平又在哪里？没有人回答过这类问题。

在美国，平衡很重要。个人的权利必须得到保护，但同时，代表所有人的政府的权利同样也要得到保护。因此，这些极端的要求在一开始的时候就不会得到批准。很显然，中国政府曾经做出过让步，而它本不该如此。

在接下来的30年内，中国将走向哪里？

这是个好问题，回答是不可预测。世界银行参加了面向中国国家主席的高层咨询会，在会上确定了中国面临的六个重大挑战，以及从现在到2030年中国需要做什么。其中一个关键领域是知识型经济，如果中国不进行转变，其经济不会是可持续的。中国的经济结构必须进行升级，让我们看看一些官方统计数据。十年前，日本经济位居世界第二，大约有15%的经济产业有着与美国一样的生产效率，而剩下的85%则只有美国生产效率的三分之二。中国现在是世界第二大经济体，但是其经济中并没有在生产效率上能与美国匹敌的。换句话说，没有一件中国制造的产品其专利是归中国人所有的：零专利，没创新，没发明。中国制造了纺织品、空调、裤子和鞋子，以及所有低科技含量的产品。以iPhone为例，世界上所有的iPhone都是在中国制造的，但中国人只从每部手机150美元的净利润中赚取了5-7美元，那是因为他们没有专利，是美国人发明了iPhone。你只是制造商，并不意味着产品就是你的，你只是劳工而已。

There is an increasing number of extreme cases in which people demand huge compensation, even threatening to kill themselves or set their houses on fire. If you were an official, what would you do? If you allow the loss of life or property, you are criticized for bad management. Or do you give in to their unjustified demand? I do believe this is a human rights issue, but I also believe that China has not yet developed an appropriate institution to deal with those extreme cases. Exorbitant compensation is also an issue of justice, justice to the taxpayers.

Is there eminent domain for the greater public good?

They do have eminent domain, but some people take extreme actions. In the United States, if you take my land to build highways, I don't kill myself. But China has hundreds of these cases. What is the correct institutional arrangement for dealing with such issues? What constitutes justice? If someone demands $5 million now, maybe the next person will demand $10 million. Where does it stop? What is fair? No one has answers to those kinds of questions.

In the U.S., balance is important. Individual rights should be protected, but at the same time the government's right on behalf of all people should also be protected. Therefore, these extreme demands should not be granted in the first place. Obviously, China's government has given in, as it should not.

Where is China headed in the next 30 years?

That's a good question; it's unpredictable. The World Bank is engaged in high-level consulting to the president of China. It identified six major challenges and what China needs to do from now until 2030. One of the key areas is a knowledge-based economy. If China cannot make the transition, its economy will not be sustainable. The economic structure has to upgrade. Let me share some official statistics. Ten years ago, Japan's economy was second in the world, about 15 percent having the same productivity as the United States. The remaining 85 percent of the economy was only two-thirds as high as in the U.S. China is now number two, but 0 percent of its economy is comparable to America's. In other words, not a single product made in China is patented by Chinese: zero patents, innovations, inventions. China makes textiles, air conditioners, pants and shoes, all low tech. Take the iPhone. Every iPhone in the world is made in China, but the Chinese earn only $5–$7 of the $150 [30-43 of 915 RMB] net profit. That's because they don't have the patent. Americans invented the iPhone. Just because you're the manufacturer doesn't make the product yours. You're just labor.

284

This is why the central government is emphasizing a knowledge-based economy and why they're doing everything to attract overseas scholars back to China. Four years of education in the United States is good but it cannot quickly convert to inventiveness. The tradition of invention has a long history in the U.S., and American educational institutions are the best in the world. These are not qualities that can simply be co-opted overnight.

Is urban planning new in China?

No. China has had urban planning for 50 or 60 years, long before communist control. But it's a different kind of urban planning. Beijing developed in rings around the Forbidden City, in a very structured way. With slow-paced or no urbanization, there were fewer questions of design. Now, under rapid urbanization, it makes the city terrible. Current urban planning focuses too much on urban design and doesn't give enough consideration to nondesign aspects. You have to remember that urban form is developed by individuals, by developers and real estate agents, and basically they don't understand the mechanism behind the way land is developed and used. They pay little attention to transportation impact. One of the things you have to remember is that cities continue to grow because of labor market integration. That's why New York is still the biggest city in the United States, even after 9/11; it continues to grow. Many thought that after 9/11, big cities were history, but that's not true.

It's basic economics. People and businesses need personal contact. It's a way of building trust. If we do not talk face-to-face, we will not do business. The importance of this city function should not be overlooked.

As an economist, I emphasize that everything has two sides, a positive and a negative. Take economic and fiscal decentralization, which provides incentive for local governments to promote growth. It is why the Chinese economy is so successful. The downside is an overheated economy, overinvestment, oversupply, and overmanufacturing capacity, all of which leads to fierce competition and instability. It also prevents the central government's policies from controlling growth. People understand one side of the story, but forget about the other side. The question is how to create balance. If you can do it, that's great. If you cannot, it's terrible. It's the same with everything.

Does China appreciate what you're saying?

It all depends. Some do and others do not, for different reasons. But it seems to me that the Chinese are far more open than before and show great interest in learning from others,

这就是为什么中央政府正在强调知识经济，也是为什么他们正努力吸引海外学者回到中国。在美国接受四年的教育是很好的，但这并不能马上转化为创造力。发明的传统在美国有着悠久的历史，而且美国的教育制度也是全世界最棒的。这些不是一夜之间就能容易补充的品质。

在中国，城市规划是不是很新颖？

不，中国的城市规划早在50或60年前就开始了，远在共产党执政中国之前。但那是一种完全不同的城市规划。北京的开发以一种结构非常良好的方式进行：呈环形围绕着紫禁城。假如发展的步伐缓慢或者根本没有城市化，就不会有太多关于设计的问题。但现在，在快速城市化的影响下，城市变得很糟糕。当前的城市规划太过于关注城市的设计，而对于非设计部分则考虑不够。你必须记住，城市形式是由个人、开发商和房地产经纪人决定的，基本上说，他们并不理解土地被开发和使用的方式背后所包含的机制，他们不怎么关注交通的影响。你必须记住的一点是，由于劳动力市场的集中整合，城市会继续增长。这就是为什么即使在911事件之后，纽约依然是美国最大的城市，因为它继续保持增长。许多人认为，在911事件之后，大城市会成为历史，但这并不准确。

这是基础的经济学。人和商业都需要私人联系，它是一种建立信任的方式。如果我们没有进行面对面的交谈，就做不了生意。城市所具有的这一功能的重要性不应当被忽视。

作为一名经济学家，我一再强调所有东西都有两面，积极的一面和消极的一面。经济和财政上的权力分散，为地方政府提供了激励政策以促进发展。这就是中国经济如此成功的原因。负面的问题则是：经济过热、投资过剩、供给太多，以及过多的制造业产能，所有这些都导致了恶性竞争和经济的不稳定，也妨碍了中央政府控制增长的政策。人们了解了故事的一面，但也忘记了另一面。问题在于：如何创造平衡！如果可以做到平衡，那简直太棒了，如果不能，就会很糟糕。其他东西同样也是如此。

中国是否欣赏你所说的东西？

这要看情况。由于不同的原因，有些欣赏有些不欣赏。但在我看来，中国人要比以前开放得多，他们也表现了对学习他人，尤其是外国人的浓厚兴趣。有许多非常聪明的人，他们理解在中国发生的一切，也知道发展需要什么。不要犯任何错误。这就是国家的希望。

在美国，许多小型农户已经被拥有大型机械的公司所取代。这种情况有没有发生在中国呢？

还只是开始阶段。许多年前，我与世界银行代表团一起访问中国。我们看到的是，大的农业公司从个体农户手中租赁土地，并将所有的农田集中起来，构成一个大约3000亩（495英亩）的大农场。农民们获得在农场中工作的机会，所以他们可以从租赁和工作中挣到双倍的收入。

另一个值得关注的问题是食品安全。确实有一些东西对人类有害，因此它们不能被食用吸收。不时就有报道指出，有人非法使用这些有害材料。中国政府颁布了严格的规定，并且对此类行为处以严厉的惩罚，但为什么还是不能阻止此类事件发生呢？我将给你举个例子。中国每年消费大约30到50亿头猪，这意味着每个人每年要消费3到5头猪。大约80%的猪都是由个体户养殖的。你可以管理一个养猪公司，但是你怎么可能每天检查成千上万个此类小企业呢？这是不可能的。实际上，和许多其他政府比起来，中国政府的工作已经做的很好了，人们并不理解这一点。在食品安全上，中国使用和美国一样的标准，有时还要高一点。例如，中国的排放标准就比美国的要好一些。

尽管说的很好，但有些人还是会认为，所说的根本没有得到执行。

在一些地方是这样的，但在有些事件中，只是因为没有足够的人力每天每处去进行检查。有些人确实在做一些非常疯狂的事情，他们甚至能为了几块钱而自杀。没有什么可以阻止这些。

所以你会在颜料中，在儿童的玩具中发现铅。

30年前，中国没有这些问题。那时的道德标准是很高的。每个人都很贫穷，但他们互相帮助。资本主义让经济得到了增长，但却使得有些人会为了赚钱而去做任何事。

particularly foreigners. There are a lot of very smart people and they understand what is going on in China and what is needed to move forward. Make no mistake. That is the nation's hope.

In America, many small farmers have been replaced by large corporations with heavy machinery. Is that happening in China?

It's starting. A couple of years ago, I visited China with the World Bank Mission. What we saw was the big farming companies renting out from individual farmers and joining all the farms together for a roughly 3,000-mu [495-acre] farm, which is huge. The farmers were given the opportunity to work on the farm so they could earn double income, from rent and work.

Another concern is food safety in China. There are certain substances that are harmful to humans, so they must not be ingested. From time to time there are reports that people have illegally used these harmful ingredients. The Chinese government has strict regulations and imposes severe penalties, so why can't they stop this from happening? I'll give you an example. China annually consumes something like 3 to 5 billion pigs, which means each person consumes 3 to 5 pigs per year. About 80 percent of the pigs are raised by individual households. You can regulate one big company, but how can you possibly check hundreds of thousands of small enterprises every day? It's just not possible. Actually, by comparison, the Chinese government has done a much better job than many other governments. People don't understand that. China uses the same standards as the U.S. for food security, sometimes higher. Emissions regulations, for instance, are better than in the United States.

Despite the talk, some would argue that there's no follow-through.

In some areas it's true, but in some cases there just isn't enough manpower to check everywhere, every day. Some people do crazy things. They'll even kill themselves for a few bucks. Nothing can stop this.

That's why you find lead in paint, and in children's toys.

China didn't have these problems 30 years ago. Moral standards were high. Everyone was poor but they helped each other. With capitalism, the economy has grown but some people will do anything to make money.

There's a lot of money to be made by everybody, on the table and below the table.

Correct. And the second thing about land revenue is this

year (2011), housing prices kind of went bust. Previously, land revenue was easily considered off budget revenue for local government, in many cases equivalent to tax revenue. Off-budget means it doesn't go through rigorous financial inspections and regulations.

Before 1993, the fiscal situation of subnational governments was better than that of the central government. The central government had been running at a small but consistent fiscal deficit while subnational governments had been running with a small but consistent fiscal surplus (they had deficits in only six out of sixteen years, from 1978 to 1993). The year of 1994, however, was the turning point. Since 1994, annual growth rates of fiscal balance for both the central and subnational governments have been extraordinary, but in different directions. The former has a rapidly rising fiscal surplus, whereas the latter has run an enormous deficit, especially starting in 2001–2002 when fiscal pressure in subnational governments began to mount.

Fiscal pressure and potential land revenues are two main drivers for the rapid increase in land transactions of urban land markets. The development of land markets in China can be divided into three periods. The first period is from the starting of the adoption of land use regulations (LURs) to 1999, labeled a rapid development of land leasing activities but not land prices. For instance, there were 52,086 lots of 5,588 hectares [13,765 acres] for a unit price of around 480,000 RMB per hectare [1,182,265 RMB, or $876,750, per acre] in 1993. They grew at annual growth rates of 28.53 percent and 55.55 percent for number of lots and total leased areas, respectively, while land prices grew at a much lower pace, at only 4.39 percent of annual growth rate from 1993 to 1999. In the second period, from 1999 to 2003, land prices skyrocketed. The annual growth rates for number of lots and total areas were pretty much similar to the previous period, but land prices rose to a 42 percent annual growth rate, ten times as big as the previous period. Unit land price increased from 660,000 RMB per hectare [1,625,615 RMB or about $256,000 per acre] in 1999 to 2.68 million RMB per hectare [6.6 million RMB or about $1.04 million per acre] in 2003. The third period is from 2004 to 2007, in which land markets were adjusted by a national cooling-off effort. The number of lots declined consistently at an average annual growth rate of minus 23 percent, while the total areas leased initially declined in 2003 to 2005 and then resumed growing in 2005 to 2007. Unit land price rose at about 18 percent of annual growth rate.

在台面上和台面下，都有很多钱正等着大家去赚。

对，关于土地收益的第二件事发生在今年（2011年），房屋的价格有点表现出崩盘的趋势。原先，土地收益很容易被认为是地方政府预算之外的收益，在许多地方相当于税收收入。预算外收益意味着，它不需要通过严格的财政审查和调控。

1993年之前，地方政府的财政情况要好于中央政府。中央政府的运作伴随着小规模但却持续的财政赤字；而地方政府则正好相反，是小规模、持续的财政盈余（在16年中，从1978年到1993年，他们的财政赤字时期只有6年时间）。然而，1994年是个转折点。从1994年开始，中央和地方政府每年财政余额的增长比率不可小觑，但方向却不尽相同。前者在财政盈余上呈增长迅速，而后者则产生巨大的赤字，特别是从2001-2002年开始，地方政府的财政压力达到了顶点。

城市土地市场中土地交易的快速增长有两个主要的推动力：一是巨大的财政压力，二是潜在的土地收益。中国土地市场的发展可以分为三个阶段：第一个阶段是从接受土地使用权利(LURs)到1999年，以土地租赁行为而非土地价格的快速增长为标志。例如，1993年，有52086块总计为5588公顷（13808英亩）的土地，均价大约是48万人民币每公顷（1185600人民币每英亩）。土地的数量和总的租赁面积以每年28.53%和55.55%的增长率持续增长，而土地价格的增长则要缓慢得多，从1993年到1999年，每年只有4.39%的增长率。第二个阶段是从1999年到2003年，土地价格大幅上扬。土地数量和总面积的年增长率大致与前一阶段持平，但土地价格的年增长率则飙升到每年42%，十倍于上个阶段。从1999年的66万人民币每公顷（1630200人民币每英亩）涨到2003年的268万人民币每公顷（662万人民币每英亩）。第三个阶段从2004年到2007年，国家通过降温行为对土地市场进行调节。土地数量以平均每年负23%的比率逐渐下降，而总的租赁面积也开始从2003年一路下降到2005年，然后从2005年到2007年重新开始上涨。这个阶段，土地价格每年的增长率大约是18%。

Traffic in Haikou City, Hainan Province, China
中国海南省海口市的交通。

Philip Enquist

SOM/Chicago

Philip Enquist, FAIA, leads the global city design practice of Skidmore, Owings & Merrill LLP (SOM), the world's most highly awarded urban planning group. Phil and his studios have improved the quality of city living on five continents by creating location-unique designs that integrate nature and urban density within a framework of future-focused public infrastructure. His work ranges from walkable, transit-enabled districts to rapidly changing urban clusters within regional ecosystems like North America's Great Lakes basin and China's Bohai Rim. Phil passionately believes that the world's explosive urbanization must be managed by humanely bold and holistically sustainable thinking at the national, regional, and metropolitan scale and that human habitat will become the alpha design science of the twenty-first century. Phil is the Charles Moore Visiting Professor at the University of Michigan's Taubman College of Architecture and Urban Planning. He was named Chicagoan of the Year in Architecture in 2009. In 2010, he received the Distinguished Alumnus Award from the Architectural Guild of the University of Southern California School of Architecture.

菲利普·恩奎斯特

SOM/芝加哥

菲利普·恩奎斯特，美国建筑师协会会员，领导斯基德莫尔、奥因斯与梅里尔建筑师事务所(SOM)的全球城市设计事务，这是世界上最享有盛誉的城市规划组织。菲利普和他的工作室已经通过创造出本地独特的设计而提升了五个大洲的城市生活质量——这样一种设计在着眼于未来公共基础设施的框架内将自然和城市人口密度整合起来。他的作品包括从可以步行、交通便利的区域到区域生态系统（像北美的五大湖盆地、中国的环渤海地区）中快速变化的城市群。菲利普热切地相信，世界爆炸式的城市化趋向必须通过在国家、地区、大都市范围内历史性地进行可持续性思考来管理，而且，人类的居住环境将成为21世纪第一位的设计科学。菲利普是密歇根大学塔伯曼建筑和城市规划学院查尔斯·摩尔的客座教授。在2009年他被称为建筑行业的年度芝加哥人。2010年他获得了南加州建筑学院建筑协会颁发的杰出校友奖。

For complete biography, please refer to the Appendix.

完整的传记，请参阅附录。

Chaoyang North District sustainable plan, Beijing CBD East Expansion Competition, China (SOM/Philip Enquist, 2009)
中国北京CBD东扩竞赛，朝阳北区可持续性计划。（SOM建筑师事务所/菲利普·恩奎斯特，2009年）

PLANNING Philip Enquist

What are the implications of global urbanization?

Urban migration has been going on for a long time but its kick-starting in Asia has been the tipping point. It was estimated in 2007 that 50 percent of the world's population was living in cities and that by 2050, it would climb to 75 percent In reality, America and Europe have been predominately urbanized places for a long time. What we've been noticing is that there's also been a dramatic shift in agriculture, so that fewer and fewer people are needed to grow food as farms become far more automated. I was recently in North Dakota and was amazed at how they farm with these massive unmanned tractors, driven by satellite, so maybe only one person per square mile is needed. So there aren't many jobs in agriculture.

Urban migration is a global trend, so there's a lot riding on cities to help solve the carbon-emissions issue, and to provide a good, healthy life for people, and the education, the jobs, and the housing they need. So the question is how can cities be the solution to a growing global population and also to an environment that's clearly in stress? The two go hand in hand, and our work in planning needs to address that.

I've been doing planning ever since I got out of architecture school at the University of Southern California in the late '70s. With the exception of Cornell and Harvard, every other school taught architecture and engineering. There was no such thing as urban design. There was regional planning, but that was studying policy, really a textbook education with no connection to design. Actually, it's still that way in many schools. When you graduate, you realize that the silo form of education is bullshit because we work with structural engineers, architects, environmentalists, landscape architects, policy planners; we work with everybody. And the more people you can get around the table to solve these complex problems, the better. The problems are really critical and too big for one discipline to solve, so you can't be educated in a silo anymore.

Are there basic frameworks that are universally applicable?

To anticipate growth and strengthen the environment is a priority and a strong guiding principle. Usually, economic growth is prioritized over environmental sustainability. Everybody is focusing on jobs, but can you really have a strong economy if you have a deteriorated environment? China is risking that right now. How long are they going to be able to keep a talented workforce with polluted air for people to breathe? I think there are other options.

About 15 years ago, we laid out for ourselves some very

全球城市化的涵义是什么?

全球移民已经持续了很长一段时间，但它在亚洲的强力推进成为了一个转折点。据估计在2007年，全球大约有50%的人住在城市里，而到2050年，这个数字将上升到75%。实际上，美国和欧洲很长一段时间以来都是主要的城市化区域。现在我们注意到，在农业之中也有一种重大的转变，随着农场变得越发机械化和自动化，需要去种植庄稼的人变得非常之少。我在北达科他州的时候，就惊讶于他们如何用大型的无人驾驶拖拉机进行耕种，这些拖拉机是由卫星控制的，所以可能每平方英里只需要一个人就够了。因此，在农业上并没有太多的就业机会。

城市移民是一个全球趋势，因此城市非常重要，它们可以帮助解决碳排放问题，为人们提供一种良好、健康的生活，以及人们所需要的教育、工作和住房。所以，问题是，面对全球不断增长的人口，以及明显压力巨大的环境，城市如何作出解决问题的方案？两个问题相辅相成，如影随形，而我们规划中的工作需要说明这一点。

在上世纪70年代晚期，当我离开南加州大学建筑学院的时候，就已经开始做这样的规划了。除了康奈尔和哈佛，所有其他学校都只教授建筑和工程，但却没有城市设计这样的东西。也有一些区域的规划，但只是学习一些政策，完全是一些毫无设计概念的书本教育。实际上，在很多学校，这种情况还在保持不变。毕业的时候你才会意识到，这种填鸭式教育太糟糕了，因为我们要和结构工程师、建筑师、环保人士、景观建筑师、政策规划者一起工作，我们是和所有人一起工作。你可以把越多的人集合起来解决这些复杂的问题，情况就会越好。要是某个学科单独解决问题，这些问题会显得非常严重，太过重大，所以你不能再接受一种填鸭式教育了。

有没有一些普遍适用的基本框架呢?

提前预计增长、加强环境，是一条优先、有力的指导原则。一般情况下，经济增长被置于环境可持续性之前。每个人都专注于就业机会，但是如果你的环境恶化了，你还真的会有强盛的经济吗？中国现在就处于危机之中。他们还能把自己的人才队伍放在受污染的空气中多长时间呢？我认为还有很多其他选项。

大约15年前，我为自己设置了一些非常简单的原则，借此我只会承担那些多功能的、基于运输的项目。最理想的情况是：重新发展那些已经被开发的土地。我们想要特别避开市郊的低密度住宅区，以及那些居住型的、依赖汽车的项目和度假村。这些东西其他人做得更好，但我们真的不太相信这么做有什么价值。我们想要专注于真正的城市工程。

对于城市居住来说，有没有一个理想的密度？

我还是谈谈简·雅各布斯吧，她做了一些最好的城市研究工作。她认为大约每英亩80单元的社区，是大街上（有"街道眼"的街道社区）真正有活力的生活需要的密度。在她的著作《美国大城市的生与死》中，提到了我十年前在旧金山住的一个社区：北沙滩社区，就是大约每英亩80个单元。并且纽约的布鲁克林，一个很大的城市社区，也大概是这个密度。所以我认为你可以说，根据她的研究，80—100单元每平方英亩（198—247单元每公顷）是我们努力奋斗争取获得的一个公交导向型社区。密度只是一个解决方案，但除非你有合适的密度，否则你就不会有很好的交通系统。在接下来的25年中，当地球人口从70亿增长到100亿的时候，密度是成功解决方案中的一个关键要素。

现在，当我们在密尔沃基工作的时候，我们看到街道上没有人，芝加哥则不同，那里的人行道上人流汹涌。两个中西部城市有两种不同的市中心。我们进行了调查，并且发现密尔沃基只有3%的人来市区上班的时候是搭乘公共交通工具的，而我认为这在美国城市中还是很典型的。芝加哥大概有60%的人乘公交车或地铁上班，在纽约则接近80%-90%。由于对公交系统的依赖，许多人在人行道上来来回回，走向或离开他们工作的场所。在密尔沃基，所有的建筑都有车库，所以人们开车去上班，没有走路的必要。

在地面上，人和建筑一般是互相影响的，像空中天桥这种在许多独立建筑之间建立广泛联系的高层连接物有什么重要意义？

就个人而言，我并非是一个多层公共基础设施的信徒，因为对于城市来说，建设这些基础设施，然后维护它们，是很困难的。我们已经参与金丝雀码头项目有20多年了。上世纪80年代，当英国政府正与伦敦爱尔兰共和军爆炸案引发的巨大安全问题

simple principles whereby we'd only take on mixed-use, transit-based projects, ideally redeveloping already developed lands. We wanted to stay away specifically from suburban low-density, all residential, car-oriented projects and resorts. Other people do that better, and we don't really believe in it. We want to focus on real urban work.

Is there an ideal density for urban living?

I still go back to Jane Jacobs, who did some of the best urban research in identifying neighborhoods of about 80 units per acre [32 units per hectare] as the density needed for really vital life on the street, eyes on the street, street neighborhoods. In her book *The Death and Life of American Cities*, she mentions North Beach, which is a San Francisco neighborhood I lived in for ten years, and North Beach is about 80 units per acre [198 units per hectare]. And Brooklyn, New York City, a great urban neighborhood, is about that density, too. So I think you could say, based on her studies, that 80 to 100 units per acre [198 to 247 units per hectare] is what we strive for to get a transit-oriented community. Density is only a piece of the solution, but you can't have good transit unless you have adequate density. Density is critical to a successful solution in going from 7 billion people to 10 billion people in the next two and a half decades.

Recently, while we were working in Milwaukee, we saw there was nobody on the street, compared to Chicago, which has a tremendous number of people on the sidewalk. Two Midwest cities with two distinct downtowns. We looked into it and found that Milwaukee has only 3 percent of its workforce coming to downtown by public transit, which I think is pretty typical of American cities. Chicago has about 60 percent arriving by bus or train. In New York, it's almost 80 or 90 percent Because of that dependence on transit, tons of people are on the sidewalks moving to and from their place of employment. In Milwaukee, every building has a garage, so people drive to work; there is no reason to walk.

People and buildings typically interact on the ground plane; what is the significance of upper-level links like skybridges for greater connection among individual buildings?

Well, personally, I am not a big believer in multilevel public infrastructure because it's so hard for cities to build and then maintain. We've been involved in Canary Wharf for over 20 years. Our firm master-planned it in the 1980s when the UK was fraught with tremendous security issues over the

Canary Wharf, London, UK, pedestrian routes and public spaces (SOM, 1992)
英国伦敦金丝雀码头，人行道和公共空间（斯基德莫尔、奥因斯与梅里尔建筑师事务所，1992年）

IRA and bombings in London. They didn't want any retail at street level. It all had to be at the lower level, while the building entrances were all highly secured and contained, and located at street level. As it has evolved, there is now retail everywhere, still predominately at these lower levels but connected and opening out onto the waterfront. It's become really vital. So that is a two-level infrastructure, but cities like Chicago really struggle when they introduce a level other than the street. Architects love to dream about new infrastructure, but who owns and builds it? Who maintains it?

Every September we are invited back to Cergy-Pontoise, which is one of the satellite cities built in the 1960s around Paris. They were all built with upper-level decks called *dalles*. The train station and parking were below, and the pedestrians and all the buildings were connected at these free-flowing upper decks. Now they're dead, grim, falling apart. No one knows how to maintain or reconstruct them, and in some places they're being dismantled. I just think it's great to keep people on the ground. Chicago and New York are prime examples of high density with everybody on the ground. Yes, you have a conflict with cars, but everybody is on the ground with the stores, food, and so on. People feel safe there. I just saw a poster today showing a 1970s photograph of people walking on a New York sidewalk. It was called "The Human Flypaper"; I don't know if that was positive or negative!

There are a number of cities that have different public levels, like Seoul, which has all this retail under the streets, and parts of Tokyo, Montreal, and even London, where you

进行斗争的时候，我们公司主导设计了码头。他们不想要任何街市层面的零售服务。所有东西都必须在建筑的低层中，而整栋建筑的入口位于街市层面上，受到严密的保护和控制。随着不断的进步，现在到处都有零售服务了，它们主要还是在建筑低层，但被连接了起来并向滨水区敞开。它现在确实变得至关重要。所以这是一种二层的基础设施，但像芝加哥这样的城市，当他们要引入一层而不是一个街区的时候，就真的要做一番斗争了。建筑师们当然喜欢幻想新的基础设施，但谁拥有它，谁建设它，谁又维护它呢？

每年九月我们都被邀请回到塞日-蓬图瓦兹，巴黎周边一座建于上世纪六十年代的卫星城。它们都是用被称为dalles（厚石板）的上层石板搭建而成的。火车站和停车场在下方，行人和所有建筑在这些自由流动的上层石板之上。现在它们已经老化、坚硬、分崩离析了。没有人知道怎么维护或重建他们，甚至有些地方已经开始拆除这些石板了。我只是认为，让人们生活在地面之上是很好的。芝加哥和纽约是每个人都在地面之上，从而保持人口高密度的最好例子。是的，你会和汽车发生冲突，但每个人都在地面上，和商店、食物等等东西在一起。人们在那里感觉很安全。我今天刚刚看到一张海报，展示了一张上世纪七十年代人们走在纽约人行道上的照片。它被称为"粘人纸"，我不知道这是正面的还是负面的！

许多城市有不同的公共层，比如首尔，在街道下面就有这些零售服务，在东京的部分地区、蒙特利尔，甚至伦敦，当你从某个地铁站出来的时候，你可以在地下通道中闲逛很长一段距离。香港也有一套运作良好的上层步行街道系统，可以让人行环道避开车流，这真是唯一解决的方式。而且在那里（香港）地形也为你提供了便利。如果你从核心区出发斜向下至海港，那你会一直在高处穿越街道。每一栋建筑都让你以有组织的方式通过。

许多中国城市已经到了重建的好时机。你是否先行开始探究现存的城市机理，并在其上增加密度的尝试？

是的，这正就是我们正在北京所做的事情，在中央商务区（CBD），一座全新的城市建立起来，之前那里还是一片重工业的工厂。这件事是关于可持续性、雨水收集和水过滤的，我们现在了解到的是，你确实无法逐栋建筑各自为战，而必须在区域的层面上寻求解决方案。

在我们的办公室里有一个机械工程小组，和一个专注于可持续性技术的小组，大家都在一个工作室，我问他们，该如何面对北京水资源匮乏的问题？这些机械工程师说，如果可以摆脱冷却塔，我们就能防止蒸发。我们探索如何利用公共公园系统提供低热策略，这样就可以制造出不依赖于冷却塔的制热、制冷系统。这就是我说"走出简仓，连结各点"背后的意思。

在天津，我们和政府合作重新开发一片四层高的围合型住宅区，它是在中国开始建设大尺度板式建筑之前修建的。大多数居住单元没有厨房。在楼道里建公共厨房，街边则有公共卫生间。这些建筑虽然简单，但其饱满的邻里空间和绿树成荫的街道，事实上是非常漂亮的。它就位于城市的中心，而这座城市正处于彻底的自我改造过程中。

我认为中国在回迁策略上必须做得更好一些。他们的某些行为非常不讲理，比如，让人们从生机勃勃的都市环境中搬走，迁到这些28层高或40层高的板楼中去。他们正在建设一些单一用途又远离城市的建筑作为安置房。

城市不停地经历着转轨和转型，即使是我们最美慕的城市也曾经从根本上发生过改变。巴黎在19世纪彻底摧毁了它的中世纪街区，盖了新房屋，然后才变成人人都爱的今日巴黎。中世纪的罗马也将自己重新装扮成一座文艺复兴的城市。纽约也在不断地翻新改变，这就是城市。

can sometimes wander for long distances in the tunnels when you come out of the tube stop. Hong Kong also has an upper-level walkway system that works very well and lets people circulate away from traffic. It's really the only way to get around. And there the topography works for you. If you head from the central district, which slopes down to the harbor, you just stay high across the streets. And every building lets you move through it in an organic way.

A lot of the Chinese cities are ripe for redevelopment. Do you anticipate going into the existing fabric of the city and actually increasing density?

Yes, that's what we are doing in Beijng, in the central business district where a whole new city is being created where there were previously heavy-industry factories. It's all about sustainability and stormwater retention and water filtration, and what we've been discovering is that you really can't combat current problems building by building; it's got to be at the district level.

We have a mechanical engineering group in our office, as well as a group that really focuses on sustainable technologies; we're all in one workshop. I asked how we could respond to Beijing's lack of water, and this mechanical engineer said that if we could get rid of the cooling towers we'd avoid evaporation. We looked at how to use the public park system for geothermal strategies so we could make heating and cooling systems that didn't rely on cooling towers. This is what I mean by getting out of silos and connecting the dots. We're also working with the government in Tianjin to redevelop a residential district of four-story block buildings that were constructed before China started building its really big slabs. Most of the residential units don't have kitchens. There are communal kitchens in the buildings and group bathrooms down the street. The buildings are simple, but the grain of the neighborhood, with its tree-lined streets, is actually very nice. It is located right in the heart of this city that is completely reinventing itself.

I just think China has to do a better job at its relocation strategies. They're doing some things that are pretty brutal, like taking people from vital urban settings and moving them way out in these 28-story or 40-story slabs. They're building single-use places really detached from the city as replacement housing.

Cities constantly go through transition and transformation, and even our most highly admired cities have transformed themselves radically. Paris tore out its medieval sections in the 1800s, built new housing, and became something

294

we all love. Medieval Rome also reinvented itself as a Renaissance city. New York reinvents itself constantly. That's what cities are all about.

We actually had an unusual project a few years ago, where we were asked by the government of Bahrain to master-plan their whole country. They had never had zoning, or any kind of planning discipline. They knew property ownership, they said, but basically the royal family owned everything. And then they were having funny problems where somebody would build a waterfront hotel and, because the water is so shallow, somebody else would put landfill in front and build a new waterfront hotel so the first hotel became a hospital because it no longer had water frontage.

We tried to teach them what we thought was right about committing to historic cities and making them healthy, and reinvesting in them instead of continuing to move out to new areas. They were in a Dubai mode, building things like Palm Island. They had a project called Durrat Al Bahrain where all these new islands were shaped like shrimp and fish. I'm serious. The city itself, Manama, which was the commercial and banking center, was left for the migrant workers and the poorest people the Bahrainis didn't want to socialize with. Nearby, the Arab city of Muharraq, filled with absolutely beautiful historic buildings, was falling apart because the building owners would just wait until they could redevelop. Nothing was conserved. Well, one of the Sheik has in the royal family was buying and restoring historic buildings and showing people that doing this added value. So we really tried to leverage that and tried to get them to invest in their cities before building sprawling new development farther out.

In a separate project, we did the master plan to preserve the historic buildings in one of Shanghai's foreign concessions, but they ended up documenting the buildings and then tearing them all down and rebuilding them.

So it's not just the new cities that are the economic drivers, but it's redevelopment in the old legacy cities as well.

In Beijing and Tianjin we're working in the existing cities, reinventing the existing places. There's one in the port of Beijing, at the Bohai Sea, which involved taking very polluted industries and relocating them into modern facilities. It's the end of a high-speed rail line linking Beijing, Tianjin and the port, and out of that strategy is a whole spider network of subway lines, a transit center, and a new city. This expansion of the central business district is taking place between the third and fourth ring roads. So in this particular case, we said you have to link to the international airport and eventually to the

我们实际上在数年之前有一个不同寻常的项目，巴林政府要求我们总体规划他们的整个国家。他们从未进行控制性规划，或者任何种类的规划导则。他们说，他们知道财产所有权，但基本上王室家庭拥有一切。而且他们还经历了一些可笑的问题，比如，有人建了水滨酒店，因为水非常浅，其他人会在前面设置垃圾填充区，然后再建立一座新的水滨酒店，这样一来，原有建筑因为不再有临水界面而改造成了医院。

我们试图教给他们我们认为正确的，关于尊重历史城市并使其保持健康的方法，同时对这些城市进行重新投资，而不是持续地迁移到新的地方去。他们走的是迪拜路线，建设了一些类似棕榈岛的东西。他们有一个项目，叫作巴林海上明珠，在那里所有岛屿的形状都像虾和鱼一样，我是说真的。麦纳麦这座城市自身，曾是商业和金融中心，却留给了外来工人和最穷的人——这是一些巴林人不愿与之打交道的人。附近穆哈拉格的阿拉伯城，满是极为漂亮的历史建筑，却正分崩离析，因为建筑的主人只是选择等待，等待着这些建筑能够被重新开发。没有什么被保留下来，王室家族的一位王妃购买并重新装修了历史建筑，也向人们展示了这么做是在增加建筑的价值。所以我们真心试图对之产生影响，试图让他们在向外扩张式的、建设新项目之前，投资他们的老城市。

在一个独立项目中，我们为上海的一处外国租界做了个总体规划，以保护那里的历史建筑，但他们最终只是将这些建筑记录在案，然后全部拆掉重建。

所以经济的推动力不仅是新城建设，对于拥有历史遗产的城市进行重新开发也可以起到相同的作用。

在北京和天津，我们就是在城市现状基础上工作，重新改造已有的东西。有个项目在北京的一个港区，位于渤海湾，它是关于搬移严重受污染的工业，并将它们重新安置于现代化工厂之中的任务。它是连结北京、天津和港口自身的高速铁路终点，由于这样一个策略，那里出现了整个蛛网结构的地铁线路，一个交通运输中心，以及一座新城市。中心商业区的扩张正发生在三环和四环的区域之内。所以在这个特殊的例子中，我们说你必须和国际机场、高速铁路站点连接在一起。

Bahrain National Plan: Selected Development Strategy Objectives, Bahrain. (SOM, 2007)
巴林国家计划：入选的发展战略目标。（斯基德莫尔、奥因斯与梅里尔建筑师事务所，2007年）

high-speed train station.

Do you typically come on board as consultants for site selection?

Not necessarily; they almost always have that decided already, and somehow, especially outside the coastal cities, projects have already been defined. We have been really fortunate that the work we've been doing, in Beijing and Tianjin especially, has been focused on very smart strategies to capitalize on increased land values around high-speed rail stations and subway lines. So usually it's a transit-based decision to increase density; we usually get involved after the decision is made.

Are your clients in China generally foreign or local developers?

In planning, almost all work is for the government. We've had a few private developers as clients, but they tend to be much more at the building scale. We usually work with the planning bureaus and land bureaus, and now we are working with mayors in both Beijing and Shanghai on redevelopment and preservation strategies.

Chinese mayors clearly put priorities on the iconic buildings. They want them identified, they want to know where they go, but they may reject buildings that they feel are too iconic because the mayor has the authority to approve or reject anything that is built. For example, Tianjin has historic districts, so there they will reject iconic contemporary architecture. They came to us last year and said, "Our city is getting too many silver spaceships and too many all-glass buildings. We are looking like everyplace else." They held forums about what is unique to their particular city, culturally and climate-wise, and looked at how to develop an architecture that is more reflective of local conditions. I think that is a real challenge with building so quickly, because a lot of this stuff looks like it could be anywhere.

Let me share with you a live site from the United Nations. I just love the series of diagrams. It begins in 1955, showing all cities with populations of 5 million or more. It's what you'd expect: New York, Chicago, Los Angeles, Buenos Aires, London, Paris, Berlin, Calcutta, Beijing, and Tokyo and Osaka in Japan. It's interesting because North America is almost all urban, around 75 percent, while Europe and South America are about 50 percent and Asia is only 25 pecent urban. Ten years later, cities of 5 million are shown in Brazil and Africa. By 1975 North America stays pretty much the same, but there's a new cluster of cities emerging on the other side of the globe. By 1985, a whole line of 5-million-population cities

你是否一般只以顾问的身份参与到选址中去?

不完全是,他们几乎总是已经做了决定了,无论如何,尤其是在沿海城市之外的地方,项目总是已经确定了。我们很幸运,我们所作的工作,尤其是在北京和天津的,都着重于制定明智的开发策略以便从高铁线及地铁线土地价格升值中获利。所以通常状态下,增加密度的决定是基于交通转运功能的,我们经常在决定确定后参与进来。

在中国,你的客户一般是外国的开发商还是本土开发商?

就规划而言,几乎都是在为政府工作。我们也有少数私人开发商客户,但他们更倾向于建筑规模上的问题。我们经常和规划部门、土地部门一起工作,我们现在正在和北京、上海的市长们一起工作制定重建和保护策略。

中国的市长们显然都把重点放在标志性建筑上。想让它们独一无二,想知道他们要去哪里,不过他们可能也会拒绝那些让人觉得太过招摇的建筑,因为市长们有权通过或拒绝任何已建项目。例如,天津有历史建筑区,所以在那里他们会拒绝标志性的当代建筑。他们去年来找我们,对我们说:"我们的城市现在有太多的银色宇宙飞船,太多的全玻璃建筑了。它看起来和其他地方没什么两样。"他们举办了一次论坛,从文化与气候的角度来看,什么是这个城市的独特之处。而且他们还想要弄明白如何开发一种更能反映当地情况的建筑。我认为这很快就会成为建筑的一个真正挑战,因为现在有很多东西好像放在哪里都可以,毫无特色。

让我和你分享联合国一个生动的网站,我非常喜欢那上面的一系列图表。它从1955年开始,展示了所有人口在500万或500万以上的城市。那里面都是你希望看到的:纽约、芝加哥、洛杉矶、布宜诺斯艾利斯、巴黎、柏林、加尔各答、北京和日本的东京、大阪。这很有趣,因为北美几乎都是城市,大约75%左右,而欧洲和南美则为50%,亚洲只有25%是城市。十年后,500万人口的城市出现在巴西和非洲。到1975年时,北美的比率依然还是差不多,但是有一大批新的城市出现在地球其他地区。到1985年,一整排500万人口的城市从伊斯坦布尔开始延伸。然后很快,在1995年、2005年,而且展望2015年,全球城市人口的拐点正在改变:北美并未有太多变化,但是南美已经变得更加城市化,而欧洲也超过50%。但是,东方发生的

Beijing CBD East Expansion Competition: Transit. Also see the related introductory rendering on page 279.
北京CBD东扩竞赛：交通。亦参见第279页相关的介绍图。

才是大变化：中国、韩国、东南亚、印度，围绕着开罗和伊斯坦布尔开始突飞猛进。所以，这里才是大场面，而且如果按人数算，那里也是人口数量最多的地方。世界的另一边正经历着一种非凡的转变，而一般的美国人并没有意识到这种转变。

那种一样的建筑产生的全球后果是惊人的。

是的，如果你不把东西做得不一样的话。随着这种惊人的都市化，真正令人恐惧的是，有一些城市尚未对这种城市增长做好准备。他们没有基础设施，没有系统，没有办法解决人口问题。他们甚至可能连资源都没有，他们有水吗？

stretches from Istanbul on. Then quickly, in 1995, 2005, and projected for 2015, the tipping point for global urban population changes: North America hasn't changed that much, but South America has become much more urban and Europe is now more than 50 percent urban. But the big change is in the East: China, Korea, Southeast Asia, India, stretching all around to Cairo and Istanbul. So the big game is here, and it's also where, just by volume, the greatest number of people are. The other side of the globe is going through a phenomenal transformation, but the average American is unaware of that change.

The global consequences of that kind of building are staggering.

Right, if you don't do things differently. With this phenomenal urbanization, the really frightening thing is that some of these cities are not ready for this kind of urban growth. They don't have the infrastructure, they don't have the systems, they don't have the way to deal with pollution. They may not even have the resources. Do they have water?

The only way to deal with it would seem to be through some kind of global initiative.

Absolutely. It is absolutely a global concern. Water is not equally distributed around the earth. We have to figure out how to do this. Why do we keep building cities in the desert with no water? That doesn't make any sense.

About 15 years ago, Richard Rogers, the London architect, published a beautiful little pink book called *Cities for a Small Planet*. The diagrams are so easy to understand: a good city needs food, energy, goods coming in—the energy is all fossil fuels—and then everything ends up as waste. The food goes out as organic waste, the energy goes out as emissions, and the goods go out as inorganic waste. Everything ends up in landfills or in the air.

What Rogers says we have got to do is get renewable energy in, and then dramatically rethink recycling food and inorganic waste. It's actually now happening in Copenhagen, which is recycling something like 95 percent of its waste, turning most of it into energy. It's like waste equals fuel, or waste equals energy. Copenhagen is one of the best examples of how you can reinvent yourself.

Urbanizing countries often look to the United States for models. What is the large and small status of sustainable development in this country?

In 1970, our firm did a state plan for California, back when nobody was looking at this scale. This led to the formation of the Coastal Commission and a number of other critical environmental groups. Also in the 1970s, we did a national plan for the United States. So we have a long history of environmental concern in our firm. We talked about the issues of more fuel. More power just turns into more demand for power. Part of the national plan was trying to organize the urban areas of the country, where you would conserve lands and where you would encourage urbanization.

We think that many current development patterns are irresponsible. I think that the standard suburban house in America has gotten so huge and is an isolating way to live; it's not sustainable. As soon as your kids go to college, your family is too small for the house. You can't retire there; it's all automobile-oriented. Older suburbs still have a sense of community, where you can walk to the commercial street or to the train station. They were building at higher density then, with smaller lots, something like a quarter acre or smaller. I

解决这个问题的方式看来只能通过某种形式的全球倡议。

绝对是，我的意思是，这绝对是一个全球关注的问题。水并非均等地分布在地球上，我们必须搞清楚怎么来做。我们为什么一直在没有水的沙漠里不停地建造城市呢？这根本不合逻辑。

大约15年前，理查德·罗杰斯，伦敦的一名建筑师，出版了一本漂亮的粉色小书，名为《小星球中的城市》。图表很好理解：一个好的城市需要食物、能源、商品货物输进来——能源都是矿物燃料——然后所有东西又作为废弃物被排出。食物作为有机废弃物排出，能源作为排放物排出，商品货物则作为无机废弃物排出。所有东西都终结于土地填埋或空气排放。

罗杰斯说，我们必须做的是获得可再生资源，然后，大幅度地重新思考可回收食物和无机废弃物。这实际上现在就在哥本哈根发生着，那里的废弃物95%都是循环利用的，大多数都转化为能源。就好像废弃物等于燃料，或废弃物等于能源一样。哥本哈根是关于你如何能够彻底改造自身的一个最好例子！

城市化的国家经常以美国为样板。这个国家可持续发展的各个方面现在情况如何呢？

在1970年，我们公司为加利福尼亚州做了个州规划，那时候还没有人预测到会有这么大的城市规模。这催生了海岸委员会以及许多其他重要环境组织的成立。在70年代，我们还为美国做了一个国家规划，所以我们公司在环境问题上有着很长的历史。我们谈论了更多的燃料引起的问题。更多的电能将会变成对电能的更大需求。国家规划的一部分是试图组织这个国家的城市区域，即在哪里你可以保留土地，在哪儿鼓励城市化。

我们认为当代的许多发展模式是不负责任的。我认为美国标准的郊区房子变得很大，是一种隔离的生活方式，它并不是可持续性的。当你的孩子上大学的时候，对于你的房子来说，你的家庭就显得太小了。你不能再用这房子养老，因为它是严重依赖汽车的。老一点的郊区依然还有一种社区的感觉，你可以走路到商业街或火车站去。那里的建筑密度更高一些，每个单元更小一些，大概有四分之一英亩或者更小。我现在明白发生了什么，我是说，芝加哥区域的最后一代城郊居民过得并不太好。他们在环境问题上不够聪明，占据了大量土地，因而唯一的选择就是开车。他们的设计就是为了让你直接把车开到你的车库里，然后在没有与其他人接触的情况下直接走进自己的房子，尤其是在这种气候下。你大概可能八个月都看不到你的邻居，直到天气开始暖和起来。每英亩两户人家，以及依赖汽车的发展，并不是最好的利用土地的方式。我认为我们已经走错路了，我真的这么觉得。

see what's going on now. I mean, the last generation of bedroom communities in the Chicago region are really not good. They're not environmentally smart, encompassing a huge use of land so that the only option is driving. They're designed so you drive right into your garage and go into your house without interacting with others, especially in this climate. You don't even see your neighbors for eight months, probably, not until it warms up. Car-oriented development at two units per acre is just not the best use of land. I think we've lost our way, really I do.

James C. Jao
J.A.O. DESIGN INTERNATIONAL

James Jao, NCARB, NABAR, SAFEA, AIA, RA, is a prominent Chinese-American architect and planner, routinely elected one of the top ten most influential people in Chinese real estate. Born in Taiwan, he relocated to New York, earned professional degrees in architecture and business, became an American citizen, and served on the New York City Planning Commission before returning to Asia in 1996. The founder of J.A.O. Design International, the largest wholly foreign-owned design enterprise in Beijing, his firm has executed more than 200 million square feet [18.5 million square meters] of residential buildings and 190 million square feet [17.6 million square meters] of commercial buildings, and has master planned land areas collectively equivalent to fifty Manhattans. Mr. Jao is a Certified Foreign Expert in Architecture and Urban Planning. He is a frequent consultant to the central government and a popular lecturer on urbanization, having addressed nearly 100,000 government officials to date. Among Mr. Jao's many publications are three Chinese books, including, in 2012, *Your City and Mine*, and also, in English, *Straight Talk*, co-authored with Dr. Janet Adams Strong.

饶及人

龙安建筑规划设计顾问公司

饶及人，美国国家建筑注册委员会，中国全国注册建筑师管理委员会，中国国家外国专家局，美国建筑师协会。他是一位著名的美籍华裔建筑师和规划师，他长期被当作是中国房地产行业最有影响力的十个人之一。饶及人生于台湾，后来迁居到纽约，获得了建筑和商业的专业学位，并成为美国公民。在1996年回到亚洲之前，他在纽约市规划委员会任职。他是龙安建筑规划设计顾问公司的创始人，这家公司是北京最大的全外资设计企业。他的公司已经为超过2亿平方英尺的住宅楼、1.9亿平方英尺的商业大楼提供过服务，总体规划过的土地面积加起来等于50个曼哈顿那么大。饶先生是建筑和城市规划方面的认证外国专家。他经常作为中央政府的顾问，他是一名很受欢迎的演讲者，迄今为止为接近十万名政府官员做过城市化方面的演讲。在饶先生出版的著作中有三本是中文的，其中包括2012年出版的《你我的城市》，以及一本与珍妮特·亚当斯·斯特朗博士合作完成的英文著作《坦言》。

For complete biography, please refer to the Appendix.

完整的传记，请参阅附录。

Master Plan for Baicheng County, Xinjiang Uygur Autonomous Region, China. (J.A.O. Design International, 2011)
中国新疆维吾尔族自治区白城县总体规划。（龙安建筑规划设计顾问公司，2011年）

What is the status of sustainable building, particularly with respect to housing, in China?

The concept is understood, but people don't want to pay extra for it. Developers say, "Yes, it's good to do, but why bother? I can't even keep up with current demands."

The rate of urbanization in China is 46 percent compared to the international average of 50 percent. A lot of people still don't have a decent place to live, so when you have a housing shortage, developers are not encouraged to innovate. Both the Chinese people and the government wrongly expected that developers should be obliged to supply low-income housing. In a highly celebrated article of 2007, I asserted that building low-income housing is the undisputed responsibility of the government. To curb rising housing prices, authorities must understand demand and supply. Providing adequate low-income housing will lower housing prices in the long run and owning property will help to stabilize society. Therefore the government's social housing program is very important.

I have been advocating to officials that when the government builds, it should at least adopt its own 3-star rating for green buildings. If the government doesn't follow its own standards, why would a developer? But it's difficult. There is a lot of confusion, a lot of conflicting information. Recently, I chaired a green-building conference in Shanghai and someone asked an official of the Ministry of Construction why the central government doesn't just mandate compliance. The official explained: "Right now the criteria keep changing so we don't have a benchmark. It is very difficult for the government to promulgate laws when there are so many variables."

Still, Chinese criteria are a lot more qualitative than in the U.S. and also a lot more rigorous in implementation phase. Green-building certification has two phases: First the government has to certify the drawings, and second, it monitors the building during and after construction. An official 3-star plaque is awarded only after a building is finished and the government has issued a permit. Even so, certification is only good for three years, after which the owner has to reapply; the government then sends out an expert to verify continued compliance. By contrast, in the United States LEED points can be earned, for example, by providing bicycle parking or using permeable paving materials. Post-construction changes are not taken into account. In China, the government is concerned with the total life cycle of a building from conception of design to completion of construction and long-term operation, even to the demolition. That's the fundamental difference between the Chinese approach and LEED in the U.S.

在中国，可持续性建筑尤其是与住房相关的建筑，是一个什么样的状况？

这个概念已经被人们理解，但他们并不愿意为此多付钱。开发商会说，"是的，这么干很好，但何必呢？当前的要求我都满足不了呢。"

中国城镇化的比例是46%，而国际平均比例为50%。许多人甚至尚未有一个体面的住处，因此在住房紧缺的局面下，开发商没有动力去创新。中国政府和人民都错误地期待，由开发商来负责提供低价的住房。在2007年一篇颇受称赞的文章中，我指出：建设低价住房是政府不容推卸的责任。要抑制不断高涨的房价，官方必须理解需求与供应的关系。提供充足的低价住房，将会在一个长远的时间内拉低房价，而人们拥有的财产将会有助于社会稳定。因此，政府的社会性住房项目非常重要。

我已经向官方建议，当政府进行建设时，至少应当采取它自己的绿色建筑三星级标准。如果政府不依照自己的标准，为什么开发商要这么做呢？但这很不容易做到。有许多令人困惑、自相矛盾的信息。最近，我在上海主持一个绿色建筑会议，有人问了建设部官员一个问题：为什么中央政府不强制执行。官员解释说："现在，标准总在改变，我们并无参照之基准。当存在如此多的变量时，对于政府来说，颁布法律是很困难的。"

和美国相比，中国标准更着重于质量，而且在工程实施阶段也更为严格。绿色建筑认证有两个步骤：第一，政府必须证实图纸符合要求；第二，在建设期间和建成之后，政府必须对建筑进行监测。只有在建筑建设完成，而且政府已经颁发许可证之后，一个官方的三星认证标志才能被授予。即使如此，认证也只在三年内有效，在那之后，所有者要重新进行申请；然后，政府派出一名专家前去核实，是否继续给予合格认证。相反，在美国LEED（能源与环境设计）的认证中，分数是可以挣到的，例如，通过提供自行车停车位或可渗透的铺设材料就可以挣分，建成之后的改变并不在考虑范围之内。在中国，政府关注的是一栋建筑的整个生命周期，从设计的概念到建设的完成，以及长期的维护，甚至还有拆除问题。这是中国方法与美国LEED标准的根本差异。

到2011年，（中国）有84栋建筑获得认证。我鼓励政府提供税收补贴，因为遵守准则是在国家的层面，而非仅仅是在独立开发项目上保护能源。试想一下：在中国，如果每个人都节约一桶水，

PLANNING James C. Jao

整个国家将节约13亿桶水。财政激励需要成为国家政策，每个人都在想这个问题，所以进步是有的，但改变需要时间，中国已有更多关注并迎头赶上。1500名专业人士参加了2011年7月在扬州举行的绿色建筑与可持续设计年度规划会议。而在这之前，如果有500名听众参加就会足以让人印象深刻了，这就是开始改变的风向标。

中国的城市规划被比作"飓风扫落叶"，政策和实践在被执行之前就开始变得陈旧落伍了。

城市规划在中国是个新鲜事物，它是西方的，而且以用户为基础，是一种民主的价值观。皇帝从来不关心为普通人做规划；因为普通人不用住得太舒服，否则他们会来挑战权威。中国已经不处于皇权统治之下了，尽管它还是一党制体系，并且也想尽可能多地进行控制。我不能说我不赞同。实际上，我认为美国民主在中国是行不通的。在美国从英国独立出来的早期，只有自由的白人男子可以进行投票，而这只占人口的一部分。女人在160年之后才有了投票权；美国黑人则还要再往后推30年。中国是一个只有60岁的国家，大部分的人口不是中等收入者，而且也不知道自己想要什么，所以很难建立一人一票的体制。

在商业活动中20/80的规则是合适的：20%的精英为全部的劳动力产生80%的利润。在一人一票的体制中，80%的普通人决定了20%精英的未来。这就是为什么民主方程式在许多新成立的国家并不起作用。这些地方需要的反而是一个强有力，自上而下的结构。只有在人们都受到良好教育，而且60%或更多的人都成为中等收入者的时候，你才可以采取公开选举的形式；在那之前，人们并不知道如何应用这个系统。

当然，腐败是存在的——在中国、印度、印度尼西亚，到处都一样。2011年的新年记者会上，中国总理温家宝表示：腐败是中国的头等大事。中国国家主席胡锦涛总书记，也在庆祝中国共产党成立90周年大会上说了同样的话，这也是史无前例的第一次。

世界人口的平均密度接近于每平方公里一万人，而在北京，上海和其他的中国大城市，平均密度达到了5万到6万人。

密度总会是一个问题。在曼哈顿，150万人密度总会是一个问题。在曼哈顿，150万人占据了23平方英里的土地，因而那里的人口密度是

By 2011, 84 buildings had been certified. I've been encouraging the government to offer tax subsidies because following the guidelines preserves energy on a national level, not just on individual development projects. Think about it: If everybody in China saved a bucket of water, the nation could save 1.3 billion buckets of water. Incentivization needs to become national policy. Everybody is talking about it, so there is progress, but change takes time. China is increasingly aware and catching up. If it's any indication, 1,500 professionals attended the Annual Planning Conference on Green Building and Sustainable Design in Yangzhou in July 2011. Previously, an audience of 500 was considered impressive.

Urban planning in China has been compared to raking leaves in a tornado, as policies and practices become obsolete even before implementation.

Urban planning is new in China. It's Western and based upon the user, which is a democratic value. The emperor never cared about planning for the ordinary people; they should not live too comfortably, otherwise they might challenge authority. China is no longer under imperial rule, although it is a one-party system, yes, and they want to keep control as much as possible. I can't say I disagree. In fact, I don't think American democracy would work in China. In the early days of U.S. independence from Britain only free white men could vote, just a fraction of the population. Women couldn't vote until 160 years later; black Americans three decades after that. China as a nation is only 60 years old. The majority of people are not middle-income and do not know what they want, so it is difficult to establish a 1-vote-per-person system.

In commercial practice the 20/80 rule applies: the elite 20 percent produce 80 percent of the profits for 100 percent of the workforce. In a 1-vote-per-person system, 80 percent of ordinary people decide the future of the elite 20 percent. That is why the democratic equation doesn't work in many newly formed nations. A strong, top-down structure is needed instead. It is not until people are well-educated and 60 percent or more are middle-income that you can have a form of open election; before that people just wouldn't know how to address this system.

And then, of course, there's corruption—everywhere, in China, India, Indonesia, it's the same. In his 2011 New Year press conference, Wen Jiabao, the prime minister of China, said corruption is the number one issue in China. Secretary General Hu Jintao, the president of China, said the same thing during the ninetieth anniversary of the Chinese Communist Party. This was again unprecedented.

Construction of high-end residential towers overlooking the Huangpu River, Shanghai, China (2011)
中国上海，俯视黄浦江的高端住宅楼群（2011年）

The world's average density is approximately ten thousand people per square kilometer [25,900 people per square mile], while in Beijing, Shanghai, and other major Chinese cities the average is fifty to sixty thousand.

Density is always a problem. In Manhattan 1.5 million people occupy 23 square miles [60 square kilometers] of land, so the density there is 65,200 people per square mile [25,175 people per square kilometer]. China uses 10,000 people per square kilometer [25,900 people per square miles] as the magic number, but I've never thought this was a good model. In the United States arable land is 80 percent of the land mass, whereas it's only 22 percent of the land mass in China. The total population of China is five times greater so the magic number should be twenty times denser than in the U.S. If you build a hundred 1-story houses and spread them out, you have no space. But if you vertically stack 100 homes, at least you've saved some open land to enjoy. I believe Le Corbusier's towers in the park with TOD (Transit Oriented Development) will be the future of China.

6.52万人每平方英里。中国将1万人每平方公里（25900平方英里）作为一个奇迹般的数值，但我从不认为这是一个多好的模数。在美国，可耕地面积是土地总面积的80%，而在中国这个数字只有22%。中国的总人口是美国的5倍，因而密度的数值应该是美国的20倍。如果你建立100栋单层楼房，并将它们铺展开来，你就没有空间了。但是如果你将100个家垂直堆叠起来，至少你会节省出一些开放空间用于休闲。我相信勒·柯布西耶(Le Corbusier)设计的、公园之中带有运输导向发展(TOD)策略的塔楼，会是中国的未来。

实际上，昆明市委记仇和听说我谈了这个话题，也公开表示，以后所有的新居民楼都不能低于100米高。这是非常有争议的，因为有许多人并不喜欢高层生活，而且抱怨说他们有恐高症，况且，楼层的垂直堆叠也消除了社区内的互动。不过，可以通过在每10到20层插入间隔的公共空间来解决上述问题，例如每10层楼为一个单元，将他们堆叠起来，每个单元都有一层用来放置公共设施。这只是一个设计的问题，但它依旧是非常有争议的。人们都被激怒了，质疑如何解决防火安全问题。事实上，我有一个主要的反对者，他公开挑战我的观点，说消防车只能达到60米高，因此我怎么能够提倡100米高的楼房呢？我回答说："喷洒系统怎么样？非可燃材料？无论建筑的高度是多少，如果你使用可燃材料，烟雾都会令每个人窒息而死。"真正需要的是一种新的建筑规范。

有许多人错误地认为，高层建筑意味着高密度。在中国，人均居住面积已经从1980年的11平方米增加到现在的32平方米。同时，人口数量也增长了10%。你如何满足这些增长的需求呢？无法逃避的结论是高层建筑。当然，你必须有一个综合的公共交通系统。车站应该被混杂在居民区中，提供便利的可达性，就像在香港一样，大型运输线就在转乘站的中心位置。

现在，人们谈论的是19个城市中的高速铁路，所以高层建筑规划将非常重要。平心而论，好像并非中国什么也听不进去。他们正试图以正确的方式去做所有事情，但规划的执行需要时间，而政府官员也不得不冒起风险。党委书记仇和由于推行百层住宅高楼而受到猛烈的批评。人们指责他增加了建设费用。实际上他是使人们花更多的钱去购买新楼房，所以他面临着众多的压力，这真的需要非凡的魄力。

每个人都喜欢花园城市的概念：低楼层，低密度，以及许多的公共空间，但这并不能容纳太多人。我还是坚持我所说的：建高楼是解决的方案，我希望这一观点可以被谈判桌的另一端所认同。

中国建筑的平均寿命如何？

日本的法律明确规定，建筑物必须维持至少79年，以获得政府的补助金。而由于低收入，他们甚至会为一座维持99年的建筑物降低按揭利率。此类想法对于中国来说是非常陌生的。即使是大型建筑，也不会维持超过50年的时间，典型的工程不会超过40年，这也是混凝土的平均寿命。中国政府官员的任期是五年，所以五年之后的任何事情都与他们无关了，他们是很短视的。我强调的是，罗马不是一天建成的，同时也主张"兴建以持存"的概念。这些在中国是新观念，它们需要扎根下来，并进行传播。

In fact, party secretary Qiu He of Kunming heard me speak about this very topic and he is now saying that all new residential buildings should be no less than 100 meters [328 feet] tall. This is very controversial because a lot of people don't like high-rise living, and complain that they have acrophobia, and also that vertical stacking eliminates community interaction. But this can be resolved by inserting interval community space every 10 to 20 floors, like a stack of 10-story buildings, each with a floor of community facilities. It's just a design issue, but still it's very controversial. People get all riled up and ask about fire safety. In fact, I have a major opponent who publicly challenges me by saying that fire engines can only reach 60 meters [197 feet] high, so how can I advocate 100-meter [328 foot]-tall buildings? I answer: "What about sprinkler systems? And noncombustible materials? No matter the height of a building, if you use combustible materials the smoke is going to choke everybody to death." What's really needed is a new building code.

A lot of people mistakenly think high-rise means high density. The quota for livable space in China has increased from about 11 square meters [118 square feet] per person in 1980 to the current 32 square meters [345 square feet] per person. Meanwhile, the population has grown 10 percent How do you accommodate the increases? The inescapable answer is high-rise buildings. Of course, you have to have an integrated mass transit system. Stations should be intermingled with residential areas for easy access, as in Hong Kong, where mass transit is right in the middle of the hub.

Right now there is talk about high-speed railroads in nineteen cities, so high-rise planning will be very important. In all fairness, it's not as if China isn't listening. They're trying to do everything right, but implementation takes time, and government officials are going to have to take risks. Party secretary Qiu He is heavily criticized for pushing 100-story residential buildings. People are accusing him of jacking up construction costs. He is actually making people pay more to buy new buildings, so he is under a lot of pressure; it takes a lot of guts.

Everybody loves the Garden City concept: low rise, low density, and a lot of open space, but this does not accommodate many people. I stand by what I've said. Building tall is the answer, and I hope this could be well received on the other side of the table.

What is the life expectancy of buildings in China?

Japanese law specifies that a structure must last a minimum of 79 years in order to receive government grants. And for

Yujiapu Residential Development, Beijing, China (J.A.O. Design International)
中国北京，于家堡住宅开发区（龙安建筑规划设计顾问公司）

low income, they will even reduce mortgage interest rates for a building that lasts for 99 years. That kind of thinking is very foreign in China. Even major buildings don't last more than 50 years, and typical projections don't exceed 40 years, the life expectancy of concrete. The term in office for Chinese government officials is five years, so anything after five years is not their job; they're very short-sighted. I stress that Rome was not built in a day and advocate the concept of built-to-last. These are new ideas in China, and they need to take root and spread.

Development in China raises serious questions about demolition and preservation.

The issue of preservation is always very difficult to judge. How do you qualify which buildings are worth saving? For instance, take the *hutongs* [neighborhoods of traditional courtyard houses lined up along narrow alleys]. A lot of Westerners say the *hutongs* should be preserved. But how can people live without indoor plumbing, and with such narrow alleyways that emergency vehicles cannot even enter? People have to be carried out on stretchers; they die because they can't get

中国的发展在拆迁和保护方面产生了严重问题。

保护的问题总是很难判定。你如何证明某些建筑值得保护？以胡同（传统的沿着狭窄的小巷而排列的四合院式住宅区）为例。许多西方人说，胡同应该受到保护。但是，人们如何在没有室内上下水系统的地方生活，而且街巷如此的狭小以至于救护车都不能驶入其中？人们必须被用担架抬出去；他们有时因为得不到救助而死去。当发生火灾时，更不用说了。不过，当胡同被拆除的时候，有许多人感到他们的遗产被摧毁了。我曾写过一篇关于这个问题的文章，阐释了保护一座城市历史面貌的价值，但也指出，新的开发是至关重要的。人们不会因为某个东西是旧的，就盲目地保护它。所以我说："如果你想摧毁所有东西，那么你是不负责任的；但如果你想要保护所有东西，你就是愚蠢的。"对于规划和发展的官员而言，获得恰当的平衡是一种挑战。意识已经有了，不过再说一次，这是一个执行的问题。在一些特殊的例子中，我曾推荐过重新安置的

Typical hutong, Beijing, China.
北京典型的胡同。

方法。你不需要在高速公路的中央保留一栋老房子，但你也不一定要失去遗产中非常珍贵的一部分。将它移动200米，那么，至少你还拥有这栋建筑，这类想法正变得流行。

现代化对于中国的社区和传统生活单元有何影响？

过去，中国有一套极有约束力的家庭登记系统，用来控制社区。1980年之前，人们并不能自由迁移。他们被安排居住在城市的特定区域内，在特定的工厂工作；人们的生活井井有条。如果你想搬到另一座城市，你必须获得允许。但在今天的自由社会中，如果人们有钱，就可以去想去的任何地方。当然，人们不希望自己所能拥有的面积大小被限定，他们想要越大越好，只要能够支付得起，而不想受到政府配给限额的制约。"哦，你结婚了？再多给你2平方米。有小孩了？再多给2平方。"所以，我不认为户口体制的限制已经被打破了；只是这些问题变得更复杂了。

help. And when there's a fire, forget it. Yet, when the *hutongs* are demolished a lot of people feel their heritage is being destroyed. I wrote an article about this, explaining the value of keeping the historical aspects of a city but also pointing out the critical importance of new development. People shouldn't blindly preserve something just because it's old. So I said, "If you want to destroy everything you are irresponsible, but if you want to preserve everything, you are foolish." Getting the right balance is the challenge for planning and development officials. The awareness is there, but again, it's a question of implementation. In certain cases I've recommended relocation. You don't have to preserve an old building in the middle of the expressway, but neither do you have to lose a precious part of your heritage. Move it 200 meters [655 feet]. Then at least you have the building. That kind of thinking is catching on.

What impact has modernization had on the Chinese community and traditional living units?

In the past, China had a very restrictive household registration system, which was used to control the community.

Before 1980, people could not move about freely. They were assigned to live in certain sections of the city, and work in certain factories; their lives were cast. If they wanted to move to another city, they had to get permission. But in today's free society, if people have money, they can go wherever they want. Of course, they don't want to be told how much area they can have. They want as big a space as they can afford, and not to be bound by government-imposed quotas. "Oh, you're getting married? Two more square meters for you [21.5 square feet]. Having a child? Another two square meters." So, I don't think the restrictions of the *hukou* system have broken down; it's just that the issues are more complex.

China will eventually become more democratic, in part because of its enormous size. The country is so large that many break it down into smaller provinces. Look at Canton. They use a different language, a different way of writing, and traditionally it's been wealthier than the other provinces. Historically, Canton always wanted to be independent. But China does not want to be a federation, so how to control the country and maintain unity is an ongoing challenge for Chinese officials. That's why they're so concerned about the stability of the political system and economic structure. Imagine you have a growing family. Your son gets married, your daughter gets married, they have children, and all of a sudden your family is much larger. Do you want to keep everybody under one roof or do you let them go out on their own? On holidays must they come back or do you let them decide for themselves? That's always an issue for parents. So the Chinese way is like being the parent. "I want everybody under one roof, but I know that kids have their own minds, and some people just don't like each other." So they are always juggling the rules. Governing 1.3 billion people is not an easy task.

The biggest challenge facing Chinese leaders is creating nine million jobs each year for the country's massive population, especially new graduates. China has a history of student rebellions; that's how Sun Yat-sen gained power during the May 4th Youth Movement in 1919. Initially I didn't understand why China had so many people collecting tolls and doing other jobs that a machine could do faster and more efficiently. But mechanization would put people out of work and that would contribute to social unrest.

In China people say: "I don't care who is in power, as long as I'm left alone to work and grow my own food." Most Chinese are not radical by nature; they respect each other. "You have your way and I have mine, and we can still live peacefully with each other." It is only when the people are pushed to the limit that they start fighting back. Westerners don't

一定程度上，由于巨大的体量，中国最终将变得更加民主。这个国家如此巨大，以至于要被分为更小的各个省份。看看广东，他们使用一种不同的语言，一套不同的书写方式，而且从传统意义上说，它比其他省份要富有一些。在历史上，广州总想要独立。但是中国并不想成为一个联邦，所以如何控制整个国家，保持统一，对于中国政府而言是一个持续的挑战。这就是为什么他们如此关注政治系统和经济结构的稳定。想象你有一个正在发展壮大的家庭，你的儿子结婚了，你的女儿也结婚了，他们有了自己的孩子，突然之间你的家庭变大了。你是想让所有人都继续待在一个屋檐下，还是让他们出去闯荡，自食其力？在周末的时候，他们是否一定要回家来，还是你让他们自己做决定？对于父母而言，这总会是一个问题。所以，中国的道路就好像扮演父母的角色。"我想要所有人在一个屋檐下，但我知道孩子们有他们自己的想法，而有一些并不互相喜欢。"所以他们总是在违反规则。管理13亿人并不是一项轻松的任务。

中国领导人面临的最大挑战，是每年要为国家的大量人口，尤其是新毕业生，创造九百万个工作机会。中国有学生反抗的历史；这也是1919年五四运动期间孙中山获得权力的方法。刚开始的时候，我不理解为什么中国有那么多人在收路费，或是做其他一些机器完全可以做的更快更有效率的工作。但是，机械化会将人们赶出工作岗位，进而引起社会动荡。

在中国，人们说："我不关心谁掌握权力，只要我可以不受干扰地工作、种植自己的食物。"大多数中国人天性并不激进，他们互相尊重。"你有你的生活方式，我也有我的；我们还是可以和平共处。"只有当人们被逼到极限，他们才开始反抗。西方人并不理解这一点，他们害怕中国，认为它将恃强凌弱，称王称霸。我告诉中国："你绝不会是第一，而且你也不应该是第一。那么，为什么你想要成为第一？"他们认为卓越是以金钱来论定的，但其实不然。汉朝时的中国是世界第一，那时人们从各个地方来这里学习文化和政治制度。现在，还有谁会选择中国的体制？

在前现代城市中，要去旅行都找不到什么理由，如果人们真的去了，也是步行或骑自行车。正如你曾讨论过的，中国人现在有更多的自由、财富和流动性，但这已经产生了新问题。

拥堵、能源、污染、人口流动总是会成为一个问题。将来的城市需要提供一种均衡的生活工作环境，让人们不至于花太多时间在往返车程上。在新疆维吾尔自治区的拜城，我们设计了一种适合步行的城市。这个城市现在规划人口为5万人，将来可能达到15万人。我们在其中设计了更短、更方便行人的城市街区，许多仅需5分钟就可步行到达的小公园。我们还提出更小型的公交汽车方案，它们可以每200米一停，因此就没有人必须跑步，干嘛要冲来冲去？等下一班公交车就好。我们的整个想法是让人们漫步其中，享受令人陶醉的景致，充分利用社区和社区中的设施。那里有许多林荫大道和公共空间，有许多住宅区小店，人们可以在这里挑选食品，将它们带回家，而不需要挤进汽车，开到几英里外的大超市。政府官员们已经表达了他们的希望，将来的发展将依照我们的原型进行。

我一直鼓励在已有城市中推行相同的做法。作为一位美籍华裔建筑师，独特的身份使我很幸运地获得了某种特权，可以和内部人士坐在一起，因此我们一起谈论了此类东西。我解释说，互动空间要比集会空间更为珍贵。给人们有绿树和长凳的公园、广场，而不是一些每年只使用一次来进行升旗仪式的大型铺砌表皮。这个观点也日益时兴，有时候我感觉自己更像是一个传教士而非一个实践者！

你做过多少个城市设计工程？

在全中国范围内，我们已经规划了面积达3000平方公里（接近于74万英亩）的新城区，它们中有许多已经建成，但那还只是总共15万平方公里（370万英亩）城市面积的一小部分。现在，市长们面临着这许多发展得压力，但我总是警告说，一个城市不能发展得过了头（一个城市的发展不能没有边界）。它必须限制容量，就像一个餐馆一样。当你有供200人使用的座位，突然有500人出现了，这时，厨房装备得再好都没有用，你并不能满足他们的需求。对于城市来说，带来挑战的是外来务工人员。

understand that, and they fear that China is going to bully big business. I told China, "You can never be number one, and you should not be number one. Why do you want to be?" They think money is the claim to preeminence, but it's not true. China was number one during the Han dynasty, when people came from all over to study culture and the political system. Who would elect the Chinese system now?

In the pre-modern city there was little reason to travel, and if people did, it was either on foot or bicycle. As you've discussed, the Chinese now have much greater freedom, wealth, and mobility, but this has generated new problems.

Congestion, energy, pollution. Moving people is always going to be a problem. Future cities need to provide a balanced live-work environment so people won't have long commutes. We designed a walkable city in Baicheng, in Xinjiang Uyur Auton-omous Region, with a projected population of 50,000, and build-out to 150,000. We've designed it with shorter, more pedestrian-friendly city blocks and many small parks within a five-minute walking radius. We've also specified smaller tran-sit buses that stop every 200 meters [650 feet] so no one has to run. What's the rush? Catch the next bus. The whole idea is for people to stroll and enjoy the amenities, to take advantage of the community and its facilities. We have lots of boulevards and communal spaces, and neighborhood shops where people can pick up groceries and carry them home instead of piling in the car and driving miles to some mega-mall. Government officials have expressed their hope that all future developments will follow our prototype.

I've been encouraging the same kind of thing in existing cities. In my unique position as a Chinese-American architect I am very fortunate to have the privilege of sitting at the table with inner circles, and so we talk about these kinds of things. Interactive space is much more precious than congregational space, I explain. Give people parks and plazas with trees and benches, not some big paved surface that is used once a year for flag-raising ceremonies. It's catching on. Sometimes I feel more like a preacher than a practitioner!

How many urban design projects have you done?

We have planned over 3,000 square kilometers [approxi-mately 740,000 acres] of new cities throughout China, many of them built, but that's just a small fraction of the 150,000-square-kilometer urban total [37.07 million acres]. Right now, mayors are under a lot of pressure to develop, but I always caution that a city cannot grow without boundaries. It has to have limited capacity, like a restaurant. When you have seating for two hundred people, and all of a sudden five hundred people appear, it doesn't matter how well equipped your kitchen is, you will not be able to meet the demand. For cities, the challenge is migrant workers.

Would satellite cities be the solution?

For how big a population? You cannot just keep building more, more, more. Satellite cities are not a solution if you can't control your population. People would be camping in open spaces because they couldn't find a place to live.

With a limited amount of space and with current technology, is it possible to create a self-sustainable city?

It's all possible, but again, you need to cap the volume. If a city is meant for two million people, it could never be self-sustainable if the population grows to three million. It would fall short of energy and other resources. This is an ongoing problem. You cannot have migrant workers coming into the city and making demands for schools, housing, and so on, only to have them move back to the country next month, next year. What will you do with all the vacant buildings? A city must have predictable growth.

So what should be done with migrant workers who come to the city for a better life?

They should be trucked in and out. Construction is not longterm. When Beijing built the Bird's Nest, for instance, all of a sudden 20,000 people were needed, but for how long? Six months? There is no magic answer to urban design problems. Finding jobs for people, making sure they are fed, providing hospitals—social issues are increasingly important.

Trucking migrant workers in and out sounds a little bit inhumane.

It's not inhumane. It's a temporary solution, a Band-Aid. Maybe just 10 percent stay, so you're building permanent facilities for 10 percent of the population. That's wasteful.

Chinese farmland is not tangible property; people lease it. But there is now talk of farmers owning land, which means they could also sell it. That would be disastrous. Right now, people from the countryside always have a place to come back to. It doesn't matter whether it's comfortable or not; the point is they have a place to go. If farmers owned the land, they would not resist the temptation to sell it for immediate profit. And after all the money was spent, they would come

卫星城会是解决的方案吗？

针对多大数量的人口？你不能只是一直建设，一直越建越多！如果你不能控制人口数量，卫星城并不是一个解决方案。人们将在空地上支起帐篷，因为他们找不到地方住。

在空间有限以及当前技术的条件下，有没有可能创造出一个"自我可持续"的城市？

当然可能，但还是那一点，你需要限制容量。如果一个城市只能容纳200万人，当人口数量到达300万的时候，它就根本不可能是"自我可持续"的。它将缺乏能源和其他资源，这是一个持续存在的问题。你不能让外来务工人员来到城市，产生对学校和住房等东西的需求，却又在下个月、下一年让他们搬回到乡村去。这样的话，所有那些空着的楼房怎么办？一个城市应该有可以预测的增长。

那么，应该如何处理那些来到城市为更好生活奋斗的外来务工人员？

应该用卡车运输他们。建设并不是长期工程。例如，当北京修建鸟巢时，突然就需要2万人来工作，但是需要多久呢？6个月？对于城市设计产生的问题，并没有一个神奇的答案。为人们找到工作，确保他们有饭吃，提供医疗设施——社会的问题正日益重要起来。

用卡车运输外来务工者听起来有点不太人道。

这不是不人道，这是暂时的解决方法，是一片创可贴。可能只有10%的人留下，而你要为这10%的人修建永久性设施，就是浪费。

中国的农田并非有形资产，人们可以租赁。但是，现在说农民拥有土地，也意味着他们可以卖掉土地，那将是灾难性的。现在，来自乡下的人总有一个他们可以再回去的地方。它舒适与否并不重要，重点是他们有地方可去。如果农民拥有土地，他们不会抵抗售卖土地获得直接利益的诱惑。在所有的钱被花掉后，他们将到城市寻找工作。但如果他们找不到工作，就不再有一个家，那么，到处都将出现贫民窟，这是在印度和其他发展中国家出现的问题。1949年之前，同样的事情也发生在上海——贫民窟，因为政府允许每个人卖掉他们的土地。现在在中国，还没有贫民窟。

Pedestrian-friendly residential district, Shenyang, China. (J.A.O. Design International, 2012)
中国沈阳方便行人的住宅小区。（龙安建筑规划设计顾问公司，2012年）

人们批评中国，说它没有人权，这并不准确。这有点像给青少年一道宵禁令："十点之前回家。"你这么做是否是在限制你孩子的人权呢？或者你实际上是帮助他们通向一个更好的未来？毛泽东鼓励每个人有五、六、七个孩子；人口数量因此暴涨。1949年，当共产党掌权时，中国有5.2亿人。这个数字现在已经变成了三倍之多。西方人认为中国的一胎制剥夺了人权，但却想要用乡村有限的可耕地养活如此多的人。西方人必须要更加理解中国面临的问题。

这就打开了一个更为根本的问题：究竟是什么赋予其他国家权利来推行它们自己的价值？有一个鲜活的例子。在1988年汉城奥运会期间，韩国人被要求不吃狗肉。狗肉恰好是一种低胆固醇、高蛋白的食物，整个韩国都在食用狗肉，在那里，狗被喂养作为食物，而不是作为宠物。一个老农民对此进行抱怨。警察非常友善地解释："狗是西方人的朋友，而且，当他们因为奥运会来到这里时，他们并不想看到你吃他们的朋友。"老农民反驳说："牛是我最好的朋友，但西方人一直都在吃牛肉！"

to the city to look for jobs. But if they couldn't find one and no longer had a home, there would be slums everywhere. That's the problem in India and other developing countries. Before 1949, the same thing happened in Shanghai—slums, because the government allowed everyone to sell their land. Currently, there are no slums in China.

People criticize China and say it doesn't have human rights. Not true. It's a little like giving teenagers a curfew. "Be home by ten o'clock." Are you limiting your child's human rights? Or are you actually helping him to a better future? Mao Zedong encouraged everyone to have five, six, seven kids; the population boomed. In 1949, when the communists took over, there were 520 million people in China. The number has tripled. Westerners think China's one-child policy usurps human rights, but try to feed that many people with the country's limited arable land. Westerners need to be more understanding of the problems China faces.

This opens a more fundamental question: What gives other countries the right to impose their own values? Here's a vivid example. During the 1988 Olympics in Seoul, Koreans were told not to eat dog meat. Dog meat happens to be a low-cholesterol, high-protein food source and is eaten throughout

Korea, where dogs are bred for meat, not as pets. An old farmer complained. The police explained very kindly that "dogs are friends of Western people and so, while they're here for the Olympics, they don't want to see you eating their friends." The farmer countered, "Cows are my best friends, and Western people eat them all the time!"

In the case of pollution, and China's environmental impact, it would seem that the world has every right to speak up.

China became the world's factory because Chinese people work hard and labor is cheap. But the result is undrinkable water, contaminated land, and deforestation. It's a dear price to pay. Awareness is growing among the elite who now say, "What we did before was wrong. We need to be more responsible, not to the international community, but to our-selves, to our families." There are two ways of planning. One is to take up a problem now, and worry about everything else later. The second way is to care for future generations, but you cannot do that until you are secure. When you are poor, you choose the expedient solution. Up until twenty years ago China was not able to feed itself, and a lot of people went hungry. You don't hear that problem anymore.

But China is importing food now, isn't it?

China is importing food, but it is also exporting rice. A sci-entist developed a new strain of rice that can be harvested three times a year. That has changed the whole landscape of China. Before, numerous rice paddies were required to produce enough food for the population. Mathematically, you now only need one-third as much land.

It's been determined that about 1.8 billion mu [approxi-mately 296 million acres] must be kept for agriculture. But developers are now challenging this. I was involved in a high-profile debate with a famous developer who argued that the government should release this controlled land for building, otherwise costs could not be cut. I said, No, the land belongs to our children and our grandchildren. If you don't protect it now, that land will be gone forever. The debate was posted on the Internet, and people voted. That the overwhelming major-ity agreed with me shows that things are changing.

How does water management impact urbanization?

Water is going to be the number one problem in the next five years. Before, people showered once a week, but now it's one, two, three times a day. Improved hygiene is encouraged, but there just isn't enough water for daily needs. Rivers are

关于污染以及中国的环境影响，看起来世界真的有权利大声地将这个问题说出来。

中国成为世界工厂，因为中国人工作非常努力而且劳动力很便宜。但结果却是不能饮用的水、受污染的土地、滥砍滥伐，这是高昂的代价。精英阶层已经不断意识到这一问题，他们现在说："我们之前做的是错的。我们需要负起更多的责任，不是为了国际社会，而是为了我们自己，为了我们的家庭。"有两种规划的方式：一种是现在先解决一个问题，以后再去担心其他别的东西；第二种是关心下一代，但在你有把握之前，你不能行动。当你是穷人的时候，你选择权宜的解决方案。上溯到20年之前，中国并不能养活自身，有许多人处于饥饿之中。你今天再也听不到这一问题了。

但中国现在正进口食物，不是吗？

中国是在进口食物，但它也在出口稻米。一位科学家发明了一种新品种的稻米，可以一年三熟。这改变了整个中国的地貌。在这之前，众多的稻田被要求为人出产足够多的食物。从数学上说，你现在只需要原先三分之一的土地。

国家已经确定，要为农业保留接近18亿亩（接近于3亿英亩）的土地。但开发商正挑战这一数字。我曾参加过一次与某家著名开发商的高层辩论，他们认为政府应该为了建筑业放松对土地的控制，否则用在土地上的费用不会下降。我说，不，土地属于我们的孩子和我们的孙辈。如果你现在不保护它们，土地将永远消失。辩论被发布在网络上，网民们进行了投票。占据压倒性数量的人同意我的看法，这也表明，事情正在发生改变。

水资源管理是如何影响在城市化进程的？

水将在接下来的五年中成为第一大问题。之前，人们每周洗一次澡，现在则是一天一次，或者一天两次，一天三次。改善卫生条件应该被鼓励，但问题是没有足够的水来满足日常的需求。河流正在干涸，星球正在改变，有些地方干旱，有些地方洪涝。当水源无法供应时，城市的末日也就来到了。有些人还坚持认为，在蓄水层有充足的地下水能够继续为中国供水一百年。如果他们是错的怎么办？如果卫星和红外照片并没有被正确地解读，该怎么办？如果所有的水都受到污染，会发生什么？

中国谈到了可持续性，但是有多少后续的行动？

在中国有一个说法："你可以骗过所有人，但你骗不了你自己。"中国人意识到资源严重匮乏。每个人都拥有一辆汽车的想法是不可思议的，他们没有足够的燃料或车库，没有足够的空间。中国现在35%的化石能源从海外进口，而这被看作是一个非常危险的缺陷。这就是为什么他们必须要和俄罗斯谈判，到南美和其他地方去购买石油。中国需要燃料来提高国家生产力，但是当石油供应为其他国家所控制时，这一点又是十分不利的。中国的主要能源是煤。煤有它自身的问题，但比核反应堆要好，特别是因为中国的核电厂已经是第四代了，而不像日本那样是第一代的。

我是一个严厉的中国政策批评者，不过，当需要的时候，也是一个重要的辩护者。在外部世界，存在许多关于中国的错误观念。如果你知道真相，那么你除了进行辩护没有其他选择。我身份特殊，中国政府鼓励我，从发展的利益出发，说出自己的想法，介绍新观念和实践活动，挑战目光狭隘的政府官员，让他们放眼世界，自我看到他们之日起，便定下了这样的基调。没有新观念，就不会有所提高。

我最近被邀请参观台湾的第二大城市高雄，那里的市长要我为一个新区域做规划。我非常直接地告诉她我的想法。在我们会议的最后，一个记者问我，高雄是否有未来。我说："我在这里的这个事实，就是希望的证明。我是一位城市医生。最坏的病人是有问题却拒绝寻求帮助的人。而认识到自己需要帮助，并遵照医生开出的康复药方行事的病人，则是大有希望的。"和大陆一样，在台湾，他们至少是在倾听——这是关键的第一步——而之前，他们甚至都不会有所在意。对病症有预判是好的。

running dry, the planet is changing, there is drought in some places, floods in others. When water supplies dry up, that's going to be the end of the city. Some maintain there are sufficient underground supplies, in aquifers, to serve China for a hundred years. What if they're wrong? What if the satellite and infrared photos are not read correctly? What happens if all the water is polluted?

China talks about sustainability, but how much follow-through is there?

There is a saying in China: "You can fool everybody, but not yourself." The Chinese realize there is a severe scarcity of resources. The idea of everyone owning a car is incomprehensible; they don't have enough fuel or garages, not enough space. China now imports 35 percent of its fossil energy from overseas, and this is seen as a very dangerous weakness. That's why they have had to negotiate with Russia, and go to South America and elsewhere to buy oil. China needs fuel to increase national productivity, but it is detrimental when petroleum supplies are controlled by other nations. China's main source of energy is coal. This has its own problems, but it's better than nuclear reactors, particularly since nuclear plants in China are fourth generation, not first generation, as in Japan.

I am a heavy critic of Chinese policy but also a great defender when called for. So many misconceptions exist about China in the outside world. If you know the truth, there is no choice but to defend it. I'm in a unique position, calling the shots as I see them since, in the interests of progress, the Chinese government has encouraged me to speak my mind, to introduce new concepts and practices, and to challenge parochial officials to open to the world. Without new ideas, there will be no advancement.

I was recently invited to visit Kaohsiung, the second-largest city in Taiwan, where the mayor wanted me to plan a new district. I was very direct and told her what was on my mind. At the end of our meeting, a reporter asked whether Kaohsiung had a future. I said, "The very fact of my being here is proof of hope. I'm an urban doctor. The worst patient is one who has a problem but refuses to seek help. There is much greater hope for the patient who realizes he needs help and follows the remedial measures the doctor prescribes." In Taiwan, as in China, they are at least listening—that's the critical first step—whereas before, they weren't even paying attention. The prognosis is good.

Jaime Lerner
JAIME LERNER ARQUITETOS ASSOCIADOS

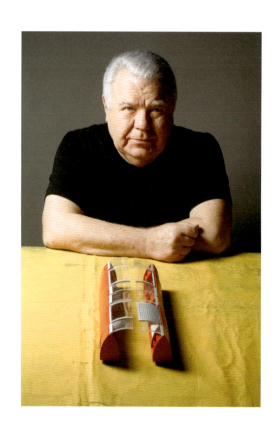

Jaime Lerner is an architect and urban planner, founder of the Instituto Jaime Lerner and chairman of Jaime Lerner Arquitetos Associados. A former president of the International Union of Architects and three-time mayor of Curitiba, Brazil, he led the urban revolution that made the city renowned for urban planning in public transportation, environment, and social programs. He twice served as governor of Brazil's Paraná State and was responsible for an urban and rural economic and social transformation. Mr. Lerner is recognized around the world as an urban visionary, and his most significant international honors include the highest United Nations Environmental Award, as well as the Child and Peace Award from UNICEF, the World Technology Award for Transportation, and the Sir Robert Mathew Prize for the Improvement of Quality of Human Settlements. In 2010, he was nominated by *Time* magazine as one of the 25 most influential thinkers in the world. Mr. Lerner's ongoing practice develops projects for the public and private sectors for cities in Brazil and abroad.

海米·勒纳
海米·勒纳建筑事务所

海米·勒纳是一位建筑师和城市规划师，也是海米·勒纳研究所的创立者、海米·勒纳建筑事务所的主席。他之前是国际建筑联合会的主席，并曾三次担任巴西库里提巴市市长，他领导的城市革命使这座城市在公共交通、环境和社会项目的城市规划上声名大噪。他曾两次担任巴西巴那拉州州长，负责城市、乡村地区的经济及社会转型。勒纳先生作为城市梦想家在世界范围内得到广泛认可，他最重要的国际荣誉包括最高的联合国环境奖，联合国儿童基金会颁布的儿童与和平奖，以及因交通运输而获得的世界技术奖，因提升人类居住质量而获得的罗伯特·马修爵士奖。2010年他被《时代》杂志提名为世界上最有影响的25位思想者之一。勒纳先生一直在巴西和全球范围内为城市中的公共部门和私人组织进行项目开发工作。

For complete biography, please refer to the Appendix.

完整的传记，请参阅附录。

Curitiba's trademark Botanical Garden opened in 1991 in prominent display of Jaime Lerner's green revolution.
在1991年海米·勒纳的绿色革命发动后才开张的库里提巴的商标植物园。

How did an architect and urban planner become mayor of a large city?

I got involved back when I was a student. At the time, we had a mayor who was destroying the city, widening streets to create big avenues; to him, the city's history was not important. So we got together, students and teachers, and started a movement against this process; we asked for a master plan. The city put out a bid—it was not a competition of ideas but a bid—and we organized ourselves to compete, but we were very young. The city selected a French company with an associate architect from São Paulo to draw up a preliminary plan. In time, they needed a local counterpart, so the city engaged some recently graduated architects, and I was one of them.

Our advisory agency evolved into an institute of urban planning. We started with just six people, and a few years later I became president of the institute. Coincidentally, I was also president of the local institute of architects. This was a time of military rule, and mayors were appointed, not elected. The government didn't want interference from politicians, so they preferred people with a technical background. That's how I became a mayor at thirty-three. It was not easy.

In the institute we always had it in mind that if, by chance, we ever got a mayor who would listen, we should have something to say. The city has to have a structure of growth, to offer job opportunities, to have this and this, so we worked up a list. When I was appointed, I asked the people I was working with, most of them my age, to assume key positions. Being appointed, I always told my team, "Look, we are not sure I am going to be mayor next week, so we have to work as fast as possible. What should we focus on? What are your ideas?"

We worked on a lot of things at once, with courage and simplicity. I realized we needed to concentrate on two separate areas: one was the needs of people and the other was the potentials. The question was how to balance the two, needs and dreams. We started to understand that it is not enough to go to a different neighborhood every week and ask what they need. In three months I had a whole mountain of needs! So I told my people that if we work only with the needs, we will never change. But at the same time, if we work only with the potentials, we'll have no impact on the large majority of people. So we have to work with needs and potentials simultaneously.

In the morning I worked on potentials and in the afternoon I went to city hall and worked on needs. Each day, I developed new insights, new motivations, and better ideas. We started to understand that you just have to start, and since

一位建筑师和城市规划师是怎样成为一个大城市的市长的?

当我还是学生时，就开始参与了。当时，我们有一位市长，他破坏城市，拓宽小巷，铺设更宽的街道。对他来说，城市的历史并不重要。因此，我们聚在一起，包括学生和教师，开始了反对城市的改造运动，并要求一个总体规划。这个城市当时进行了招标，但那只是招标，并不是观点的竞争。于是，我们组织起来为此竞争，但当时我们都非常年轻。最后，一家法国公司中标了，资深建筑师来自圣保罗，他负责拟定初步规划方案。他们需要一个当地的合作搭档，所以政府招聘了一些刚刚毕业的建筑师，我就是其中之一。

我们的顾问机构演变为一个城市规划研究所。最初只有六个人，几年后，我成为了该研究所的所长。巧合的是，我那时也是当地建筑师学会的会长。我们的国家当时由军方统治，市长直接被任命，而不是选举。政府不希望受到政治家的干扰，因此他们更倾向于具有技术背景的人。这就是我如何在33岁时就成为市长的原因，的确不容易。

在研究所工作时，我们一直在想，如果碰巧有市长愿意听取意见的话，我们一定会提出一些想法。这座城市需要有一个利于增长的结构，来提供就业机会，对此，我们制定了一个清单。当我被任命为市长时，我要求那些曾经与我一起工作的人来扮演关键岗位上的重要角色，他们当中的大部分人与我年龄相仿。在任命之后，我总是告诉团队，"你们看，我们并不知道我下周还是不是市长，因此我们必须尽快开始工作。我们应该关注什么?你们有什么想法?"

我们立即开始着手各种工作，非常大胆，也十分真诚。我意识到，我们需要专注于两个独立的方面:一个是人民的需求，另外一个是可能性。问题是如何平衡这两个方面，即需求和愿望。我们开始明白，每周去拜访不同的居住区并询问市民需要什么，这是远远不足的。在3个月的时间里，我们了解到大量的需求!所以我告诉我的工作人员，如果我们只关注于需求，我们将永远无法改变。但同时，如果只关注于可能性，那么我们将不会为大多数人带来影响。因此，我们必须同时专注于需求和可能性。

上午，我专注于可能性方面的工作;下午，我去市政厅关心需求方面的工作。每天，我会形成新的见解、新的动机和更好的想法。我们渐渐明白，你只需要行动，因为你无法获得所有问题的答案，所以要留一些余地，以备纠正。重要的是，要意识到一个城市就是一个结构，它涵盖生活、工作、交通各个方面。我将城市比喻为乌龟。它的所有功能都结合在一起，它的胳膊、腿和头一起活动，并且在同一个地方生活和工作。但是想象一下，如果你将乌龟的壳切成数片，那么它就会死掉。这正是世界上大多数城市正在做的：将城市功能分离，按照收入、年龄和宗教进行分离；其实这无异于是在谋杀城市。

以前担任市长时，我每天晚上都让司机带我到不同的住宅区和社区，听取公众的讨论。当时，并没有PowerPoint或类似的东西，所以我们不得不带上城市模型。几年后，当我成为州长时，我让那位司机和我一起访问巴西利亚。他从来没有乘坐过飞机。我们起飞时，他看了看城市，说道："哦！它真的就像当初的那个城市模型！"我当时感到非常欣慰和满足，因为库里提巴的设计和增长结构是如此明确。

库里提巴并不是天堂。但让我感到自豪的是大约70%的人生活在多样的住宅区。此外，人们尊重这个城市。当然，我们也存在问题，就像所有其他城市一样，但在20年至30年内，我认为在城市成果的数量和范围方面，我们会处于领先地位。我说的不仅包括公共交通，还包括医疗保健、儿童保健和生活质量这些问题。你知道我们有多少个儿童日托所吗？260个！我们还有110个卫生保健中心，不是医院，是卫生保健中心。

是什么给了你去创新公共交通系统的灵感？

需求。库里提巴在1971年大约有60万人口。当时有人说，每一个拥有一百万人口的城市都应该建设地铁。我们试图搞清楚良好的公共交通应具有哪些要素。它必须具有速度、可靠性和合适的频次。于是，我们开始考虑，为什么不发展地面交通呢？我们不能花大量时间或金钱征用土地，去修建非常宽阔的街道，所以我们决定利用现有的街道。我们并没有修建70米宽的大道（230英尺），而是采用了三路道。在密度较高的地方，我们提高了在公共交通中的班次和质量。我们开始计划的是每天2.5万的乘客流量，然后不断改进和完善，增加区际线和支线。现在，该系统每天可运送240万人次。相比之下，伦敦地铁每天运送300万人次，但伦敦地铁位于一个更大的城市，并且是世界上最古老的地铁。

you can't have all the answers, you need to leave some room for corrections. It is important to understand that a city is a structure living, working, and moving all together. My metaphor for the city is a turtle. All of its functions are integrated so that its arms and legs and head all move together, and it lives and works in the same place. But imagine if you cut the turtle's shell into different pieces—the turtle would die. That is exactly what most of the cities of the world are doing: separating urban function, separating by income, by age, by religion; you're killing the city.

Back when I was mayor, I would take my driver to different neighborhoods and communities for public discussions every night. At that time there was no PowerPoint or anything like that, so we had to carry the model of the city with us. A few years later, when I became governor, I invited the driver to visit Brasilia. He had never been on a plane. As we were taking off, he watched the city and said, "Oh! It's exactly like the model!" It was a moment of great gratification to me that Curitiba's design and structure of growth were so apparent.

Curitiba is not a paradise. But what makes me proud is the fact that around 70 percent of the people are living in diverse neighborhoods. Also, the people respect the city. We have problems, of course, just like every other city, but in twenty to thirty years, I think we will probably lead in the number and scope of our urban achievements. I am not only talking about public transport, but health care, child care, quality-of-life issues. Do you know how many day care centers we have? Two hundred sixty! And one hundred and ten health care centers, not hospitals, health care.

What inspired your innovative public transport system?

The needs. Curitiba had around 600,000 people in 1971. At that time, it was said that every city of about one million people should have a subway. We tried to understand what makes for good public transport. It has to have speed, reliability, good frequency. We started to wonder, why not on the surface? We couldn't spend time or money on a big expropriation for very wide streets so we used existing streets. Instead of one 70-meter-wide avenue (230 feet), we took three streets, a trimary. Where we have more density is where we have a better headway in public transport, better quality. We started with 25,000 passengers a day and kept improving, improving, improving, adding interborough lines, feeder lines. Now the system transports 2.4 million passengers a day. By comparison, the Underground in London, which is in a much bigger city and is the oldest subway in the world, transports three million a day.

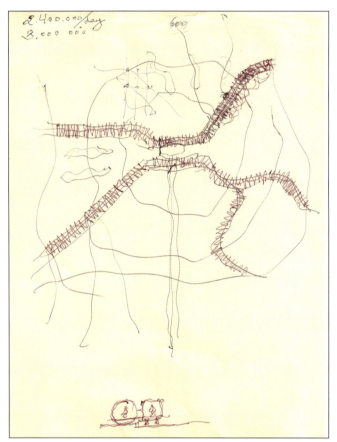

Left: Jaime Lerner sketch of the trimary transit arteries he established in downtown Curitiba. Right: Express Buses in dedicated traffic lanes.

左：海米·勒纳在库里提巴市中心建立的三角形交通干道的草图。右：特定公交线路上的快速公交。

Significantly, our system pays for itself; there is no subsidy because the money is needed for social programs. We started to understand that every problem in the city has to have a good equation of co-responsibility. The fleet cost three hundred million dollars. If we didn't have the funds, what was the solution? We posed the question to private initiatives. "We are going to invest in the itinerary as long you invest in the fleet," we said, "and we will pay you by kilometer." That's all. Since then, almost forty years, the system has worked.

Four years ago, I was interviewed by a reporter from the *New York Times* for a big issue they were doing on Curitiba in the weekly magazine. He asked me about the key elements of our public transport. I told him it's dedicated lanes, boarding fast because you pay before getting on the bus, and boarding at the same level, and also you don't have to wait more than a minute. My wife, who was listening, said, "No,

值得注意的是，我们的公共交通系统是自给自足的，我们无法得到补贴，因为各种社会项目需要钱。我们开始明白，城市中每一个问题的解决都需要责任共担。公共交通车队需要耗资3亿美元，如果没有足够的资金，那么有什么解决方案呢？我们把问题提交给私人项目。"只要你们投资车队，我们将会对行车路线进行投资，并按公里付款。"事情就是这样。从那时开始，我们的公共交通系统一直运行良好，到现在已将近40年。

4年前，《纽约时报》的一位记者对我进行了采访，他们准备在周刊杂志上对库里提巴进行专题报道。他问道，库里提巴公共交通的关键是什么？我告诉他，是专用车道、快速登车，因为你是在登车前买票。你在同一位置上登车，并且等车时间不会超过一分钟。我的妻子当时也在旁边，她说："不，这是做不到的。"我当时感到有些

Typical tube station in Curitiba's innovative and efficient transit system.
库里提巴革新性、高效率交通系统中典型的地铁站。

难堪，因为在《纽约时报》记者面前，我像是骗子！所以我们来到站台，开始对公交车的经过时间进行计时，结果每40秒就会有一辆公交车到达。因此，我的声誉得到了恢复，至少在一段时间内我成了家里的主人。

这个系统在刚开始就具备了这些要素？还是说它是一个逐渐发展的过程？

最初，我们由公共汽车专用车道，乘客正常登车，大约每隔500米设停车站。但直到我们了解登车过程之后，事情才真正得到改善。我们意识到：公交车与地铁在需求上具有有相同的性能要求，登车可以在同一位置上，你必须提前支付，这样就不会因为某个人翻找钱包或费力爬上阶梯而浪费时间了。这种模式下我们就可以增加频率，提高公交车的利用率。这是一个综合的生活工作系统，而不仅仅是一个交通系统。

it's not possible." I was so embarrassed, being called a liar in front of the *New York Times*! So we went to the terrace and started to clock the buses, and every 40 seconds one arrived. So my credibility was restored, and for a while, at least, I was lord of the manor.

Did you begin with all elements of the system in place or did they evolve?

At first we had normal boarding, with dedicated bus lanes and stops about every 500 meters. But it wasn't until we understood the boarding process that things really improved. We realized that buses needed to have the same performance as subways. Boarding has to be on the same level and you have to pay in advance, so time isn't lost while someone searches for their wallet or struggles to climb the stairs. By doing this, we could increase frequency and refill the bus more often. It's

库里提巴有16个大公园，14座森林和超过1000处的绿地，为每人提供平均50平方米的（540平方英尺）公园绿地。

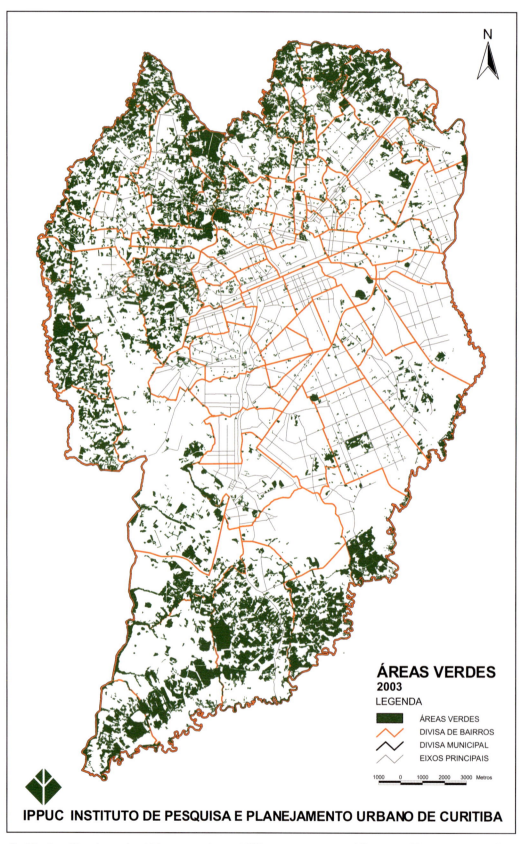

N

ÁREAS VERDES
2003
LEGENDA

ÁREAS VERDES
DIVISA DE BAIRROS
DIVISA MUNICIPAL
EIXOS PRINCIPAIS

1000 0 1000 2000 3000 Metros

IPPUC INSTITUTO DE PESQUISA E PLANEJAMENTO URBANO DE CURITIBA

Curitiba has 16 major parks, 14 forests, and over 1,000 green spaces, providing some 50 square meters [540 square feet] of parkland per person.

库里提巴有16个大公园，14座森林和超过1000处的绿地，为每人提供平均50平方米的（540平方英尺）公园绿地。

交通系统被规划为每天最多使用两次，这是不可行的，人们需要的是一种可以更好地与日常生活相结合的运输系统。你并不是一位交通工程师，对吗？不是，太好了，感谢上帝，因为交通工程师可以扼杀一个城市，他们是连环杀手，总是将重点放在交通工具上面，而不是城市。城市远不止于此。

你曾经说过，每一个城市都能够在三年内得到改善，而不论其规模或预算。

我是指某些特定方面，并不是说全部，排水系统是一个好的例子。我们不需要另外挖掘、拓宽河流，只是根据我们的公园网络、可渗水走道以及可自然收集雨水的河流和湖泊情况，我们就对库里提巴的水源和排水进行了良好的控制。

你是如何发展公园系统的？是否需要拆除建筑物以创造空间呢？

这源于对城市的讨论。我记得有一个经验法则，每隔0.3英里（0.5公里），应该有一个广场，或每隔2/3英里（1公里），应该有一个小公园。因此，我们开始建设小公园和广场网络，但人们对此提出质疑，为什么要花费这笔钱，为什么不保存现有的树林呢？新的树木长出来，将需要二三十年的时间，所以，为什么不保护现存的树林呢？1971年，库里提巴的人均绿地面积仅0.5平方码多一点。现在，我们的人均绿地面积为62平方码，并且人口已经翻了三番，这意味着绿地增长了30万倍之多。

让我来解释一下。假设你的家庭（King一家人）有大约25英亩的土地，你们已经拥有这笔财产很多年了。你在这里盖起了房子，养育子孙并缴纳税款，但现在25英亩土地的税收是一个问题，也许你并不需要这么多土地，你可以将其出售，以便进行开发。我们谈判达成了一项协议，我建议这个家庭保留25%的土地，然后我们会将其他土地改造成一个公园。价格一定会非常合理。为什么呢？因为通过维护和培育园区，土地价值将会增加，因此，对于这个家庭和城市都是一件好事。为了对这些多年来拥有土地的家庭表示敬意，我们以这些家庭的姓氏为公园命名，所以，这个公园就叫"King公园"，同时，我们取消这个家庭的赋税。这就是98%的人都会同意的原因。很快，我们的公共树林和绿地面积大量增加。在这之前，土地是否开发，

an integrated living-working system, not a commuting system.

Commuting systems are planned for maximum use twice a day. It's not feasible; you need a system for moving that is more integrated with daily life. You're not a specialist in traffic engineering, are you? No. Thank God, because traffic engineers can kill a city. They're serial killers, because their focus is the vehicle and not the city. The city is much more than that.

You have said that every city could be improved in three years, irrespective of its size or budget.

I'm not saying the whole, but in targeted areas. A good demonstration is drainage. We have good control of water and drainage in Curitiba due to our network of parks, permeable walkways, and ponds and lakes that collect rainwater naturally rather than having to channelize our rivers.

How did your park system come about? Did you have to demolish buildings to create space?

It grew out of discussions with the city. I remember there was a rule of thumb that every three-tenths of a mile (.5 km) there should be a plaza, or every two-thirds of a mile (1 km) a small park. So we started to develop a grid of small parks and plazas, but people started questioning why we would spend the money, why not just save the existing woods? It would take twenty or thirty years for new trees to grow, so why not just save the ones that already exist? In 1971, Curitiba had a little more than half a square yard of green area per inhabitant. Now we have 62 square yards of area per capita and the population has tripled, which means 300,000 times more.

Let me explain. Suppose your family, the King family, owns roughly 25 acres and it has kept the property for many years. You've built a house here, raised children and grandchildren, paid taxes, but now the taxes on 25 acres are a problem, maybe you don't need all that land and you might be tempted to sell it for development. We negotiated an agreement. I proposed that the family keep 25 percent of the area and that we would transform the rest into a public park. The price had to be very reasonable. Why? Because by maintaining and cultivating the park, land values would increase, so it was good for the family and good for the city. We paid homage to the families who had preserved the land for so many years by naming the parks after them—in this case it would be King Park—and we also repealed their taxes. That's why 98 percent of people agreed. All of sudden, we had an increase of public woods and green areas everywhere. Previously, it was up to the individual owners whether to develop the land or not, but now that it is public parkland they cannot. It is forbidden.

What I'm trying to say is you can do a lot in three years as long as you don't limit yourself by thinking that everything has to come out of your budget. Take waste management. We didn't have enough landfills, so we had to transform the problem of garbage, We determined that one-third is recyclable, one-third is organic and can be transformed for reuse, and one-third is really garbage. So if people could separate their garbage, it would decrease the problem by almost 70 percent. We started by teaching schoolchildren. In six months they learned how to separate, and then they taught their parents. That's how we started, and that's why, since 1989, Curitiba has the highest rate of separation of garbage in the world. There is no fine or other penalty for nonparticipation, but 60 to 70 percent of people do it voluntarily.

All of the improvements in this city are examples of what can be done without much money. Too much money is not good. If you want creativity, cut one zero off your budget; if you want sustainability, cut two zeros. And if you want solidarity, you need to keep your own identity while respecting diversity. But it's only with time that I've realized how true this is.

I am shocked by how much waste, and wasteful mistakes, I see in cities all over the world. I always talk to the mayor and to the office of management about their city's structure of growth. They put some dots on some vectors. They are afraid of designing. If I ask a taxi driver or a barber, they know what is working and what is not, but the authorities, with all their resources and sophistication, do not. If you don't know the structure of growth in a city, you have no priority. Cities are big, so you cannot invest in the whole unilaterally; you have to determine priorities. Once you do that, you can build multiple good equations of co-responsibility in every part of the city. And the people are happy because you are investing in what is important to them.

It's logical and simple. I came to realize that most decision makers don't like simplicity. Why? Because they say to themselves, "If the solution is so simple, why didn't I think of it before? It must be wrong."

You have said that cities are not the problem; they are the solutions.

Every time we separate the economy from the people, it's a disaster. Political decision makers think they own the economy, and they forget that quality-of-life issues can provide more jobs than anything else. The whole world invested $3 trillion to stem the economic crisis. Invested in what? Banks

由个体所有者决定，但现在，这些土地成为了公共绿地，并禁止开发。

我想说的是，你在三年的时间内，可以做很多事情，只要你不去想每件事都要占用你的预算并以此来限制自己。以废物管理为例，我们没有足够的垃圾填埋场，所以不得不设法改变垃圾处理的问题。我们认为，三分之一的垃圾应具有可回收性，三分之一的垃圾为有机垃圾，这些垃圾可以转化并重新利用，还有三分之一的垃圾是真正的垃圾。因此，如果人们能够将垃圾分类，那么可以将该问题解决近70%。我们首先对学生进行宣传教育，在6个月的时间内，他们学会了如何为垃圾分类，然后他们会将垃圾分类方法告诉他们的父母。最初我们就是这样开始的，这也是自1989年以来，库里提巴有全世界最高垃圾分离率的原因。对于不参与垃圾分类的人们，我们并没有采取罚款或其他处罚措施，但60%至70%的人们自愿这么去做。

这个城市的各种改善都是在没有太多钱的情况下实现的。钱太多并不一定就好。如果你希望创新，那么就将你的预算减掉一个零；如果你希望可持续性，那么就将预算减掉两个零。如果你希望团结，那么在保持自身特性的同时，尊重多样性。随着时间的推移，我才意识到这是多么的正确。

对于浪费程度和浪费的错误行为，我感到震惊，我在世界各地的城市都看到了这种情况。我总是跟市长们和管理办公室的工作人员谈论城市的增长结构。他们将眼光放在一些固定数量上，畏惧设计。如果我问一位出租车司机或理发师，什么在起作用，什么没起作用，他们是知道的；但手上握有资源，思维精明的当局者却不知道。如果不知道一个城市的增长结构，那么就无法确定优先事项。城市是非常大的，所以不可能片面地投资全部事项，你必须确定优先事项。如果你这样做了，就可以在城市的每一个部分平衡多重有益的共同责任。人们会感到幸福，因为你投资的是他们认为重要的东西。

它合乎逻辑并且很简单。我认识到，大多数决策者并不喜欢简单。为什么呢？因为他们对自己说，"如果解决方案这么简单，为什么我以前没想到？它一定是错的。"

With its "Garbage that is not Garbage" program, Curitiba pioneered recycling and retains the highest recycling rates in the world.
"垃圾不是废物"的项目，使库里提巴成为废物回收方面的先行者，它也保持了世界上最高的回收率。

你曾经说过城市并不是问题所在，而是解决方案。

每次我们抛开人民谈论经济，都会是一场灾难。政治决策者们认为他们掌握经济，但他们忘记了生活质量可以带来更多的就业机会和其他任何事情。整个世界投入了3万亿美元来遏制经济危机，这些钱都投到了哪里呢？答案是银行和汽车行业。如果他们仅将这笔钱的三分之一投入到世界各地的城市生活质量上，就会创造出大量的就业机会。有时，我们需要打破条条框框。如果你正在试图解决城市化问题，那么提供更多的就业机会，才能真正帮助城市和整个国家。

我喜欢与那些有技术或哲学背景的经济学家们一起工作，因为他们不会注重抽象化的东西。这种不抽象，不是关于减息，而是关于人民。这就是为什么当我与市长或规划者谈话时，我的第一个问题永远是："你有什么难题吗？"答案有所不

and the car industry. If they had invested even a third of this money in quality of life in cities all over the world, there would have been a huge number of jobs created. Sometimes we have to think outside the box. If you're trying to solve the problem of urbanization, provide more jobs. In this way, you can really help cities, entire countries.

I like to work with economists who have a technical or philosophical background because they don't focus on the abstract. It's not abstract. It's not about cutting interest, it's about people. So that's why, when I speak to mayors or planners, my first question is always, "What is your problem?" The answers vary—traffic, safety, or education, pollution— every city has its own issues, but mostly they have a lot of common problems. My second question is, "What is your dream?" It is not the dream that will solve your problems, I explain, but a higher dream that I'm talking about. It's hard to extract, like birthing, but I've heard so many smart and sustaining answers.

When I was in Russia I spoke with a government official, a very wise man. He showed me a city of a million people and explained that it had formerly been a supplier of army rockets, but now there was no need. The city, which had grown on both sides of the river, could not afford to build bridges. It had problems with public transport, with providing access to parks on one side of the river to the neighborhoods on the other side. "What is your dream?" I asked. He answered, "For our young people not to move to Moscow," and he showed me a program to transform this city into the cultural capital of Russia. He had some ideas, great ideas, and if he acted on half, or even a third of them, he would have a good city where the young people would want to stay.

In another instance, I had a visit from the mayor of a very poor municipality in the state of São Paulo. The government had created an affluent new municipality, leaving behind all the slums in this mayor's precinct. When I asked about his dream and what he was proposing, he explained that they were building glass houses. The area had sugar cane and also fruits, so they started to teach the children how to make sweets. At first they made jellies in their houses, but then, on the road that divided the slum from the new municipality, they started building shops with big glass windows to show their products. Cars driving through started to stop, and all of a sudden this very, very poor community had a lively business to increase the income of its people. This guy was a genius!

Backtracking a little, you served for three terms as mayor of Curitiba before being elected governor. Were you able to implement the same kind of thinking on both the local and state levels?

Yes. It was interesting because when I started running for state office, I had just a 6 percent rating in the polls. My opponent, who had 52 percent, was the incumbent, a former soap-opera actor, young and very good-looking and, in truth, not a bad governor. About two months before the elections we had a half-hour televised debate, and it was assumed that, being so handsome, he would destroy me. But I won because I tried to make the people understand their state, which is like a jigsaw puzzle and which they usually don't pay attention to but instead just concentrate on their own particular city or part of the country. I pointed out on a map the main roads, rivers, and cities, and I tried to explain how they were interrelated, and the importance of my plan to create a vital pole of universities, hospitals, and so on close to every city. That way, no one would be more than an hour from such services; we would organize our infrastructure for support. The people understood

同——交通、安全、教育或污染，每个城市都有它自己的问题，但大多数情况下，它们具有很多相同的问题。我的第二个问题是："你有什么梦想？"这不是你想要解决具体问题的愿望，我解释道，我谈的是更高层作为梦想的愿望。这就像分娩一样并不容易，但我也听说过许多明智和有效的答案。

在俄罗斯，我与一位政府官员谈过话，他是一个非常聪明的人。他向我展示了一个有百万人口的城市，并解释说，这里以前是军队的火箭基地，但现在已经不需要了。这座城市在一条河的两岸发展，但却没有钱搭建桥梁。它存在着公共交通的问题，以及分属河岸两侧的公司和社区的可达性问题。"你有什么梦想？"我问道。他回答说："让我们的年轻人不再向莫斯科迁移。"他给我看了一个方案，可以将这座城市改造成俄罗斯的文化之都。他有一些想法，非常好的想法，如果能够将这些想法的一半甚至三分之一付诸实践，那么他将会打造出一个很好的城市，这样年轻人就愿意留在这里。

还有一个例子。圣保罗州一个非常贫穷的自治市市长访问了我。政府建设了一个全新的自治市，而将所有的贫民窟留给了这位市长的辖区。当我问到他的梦想以及他有什么提议时，他解释说，他们正在建造玻璃房子。该地区盛产甘蔗和水果，所以他们开始教孩子们如何做甜食。起初，他们在自己的房子里做果冻，然后在新自治市与贫民窟之间的道路旁边，修建带有大玻璃窗的商店，展示他们的产品。路过的汽车开始停下来，突然间，这个非常贫困的社区有了新的生计，人们收入有所提高。这家伙简直是个天才！

在当选为州长之前，你曾经连任了三届库里提巴市长。你能够在地方和州实践类似的想法吗？

是的。这很有趣，因为开始竞选州长时，我在民调查中仅获得6%的选票。我的对手，获得了52%，他是现任州长，以前是一位肥皂剧演员，年轻帅气，说实话，他是一位不错的州长。大约在选举的两个月前，我们进行了半小时的电视辩论，人们认为他是如此英俊，一定会打败我。但我赢得了辩论，因为我努力让人们了解他们自身的状态，这就像一个拼图游戏，他们通常并不重视这个问题，而只是专注于自己特定的城市或国家的一部分。我在地图上指出主要的道路、河流和城市，并努力解释它们之间的相互关联性，以及我所计划的重要性，即努力让大学、医院等重要设施靠近每一个城市。让每个人都可以方便享受这些服务，并且车程不会超过一个小时；我们将会组织基础设施，以支持这些服务。人们理解了这种情况，并给我投赞成票，而我的对手只提出了减税和提供更多的就业机会。他这种习以为常的提议显得过于遥远，并且过于抽象，他的主张究竟是什么呢？人们并不知道。

对于一个州，你可以是一个优秀的政治家，而不是一个好的行政官，在这里，政策非常重要。对于一个城市，你要清楚各种情况。如果要改善现状，就必须付诸行动，这样情况才能变得越来越好。在墨西哥城的一个小广场上，有一块牌匾，上面写着："宁可要不完美的优雅，也不要完美的不优雅"。你无法获得所有的答案，但是这并不意味着不可以开始。

您是如何应对那些受现有规章制度束缚的人？

人们总是可以找到一个很好的借口。我可以举一个例子。当我还是一个22岁的学生时，我设计了这个房子，并在毕业时获得了一笔奖金，我用这笔奖金建了这个房子。我设计的混凝土壁炉很难塑形，于是我告诉了施工队长。他向我保证他可以做到，他的确出色地完成了这项工作。"有两种工人，"他告诉我，"一种工人会告诉你为什么一些事情无法去做，另一种工人会告诉你如何做到这一点。"这同样适用于城市。出于种种原因，人们有时会将情况复杂化，并作出详尽说明。20分钟后，他们会说："你看，我可以完成这项工作，不过它非常难，你需要花不少的钱。"

这并不是一个规模或金钱的问题，而是你如何去处理的问题。我任州长期间，有一个海湾受到了严重污染。里约热内卢曾获得了8亿美元的贷款，以清理它的海湾，但我们却没有这么多钱。那么有什么解决方案吗？当然，您可以创建公共工程，但只要人们继续向水中扔垃圾，污染就不可避免。因此，我们与渔民达成一项协议：如果你们逮鱼，那么逮到的鱼归你们。如果你们捞垃圾，那么我们会买这些垃圾。如果某一天不适合逮鱼，那么他们可以捞垃圾。他们捞的垃圾越多，能得到的钱也就越多，同时，海湾也会变得更清洁。海湾越清洁，就会有更多的鱼。这是一个双赢的解决方案。对于每个州和每一个城市，你需要有一个好的州长和市长，他们能够打破常规思维，另辟蹊径。

the scenario and they voted for me, while my opponent had offered only to reduce taxes and supply more jobs. His familiar message was too remote, too abstract. What exactly was his proposal? The people did not know.

On a state level, you can be a good politician but not a good executive; policies are important here. But at the city level, you have to be very clear. You have to make things happen if you are going to improve the diagnostic, so that it always gets better and better. In a small square in Mexico City there is a plaque that reads "Better the grace of imperfection than perfection with no grace." You cannot have all the answers but this doesn't mean you should not begin.

How do you overcome people tied down by existing rules and regulations?

People can always find a good excuse. I'll give you an example. I designed this house when I was a twenty-two-year-old student, so when I graduated and won an award, I used the prize money to actually build it. The concrete fireplace I designed was difficult to shape, so I spoke to the job captain. He assured me he could and did a beautiful job. "There are two kinds of workers," he told me, "one who tells you why something can't be done, and one who tells you how to do it." The same is true for cities. For all sorts of reasons people will sometimes complicate situations and make elaborate presentations. After twenty minutes, they say, "Look, I can make this work, but it will be very difficult. It will cost you."

It is not a question of scale or money but a question of how you handle it. When I was governor, there was a bay that was very polluted. Rio de Janeiro got an $800 million loan to clean its bay, but we didn't have that kind of money. So what was the solution? You can create public works, of course, but as long as people are throwing garbage in the water there will always be pollution. So we made an agreement with the fishermen: If you catch fish, they're yours. If you catch garbage, we'll buy it. So if a particular day was not good for fishing, they caught garbage. The more garbage they caught, the more money they'd have and the cleaner the bay would be. The cleaner the bay, the more fish would grow. It's a win-win solution. In every city, in every state, you need good mayors, good governors, who can think outside the box.

Liu Thai Ker
RSP ARCHITECTS
PLANNERS & ENGINEERS

Since 1992, *Liu Thai Ker, M. Arch., Hon. Sc.D.,* has been director of RSP Architects Planners & Engineers Pte Ltd., a 1,000-person consulting firm with 11 overseas offices and projects in 17 countries. He is also the founding chairman of Centre for Liveable Cities (2008). As architect-planner and chief executive officer of Singapore's Housing & Development Board (1969–1989), he oversaw the completion of over a half million dwelling units. As CEO and chief planner of the Urban Redevelopment Authority (1989–1992), he spearheaded the major revision of Singapore's Concept Plan and key direction for heritage conservation. Dr. Liu earned a Bachelor of Architecture from the University of New South Wales and a Master of City Planning from Yale University. He received a Doctor of Science honoris causa from the University of NSW in 1995. A long-standing professor and leader in the cultural arena, he also serves on several governmental bodies in Singapore, and as planning adviser to over 20 Chinese cities. Among his many honors is the prestigious Asian Achievement Award for Outstanding Contributions to Architecture.

刘太格

雅思柏设计事务所

从1992年开始，刘太格就担任雅思柏设计事务所的主管，这是一个有着1000名职员，11个海外办公室，并在17个国家拥有项目的咨询公司。他也是宜居城市研发中心（2008年）的创始主席。作为新加坡建屋发展局的建筑规划师和首席执行官（1969年-1989年），在他的监督指导下建成了超过50万座住宅单元。作为新加坡市区重建局的首席执行官和主要规划者（1989年-1992年），他发起了对于新加坡概念规划的重大修正，并且重新确定了文物保护的关键方向。刘博士在新南威尔士大学获得了建筑学学士学位，并在耶鲁大学获得城市规划方面的硕士学位。1995年，他获得了新南威尔士大学授予的荣誉科学博士学位。除了长期担任教职并作为文化领域的领袖，他也为新加坡许多政府机构服务，同时在超过20个中国城市担任规划顾问。在他的诸多荣誉中有一项是专门颁给在建筑行业做出突出贡献的，著名的亚洲成就奖。

For complete biography, please refer to the Appendix.

完整的传记，请参阅附录。

Bishan Public Housing New Town, developed in the 1980s and 1990s.
璧山公屋新城，开发于上世纪80年代和90年代。

Did the United Nations serve as planning adviser to the government of Singapore shortly after the nation became independent?

Yes. In those days we did not have planning expertise, so we engaged the UNDP to do the first concept plan for Singapore (1968–72). It was really very contextual, not very detailed, just broadly mapping out the urbanized area and the natural area for protection. Singapore has been built based on the plan, with minor adjustment and further detail, of course.

Were Singapore's early new towns located along main transportation arteries?

Definitely. One of the key things that the UN proposed was the metro lines. We have basically followed that proposal, but with refinements and additions. It laid the foundation for our current very effective public transportation system in Singapore.

In addition to early efforts to create urban public housing, was there a parallel effort away from the city?

I came back from the U.S. in 1969 while Singapore's concept plan was still being developed, so I dove straight away into the preparation of new towns. By then, we had already built Queenstown and we were beginning to build Toa Payoh. I got my superior's permission to give definition to what we intended for a new town in terms of population, kinds of facilities, and so on, in order to achieve a high degree of self-sufficiency. That took a year, after which I went on to plan the rest of Toa Payoh and other new towns. I tried to blend the new town planning with the original concept plan.

In the development of new towns, there was hardly any demolition of old buildings of merit. There were very, very few, and that actually gave me a chance, when I became chief planner in 1989, to work out an ambitious plan to conserve a vast number of historical buildings. As chief planner, I gave the Concept Plan its first amendment and upgrading since 1972. I also directed my staff to spend a year combing through all the old buildings in Singapore, and recommended to my minister that some 6,000 be preserved according to internationally recognized criteria. I am happy to say that today we have preserved over 7,000, and there are still as many shophouses [2- or 3-story buildings with retail space on the ground floor and residences above], although not quite as good, existing in Singapore. So we do have a good preservation and conservation record.

在新加坡独立之后的短暂时间内，联合国是否曾经扮演过其规划顾问的角色？

是的，在那些日子里，我们并没有规划专家，因此我们请联合国开发计划署(UNDP)为新加坡做了第一个概念蓝图(1968年－1972年)。它真的非常因地制宜，不是太详细，只是大概勾勒出城市区域和要保护的自然区域。新加坡就是在这个蓝图的基础上建立起来的，当然，我们也做了一些细微的调整和进一步的细化。

新加坡早期的新城镇是不是都在主要交通动脉的附近？

当然，联合国提出的一个关键就是地铁线路问题。我们大致上遵循了那个建议，同时也进行了改进和补充。这一点为新加坡当前非常高效的公共交通系统打下了基础。

除了早期建设城市公共住房项目的努力之外，有没有相应在城市之外做一些工作？

我1969年从美国回到新加坡，那时新加坡的概念蓝图依然处于发展阶段，因此我直接进入到新城镇的准备工作中。那个时候，我们已经建成了昆士城，并且开始建设大巴窑。我得到了上级的允许，依照人口、设施类型等等因素，提出自己关于一座新城镇需要什么的观点，以便能够获得高度的自给自足。这花了一年的时间，在那之后我才继续规划大巴窑的其他部分，以及另外的新城镇。我也尝试着将新城镇的规划与原先的概念蓝图融合在一起，来制定建设一座新城所需的。

在新城镇的开发发展中，几乎没对有价值的旧建筑进行破坏。破坏真的非常非常少，1989年，我成为了总规划师，这实际上给了我一个机会去制定一个雄心勃勃的计划，以保护大量的历史建筑。作为总规划师，我对概念蓝图作出了自1972年以来的第一次修改和升级。我也让我的员工花了一年的时间，将新加坡所有的旧建筑梳理了一遍，并将此推荐给我们的部长，最终根据国际认可的标准，有6000栋建筑得到了保护。我可以很高兴地说，今天我们所保护的建筑超过了7000栋，现在依然有很多的店屋（2层或3层建筑，底下一层用来经营商铺，上面的则用来居住），虽然可能不是那么完好，但依然存在于新加坡。所以我们确实有着非常好的保存和保护记录。

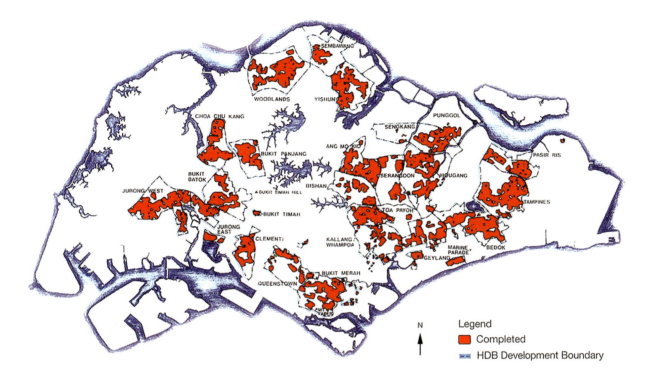

Location of Housing and Development Board (HDB) New Towns. Each has a population of 200,000, and comprises a basic building block of Singapore's Master Plan.
新加坡建屋发展局(HDB)新城的位置。每个新城都有20万人口，包含一个新加坡总体规划的基本构件。

新城镇仅仅是用于居住，还是说它们也包含了一些商业/工作的构成部分?

在几乎所有的新城镇中，只有45%的土地用于住房建设。剩下的则用于道路、公园、体育设施和商业设施，以及8%-10%的清洁工业，这就使很多人不用到社区之外去另找工作。

你的意思是人们可以在新城镇中找到工作，还是说他们依然要到一些中心地区或工厂区工作?

我们在新城镇提供了很多的就业岗位，但不是100%。企业高管和办公室工作人员依然需要往返中心商业区进行工作，但在新城镇中，有为教师、护士、店主和园丁等等提供的工作岗位。而且，10-12%的清洁工业雇佣了许多妇女，她们可以在午饭的时候回家照看自己的孩子。

我并不知道具体的就业率，但我们确实在新城镇中创造了大量的新工作，所以，它当然不仅仅是一个住宅区。通过这种方式，我们帮助缓解了中心商业区的交通拥堵现象。人们可以在新城镇内满

Were the new towns exclusively residential or did they also include commercial/work components?

In almost all new towns, only about 45 percent of the land is used for housing. The rest is used for roads, parks, sports facilities, and commercial facilities, as well as about 8 to 10 percent clean industry so some people don't have to go outside the community to find work.

Are you implying that people can find employment in the new towns, or do they still have to commute to some central area or factory district to work?

We provide a lot of employment in the new towns but not 100 percent. Corporate executives and office workers still have to commute to a central commercial district, but within the new town there is employment for teachers, clinical nurses, shopkeepers, gardeners, etc. Also, the 10 to 12 percent clean industry employs lots of women, who can return home at lunchtime to look after their children's needs.

I don't know the percentage for employment, but certainly we create a lot of new jobs within the new town. It's certainly not just a dormitory community. By doing so, we help to relieve traffic congestion in the central business district. People can meet a lot of their shopping needs within the new town without having to go to the central business district.

In consequence of my year-long study, I established 200,000 to 250,000 as the population of new towns, large enough to attract and sustain supermarkets, department stores, and, of course, lots of shops, as well as sports facilities, and a restaurant large enough to hold weddings, birthday parties, or other celebrations. There are primary schools, which we tried to locate within about a five-minute walk from homes, as well as secondary and technical schools. One hospital might serve several of the new towns, while residents rely more locally on clinics. Of course, this list is not exhaustive. The new towns have many other facilities as well, so it's very comprehensive.

We wanted a critical mass to support all of the facilities so that people could live in new towns and find them self-sufficient. On the other hand, we integrate the town into the overall Concept Plan and, in order to fit them into the rail-bus public transportation system a mass-transit line cuts through the new town. One of the stops is at the town center, where there is a bus depot so people can get off the train and transfer to a bus to different parts of town. The public transportation system is actually quite convenient.

We developed our new towns progressively, meaning that after finishing one, we moved to the next. We started with a neighborhood equipped with schools and shops. If the population was too small to sustain a metro line, we would extend the bus services and build a neighborhood shopping center. This ensured that early residents of new towns would still have essential facilities.

Basically, we divided the island of Singapore into five zones. Each zone would have four to six new towns, with a total population of 1 to 1.2 million; each zone is like a small city. In each zone, we give one new town the most convenient train service system to create a regional center. If people cannot get what they need in the town center, they can always look for it in the regional center before even considering going into the central business district. In fact, following Singapore's example, China is trying to establish regional centers as a kind of back-of-house office development so that companies won't need to bring everyone into the central business district, just the people who really need to be there. Among the many advantages are relieving traffic congestion in the central area, reducing commuting time, and saving energy.

足他们的许多购物需求，而不需要去中心商业区。

基于多年的研究，我为这些新城镇安置了20万到25万的人口，这样的新城镇足够大，也足够吸引和支撑超市、百货公司，当然还有许许多多的商店以及体育设施，一家足以举办婚礼、生日派对和其他庆祝活动的大酒店。那里也有许多小学——我们努力将它们安置在离家大约五分钟的路程之内，以及一些中学和技术学校。会有一家医院可以为许多的新城镇服务，而实际上居民更依赖于当地的诊所。当然，这个清单并不详尽，新城镇还有许多其他的设施，因而它是非常全面综合的。

我们想要占有决定性的多数人来支持所有这些设施，这样人们就可以生活在新城镇中，并且发现它们是自给自足的。另一方面，我们也将这些城镇融合到整个的概念蓝图中，以便它们能够适应铁路——公交车的公共交通系统。一条公共交通线路穿过新城镇，其中的一个停靠站点就是在城镇中心，那里有一个公交站点，所以人们可以从火车上下来然后换乘公交车到城镇的不同地方去。公共交通系统实际上非常便利。

我们以渐进的方式开发新城镇，这意味着，只有在完成了一个之后，才会转向下一个。我们从一个配备了学校和商店的街区开始，如果人口数量太少以至于不能支撑一条地铁线路，我们就将拓展公交车服务，建立一个街区购物中心。这就保证了新城镇的早期居民依然享有基础设施。

大致上，我们将新加坡岛分为五个区域。每个区域将有4-6个新城镇，人口总数将达到100万至120万之多，每个区域都像一个小城市。在区域之内，我们给予一个新城镇最便利的火车服务系统，以便创造出一个区域的中心。如果人们在城镇中心得不到自己想要的东西，在考虑要不要去中心商业区之前，他们总是可以到区域中心去找找看。实际上，依照新加坡的范例，随着一种居家办公方式的发展，中国现在也试图建立区域中心，公司将不需要让所有人都到中心商业街区——除了那些确实需要去的人之外。在这种方案的许多益处中，还包括缓解中心区域的交通拥堵，减少上下班花费在路上的时间，以及节约了能源。

新加坡的新城镇是规划成多种用途还是只作为单一目的的住宅小区？

让我来解释一下。新加坡的建设从接受基本的英国惯例开始，但是在英国，一个典型的新城镇可能有6万到12万人，这对于我们的住房需求来说是不够的；它也不能满足我们想要的那些设施。所以我们将新城镇的人口定在了20万到25万人，并将他们打散，分布在人口为1.5万到2万之间的各个街区，这些街区的人口数足够支撑街区的各种设施。然后我们将这些街区分为面积大约在2到3公顷或5到8英亩的区块。早些时候，在这些区块内，我们安置了一些街角小店。但是居民们表示反对，因为一个区块就是一个非常独立的社区，他们并不希望有外来人进入社区。另外，这些店铺也相对较小，并没有足够的客源，所以我们放弃了这个想法。这些区块现在纯粹就是居民区。但是在街区和城镇中心，我们当然有商业和市政设施的混合搭配。

住宅楼有多高？这些年高度有没有改变？

在20年里，我已经对公共住房做了许多研究。在最初设定新城镇的努力之后，实际上几乎是在同时，我们就意识到需要确定合适的密度。在开始的时候，政府的公共住房政策只是向寮屋居民提供基本的庇护所。然后，大概在1964年到1965年，新加坡政府决定，它不想和英国或中国香港政府一样，不想只是简单地提供租赁屋。相反，它想让大多数的人拥有自己的公寓。实际上，在今天，93%的新加坡人拥有自己的房子，包括那些并非居住在公共房屋中的人。这在整个世界都是最高的拥有比率。在美国，受当前经济衰退的影响，这个比率低于50%。

我们知道在一个新城镇中，需要有一定的密度，以保证有足够的土地用于长远的发展。在那个时候，有大约12位社会学家、博士帮助我研究理想的建筑高度和个人偏好。我们发现，大多数人想要住在第2层，靠近底层但同时有更多的私人空间。因此，我们将底层用于公共用途：婚礼、葬礼、派对等等，甚至于用作老人之家和幼儿园。其次最受欢迎的是第3层到第6层，然后是第7层到第12层。大多数人不愿意住得更高，因为一旦电梯出现故障，爬上爬下是很困难的。为了克服这些问题，我们在电梯机制和监控上做了很多研究。

过去30年或者更长一段时间，新加坡建屋发展局：(HDB)对每部电梯的监控频率接近每五秒钟一次。我们的工程师会在居民之前知道电梯故障，这些故障分为五到六种类型。将一到两种类型结合起来分析，就会告诉我们问题的性质，可能是某种故意损坏，之后我们会优先派出抢修队伍，基本上可以解决全部的问题。公寓的高度现在并不是一个

Were Singapore's new towns planned for mixed use or as single-purpose residential communities?

Let me explain. We began by looking at basic British practice. But in Britain a typical new town might have 60,000 to 120,000 people, which was inadequate for our housing needs; nor could it sustain the facilities we wanted. So we set our new town population at 200,000 to 250,000 people, and broke them down into neighborhoods of about 15,000 to 20,000, a sufficient number to sustain neighborhood facilities. We then broke the neighborhoods into precincts of about 2 to 3 hectares, or 5 to 8 acres. Within each precinct, in the early days, we put in corner shops. But the residents objected because a precinct is quite an exclusive community, and they didn't like outsiders coming in. Besides, the shops were relatively small and they didn't have adequate clientele, so we dropped that idea. The precincts are now purely residential. But in the neighborhood and town centers, of course, we have a mixture of commercial and civic facilities.

How tall were the residential buildings? Has the height changed over the years?

I've done a lot of research in my 20 years in public housing. After the initial effort to define new towns, actually almost simultaneously, we realized that we needed to determine appropriate density. In the beginning, the government's public-housing policy was just to provide basic shelter to the squatters. Then later, around 1964–65, the government decided that, unlike Britain or Hong Kong, it didn't want simply to supply rental housing. Rather, it wanted most of the population to own their apartments. In fact, today, about 93 percent of Singaporeans own their own flats, including the non–public housing population. That's the highest ownership rate in the world. In the U.S., affected by the recent economic downturn, the rate is below 50 percent.

We knew that within a new town, we needed to package a certain density to make sure we would have enough land for the long, long, long haul. At that point, I had about twelve sociologists, Ph.D.s, to help me research ideal building heights and personal preferences. We found that most people wanted to be on the second floor, near the ground but with greater privacy. So we use the ground floor for community purposes: weddings, funerals, parties and so on, even homes for the aged and kindergartens. The next most popular are from the 3rd to 6th floors, then from floors 7 to 12. Most people didn't want to go higher because of the difficulty climbing up and down in the event of elevator breakdowns. To overcome that, we did a lot of research into elevator mechanisms and monitoring.

Tiong Bahru Public Housing Estate by Singapore Improvement Trust, 1950s and earlier.
20世纪50年代及更早的时候，新加坡改良信托局在中峇鲁的公屋。

For the last three decades or more, HDB monitored each elevator almost every five seconds. Our engineer would know about a breakdown before the residents did. There are five or six categories of breakdowns. A combination of one or two categories would tell us the nature of the problem, maybe vandalism of some sort, so we'd send our rescue team on a priority basis, and that reduced the problem nearly totally. The height of apartments is now not a serious issue because people are no longer afraid of elevator breakdowns. I discovered that good maintenance equals good high-rise housing. We put in a huge amount of work to tackle that.

What would you say is comfortable density?

In the new town, the density is about 20,000 to 22,000 persons per square kilometer [51,000 to 57,000 per square mile] which is very, very high. But if you take the whole of

严重的问题，因为人们不再害怕电梯故障。我发现好的电梯维护就等于好的高层住宅。我们做了大量工作来处理这个问题。

你认为什么才是舒适的人口密度？

在新城镇中，人口密度大约是每平方公里2万到2.2万人（51000到57000人每平方英里），这已经非常非常高了。不过，如果你纵观整个新加坡，由于我们有很大的公园和低层建筑等，人口密度大约在7500人每平方公里（2925人每平方英里）。别无选择：我们必须建高层建筑，否则就没有足够的土地为每个人提供住房。

同时对整个新城镇进行规划是非常重要的。所以我们建设高层住宅楼，但利用学校、操场甚至购物中心进行缓冲。换句话说，我们将土地使用的高密度部分和低密度部分结合起来，因此，人们依然可以从他们的建筑中获得开阔的视野。但如果你去上海或者香港，那里没有缓冲，只有从这头到另一头的高层、高密度建筑。我们实际上通过不同的土地使用来平衡不同的楼层高度，因此，当人们来到新加坡的时候，他们并不会感受到高密度建筑带来的压迫感。

新加坡拥有的一个很大优势是气候。作为一个热带国家，每块空闲的土地都可以用于种树和其他绿色植物。

我们确实非常充分地利用了我们的气候，以及我们大片高耸的树木来缓解新加坡的建筑环境。但我不认为温和气候是我们不去创造一个无压迫高密度环境的好理由。这真的只关乎于细心的规划。想象一下如果新加坡没有树，只有足球场和学校操场等等设施。我们依然不会有你在其他高密度城市感受到的、相同程度的压迫感。实际上，当我是首席规划师的时候，我就非常注意改变从街道中某一段到另一段的密度，即使是在非公共住宅区。

所以你不会感到自己好像正在穿越一大片高密度建筑。实际上，我有许多的中国参观者，他们中的一些人是我的客户，一些人是城市官员，在新加坡呆了一两天之后，他们会说，新加坡这么小的岛屿，城市的感觉却比实际的面积要大得多，这令他们感到惊讶。在城市规划中，很多很多的理论，我大量借鉴的一种手法，就是营造一种幻觉。换句话说，我们交替穿插不同的密度，所以你不会产生压迫感。而且，通过极具策略地安排绿色区，即使我们没办法为公园腾出一大块的土地，但你依然产生新加坡是一座花园城市的感觉。在规划新加坡的时候，营造这样一种面积、密度和绿化所产生的幻觉，在我的想法中占据着很大的比重。

新加坡是否已经成为其他国家效仿的模板？

是的，不断有人从非洲、东欧、南美以及亚洲来新加坡参观。实际上，单是中国每年就派遣大约2万人来到新加坡学习我们如何进行城市规划，其中包括高级官员，甚至还有市长。他们也学习其他方面的经验，比如警察、财政和中央公积金体系等。新加坡国立大学、南洋理工大学以及新加坡政府部门会负责安排这些课程。

过去3年我都担任宜居城市中心的主席，这个中心是由国家发展部和环境及水资源部联合成立

Singapore, because we have big parks and low-rise buildings, etc., it's around 7,500 per square kilometer [19,425 per square mile]. There was no alternative; we had to build high-rises, otherwise there would not be enough land to house everybody.

It's very important to plan the whole new town at the same time. So we built tall residential buildings, but used the schools, the playgrounds, and even shopping centers as relief. In other words, we interlocked high- and low-density land parcels so that people can still have a long view from their apartment. Whereas, if you go to Shanghai or Hong Kong, there is no relief, just high-rise, high-density buildings marching from one end to the other. We actually juggle the different building heights by virtue of the different land uses so when people come to Singapore, they do not feel the oppressiveness of high-density buildings.

The one big advantage Singapore has is weather. As a tropical country, every piece of available land can be used for growing trees and greenery.

We definitely make full use of our weather, and our large and tall trees to soften it. But I don't think that is a good enough excuse in a temperate climate for not creating an unoppressive high-density environment. It's really a matter of careful planning. Imagine if Singapore didn't have the trees, but just football fields and school playing fields and so on. We still wouldn't have the same degree of oppressiveness you get in other high-density cities. In fact, when I was chief planner, I took care of varying the densities from one stretch of streets to another, even in the non–public housing area.

So you don't feel as if you're going through a wall of high-density buildings. In fact, I have many Chinese visitors, some of them are my clients, some are city officials, and so on, and after a day or two in Singapore they say they are surprised that for such a small island, the feeling of the city is much larger than actual size. Among urban planning's many, many theories, the one I used—a lot!—is creating a sense of illusion. In other words, we alternate densities so you don't get a sense of oppressiveness. And also, by placing green areas very strategically, you get a sense of a garden city even though we can't spare a huge amount of land for gardens. When I was planning Singapore, creating a sense of the illusion of size, density, and greenness was very much in my thoughts.

Has Singapore become a model for other countries?

Yes. People from Africa, Eastern Europe, South America, and Asia are coming in increasing numbers. In fact, China alone sends something like 20,000 people annually, including high officials, even mayors, to study how we plan. They also study

MacPherson Public Housing Estate, built in the 1960s and 1970s, was among the earliest in Singapore.
麦波申公屋，建于20世纪60年代和70年代，是新加坡最早的公屋。

other aspects, like police, financing, and Central Provident Fund, etc. The classes are conducted by the National University of Singapore, Nanyang Technological University, and Singapore government agencies.

For the past three years I have been chairman of the Center for Livable Cities, which was set up jointly by the Ministry of National Development and the Ministry of the Environment and Water Resources. CLC also conducts courses. Whereas other universities offer classes in a broad spectrum of governance, CLC focuses mainly on urban development issues. In fact, the World Bank, the Asian Development Bank, and also the Urban Land Institute are working out some form of collaboration with us in order to better know Singapore's urban story.

What is the biggest challenge that remains for Singapore?

Once you leave government service, a lot of your ideas are no longer followed, but I wish I could develop the regional centers more comprehensively, having not only offices and

的。宜居城市中心也安排一些课程。尽管其他大学提供更为广泛的政府管理课程，我们中心则主要专注于城市开发问题。实际上，世界银行、亚洲开发银行以及城市土地协会正在与我们展开某种形式的合作，以便宣传推广新加坡的经验。

对于新加坡来说，接下来最大的挑战是什么？

一旦你离开了政府服务部门，你的许多观点都不会被付诸实践，但我希望我可以让区域中心发展得更全面，不只是有办公室和酒店，而且有文化设施。如果我们可以做到这些，就能更好地解决中心地区的交通拥堵问题，同时更进一步加强人们与文化的接触。

说一下你现在在俄罗斯的工作。

我的工作是在俄罗斯联邦鞑靼斯坦共和国首都喀山市进行城市规划——喀山在莫斯科东边，乘坐飞机大约需要一个半小时的时间。他们有一位非常有活力的总统，才53岁，身边尽是一群30到40岁的部长、首相。他们想要把鞑靼斯坦变成另一个新加坡。我正在帮助他们开发信息技术村，以及安置相应的人口来支撑这个地方。

俄罗斯和新加坡在气候条件上非常不同。这对开发发展有什么影响吗？

规划的基本原则可以在世界上任何地方适用。我认为环境因素可以在建筑设计的层面上予以解决，但是为了表明我认真地考虑过这个问题，我让我的工作人员研究了建筑布局、太阳倾角等等要素之间的关系。我们也依赖于鞑靼斯坦的建筑师对我们的指导，因为显然我们对于在冬天如何最大限度地接收太阳光并没有清晰的实践经验。但是我们并未放弃。有趣的是，他们可能并不想让建筑面朝南方，因为他们不愿意在漫长的夏天，整夜整夜地忍受无穷无尽的阳光。我确实尝试着去理解这个问题，以便为总体规划安排恰当的人口密度。不过，严格来说，这并不是一个规划的问题，而是一个建筑设计的问题。

回到早期的住宅设计，标准的公寓面积是多少？

在我为公共住房项目工作的最后一段时间里——即我离开之前的三四年，可能是1984年到1985年——我要求我的工作人员做了一个调查研究。在早些时候，新加坡很穷，所以政府只能负担小公寓。实际上，许多小公寓被用于重新安置。1960年，当新加坡第一次从英国独立出来，我估计大约有130万到190万人住在寮屋中。那就是我们所面临的面积上的挑战，这还不算上贫民窟以及其他一些地方，我们能做的就是建设更小的公寓。

但是在上世纪80年代左右，我感到不能再继续这一趋势了，因为随着财富的增加，人们会希望升级到更大的公寓，然后小公寓将会变得过时而被淘汰。我们可能需要拆掉它们，在新加坡，这是对我们有限资源的一种浪费。所以我们对家庭的生命周期进行研究，并得出结论：三个卧室的公寓，包括一个功能共享的起居室和餐厅，对于家庭的长期使用来说是比较理想的。当家庭还很年轻的时候，父母占据了主卧，儿子和女儿则占用了另外的卧室。当家庭变老的时候，可能孩子们会购买他们自己的公寓，然后祖父母们就可以搬进来。

hotels but also cultural facilities. If we could do that, we could resolve traffic congestion in the central area even better, while simultaneously advancing people's exposure to culture even more.

Tell us about your recent work in Russia.

It's urban planning, in the Republic of Tatarstan. Kazan is the capital, east of Moscow about one and a half hours by plane. They have a very dynamic president, 53 years old, surrounded by ministers and prime ministers all around 30 to 40 years old. And they want to turn Tatarstan into another Singapore. I'm helping them to develop an Information Technology Village, with a corresponding population to support the Village.

Russia and Singapore are climatically very different. What impact does that have on development?

The basic principles of planning can be applied anywhere in the world. I think climate issues can be resolved at the architectural design level, but to show that I take such concerns seriously, I had my staff study the relationship of the building layout, the sun angle, and so on. We rely on the Tatarstan architects for guidance because there is apparently no clear-cut practice on how to maximize exposure to the sun in winter. But we're not giving up. The funny thing is that they may not want buildings to face south, because during the long summer days, they don't want endless sun all night long. I do try to understand the issue in order to establish the appropriate density for the master plan. But this concern is, strictly speaking, not a planning issue but an architectural design issue.

Going back to early housing, what was the standard apartment size?

Toward the end of my stay in public housing, three or four years before I left, maybe 1984–85, I asked my staff to do a study. In the early days, Singapore was so poor that the government could only afford small apartments. In fact, a lot of the small apartments were used for resettlement purposes. In 1960, when we first became independent of the British, I estimate that around 1.3 million out of 1.9 million people lived in squatter huts. That was the size of the challenge we faced. And that's not counting the slums, and so on. All we could do was build smaller apartments.

But around the early 1980s, I felt that we could not continue that trend, because with increased prosperity, people would wish to upgrade to larger apartments, and then the smaller ones would become obsolete. We might have to demolish them, and that is a waste of the limited resources we have in Singapore. So we did a study of family life cycles and concluded that the three-bedroom apartment, including a combined living room and dining room, would be ideal for long-term occupancy. When the family is young, the parents occupy the master bedroom, with sons and daughters in the other bedrooms. When the family gets older, maybe the children will buy their own apartment, and then the grandparents can move in.

This kind of apartment would have a floor area of around 100 square meters net [about 1,100 square feet], encompassing the area from the front door in, not counting the corridors, the elevators, and so on. We started to promote and increase the supply of these four-room flats, so now it's the single largest type of apartment we have. The second-largest is the three-room, in other words, two bedrooms and a living room, about 85 square meters [915 square feet]. And then there is the two-room apartment, 60 square meters [645 square feet]. There is also a five-room apartment, which has three bedrooms, a living room, and a dining room. It's the only one with a separate dining room. We stopped building one-room studio apartments, 36 square meters [390 square feet]. But recently the government realized the need for more one-room apartments, partly for the aging population.

Later, the government felt, quite correctly, that there was a gap between the five-room apartments and the private housing, meaning that there are some people whose income is too high to qualify for a five-room apartment but not high enough to be able to afford private housing. So the government introduced an Executive Apartment to cater to that, so now we have six types.

Does public housing compete with the private sector?

It's not competing for customers, because the government's stand is that whatever the private developers can do, the government should not do. The Singapore government needs the private sector to maintain the working of market forces, so were the government to take over everything, it would have no idea of real market forces, no idea of the real value of apartments. So the government's attitude is to do only what is necessary, filling the gap that the private sector cannot. So in that sense, there has never been any competition. In fact, the government doesn't want that competition. In terms of architectural appearance, yes, they are trying to make it look a little bit more sexy.

Personally, I think that is a misplaced enthusiasm. While we want the buildings to look handsome, that doesn't mean you need to make them look "vibrant and iconic." I don't know what those two words mean. The government spends money on appearance, whereas during my time, the policy was to

这种公寓每层的净面积大概在100平方米左右（大约1100平方英尺），它包含了从前门进入到房间的面积，而不算走廊、电梯等其他部分的面积。之后我们就开始推动和增加四室公寓的供应，所以现在，这是我们所拥有的唯一最大的公寓类型。第二大类型是三室公寓，换句话说，两个卧室和一个起居室，大约在85平方米左右。然后是两室公寓，大约是60平方米。当然也有一些五室公寓，有三个卧室，一个起居室和一个餐室，这是唯一一个有独立餐室的公寓。我们不再建设单间公寓，它一般是36平方米（390平方英尺）。但是现在政府意识到人们对于单间公寓有更多的需求，这一定程度上是因为人口老龄化的缘故。

后来政府感到，在五室公寓与私人豪宅之间存在一个缺口，也就是说，有一些人他们的收入很高，因而没有资格获得一套五室公寓，但他们的收入也没有高到可以买得起一栋私人豪宅。因此，政府引入了服务式公寓来满足这类人的需求。所以现在我们有六种公寓类型。

公共住房能否与私营机构相竞争？

他们不会争夺客源，因为以政府的立场，他不会去做私人开发商可以做的事情。新加坡政府需要私人机构来保持市场力量的持续运作，政府如果接管一切，他便完全不知道真正的市场力量，不知道公寓的真实价值。因此政府的态度是，只去做需要的事情，填上私人机构无法填上的空缺。所以在这个意义上，并不存在任何的竞争。实际上，政府并不想要这样的竞争，是的，从建筑的外形上看，他们确实想要让公共住房显得更迷人一些。

从我个人的角度，我认为这是一种错位的热情。我们确实想要让建筑看起来美观大方，但这不意味着你需要使它们看起来"充满活力且成为标志性建筑"。我不知道那两个词意味着什么。政府现在将钱花在外观上，而在我的时代，政府的政策是把大部分钱用于为人们建造尽可能大的楼房，当然，之后也会用最少的装饰营造出一种好看的外观。以这样的方式，当市民们变得更加富有并想要进行装修的时候，他们可以采取措施，但却不能再增加楼层面积了。因此，我们所关注的是，一开始就给他们尽可能大的楼层面积。我认为这个观念现在并没有被很好地理解。

Town Centre, Ang Mo Kio Public Housing New Town, 1970s and 1980s.
20世纪七八十年代，宏茂桥公屋新城城区中心。

由于对自然通风的需求，新加坡的建筑形式变得更加丰富了。空调系统产生了哪些影响？

自然通风的观念当然还在我们的思想中。建屋发展局总是将自然通风放在非常优先的地位，但并没有完全认识到，自然通风使得住宅区更加绿色环保，同时也降低了空调的使用率。实际上，我们从一项调查中学到了很多东西，这项调查指出，3—5%的人抱怨说，我们的一些公寓"太冷了"。你可以想象得到吗？

　　那是因为我们非常注意自然通风。现在，为了使我们的建筑看起来更迷人，更像私人公寓，对流通风的情况据我所知并不太好。不过，保持建筑自然通风的观念依然存在。

use the bulk of money to create the largest possible floor area for people and then, of course, to create a good-looking exterior with minimum frills. In this way, when citizens become richer and want to decorate, they can do that, but they would not be able to add floor area. So we focused on giving them the largest floor area possible right in the beginning. I think that concept is currently somehow not well understood.

Building forms in Singapore are enriched by the requirement for natural ventilation. What has been the impact of air-conditioning?

Natural ventilation is definitely still in our minds. HDB always put very high premium on natural ventilation, without quite realizing that it made the estates very green and minimized the use of air-conditioning. In fact, we learned from surveys that 3 to 5 percent of the population complained that some of our apartments were "too cold." Can you imagine?

That's because we paid a lot of attention to natural ventilation. Nowadays, in order to make our buildings look sexy and look more like private condominiums, the cross ventilation, I was told, is not as good. But the concept of keeping the buildings naturally ventilated is still there.

What, to your mind, constitutes a reasonable height limit for residential buildings?

We at RFP just did seven blocks of 52-story buildings. Personally, I don't wish to see this replicated too much as public housing, because when a person lives too far from the ground, he loses his sense of community. This is not just my view, but also my sociologists' recommendation to me. We don't want the people living in a new town to feel that they have nothing to do with the ground, as if they're just guests.

Have the Chinese been skeptical about trying to apply the Singapore solution to China?

Well, first of all, they love what they see in Singapore. But now fewer and fewer would say, as they did before, that the situation in China is different. In fact, when the Chinese come to Singapore, I say "Welcome to a truly communist country!" When they go to a new town, they see that we are truly practicing looking after the masses. So the problem is not that they reject our experience. The problem is that their political and administrative system makes it difficult. Singapore has an advantage in that we have a highly disciplined one-tier government. It only formulates strategies; it doesn't interfere with the professional world, whereas in China, there is a three-tier or four-tier government, and of course, many leaders go beyond laying down strategic decisions and interfere with operational issues. It's very hard for talented officials in China to get the job done properly.

The Chinese government has done two things that, I believe, are not only good conceptually but also are very well implemented. One is that most cities must adhere to a density of a million people per 100 square kilometers [39 square miles]. My concern about that is that they tend to apply the same rule to smaller cities, to counties, and so on. If they can vary that a bit, I think this would be an excellent rule. Second, they enforce this policy very diligently. If a local government wants to take over existing farmland, it must first prove to the central government that it has a replacement to create new farmland. These two rules are the big saving grace of China. Where they've not done well is that few of them really understand planning, don't know what constitutes good planning. And they tend to make very arbitrary decisions. This is where we could help.

在你看来，是什么因素构成了一种对住宅楼高度的合理限制？

我们最近在RFP公司刚刚完成了7栋52层的高楼。从个人角度说，我并不希望这种现象在公共住房中被大量复制，因为当一个人住得离地面太远的时候，他就失去了社团感。这不仅仅是我的观点，也是我们的社会学家向我提出的建议。我们并不想让住在新城镇中的人感到，他们仿佛是客人一般，与土地没有任何关联。

中国人对于试图将新加坡的解决方案应用到中国这一点，有没有怀疑？

首先，他们很喜欢在新加坡看到的东西。但是现在，很少有人会像过去那样，说中国的情况是不同的。实际上，当中国人来到新加坡的时候，我向他们说，"欢迎来到真正的共产主义国家！"当他们到一座新城镇去的时候，他们看到我们真的在执行为人民服务的宗旨。因此，问题并不是他们拒斥我们的经验。问题是，他们的政治和管理体制使这些变得困难。新加坡有一个优势就是，我们有高度自律的一级政府。他只进行战略规划，并不干涉实际工作。而在中国，政府分为三级或四级，当然，许多的领导人超出了制定战略决策的职责，转而去干涉操作问题。这让中国一些才华横溢的官员很难合理地去完成工作。

我相信，中国政府已经做的两件事，不仅在概念上是很好的，而且也得到了很好的贯彻实行。一是大多数城市的人口密度必须控制在每100平方公里（39平方英里）100万人。我所关心的是，他们倾向于将同样的规则应用到更小的城市，应用到乡村去。如果他们能够做一些小调整，我认为这将是一个非常出色的规则。第二，他们非常努力地强制推行这一规则。如果一个地方政府想要占用现有农田，它必须首先向中央政府证明它有一块可替代土地能变成新的农田。这两项规则是中国很大的长处。但他们也有做得不足的地方，比如：很少有人真的理解规划，也不知道什么可以构成好的规划。他们总是会做出非常武断的决定，这是我们可以提供帮助的地方。

和欧洲一样，在中国，获奖的项目都是以一场概念竞赛为基础。关于这一问题，你的哲学是什么？

首先，我拒绝参加任何竞赛，特别是在规划工作上。我告诉他们我实际上是一位城市医生。如果我没有看到病人，感受到病人的脉搏，我怎么为

Lineal Park, Bishan Public Housing New Town, 1980s and 1990s.
20世纪八九十年代璧山公屋新城线形公园。

他/她做规划呢？那样我无法去规划了。所以如果你想要举行一场竞赛，去找其他人。如果你需要我，直接找我就好。这个比喻非常令人信服，所以我在中国的每一份工作都是他们直接委托我做的。通过理智的推理，我反对竞赛。

As in Europe, projects in China are all awarded based on conceptual competitions. What is your philosophy regarding that?

First of all, I refuse to participate in any competition, especially in planning work. I tell them that I'm actually an urban doctor. How can I plan for you if I don't look at the patient, feel your pulse, and so on. I wouldn't be able to plan. So if you want to stage a competition, go elsewhere. If you want me, appoint me directly. The metaphor is very, very convincing, so every job I do in China is by direct appointment. By intellectual reasoning, I'm against competition.

Bruno Padovano
FAUUSP

340

Bruno Roberto Padovano, M. Arch, Ph.D., was born in Milan, Italy, and relocated to São Paulo after an eight-year stay in South Africa, where he attended the School of Architecture of Witwatersrand University, in Johannesburg. He received a professional degree in Architecture and Urbanism from Faculdade de Arquitetura e Urbanismo da Universidade de São Paulo (FAUUSP), two master's degrees from the Harvard Graduate School of Design, and a Ph.D. from FAUUSP. An associate professor at FAUUSP since 2007, he is vice head of the Design Department and co-editor of *Projetos* magazine. Mr. Padovano has practiced planning and design in Brazil and other countries, notably China. He has published books and articles on these subjects, and was the curator of the 8th International Biennial of Architecture of São Paulo (2009). He is currently Scientific Coordinator at Nucleo de Pesquisa em Tecnologia da Arquitetura e Urbanisatmo da Universidade de São Paulo, and works as consultant to public agencies and private groups.

布鲁诺·帕多瓦诺
FAUUSP

布鲁诺·帕多瓦诺生于意大利米兰，在南非待了8年之后（在那里他进入了约翰内斯堡的金山大学建筑学院）他迁居到了圣保罗。他在圣保罗大学建筑与城市规划学院(FAUUSP)获得了建筑与城市规划的专业学位，并在哈佛大学设计研究院获得两个硕士学位，然后又在FAUUSP得到了博士学位。从2007年开始他成为了圣保罗大学建筑与城市规划学院的副教授，并且是设计系的副主任，以及《项目》杂志的主编。帕多瓦诺先生在巴西和其他国家，尤其是中国，进行过规划和设计的具体工作。他出版发表了一系列关于这一主题的书籍和文章，他也是圣保罗第八届国际建筑双年展的策展人（2009年）。他现在是圣保罗大学建筑与城市规划技术研究中心的科技协调员，并作为顾问为公共部门和私人团体工作。

For complete biography, please refer to the Appendix.

完整的传记，请参阅附录。

Panorama of the city of São Paulo, 2009.
2009年，圣保罗市全貌。

You have a unique perspective on urbanization resulting from direct experience in both Brazil and China.

The first time I visited China was in 2000, when I participated in an international conference on megacities that was held in Hong Kong by Professor Steven Lau from the University of Hong Kong. I subsequently traveled and lectured in China and actually worked there from 2003 to 2006 in association with the office AMP—Architecture of Metropolitan Post in Hangzhou, in China, and by Internet from São Paulo. I recently returned to China on a consultancy in 2008 for the mainly Muslim city of Yinchuan, in an autonomous region just south of Inner Mongolia (a study for a huge mixed-use complex for the elderly).

One thing about China that took me by surprise was the highly conceptual nature of public competitions. Real numbers were not assigned and everything was very loose. "Where is the program?" I asked. "What do you want me to do?" Except for some very generic guidelines, they said, "We don't know; make suggestions; tell us what we should do." It was a bit difficult to be on a new continent, in a completely different cultural milieu, and be asked to determine what was best for the native people and conditions. I had some trouble with that, but at the same time, it was very interesting because you're very free to do what you want to do. It was a unique professional challenge unlike anything I'd ever experienced before, or since.

How does urbanization compare in China and Brazil?

São Paulo and Hong Kong share a bit of the same rhythm and also the same kind of explosive growth pattern. Hong Kong went through it in a different way but with the same kind of intensity, growing quickly in just a few decades from a relatively small city of maybe 300,000 to 400,000 in the 1960s to over 8 million now. Similarly, when I arrived in São Paulo in the late sixties, the metropolitan region had a population of 8 million; now we're up to almost 20 million, in 40 years. Shenzen in China has displayed very rapid urban growth, too, in the last two decades.

Generally speaking, the Chinese have been paying more attention to urban planning issues, whereas growth in Brazilian cities has been more spontaneous. In the past, public authorities weren't really able to meet the intensified demands of great numbers of people moving from rural areas. So what we call the "peripheral ring" that developed around São Paulo on cheaper land occurred pretty much without government involvement. Settlements—some legal, some illegal—started to develop into a sort of self-built metropolis. An amazing number of people live in that outer ring. In this aspect, São Paulo is quite different from what has gone on in other large

您对城市化的独特见解来自在巴西和中国的亲身经历。

2000年，我第一次访问中国，当时是参加香港大学史蒂文·刘（Steven Lau）教授举办的一个关于大城市的国际性会议。随后，我在中国旅游、讲学，而真正的工作是2003-2006年期间，在圣保罗通过互联网，就职于AMP（美波登都市建设）杭州有限公司。我最近一次回到中国是2008年，当时要为银川市做咨询工作。银川是以穆斯林为主的城市，位于内蒙古南部的一个自治区（即宁夏）内（为一个老年人用的大型混用建筑综合体做研究）。

关于中国，有一件事让我非常吃惊，那就是公开竞争这件事的高度概念化。他们没有划分确切的数字，一切都非常宽松。我问："项目在哪儿？你们想要我做什么？"除了一些非常笼统的条条框框，他们会说："我们不知道；请给出一些建议；告诉我们应该做些什么。"在一个全新的大陆，一个完全不同的文化环境内，人们要你来决定什么才是对当地人和本国国情最好的措施，这有点难度。我对此感到有些困扰，但也非常有意思，因为你很自由，想做什么就做什么。这是我之前从未遇到过的专业挑战。

如何比较中国和巴西的城市化？

圣保罗和香港都经历了同样节奏和同类爆发型的增长。香港的增长方式不太一样，不过速度相差无几：都是从上世纪60年代人口三四十万的小城市发展到今天人口超过800万的大城市。同样的，60年代晚期我到圣保罗的时候，这个大都市市区的人口只有800万；40年后的现在，已将近2000万了。中国的深圳在过去的20年间也经历了相当快速的城市化发展。

一般来说，中国一直都很重视城市规划，而巴西的城市化进程则更加自发性。过去，大批人口从农村转移到城市，公共机构不能真正地满足他们的需求，因此，我们所谓的"边缘地带"（peripheral ring）就在圣保罗周边更便宜的土地上出现了，政府并没有参与这一进程。聚居地——有些合法，有些非法——开始发展成一种自我建设类的大都市，居住在外环的人数令人震惊。在这方面，圣保罗与中国其他的大城市和大都市所经历的相当不同，因为中国一直有更多的政府参与。

Favela do Moinho, São Paulo, Brazil (2011). Slums with adjacent modern high-rises are common in Brazilian cities.
巴西圣保罗，莫英豪贫民窟（2011年）。贫民窟和现代高层建筑在一起，这在巴西城市中很普遍。

巴西的增长模式是什么？

上个世纪以来，巴西的城市增长过程持久而快速，其人口增长了8倍，一个世纪就从2000万增加到了1.6亿。与此同时相对贫穷的农业地区（尤其是东北部各州）有大量人口迁入城市；同时也有很多来自欧洲和亚洲的移民。20世纪60年代，每个妇女平均生育六个孩子，现在这个数字在全国范围内下降到不足两个。未来20年，这种趋势仍将继续，到本世纪30年代，巴西的人口应该稳定在2.3亿左右。这是一个大数目，但相对于这个大陆国家的规模——800万平方公里（310万平方英里）的国土面积而言，它实际上只是个小数字，所以有可能巴西会开始从其他国家输入人口以维持可持续的增长模式。因此，当本国人口正趋向老龄化，且不再繁衍时，欧洲国家也采取了同样的做法：大量引入移民务工；而且俄罗斯显然现在也在这么干。

最初，移民起因于欧亚严重的经济问题。巴西现在正处于上升期，也被人们认为是一片充满机遇

cities and megacities in China, where there has been more governmental participation.

What has been Brazil's growth pattern?

Over the last century, Brazil's urban growth has been very sustained and intense. It increased eightfold, from 20 million people to 160 million people in a century, as people migrated from the poorer agricultural regions (especially from the northeastern states); there was also a lot of immigration from Europe and Asia. In the 1960s the average fertility rate was six children for each woman. Now it's dropped nationally to less than two. The trend will likely continue in the next two decades, and in the 2030s Brazil should stabilize at a population of approximately 230 million. That's a large number, but for the size of the country, continental, with its 8 million square kilometers [3.1 million square miles], it's actually small, so probably Brazil will start bringing in people from other countries to continue a sustainable growth pattern. When its population was aging and not reproducing itself anymore, Europe did the same thing, bringing in immigrants to work, and apparently Russia is now doing the same.

Housing complex that was once a slum, located in the Lemon neighborhood of São Paulo, 2011.
2011年，这个宿舍曾经是个贫民窟，位于圣保罗的柠檬社区。

Originally, immigrants came because of serious economic problems in Europe and Asia. Brazil is now on the upsurge and people are seeing it as a land of opportunity. The people now thinking about coming here are usually more educated than the original immigrants—for example, those who came after the Italian wine crisis. At the turn of nineteenth century, Italy's grapes were blighted by disease, so a lot of people moved to Brazil. They really helped to build this country, and you have to this day a huge population of Italian origin in São Paulo.

This megacity grew due to its location close to the harbor of Santos and the huge coffee commerce at the turn of the twentieth century; development started there and intensified over the last one hundred years. The Tietê River flowing inland made it easy to colonize the interior of the country, and railroads built by the British accelerated an agriculture based on coffee and the implementation of the first factories in the blossoming metropolis. São Paulo was very strategic for the development of this region of Brazil. Basically, it is a city of rings; the further out you go, the poorer and more self-built—with some exceptions, such as the Alphaville-Tamboré high-income suburb, where I live, and new middle-class centralities on its east-west axis, such as Guarulhos (east) and Osasco and Barueri (west).

的土地。如今想来到这里的人，通常比最初的移民受教育程度更高——例如，意大利葡萄酒危机之后的一批人。19世纪初，意大利葡萄饱受疾病摧残，因此很多人迁移到巴西。他们对这个国家的建设确实有很大帮助，直到今天，你依然还可以看到许多圣保罗人拥有意大利血统。

这座大城市的发展得益于靠近桑托斯港口的地理优势，以及19世纪末20世纪初蓬勃的咖啡贸易；发展的进程由此开启并在过去一百年间不断加快。流经国内的提亚特河很容易地就使巴西内陆成为殖民地，英国人建造的铁路也加速了以咖啡为基础的农业发展，并让繁荣的都市有了第一批工厂。圣保罗对于巴西的发展具有非常重要的战略意义，它基本上是一个环形的城市，越往外沿，就越贫穷，自建的聚居地就越多——不过也有一些例外，像我所居住的阿尔伐城(Alphaville-Tamboré)高收入郊区，以及新中产阶级集中的东西轴线：如瓜鲁柳斯(Guarulhos)(东)、奥萨斯库(Osasco)和巴洛艾利(Barueri)(西)。

政府在现代城市发展中扮演什么角色？

巴西的基础设施已有很大改善，政府资金较之以往更为雄厚，因此也倾向于处理过去一直没有解决的问题，例如住房、污水处理、防洪和公共交通等等。以前，人们习惯自己解决问题，而现在，政府显然能够提供更好的帮助。现在有这样一种趋势：棚户区（贫民区）将逐渐消失，自建房屋也将由业主进行改善。边缘地带正变得丰富多彩：人们为自己的房屋添彩加色，购买现代化电器（在巴西，60%的家庭拥有电脑）。这就是所谓的"新中产阶级"。

在上世纪80年代之前，基础设施一般而言还只是为中高阶层中的个体提供便利。之后，政府开始致力于改善道路、街道系统和机场建设。巴西最初开拓了全国铁路系统，但很快，相对于公路，铁路建设就失去了优先权。人们的公共卫生意识日益增强，因而在这方面投入了大量的资金。这是一个持续的过程，对社会发展很重要，显然与较发达的国家相比，我们仍然是落后的，仍有很多事情要去做，但在联邦、州和城市的层面上我们正在做。我认为评估这些项目很有趣，大学也可以参与其中：跟进、评估和监测这些项目，看看它们的有效性如何。此外，军事政权以后，在过去的20年，人们更多地参与并了解自身的权利和需求，也更具组织性。没有社会大众一定程度的参与，什么也做不成，这一点与美国上世纪60年代的情形类似。

巴西如何解决可持续性问题？

我正在写一本书，评价圣保罗市政府一直在做的事情。这本书对可持续性提及较多，但是至于所有这些行动如何协调起来，哪些方面花费了多少钱，如何影响人们的生活质量，还不是非常清楚。这是一个运用多学科进行分析的方法，包括社会科学、工程经济学和公共政策本身，而不仅仅是空间组织（建筑与都市化）。

在不同的部门，大量种类各异的行动都有很好的人才和丰富的资源，但是由于目前的市政组织仍然非常集权，所以他们之间少有交流。例如，工程部也许不与街道部门联系，住房（部门）可能不与公共交通（部门）联系。依照可持续性模式，很重要的一点就是更多地关注这些部门间的协调互动，并尽可能建立一个能够促进信息交叉的机构，以关注可持续性相关的事项。作为一个中立机构，大学能够在这种情况下提供很大的帮助，尤其是

What is the government's role in the development of the modern city?

Brazil's infrastructure has improved greatly and the government has more money than before, so there is a tendency to tackle problems that weren't being addressed in the past, such as housing, sewage treatment, flooding control, and mass transportation, etc. Previously, people became accustomed to solving their own problems, but now the government is in a better position to help. There's going to be a tendency for the favelas (ghettos) to gradually disappear and for the self-built houses to be improved by the owners. The periphery is becoming colorful; people are painting their houses and buying modern electrical appliances (60 percent of households in Brazil own a computer). It's what is being called "the new middle class".

Until the 1980s, infrastructure mostly meant individual mobility for the higher and middle classes. The government was interested in improving roadways, street systems, airports. Brazil had the beginning of a national railway system, but most of it was lost to the priority given to highway systems. Then there was a growing awareness of the need for sanitation, so there was a big investment in that area. It's an ongoing process, fundamental for social development, but we are still lagging behind compared to more developed nations. Much remains to be done, but it is being done, on federal, state, and municipal levels. It is interesting, I think, to assess these programs; that's where the university comes in: following up, evaluating, and monitoring these programs to see how efficient they are. Also, in the last 20 years, after the military regime, people are much more involved and knowledgeable about their rights and needs, much better organized. Nothing is done without some level of community participation, similar to what went on in the U.S. back in the sixties.

How is sustainability addressed in Brazil?

I'm writing a book to assess what the municipality of São Paulo has been doing. It's been talking a lot about sustainability but it's not very clear how all these actions come together, how much is being spent on what, and how this affects people's quality of life. It's a multidisciplinary approach that includes social sciences, engineering economics, and the public policies themselves, not just spatial organization (architecture and urbanism).

There is a lot of uncoordinated action with very good people and much more resources in different sectors, but because of the current municipal organization, still very centralized, they don't always talk to each other. So the engineering department, for example, might not interface with the street department, or housing with mass transit. In view of the sustainability paradigm, it is important to give more attention to these interactions and maybe to create an agency that would promote cross-information and help to focus on sustainable issues. The university, as a neutral body, could be very helpful in this context, especially on state and municipal levels, because we are not directly involved in the actions and maintain a critical view, which can help governments to improve their actions, if they are prepared to listen and involve the university in their programs. I've personally been in meetings to talk with the mayors and explore possibilities, even for some pro bono consulting work. Sometimes they may just want to speak to an expert, a transportation specialist for example, but may not have one in their municipality, so I have tried to make intermediate connections. We are talking about large-scale problems that have to be treated much more collectively.

There's a lot of environmental education going on right now, but, of course, not enough. Compared to what was going on 20 years ago, there's been a tremendous improvement, but much remains to be done. The next step is the coordination for more sustainable development and urban regeneration in the old industrial brownfields downtown, which have been largely abandoned as the megacity changed its economic nature in the last forty or so years (from industrial to services).

A lot of local architects and urbanists are concerned with this issue, working in both the public and private sectors. I would even call it a "sustainability industry" in Brazil. If you're not sustainable, you're out. Owners require it. There's a new green trend and everyone seems interested in it. You might think that private developers might be opposed because of higher costs, but that's not the case. The private sector is extremely interested in sustainability. They run ads saying "our building is more sustainable" and people prefer green urban developments, though they can be misguided, especially with regard to car ownership. A lot of new urban settlements in the outskirts of São Paulo are being sold as sustainable because they adopt recycling (garbage, water, etc.), more green areas and biodiversity on their sites, but in order to live out there, house owners drive more and pollute the air, facing higher living costs and traffic jams. This is my experience in living in Alphaville, where we are now expecting a monorail system to link us up to the metropolitan train and subway systems, thus reducing our dependency on car use.

I don't think sustainability can be obtained fully, but if you make things work in a more sustainable way, it can be very successful in balancing urban development. In the early 1900s Brazil's population was roughly 30 million, 80 percent of which

在州和自治市层面，因为我们并不直接参与行动，却可以坚持重要的观点，如果政府准备听取这些观点并让大学参与到这些方案中来，就能极大改善效率。我甚至做过无偿咨询工作，亲自参加一些会议，与市长交谈，探究各种可能性。有时候，他们可能只想与专家交谈，譬如一位交通专家，但他们的自治市却没有这样的专家，所以我设法帮助他们建立联系。我们现在讨论的是规模较大的问题，处理这些问题必需更加具备集体意识。

目前，很多环境教育正在进行之中，当然这还远远不够。虽然与20年前相比，已有了极大的改善，但要做的事情仍然很多。下一步，是在更可持续的发展与老城区工业棕色地带的复兴之间进行协调，在过去的40年左右，由于大城市经济特性的改变（从工业到服务），大部分老城区的工业地带已经废弃了。

许多本地的建筑师和城市规划者也在关注这个问题，他们都在公共部门和私立部门工作。在巴西，我甚至会称之为一种"可持续性产业"。如果你不是可持续性的，你就出局了。业主要求可持续性，似乎每个人都对这种绿色趋向感兴趣。你可能会认为私人开发商由于较高的成本会反对，但情况并非如此。私立部门对可持续性极其感兴趣。他们投放广告说"我们的建筑更具可持续性"，而人们偏爱环保城市的发展，尽管他们可能会被误导，特别是在拥有汽车的问题上。许多圣保罗市郊的新城市聚居地正以可持续为卖点，因为采用了再循环（垃圾、水等）系统，那些地方有更多的绿地和生物多样性，但是为了住在那里，房屋业主们需要驱车更远，并由此造成空气污染，面临更高的生活成本和交通堵塞。这就是我生活在阿尔伐城(Alphaville)的经历，在那里，我们正期待一个单轨系统来将我们与城市列车和地铁系统联系起来，从而减少对汽车使用的依赖。

我认为可持续性不能完全实现，但是，如果你用一种更可持续的方式让事情运作起来，那么你就可以成功地平衡城市发展带来的问题。20世纪初，巴西的人口大约有3000万，其中有80%在农村，这3000万人现在仍生活在这个国家。但城市人口从1000万猛增到1.5亿。巴西基本上是一个城市化国家，有87%的人居住在城市。中国也正朝着同样的方向迈进，但是还有很长的路要走。另外，农村的环境也不像以前那样落后，人们有电视、互联网，还开着小汽车。这只是整个城市化现实的一部分。如果城市变得更加可持续，国家就会变得可持续。

巴西正处于上升期，但是从许多城市的角度看，圣保罗似乎濒临崩溃的边缘。

圣保罗是当今世界上最不可持续的城市之一，但采取了许多措施进行改变。例如，政府一直尽力改善公共交通。我们的地铁规模很小——地铁线路大约70公里（44英里），考虑到圣保罗的城市规模，这算不上什么——但是我们能够将城市列车、部分地铁路线，变成公共交通系统的一部分，这是令人吃惊的，因为我们建成了260公里（162英里）和将近100个站点。现在，这300公里网络(186英里)很大程度上将城市不同区域联系起来；在私立部门的帮助下，新的地铁线路（包括单轨铁路）正在兴建，每个月都有一个新的交通连接点。本来旧地铁系统的用户正在流失，但是现在，随着新的地铁/城市列车/公交综合系统的出现，载客量飞速增长，人们能够找到私家车的替代物。我的两个儿子也因此搬回到市中心，不再使用小汽车了。

研究表明，生活在圣保罗的人平均每天出行两次，那么2000万人口，每天就有4000万次出行。三分之一的人步行，三分之一的人乘坐小汽车或出租车，三分之一的人乘公共交通。如果条件继续改善，公共交通所占份额将会增加至50%，甚至60%，这对大城市的可持续性将是一种巨大的提升。而且，在未来的十年，如果汽车的消耗从化工燃料过渡到电力，空气质量将会得到极大改善。因为圣保罗的主要污染源是汽车。

我认为，在未来的30年，圣保罗即使不是可持续的，至少也更适宜居住。这取决于这些行动的效率和时间安排，以及整合行动的方式。以自行车道为例，圣保罗是丘陵地带，因此不是一个易于骑车的地方；只有河流附近的低洼地区真正适合骑自行车。城市的各处已建有自行车道，却没有总体规划。还有成本和维护的问题。由于维护不够，许多自行车道散落着垃圾。这些事情必须予以认真对待。

圣保罗附近有38个自治市，构成了这个都会区，但是城市本身不再显著扩大。即使你有一千多人，也可以安置在运输站附近的新混合型住房中，这些地区可以变成带状公园、休闲区。这些已经开始发生，并且更多的改变即将到来。如今待在圣保罗是一个好时机，因为你可以看到一个市容非常糟糕的大城市正处于改善的高潮中。

was rural. Those 30 million are still living in the country. But the urban population has jumped from 10 million to 150 million. With 87 percent of its people living in cities, Brazil is basically an urban nation. China is moving in the same direction but has a long way to go. Also, the rural environment is not as backward as it was before; people have TV, Internet, and drive cars. It's part of the overall urban reality. If the city becomes more sustainable, the nation becomes more sustainable.

Brazil is on an upsurge, but from many urban perspectives São Paulo seems to teeter on the verge of collapse.

It's one of the most unsustainable cities in the world right now, but a lot of changes are taking place. There's been a tremendous effort to improve public transport, for example. Our subway is still small—we have something like 70 kilometers [44 miles] of subway lines, which is nothing, given the size of São Paulo—but we were able to get the metropolitan trains, sort of subway trains, to become part of the public transportation system. This was amazing because we gained 260 kilometers [162 miles] and almost 100 stations. So now this 300-kilometer [186-mile] network is helping tremendously to link different parts of the city; every month there's a new transport connection, and new subway lines (including monorails) are being implemented, with the help of the private sector. The old subway system was losing users, but now, with the new integrated subway/metropolitan train/bus system, ridership is growing very quickly and people can find an alternative to the use of the private car. My two sons, who moved back into downtown, don't use cars anymore.

Studies show that people in São Paulo average two trips per day, so with 20 million people, that's 40 million daily trips. A third are done by foot, a third by car and taxi, and a third by public transportation. If improvements continue, public transport will increase to maybe 50 or even 60 percent. That would be a tremendous improvement toward metropolitan sustainability. And if, in the next decade, cars move from fossil fuels to electricity, air quality would improve tremendously, because cars are the main source of pollution in São Paulo.

I think in the next 30 years São Paulo will be, if not sustainable, at least much more livable. This depends on the efficiency and timing of these actions and the way they are brought together. Take bicycle lanes, for example. São Paulo is very hilly so it's not an easy place to ride a bike; only the low areas near the rivers are really suitable. Bicycle lanes have started to appear here and there but there's no general plan. There's also the question of cost and maintenance. A lot of bicycle lanes are littered with garbage because there's insufficient maintenance. These things have to be looked at carefully.

There are 38 municipalities around São Paulo that constitute the metropolitan region, but the city itself is not growing significantly anymore. Even if you have more than one thousand, these people can be housed in new mixed-use housing developments near the transport stations and these areas can be turned into linear parks, leisure areas. This is already happening but more of it should occur. It's a nice time to be in São Paulo because you get to see a mega-city that was in very bad shape but now is on the upsurge.

How is urban planning handled in Brazil, and São Paulo in particular?

Federal law requires every city with more than 20,000 people to have a master plan. If the mayor doesn't do it, he can be prosecuted. The paradigm is participative, not top-down planning, but listening to people, to commercial and industrial associations, to the different suburbs. In São Paulo—which is very large—the city is divided into subprefeituras or submunicipalities and these have a great deal of local involvement. We have a problem with garbage disposal, so now some of our garbage dumps are being turned into a source of energy, and there are initiatives for saving water. Even though we have a lot of rain in the summer, our drainage system can't handle the runoff. Twenty million people consume a lot of water, so we need more and bring it from very far away. There is a lot to be done with water management.

What is the density of São Paulo?

The general density of the metropolitan region is about 80 persons per hectare [2.5 acres] but there are a lot of open spaces and natural preserves. I believe that the average density is closer to 100 to 150 persons per hectare [2.5 acres] in the urbanized areas, which is pretty high. The densest places are the favelas, which can go up to 1,800 people per hectare [2.5 acres].

What is the tallest building in São Paulo?

We don't really have very tall buildings; the highest is maybe 40 stories.

What accounts for the uniform density?

There are restrictions because of helicopters, but basically it's the FAR coefficients that really determine building heights. Our top FAR is 2.5, which is very low compared to Manhattan's, which permits very tall buildings. I'm very much in favor of higher densities near transit stations. I've

巴西，尤其圣保罗，是如何进行城市规划的？

联邦法律要求每座超过2万人口的城市都要有一个总体规划。如果市长没有做到，他就可能被起诉。这种模式是参与性的，不是自上而下的规划，而是听取人民、工商业协会和不同郊区的声音。圣保罗是一个非常大的城市，这个城市分为自治区(sub-prefeituras)或次级自治市(submunicipality)，有大量的地方参与。我们有垃圾处理问题，因此现在一些垃圾场正被改造成能量来源，同时也有一些节约用水的倡议。虽然夏天雨水很多，但是城市排水系统却不能处理降水径流。2000万人口要消耗大量的水，因此我们需要得更多，也不得不从非常远的地方取水。关于用水管理，还有许多事情要做。

圣保罗的人口密度如何？

都会区的一般密度为每公顷80人左右，但还有很多开放空间和自然保护区。我认为，市区的平均密度每公顷大约100人至150人，这是一个相当高的数字。最密集的地方是棚户区，可以高达每公顷1800人。

圣保罗最高的建筑物是什么？

我们没有真正非常高的建筑物；最高的可能只有40层。

均衡密度的原因是什么？

因为有直升机，所以对建筑会有些限制，但基本上说，真正决定建筑物高度的是容积率(FAR)系数。我们最高的容积率是2.5，与曼哈顿相比，这是一个非常低的数字，而曼哈顿允许建造非常高的大楼。我非常赞成中转站附近的密度可稍高一些。我和我的学生一直在做一些工作，在地铁站附近，我们提供的密度为每平方公顷1,500人，这差不多与香港接近。我们并没有像在其他国家那样租用公共住房，在这里，国家进行补贴，让人们成为自己公寓的主人。

如今，社会住房几乎全部用于居住，没有用于商业或服务。我们一直主张多种用途，差不多应实行70/20/10公式，即：70%的空间用于居住，20%用于商业，10%用于服务。我认为，在公共交通附近，保持高密度并提供社会住房是一种可以尝试的做法。这里不仅有住房，而且有完整的街

The Santo Amaro Station of São Paulo Metro (2010)
圣保罗地铁的圣阿玛鲁站（2010年）

区，离哪儿都很近，因此没有必要使用汽车。

直升机是如何影响这座城市的？

圣保罗拥有世界第三大直升机机队。但由于进出城市不容易，因此直升机也由公共部门使用。例如，市长使用直升机作为交通工具。但基本上，商人，尤其是银行家，使用直升机较多。我认识一位银行家，他乘坐直升机从一栋楼到另外一栋，穿过同一条大街，从一个屋顶到另一个屋顶。为了安全起见，他不会走到街上去。直升机噪音很大，而且没有真正地解决群众的问题。直升机应该真正服务于集体利益，而不是为我们这样密集环境中的私人利益服务，它应该受可持续性的限制，就像小汽车那样（我们一直在讨论越来越多的过路费）。

been doing some work with my students where we give them a 1,500-person-per-hectare [2.5 acres] density near a subway station, moving more or less toward a Hong Kong paradigm. We don't have rented public housing as in other countries. Here, people become owners of their own apartments, subsidized by the state.

Nowadays social housing is practically just 100 percent housing, with no commerce or services. We have been arguing in favor of more mixed use, working more or less with a 70/20/10 formula, which is 70 percent space for housing, 20 percent for commerce, and 10 percent services. To maintain high densities and offer social housing near public transportation, I think, could be a way to go. Not just housing but full neighborhoods, with everything close by, so there's no need for the use of cars.

How do helicopters affect the city?

São Paulo has the third-largest helicopter fleet in the world. It's very difficult to move in the city so they're also used by the public sector. The mayor, for example, uses a helicopter to get around. But basically they're used by businessmen, bankers especially. I know a banker who helicopters from one building to another, across the same avenue, rooftop to rooftop. He doesn't go down to the street for safety reasons. The problem with helicopters is that they are very noisy and don't really solve a mass problem. They should really be used for collective interests, not private interests in a dense environment like ours, and should be subject to sustainable restrictions, just like cars (we've been talking more and more about tolls).

Progress is often prevented by existing regulations. It is difficult to convince the authorities, for example, that more FAR is needed to save the countryside. They have a job to do and follow the book.

You have to rethink the book. If you think about things and discuss them in a new way, maybe new concepts can be introduced.

There was this very important Italian legislator, Norberto Bobbio, who talked about promotional legal systems. Rather than just prohibit, give incentives for positive actions; that could be a major change in law-making. Rather than state what you are not allowed to do, go in the opposite direction. For instance, if you do this, then a higher FAR will be permitted. The Americans were actually the ones who started this. While I was at Harvard in the 1970s, Jonathan Barnett, responsible for the Urban Design office of New York with Mayor Lindsay, was working toward exactly that, a more promotional urban regulatory system.

Another concern is that you may have general planning regulations that are sound enough in concept but that need greater scrutiny in their local applications. This is not to stunt creativity, but to give incentives for more private gains with public benefits and see how much of the law can be used and which parts should be made flexible, because there can be times when the law prohibits good things from occurring (such as more mixed-uses in buildings). So we have an urban design tool called AIUs—Areas for Urban Interventions, a recent part of national urban policy, which allows for various locations to be studied with more care. This is something that has been going on in the States and in Europe for a long time, but in Brazil it's new. Traditionally, you either have that very generic planning law or some very generic urban design, like Brasilia, and there is very little space for these sinterventions. But now they are appearing. For example, we have one in São Paulo in Area Luz (luz means 'light' in Portuguese) because

现行法规常常是进步的阻碍。比如说，很难说服当局，需要更高的容积率来保护乡村。他们通常按照书本行事，完成自己的工作。

你必须反思这本书。如果你思考一些事情，并用一种新的方法讨论，也许可以发现新的概念。

有一位非常重要的意大利立法委员名叫诺伯·博比欧(Norberto Bobbio)，他谈到推广法律制度时认为：与其只是禁止，不如对积极的行为给予奖励。在制定法律的过程中这可能是一个重大的转变：与其规定不允许你做的事情，不如朝着反方向走。比如说，如果你这么做，更高的容积率就会被允许。美国人实际上开始就是这么做的。20世纪70年代我在哈佛的时候，乔纳森·巴尼特(Jonathan Barnett)与市长林赛(Lindsay)一起负责纽约城市设计办公室，他们当时正是朝着一个更容易推广的城市管理体系迈进的。

另外一个值得关注的问题是，你可能有总体的规划章程，这些章程在概念上是足够完善的，但是在本地应用时却需要更多的周密调查。这不是要抑制创新，而是要更多地激励个体在公共好处中有所获得。弄清楚可以运用哪些法律，哪些部分可以弹性化，因为有时候法律会阻碍有益事件的发生（例如，建筑物多功能化）。因此，我们有一个城市设计工具叫作AIUs——城市干预区域(Areas for Urban Interventions)，它是近期全国城市政策的一部分，可以更细心地研究不同的地点。美国和欧洲已经在这方面进展了很长一段时间，但在巴西它却是新的。从传统上看，你要么有非常通用的规划法律，要么有某种非常通用的城市设计，像巴西利亚，在这里，干预的空间非常狭小。但是现在，这些干预出现了。例如，圣保罗的卢斯区(Area Luz)(luz在葡萄牙语中是"电灯"的意思)，因为英国电灯公司曾在那里盛极一时。它位于这个中心城市衰落的一角，卢斯区有一些非常漂亮的旧式公共建筑。因此，现在全市正在研究如何改善这个地区，应该移除哪些建筑物，应该落成哪些新大楼，如何处理所有那些公共区域。由于店主的反对，他们正经历一段艰难的时期，当然，在这种关注中，所有本地的利益都被提出来，但也可能会发生一些有趣的事情。

这就是大学可以起重要作用的地方，因为公共部门有时间和费用的限制。为了解决这样一个问题，可能需要花费六个月以上的时间。与社区打交道通常很费时间和金钱，但是你必须让所有人共同努力以想出一种更民主的城市设计方案。

PLANNING Bruno Padovano

Panorama of the city of São Paulo, 2006.
2006年，圣保罗市全貌。

我认为，现在私营部门和公共部门需要更多的共同协作。你不能撇开公共利益来看待私人利益。西方社会赋予私营部门很多权力，比如：个性、实现你自己的目标、增加你的财富等等。但有时候要以集体的付出作为代价。集体社会以相反的方式遇到类似的问题：他们忘记了人们需要追求自己的梦想，寻求财富、个人和家庭的幸福。因此，你必须用一种平衡的方式将集体和个人结合起来。大家互相帮助，但必须进行充分的交流。另外，这也是大学可以参与的地方，因为你可以把私人和公共部门都请来，让他们聊天和讨论，在大学校园中，通常一切都进行得相当文明。

the British Light electric company used to be active there. It's in a degraded part of the center city that has some very beautiful old public buildings. So now the municipality is studying how the area could be changed, which buildings should be removed, which new buildings should be introduced, what to do with all those public spaces. They are having a tough time because the shopkeepers don't agree, and of course, all the local interests come up in this kind of focus, but some interesting things may happen.

This is where the university can be very effective, because the public sector has limitations of fees and time. To solve such a problem might take more than six months. Dealing with the community often takes a lot of time and is very expensive, but you have to get everyone to work together to come up with a more democratic urban design.

I think private and public sectors nowadays need to work together more and more. You can't look at private interests divorced from public interest. Western society has given a lot of power to the private sector—individuality, achieving your own goals, expanding your fortune—but sometimes at collective expense. Collective societies go through similar problems the other way around; they forget about people needing to follow their dreams, to seek wealth, and personal and family well-being. So you have to merge the collective with the individual in a balanced way. Each one helps the other, but they have to talk to one another—a lot. And again, that's where universities come in, because you can bring in the private and bring in the public and have them chat and talk, and within the university walls, usually everything is rather civilized.

SECTION THREE / FORUM OF EXPERTS

Sustainability

Ajmal Aqtash, FORM-ULA, CORE.FORM-ULA
Dickson Despommier, Vertical Farm
Gregory Kiss, Kiss + Cathcart Architects
Ashok Raiji, ARUP
Richard Register, Ecocity Builders
Maria Sevely, FORM | Proforma

第三篇/专家论坛

可持续性

阿杰马勒·阿科塔什，形—式建筑事务所

迪克森·德波米耶，垂直农场

格里高利·奇斯，奇斯 + 凯斯卡特建筑事务所

阿肖克·瑞吉，奥雅纳公司

理查德·瑞杰斯特，生态城市建设理事会

玛利亚·塞弗莉，FORM | Proforma 设计公司

Ajmal Aqtash
FORM-ULA
CORE.FORM-ULA

Ajmal Aqtash, AIA, is co-founder of form-ula, an award winning multidisciplinary practice that seeks to understand the intersection of design and engineering to produce culturally rich, high-performance architecture. He also co-founded core.form-ula for research and development. Ajmal has an extensive background in morphology. He is assistant director of the Center for Experimental Structures (CES) at Pratt Institute, and is exploring folded structures, particularly on the "AlgoRythms" technology and X-Surf (expanded metal) with Milgo-Bufkin. He previously worked at Skidmore, Ownings & Merrill on projects ranging from supertowers to large complex proposals for airports and universities. Mr. Aqtash worked within the Design, Technical, and Digital Design groups, and founded the Advanced Design Group and the Performative Design Group. In each case his goal was to push advanced design methods, computational techniques, and sustainable design. Mr. Aqtash is an adjunct professor at the Pratt School of Architecture, and has taught and served as juror at other leading universities.

阿杰马勒·阿科塔什

形—式建筑事务所
核心·形—式建筑事务所

阿杰马勒·阿科塔什是美国建筑师协会会员，形—式建筑事务所的创始人之一，一位成功的多学科协作实践者，他想要理解设计和工程的关系以便能创造出富有文化气息、高性能的建筑。他也是公司致力于研究、开发的学术分支"核心.形—式建筑事务所"的创始人之一。阿杰马勒有广泛的形态学背景，他是普拉特学院实验结构中心(CES)的助理主任，他正在同美乐高-巴伏金公司一起探索褶皱结构，特别是"算法"技术和X波浪形多孔金属网。他原先在斯基德莫尔、奥因斯与梅里尔建筑师事务所负责从高层楼房到为机场、大学做大型复杂策划的各种项目。阿科塔什先生在设计、技术和数字技术部门工作，并创立了高级设计部以及功能设计部。在每个领域，他的目标都是推进设计方法、计算技术和可持续性设计。阿科塔什先生是普拉特建筑学院的兼职教授，同时在其他一些顶尖学校教书并担任评定人员。

For complete biography, please refer to the Appendix.
完整的传记，请参阅附录。

"Electric Plant" retrofit proposal, exterior rendering, Los Angeles. (form-ula, 2012)
洛杉矶"电力工厂"加装减排装置的体案，外观（形—式建筑事务所，2012年）

How do you make a high-rise sustainable?

Skyscrapers began with the invention of beams and columns, using steel as the material of choice. Now, with the recent surge of building super tall, steel may no longer be the most efficient means of construction in certain parts of the world. You are able to achieve a great deal with steel, but it has limitations. For example, the construction sequence of erecting a super-tall tower can present problems, specifically when you have to transport the steel to the site, hoist it up, position it in place, and finally secure the steel mechanically, and then repeat this process over and over again.

In the case of Burj Khalifa, Samsung High Rise Construction was charged with the task of erecting the world's tallest building. Instead of implementing typical construction methods, they engineered a concrete mixture that was fluid enough to be mixed on site quickly, pumped up to great heights, and then poured in place. This was the only way for the super-tall tower to be constructed in that region. For Samsung, being innovative and concerned with sustainability was not just how a building behaves during and after construction but how the tower is constructed efficiently.

It also has a great deal to do with the idea of pure mass. As you build taller, you need more mass, in other words, more weight to sustain the greater wind loads. In the case of Burj Khalifa, it's a series of tiers that are configured in plan and gradually move upward. The design is not just purely gestural; its geometry is actually logical. The seemingly random tiers are designed to confuse the wind flow, meaning the geometry is designed to disrupt the wind patterns, which in turn stabilizes the tower. When the wind is flowing in cohesive, successive, repetitive motions, it causes the tower to sway back and forth. The Burj is designed to cut into the wind with its helical strakes at different heights.

Such innovative ideas do not develop within the project schedule alone but over the life span of multiple projects, and are perfected from one project to the next. For example, SOM developed the Buttressed Core scheme of Burj Khalifa from the Bundled Tubes scheme used in the Sears [now the Willis] Tower. Another example is Mies van der Rohe who started prototyping his ideas of a facade on Lakeshore Drive, and later perfected the facade on the Seagram Building. A comparison of the drawings for both projects reveals how the details developed into more compact, more efficient, more refined design. Perfection of ideas depends on a conscious agenda. I believe that within our competitive profession, we must participate in a design discourse that challenges new ways to produce architecture through research and development, and thereby ensure innovative architecture.

你如何让一栋高层建筑具有可持续性？

摩天大楼开始于梁柱体系的发明选择钢作为主材。现在，随着超高层建筑的兴起，在世界的一些地方，钢材可能已不再是最有效的建筑方式了。你能用钢材实现很多，但它是有局限的。例如，竖立起一座超高层建筑的施工步骤可能会呈现一些问题，特别是你需要将钢材运送到（施工）现场，升起它们，放到特定位置，最终以机械的方式固定住钢材，然后继续一遍又一遍地重复这个过程。

例如哈利法塔，三星高层建筑公司受命建设这一世界上最高的建筑。他们没有采用传统的建筑方法，而是采用了一种混凝土混合材料，它足够液态化，可以迅速进行当场搅拌，并用泵输送到极高的地方，然后倒在需要的位置上。这是在那一地区建设超高层建筑的唯一方法。对于三星公司来说，创新和对于可持续性的关注，并不仅仅是一座建筑在建设期间以及在建设之后的表现，还有这一建筑如何有效地被建设完成。这与纯粹体积的观点也有很大的关系。你建得越高，就需要越多的体积，换句话说，需要重量来抵抗更大的风力载荷。就哈利法塔而言，它是一系列的阶梯，层级在规划中就已经设置好了，并且它们是逐步上升的。这种设计并非纯然是装腔作势，其几何学原理是非常合逻辑的。看起来随意的层级堆叠，实际上是设计来搅乱气流的，这意味着几何学被用来扰乱风的型态，相应地让高楼更加稳固。当风以内聚的、连续的、重复的方式流动时，它使得建筑前后摇晃。哈利法塔的设计使它可以利用不同高度上的螺旋侧板减小风的作用。

这些创新的想法，并不是在项目安排之内产生的，而是在多个工程的完整周期内，从一个工程到另一个工程不断完善进步的结果。例如，SOM建筑设计事务所从希尔斯（现在的威利斯）大楼上的管束方案中开发出了适合哈利法塔的核心支撑方案。另一个例子是密斯·凡·德罗，他开始在湖滨公寓勾勒出他的"外观模式"观念，然后在希格拉姆大厦对其进行了完善。对照这两个项目的图纸可以发现，这些细节如何变成更加紧凑、更加有效、更加精妙的设计。想法的完善依赖于有意识的日程安排。我相信在我们这一竞争性的职业中，我们必须参与到一种设计讨论中，即通过研究和进化的方法对新型建筑建造提出挑战，从而也确保了创新建筑的产生。

如何能从过往的传统中跳出来？

我会考虑摩天大楼的文化，这点甚至要先于讨论摩天大楼的经济情况。建立摩天大楼是文化上和经济上都需要的，特别是在大城市——那里建筑空间有限或者往往有强大的政治资本。当前，人们似乎对拥有世界最高建筑很感兴趣。这可能还会继续。

当前摩天大楼的个性化是否给建筑类型赋予了新的生命？

在塔楼设计中，品牌总是非常基础的一个因素。一个客户可能会说："我需要这个，能不能尽量有效地做出来？"这样的机会非常少，所以你要充分利用，但在这个方向上的某个地方，我们失去了一些必要的逻辑来推动摩天大楼。

例如，SOM被邀请参与（并且最终赢得了）韩国首尔超级摩天大楼的竞赛。如果这座高850米的大楼建成，那么它将是当时世界上的最高建筑。客户想要一个企业核心概念，要某种"高并且神奇"的东西。除了避免挡住航空路线，并无其他别的限制，所以问题归根结底就只有一个，如何将多种功能结合到一个高楼中。酒店、居住、商业区和零售区各自有不同的楼层，不同的规划模式，不同的核心筒布局和不同的交通动线，因此，大楼不能被理解成仅仅是在复制的楼层平面中重新规划功能。实际上，先规划出一系列有效的平面布局，然后将它们重叠起来用以发现它们的相似之处。在这样的过程中，你可以开始在设计中就有效地进行发明和创新。在这个例子里，客户要求一栋高层建筑，并且要求建筑师为之进行论证。

你是如何应用可持续性技术的？

洛杉矶是一个相对干热的地方。我们为60年代后期的政府建筑设计出一个方案，这建筑本身包裹在四面都装上相对较小的遮阳玻璃窗的外壳里，建筑的采光因此很糟糕，另外有相对较大尺度的空间需要大型屋顶机械系统进行制冷。我们的建议是，将这些建筑当作正要收获的植物进行翻新，寻找多种产生能量的方式。

我们的第一个干预措施是加入三个大型的挑空空间(voids)，并将其命名为"电力工厂"(EP)。这些挑空空间让我们有机会将室外立面的情况带入到建筑主体的核心部分去。这些空间被整合进去，以产生一种烟囱效应，通过流动的侧风将建筑内的

How can the leap be made from the conventional past?

Before even discussing the economies of skyscrapers, I would consider the culture of skyscrapers. Skyscrapers were brought about by necessity, culturally and economically, especially in major cities with limited building space or strong political capital. Currently, it seems there is a great deal of interest in having the world's tallest tower. This will likely continue.

Has the recent personalization of skyscrapers given the building type new life?

Branding has always been a fundamental within tower design. A client might say, "I need this, and can you also make it as efficient as possible?" Those opportunities are rare, so you take full advantage of them, but somewhere along this trajectory, we lost some of the basic logical necessities that drove skyscrapers.

For example, SOM was invited to participate in (and eventually won) a competition for the Super Lotte Tower, in Seoul, South Korea. If the 2,790-foot [850-meter] tower had been constructed, it would have been one of the tallest in the world. The client wanted a corporate centerpiece, something "tall and mystical." There were no restrictions other than avoiding the flight paths, so it came down to a question of how to stack multiple programs into one tower. Hotel, residential, commercial, and retail all have different floor plates, different planning modules, different core layout, and different circulation paths, so the tower could not be understood as merely reprogramming a repetitive floor plan. Rather, it was taking a series of efficient layouts and overlapping them to discover the specific moments where they share similarities. Within these moments, you can begin to invent and innovate efficiency in design.

How have you applied sustainable technologies?

In Los Angeles, which is relatively hot and dry, we developed a proposal for a late-1960s government building wrapped in ribbons of relatively small tinted windows on all four sides. The building consequently has poor daylighting and relatively large spaces that need to be cooled by large rooftop mechanical systems. Our proposal was to retrofit the building as a harvesting plant and to seek out multiple ways of generating energy.

Our first intervention involved introducing three large voids within what we rebranded as the "Electric Plant" (EP). The voids allowed us to bring exterior facade conditions deep

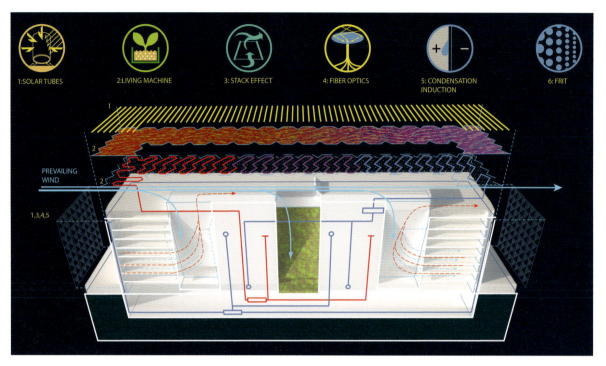

1:SOLAR TUBES   2:LIVING MACHINE   3: STACK EFFECT   4: FIBER OPTICS   5: CONDENSATION INDUCTION   6: FRIT

PREVAILING WIND

1,3,4,5

"Electric Plant" performance section.
"电力工厂"性能部分。

into the center of the mass. The voids were integrated in order to create a stack effect, pulling hot air out of the building by prevailing crosswinds, and drawing cooler air into the building. We were then able to introduce an all-glass facade along the perimeter of the existing building, which is radically different from most conventional facades in the southwest, mainly because of the issue of heat gain.

In addition to the all-glass facade, we proposed the idea of reclaiming the roofscape as a living organism for water retention. In this process, gray and black water is pumped to the roof and pushed through a series of stages / plant species for purification. This water is then pumped back into the building for day-to-day use and gardening.

What kind of maintenance does the living machine require?

Maintenance is minimal, similar to the maintenance required by a typical house garden. Some of the water is absorbed by the plants, but the excess, once it's cleaned, is recycled back into the building. The green roof offers a scenic amenity and a layer of insulation that draws down interior temperatures.

The first step in this process, obviously, is to address the

热空气排放出去，同时将冷空气输送到建筑中。然后，我们可以沿着现存建筑的边界，引入一套全玻璃的外观，这与西南方大多数传统的建筑外观有着根本的差别，主要是因为热增量的问题。

除了全玻璃外观，我们还提出了一个观点，改造利用屋顶空间，把它当作一个保持水资源的有机体。在这个过程中，灰水和黑水被抽到屋顶，然后通过一系列净水流程和特殊植物品种完成水的净化。这种水随后被抽回到建筑中供日常使用和园林浇灌。

这种有生命的机器需要怎样的维护？

这种维护是最低限度的，类似于一个典型室内花园所需要的维护工作。有些水被植物吸收，而那些剩余的部分一旦被净化，便会循环回到建筑中。绿色屋顶不仅是一种对风景的美化，还可降低室内温度的隔离层。

很显然，这个过程中的第一步，处理引入这种系统所需要的管道问题，特别是当现存的管道系统需要重新定向、组织时。我已经花了很多时间研究NASA循环利用水以及过滤脏空气的技术。有一

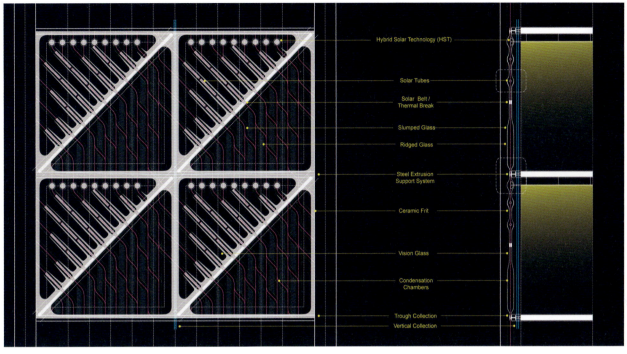

"Electric Plant" partial curtain wall elevation and section.
"电力工厂"局部幕墙的立体结构、横截面。

些特殊的植物，只要把它们置于室内空间之中，就可以自然地净化空气。

太阳能光伏呢?

很多人相信PV（太阳能光伏）板是最有效的收集太阳光能的方式，但是在过去五年中，这一趋向已经缓慢地转向了另外一些方式。其中有一个例子是太阳管，其直径为1到2英寸，长为3到4英尺。这些管依照队列排序安置而产生围幕，围幕允许光穿过，所以如果你的绿色屋顶有一些植物还在土里孕育，光合作用还是能够发生的。从地上反射回来的太阳光也被下方的太阳管吸收，使得它特别高效。当你需要接近底面时，这帷幕可轻松旋转。

它们是否生产热水或电流?

都有。你可以在太阳能光伏管和真空管之间进行选择。可供选择的有三四个不同的类型。在洛杉矶的"电力工厂"中，对热水的需求不如收集能量的

plumbing issues involved in introducing a system like this, especially in a retrofit where existing plumbing needs to be redirected and reorganized. I have spent a great deal of my time looking at NASA technologies for recycling water and filtering dirty air. There are specific plant species that, just by placing them within the interior spaces, clean the air naturally.

What about photovoltaics?

Most people believe PV panels are the most efficient way of harvesting solar energy, but within the past 5 years, the trend has slowly moved toward other means. One example is solar tubes, which are 1 to 2 inches [2.54 to 5.08 centimeters] in diameter and 3 to 4 feet [1 to 1.3 meters] long. The tubes are arranged in rows that are spaced to produce screens. The screens allow light to come through, so if you were to have vegetation underneath for a green roof, photosynthesis can still occur. The solar rays bouncing off the ground also get absorbed by the underside of the solar tube, making it extremely efficient. The screens are easily rotated if you need to access the underside.

"Electric Plant" interior view (model).
"电力工厂"内部结构图（模型）。

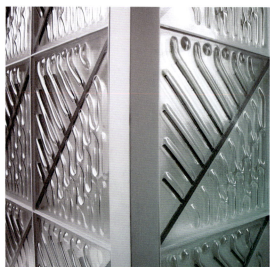

"Electric Plant": exterior glass panels (model).
"电力工厂"外部玻璃层（模型）。

Do they generate hot water or electricity?

Both. You have a choice between photovoltaic tubes and evacuative tubes. There are three or four different types from which you can choose. In the case of the "Electric Plant" in L.A., hot water was not desired as much as harvesting energy. We chose to go with solar tubes, a water generator, and daylight receptors. You can do some interesting things when you co-generate using different technologies together; in fact, we like to hybridize everything we do. When we propose a facade, we're actually inflating the skin and hybridizing multiple technologies into one cross section. We like to think that we're not layering the skin but rather making it wider and fatter; fat is a great insulator.

Do photovoltaic tubes have the same 15 to 20 year life span as photovoltaic panels?

Yes, and the solar tubes are very easy to service because they're standardized. In addition, there are geographical maps available that indicate where certain sustainable technologies would be most effective. Understanding the proper coordinates, the rotational angles in plan and section, can assist in identifying specific technologies for harvesting energy. The horizontal orientation is obviously an ideal condition for positioning a solar collector, but if you were to move to the vertical surface, what would be the optimal angle for collection?

需求来得多。我们选择搭配上太阳管，一种水能发电机，以及日光接收器。当你将不同技术一起使用的时候，你可以做出非常有意思的东西，实际上，我们想将所有东西都混合起来。当我们提出一个立面设计时，实际上是扩张了表层，同时将多种技术混合到一个横截面中去。我们常常认为，我们不是对表面进行分层，而是使它变得更宽更厚，厚度是一种很好的绝缘体。

太阳能光伏管是否与光伏板一样有15到20年的使用寿命？

是的，太阳管非常容易发挥作用，因为它们是标准化的。另外，也有一些可以利用的地图，它们会指出特定的可持续技术将在哪里发挥最大效用。了解了正确的坐标，平面图和结构图中的转动角，可以有助于确定某种特殊的能量收集技术。水平方向很显然是一个放置太阳能集热器的理想位置，但如果你想移到垂直表面上，那什么才是最佳的收集角度呢？

你不仅要理解收集太阳能的几何关系，也需要知道对于一个地区而言最好的技术是什么。而对于太阳管来说，越是靠近赤道，它们的效用就会越低。太多的太阳照射和热量会破坏光伏电池，使它们变得无效。了解一个特定地区的气候，以便更好地确定在一段长时期内，最好地适应这种条件的技术，这一点也是非常重要的。

在开发电力工厂项目的过程中，我们找到了一些问题的答案。在一道传统的幕墙中，你一般会看到一个5英尺的规划模板，它带有一种楼面到楼面高度为10英尺的玻璃面板。在我们的电力工厂设计中，我们将这种玻璃面板最大化，扩展为13英尺的方形面板。你可能会好奇，玻璃如何做的这么大，而且在运输的时候不会损坏。

利用当前在表层建设上的方法（不过这些对于玻璃行业来说却算不上新颖），"热弯"技术已经为设计打开了新的领域。热弯是一种工艺流程：你将钢化玻璃放到陶瓷模型上，同时它通过上面的铜加热线圈暴露在高温之下。当玻璃达到一定温度的时候，玻璃会软化，变成了陶瓷模型的形状，因此，如果陶瓷模型下垂，受热的玻璃也会向下垂，这就产生了一种自然的弯曲，这种弯曲使大块的玻璃更加坚固。通过在玻璃几何学中引入结构折叠，我们可以获得大块的玻璃。

热弯过程也减少了通常在框架设计时，对于金属模型的大量需求，因为玻璃现在可以支撑自身。另外，从相反方向对双层玻璃进行热弯处理的想法，在两层玻璃之间创造了大约6到8英寸的空间，这就产生了一个其他可持续性技术可以占据的小区域。例如，在厚玻璃板的一个区域内，我们放置了太阳管来收集直接和间接的太阳能量。在其他区域，我们放置采用混合太阳能技术(HST)的接收盘以收集自然光。利用这种技术，反射盘吸收阳光，并将其反射到与一堆光纤电缆连在一起的透镜上。这种电缆进入到50英尺内的区域中，而这些区域原来并没有获得自然光的途径。

在EP项目上，我发现最有趣的是如何收集水的问题。一般城市每年有40到50英寸的降雨量，与它们相比，洛杉矶每年的降雨量只有10到20英寸。尽管降雨有限，洛杉矶的空气中还是有可以收集的水分。想象一下你正困在一座岛上，你如何得到新鲜的饮用水？有一个办法是，收集一碗海水，将一张塑料薄膜放在碗上，让阳光照射。最后，挥发作用发生，盐分被分离，剩下的就是新鲜的饮用水。如果你打算采用这个办法，并将这套系统装进建筑的外层，就会产生至关重要的机会。这就是我们想要对电力工厂表层中的下半部分进行挑战的地方。

通过使用一个小型的对流器，建筑的下层部分得以降温。内外环境在温度上的差异足以令玻璃面

You not only must understand the geometric relationships for collecting solar energy, you also need to know the best technology for a given region. With solar tubes, the closer to the equator, the less efficient they become. Too much solar exposure and heat can derogate the PV cells and render them useless. It's also important to understand the climate for a particular region in order to better determine the technology best suited for its conditions in the long term.

We learned the answers to some of these questions while developing the Electric Plant project. In a conventional curtain wall, you're typically looking at a 5-foot [1.5-meter] planning module with a 10-foot [3-meter] floor-to-floor height for glazing panels. In our EP design, we maximized the glazing to 13-foot square [4-meter] panels. You might wonder how glass can be made that big and shipped without breaking.

With recent methods in facade construction (but not new to the glass industry), slumping has opened up new territories for design. Slumping is a process in which you lay treated glass over a ceramic mold that is exposed to high temperatures via a copper heating coil directly above. When the glass reaches certain temperatures, the glass softens and takes the shape of the ceramic mold so that, for instance, where the ceramic mold dips down, the heat-treated glass dips down into the void, causing a natural inflection that makes the large sheet of glass stronger. By introducing structural folds into the geometry of the glass, larger sheets of glass can be achieved.

The slumping process also eliminates the need for the usual large amounts of extruded metals for framing because the glass is able to support itself. In addition, the idea of slumping two sheets of glass in opposite directions creates spacing between the two sheets, roughly 6 to 8 inches [15 to 20 centimeters], producing a pocket for other sustainable technologies to occupy. For example, in one region within the fattened glazed panel, we placed solar tubes to collect direct and indirect solar energy. In another region, we placed Hybrid Solar Technologies (HST) receptor dishes to harvest natural daylight. Utilizing the HST, a reflective dish absorbs daylight rays and redirects them into a lens connected to a bundle of fiber-optic cables. The fiber-optic cable runs 50 feet [15 meters] into areas that otherwise do not have access to natural daylight.

What I found most interesting in the EP project was how to harvest water. Los Angeles gets 10 to 20 inches [25.5 to 51 centimeters] of rainfall annually, compared to other cities, which typically get 40 to 50 inches [102 to 127 centimeters] each year. Despite its limited rainfall, L.A. has moisture in

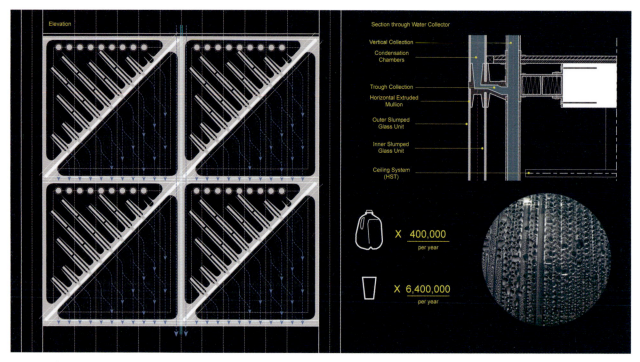

"Electric Plant" partial curtain wall elevation and section through water collector.
"电力工厂"中通过集水器的局部幕墙立体结构和横截面。

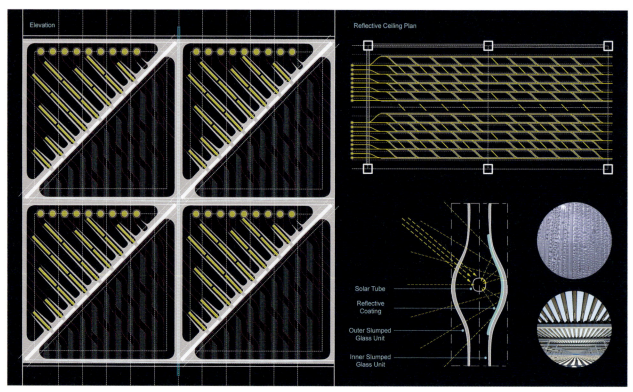

"Electric Plant" light energy: partial curtain wall elevation, reflective ceiling plan, section.
"电力工厂"光能：局部幕墙的立体结构，反光顶棚平面图，横截面。

板内面发生凝结，如此你可以设计出一套系统来收集水分供建筑使用。我们做了一些基本计算，来弄清楚基于建筑的总体面积，可以收集到多少水分：每年可能会产生4000加仑或640万杯的饮用水。

在电力工厂的例子中，我们想要对玻璃面板进行挑战。表层并不只是一种特殊阳光收集方式，而是涉及到大量收集阳光，同时将它们集中起来，重新导向以供建筑使用。混合型表层对于我们来说是一种方法，有助于探寻崭新的水收集方式，而这一点对于这个地方和这个项目是再合适不过了。它干净、简单、出色，并且都包含在一种新的建筑经验形式中。

这建筑就像一个活的有机体。

这是设计理念。电力工厂表层最好的地方是从系统内产生出水来。另外，由于凝结发生在表层之内，系统也运作得如同一个绝缘体。由于水从内部表面产生出来，房间内的热量被凝结效果所吸收。这种水最终排入位于玻璃面板底部的一个水槽中。因此，随着水被排出，热量也一起被排出。唯一的问题是让客户相信，建成这些相对而言并不昂贵。

除了玻璃，这是一个非常昂贵的方案，因为它融合了多种技术。

是的。与玻璃相比，我更担心技术部件，因为你必须说服可持续元件的制造商定制他们的产品以达到设计的要求。

有了光纤对主要光源的补充，你们能否增加玻璃到大楼核心的净距离？

并不需要。我相信技术会被理解为一种提高空间质量的方式——例如，在自然光受到限制的环境中就是如此。

通过光纤传输的光，其质量非常近似于自然光，因为此种技术只是简单地改变了光源的方向。大多数人造光并不具有和自然光相同的质量，这就是为什么一些光源让人疲乏的原因。

推动我们设计实践的一个根本问题是："如何将居住者和外在世界之间的空间最小化？"在我看来，这一空间拉得越开，居住者的境况就变得越糟糕。就电力工厂项目来说，我最爱的是，我们正在推进居住者与外在环境之间的空间关系，这一点在大多数例子中都很难达到。弗兰克·劳埃德·赖特(Frank Lloyd Wright)理解与自然联系的重要性，

the air that can be harvested. Imagine that you are stranded on an island; how do you get fresh drinking water? One way is to gather a bowl of seawater, place a plastic membrane over the bowl, and allow for solar exposure. In time, evaporation occurs, so that the salt is separated out, leaving fresh potable water. If you were to take this idea and package this system within a facade, radical opportunities can happen. This is where we wanted to challenge the lower half of the EP facade.

The lower chamber is cooled utilizing a small convector. The temperature difference between inside and outside conditions is enough for condensation to occur on the interior surface of the glazing panel, from which you can devise a system to collect water for building use. We ran some basic calculations to understand how much water can be harvested based on the total surface area of the building: potentially 4,000 gallons [15,160 liters] or 6.4 million glasses of fresh drinking water can be produced per year.

In the case of the EP, we wanted to challenge the glazing panel. The skin was not just about one particular type of ray to collect, but rather looking at the vast collection of rays that can be harvested and redirected for building use. Hybridizing the skin was a way for us to explore new ways of harvesting water, which was appropriate for this site and project. It's clean, simple, and elegant, all packaged within a new type of architectural experience.

The building is like a living organism.

That is the idea. The best part of the EP facade is the production of water from within the system. Additionally, as condensation occurs within the facade, the system also behaves like an insulator. As the water is generated on the inside surface, heat within the chamber is being absorbed by the condensation. This water eventually drains into a trough located at the base of each glazing panel. Thus, as the water is drained out, so too is the heat. The only problem is convincing clients that it's relatively inexpensive to build.

Glass aside, this is an expensive proposition because it incorporates multiple integral technologies.

Correct. I'd be more worried about the technology component than the glass because you have to convince the manufacturers of the sustainable componentry to customize their products to meet design intent.

With fiber optics supplementing the main source of light, are you able to increase the net glass-to-core distance?

Not necessarily. I believe the technology should be understood as a way to enhance the quality of the space—for example, in conditions where natural daylight is limited.

The quality of light through fiber optics is very similar to that of natural daylight because the technology simply redirects the light source. Most artificial lights do not have the same quality as natural daylight, which is why some light sources cause fatigue.

Our design practice is driven by the fundamental question: "How can we minimize the space between the occupant and the outside world?" In my opinion, the further the separation, the worse the condition for occupants becomes. What I love most about the EP project is that we are forcing the issue of the occupant's spatial relationship with the outside environment; that is difficult to achieve in most cases. Frank Lloyd Wright understood the importance of connecting with nature, and Buckminster Fuller talked about the synergy between humans and nature. I openly embrace this relationship, but I would also like to seek out new and interesting spatial opportunities that garner a different kind of experience. The notion of minimizing, compressing, and fattening would be some of the things that can open up new worlds.

What is the visual experience of looking out from within the EP project?

Because of the sheer size of the glass panels, you might look past the technological components and see the world in a new way, similar to looking through filters that reveal a different shade of nature. For some that may be exciting; for others, it might not be acceptable. I arrived at this point through my reexamination of Mies van der Rohe and his fascination with transparent glass skyscrapers. That idea has been flipped, redirected, and reinjected into the cross section of the glass skin for the EP project. If you were to replicate the EP skin within a vertical tower, it would be just glass, where you would read only the core and floor plates. That becomes the formula, not the silhouette.

What I am proposing is to completely wrap a tower with only glass. It has its moments of being 100 percent transparent and moments of being opaque. With enough time, prototyping, and some value engineering, there is no reason why there couldn't also be a clear vision band for greater balance. This is a radical challenge to conventional thinking. If we're going to move into the future and have a dialogue with engineers and manufacturers, we have to think unconventionally, think about the things they haven't done. Glass is a delicate material within the architecture and engineering professions, refined over hundreds of years. I am very sensitive to that

巴克敏斯特·富勒(Buckminster Fuller)也谈到了人与自然的协同作用。我公开欢迎这种关系，但我也想要寻找新的、有趣的空间关系，从而能够获得一种不同的经验形式。最小化、压缩和厚重的观点将是一些可以打开崭新世界的东西。

从这个电力工厂项目向外看，会有什么样的视觉经验？

由于玻璃面板的绝对面积，你可以透过技术元件，以一种新的方式看待世界，类似于通过滤镜进行观看，这种滤镜展现出不同的自然影像。对有些人来说，这可能是非常让人激动的一件事，对于另外一些人来说，它可能是不被接受的。我通过重新审视密斯·凡·德罗和他对透明玻璃摩天大楼的痴迷而了解到这些。这种观点已经被激发，并重新定向，重新注入到电力工厂项目的玻璃外层横截面中。如果你想在一栋垂直高楼中复制电力工厂的表层，那只能是采用玻璃，通过它可以看清核心和楼面板。这已经成为一种惯例而不仅仅是做做样子。

我的建议是，只用玻璃来完全包装一座大楼。它既有100%透明的时候，也有模糊不透明的时候。有了足够的时间、原型设计和一些价值工程学知识，它没有理由不成为一个清晰而又更加平衡的视觉带。这对于传统思维是一个根本性的挑战。如果我们打算进入未来，并且与工程师、制造商展开对话，我们必须不循规蹈矩地进行思考，想想那些我们尚未做过的事情。在建筑和工程行业，玻璃是一种精致材料，数百年来都不断进行改进。我对于这种文化极有感触，但我的兴趣在于改进和利用玻璃。从我的建筑生涯一开始，即在SOM的时候，这种想法就已经开始酝酿了。我那时已经通过了SOM的新手训练营，也就是说，理解了如何设计一个完美的玻璃盒。我感兴趣的是修改系统，而不是破坏系统，我也想去理解那些失败和断裂点。之后我才能够稍稍对这一系统重新做出调整，以便引入一种新模式。我认为继承密斯的精神，建造一座全玻璃的建筑会非常有趣。

建筑这一职业的本质越来越紧密地与工程联系在一起。

为了进步，有许多东西需要你去理解，特别是与可持续性相关的东西。我不喜欢可持续性这个词，我更喜欢用平衡这个词，它更直接也更清晰，反映了我们现在的位置以及我们将来想要去的位置。

"Electric Plant" rendered interior 11:00 a.m.
"电力工厂"上午11点时内部的样貌。

"Electric Plant" rendered interior 5:00 p.m.
"电力工厂"下午5点时内部的样貌。

culture, yet I am interested in furthering the implementation of glass. These ideas have been incubating since the first part of my architecture career, at SOM. I have been through the SOM Boot Camp, so to speak, and understand how to design a flawless glass box. I am interested in tweaking, not undermining, the system, to understand the failures or breaking points. Only then can I recalibrate the system slightly to introduce a new mode. I thought it would be interesting to do an all-glass tower in the spirit of Mies.

The nature of the architectural profession has become closely integrated with engineering.

To be progressive, there are things you need to understand, especially relative to sustainability. I dislike the word sustainability. I prefer to use the word balance, which is a more direct and clear term that reflects where we are and where we want to be in the future.

Is net-zero building achievable?

I would like to say yes, but I cannot be certain. The issue is that after completion, the building should be monitored and tracked over the course of two to three years. Once the data is collected, organized, and analyzed, net zero can be determined. The different types of analysis required include energy modeling, solar radiation modeling, and computational fluid dynamics modeling. Volumetric modeling is more important than surface modeling, because this is where you can identify energy exchange, heat gain, and heat loss.

For Pearl River Tower, "net zero" was determined mainly by energy modeling. Energy modeling can compute how much energy is consumed in maintaining heating and cooling conditions, among other things. Pearl River utilized many sustainable technologies, but the building was tested with an assortment of analytical software in advance. For example, if I design a facade, I would first understand the total solar radiation the tower will be exposed to over the course of a year. I can analyze the extent to which the facade can be shaded through solar modeling. Ultimately, energy modeling is going to tell me the ratio of opacity to transparency for the facade. This information informs the early stages of the design process. What I am proposing is that testing should be done not only before the building is constructed but also after the building is completed and fully occupied.

Achieving a net-zero building has also a great deal to do with the location. The climate can change and shift certain design strategies, and at times can render them useless. The design team and engineering team need to work very closely with all of the building components to understand how they

零能耗建筑能被实现吗?

我想说是的，但我不能确定。问题是，在施工完成之后，建筑应当被监视和跟踪两到三年时间。只有完成数据的收集、组合、分析，零能耗才能被确定。不同的分析类型需要包括能量建模、太阳能光照建模、计算流体动力学建模。体积建模要比表面建模重要很多，因为这是你能确定能量交换、热量获得和热量损失的地方。

对于珠江大厦来说，"零能耗"主要由能源建模决定。能源建模可以计算出在保持加热和冷却的条件下，需要耗费多少能源。珠江大厦采用了许多可持续性技术，但是该建筑提前就通过一套分析软件进行了测试。例如，如果我设计一个建筑表皮，我首先会了解建筑在一年的时间中将会收到的太阳光照总量。通过阳光建模，我能够分析表皮可遮蔽的范围。最后，能量建模将告诉我表层中不透明度与透明度之比。这些信息会在设计过程的早期阶段给出来。我想说的是，应该做测试，不仅是在建筑建设之前，而且在建筑完工并且正式使用之后。获得一栋零能耗建筑也与地点有着极大的关系。气候可以改变、转化特定的设计策略，甚至有时会让这些策略失去效用。设计团队和工程团队需要非常仔细地利用所有这些建筑要素，以便理解他们如何在特定位置被实施。

你会不会预测有这么一个时候，那时建筑将会定期更换表层?

很有趣的是，政治在建筑中扮演着这个角色。奥巴马政府就极力主张可持续性技术。奥巴马总统不是通过石油钻井，而是借助于现有的、崭新的技术，通过推动其他手段来获取能源。他现已连任成功，我们可以看到他会颁布一项命令，要求改造现有建筑以达到政府颁布的能源标准，而这将开启一系列对现有建筑的整修改造工作。当然也包括重新更换建筑表皮，使它运行得更好。

这些能源法案——在过去没有受到应有的重视——有可能要求所有建筑达到新的节能标准。要求在两到三年时间内监控建筑表现的技术，将会得到实施，各种分数决定了到底是对建筑进行彻底改造还是小修小补。在那之后，改造过的建筑将会定期进行测试，就像服从汽车年检的规定一样。

主要的大学已经开始修复进程了，因为他们知道能源法案马上就要颁布了。也有一些成为第三方托管的工程小组，他们会对建筑改造进行评估，然后提前给出总的费用。在这些评估中，在尚未付费的情况下，他们会帮助指出所有需要改造的地方。在5到10年的时间中，学校从改造中节约的钱会变成工程费，因此，这项服务是靠自己支付的。

学校里教授性能建模吗？

并不像人们期待的那样多。我现在就在普拉特学院建筑系教书。我所教授的课程中，有一门叫作"样板性能"，其中涵盖了性能建模的许多方面，包括太阳光照建模、能量建模、计算流体动力学建模。这个课程倡导可持续设计的主题，以及一些更具体的，比如高性能设计策略如何融入到建筑设计的过程中去。在这个框架内，我利用我关于幕墙的知识，加上性能建模，开发出了更先进的建筑表皮。

can be implemented for that specific location.

Do you foresee a time when buildings will be routinely reskinned?

Interestingly, politics play a role in architecture. The Obama administration is a huge advocate of sustainable technologies. Instead of drilling for oil, President Obama promotes other means to harvest energy via existing and new technologies. Now that he has been re-elected, we could see him roll out a mandate for retrofitting all existing buildings to meet government-issued energy benchmarks, which will open up an assortment of retrofits for existing buildings. This would also include reskinning buildings to perform at higher thresholds.

This energy bill, which has not yet received proper attention, can potentially require every building to meet new energy efficiency standards. The technologies required to monitor building performance over two to three years will be implemented, the various scores determining whether a major or minor overhaul is required. Thereafter retrofitted buildings will be tested periodically for compliance much the way automobiles are annually inspected.

Major universities have already started the remediation process because they know the energy bill is coming. There are engineering groups called escrows that will assume the up-front costs in evaluating a portfolio of buildings. From this evaluation, they will make all of the necessary retrofits, still without payment. Over the course of 5 to 10 years, the money the school saves from the retrofit goes towards the engineering fees, so the service pays for itself.

Is performance modeling taught in schools?

Not as much as one would hope. I am currently teaching at Pratt Institute's School of Architecture. Among the courses I teach is "Prototyping Performance," which covers many aspects of performance modeling, including solar radiation modeling, energy modeling, and computational fluid dynamic modeling. The course addresses sustainable design issues, and more specifically how high-performance design strategies are integrated into the architectural design process. Within this framework, I am taking my knowledge of curtain walls and overlaying it with performative modeling to develop advanced skins.

Dickson Despommier
Professor Emeritus at
Columbia University

Dickson Despommier has a doctorate in microbiology from the University of Notre Dame. After 27 years of laboratory-based biomedical research at Columbia University with NIH sponsorship, Dr. Despommier, now professor emeritus, is engaged in a mission to produce significant amounts of food crops in tall buildings situated in densely populated urban centers. His recent book, *The Vertical Farm: Feeding the World in the 21st Century*, and website (www.verticalfarm.com) have stimulated progressive planners and developers to envision vertical farms in future cities. Six are already operational; many more are planned. Professor Despommier is the recipient of eight Best Teacher of the Year awards and of the national American Medical Student Golden Apple Award for Teaching Excellence (2003). Extensively published on a variety of subjects, he has delivered numerous lectures about vertical farming at leading universities, government agencies, professional conferences, and organizations, and frequently appears on nonprofit television devoted to ideas worth spreading.

迪克森·德波米耶
哥伦比亚大学教授

迪克森·德波米耶在圣母大学获得了微生物学博士学位。在哥伦比亚大学进行了27年由国立卫生研究院资助的基于实验室的生物医学研究之后，现在荣誉退休，投入到一项在密集分布于城市中心区域的高层建筑上生产大量粮食作物的工作中去。他最新的著作《垂直农场：在21世纪为世界提供粮食》以及网站(www.verticalfarm.com)，都对一些前沿的规划者和开发商产生了刺激，让他们开始想象将来城市中的垂直农场。目前，已经有6座垂直农场在运作了，更多的还在规划中。德波米耶教授是8位获得年度最佳教师奖项中的一员，也因出色的教学获得了美国医学学生金苹果奖（2003年）。他就很多主题发表了大量作品，在许多顶尖大学、政府部门、专业会议和组织做过关于垂直农场的讲座，也经常出现在非营利性电视台中，贡献一些值得推广的观点。

For complete biography, please refer to the Appendix.
完整的传记，请参阅附录。

Sky Farm proposal for Chicago. (Wonwoo Park, 2010)
芝加哥的空中农场方案。（朴勇友，2010年）

What do you say to pragmatic architects who see vertical farming as just some romantic notion?

There are now many functional vertical farms up and running throughout the world: Korea (a 3-story experimental VF), Singapore (a 2-story prototype), Chicago, (a 3-story retrofitted VF), England (an 8-story VF in a retrofitted office building), and a 17-story commercial VF being constructed in Sweden by the Plantagon Corporation. PlantLab in Holland has plans for a 3-story VF that will be built underground.

What is the definition of a vertical farm?

It has to be more than 2 stories. That's my definition of vertical.

How did your interest in vertical farming begin?

My passion is for ecology, for the natural world, because I lost a trout stream when I was about 16 years old. They built right down to the bank of the stream and put in storm sewers and completely destroyed everything that could possibly live there, except people. I determined then to become a biologist, I just didn't know what kind. After college, I got a job in a hospital doing research on malaria in a parasitology lab, and went on to get a Ph.D.

Three years later, I got a scholarship to study at Rockefeller University. That's where I met René Dubos, the great microbiologist and Pulitzer Prize–winning author of *So Human an Animal*. The book is all about why people resist change in their own world. This wonderful man, who coined the phrase "Think globally, act locally," profoundly influenced my life, but I didn't know it for another 20 years.

I was hired by Columbia University and worked in a laboratory on Trichinella spiralis, the worm you get when you eat raw pork. Then one day in 1999, a letter arrived withdrawing the grant that had supported my work for 27 years. That was the year of the West Nile virus, so I started paying attention to the ecology of that, and started teaching more. One of my courses was Medical Ecology, a term which, by the way, was introduced by René Dubos. He meant that good things come from the Earth, if you pay attention. Dubos actually discovered the world's first antibiotic, gramicidin, which is produced by a soil bacterium. He was epic, so wonderful, so caring, so nurturing, and so well thought out you couldn't ignore his work.

My medical ecology class was basically what goes on in the stratosphere. On summer days like this there is too much ozone on the surface, for instance, so asthmatics have a real problem. We do this to ourselves, with or without

你会对认为垂直农场只是浪漫的概念的使用主义建筑师说些什么？

现在，全球已经有很多实用有效的垂直农场投入运作：韩国（一栋三层的实验垂直农场）、新加坡（一座两层楼的模型）、芝加哥（一座三层，改装过的垂直农场）、英国（在改装的办公楼内有一座八层高的垂直农场），以及 Plantagon 公司在瑞典建设的一座高达17层楼的商业垂直农场。荷兰的"植物实验室"团队，则打算在地下建造一座三层高的垂直农场。

垂直农场是如何定义的？

它必须超过两层，这是我对垂直的定义。

你是如何开始对垂直农场产生兴趣的？

我对生态学和自然世界充满热情。因为16岁的时候，我失去了一群鲑鱼。人们在河岸正下方施工，置入雨水管，最后完全毁灭了可能生活在那里的一切——除了人类。从那以后，我决心成为一名生物学家，但我并不知道要成为哪种类型的生物学家。大学毕业后，我在医院找了份工作，在寄生虫实验室从事疟疾研究，然后继续求学，获得了博士学位。

三年后，我得到了一份奖学金，资助我在洛克菲勒大学学习。在那里，我遇到了瑞尼·杜波斯（René Taliesin Dubos），伟大的微生物学家，他是《因而人是动物》一书的作者，并因这本书获得了普利策奖。该书讨论了为什么人在自身世界中拒绝改变。这位出色的作家，提出了"以全球视野进行思考，立足本土展开行动"的警句，他深刻影响了我的生活，但我在过去的20年时间里却没有意识到这一点。

我受雇于哥伦比亚大学，在一个研究旋毛虫（当你吃生猪肉的时候，就会感染它）的实验室工作。然后在1999年的某一天，我收到一封信，上面宣布撤销已经支持我工作长达27年之久的基金，那是西尼罗病毒猖獗的一年。因此我开始专注于生态学，并开始做更多的教学工作。我的一个课程是医疗生态学，顺便说一下，这个提法是由瑞尼·杜波斯介绍的。他的意思是：如果你留心就会发现，好的东西都来自于土地。杜波斯实际上发现了世界上第一种多肽抗生素，短杆菌肽，它是从土壤微生物中生产出来的。杜波斯是一个传奇，如此

Sky Farm detail within building skin (Wonwoo Park, 2010).
空中农场细节建筑表皮内。（朴勇友，2010年）

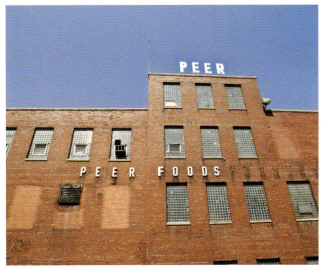

"The Plant:" industrial building converted to urban farm, Chicago.
植物转化为城市农场，芝加哥的工业大厦。

善良，如此关怀，如此扶持，如此深思熟虑以至于你绝不能忽略他的工作。

　　我的医疗生态学课程大体上关注的是发生在对流层的各种情况。例如，像今天这样的夏日，大气表层有太多的臭氧，就会引发哮喘病。不管有没有意识到，这都是我们自己造成的，它取决于是否控制像燃煤发电厂这样的产业。问题是，我的学生在课程进展到一半时，就已经厌倦听这些东西了，他们想先做些其他的事情。我说："好的，我将做你们的导师，支持你们并推荐一些资源，但是你们要自己完成所有的工作。"过了一周他们回来，提出了一种屋顶花园计划：在那，人们可以种植一些食用的蔬果。我将他们带到地图室——那时我们还没有谷歌地图（Google Earth）——他们坐下，手里拿着尺子，在曼哈顿街区的地图上计算着面积大小。

　　在那一天结束时，他们发现，如果在所有的屋顶都种植稻米，这只能为2%的曼哈顿人提供足够的食物，学生们对结果感到很沮丧。我说："不要灰心失望，你们已经做的很棒了，只是得出了错误的答案。你们应该得到的回答是：不可能从屋顶上收获足够供给整个城市的粮食。"我问道："如果换一个问题，会发生什么呢？比如将你们的想法应用在布朗克斯（Bronx）那些被遗弃的楼房里，会发生什么呢？"他们看着我大喊：对呀！

　　很多学生听说了这个项目，第二年登记参加人数增加到12人，然后是30人，40人。我在自己

knowing, depending on whether you control an industry, like coal-burning power plants. The point is that my students got sick of hearing it about halfway through; they wanted to work on something else. I said, "Alright. I'll be your mentor. I'll support you and recommend resources, but you'll do all the work." They came back the next week and proposed rooftop gardening, plants you can eat. I sent them to the map room—we didn't have Google Earth in those days—and they sat with their rulers and calculated the square footage from Block & Lot maps of Manhattan.

At the end of the day they found that if they planted one crop of rice on all the rooftops, they could raise enough food for 2 percent of Manhattan. They were despondent with the result. I said, "Don't get so upset. You guys did a great job. You just got the wrong answer. You got the answer you should have gotten, and that is that it's impossible to grow enough food on the rooftops to feed the city. What would happen," I asked, "if you asked a different question? What if we took your idea and put it inside all those abandoned apartment houses over in the Bronx?" They looked at me, YEAH!

Students heard about the project and the next year enrollment increased to twelve, then thirty, and forty. I list all 106 of them in the back of my book. They are the inventors of vertical farming; I was their mentor. That's how the idea started. Around 2004, I started getting a lot of unsolicited ideas from architects. I even gave a talk in India at the annual meeting of their national institute of architecture; I think I was the only nonarchitect there. So I began to preach—I guess you can call it preaching—to architecture groups, the Council on Tall Buildings, the Center for Architecture in New York, and so on. I designed a course called Ecology for Architects and Engineers and gave a summary at Taliesin West. The director thought it was great. George Ferguson, president of the British Institute of Architecture, loved it, too.

I also gave a talk at the annual NGO meeting at the United Nations. They wanted to hear about this because they thought we could maybe generate funds for a prototype. Make the prototype work, then maybe do it in modular form, and take it to a place like Chad or Mali, places that are really in trouble because they have no farming to begin with. Each year, they would lose their crops. One year, they actually had a good harvest but the locusts ate everything, so they ended up eating locusts. That's the wrong way to go about things.

Can vertical farming be integrated in a way that makes sense in tall buildings?

Yes. People laughed at the Wright brothers and said that thing won't fly. As far as vertical farming goes, it's already flying. Pragmatists have a real job, while just around the corner there are a lot of people who are going to invent the future, and the pragmatists are going to make it happen. We're about two fingers away from making it work altogether. Korea is willing to invest probably $100 million dollars in their prototype vertical farm and an agro-bio-diversity building that houses seeds from every single plant that grows there. Every country needs one of these. It's as pragmatic as you can get.

Is there a limit to how high a vertical farm can go?

There shouldn't be any. How high do you want to go? How many people do you want to feed? With what? You have to answer those questions first. Every situation will be different. In my view, Iceland is the best country for vertical farming to start with because they have unlimited geothermal energy. Even low-tech grow lights would work really well for Iceland all year. Alternatively, if you go to Australia, they have so much solar and so little realized energy from it, they're embarrassed. They are in the process of figuring it out.

They need to farm indoors. That country was nearly flooded out of existence; six months ago it nearly burned to the ground. Try living like that as a farmer! You know what the suicide rate is among farmers in Australia? It's the highest in the world, second only to India. Why are they committing suicide? Their farms are failing, their families are dying, they

的书本最后列出了所有这106个人的名字，他们是垂直农业的发明者，我是他们的导师，这个想法就这样开始了。2004年，我开始得到一些建筑师主动提供的建议，甚至在印度的国家建筑机构年会上做了个演讲，我想我是唯一一个非建筑师参会的人。此后，我开始为建筑团体、高层建筑委员会、纽约的建筑中心等机构做宣讲——也许你们会称之为讲道。我为建筑师和工程师设计了一门生态学课程，并在西塔里耶森(Taliesin West)做了一次概述。系主任认为课程很不错，英国建筑协会的主席乔治·福格森(George Ferguson)也很喜欢。我也在联合国的非政府组织(NGO)年会上发表演讲。他们想听这个演讲，想通过一个原型找到资金，运作这个原型，然后将其制作成模块的形式，并带到乍得或马里——那些由于无地可种而陷入困境的地区。每年这些地区的农作物都遭受损失。有一年，他们实际收成不错，但却被蝗虫吃光了一切，最后人们吃掉了蝗虫，事情本不应当如此发展。

垂直农场可以应用到高层建筑上吗？

可以。人们嘲笑怀特兄弟，并说那玩意儿飞不起来。就目前垂直农场的发展来看，它实际上已经飞了起来。实用主义者有着实际的工作，不久后，会有许多人试图发明未来，而实用主义者就会努力使这一理想成为现实，我们已经十分接近于将理想与现实协同工作了。韩国打算投资约1亿美元建造垂直农场的原型，以及农业—生物—多样性建筑（这种建筑储藏着生长在那的每一类植物种子）。每个国家都需要这些，它是你可以获得的最实用的东西。

垂直农场在高度上有限制吗？

不会有任何限制。你想要多高呢？要供养多少人？以什么来供养他们？你必须先回答这些问题。每一种情形都不相同，在我看来，冰岛是最适宜开展垂直农场的国家，因为他们有无限的地热资源。即使是低技术水平的植物生长灯，也会在冰岛常年正常工作。相比，如果你去澳大利亚，那有太多阳光，却很少能够将其转化为能量，这着实有些尴尬。他们还处在理解这一项目的阶段。

澳大利亚需要在室内种植，这个国家曾经几乎被洪水淹没，六个月前又几乎被大火焚毁。试试作为农民在那里生活！你知道澳大利亚农民的自杀率有多高吗？全世界第二，仅次于印度。他们为什么自杀？他们的农庄失败了，家庭正在灭亡，自己饱尝饥荒，他们不想搬到城市去。砰！砰！砰！砰！不开玩笑，事实就是这样。

你可以在火星上，在太空飞船上，在任何出乎意料的气候中做垂直农场，这看起来非常合理。

现在在你能发现的任何地方做垂直农场都是合理的。

有一些实践和产品被说成是可持续的，但由于在其生产过程中消耗的资源，实际上并非如此。也许你可以生产出许多食物，但代价是什么呢？能量是否保持平衡？

我这样解释一下，生态学在光谱的一端，而建筑环境在光谱的另一端，它们没有共同之处。实际上，它们是相互间的镜像，城市占据了地球上2%的土地面积，却产生了70%的二氧化碳。这是一个生态学的解决方案吗？这是一个生态学的悲剧！

一个生态系统如何运作？与生态系统联系在一起的词语是"弹性"和"生物多样性"。它有混合和后备系统，像一架飞机一样，为了预防一些危害生命安全事件的发生，它装配了三个不同的后备系统，这些系统必须协同工作。今天，人们说必须处理能源浪费问题，但人们并不真的这么想。

仅仅是去处理我们的排泄物，你知道就要花多少钱吗？是数以亿计的。但如果你想要对自然说"我们打算销毁所有动植物的排泄物"，你知道会发生什么吗？你将不花一分钱。所以，我们为什么不如此处理废弃物呢？实际上，现在有些地方正在这样做。加利福尼亚的圣安娜（Santa Ana），就对自身的污水进行处理：从尿液中产生了饮用水。它是完全可以喝的，但人们却拒绝饮用。因此，圣安娜市将这种处理过的水重新灌到地下蓄水层中，然后人们才开始饮用它。

所以，这是一个认识的问题。

没错。整个投资耗时8年，花费5亿美金。现在，水源不会限制生活在圣安娜居民的数量，因为这些人制造水的同时又重新饮用水。很简单，对吧？很多其他的地方也如此，像瑞典的生态城市马尔默(Malmo)，但是我们如何处理固体呢？

纽约市的例子，它产生的粪便，每人每天200克（7盎司），365天，8万人，对吗？如果我们

are starving, they don't want to move to the city. Bang, bang, bang, bang. That's it, no joke.

You could do this on Mars, on a spaceship, anywhere the climate is unpredictable, which seems very reasonable. It's reasonable right now to do it wherever you can.

There are practices and products presented as sustainable, but because of the resources consumed in their production, they end up anything but. Perhaps you can produce a lot of food, but at what cost? Is the energy balanced?

Let me explain it this way. Ecology is on one end of the spectrum and the built environment is on the other. They have nothing in common. In fact, they are mirror images of each other. Cities take up 2 percent of the world's land mass and give off 70 percent of the CO_2. Is that an ecological solution? It's an ecological tragedy.

How does an ecosystem work? It has words associated with it like "resiliency" and "biodiversity." It has mixtures and backup systems. Like an airplane, it has three different backup systems for the same life-threatening event. And it all has to work together. Nowadays people say we have to process waste energy, but we don't really mean it.

How much money do you think we spend annually just trying to get rid of our own feces? It's in the billions. If you were to say to nature, "We're going to destroy all of the waste products of the animals and plants," you know what happens? You get a big zero. So why don't we do that with our waste? Actually, there are a few places that do it. Santa Ana, California, processes its own water, i.e., generates drinking water from urine. It's totally drinkable but people refused. So they recharge the aquifer with that water, and then people drink that.

So it's a question of perception.

Correct. The entire investment was eight years and $500 million. And now water poses no limit to the number of people who can live in Santa Ana because they make the water and redrink it. Pretty simple, right? There are plenty of other places that do this, like Malmö, an ecocity in Sweden. But what do we do with the solids?

Take the example of New York City, which produces 200 grams (7 ounces) of feces per person per day; that's 365 days times 8 million people, right? If we took this human waste and dried and incinerated it, and did this with all of New York City, we would generate 900 million kilowatt-hours of electricity per year, just with our own feces. You can run about 80 vertical farms on that. Why hasn't anybody done it?

374

Why? It's because it's too goddamn easy, that's why. They'd rather spend billions to make infrastructure to get rid of it than to use it for something purposeful. Tell that to nature and nature will say, "You guys are doomed because you don't know how to recycle. You're doomed." What people mean by recycling is taking their aluminum cans, crushing them, and reprocessing them. Nature is sense, and this is nonsense. So let's be sensible. I'm not accusing people of anything except a lack of information. I'm an ecologist. This is my specialty.

Doesn't human feces contain toxins that have to be gotten rid of?

You can go around the country and find all the trace elements that exceed EPA limits. Don't go to Toledo, because mercury is in the soil and it comes into the drinking water, so Toledo's feces can't be used for food production. But you can use it for landfill. The EPA has a list of cities and whether their feces would be good for farming or for something else. That's all it takes!

But wait! There's a process called plasma arc gasification, or pyrolysis. Imagine sitting on the sun. How long would you last? You'd vaporize before you even got close. Vaporize! What if I could make a temperature hotter than the sun, 7,000 degrees Centigrade [12,632 degrees Fahrenheit], right here? What do you think I could do with waste? I could vaporize it. What do you think you could do with all that energy that is released when you break those chemical bonds? You could power something, like the machine that pyrolyzed everything to begin with. If you don't want a landfill, this is the perfect solution.

About three-quarters of Japanese cities run on plasma arc gasification. They generate their energy from it and have no landfills. None. The only concern is toxic materials like heavy metals because everything else is reduced to its elements. In other words, if you crushed up this table and chair and made a powder out of it, and ran it through this big electrical arc that forms a plasma in the middle, out the other side would come the elements. It's like a miracle, like sitting on the sun.

The more carbon you have, the better to produce more energy because there is another process after that. Carbon and hydrogen: we're largely made of that: carbohydrates. So plasma arc gasification gives you pure carbon, and pure oxygen and pure nitrogen, which you release back into the atmosphere. You want to keep the hydrogen and the carbon because you want to recombine them some way; you need a little oxygen, too. If you knock off hydrogen, you have something that almost looks like methane. When you go home

将整个纽约城中的这些排泄物收集起来烘干焚烧，那么，仅仅是利用我们自己的排泄物，每年就将产生9亿千瓦时的电能，这可以为80座垂直农场提供能源。但为什么没有人这么做呢？

为什么？因为它太简单了，这就是原因。他们更愿意花费数十亿美元兴建基础设施，摆脱这些排泄物，而不是有目的地利用它们。将这些告诉大自然，大自然会说，"你们这些人无可救药，你们不懂得如何循环利用，着实无可救药。"人们所谓的循环利用，往往就是收回那些铝罐，压碎它们，然后进行再加工。大自然是有意义的，人们做的循环利用是没有意义的。所以，让我们清醒一些。除了认为在信息上有所匮乏之外，我没有对人们提出任何其他的责难，我是一名生态学家，这是我的专业领域。

人体排泄物不是包含一些必须被处理掉的毒素吗？

你可以走遍全国，找到所有超出环境保护署规定的微量元素。别去托莱多(Toledo)，因为那里的土壤含有水银，而且已经进入到饮用水中，因此托莱多排泄物不用于食物生产。但是你可以用它来为垃圾填埋。环境保护署有一个城市的名单，上面说明哪些城市的排泄物有益于农业耕种，或有利于其他什么东西，要做的就只是这些！

但是等等！有一道工序称为等离子电弧汽化法或高温分解法。想象一下坐在太阳上，你可以待多久呢？其实在接近它之前，你就会蒸发。蒸发！如果可以产生出比太阳表层还要高的温度，7000摄氏度（华氏12632度），那么这里将会怎样？你觉得我会如何处理垃圾呢？我可以蒸发它，当切断所有这些物质的化学链时，其中释放出来的能量可以作何处理？你可以给一些东西供电，就像那种高温分解物质的机器一样。如果你不想要垃圾填埋地，那么这是绝妙的解决方案。

大约四分之三的日本城市采用等离子电弧气化法。他们从中开发出能源，而且不用垃圾填埋地。唯一的问题是类似于重金属的有毒物质，因为所有其他东西最终都分解为元素。换句话说，如果你毁掉这张桌子和这把椅子，并将其变成粉末，使其通过中间形成等离子的巨大电弧，那么，当它们在另一端出来的时候，将呈现为元素的形式，这就像是一个奇迹，像坐在太阳之上。

你有越多的碳，就能产生越多的能量，因为在那之后还有个反应过程。碳和氢，大部分物质

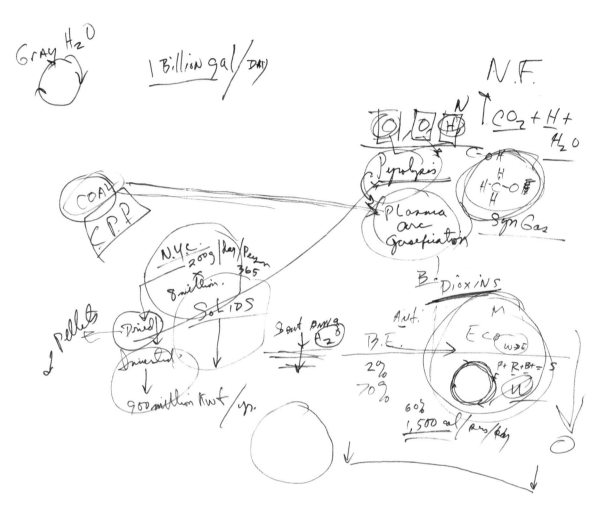

Closed system for New York City water and waste energy recycling. (Dickson Despommier)
纽约市的水资源和浪费能源回收封闭系统。 （迪克森·德波米耶）

是由它们构成的：碳水化合物。因此，等离子电弧气化法产生纯粹的碳，纯粹的氧和纯粹的氮，而我们会将这些释放回大气中。你可能想要保留氢和碳，因为你想以某种方式重组它们；你也需要一点氧。如果你移除氢，你就有了某种看起来像甲烷一样的东西。当你今晚回家，打开火炉，那就是用来煮饭的东西，它被称为合成气，因为它就是一种合成气体。人们可以从高温分解的副产品中得到它们，产生的热量被用来推动涡轮，为电弧提供电源。放入的碳水化合物越少，通过机器消耗的能量就越多，而你并不会收回太多的合成气体。但是如果你把人的排泄物放入高温分解装置之中，你就可以获得大量的合成气体——

tonight and turn on your stove, that's what you're cooking with. It's called syn-gas because it's synthetic. They can make it from the byproducts of pyrolysis. The heat generated is used to drive a turbine to make electricity for the arc. The fewer carbohydrates put in, the more energy is consumed by the machine and you don't get a lot of syn-gas back. But if you put human feces through the pyrolysis machine, you could make lots of syn-gas, which you can then burn, cleanly, like methane, and the only byproducts are CO_2 + heat + water. Plants need all of this, so if you put those products into a vertical farm, you've got loops within loops within loops. A city can take its waste, pyrolyze it and recirculate the energy back into a vertical farm, which is what should be done. You could do the same with all waste. It's a known technology, used in Japan, Germany, Holland, in thinking countries with no agenda. It's going to replace nuclear power plants in Japan. I guarantee it. And you can make the plants big, small, medium size. People make a lot of waste, so you might as well do something with it.

Why isn't it used here?

Good question. You've heard of coal-powered power plants? They're what runs America. China is putting up one a week. One a week! Every time someone suggests pyrolysis, the coal industry says, "Oh, you don't want to do that because it produces dioxins. Coal doesn't do that." So they send out a stream of misinformation. Port St. Lucy wanted it, Newark and Chicago wanted it, but every time, the coal industry would say, "No. We have no vested interest in this, of course. We're just trying to do the right thing." You can read the arguments online. Talk to Dr. Cicero from Georgia Tech. He used to have hair, but he's pulled it all out over this!

As far as vertical farms go, they're up and running so I don't have to push too much in that direction anymore. The snowball is already rolling down the hill, getting bigger all the time. Everyone will figure it out.

The Korean national government's vertical farm is like the vertical farm of the future, the future being two to three years away. All these questions that we're now struggling with, someone very smart at MIT or Caltech will sit there and go click, click, click. Done. You want to take a ride? Sure. That's how confident I am that this is going to work.

If a city is going to be like an ecosystem, then you have to do a lot more of this. On a linear scale where the most extreme bad behavior is at one end and the most extreme good behavior is at the other, where is the world right now? Very close to the wrong side. So your and my job is to push as hard as we can to restore the balance, any way that we can, on any level. Otherwise, some other species will be reading about our extinction pattern.

What requirements does vertical farming impose on mixed-use buildings?

You must reuse water. It's against the law to throw it away. New York City uses over a billion gallons [3,785,411,780 liters] a day. What is the biggest problem in the world today? Water. There's only so much of it and it's maldistributed. That's why farming in India and China is failing; that's why farming in the United States is failing. You either have too much or too little.

这是一些类似于甲烷的洁净可燃气体；而副产品也仅仅是二氧化碳+热能+水。植物需要这些副产品，所以如果你将它们放到垂直农场中，你就可以获得能量物质的不断循环。一个城市可以收集排泄物，高温分解，使能量循环回到垂直农场，这事值得去做。你也可以同样处理所有废弃物，这是一种大家熟悉的技术，被用在日本、德国、荷兰，对比想想没有此议项的国家，我可以保证，它将在日本取代核能。你可以建立大中小型的工厂。人们制造了许多垃圾，所以应该想想如何处理它们。

为什么在我们这里并没有应用此种技术？

好问题。你有没有听过燃煤发电厂呢？是它们推动着美国发展。中国一周之内就会兴建一座燃煤发电厂。一周之内！每次有人提出高温分解的建议，煤炭行业就会说："哦，你不会真想那么做的，因为那会产生二恶英，而煤则不会。"因而他们传达出一连串的错误信息。圣露西港、纽瓦克和芝加哥都想要这种技术，但是每次煤炭行业就会说："不，我们当然对这没有兴趣，我们只想做正确的事情。"在网上能够读到这些观点，你可以和乔治亚理工学院的西塞罗（Cicero）博士谈谈，他过去是有头发的，但因为这个问题，他的头发都掉光了！

随着垂直农场的出现，它们必定会兴盛起来，因此我并不需要往那个方向用力推了。雪球已经从山顶滚下，越变越大，每个人都明白这一点。

韩国政府的垂直农场就像是未来两三年之后的。所有那些我们现在面临的问题都将迎刃而解，一定有某个麻省理工学院或加州理工学院的聪明人坐在某个地方，不断地点击鼠标，然后说：成了！你想要去兜风吗？当然，我就是这么有信心。

如果一个城市想要有一个生态系统，那么需要做的比这还要多。在一个水平的天平上，最坏的行为和最好的行为各居两端，那么，现在世界处于哪个位置上？非常接近于错误的那端。所以你和我的工作，是以任何可能的方式，在任何层面上尽力推动，去恢复那种平衡。否则，一些其他的物种将会看到我们灭绝的方式。

垂直农场对混合用途的建筑有哪些要求？

你必须重复使用水。把它扔掉是违法的。纽约市

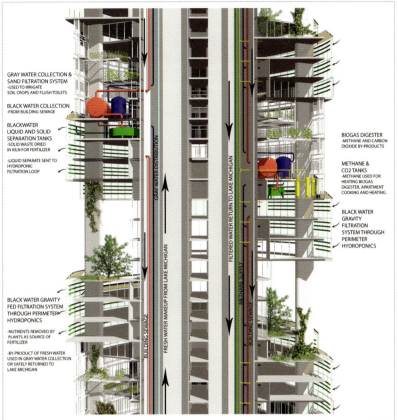

GRAY WATER COLLECTION &
SAND FILTRATION SYSTEM
-USED TO IRRIGATE
SOIL CROPS AND FLUSH TOILETS

BLACK WATER COLLECTION
-FROM BUILDING SEWAGE

BLACKWATER
LIQUID AND SOLID
SEPARATION TANKS
-SOLID WASTE DRIED
IN KILN FOR FERTILIZER

-LIQUID SEPARATE SENT TO
HYDROPONIC
FILTRATION LOOP

BLACK WATER GRAVITY
FED FILTRATION SYSTEM
THROUGH PERIMETER
HYDROPONICS

-NUTRIENTS REMOVED BY
PLANTS AS SOURCE OF
FERTILIZER

-BY-PRODUCT OF FRESH WATER
USED IN GRAY WATER COLLECTION
OR SAFELY RETURNED TO
LAKE MICHIGAN

BIOGAS DIGESTER
-METHANE AND CARBON
DIOXIDE BY-PRODUCTS

METHANE &
CO2 TANKS
-METHANE USED FOR
HEATING BIOGAS
DIGESTER, APARTMENT
COOKING AND HEATING.

BLACK WATER
GRAVITY
FILTRATION
SYSTEM THROUGH
PERIMETER
HYDROPONICS

GRAY WATER DISTRIBUTION

FRESH WATER MAKEUP FROM LAKE MICHIGAN

BUILDING SEWAGE

FILTERED WATER RETURN TO LAKE MICHIGAN

METHANE SUPPLY

BUILDING SEWAGE

Proposal for a Sky Farm in Lake Michigan. (Blake Kurasek, 2010)
密歇根湖的空中农场方案。（布莱克·库拉塞克，2010年）

So every building must recycle its gray water. That's number one. Number two: do an energy calculation: 1,500 calories per person per day, that's what everybody eats. Your vertical farm has to produce that many calories, or even 60 percent of that number, for the building. That would make a huge difference. Then you can start to be a little more biodiverse.

One of the characteristics of the ecosystem is diversity. That's what gives it resiliency and ultimately sustainability. Notwithstanding climate change, different weather patterns, rains, volcanoes, tidal waves, tornadoes, hailstorms, you name it, we're still here! We rose out of that system and have everything in our genome to become as resilient as any other species in nature. We even create our own environments to resist those changes. That's what architecture is all about.

An ideal mixed-use building should supply most of its water and most of the food, and have a built-in waste energy recycling system. Plasma arc gasification is perfect because it comes in various sizes and it is very, very safe. When you want it to go off or on, you flick a switch. There's no nuclear meltdown possibilities, no exhaust to pollute the environment. You can collect the heavy metals, by the way, in the form of a slag and actually sell it. For instance, if you were to process all the abandoned electronic technologies, all the cell phones, all the computers, forget about it. Power up, throw the stuff through the plasma arc gasifier and the heavy metals—platinum, gold, silver—will collect in the slag. You can send that off to a smelter and they'll recover it 100% because they're solids, not vapors. It's worth a lot of money so this is how you get it all back. In the building you need circles. Think of any loop within a loop within a loop within a loop. You have to make loops. That's the only rule, but how do you do that? I don't want to sound elitist, but I know that in school, architects don't get any ecological training whatsoever, no biology. How can you leave out biology from the human condition? How can you do that?

If you want to recapitulate nature, which is what you should be doing, then cities need to take a lesson from nature. An ecologist once said that nature has all the answers. It doesn't matter what the question is. Find the natural system that does that function and technologically imitate it. The best example I know is the introduction of the zebra mussel, a horrible little pest, to this country. It's tiny, about the size of your pinky fingernail, and has zebralike stripes. What does it do? It reproduces. Imagine what Lake Erie looked like 20 years ago when it was green, and compare it with the clear water of Lake Erie today. Who did that? The zebra mussels did. Nature has all the answers.

每天使用超过一个亿加仑（37亿）升的水。当今世界的最大问题是什么？水！水只有这么多，而且分布不均。所以印度和中国的农业正在衰败；美国的农业也正在衰败。我们要么有太多水，要么就是严重缺乏。所以每栋建筑都必须循环利用洗漱污水，这是第一条。第二，做一个能量计算：每人每天会吸收1500卡路里的热量。垂直农场必须为建筑生产出这个数量的卡路里，甚至60%也可以，这将令一切大为不同。然后，你可以开始变得更加富有生物多样性一些。生态系统的另一个特征就是多样性，这使它富有弹性和可持续性。尽管气候改变，出现不同的天气形态：雨水、火山、潮汐、龙卷风、冰雹……凡是你能说出来的种种，但我们依然还在这里！人类从那个系统中脱颖而出，和自然界中其他物种一样，也使自身的基因组变得富有弹性。我们甚至创造了自己的环境，以抵御那些改变，这就是建筑的本义。

一座理想的混合用途建筑，应该能够为自身提供大部分的水和食物，并有一个内部构建的废弃物能量循环系统。等离子电弧汽化法是一个非常好的方案，因为它的尺度灵活多变，而且非常安全。当你要起动或暂停它的时候，只需要按下开关，不存在核泄漏的可能，也不会因为废气而污染环境。你可以收集重金属，而且将其作为一种溶渣予以出售。例如，如果想要处理所有废弃的电子设备：手机，电脑，别再想了！直接通上电，将那些东西进行等离子电弧汽化处理，然后可以在溶渣中收集到重金属——铂、金、银，这些被送到冶金厂，可以百分之百地复原，因为它们是固体不是蒸汽，这些东西很值钱。回收过程就是这样。在建筑中，需要循环圈，想想那些重复循环的系统，你必须创造出许多循环来。这是唯一的规则，但怎么做呢？我不想说得自己好像精英人士一样，但我知道，在学校里，建筑师没有受到过生态学方面的训练，当然，也没有生物学的。你怎么能够在人类的处境之中遗漏生物学呢？这怎么可能！

现在应该做的是重述大自然，城市需要从自然中吸取教训。一位生态学家曾说过：不管问题是什么，自然拥有所有答案。发现具有这种功能的自然系统，并在技术上对其进行仿效。我知道一个最好例子：有种可怕的小害虫，叫斑马贝，被引入到美国。它只有指甲那么小，带有斑马的条纹。用它做什么呢？繁殖。20年前，伊利湖（Lake Erie）还是绿色的，而现在湖水清澈无比。这是谁的功劳呢？是斑马贝！所以说，自然拥有所有的答案。

　　想要将洗漱污水转化为饮用水？看起来像希腊竞技场的沃兹岛，为什么我选沃兹岛？那是纽约市所有废液的堆积处。在加热杀菌之后，人们以离心机分离并固化，再把加热杀死的所有微生物以肥料形式出售。然后，人们带走洗漱污水，放回到运来的船上，并在氯化之后重新倒回到河口处。你说的是破坏臭氧层的氯？倒回河口？布隆伯格(Bloomberg)究竟怎么了？他是一个工程师，他给出的理由是：我们是一个有着18世纪基础设施的21世纪城市。那么，把那该死的设施修好！斑马贝现在在哪？他们正在过滤哈德逊河，以使之达到咸水标记（它是一种淡水物种）。

　　如果自然拥有所有答案，那么你的问题是什么？该如何处理洗漱污水？该如何处理固体来获得能量？该如何建造一座大楼，为我生产食物？这些都是正确的问题。建筑师需要以此为导向，我并没有直接的答案。

　　中东地区的光照充足，你想用多少就有多少。如果这些来自荷兰"植物实验室"的人是对的，你便可以利用窗户过滤掉所有不想要的阳光，而只让想要的阳光照射进来。使用日光灯在能源上太没效率，你只需要为叶绿素A和B提供两种东西，显然会节省很多能量。明白我的意思吗？

　　有许多好事将会继续发生。在我的生态城市中需要的所有东西，现今都在全球范围内发生，只是并非同时发生。现在的问题不再是要不要运作，而是这种东西如何运作，或那种东西如何运作。我们需要一个年会，需要设计一个竞赛，找出2012年最优秀的垂直农场设计方案。联合国已经就这个问题和我交流过了。

Want to turn gray water into drinking water? There's a water feature that looks like a Greek amphitheater the size of Wards Island. Why did I pick Wards Island? Well that's where New York City brings all of its liquid waste. They centrifuge and solidify it and then, after they heat it up to kill all the microbes, sell it for fertilizer. Then they take the gray water and put it back into the boats that brought it there, and dump it into the estuaries after they treat it with chlorine. Did you say ozone-depleting chlorine? Into the estuary? Bloomberg is an engineer. What's wrong with him? He gives the excuse that we're a 21st-century city with an 18th-century infrastructure. Well then, fix the damn thing! And where is the zebra mussel now? Filtering the Hudson River up to the saltwater mark (it's a freshwater species).

If nature has all the answers, what's your question? How do I process gray water? How do I process solids to get energy? How can I build a building that will make food for me? These are all the right questions. Architects need to run with it. I don't have any direct answers.

The Middle East has as much sunlight as you could possibly use. If these guys from Plantlab in Holland are correct, you can make the windows filter out all the unwanted solar and just let in the light you want. Using fluorescent bulbs is too energy inefficient. If you just use the two narrow ones that you need for chlorophyll A and B, that saves a lot of energy. Get my point?

There is a lot of good stuff going on. Everything I need in my ecocity is already happening around the world, just not all together. It's no longer a question of whether it will work. It's, How does this work? Or how does that work? We need an annual meeting. We need a design competition for the best vertical farm designs for the year 2012. The United Nations has already talked to me about it.

Gregory Kiss
KISS + CATHCART ARCHITECTS

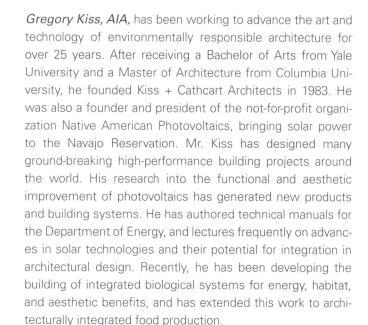

Gregory Kiss, AIA, has been working to advance the art and technology of environmentally responsible architecture for over 25 years. After receiving a Bachelor of Arts from Yale University and a Master of Architecture from Columbia University, he founded Kiss + Cathcart Architects in 1983. He was also a founder and president of the not-for-profit organization Native American Photovoltaics, bringing solar power to the Navajo Reservation. Mr. Kiss has designed many ground-breaking high-performance building projects around the world. His research into the functional and aesthetic improvement of photovoltaics has generated new products and building systems. He has authored technical manuals for the Department of Energy, and lectures frequently on advances in solar technologies and their potential for integration in architectural design. Recently, he has been developing the building of integrated biological systems for energy, habitat, and aesthetic benefits, and has extended this work to architecturally integrated food production.

格里高利·奇斯
奇斯+凯斯卡特建筑事务所

格里高利·奇斯，美国建筑师协会会员，他已经为改进环保建筑的艺术和技术工作了超过25年的时间。在从耶鲁大学获得艺术学学士，并在哥伦比亚大学获得建筑学硕士之后，他在1983年创立了奇斯+凯斯卡特建筑事务所。他也是非营利性组织美国土著居民光伏协会的创始人和主席，这个组织为纳瓦荷保留区带来了太阳能。奇斯先生在全世界设计了许多开创性的高性能建筑项目。他对太阳能光伏功能和美学改进的研究，已经产生了新的产品和建筑体系。他为能源部门编写技术手册，并经常就提升太阳能技术，以及将其整合进建筑中的潜力发表演讲。为了能源、居住和美学上的益处，最近他开发了一栋整合生态系统的建筑，并将这一作品延伸到将建筑与食物生产整合在一起。

For complete biography, please refer to the Appendix.

完整的传记，请参阅附录。

2020 Tower, rendered distant night view, proposal for New York. (Kiss + Cathcart Architects, 2001)
渲染的夜间远景透视，纽约2020大厦设计方案。（奇斯+凯斯卡特建筑事务所，2001年）

How did you become a leader in sustainable architecture?

Some of our very first projects were pioneering solar buildings, including the world's first factory that produced thin film photovoltaics, in Port Jervis, New York (1983–85). And in 1988–89, in Fairfield, Ct. we did the first building-integrated photovoltaic curtain wall (BIPV) building in this country. Most of the glass in the control room's curtain wall is photovoltaic, so it produces electricity and saves the cost of traditional spandrel glass.

Back in 1990, this kind of photovoltaic glass cost about $15 per square foot [$160 per square meter]. The wiring and inverters cost about the same, so the solar facade cost about $30 per square foot [$325 per square meter], in other words, an energy-generating facade that costs less than some conventional ones. Imagine, in the right project, a net zero cost for the solar curtain wall. Free renewable energy! This is the big promise of photovoltaics in architecture. Any building that the sun shines on wastes income if the solar energy isn't used. So why wouldn't you do it? It's a no-brainer.

The Fairfield project was designed for the headquarters of APS, the solar panel manufacturer, with minimal cost premium. Why hasn't this been duplicated? Why don't you see BIPV everywhere?

There are plenty of BIPV buildings, but instead of being economical, they have tended to be enormously expensive. Most of the "green" community has no idea that economical BIPV is even possible. There is a sense that solar is nice but just too expensive now—worthwhile, maybe, but not for economical reasons. This is a shame, because I know it's possible, we've done it.

Why isn't everyone using this technology?

There's no single reason but lots of small issues that collectively make it very difficult. The first is lack of awareness that this technology exists. Second, it's the industry. Glass is typically ordered in custom sizes and the price per square foot is pretty much the same. That's the way the building industry works but not the way the solar industry works. Although you can now get pretty much any size and shape of solar panel, the cost of customizing is very high and the PV industry isn't set up to do that on a routine basis. They don't really need to because they're selling all the standard product they produce. Most BIPV projects are done as demonstrations, so cost isn't the primary concern. Instead of $15 per square foot [$160 per square meter], custom BIPV modules can cost 20 times as much, considerably more than any standard building

你是如何成为可持续发展建筑领域的一位领先者的？

我们开始的一些项目是非常前沿的太阳能建筑，包括位于纽约杰维斯港口的、世界上第一个生产光电薄膜的工厂（1983年-1985年）。而在1988年至1989年，我们在加利福尼亚的费尔菲尔德，建造了这个国家第一座建筑一体化光伏幕墙(BIPV)建筑。控制室幕墙中的大多数玻璃都是光伏的，所以它会生产电能，并节约传统拱肩玻璃的费用。

早在1990年的时候，这类光伏玻璃的费用大约是15美元每平方英尺（162美元每平方米），电缆加上换流器的费用大概与此相当，因此太阳能表皮的花费约为30美元每平方英尺（324美元每平方米）。换句话说，一个能够产生能量的表皮，它的成本有些比传统表皮要小。想象一下，在某个合适的项目中，太阳能幕墙的成本为零——免费的可再生能源！这是太阳能光伏在建筑行业的一个巨大希望。如果不利用太阳能，任何被太阳照射到的建筑都是在浪费金钱。那为什么不这么做呢？这是个不需要动脑的判断。

费尔菲尔德的项目是为APS（太阳能面板制造商）总部设计的，其额外花费是最低的。这个项目为什么没有被复制推广呢？为什么你没有在所有地方都看到这种建筑一体化光伏幕墙呢(BIPV)？

其实，有很多的BIPV建筑，但他们趋向于极其昂贵而非经济型。大多数的"绿色"团体甚至都不知道，这种经济的BIPV是有可能实现的。人们有一种感觉，太阳能很棒，只是现在还太贵——它可能值得去做，但不是出于经济上的理由。真是令人遗憾啊，因为我知道这是可能的，而且我们早就实践过了。

为什么不是所有人都采用它呢？

不是单一的一个原因，而是有很多的小问题积累起来，使得这件事变得困难。首先是没有意识到这种技术的存在，其次是行业的问题。一般来说，玻璃的尺寸是可以定制的，而每平方英尺玻璃的价格则几乎是一样的。这是建筑行业的运作方式，但却不是太阳能行业的。尽管你现在可以得到几乎任意尺寸和形状的太阳能板，但定制的费用非常高，而且光伏行业并未将此作为常规基本原则，他们也不需要这么做，因为他们正在卖自己生产的所有标准产品。大多数BIPV项目都只作为一个展示项目，因此费用并非是主要的关注点。定制的BIPV模型，

Left: Port Jervis Manufacturing Facility, Port Jervis, NY (Kiss + Cathcart Architects, 1984)
Right: APS Fairfield Photovoltaic Facility, Fairfield, CT (Kiss + Cathcart Architects, 1991)
左：纽约州杰维斯港的生产设施（奇斯+凯斯卡特建筑事务所，1984年）
右：APS公司在康涅狄格州费尔菲尔德的光伏设施（奇斯+凯斯卡特建筑事务所，1991年）

其费用可不是15美元每平方英尺（162美元每平方米），而可能是这个价格的20倍，大大超过了任何一种常规的建筑材料。第三，依然有许多需要说明的规则和标准问题。

为了产生能量，太阳能板牺牲了自然光的进入。如何进行这种调节？

在一栋可持续建筑中，你不想要超过20—40%的玻璃，即使是为最优化的日光照射，因此60—80%的表皮都没有开窗。你当然可以把太阳能材料用来遮阳，或者用太阳能窗产品，因为他们已经生产，出来了。

但在建筑的垂直表面上，太阳光的照射并不是非常高效的。

对的，但如果你认识到净成本很低甚至为零，那上面的情况就没有什么关系了。如果有人在建设中想要使用价格为30美元到40美元每平方英尺（324美元到432美元每平方米）的涂层材料，而你可以使用同样成本的太阳能材料，那么你在电力上的收益是非常巨大的。

在费尔菲尔德的"太阳能光伏体"中，我们把BIPV的参观者中心和控制室置于一个呼吸幕墙之

materials. Third, there are still a number of code and regulatory issues that need to be addressed.

Solar panels compromise the entrance of natural light for the sake of generating energy. How is this reconciled?

In a sustainable building, you don't want more than 20 to 40% glazing, even for optimized daylight, so 60 to 80% of the facade is not even glazed. You could certainly use solar material as a sun shade, or use solar window products as they become available.

But solar glazing isn't very effective on a vertical face.

Right, but it doesn't matter if you accept the notion that the net cost is low or even zero. If somebody were willing to build with a cladding material at $30 to $40 per square foot [$325 to $430 per square meter], and you can use a solar material for the same cost, then any electricity gain is gravy.

In Fairfield's "PV Cube," we set the BIPV Visitor's Center and Control Room in a ventilated curtain wall. Thin film (amorphous silicon) has the excellent characteristics for BIPV, such as good generation performance at high temperatures and low light levels, visual uniformity, and low cost (down to less than $10 per square foot [$110 per square meter] today). Energy from the PV Cube's curtain wall and skylight, together with the awning that shades the south side of the adjacent factory, generates approximately 20% more energy than the Cube consumes, making it a prototype of energy-positive buildings. Even now, 22 years later, few buildings have achieved this level of economical sustainable performance.

A big skylight has conventional tinted skylight glass in the center and PV in the corners. I originally thought that the 4 to 5% transparent PVs would be too opaque, but actually it turned out fine. Normally, you don't want big skylights in a workplace because it's too hot and contrasty, but the light coming through these PVs is very diffused, so you can cover the whole roof of an office or classroom, for example, and get the daylight benefit at the same time that you're generating energy.

Could this be used for vertical farming?

Not really. Most plants want all the light they can get. They compete with solar panels for the sun. This is particularly true of most kinds of produce—lettuce, tomatoes, squash, and so on. Having gotten involved in solar so early, we did a series of technical studies for the Department of Energy's National Renewable Energy Lab. It started us thinking about sustainable green buildings. Back in the 1980s there was no U.S Green Building Council.

My partner and I started working together directly out of Columbia architecture school in 1983. We became interested in how you could make environmentally productive elements architecturally significant parts of a building. We did some low-cost housing in Holland in the early '90s, with different kinds of solar technology, including solar thermal panels for hot water and customized PVs with extra transparency.

We later did a project in Panama for the Smithsonian Tropical Research Institute (2003), which really got us thinking about high-performance buildings as a pragmatic problem. We weren't getting into the green thing for moralistic reasons but found it interesting, and challenging, and it created opportunities. The Research Institute was the first one that brought a lot of these subjects together, resulting in a near-zero-impact building. We went to great lengths to design the classrooms and labs for natural ventilation, but in practice they run the air-conditioning all the time, so the building has a very big energy load. But, as designed, it was zero energy.

All of the energy was generated by the solar roof, which is also the water source for the building. It rains a lot in Panama, so we collect the runoff. In fact, the roof supplies not only this building but the whole campus. There is no issue with ground water and aquifers, and because you're generating all your energy from the sun, there's no pollution of carbon emissions.

中。薄膜（非晶硅层）在BIPV中有出色的性能，比如，在高温和低光条件下出色的产能表现，视觉上的一致性，以及较低的成本（今天已经降到小于10美元每平方英尺〈108美元每平方米〉）。从太阳能光伏体的幕墙、天窗中产生的能量，加上遮蔽相邻工厂南面的遮阳棚，总共产生的能量比太阳能光伏体所消耗的还多出近20%，这使它成为一座典型的正能量建筑。不仅是现在，22年之后，也很少有建筑在可持续发展的经济表现上达到这一个程度。

一个大型天窗，中间是常规玻璃，角落则是太阳能光伏板。我以前认为，4—5%透明度的光伏板会太不透光了，但实际上它是合适的。一般来说，在工作场合，你并不想要一个很大的天窗，因为太热了，而且光线反差过大，但经过这种光伏板照射进来的阳光是很分散的，所以你可以整个覆盖办公室或者教室的屋顶，在生产能量的同时，还能得到阳光。

这可以应用在垂直农场上吗？

不见得。大多数植物需要天窗以获得的全部阳光，它们会与太阳能板争夺阳光。在很多作物身上尤其如此，比如：生菜、西红柿、瓜类等等。我们很早就已经进入了太阳能行业，并为能源部的"国家可再生能源实验室"做了一系列技术研究。这让我们开始思考可持续性的绿色建筑，上世纪80年代，美国还没有绿色建筑委员会。

1983年，我和我的伙伴离开哥伦比亚建筑学院，直接在一起合作研究。我们开始对如何能让那些对环境有产出效应的因素，成为一栋建筑中的重要组成部分这件事产生了兴趣。在90年代早期，我们在荷兰建了一些低成本的房子，采用了各种不同的太阳能技术，包括用于生产热水的太阳能热板，以及具有很大透明度的定制太阳能光伏板。

后来我们在巴拿马为史密森尼热带研究所做了一个项目（2003年），它让我们开始认真地将高性能建筑当作一个实际问题来进行思考。我们现在从事绿色工作，不是源于道德说教，而是发现它很有趣，很有挑战性，而且可以创造机会。史密森尼热带研究所第一次将这些论题整合起来，实现了几乎零消耗的建筑。我们竭尽全力设计教室和实验室的通风设备，但实际上，人们一直都开着空调，所以整栋建筑有很大的能源负载。但像原本设计的一样，它是零能耗的。

所有的能源都是通过太阳能屋顶产生的，它也是建筑的供水源。巴拿马降雨量很大，所以我们收集这些水分。实际上，屋顶不仅为这栋建筑，还为

Smithsonian Tropical Research Institute, Bocas, Panama. (Kiss + Cathcart Architects, 2003)
巴拿马博卡斯史密森尼热带研究所。（奇斯+凯斯卡特建筑事务所，2003年）

整座大学提供水分。这里不存在地下水和地下蓄水层的问题，而且由于你的能量是来自太阳的，所以也没有碳排放的污染问题。

在2003年，我们能够做出零能耗的建筑。这就已经是一个启发，令人惊讶的是，迄今为止，依然很少有项目以此为目标。人们开发了更多零能耗的项目，但是几乎没有项目尝试真正经济的BIPV。

什么时候开始，可持续性建筑获得了主流的关注？

我们注意到，2004年在纽约发生了一次重大的转变，那时环境建筑的理念，甚至于那些非常激进的理念，也忽然变得大受欢迎。在此之前，我们的高性能项目对于客户来说显得那么不寻常，并且非常陌生遥远。现在，我们正在纽约做很多非常有意思、有雄心的项目。1999年的康尼岛史迪威大道地铁站是一个开始。根据纽约市的交通情况，它实际上是最繁忙的公共运输站点。

这个站点一百多年来一直忍受着维护不及时的痛苦，基本上说是在不断败坏的过程中，所以他们重建整个车站。我们的工作是在轨道上方设计一个全新顶棚。主要规划组内部决定，因为是对整站

In 2003, it was a revelation that we could do real zero-impact buildings. The surprise is that so few projects have attempted it since. There are more zero-energy projects being developed, but almost none try for the true economy of BIPV.

When did sustainable building command mainstream attention?

We noticed a big change around 2004 in New York City when environmental building ideas, even pretty radical ones, suddenly were welcome. Before then, our high-performance projects were very unusual for clients, and far away from home. Now we're doing a lot of really interesting, ambitious projects here in New York. It started in 1999 with the Stillwell Avenue Terminal in Coney Island. According to New York City Transit, it's the world's busiest mass transit terminal.

The terminal was suffering from almost 100 years of deferred maintenance—falling down, basically—so they rebuilt the entire thing. Our part was to design a new roof over the above-ground tracks. Within the capital planning group they decided that since they were rebuilding the station, we should consider a glass roof to maximize daylighting, so why

Stillwell Avenue Terminal, Coney Island, NY, U.S.A. (Kiss + Cathcart Architects, 2005)
纽约州康尼岛史迪威大道地铁站。（奇斯+凯斯卡特建筑事务所，2005年）

not solar glass? It was an amazing decision on New York City Transit's part. If you get the right person in the right place, anything is possible. The station operates constantly and has 30-year intervals between significant maintenance. We had to make sure that maintenance on the roof would never interrupt the train service, so everything had to be serviceable from above. The structure has a nostalgic aspect, recalling Coney Island's 19th-century amusement park structures just beyond. We worked hard to get the right kind of dappled daylight through the integrated solar roof.

What do you mean by "productive architecture"?

Green building has become such a ubiquitous topic, but its meaning has been diluted. What do people mean by green? A LEED Platinum building may seem like a pretty exotic thing. But in terms of energy, which is the single biggest part of LEED, as it should be, a Platinum building is unlikely to be more than 60% better than business as usual; doing any

的重建，我们应当考虑一种玻璃顶棚，使日光最大化，所以，为什么不用太阳能玻璃呢？在纽约市的交通部门，这是一个惊人的决定。如果你让正确的人处在正确的位置上，什么都是有可能的。这个站点持续运作，每隔30年进行一次重大维护。我们必须保证，对顶棚的维护不会干扰到地铁的正常服务，所以所有构建的维护都需能从上部被服务到。这个结构还有一种怀旧的味道，让人想起不远处19世纪康尼岛的休闲公园。我们很努力工作，通过综合集成的太阳能屋面，来获得合理的洒落的太阳光。

你所谓的"生产性建筑"是什么意思？

绿色建筑已经成为了一个大家都在谈论的问题，但它的含义却被削弱了。人们说的绿色是什么意思？一座LEED白金认证的建筑可能看起来是非常奇特。但是从能源角度说（能源是LEED中最大的构成部分），一座白金建筑不太可能在能源方面比普通的商业建筑要好60%以上（它按规定本应如此）。如果做得好没有奖励，那就只能做得不那么坏了。这样的心态从根本上是有问题的。

对我们来说，我们试图在建筑上和技术上获得最佳方案。至少，你绝对不能损害环境，对吧？这是一个很简单的概念。如果你想到这一点，那其实就意味着一种零能耗建筑。今天，在大多数绿色建筑专业人员中，这看起来是一个乌托邦式的目标。不过我相信，在不太遥远的未来，这将成为最低标准。它就是所谓的可持续性。然后问题变成了：你能做得更好吗？有没有可能做一些对环境真正起积极作用的项目呢？这就是我们所说的生产性建筑，它是真正的目标。

你们在纽约的其他生产性项目怎么样呢？

有一个非常有趣的项目是，布朗克斯河园林路的河边住宅，其外围包裹的是镀锌钢和网格幕布，还覆盖着藤本攀爬植物。你可以使用太阳能板来产生电能，但植物乃是终极的太阳能装置，所以我们找了一些方法，将它们以建筑的方式，合乎功能地融合到建筑中去。

如果你在炎热的夏天穿过古老的森林，那里的温度可能会比周边低华氏52度（摄氏11度），因为树叶、蕨类、苔藓等会层层堆叠起来。我们的想法就是在系统中产生这种效果。我们有一个离建筑4英尺远的幕布墙，上面生长着藤本攀爬植物。第二层是一种雨幕，上面长着苔藓，同时还有一个滴流灌溉装置以保证苔藓的湿润。所有这些都来自屋顶的雨水，以及一个蓄水系统。这比植物对水的最低需求量要多的多，所以，特别是在夏天，他们会将多余的水蒸发掉，以创造一个比温度颇高的周边环境要凉爽很多的微气候。绿墙也会进行空气过滤和氧气生产。这是一个节约能源的好方法，在遮阳的同时改善了空气温度和空气质量。这栋建筑是公园的一个活跃部分，季节性的攀爬植物在夏天抽拔出了厚厚的绿叶，秋天时变得红艳动人，而在冬天叶落之后，你就可以很高兴地享受那透过窗户的阳光了。

这需要什么样的维护措施呢？

必须有一些措施保证攀爬植物在遮幕之上生长，但更有必要的是，要有些人去修理它们。如果它们在整座建筑之上生长，那实际上是没问题的。叶面积系数和植物的容积率是类似的，也就是在森林的地

better brings no reward, it's about doing less bad. That mindset is fundamentally flawed.

For us, we try to achieve the best possible solution technically as well as architecturally. At minimum, you should do no damage at all to the environment, right? It's a pretty simple concept. If you think about it, that means a zero-impact building. Today that's seen as a utopian goal among most green building professionals. But I am convinced that in the not too distant future it will be code minimum. That's what sustainability is. The question then becomes: can you do better? Is it possible to do projects that produce a real positive benefit to the environment? That's what we mean by productive architecture, that's really the goal.

What about your other productive projects in New York City?

One of the more interesting is the Bronx River Greenway River House, which is wrapped in a galvanized steel and mesh screen wall and covered in vines. You can use solar panels to make electricity, but plants are really the ultimate solar devices, so we looked for ways to integrate them architecturally and functionally into the building.

If you walk on a hot summer day through an old forest, it can be 52 degrees Fahrenheit [11 degrees Celsius] cooler than the ambient temperature because of the various layers of leaves on the trees, ferns, moss, etc. The idea was to turn that system on its side. We have a screen wall four feet away from the building (important for security) with vines growing on it, and a second layer, a kind of rain screen with moss growing on it, with drip irrigation to keep the moss moist. All of this is fed by rainwater coming from the roof, and a storage system. This is much more water than the plants minimally need, so, in summer in particular, they will transpire the extra water, creating a microclimate that should be significantly cooler than the ambient temperature. The green wall will also do some air filtration and oxygenation. It's a way of saving energy that improves air temperature and air quality while shading. The building is meant to be a living part of the park; the seasonal vines produce thick green summer foliage that turns brilliant red in the fall, and then the leaves fall off in the winter when you're happy to have the sun streaming in the windows.

What kind of maintenance is required?

Some to ensure that the vines grow on the screen, but more likely somebody will have to keep chopping them back. It's actually OK if they grow over the whole building. The leaf area index is like the FAR for plants. It's the leaf area per square foot of the ground area in a forest. We will end up with a net leaf area index that may be almost the same as if it were an old-growth forest.

What is the ultimate application of plants in a building?

To grow food. We've been collaborating with Dickson Despommier for a while. He introduced us to the Science Barge by New York Sun Works, an amazing floating educational facility that runs entirely on renewable energy: solar and wind turbines. It uses rainwater and can also purify river water to operate its hydroponic greenhouse. Working with NY Sun Works, Dickson, and Arup, we entered the Buckminster Fuller competition a few years ago to adapt the hydroponic process to integrated architectural systems. The NFT (nutrient film technique) system is a common way to grow vegetables hydroponically.

We decided to adapt this to a vertical system in a double-skin facade about a meter and a half deep. We were thinking about places like China and the Middle East, or very dense cities with more vertical than horizontal surface.

You don't want to have farm workers running up and down the facade weeding and picking lettuce in front of people's offices or bedrooms, so the system rotates vertically on a very slow cycle to a processing area at the bottom. It takes about thirty days to come around, and by then the plants are grown and can be replaced by more seedlings. We figured this could have other symbiotic benefits and worked with Arup to develop its function as a light-control device, a big venetian blind, essentially. The plants want as much light as they can get, but the building really wants very little—about 5% of sunlight is about right for indoor use—so you can adjust this daily or seasonally to optimize lighting for both the plants and the interior.

The depth of the double skin provides an intermediate zone between inside and outside where you can supply waste heat. Carbon dioxide–rich air exhausts from the building to the plants, and they give off oxygen in exchange. In hot weather, this greenhouse-plant system provides evaporative cooling. Double-skin facades have become a bit of a cliché in green architecture. This is a way to actually make them do something positive. Beyond growing plants, the biggest benefit, I think, is not the value of the crops or the energy. In a place like Dubai or many of the other fast-growing parts of the world where there is no water and little natural green, making plants part of our living or working space improves health, happiness, and productivity.

We did a proposal for the UAE to create an urban market with a double-skin envelope, both vertically and horizontally, to permit growing all around the market space. Building a

面上每平方英尺的落叶面积。我们最终会得出一个净叶面积系数，从这个系数看，它几乎就是一片原始森林中的一部分。

在一栋建筑中，植物的终极应用是什么？

生产食物。我们已经和迪克森·德波米耶合作了一段时间。他向我们介绍了"纽约太阳能工厂"的科学之舟，一座惊人的、完全使用可再生能源的浮动教育设施：太阳能和风力涡轮机。它利用雨水，同时也对河水进行净化处理，以便让水培温室可以工作。我们与纽约太阳能工厂、迪克森、奥雅纳一起工作，并在几年前参加了巴克敏斯特·富勒挑战赛，以改造水培过程和整合建筑系统。NFT（营养液膜技术）系统是以水培方式种植蔬菜的一般方法。

我们决定将这项技术改进应用到一个1.5米厚的以双层玻璃幕墙为表皮的垂直系统中去。我们考虑的是像中国和中东这样的地方，以及人口非常密集、建筑表面更多是垂直而非水平的城市。

你不会想让农场工作人员在人们的办公室或卧室前面上上下下除草和采摘生菜，所以这个系统非常缓慢地在垂直方向上循环旋转，以到达底部的处理区域。旋转一个来回用30天，这样的话，植物就开始生长，而且能够被更多的幼苗所替代。我们发现这可以有共生好处，然后我们与奥雅纳合作，将其功能发展成有控制光线功能的设备，当然，其本质是一种很大的百叶窗帘。植物想要尽可能多地获得阳光，但是建筑想要的很少——室内的阳光使用需求大概只有5%左右——所以你可以每天对此进行调节，或者季度性地为植物和建筑内部提供阳光上的优化。

双重表层的厚度在建筑内外之间形成了一个中间区域，你可以以此提供余热。从建筑中向植物排除丰富的空气废弃物——二氧化碳，而植物则生产出氧气作为交换。在炎热的天气，这种温室植物系统提供了蒸发制冷。双重皮肤的表层已经成为了绿色建筑中的陈词滥调。这是一种使它们实际上能够发挥一些积极作用的方式。除了种植植物，最大的益处我认为并不是农作物或能量。在像迪拜或世界上许多其他快速发展的地区，几乎没有什么水，也没有天然的绿色，植物成为我们生活和工作场所的一部分，会促进健康、幸福和生产效率。

我们为阿拉伯联合酋长国做了一项提案，建造一个有双重表皮外壳的城市市场，垂直和水平的方向都有，这样就允许在市场空间的周围栽种植物。在炎热的气候中为人们建造一个温室看起来非常荒唐可笑，但实际上它可以运作得很好，因为植物有它自身感觉舒适的气候范围，这个范围比人的可要大多了。它们在更加炎热、更加潮湿的气候中感觉十分良好。你可以利用蒸发冷却来做很多事情，这会让能源需求最小化。你也可以处理内层和外层的玻璃，并在各自独立的玻璃层上单独进行阳光控制。这个中间区域，是一种气候条件，而内层则设计得让人感到舒适。食物的运输和冷冻都不是问题了，因为它们就在那里。

这些原则和技术有没有用于高层建筑？

在2001年，我们应国家建筑博物馆的委托，为"大+绿"展览设计了一个假想办公楼。在与奥雅纳的合作中，我们为所有建筑物确定了新的生态、都市规划和生命质量标准。当技术改进的时候，这种建筑就不是一个乌托邦的景象，而是我们精心设计，使之到2020年时可以变成现实，变得经济的建筑。

2020年大楼会有150层高——800米（870码）——这个高度在目前看起来几乎难以想象，但是奥雅纳的结构小组说这是可能的。当然，现在有更高的建筑了，甚至一些还要高的建筑也在规划之中。我们在那时并不知道这一切会发生得如此之快。

我们给那栋建筑设计了很薄的外层封装，以获得更好的阳光、视野和自然通风。高层建筑一般来说都有一个中间核心，那样是有效且廉价的。建筑周长被最小化，因为这一项耗资昂贵，不管在夏天还是冬天，你都会经过外皮获得和失去热量。但如果有更高端的技术和材料，你会不会只从表层获得能量而不是失去能量呢？你肯定需要尽可能大的表层。

通过拓展这一紧凑的规划，我们将核心分为几块。这是一个根本性的安全问题。看到911中世贸大厦的倒塌，我再也不会设计只有一个出口的高层建筑了。我们在建筑的外围边缘安装了楼梯，所以你实际上可以从那里看到外边。这些楼梯被仔细设计，成为建筑中最精致的地方之一，因而人们将会想要使用它们。当然，这花费不菲，但却很值得。这是建筑周长很长的一个例子，因为值得这么做。

在我们的技术分析中，我们试图真的严格起来，因为在一般所谓的绿色世界中有太多的欺骗行为，有很多说自己是零能耗的建筑甚至连接近这一标准都做不到。我们得出结论认为，一栋多种用途的大楼，其平均的能耗大概会是每年55千瓦时每平方米（66千瓦时每平方码），大约是今天一栋典型办公楼的四分之一。住宅的能耗会小一些，这

greenhouse for people in a hot climate seems completely ludicrous, but it could actually worked pretty well because plants have their own climate comfort range, which is wider than people's. They're happy in hotter, more humid conditions. You can do a lot with evaporative cooling, which minimizes energy demand. You can also deal with both the outer and inner glass layers and do solar control individually on each. You have this intermediate zone, which is one climate condition, and then the interior, which is designed for human comfort. Food miles and refrigeration are eliminated because it's all right there.

Do these principles and technologies apply to tall buildings?

In 2001, we were commissioned by the National Building Museum to design a hypothetical office building for the "Big + Green" exhibition. In collaboration with Arup, we targeted new ecological, urbanistic, quality-of-life standards for tall buildings. While technologically advanced, the design is not a utopian vision but a building we carefully engineered to be practical and economical by the year 2020.

The 2020 Tower is 150 stories high—800 meters [2,616 feet]—which seemed unimaginably tall at the time, but Arup's structural group said it was possible. Now, of course, taller buildings exist and even taller ones are being planned. We didn't know then that this would happen so quickly.

We gave the building a thin footprint for good daylight, views, and the possibility of natural ventilation. Tall buildings typically have a single central core because it's efficient and cheap. Their perimeter is minimized because it's expensive, and you gain and lose heat through the envelope in both summer and winter. But what if, with advanced technology and materials, you gain energy from the skin instead of losing it? You'd want as much skin as possible.

By spreading out the compact plan, we break the core into pieces. It's a basic safety issue. After watching the World Trade Center fall on 9/11 I would never design a tall building with only one way out. We put the stairs on the perimeter so you can actually see out. The stairs are deliberately some of the nicest places in the building, so people would want to use them. This costs more, of course, but there is value to it. This is an example of building a lot of perimeter because it pays to do so.

In our technical analysis, we tried to be really rigorous because there is a lot of smoke and mirrors in the green world generally, and lots of projects that claim to be zero energy but aren't even close. We concluded the average energy use of the mixed-use building would be about 55 kilowatt-hours per square meter [66 kilowatt-hours per square yard] per year, about a quarter of the typical office building today. Residential energy is somewhat less, so that factors in.

The building's integrated photovoltaic skin uses advanced thin-film modules, more than 15% efficient and similar in cost to coated laminated glass. The output from the opaque BIPV spandrels, and transparent BIPV windows (about one-third the efficiency of the opaque units) provide two-thirds of all the energy consumed by the building. We assumed it would be surrounded by 50-story structures—these are New York City conditions— and given the fact that the tower shades itself with its wings, wind turbines were used to provide the remaining one-third of the building's energy.

Building-integrated wind is problematic and most of the time won't work well. I'm more convinced of that now than I was back in 2001, but in this case we figured that at the top of such a very tall structure surely there would be good wind. This is assuming year 2020 technology. If you wanted to build the 2020 Tower today, you could make it zero energy by putting additional PVs above a 4-story parking garage big enough to accommodate the building's required parking.

The tower is divided into 30-floor units vertically. Every 30 floors, there is a major public level, with some special event like a hotel or theater complex, a university, or botanical garden, and also smaller-scale services and retail. There are express elevators to each public floor. Also, every 30 floors there would be outdoor parks in the sky, with wind shielding and, with 9/11 in mind, space for helicopter landings. You'd never be more than 15 floors from rescue.

Below each public level is a mechanical and hydraulic floor ringed by greenhouses containing biological waste–treatment systems; this allows reuse of more than 90% of the building's water. Waste is a source of energy as well. Systems like plasma gasification can cleanly extract a lot of energy from waste.

Aquaponics, which is growing fish in conjunction with vegetables, would work well in this building. Plant nutrients drain into the fish tank and the fish wastewater fertilizes the plants in a self-sustaining system. Fish don't need much light, so they occupy otherwise undesirable areas.

The daytime population of the 2020 Tower is about 50,000. Close to the bottom, the building gets deeper in plan. In the commercial market for office space, you always hear people insisting that the bigger and deeper the floor plate, the better. I think that is not true. Most operations would benefit from a daylight environment.

Most of the building is no more then 15 meters [49 feet] deep, which means it can be easily daylit and easily reconfigured: all residential, all office, or any mix of the two you want.

被纳入重要因素之一。

建筑的集成太阳能光伏外层使用了现今的薄膜模块，它和涂层玻璃在成本上相差不多，但性能上高出15%。整座建筑所消耗能量的三分之二，都是不透明的BIPV拱肩以及透明的BIPV窗户（它的功效大约是模糊部分的三分之一）产生的能量提供的。我们是假定了建筑将会被50层高的构筑物所包围——这些就是纽约的环境——而且，鉴于这座建筑的侧翼遮蔽了自身这一事实考虑，风力涡轮机被用来提供剩下的，建筑消耗总数三分之一的能量。

风势与建筑融为一体时是问题的，而且大多数情况下效果不好，比起2001年，我现在更加确信这一点。但在这个案例中，我们认为在如此高的结构顶部，当然会有很好的风势。这可以假定是2020年的技术，如果你想要在今天建2020大厦，你可以通过放置额外的太阳能光伏板在四层高的足以满足整栋楼停车需求的停车楼上，就可以实现零能耗。

在垂直方向上，大楼被分为多个30层构成的单元。每隔30层就都有一个大的公共层，里面有一些特殊的东西，像酒店或者影院的综合设施，一个大学，或植物园，以及规模较小的服务和零售中心。有一些特快电梯可以直达每个公共层。而且每隔30层还会有空中的室外公园，由于一直想着风阻和911事件，所以还为直升机的降落留下了空间。救援距离绝对不会再超过15层了。

在每个公共层下面，是一个机械和液压层，环绕着具有生物废弃物处理系统的温室，这将会重新利用超过90%的大楼水源。废弃物同样也是能量的来源，像等离子气化这样的系统，可以无污染地从废弃物中提取到大量的能量。

鱼菜共生，即成长中的鱼同蔬菜在一起，将会在这种建筑中良好运作。在一个自给自足的系统中，植物养分沉入到鱼池中，鱼产生的废水则用于给植物施肥。鱼不需要太多的阳光，所以它们占据了其他一些不太需要的空间。

2020大厦每天的人流量大约是5万人。在接近底部的地方，建筑平面就更深些。在办公空间的商业市场上，你总是听到人们坚持，楼层越大、越深就越好，我认为这并不对，大部分的运作都会从日光的环境中获益。

大多数建筑在地下的部分都没有超过15米深（16.5码），这意味着它能够很容易获得阳光，也很容易被重新配置：全部是民居，全部是办公室，抑或依照你的要求混合搭配。典型的多用途建筑都是酒店、办公室、公寓混合的，每个功能都被放在独立的区域中，既不生动，也无趣味。在这里我们试图创造一个将功能和体验相结合的组合体来实现更好的社会可持续性。

我们预测，在2020年，能源消耗将少于60千瓦时每平方米。在今天，一栋典型的建筑要消耗200-250千瓦时，而非常节能的建筑则接近于100千瓦时。所以我们正朝那个目标迈进。我们也预测，就生产方面来说，和今天相比，我们将会有功效高出18%，每平方米的价格却要便宜得多的薄膜太阳能光伏板，现在这种技术只是提升了5—12%的功效。你现在可以买到功效提高20%的太阳能光伏板，但它们的价格要贵很多。我们也为住宅、办公室、学校做了平均使用量的假定。关键在于，有了这样的假定，我们就可以只通过大楼的表皮和风力涡轮机，使2020年的建筑变成零能耗的；今天如果我们有额外的场地来建设太阳能板覆盖的停车场，也可以实现零能耗。

都市农业在建筑中扮演什么角色？

在2001年，没有人谈论与建筑融合在一起的农业。从那之后，我们已经意识到食物大约占据了城市环境足迹的40%，这是个大问题。所以我们增加了3英寸厚的双重表层（它有可能进一步优化变得更纤细），然后将生长系统置入其中。你总是不时地需要一大块的加工区域，所以每隔10层都有一个专属的楼面空间用于储存和加工食物。

营养指南建议每人每年消耗100千克（220磅）的新鲜果蔬——平均下来美国人没有吃这么多。我们的垂直温室系统将生产出楼层中的办公室工作人员所要消耗的全部蔬菜。而对于一栋居民楼而言，你所生产的将是居民所需总量的5倍之多。当然，无论哪一层被用于生产食物，你都至少是放弃了一些太阳能的产品。

可能会有一些很有趣的方式来鼓励这种发展。纽约市的环保法规已经使额外增加的墙（为了高度绝缘的关系）所产生的厚度不算在楼层面积之中。同样的规定也可以被用到垂直温室上，它已经应用在了水平的顶棚温室中了。如果你那么想的话，你就没有因为双层墙而失去任何空间，它也没有限制墙后面的用途，不管是用于办公室、学校还是公寓。

我们从未停止在这个项目上的工作。它已经成为了一种内部的研发设计项目，一个实验新想法的载体，就像垂直温室一样。这个项目为两个不同的美国国家科学基金课题提供了灵感——这两个课题探究建筑中独立自足的基础设施的涵义。当然，我也曾试图

The typical multi-use hotel/office/apartment building mix, in which each function is stacked in discrete zones, is not very dynamic or interesting. Here we were trying to create a combination of uses and experience for better social sustainability.

We assumed that in 2020, energy consumption would be less than 60 kilowatt-hours per square meter [646 square feet]. Today, a typical building consumes 200 to 250 kilowatt-hours, and very efficient buildings are approaching 100. So we're getting there. We also assumed, on the production side, that we'd have 18% efficient thin-film PVs at much lower cost per square meter than today; now this technology is only about 5 to 12% percent efficient. You can now buy PV panels that are about 20% efficient, but they're much more expensive. We also made average-usage assumptions for residential, office, schools. The point is that with these assumptions, we can make the 2020 Tower zero energy just from its facade and wind turbines; and today we could do it if we had some extra site area for that solar-covered parking garage.

What role does urban agriculture play in the building?

Nobody was talking about building-integrated agriculture in 2001. Since then, we've realized that food is about 40% of the environmental footprint of the city; it's a huge issue. So we added a 3-foot [.9-meter]-deep double skin (which could probably be optimized thinner), and the growing systems fit inside. You'd need a larger processing area every now and then, so every 10 stories there's dedicated floor space to store and process the food.

Nutritional guidelines recommend 100 kilograms [220 pounds] of fresh produce per person per year—not that average Americans eat that much. Our vertical-greenhouse system would produce all the vegetables consumed by the office workers on that floor. For a residential floor, you produce about 5 times as much as the residents need. Of course, for whatever area is dedicated to growing food, you're giving up at least some of the solar-energy production.

There might be interesting ways to encourage this kind of development. Already, green codes in New York City exempt extra wall thickness for very high insulation from floor area. The same provision might apply to vertical greenhouses. It already does for horizontal rooftop greenhouses. If you think that way, you're not losing any space from the double wall, and it doesn't limit the uses behind, whether for offices, schools, or apartments.

We've never stopped working on this project. It has become a kind of in-house R&D program, a vehicle to try out new ideas, like the vertical greenhouses. This project became the inspiration for two different National Science Foundation grants looking into implications of self-sustaining infrastructure in buildings. And, of course, I've been trying to persuade real people to actually build one.

What is your relationship with Dickson Despommier?

We're architects, so we come at this from a particular point of view. Dickson comes at it from the perspective of public health, epidemiology, and global ecology. His vision is a big one: you stop industrial farming as we know it, grow food in dense, controlled environments, and let all that farmland revert to nature. That is the best carbon sequestration strategy possible, with incredible benefits for the environment generally, eliminating the poisoning of the ocean, desertification, etc. I am skeptical of the idea of building towers to grow every type of food in a single building, in prime midtown sites. But that's not to say the basic principle will not work well in the urban environment. What appeals to us is that integrated agriculture is a multifunctional form of renewable energy. Multifunction is a principle we try to apply to everything we do; we find that if something is multifunctional, it works better and is more elegant as a design, from the shape and the footprint of a building to onsite energy generation, each reinforcing the other. In this case, agriculture improves the building, not only in terms of energy efficiency and food production but psychologically for the people inside who are removed from the earth.

It's important for these principles to be on good trajectories. So even if some extreme visions of vertical farm towers don't make much sense, integrated agriculture brings so many other benefits to the table. If natural elements in buildings means a 1% improvement in education performance, or a 1% reduction in office absenteeism, or a 1% sales increase in retail environments, the value gained is 10 times as much as the value of energy saved. That's why we're integrating it into our fabric in a visible way. It has real value.

Have there been significant innovations in the design of tall buildings in consequence of 9/11?

I have to say that the lack of innovation that we've seen is disappointing. 9/11 was an opportunity in so many ways that was not embraced, least of all in building design. There have been a lot of misleading and overhyped claims about sustainability in tall buildings, incremental improvements in safety that don't go far enough, and not much innovation in terms of social sustainability. The building sector is very conservative!

说服人们真的去建设一栋这样的建筑。

你和迪克森·德波米耶是怎样的关系?

我们都是建筑师，所以我们从各自独特的角度明白了这一点（对于建筑的看法）。迪克森是从公共卫生、流行病学和全球生态出发的，他的视野是很大的。你不再像我们知道的那样采用工业化的农业生产模式（在控制的环境中密集地种植粮食），而是让所有的农田退耕，还给自然。这是有可能的，最好的固碳策略对总体环境有着巨大的共享，比如消除海洋污染、荒漠化等等。在市中心地区，建造一栋建筑，然后在其中种植所有类型的食物，我对这一点是有所怀疑的。但这并不意味着在城市环境中，这条基本原则是无效的。吸引我的是，这种综合农业乃是可再生能源的一种多功能形式，多功能是我们想要应用在一切事情中的一条原则，我们发现如果某种事物是多功能的，它会运作的更好，而且作为一种设计也更优秀，从建筑的形状、轮廓到就地的能源生产，都会互相增进。在这样的情况下中，农业提升了建筑，不只是在能源效率和食物生产方面，而且对那些离开地面、高层中人们的心理也是有很大帮助的。

这样一些原则在正确的轨道上运作是非常重要的。所以即使有一些极端的现象，比如，垂直农场的大楼没有发挥太大作用，综合农业还是明显带来了很多其他的好处。如果建筑中的自然因素意味着在教育效率上获得了1%的提升，或者在办公缺席率上减少了1%，或者在零售环境销量上升了1%，那么它所获得的价值，就是节能所获得价值的10倍。这就是为什么我们以一种可见的方式将它整合到我们的建筑中去。它有真实的价值。

伴随着911事件的后果，高层建筑设计上有没有重要的创新?

我必须说，创新的缺乏非常令人失望。在很多方面，911都是一个机会，但却没有把握住，至少在建筑设计领域如此。出现了许多误导性的、过度炒作的关于高层建筑中可持续性的主张，而在安全性上的提高却远远不够，在社会的可持续性方面也没有太多的创新。建筑部门还是相当保守!

2020 Tower: university exterior (top), botanical garden interior; right: 2020 lower wall section.
左上：2020大厦，大学外部（顶部）；左下：内部的植物园；右：2020大厦低处墙面。

可持续性建筑或者生产性建筑是未来的希望么？

最终必定是。所有这些生产性技术变得越来越便宜，而传统方法的费用则不断攀升。所以我认为，市场因素结合了城市化的压力、不断增长的气候问题，所有这些将会让它成为未来的希望。我只是尝试着从设计策略的角度进行思考，这些策略我认为根本上是有意义的，而且在技术上是可行的。如果我们不相信一种技术会在不久的将来变得更经济，那么在它身上花时间就是一种浪费。

Is sustainable or productive architecture the promise of the future?

Ultimately it has to be. All these productive technologies are becoming less expensive, as the cost of conventional approaches rise. So I think the market, combined with the pressures of urbanization and increasingly obvious climate issues, will make it happen. We try to consider only design strategies that we think make sense fundamentally and are technologically realistic. If we didn't believe a technology would become economical in the near future, it would be wasteful to spend time on it.

Ashok Raiji
ARUP

Ashok Raiji, P.E., LEED AP, is a principal in the New York office of Arup and has led the planning of city-scale developments, including the master planning of new eco-cities in Asia. He is Arup's commercial, residential and retail business leader in the Americas, and leads Arup's global residential business. Mr. Raiji has led the design of many high-performance buildings around the world, including, most recently, the Songdo Convention Center and the 1,050-foot-high [320m] Northeast Asia Trade Tower, both in Korea, as well as the Boston Seaport Sustainability Master Plan, and the Kresge Foundation Headquarters (LEED Platinum rated). Mr. Raiji lectures frequently on Sustainable Design and has served as technical adviser on sustainability for the National Building Museum and the Museum of the City of New York. He is a LEED Accredited Professional and a licensed professional engineer in 25 states. Mr. Raiji is a visiting professor of architecture at the Irwin S. Chanin School of Architecture at The Cooper Union in New York.

阿肖克·瑞吉
奥雅纳

阿肖克·瑞吉，职业工程师，能源与环境设计指导认证专家，他是奥雅纳纽约办公室的负责人，主持过城市发展规划，包括在亚洲的新型生态城市的总体规划。他是奥雅纳在美国的商业、住宅和零售业务方面的领导者，也主管奥雅纳的全球住宅业务。瑞吉先生在全球范围内主持了许多高性能建筑的设计，包括最新的仁川会展中心和高1050英尺（320米）的东北亚州贸易大厦（这两座建筑都在韩国），还有波士顿海港可持续发展总体规划，斯吉基金会总部（获得LEED白金认证）。瑞吉先生经常就可持续性设计发表演讲，并担任国家建筑博物馆和纽约市博物馆可持续性问题方面的技术顾问。他是能源与环境设计指导认证专家，在25个州获得职业工程师执照，同时是纽约库伯高等科学艺术联合大学建筑学院的访问教授。

For complete biography, please refer to the Appendix.

完整的传记，请参阅附录。

Drivers of Change, first printed by Arup in 2006, is an ongoing series that typologically addresses key issues affecting the built environment.
《改变的动力》由奥雅纳于2006年第一次印刷出版。它是一系列还在进行的，分类阐述影响建设环境关键问题的论述的书籍。

What role does sustainability play in modern architecture and engineering?

People are beginning to assume that it's just part of good design practice. There have been times when I would sit in a meeting and try to count how many times the word sustainability was used. There are fewer and fewer now. In fact, I was in a meeting not too long ago where we didn't use the word at all. When you have like-minded people who are in tune with the environment and are being responsible, you don't have to mention it; it's just part of what we do

Is this a relatively new phenomenon?

Well, the notion of sustainability has actually been around a long time; it's just that the media has gotten hold of it over the last 15 years or so. It started off with a general concern about the environment, spurred by two milestone events. One was in the late 1990s, when the IPCC [Intergovernmental Panel on Climate Change] concluded that what we were doing on our planet was impacting global warming and climate change. That brought a lot of awareness to sustainable design. The second catalyst was Al Gore. He had a high profile—everyone knows Gore—and he had a very good message and presented it well. What's pushing it even more now is the price of energy. In the U.S., energy is still about a third of what it is in Europe. Let's use gasoline as a baseline since everyone can relate. When gasoline was $2 per gallon [3.8 liters] in the U.S., it didn't matter, but now it's approaching $5 in some places. So people are paying attention. Also, in terms of building design, a lot of our professional organizations, like the American Institute of Architects, American Consulting Engineers Council, etc., have put it at the top of their agendas. Just about every professional organization that deals with the built environment has a "plan" that addresses greenhouse gas emissions and climate change.

There was actually one more trigger: the Kyoto Protocol. The U.S. signed it but Congress didn't ratify it, which for people in the industry—certainly for me—was a big source of embarrassment. A lot of cities decided to enact legislation and essentially ratify on their own. It just kind of gained a lot of momentum and ultimately over 100 cities got involved. It started off with cities like Seattle, Portland, and New York, and the desire of the leadership (mayors) of American cities to do what our federal government refused to do has led to over 100 cities putting in local laws and regulations for reduction of greenhouse gas emissions. New York is actually a very sustainable city. From the standpoint of transportation, it can't be

在现代建筑学和工程学中，可持续性扮演着什么样的角色？

人们现在正开始假定，它只是良好设计实践的一部分。有好多次当我坐下来参加会议时，我都试图数清楚"可持续性"这个词到底被使用了几次。现在，这个词的使用是越来越少了。实际上，在不久之前我参加了一次会议，会上我们根本就没有提到这个词。当你与一些志同道合的人合作时，他们总强调与环境相协调，而且勇于担负责任，那么你就没有必要专门提到这个词；可持续性就是我们工作的一部分。

这是不是一个相对较新的现象？

实际上，可持续性的概念已经出现很长时间了；只是媒体大约在15年前才开始理解这个词。它开始于对环境的普遍关切，而两起里程碑式的事件则起到了刺激推动的作用。一是发生在上世纪90年代后期，IPCC（跨政府气候变迁委员会）得出结论说，在地球上，我们当前的所作所为正在导致全球变暖和气候改变。这使得很多人开始了解、认识可持续性设计；第二个刺激因素是阿尔·戈尔(Al Gore)。他有着很高大的形象，每个人都知道戈尔，他有一个非常好的信息，而且将其表述的很好。今天，对这个信息更大的推动来自于能源的价格。在美国，能源的价格依然是欧洲的三分之一。让我们以汽油作为一个基准——因为每个人都与之相关。当美国的汽油为2美元每加仑的时候，问题不大，但是现在它在某些地方已经接近于5美元每加仑。于是人们便开始关注这些问题了。而且，从建筑设计的角度说，我们许多专业的组织，像美国建筑师学会，美国工程咨询委员会等等，都将这一问题作为他们日程中的头等大事。几乎所有处理建筑环境的专业机构，都有一个针对温室气体排放和气候变化的"规划"。

实际上，还有另外一个促发因素：京都议定书。美国在议定书上签了字，但国会并未正式批准它，对于这个行业中的人来说（当然对我也是如此），这是个非常大的、引起尴尬的原因。许多城市已经决定立法，并从根本上批准了这一法律。这确实产生了很大的推动力，最终超过100个城市投入到这一运动中。它从西雅图、波特兰和纽约这样的城市开始，美国城市领导者们（市长们）想要做那些我们联邦政府拒绝做的事情，这已经使得超过100座城市通过地方法律、法规，来减少温室气体排放。纽约实际上是一个非常具有可持续性的城市。从交通运输的角度上说，它不能再好了；公共客运系统非常棒。幸运的是，我们有很好的水源供给，对土地的使用也很聪明。城市密度很大，因此也产生了许多好的事情。

奥雅纳如何顺应不断改变的需要？

当奥韦·奥雅纳（Ove Arup）于大约70年前成立这家公司的时候，它主要关注的是结构设计。在上世纪60年代后期，奥韦开始有了这样的想法：好的设计是整体设计，也就是说，这不是关于一个单一学科的问题，而是不同学科的综合。正是在那之后，我们拓展到了机械、电气、水暖、消防、声学等领域。我们也规划和设计了许多基础设施，比如道路、桥梁、高速公路、机场、铁路系统等。例如在纽约，我们就设计了地铁第二大道线。与此同时，我们也做城市规划。根本上说，我们是处于建筑环境中的设计师，项目是什么并不重要，不管是一栋楼房、桥梁还是工厂，总是有许多的协同合作。我们的结构工程师向桥梁设计师学习，桥梁设计师也向结构工程师学习，拥有不同技术能力的人之间相互影响，创造出了一些思虑周详的项目。当然，在经济环境不佳的时候，多样化也是好的。我们实际上比大部分人知道的要更加多样。现在，我们在全球至少35个国家有大约100个办公室，这些办公室的员工人数差不多有10000人。

奥雅纳有没有一个专门的可持续性部门，抑或这是一种普遍认同的企业文化？

我们倾向于以团队的形式工作，团队提供了一个组织结构，因而人们可以得到监督和指导。但是，每个人对于可持续性设计都完全负责，因而每个机械工程师或电气工程师都可以谈论、设计各种类型的环保系统。在某种意义上，我们的"可持续性顾问们"更多的参与是可持续的理智层面内容，像企业的社会责任等。他们也已经研究了一些我们称之为"变化驱动力"的东西，比如说，那些能影响我们生活方式、工作方式、将来我们开展自己业务方式的东西。我们也尝试去期望、去思考像将来的宾馆会是什么样的这类问题。它们会和现在的一样还是不同？是什么让它们发生了改变？

better; the mass transit system is amazingly good. Fortunately we have a good water supply, and land use is brilliant. Urban density is high, so that gives rise to a lot of good things.

How has Arup responded to changing needs?

When Ove Arup founded the firm almost 70 years ago, it basically focused on structural design. It was in the late 1960s that Ove embraced the idea that good design is total design, i.e., that it isn't about a single discipline but about different disciplines coming together. It was then that we expanded into mechanical, electrical, plumbing, fire protection, acoustics, and so on. We also plan and design lots of infrastructure, i.e., roads, bridges, highways, airports, rail systems. We have designed the Second Avenue subway, for example, here in New York. We also do urban planning. Ultimately we are designers within the built environment. It doesn't matter what the project is, whether a building, bridge, or factory, since there are lots of synergies. Our structural engineers learn from our bridge designers and vice versa, and the interactions between people of different technical skills create projects that are holistically thought out. Of course, diversifying is also good in a bad economy. We are actually much more diverse than most people know. Right now, we have roughly 10,000 people in about 100 offices in at least 35 countries around the world.

Does Arup have a sustainability department or is it a widespread corporate culture?

We tend to work in teams, which provides an organizational structure so that people can be supervised and mentored. But everybody is completely up on sustainable design, so any mechanical engineer or electrical engineer can talk about and design all kinds of environmentally responsible systems. What our "sustainability consultants" get involved in is, in a sense, more of the intellectual side of sustainability, like corporate social responsibility. They have also been researching things that we call "Drivers of Change," i.e., things that will affect the way we live, work, and conduct our business in the future. We also try to look forward and think about, for example, what hotels are going to look like. Are they going to be the same as now or different, and what's going to drive them there?

What is the outlook for future cities?

The city is in real crisis mode, not so much in the U.S., where about 70 percent of the population lives in urban areas, but in the developing world, particularly China and India. They have some really serious problems. There are many different statistics, but effectively what it boils down to is a staggering increase in urban population. Over the next 15 to 30 years, China will have to build 40 cities with populations of 5 million or more. Think about that. And India is about the same, envisioning something like 50 cities in the next 20 years. It's a very serious problem.

How do you distinguish India's problems from those in China?

The problems are similar. Although the two economies are actually quite different, both are very successful. Right now China's economy depends on exports; more and more products are being manufactured in China and then sent all over the world, while India's GDP increase is actually due to domestic consumption. That's a significant difference. In India, particularly in the big cities, pressure on land is so great that prices are completely out of whack. The rich get richer and the poor get poorer. In Bombay, where I grew up, land prices are close to the high end of Manhattan real estate. I believe there will be a bust pretty soon; it just cannot continue.

Another difference is that the middle class in India has really improved its lifestyle over the last 20 years. But having said that, I think China's top-down government offers much greater potential to solve problems, whereas India is a democracy; things take time, they have to build consensus. To give you an example: About two years ago we were planning a new city of about 300,000 to 400,000 people in China's Hunan Province, and we got to talking about how we were going to get electric power out there. I sat down with the local providers, and they said it wasn't a problem, that they could essentially design, build, and commission a power plant in about 18 months. In 18 months, we can't even finish the paperwork in the India! China's way of government allows things to happen very, very quickly. So I think China will solve its organizational problems a lot quicker than India will.

Doing things quickly doesn't necessarily mean doing them well.

I don't think that applies in China, because there is a lot of good planning going on. In fact, planning in China is arguably at a much more advanced level than in India. In India there's a kind of business-as-usual scenario, whereas in China they definitely want their new cities to be better than their legacy

将来城市的前景如何？

城市确实是一种处于危机之中的模式，这一点在美国可能不是那么明显，因为这里70%的人口生活在城市地区；而在发展中世界，特别是在中国和印度，它们已经产生了一些非常严重的问题。有许多不同的统计数据，但是实际上归根结底就是一点，城市人口在令人吃惊的速度增长。在将来的15-30年间，中国必须要建立40座人口达500万甚至更多的城市，想想这些吧。印度的情况也差不多，正规划着在接下来20年中建立50座新城市，这真是一个非常严重的问题。

你如何区分印度的问题与中国的问题？

二者的问题非常相似。尽管两个经济体实际上非常不同，但都很成功。现在，中国的经济依赖于出口；越来越多的产品在中国制造，然后在全世界范围内销售，而印度GDP的增长实际上得益于国内的消费。这是一个重大的差别。在印度，尤其是在大城市，土地的压力如此巨大以致于其价格已经完全不正常了。富人越发富有，穷人越发贫穷，在我从小长大的孟买，土地的价格已经接近曼哈顿高档地区的房地产价格。我相信很快就会有一次崩盘，这并不能持久。

另一个差别是，印度的中产阶级在过去20年间确实已经提升了他们的生活方式。但尽管如此，我认为中国自上而下的政府管理机制提供了更大的解决问题的潜力，而印度是一个民主社会，做什么都要时间，他们必须达成共识。举个例子：大概两年前，我们在中国的湖南省规划一座约有30到40万人口的新城市，我们必须说清楚准备如何给新城市提供电能，我与当地的供应商坐在一起，他们说这不是问题，他们大致可以在18个月内完成一座发电厂的设计、建设、委托施工。18个月，我们在美国连文书工作都做不完！中国政府的办事方式允许事情发生的非常非常快。所以我认为，中国解决它的组织问题将要比印度快很多。

但做的快并不就等于做得好！

我不认为这适用于中国，因为有许多规划都非常出色。实际上，和印度相比，中国的规划可能比印度要处于一个高很多的水平上。在印度，有一种照常营业的方案，而在中国，他们确实想让新城市比老城市来得更好。对于再生能源、温室气

Smart Solution For Cities: An iPhone uses augmented reality to indicate the distance to the nearest rail and metro stops (one of the Smart City examples presented by Arup to the C40 and City of Melbourne workshop on Smart Cities).

城市智能解决方案：一部iphone手机使用增强现实技术来指出到最近铁路和地铁站的距离（这个关于智能城市的例子，是奥雅纳向C40城市和墨尔本智能城市工作组陈述的）。

体排放、能源效率等问题，政府有非常强有力的政策。在这一点上，印度和中国，还有印度尼西亚、泰国、马来西亚和美国的发展中部分——实际上都见证了智能城市的诞生。这不仅仅是市场营销，实际上它已经发生了，人们对这个问题是很认真的。

　　例如，思科系统已经在韩国迈出了第一步。他们正在实施所谓的智能连结城市。我喜欢称之为城市信息基础设施，从根本上说，它将可获得的现有电子信息收集起来，然后以某种方式利用这些信息，来帮你做出明智的决定，让你能够做很多高效率的事情，做很多使人们生活得更容易的事情。

　　简单地以等待一辆公交车为例。在纽约，你站

cities. The government has very strong policies on things like renewable energy, greenhouse gas emissions, energy efficiency, etc. Both India and China—and, for that matter, Indonesia, Thailand, Malaysia, and developing parts of the U.S.— are actually seeing the birth of the intelligent, or smart, city. It's not just marketing; it's actually happening. They are very serious about it.

Cisco Systems, for example, has taken a big first step in Korea. They are putting in what they call smart connected cities. I like to call it urban information infrastructure, which is basically taking all the electronic information now available and somehow using it to help you make informed decisions. It lets you do lots of good things in terms of efficiency, in terms of making people's lives easier.

Take something simple like waiting for a bus. In the past, you would stand at the bus stop with no clue as to when the bus would arrive. But now, in cities like metropolitan Seoul, the entire bus system is tracked by computer. At every stop, there is a display that posts when the next bus will arrive. You can also get that information pushed to your cell phone so you'll know that the bus is 5 minutes away. You no longer have to stand there waiting. It's all about making life easier.

The same kind of technology applies to just about everything. Some of it is a little gimmicky, arguably. But for elevators, for example, there is a dispatch system that is being used to improve the efficiency of elevators. Instead of pushing a button to go up, there is a keypad where you key in your floor. The elevator system is able to see how many people are going to, say, the 8th. If a large number of people want to go to the 8th floor, the system will dispatch an express elevator. Anybody going to the 8th floor will be sent to elevator number 3, so instead of stopping at floors 5, 6, 7, it just goes directly to 8; it improves efficiency. That system has existed for the last 5, 7, maybe even 10 years. The information infrastructure can take information from a security system and integrate that with the elevator system remotely. So you walk into your office building and swipe your ID card—in fact you don't even have to swipe it anymore; it's read remotely. Now the system knows who you are and knows that you work on the 8th floor. It will tell you as soon as you walk by a security device to go to elevator number 3.

What is the importance of tall buildings in the future?

I think they are the future for many of these cities because with the exception of the U.S. and Canada, land is in very short supply. You can't get more efficient than the skyscraper in terms of land use or numbers of people per acre. But it's an expensive building and it comes with challenges, like life safety and elevatoring and things like that. These are challenges that the industry has solved very, very well. When the Burj Khalifa opened three years ago at 800 meters [2,625 feet] tall, that was a step change. Today there is a building being designed in Saudi Arabia that is a kilometer [3,281 feet] tall. We at Arup have been through the exercise; there is really no notional limit; we can design buildings that are a mile [1.61 kilometers] high if we wanted to.

When you think of it, the construction industry is really still quite primitive. We are using the same materials: steel and concrete. We have not started to look into the real space age materials like carbon fiber, and things like that, which are much lighter, much stronger, much more flexible. That's one

在公交车站，对于公交车将何时抵达毫不知情。相反，在首尔都市圈，整个公交系统都由电脑跟踪。在每个停靠站都有一块显示屏，将会显示下一班公交车何时会抵达。你也可以让信息推送到你的手机中，进而迅速获得信息，这样你就会知道公交车五分钟后就会到达，你不需要站在站台继续等待，所有这些都让人们的生活变得更便捷。

同样类型的技术被应用到几乎所有东西上，其中有些技术可能略有"噱头"之意。例如电梯，有一套可用来提升电梯效率的调度系统。它有一个键盘，你按一下按键就可以到自己的楼层去，不用再像以前那样按按钮。电梯系统可以知道有多少人想要到某个地方，比如到八楼去。如果有很多人想去八楼，系统将会调来一部直达电梯。所有想去八层的人都会被送到3号电梯，所以它不用停在五、六、七层，可以直接去八层，这提升了效率。这个系统已经存在了至少5或7年，甚至可能是10年了。信息基础设施可以从一套安全系统中调取信息，并远程与电梯系统结合起来。所以，你走进办公楼，刷了你的ID卡——实际上你甚至都不用刷卡，信息就在远程被读取了。现在系统知道你是谁，也知道你在八楼工作。你一路过一套安全装置，它就会告诉你，去3号电梯。

在将来，高层建筑有哪些重要意义？

我认为它们是许多城市的未来，因为除了美国和加拿大，土地的供应非常紧缺。就土地利用或每平方英亩上的人口数量而言，没有什么比摩天大楼更有效率的了。但高层建筑是非常昂贵的，而且总会带来挑战，比如生命安全以及像电梯升降之类的东西。这些挑战已经被工业解决得非常好了。当三年前哈利法塔（Burj Khalifa）在800米高的地方开张时，那简直是一个巨大的转变。现在，沙特阿拉伯正在设计一栋建筑，有1000米之高。我们在奥雅纳已经通过了测试，想象没有限制，如果我们想，我们可以设计出高达1英里的建筑物。

当你想到这一点，会发现建筑行业真的是非常原始。我们正在用着同样的材料：钢铁和水泥。我们还没有开始研究真正空间时代的材料，比如碳化纤维以及此类的东西，它们往往更轻，更牢固，也更柔韧。这是行业必将非常认真去研究的一个东西。

你能否预见超高建筑，2英里高的那种？

不，我不这么认为。这些城市的生存方式将是60到70层的建筑物，如果加上建筑上的雄心，它有可能达到100到120层高。任何超过这些的建筑，至少在现在，实现起来一点都不经济。建筑的楼芯变得很大，只为每层留下很小的可用（可租赁的）空间。所以，除非你真的有很大的雄心，以及很多的金钱，否则超高建筑看起来将是无法建成的设计练习。

建得高最大的挑战是什么？

依据你在哪里，挑战并不相同。电梯是个大问题，但我再说一次，建筑师、工程师和电梯顾问真的已经解决了这个问题。这并非刻板的公式，但确实有三到四个不同的排列、组合方案。有像世贸中心那样的空中大堂，或者你也可以有低层、中层、高层的银行大楼，或者是一套调度系统，一种双层结构的电梯。这些都是可选项，而对它们的分析则是非常直接明确的，这是关于究竟什么适合一种特殊建筑的事情。

有没有一个系统，允许一个电梯间上升到空中大堂，然后在同一个运输体系中横向移动，然后再垂直移动？

我不知道有这种东西存在。奥的斯已经探讨，电梯不只垂直运动，而且横向运动。当我们在设计北京的中央电视台(CCTV)大楼时，我们已经就此问题进行了认真的讨论。CCTV是极不寻常的建筑，它既是垂直的又是水平的，但最终没有人想要做这个研究项目。

回到你的问题，500到600米（1640到1968英尺）超高建筑的另外一个巨大挑战——尤其在世界上的一些特殊地区，比如环太平洋沿岸，就要考虑如何处理风和地震的影响。但是在超高建筑中，结构设计通常受制于房顶高速刮过的风产生的风压，而不是地震的影响。以香港为例，即使它并不处于地震带，但如果香港地震，风也将是决定性的标准，因为在气旋影响下风压将非常巨大。风将带来巨大的挑战，特别是对结构工程师而言，所以我们需要考虑的是风、地震和电梯。我确定存在建筑上的挑战，不过再提一次，韩国和日本的公司已经走在前沿，他们已经发现了如何将水泥用泵抽到500-600米的高度。

thing the industry is going to have to look into quite seriously.

Do you foresee ultra-high buildings, say 2 miles [3.22 kilometers] high?

No, I don't think so. The bread and butter of these dense cities is going to be the 60- to 70-story building. When egos are involved, it will go up to 100 to 120. Anything beyond that, at least right now, just doesn't work out economically. The building cores get very large, leaving a small amount of usable (rentable) space on each floor. So unless you have a real big ego and a lot of money, ultra-tall buildings would seem to be unbuilt design exercises.

What is the biggest challenge to building tall?

It's a different challenge depending on where you are. Elevatoring is a big issue, but again, the architects, engineers, and elevator consultants have really figured it out. It's not formulaic, but there are three or four different permutations and combinations available. There's the sky lobby, like in the World Trade Center, or you can have low-rise, mid-rise, high-rise banks, or a dispatch system, or a double-decker elevator. These are the options, and the analysis is quite straightforward. It becomes a matter of what is appropriate for a particular building.

Is there a system that allows a cab to rise to a sky lobby and then, in the same carrier, to move horizontally across, before continuing vertically again?

I'm not aware that exists. Otis has been talking about elevators going not just vertically, but also horizontally. When we worked on the CCTV building in Beijing, we had some very serious discussions about this. CCTV is a very unusual building; it's both vertical and horizontal. But ultimately, nobody wants to be the research project.

To go back to your question, the other big challenge in super-tall buildings of 500 to 600 meters [1,640 to 1,968 feet], particularly in certain parts of the world like the Pacific Rim, is how you deal with wind and seismic effects. But in super-tall buildings, the structural design is usually controlled by wind pressures resulting from very high wind speeds at the top, rather than seismic effects. Take Hong Kong, for example, even though it is not seismic. But if Hong Kong were seismic, wind would be the defining criterion because the wind pressure during cyclonic conditions is amazingly high. Wind can be a big challenge, especially for structural engineers. So it's wind, seismic, elevatoring. I'm sure there are construction challenges, but again, the Korean and Japanese companies

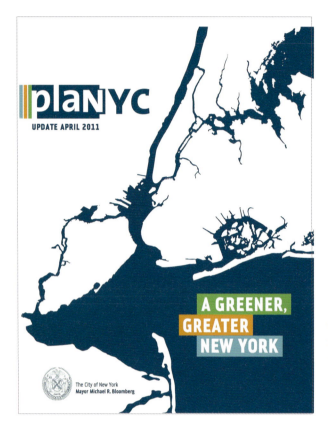

PlaNYC, a bold strategy announced by Mayor Michael Bloomberg in 2007 to make New York the first environmentally sustainable city in the United States.

"规划纽约"是迈克尔·布隆伯格市长2007年提出的一项大胆策略，旨在使纽约成为美国第一个环境可持续性的城市。

come to the forefront, having figured out how to pump concrete up 500 to 600 meters [1,640 to 1,968 feet].

What about new curtain wall developments, particularly with regard to solar energy?

We are beginning to see buildings with building-integrated photovoltaics (BIPVs). They are still very expensive, but that's getting better. From the U.S. perspective and the cost of energy here, payback is in 16 to 18 years. For a corporation, that's not a good return; they can get much better elsewhere.

The life span for PVs might be 20 years, but it's like so many other things in buildings, like air-conditioning units, which need to be replaced. In 20 years, the new generation of PV cells will have much higher efficiencies. You would already have the infrastructure in place so you'd just be replacing the tail end of the system.

Reskinning every 20 years is a new way of thinking about buildings, no?

Correct. In other parts of the world, where there is much

新式幕墙，尤其是与太阳能相关的幕墙，其发展情况如何？

我们开始看见有着光伏建筑一体化的(BIPVs)建筑。它们还是很贵，但情况已经越来越好。从美国的视角，以及能源价格来看，投资的回报在16-18年之内才有可能产生。对于一个公司来说，这绝不是好的回报，他们可以在其他地方得到更好的收益。

光伏系统的生命周期可能是20年，但是和建筑中许多其他东西比如空调组相似，光伏系统也是需要替换的。在20年间，新一代的光伏电池将有更高的效能。你应当将这些基础设施放在特定的位置上，然后当有需要的时候，你可以只替换系统的末端。

每20年更新一次建筑的外层，这是思考建筑的一种新方式，不是吗？

对，在世界的其他地区，人们对温室气体更为关注，因此也希望拥有可再生能源，政府为此提供了许多良好的激励机制。例如，韩国政府就签订了京都议定书和克林顿气候行动计划，并设定了接下来15或20年内利用可再生能源的目标。但他们也意识到，市场不会推动可再生能源的发展。这些能源系统非常昂贵，而韩国的能源价格相对较低，因此并没有吸引人的回报。所以，韩国政府通过公用事业公司，提供一种激励机制。举个例子，如果你在建筑里安装光伏系统，公用事业公司将以35分每度的价格从你那里买回电能，而它自己在出售时候的价格只有9分每度。实际上，你正在生产和出售的电能，是你买电时支出费用的四倍左右。所以，将这些电能卖出去而不是使用它们，是合乎情理的。

想想传统的光伏系统，它作为一种电能生产设备，通过电气设备（包括仪表等），与公用电网相连结。当你使用电能时，仪表开始转动，并增加读数。如果你有光伏板，将电能输回电网，仪表就会往回走。这被称为电力回馈。但是在韩国，他们使用两个仪表，一种是用于计算消耗的，另一种则是用来计算将电能卖回公用系统的。实际上，财政激励措施使得光伏系统更加可行，所有这些最终有易于减少温室气体的排放。对于风力涡轮发电机，也有类似的激励政策，尽管它的吸引力要小一些，因为风力涡轮发电机的成本与光伏系统相比，不是一个数量级的，前者要高出许多。也有针对燃料电池的激励政策。因此，他们正通过财政激励和立法来推进可再生能源事业。

中国或印度有没有相似的东西？

我还不知道是否有。大多数美国公用事业公司都有一个回购项目，但回购价格只是略比电能售出价格高一点，绝对不是4:1的差异。这些公用事业公司备受威胁，所以他们不想让你成为电能生产者。从法律上说，他们很难拒绝，但可以提出各种基础设施上的条件要求来让你很难共同参与，这些设施让你能够与电力网络连结起来，但使得你的设备安装费用往上攀升。

more concern about greenhouse gases, and therefore the desire to have reusable energy, governments provide lots of good incentives to do it. The South Korean government, for example, signed the Kyoto Protocol and also the Clinton Climate Initiative, and has set targets for renewable energy use over the next 15 or 20 years. But they realize the market can't drive renewable energy. Renewable energy systems are expensive and energy costs are fairly low in South Korea. Therefore, there's no attractive payback. So what the government has done through the utility company is to provide an incentive so that if you install PVs, for example, in your building, the utility company will buy that power back for roughly 35 cents per kilowatt-hour, while selling it for 9 cents per kilowatt-hour. Effectively, you're producing and selling at a rate four times more than what it costs to buy. So it makes sense to sell it, not use it.

Think of the conventional PV system as a power-generating device with a connection to the utility grid via electrical hardware that includes a meter. When you use power, the meter rotates and increases the reading. If you have PV panels and you are pushing power back into the grid, think of the meter as going backward. It's called net metering. But in South Korea they use two meters, one for consumption, the other for selling power back to the utility. Effectively, the financial incentive makes PV systems more viable, and all this eventually results in a reduction of greenhouse gas emissions. There's a similar incentive for wind turbines, although it's less attractive because wind turbines don't cost as much as PVs on a dollars-per-kilowatt basis. There's also an incentive for fuel cells. So they're pushing their renewable energy agenda by both financial incentives and legislation.

Is there anything similar in China or India?

I'm not aware of it. Most U.S. utility companies have a buy-back program, but the rate is just slightly more than the selling price of electric power, certainly not 4 to 1. Utility companies here feel threatened, so they make it hard for you to be a co-generator. By law they can't refuse, but by putting in all kinds of infrastructure requirements in order to connect to the electricity grid, it pushes the price of your installation higher.

Is it a failure of government initiative?

We have always been a market-driven economy, and the market is not going to drive renewable energy for a long, long time. If we are serious about emissions and the future of

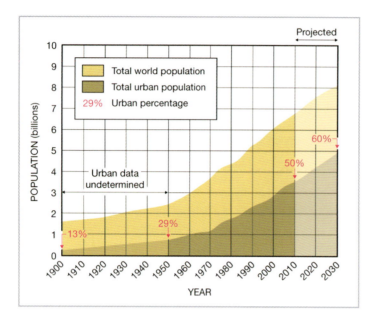

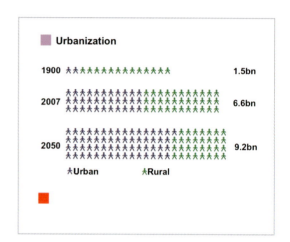

World Urban Population and Urbanization Growth.
世界城市人口和城镇化的发展。

fossils fuel, etc., we need legislation. There is no other choice since very few people will do it voluntarily. We have no carbon tax so nothing is helping renewable energy along. We have a very sophisticated client—I can't tell you who—that believe very strongly in the environment, and actually require us to include an artificial "carbon penalty" of x dollars per kilogram of carbon used. They're the impressive exception, not the rule. We need more organizations like that, but the only way is for government to legislate it. So that's where we are with renewable energy. I hope it's not too little too late, but certainly it seems that way.

What are the potential consequences?

The amount of carbon dioxide that we're putting into the atmosphere will increase, which means the impact of climate change will become more severe. Even without looking at statistics you can just tell that over the last 10 years, anytime there's been a natural event, whether a hurricane, tornado, flooding, or drought, it just seems to be more serious and happening more often. There will be an acceleration of that. I always imagined that we'd eventually sort out renewable energy. I thought nuclear energy would be the stopgap until renewable energy came into its own. Recent events in Japan have caused a bit of a setback for nuclear energy.

But there are so-called killer technologies on the horizon. Concentrated solar thermal is something venture capitalists in California are betting on strongly. They tend to be smart

这是否是政府措施的失败呢?

我们的经济大致都是由市场驱动的，市场不会很长时间地推动可再生能源。如果我们真的在意气体排放和矿物燃料的未来等问题，就需要立法。没有其他选择，因为很少有人会自愿做这些事情，我们没有碳税，所以没有什么可以一直推进可再生能源。我们有一位富有经验的客户，我不能说是谁，他非常强烈地相信环境的重要性，并且在实际中也要求我们加入一种人为的"碳惩罚"，即：在使用每千克碳时要交数美元的费用。这是一些令人印象深刻的特例，但却不是规则。我们需要更有组织，而唯一的解决办法是政府通过立法来规定这些，所以这就是我们和可再生能源的现状。我希望它不是太少太迟，但看起来它确实如此。

潜在的后果是什么?

向空气中排入的二氧化碳数量增加，意味着气候变化的影响将会更加严重。即使不看统计数据，你也可以看出，过去十年，不管在什么时候，当发生自然灾害时，无论是飓风、龙卷风、洪水还是干旱，看起来都越来越频繁，造成的后果也越发严重了。这些现象将会有一个加速运动。我总是在想象我们最终会挑选出合适的可再生能源。我认为在可再生能源出现之前，核能将会被当作是权宜之计，日本当前发生的事故，对于核能使用来说，确实是产生了一点小挫折。

但是有一些所谓的杀手级技术正在出现。集光型太阳能热，现在是加利福尼亚的风险资本家压上重注的行业，他们倾向于聪明人，所以他们可能是对的。如果能现实一点，我可能会说，我们在可再生能源上已经做的有点迟了，美国太大了。像丹麦这样的国家很有可能解决他们自己的问题，因为国家很小，不仅需求不多而且风能充沛。我们也在某些地方拥有这种能源，但不是遍地都有。从根本上说，如果我们打算解决气候问题，那么给出的解决方案应该是个综合体，既要有新的可再生能源生产技术，又要减少我们的能源消耗。

美国人总是深信，科学将会解决问题。它确实会解决问题，但需要付出很多的努力和投资，而且不幸的是，这些看起来都还很遥远。我将很乐于看到奥巴马的经济刺激计划更多地倾向于可再生能源和通常的基础设施，但实际情况却非常让人失望。以高速铁路为例，本来我以为在美国它会非常成功，但我们的政客只想着人为地制造障碍。例如，新泽西州州长克利斯蒂(Christie)就叫停了穿越哈德逊河的隧道建设，我理解他所顾虑的东西，但那真是毫无远见的做法。与此类似的还有佛罗里达州的高速铁路：政府打算支付25亿美元，建造从坦帕市到奥兰多市的高速铁路，但州政府却不答应。

在中国，国民生产总值依赖于建设。是不是即使没有市场需求，他们也还是要建设，最终让很多楼房空闲着无人居住？

有一些城市深受其苦。在某种程度上，这是一种2008年奥运会现象；他们想要向世界展示中国的能力，但却建设过度了。不过，次级城市的需求却是真实的。我不认为你会在天津或重庆这样的地方找到空荡荡的住宅楼。有许多人对中国未来的潜力信心满满。如果你回到基础数据，那么，在接下来的15到20年间建设40座能容纳500万人的城市，这是个大事业。

现在，中国已经很开放了，人们也看到了在世界其他地区正在发生的事情，他们想要追上来。印度的情况不太一样，当然，因为它曾处于英国的统治之下，而且并未闭关锁国。这两个国家未来的前景会有什么不同？

我不能从政策层面上，而只能从一种私人层面上进行阐述。我不太喜欢说这些，因为印度是我出生的地方，但是在中国工作要容易很多。他们看起来有一些确定的规章程序，这些可能并不一定

people, so maybe they're right. If I had to put on my reality hat, I think I would say we're a bit too late with renewable energy. The U.S. is far too big. Countries like Denmark can likely solve their problems because they are small, with small demands and huge wind resources. We have it regionally but not all over the place. Ultimately, if we are to solve the climate problem, it will be a combination of new renewable technologies for energy production, along with a reduction in consumption.

Americans are convinced that science will solve the problem. It will, but it takes a lot of effort and investment and, unfortunately, that doesn't seem to be coming. I would have liked to see Obama's stimulus program much more geared toward renewable energy and infrastructure in general. It's actually been quite disappointing. For example, take high-speed rail, which I think would be quite successful in the U.S., yet we have politicians who just want to create obstructions. New Jersey's Governor Christie, for example, canceled the trans-Hudson tunnel. I understand his concerns, but that was very shortsighted. Similarly, high-speed rail in Florida: the government was going to underwrite something like $2.5 billion to create high-speed rail between Tampa and Orlando, but the state said no.

In China the gross national product depends on construction. Is it true they build even though there is no market and that buildings remain vacant?

Some of the big cities suffer from that. To some extent it's a 2008 Olympics phenomenon; they wanted to show the world what China can do, and they overbuilt. But the demand in secondary cities is real. I don't think you'll find empty residential buildings in places like Tianjin or Chongquing. There are a lot of people who believe very strongly in China's potential for the future. If you go back to the basic numbers, to build 40 cities of 5 million each in the next 15 to 20 years, that's a huge undertaking.

Now that China is open and the people see what's going on in the rest of the world, they want to catch up. India's situation is different, of course, since it was under British rule and not closed. How do the two countries differ in outlook?

I can't speak from a policy level but only from a personal level. I don't like to say this because India is the place of my birth, but it's a lot easier to work in China. They seem to have procedures nailed down. They may not necessarily be state of the art, but at least there is a process. In India, it's very, very chaotic. You never know what to expect and everything takes much, much longer. The major difficulty in China is language, whereas in India most people, certainly in the business world, speak English.

About 10 years ago, India started designating Special Economic Zones around big cities. The idea was basically to develop these SEZs into mixed-use business, educational, and residential communities but with a very strong focus on information technology (IT). So whereas China is very strong in manufacturing, India—which, by the way, is not as strong in manufacturing—saw its way out into the world with IT. That's been the push and their focus of its development.

A lot of the big blue-chip companies signed up for the SEZs. They have the money, the influence, and the connections to get things done, and certainly they see big returns. When you consider the Manhattan-like prices of residential real estate but with construction at about one-third the cost, taken all together, the revenue stream is very, very attractive. And there's a huge market.

For a long time, India suffered from the so-called brain drain. Smart young people would leave and not go back. But in the late 1990s, that trend reversed so that lots of kids studying abroad started heading back home because of all the opportunities India now offers.

In a democracy, individuals or groups have a much louder voice. They can oppose the development process, or get involved and complicate it. Has this been a problem in India?

You would think that happens a lot, but my observation is that it doesn't happen to that extent. This comes from India's culture, which is based on a great deal of tolerance. I remember growing up, my grandmother and grandfather always telling me to be tolerant and flexible. Go with the flow and find ways to accommodate people. So to create a bit of difficulty for others is cultural taboo. The way this manifests itself in government is through the electoral process. If the government is not doing its job to the satisfaction of the people, it is elected out. It's a parliamentary system, so the opposition party can put in what's called a no-confidence motion, which means the government essentially has to resign. This is overused and abused, but nonetheless that's the power the people have: they can peacefully bring the government to its knees and elect a replacement.

How much of a problem is corruption?

Corruption is really a big problem in India. It's not that you

是某种艺术的表达，但至少确实存在一个这样的程序。在印度，情况则是非常混乱无序。你从来不知道该期待什么，而且所有东西都要花费很多的时间。在中国，最大的困难是语言，而在印度，大多数人，尤其是在商业领域，都说英语。

大约十年之前，印度开始在大城市周边设计特殊经济区。印度的想法大致上是要把这些特殊经济区发展成具有多种用途（商业、教育、居民社区）、但同时也特别关注信息技术(IT)的开发区。因此，尽管中国在制造业上非常强大，但印度（顺便说一下，印度在制造业上要比中国弱很多）将IT行业看作是进入世界的一个渠道。这已经成为印度发展的推动力和关注焦点。

许多大型的蓝筹公司与这些经济区签订了协议。他们有资金、影响力，以及把事情做成功的关系，当然，他们也看到了巨大的回报，那些和曼哈顿类似的居民房地产价格，其建设只需要三分之一的费用，综合来说，收入来源是非常吸引人的。这是一个巨大的市场。

很长一段时间以来，印度深受所谓人才流失之苦。聪明的人纷纷离开印度，不再回来。但在上世纪90年代后期，这一趋势有所逆转：由于印度现在所能提供的机会，有许多在国外学习的孩子开始回到家乡。

在民主社会，个人或团体有更大的声音。他们可以反对开发的进程，或者卷入其中，令过程变得复杂。在印度存在这样的问题吗？

你可能会认为，印度肯定发生了很多此类事件，但据我观察，情况还没有到那个境地。而这是因为印度的文化，这种文化奠基在大量的忍耐之上。我记得当我长大时，我的爷爷奶奶总是告诉我要忍让和灵活，要顺其自然，找到包容他人的方法。所以，给他人制造一丁点麻烦都是文化的禁忌。这些文化在政府中通过选举过程得以反映出来。如果政府没有尽到自己的职责，满足人们的需求，它就会被迫下台。这是一套议会体系，因此反对党可以提出所谓的不信任议案，这意味着政府基本上需要辞职。虽然是一种过度滥用，但也正是人们所拥有的权力：他们可以和平地罢免政府，同时选出一个替代者来。

腐败问题有多严重呢?

在印度,腐败当然是个大问题。不是说你必须为日常工作向人行贿,而是,如果你想要让事情做的快一点,那么你必须知道谁是正确的人,然后送点礼给他,以让事情得以进行。从长远看,这将对印度造成伤害。

作为一个公司,我们依照非常高的道德和伦理标准待人处事。我们赢得工作,乃是基于自身的能力以及过去的成绩,我们没有也不会参与任何可能引起质疑的行动。从长远看,但愿这对我们会有所帮助。那些向华尔街股东汇报的公司,可能会发现一种非常不同的反应,改变总是需要等待新的一代人,但我在印度没有看到这些。新的一代几乎和正被扫地出门的老一代人一样腐败堕落。更年轻的一代人以西方为目标,在我的成长过程中,很少有朋友出国,现在,有一半的中产阶级都会出国旅行。他们看到了不同的东西,而且想要这些东西。电视和因特网也有着重大的影响,新的一代对于世界正在发生什么更加了解。他们有机会让印度变得更好,但我不认为他们有老一代人所具有的毅力或动力。我希望我是错的。

我非常欣赏的国家是新加坡。我记得在上世纪60年代后期,我曾经去拜访过我在新加坡的祖父母,那时,城市有一半的地方没有自来水。今天的新加坡却令人难以置信。尽管存在着土地不足,人口相对过多的问题,新加坡是世界上城市人口密度最高的地方之一。如今他们已经将自身治理成为这样一个地方:人们真正乐于生活在其中,也乐于在那里从事商业活动。新加坡有一个强势的政府,但我认为它是合适的政府。这也是为什么印度在1948年独立之后徘徊不前的部分原因。对于政府来说,在起步阶段,民主可能并不是正确的形式。它需要一只强有力的手,可能有点类似于社会主义,让政府不必总是去建立共识。

have to pay people off for routine operations, but if you need things done urgently then you have to know the right people or grease a few palms to make it happen. That will hurt India in the long run.

As a firm, we operate with very high moral and ethical standards. We win work based on our capabilities and track record and have not and will not partake in activities that would be questionable. Hopefully in the long run it will serve us well. Companies who report to Wall Street shareholders might find a very different reaction. Change often has to wait for a new generation. But I don't see that in India. The new generation is just about as corrupt as the one being pushed out the door. The younger generation is very Western-focused. When I was growing up very few of my friends went abroad; now half the middle-class population travels. They see things that are different and want them. Televisions and the Internet are also big influences. The new generation is much more informed about what is happening in the world. They have an opportunity to make India better, but I don't think they have the stamina or the drive that older people had. I hope I'm wrong.

The country I have tremendous admiration for is Singapore. I remember visiting my grandparents in Singapore in the late 1960s and half the city didn't have indoor plumbing. Singapore today is incredible. Despite problems of not having enough land and having a fairly big population—Singapore has one of the highest urban densities in the world—they have managed to position themselves as a really desirable place to live and to conduct business. It has strong but, I think, appropriate government. That's part of the reason why India has floundered since her independence in 1949. Democracy was probably not the right form of government during the early days. It needed a strong hand, maybe a little bit of socialism so the government didn't have to build consensus all the time.

Richard Register
ECOCITY BUILDERS

Richard Register, founder and president of Ecocity Builders, is one of the world's great theorists and authors in ecological city design and planning, having advocated the building of better cities for more than four decades. Among his many firsts was convening the inaugural Ecocity International Conference (Berkeley, California, in 1990). He is founding president of Urban Ecology (1975) and founder and current president of Ecocity Builders (1992), both not-for-profit educational organizations. Mr. Register's acclaimed self-illustrated books include *EcoCities: Building Cities in Balance with Nature* (2002), *Ecocity Berkeley: Building Cities for a Healthy Future* (1987), and *Another Beginning* (1978). He is also the editor of *Village Wisdom: Future Cities* (1997). Mr. Register has lectured frequently in some 29 countries and 29 U.S. states. He was recently appointed to the International Scientific Advisory Committee on Active Ecological Urban Development to the Scientific Committee on Problems in the Environment (SCOPE), an international association of nation states and two dozen major scientific associations.

理查德·瑞杰斯特
生态城市建设理事会

理查德·瑞杰斯特是生态城市建设理事会的创始人和会长，是世界生态城市设计和规划方面重要的理论家和作家，他倡议建设更好的城市这一主张已经有40多年了。在他的许多首创中包括召集并成功举办了生态城市国际会议（1990年在加州伯克利）。他是城市生态组织的创始人（1975年），也是生态城市建设理事会（1992年）的创始人和现任会长，这两个组织都是非营利性的教育组织。瑞杰斯特先生广受赞誉的书籍包括《生态城市：在保持自然平衡中重建城市》（2002年）、《生态城市伯克利：为一个健康的未来建设城市》（1987年）以及《另一个开端》（1978年）。他也是《乡村智慧：未来的城市》（1997年）一书的编者。瑞杰斯特先生频繁的在29个国家，以及美国的29个州进行演讲。他最近任职于国际环境问题科学委员会的国际生态城市发展科学咨询委员会，这是一个各个国家合作的国际组织，他同时也在20多个大型的科学组织中任职。

All drawings by Richard Register.
皆为理查德·瑞杰斯特绘制。

For complete biography, please refer to the Appendix.
完整的传记，请参阅附录。

Union Square, San Francisco, rethought
联合广场，旧金山新思路

410

Having coined the word "ecocities" in your 1987 book *Ecocity Berkeley: Building Cities for a Healthy Future* and developing the idea in subsequent books like *EcoCities: Rebuilding Cities in Balance with Nature* (2002), you have argued compellingly that cities are man's largest creation and that the quality of our lives depends largely on how we build them. There has been increased interest in sustainability and green building practices recently, but why has it taken so long?

People have their habits and most of us very much resist changes that are uncomfortable. After many decades of building automobile-oriented sprawl, and expectations of cheap energy from oil and other sources, we've gotten used to a lifestyle and comfortable jobs and everything that makes it difficult to open minds to rearranging things more sensibly, even though a very large majority of the world lives in apartments and doesn't have cars, and gets around on foot. But very few people seriously consider the implications of the basic shift from two dimensions to three dimensions in designing and laying out cities.

We complex living organisms are all designed in three dimensions, not flat like tortillas. Cities are complex living organisms we make and live in, and, like all complex organisms, they need to be essentially three-dimensional too, not flat and scattered about in two dimensions held together by cars, asphalt paving the world, parking structures, freeway interchanges, floods of gasoline—about 500 gallons [1,900 liters] per year per person for the average car driver in the U.S.

Building compact cities, towns, and villages with their functions close together saves land, energy, time, money—pretty much everything, including, if we chose ecocity design, over a million lives lost around the world in violent crashes every year. It is pretty logical, really: organizing complex things for many functions works best when lines of connection are as short as possible, overlapping one another. Think of an animal's circulatory, lymphatic, and nervous systems, for example: distances are short; linkages overlap one another. In cities, the flatter and the lower the density, the longer the connections of paved streets, miles of pipes and wires for subway systems, water, gas, electricity, phone, routes for emergency vehicles and land consumed, farms pushed off the best soils. And of course the arrangement means people have to move faster to get anywhere in reasonable amounts of time—which really turn out to be not very reasonable after all with all the collateral damage speed causes in environments that are supposed to be for people. In the meantime, immense amounts of energy are consumed

你在1987年的著作《生态城市伯克利：为一个健康的未来建设城市》中，引入了"生态城市"这个词，并在你后来的作品，比如2002年的《生态城市：在保持自然平衡中重建城市》中，发展了这个概念。你令人信服地指出，城市是人类最大的发明，我们的生活质量很大程度上依赖于如何建设城市。近来，人们对可持续性以及绿色建筑的兴趣与日俱增，但为什么发展如此之缓慢？

人们有自己的习惯，大多数人会抵制那些不舒服的改变。多年来，我们以汽车为导向去扩张，并期待在石油和其他资源中找到更便宜的能源，我们已经习惯了一种生活方式，一种舒适的工作，以及所有那些令人很难展开想象去重新安置的事物，虽然世界上有很大一部分地区的人都住在公寓中，且没有汽车，完全靠自己的双脚出行。但是，在设计和规划城市时，很少有人认真考虑从二维到三维空间这一根本转换的实施措施。

我们复杂的有机体都是以三维方式设计的，而非像玉米饼那样是平面的。城市是我们创造并生活于其中的复杂有机体，而且和所有有机体一样，它在本质上也应该是三维的，而不能是平面式的散落在二维空间中——这种二维空间通过汽车、全世界的柏油路、停车位、高速公路交通枢纽以及大量的汽油（在美国行驶的普通汽车，每人每年要消耗大约500加仑的汽油）组合而成。

建设集体的、功能相似的城市、小镇和乡村会节约土地、能源、时间和金钱几乎所有事，包括如果我们选择生态城市设计可以每年在全世界的严重车祸中丧命的一百多万人口。这真的很符合逻辑：当联系线路尽可能短，相互重叠的时候，组织具有多种功能的复杂东西才会工作得最好。例如，我们设想一只动物的循环、淋巴和神经系统：若距离很短，连结就会相互重合。相反，在城市中，越呈平面化，密度就越低，铺设的街道、地铁系统、水、气、电、电话需要的线路和管道、急救车和土地开发利用的线路就越长，农业也将被推到离最佳土壤更远的地方去。当然，这种安排意味着：人们必须在合适的时间内，更加快速地到达目的地——但实际上这并不合理，因为在人类环境中会由速度引发许多附属损害。同时，大量的能源被消耗，产生大气污染、气候变化、海平面上升等等问题。这真是荒诞不经，并足以让人觉得愧疚，这也是很多人不

Downtown San Francisco linked by bridges.
由桥梁连接起来的旧金山市区。

愿思考这个问题的另一个原因，即使是思考城市设计或者考虑替代物。

什么是单维度的发展？

三维的开发是有一定体量的，它具有我刚刚提到的相互交叉连接的可能。想象一下地下的地铁，街道上的步行交通，以及在市镇更密集部分横跨街区的桥梁。二维的开发效率没那么高，就像一座面积巨大的城市从山间斜下，延伸到平地上。而在废弃物、能源、时间和相对土地"服务"人数方面，单维的城市是最糟糕的。

直线性的另一个说法就是单维度，它的发展可以变得很长，并以很低的效率在一些有限的方向上阻止发展。像过去拉斯维加斯那样：线条状在其主路以及少数的辅路被勾画出来，以二维的方式进行填充，从而盖住那贫瘠干硬的荒漠——伴随着向另一维度上少量的延伸，其本质上是一种单维度的线条。

我曾在匈牙利从空中看过同样类型的开发形式，在那里，乡村沿着乡间小道分布，这些小道无限制的延伸，没有形成任何的乡村中心。在小房子的沿街墙壁后面，你可以看到灌水渠垂直流向街道，土地一代一代地细分给后辈，直到进一步的细分不再可能为止。他们并没有一套长子继

producing atmospheric waste, climate change, rising seas, and on and on. It's really quite surrealistic. It is also enough to make one feel guilty—and does—which is another reason so many people don't want to "go there," that is, even think about city design or look at the alternatives.

What is one-dimensional development?

Well, three-dimensional development has some bulk to it and the possibility of linkages crossing over one another as I just mentioned. Think of subways underground, foot traffic on the street, and, say, bridges spanning streets in the more compact parts of town. Two-dimensional development is much less efficient, like a city of vast acreage pouring over hills and spreading to the horizon. One dimension is worst of all in terms of wasted material, energy, time, and land area relative to the number of people "served."

Linear—another term for one-dimensional—development can get very long and can head off in a limited number of directions at extremely high inefficiency. A linear strip like Las Vegas in the old days—before its Strip and few side streets stretched and filled in two-dimensionally to cover its barren, baking desert—was essentially a one-dimensional line with minimal extension to either side.

I have seen this same development pattern from the air, by the way, in Hungary, where the villages are laid out along

Downtown, car city.
市区，汽车城。

a rural road that goes on forever through the countryside, with no village center. Behind the street wall of small houses you can see irrigation ditches taking off perpendicular to the street, with land subdivided down through generations of children to the point that further subdivision is impossible. They didn't have a primogeniture system where the older son or another individual inherited everything and held it all together. Instead, they divided and divided and divided into these thin strips running generally from the road outward. The result looks from the air like centipedes crawling all over the landscape, the street being the body of the centipede and the thin strip of property between fences stretching out from the body looking like the legs.

The solid street wall has only a few scattered shops. I was informed by a friend in Austria who had been through such landscapes many times that it was the most barren cultural environment he'd ever been in. So you have to go

承权系统，不是由长子或其他人继承一切东西并完整地保存。相反，他们将所有东西不停地再分配，变成了这些细带，渐渐远离了通向外界的道路。从空中看，情景宛如蜈蚣般蜿蜒盘旋于大地之上，街道成了蜈蚣的身体，连接围墙之间房屋的细带就像从蜈蚣身体中延伸出来的脚一般。

封闭的街墙只有一些散落的商店。一位曾多次穿越这片土地的奥地利朋友告诉我，这是他所到过的最无趣的文化环境。所以，如果你想找到什么值得去的地方，就要走出很远很远的距离，这是可以想象到的最坏的发展形式。

从广义上说，这就像是上世纪50年代以后，美国沿着公路发展的条状方式（此时期，环形公路还未包围城市并摧毁其核心）。非常高、非常细的大楼像绳索一样冲向天空，这就是单维度的，尽管它们增加了密度，对于整座城市而言也增加了一些功效——但是为了将大

Car city (facing page) transformed into downtown ecocity.
汽车城（上一页）转变为生态城市区。

楼固定住，在其底部无效率地使用那些厚重、大功率、浪费资源的建筑材料，这种方法甚至消除了高楼的功效。建筑的结果是在内部大量应用电梯。在线性发展中，即使它是垂直的，巨大的投资也将耗费在运输上。这样的极高建筑，根本上就像商会风格的线状广告版，冲击着所有人的眼球，并宣称："嘿，我们是一个伟大的城市，来参观、购物、交易和投资吧！"在美国西南部地区，有许多二维风格的城市，像新墨西哥州的阿尔伯克基和亚利桑那州的菲尼克斯。美国东部的城市一开始就作为殖民地、旧工业区及类似的区域，它们在被汽车占领之前就已经建立起来了，因而显得更加紧凑。但是新兴的美国城市，或多或少从一个核心区开始的——当然，这些核心区绝不像欧洲的那样紧凑——然后它们向东南西北四个方向辐射出去，成为了一些区域带。

miles and miles and miles to get to anything. That's the worst development form imaginable.

On a larger scale it's like strip development in the United States, which was happening in the 1950s and after, before ring roads went around cities and destroyed their cores. Very tall, thin buildings rising like wires into the sky are similarly one-dimensional, though they add density and hence some efficiency to the total city—which they turn around and largely nullify by enormously inefficient heroic use of thick, powerful, and energy-wasting building materials at the base to hold it all up. Such spectacle buildings end up having much of the interior used up in elevators. In linear development, even when it's vertical, enormous investment has to be made in transport. Such extremely tall buildings are essentially Chamber of Commerce–style linear billboards rising into everyone's view, announcing, "Hey, we're a great city. Come visit, buy, trade, and invest!"

There are lots of the two-dimensional–type cities in the southwest, like Albuquerque, New Mexico, and Phoenix, Arizona. Eastern cities in the United States are more compact, beginning as colonial settlements, old mill towns, and the like, which were built before the takeover by cars. But the newer U.S. cities began with a core, more or less—never compact like European cores, of course—and then they radiated out in four directions, north, south, east, and west, which became strips.

Some like Las Vegas became sinfully wasteful of water, energy, and money just to make more money, kind of like Wall Street going into the Crash of 2008, with no discernable physical product or service other than time-wasting entertainment, while the real crises grows worse for lack of attention. By the way, there's a casino and a whole Italian-themed town on its eastern fringe built on a great big artificial lake called Lake Las Vegas, which gets its water from fossil deposits millions of years old deep under the desert and evaporates something like a billion gallons [3.79 billion liters] into the hot sun and dry sky every year. The lake is big enough for ocean-going yachts and has some floating about. It's an absolutely insane design and location for a big city, sprawling suburbs or not, but there it is. There is a famous book by the architect Robert Venturi called *Learning from Las Vegas*. If you don't learn it's a disaster to build that way you must be learning how to destroy a planet. *Star Wars* has nothing on *Learning from Las Vegas*. This kind of city is stunningly destructive to the whole climate system of our now space-age planet.

Have any cities actually been built according to your principles?

No. None that I know of, and I actually advise one called Tianjin Eco-city, now being built for 350,000 people in China. The Chinese are trying and having some success, but are still not very close to what I would think of as a full-spectrum ecocity. But there are places where pieces are coming together more and more.

Tianjin Eco-city is one of them. Another is Changwon, South Korea, where I'm on the mayor's International Advisory Council. They have a project there called City 7 that has 30-story residential towers, a hotel and conference center, and six floors of commercial development with big atrium gardens; rooftop terraces; sculpture gardens; bars, restaurants, and cafes; a movie theater; clothing stores; convenience stores like your usual corner store; ATMs, of course; and currency exchange at the hotel. All of this is linked together on the ground in a discontinuous garden and above by multiple

有一些像拉斯维加斯一样的城市，为了赚取更多的钱，可耻地浪费水、能源和金钱；它们也像2008年陷入崩溃的华尔街一样，没有看得见摸得着的真实产品或服务，只有浪费时间的娱乐业；由于缺乏重视，危机的真实情形越来越糟。另外，在其东部有一个赌场和一个完全意大利风格的市镇，建在被称为拉斯维加斯湖的一个巨大人工湖上，这个湖从沙漠深处数百万年的化石沉积层获取供水，且每年向炙热的太阳和干燥的天空蒸发掉数十亿加仑的水分。这个湖对于航海的游艇来说都足够大了，而且湖里还是流动的活水。对于大城市来说，无论有没有向外蔓延的郊区，这无疑都是个疯狂的设计和地点，但它就屹立在那里。建筑师罗伯特·文丘里（Robert Venturi）写了一本著名的书：《向拉斯维加斯学习》。如果你不认为建成那个样子是一场灾难，那你一定知道如何摧毁一颗行星。《星球大战》都比不上《向拉斯维加斯学习》。在当下的太空时代，这种城市简直是对整个星球气候系统的严重破坏。

有没有哪个城市实际上是依照你的原则建立起来的？

没有，据我所知，事实上并没有这种城市。而我实际上对现在一座名为"天津生态城"的城市提出了自己的建议，它位于中国，可以容纳35万人居住。中国人正在做出尝试，并取得了一些成效，但还不是很接近我所认为的那种全方位的生态城市。不过，现在已经有一些地方，各个部分越来越多地集中到一起。

天津的生态城是其中之一。另外一个是在韩国的昌原，我在那里参加了市长主持的国际咨询委员会。他们有个被称为7号城市的项目，其中有一些30层高的住宅楼，有一间宾馆和会议中心，用于商业开发的带中庭花园的六层楼；有屋顶露台、雕塑公园、酒吧、餐厅和咖啡馆，有一家电影院，服装店以及像普通街头小店那样的便利店，当然还有自动取款机，宾馆中还可以进行外汇兑换。所有这些都在地面上，在一座不连续的花园中连接起来，而在楼层上，则通过许多桥，连通了屋顶和楼层。另外，该城正在疯狂地提倡自行车。这是我看过最接近于生态城市片段的一个例子。所谓的"片段"，我指的是作为整个城市的一部分，它现在具有了基本的构成要素：住房、工作、商业等等，并组合的很好，而且还结合了最好的阳光角度以获得被动式太阳能、美食花园、艺术和教育等等。

我们需要思考的一条基本原则是均衡性：汽车在塑造当代城市上扮演着关键角色。问题是，汽车的平均重量是人类的30倍，当它静止的时候所占的空间是人类的60倍，而运动的时候显然还要更多。这意味着，在汽车的"需求"和人的需求之间有着巨大的不匹配。想象一下在你的客厅装进一只大象，或者相反，建一个像体育馆那么大的客厅。

你想如何应用那些纬度设计出一个得体的人类环境？先认识到比例的重要性，然后按照优先顺序依次使用最佳方案处理问题，这一点非常关键。城市正在破坏全球环境，而我们要解决许多最为棘手的问题，可能还有很长一段路要走。但是我们要先想出对的路，才能建设出合理的城市。

没有人认真对待这一点。关于气候变化，人们并不谈论城市形态——这一人类所建立的最大的东西。实际上，这就是对不能均衡化处理问题，不善于按顺序安排事情的最好范例。人们谈论的是控制总体经济，意思是你修改已经存在的事物或模式，以使他们更加有效一点。但是，假如你让某些不正常的东西变得更有效率，你就创造了一个更坏的情形。也就是说，实际上你创造了一种更健康的疾病，你不是想要那种疾病变得健康，你是想要自己变得健康。

现存的城市就像一些不幸的、有着可怕残障的动物，对于城市而言，这种残障就是"为汽车修建的柏油路"和"建筑使用的水泥"；像悬在空中的木杖，根本上是单维度的而不是三维的。而且，还有巨大数目的材料，从基底上将它们整个支撑起来。这些东西就是人们尚未面对的均衡问题。如果人们面对了这些问题，就会开始提出生态城市的设计，这种设计涉及到主宰生态系统的原则：在所有系统中，器官必须很接近，连接必须很短，以及其他一些生态原则，否则你就不会有一个健康复杂的有机体。

有这么多关于可持续发展的讨论，为什么并没有兴建起更多的生态城市？

我想它是被抵制的。我已经试图在我的所写所画中，开发出一种关于可持续性的语言，但是我发现我总是被阻碍，主题已经提出，但人们并不想要谈论它。如果我们不谈论发生在你眼前最大最重要的事情，那他们该怎么学习呢？实际上，我们就生活在其中，但却没有看到它实质上是一个

bridges, roof to roof and level to level. Plus the city is promoting bicycles like crazy. It's the closest thing I've ever seen to an ecocity fractal. By "fractal" I mean a fraction of the whole city that has all the basic components present and well organized: housing, jobs, commerce, and so on, but also integrating best sun angles for passive solar heating, native and food gardens, art and education, and so on.

One of the basic principles we really need to think about is proportionality: the car plays a key role in shaping contemporary cities. The problem is that the car is 30 times as heavy as the average human being, takes up about 60 times the volume standing still, and much more space when it is moving. This represents an enormous mismatch between the "needs" of the car and the needs of the human. Imagine stuffing an elephant into your living room or, conversely, building a living room the size of a gymnasium.

How are you going to design a decent human environment around something with those dimensions? Getting a sense of proportion, then prioritizing to deal with the problem with best solutions, is crucial here. The city is wrecking the global environment while it could be taking us a long way toward solving many of our most intractable problems. But we have to build it right—after thinking it right.

No one takes this seriously. The climate-change people don't talk at all about urban form, the largest thing human beings build. That's practically the definition of failing to proportionalize and prioritize. They talk about manipulating the general economy, meaning you modify things and patterns that already exist to make them a little more efficient. But if you make something dysfunctional more efficient, you create a much worse situation. You are actually creating a healthier disease. You don't want the *disease* to be healthy, *you* want to be healthy.

The existing city is like some unfortunate animal that has a terrible handicap, which is asphalt and concrete for automobiles and buildings, like sticks in the air, that are essentially one-dimensional, not three. And they have an enormous amount of material at the base to hold it all up. These things are proportionality problems that people aren't facing. And if they did, they would start coming up with ecological city design, which runs on principles that govern living systems in which organs have to be close together and linkages have to be short, and other ecological principles, otherwise you don't have a healthy complex organism.

With all the talk about sustainability, why aren't more ecocities being built?

I think it's denial. I have tried to develop a language about it in what I write and draw, but I find myself stonewalled constantly. The subject comes up and people don't want to talk about it. How are they going to learn if we don't talk about the biggest most consequential thing going on right in front of our eyes? In fact, we are living right in the middle of it and not seeing it for what it is: the seriously dysfunctional city.

Who is "they"?

Everyone from conventional architects to people who don't want to see their cities change; they're comfortable with the way things are. A really good example is Berkeley, where I spent 35 years trying to do good projects. A number of friends and I did manage to get some creek restoration projects created, urban orchards planted, and some buildings influenced in a very good way, but when it came to serious reshaping for real changes, little happened. The city refuses to establish even a single block of pedestrian street, and even most of the people who say they love their creeks oppose zoning changes to make it possible to remove a decrepit house in a willing seller deal to open a stretch of buried creek. Eighty-five percent of the creek system of Berkeley is buried in concrete pipes, and the friends of the creeks don't want to allow the removal of a single house to open up the system more. They just want to plant native plants on their banks and count fish. At a couple of city zoning board meetings where I've spoken, people said, "Richard, get a horse."

What does that mean?

You're out of place. Forget it. Change the subject. People don't understand and don't want to understand because they might have to work hard, they might have to see a change in their neighborhood; their housing value could go down. A lot of this is pure fantasy and unfounded fear, but it put a stop to what could have been some extremely positive changes.

In *EcoCities*, published in 2002, you discuss the disturbing conspiracy of oil companies, auto and oil manufacturers, and bus lines to destroy rail competition in the United States.

That's right, this took place over nearly three decades during which the country's urban rail system almost entirely disappeared. Finally in 1955, the conspirators were found guilty of violating antitrust laws but only received a slap on the wrist. Car dependency was intentionally designed into the American lifestyle to maximize corporate profits. Urban sprawl was

功能严重失衡的城市。

"他们"是谁?

从传统的建筑师到不想看到自己城市发生改变的所有人,他们对现有的状况感到舒适。一个真实的例子是伯克利,我花了35年时间试图在这里做一些好的项目。一些朋友和我尝试并实现一些溪流生态系统修复项目,城市果园种植项目,和一些影响很好的建筑,但当严肃认真的为了改变而对城市进行重塑时,基本上什么也没有发生。这个城市拒绝建立哪怕一个带人行道路的街区,甚至大部分热爱他们的溪流的人也反对区域改革。即使这些改变可以通过意愿达成交易,拆除老旧房屋,来打开一片被掩埋的溪流。伯克利85%的溪流系统被埋在水泥管道之下,而溪流之友们并不打算拆除某一栋房子以更多地开拓整个系统。他们只是想在岸边种一些本地植物,然后欣赏水中的鱼儿。在我宣讲的许多城市分区理事会上,人们说:"理查德(Richard),别着急啊。"

那是什么意思?

你不了解实际情况,算了吧,换一个话题。人们不理解也不想理解,因为他们可能要努力工作,要看到自己的社区发生改变,他们的房子价值可能会下降。这些大多数是纯粹的幻想和没有根据的害怕,但它确实阻止了原本可能发生的一些积极改变。

在2002年出版的《生态城市》一书中,你讨论了美国石油公司、汽车和石油制造商以及公交路线摧毁铁路竞争的阴谋。

是的,这发生在大约30年的时间里,在这期间,国家的城市铁路系统几乎全部消失。最后在1955年,阴谋者由于违反反垄断法被判有罪,但也只是受到了轻微的处罚。汽车依赖症被有意地植入美国人的生活,从而让那些公司的利润实现最大化。城市的延伸并不只是放任公众的欲求,它必须被规划。这种事情是典型的资本主义,而且只是许多问题事件中的一桩。

我现在正在写一本书,这本书必定要讨论到我们所建设的经济。人们更喜欢谈论他们如何从一笔投资中获得最大回报。那么,所有这些都是从哪里来的?来自树林对太阳能的汲取,来自农耕土地,它们作为能量源泉能够融化金属、启动机器，还可以激活我们的身体。所有这些东西都来自大自然，而人们并未认真地思考这些关联，所以实际上是在排斥。许多开发商在城市扩张中大发其财，所以他们想要继续这么干下去。你必须观察人们的动机，为什么他们不愿意着眼于更大的图景，就像那个特别的阴谋（摧毁铁路的阴谋）。还有一种私人的阴谋：不想看到任何东西带来改变。人们说："我要保证我的收入，让我的孩子待在大学里。"情势十分艰难。

但是，最后有一样东西被塞拉俱乐部这样的主流环保组织证明，那就是，可持续发展的道路会带来更大回报，最终也有更好的工作机会。这仅仅是个开始，因为如果你理解了生态城市的规划与设计，就会发现，当地咖啡店的女服务员给那些从街区走进店里的人送咖啡和三明治是一份绿色工作，因为这些人就在街区里居住、工作、购物、享受生活，女服务员的工作就像资源回收工人和绝缘材料的安装工一样。同样地，钢铁工人在工作地点和交通站点附近建立了一座新的公寓大楼，"平衡"了发展方案，缩短了交通运输、资源供给和回收的距离，阻止了城市过分扩张，也防止它侵占溪流、溪谷、群山和山谷。相反，零售店中的女服务员；钢铁工人在农田上建立的仓库，离人们居住工作的地方有20英里远；装配工人在边远郊区，为管道装上太阳能系统，这些都不算绿色工作。地区意识，以及距离、能源、物质和时间在地区之间的关系如何规划和设计联系在一起，这在经济上至关重要。之后你才会知道投资什么，才会有个很长、很丰富的绿色工作列表，它与一个健康的将来是相契合的。

非专业人员有何回应？

这对他们而言太复杂了。对于没有受过专业训练的人来说，这真是一个困难的话题，他们对于分区和改造城市的语言并不熟悉，也没有足够冷静到去市政厅，为一些东西而斗争。通常来说，只有在某些东西威胁到他们的家庭，或者提高税收的情况下，他们才会起来反抗。他们并不想要一个更好的世界，他们想要的是更好的房子，孩子能获得更好的教育机会。在短时期内，他们是完全理性的；但你不能指望在短时期内施加影响，获得一个健康的未来。你必须依照基本原则行事，这会提示你一个健康未来，实际上需要什么，以及如何去设计。无论好坏，但长期规划需要的是：与短期在本质上不同的方法。我们可

not just the drift of public desire. It was planned. This kind of thing is typical of capitalism and only one of many things that are problematic.

I'm working on a book now that has to do with the economics of what we build. People prefer to talk about how they can make an investment for best return. Well, where does it all come from? It comes from solar energy pumping through forests and from agricultural land coming out as energy sources to melt metal and power machines and animate our bodies. There are all these things that come from nature, and people are not seriously considering that link, so in effect they're denying it. A lot of developers are making money on sprawl, so they want to keep doing the same thing. You have to look at people's motives about why they don't want to look at the bigger picture, like that particular conspiracy. There is a private conspiracy of not wanting to look at anything that might bring about change. People say, "I have to keep up my income so I can keep my kids in college." Things are tough.

But if there is one thing that's been demonstrated conclusively, and even by mainstream environmental groups like the Sierra Club, it is that there are more good-paying and eventually better jobs down the sustainability road. And that's just the beginning, because if you understand ecocity layout and design, it turns out that the local waitress serving coffee and sandwiches to people who just walk in from the neighborhood where people live, work, shop, and enjoy themselves has a green job just like the recycler or installer of insulation. So does the iron worker building a new condominium near jobs and transit, "balancing" the development pattern, shortening those lines of transportation and supply and recycling, keeping the city from spreading out and covering creeks, valleys, hills, and dales. The waitress at the stripmall, the iron worker building a warehouse out on farmland 20 miles [32 kilometers] away from where people live and work, and the installer putting a solar electric system on a hose in the distant suburbs do not have green jobs. Place consciousness, as well as understanding how places relate to one another in terms of distances, energy, material, and time in making the connections required by layout and design, is economically crucial. Then you know what to invest in. Then you have a long, varied green job list and it fits a healthy future.

How do laymen respond?

It's too complex for them. It's really a tough subject for people who aren't trained in it, who aren't familiar with the language of zoning and changing the city, who aren't hard-headed enough to go down to city hall and fight for something. Usually they'll only fight if something endangers their family or raises their taxes. They don't want to have a better world; they want to have a better house, a better education for their kids. They're completely rational in the short term, but you can't leverage the short term and come up with a healthy future. You have to work on basic principles that give you some indication of what a healthy future actually needs, and how it needs to be designed. The long term, for better or worse, requires a substantially different approach from the short term. We can figure out a better future, but we have to try, work, and persevere to change some things.

Beyond Berkeley, is the basic principle applicable elsewhere?

You can apply it everywhere in the whole world when people decide they want to protect their children's future and really commit to it. How can you protect a distant future? People talk about climate change. It's here. Species are becoming extinct, the climate is changing, economics are gyrating all over the place. Sea level is even rising.

To get a grip on the kind of good changes we need, as I mentioned a moment ago, we need to proportionalize, see things in the order of scale and importance, and get busy.

I see five big crisis areas that need immediate attention. One is the built environment, which is disastrously structured, when, in fact, our cities, towns, and villages could actually build soil and enhance biodiversity, be a net contributor to health in the biosphere. It's a matter of imagining it, designing it, building it, and staying committed to the vision.

Another "Big One," as I think of these essential issues we have to confront, is overpopulation. There are so many people bringing down so many resources, just the sheer amount of food we have to eat, the clothing, the basics for shelter are overwhelming. And then there's the agricultural nexus, which takes up immense areas of landscape and is largely built around a heavy meat diet, lots of chemicals, lots of machines, lots of refrigeration, and is global, not local at all. The fourth Big One is what I call "generosity." Everyone is hoarding for him- or herself or expanding mainly only as far as the family. It's assumed that humans aren't smart enough to figure out systems for themselves so they're always on the defensive, and think reactively. It's always the other guy that's like that, of course.

The big term these days is resilience. You want a climate-resilient city, economy, lifestyle. False! That's a bad idea— well

以设想出一个更好的将来，但我们必须尝试、工作、坚持着去改变一些东西。

除了伯克利，这一基本原则是否可以应用在其他地方？

当人们决定要保护他们孩子的未来，并真的致力于此时，你可以将这一原则应用到世界各地去。如何保护一个遥远的未来呢？人们谈到了气候改变，它就在我们身边，物种逐渐灭绝，气候正在改变，经济统治一切，海平面也在上升。

正如我前面提到的，要把握住那些好的改变，我们需要均衡，按照比例和重要性的次序来看待事物，并且忙碌起来。

我看到五个重大、需要被迅速关注的危机领域。一是已建环境，它被灾难性地建构而成；实际上，我们的城市、城镇和乡村可以真的创造土壤，提升生物多样性，从而，成为确保生物圈健康的纯粹贡献者。关键就在于想象之、设计之、建设之，并保证对这种愿景的承诺。

我认为我们不得不面对这些本质问题，另一个大危机，是人口过剩。有如此多的人，消耗了如此多的资源，仅仅算上我们必须吃掉的食物、消费的衣物、基本避寒的房屋数量，就已经是非常巨大的了。第三个是占据了大片土地的农业关系网，它在很大程度上是环绕着肉食食谱、诸多化学品、机械、冷藏设备而建立的，它是全球性的，不再是地方性的了。第四个重大危机，我称之为"慷慨大方"。每个人都为自身，或者推而广之主要也就是为了家人而囤积储藏。人类没有聪明到能为自身构思出系统，所以他们总是处于防御状态，总是被动地思考。当然，人们总是认为其他人才是这样的。

今天的一个大词汇是弹性。你需要一个在气候上有弹性的城市、经济和生活方式。错！这是一个坏主意——并不一定。弹性是好的，但它不能是首要的，因为它有反作用。弹性意味着如果发生了某件事，你已经做好准备以一种有效的方式来处理它。所以弹性是等待着，做好准备做出反应，而创造性才是首要因素。

教育，包括学习和为创造性做准备，乃是第五个领域。教育受到绝对必要的关注，它用来设计出另一个系统，使得我们有弹性去应对那些不好的事情，与之相反，可战而胜之的东西不会一下子就出现。这需要我们直面"五个重大危机"：人口，农业、饮食的关系，城市、城镇、乡村的建设环

Roll back sprawl.
击退扩张。

not really. Resilience is okay, but it cannot be primary because it's reactive. Resilience means that if something happens, you're ready to deal with it in an effective way. Creativity takes the initiative while resilience waits, well prepared to react.

Education, which involves learning and prepares for creativity, is the fifth area. Education is of indispensable concern, and has to come to designing another system so that the nasty things that happen that we have to be resilient to, counter, and triumph over don't come up in the first place. That in turn requires that we forthrightly confront the "Five Big Ones": population; the agriculture/diet nexus; the built environment of cities, towns, and village; being generous and educating with strong, even emergency, emphasis on the above four, or we're not going to make it.

The attitude frequently is that the problem is too complex for me; science will solve it.

No. That's not true. Complexity is comprehensible if it's organized well. We can all understand it, not just scientists, but you have to do the proportionalizing and prioritizing to get a grip on the overall pattern, the important things that relate to one another. Then the task is not so difficult for anyone. For example, you look out across the sprawling city and just have no sense about how to get this thing together. It is complicated, overly contradictory, confusing, not genuinely complex. It lacks the order and intelligence of complexity. But look at a bird. It has a beak for eating, legs for standing, wings for flying, feathers for giving the body form so it slides through the air. It eats insects and lives in the trees. It is an amazingly complex thing down to the DNA-encoded instructions deep in its eukaryotic cells. But it is comprehensible if we look carefully to see the basic principles in its functioning. All these pieces come together and it's very comprehensible, if the complexity is ordered well. If not, it's incomprehensible—or dead or dying or a phantom of the imagination, so strict are the rules of biological order in complexity.

Where does the change need to happen?

People need to be willing to work, to think, to study, to stop watching TV, stop accruing money to buy a better sound system for their automobile, start ignoring ads, and start paying attention to the larger problems and start thinking about cultivating their own creativity, that can-do attitude. People have to decide in favor of what's useful and healthy, and then actually do it seriously, and stop doing all these things that are wasting time, energy, money, and our very

境，"慷慨大方"和主要对上述四种危机进行有力甚至紧迫说明的教育，否则我们就不能够成功。

人们的一般态度是：问题对我而言太过复杂，科学会解决它的。

不，不是这样的。如果组织有序的话，复杂性是可以理解的。不只是科学家，我们都能够理解它，但你必须进行均衡化处理和优先处理，以掌握整体的形式和那些互相联系的重要东西。这样任务对任何人而言就不那么难了。例如，你放眼远眺向外扩张的城市，却不明白如何将所有这些东西结合起来。这是混乱、矛盾、令人困惑的，而不是真的错综难解，它缺少秩序和对复杂性的认识。但是看看一只鸟，它有用于进食的嘴，用于站立的腿，用于飞行的翅膀，有能够让身体有形并在天空滑行的羽毛，它以昆虫为食并住在树上。这是一种具有惊人复杂性的东西，这种复杂性可归于DNA封闭指令，这种指令深藏于真核细胞之中。但是如果我们仔细观察其功能中所具有的基本原则，就可以理解。如果复杂性组织有序的话，所有这些片段合在一起，就完全可以被理解。如果无序，那就是不可理解的——或是已死的、垂死的或是一种虚幻的想象力，在复杂性中，生物的秩序规则是如此的严格。

什么地方需要改变？

人们需要乐于工作、乐于思考、乐于学习、乐于停止看电视、停止不断挣钱去为他们的汽车买一套更好的音响设备，可以开始忽视广告、开始将注意力放到更大的问题上、开始思考如何培养自己的创造性，这才是乐观进取的态度。人们必须决定，支持有益健康的东西，然后去认真地进行实际工作，停止做所有那些浪费时间、能量、金钱和生命的事情。在这个过程中，关键是要确定：我们应该多做需要做的事，少做可能想做的事，而不是简单冲动地想做什么就做什么。想想这一点！

2012年6月，联合国可持续发展大会里约+20峰会的主题，或者说主题句就是："我们期望的未来。"我们需要的未来是什么？许多人想要更多，但增长的时间已经超越了地球生态环境的极限，如果我们的后代要兴旺繁荣或者甚至有许多野生动物陪伴他们（可能任意一些都行），这些肯定就没有意义。我们需要为一个健康未来做点什么，而且为了保证这一点，我们将会发现，就像我们熟悉的情况那样，我们要做的事情，是一些不同于刚开始必须要做的事情。

在生活中不仅有权利，也有义务。我们依然有时间享受，最后获得更高的满足，但这种享受会和未来有价值的东西相关联，是我们给予孩子的礼物。许多投身"可持续运动"的人说，我们认为自己理解了设计事物的要求，因而似乎人们也应该可以感受到这些东西是容易的、简单的，也就是我们常听到的，他们也可以"摘到低垂的果实"。我却说，要为子孙后代保留低垂的果实，我们要做一些困难的、成年人的事情，这些需要真实的工作、需要改变、需要经典的目标驱使来成就人们的共同利益。没有人能在战斗的口号"放松"中获胜。

在你看来，中国对环境的承诺保证有多认真？

一些人做出许多重大的承诺，但同时存在着一种创造工作、执行项目的强烈愿望，它不可避免地会损害生态环境。我参与了淮北市的项目，这个城市有200万人口，在历史上曾是一个能源城市，它在上海西南方向200英里左右。人们在大运河上开采出煤矿，并挖了一个巨大的湖来获取煤矿。这几乎是一幕超现实的场景，非常令人迷茫，因为土地正在坍塌，所有的郊区和城市周边都向这个湖倾倒。实际上，你可以划着船，在水下俯瞰所有的楼房。

在淮北，人们对生态城市规划特别感兴趣，或者人们说自己很感兴趣。现在距我去那里已经有四年了，所有的解决方案都在供应的层面：更多的太阳能、更多的风能用以替代石油、更多机器、更多的所有东西……更快、更多。我则站在需求的一边，提倡造一座不需要太多能源的城市。它将会是世界的领导者，将会是这样一个极佳案例：一个煤矿能源城市如何通过城市自身的设计，在能源转化领域占据领导位置。可能在提出策略的一边看来，在生态城市的模式中，获得太阳能有一点贵，但如果你对城市进行设计，会发现并不需要这么多的能源，不管是开始还是它在数年之内改变。我为该城市的一个模范工程做了很多设计，我认为这个城市是完整的，一个健康城镇需要的所有要素都集中在一起，主要的步行区域与临近的自然恢复区相互协调。在我的设计中，基本上是采取我之前提到的生态城市的"片段"理念，其中有三维的联系，大楼之间有连桥，外部的玻璃电

lives. A crucial element here is in deciding that we maybe should do more of what is needed and less of what we might think, rather impulsively, we simply want. Think about that.

The theme for the Rio+20 United Nations Conference on Sustainable Development, in June 2012, carried the theme, or tag line, "The Future We Want." What about the future we need? A vast number just want more, but the time of growth is well into ecological overshoot for our planet, and certainly that doesn't make sense if our children are going to thrive or even have many—maybe any—of the wild species to keep them company. We need to do what's needed for a healthy future, and in making a commitment to it we will discover, as we learn the ropes, that we will end up wanting something a little different when we start facing what has to be done.

There are duties as well as rights in life. We can still have time for pleasures and eventually higher satisfaction too, but it will be relevant to something of value to the future, a gift to our children. Many in the "sustainability movement," such as it is, say we who think we understand need to design things so people can feel that things are easy, simple, as I often hear, so they can "pick the low-hanging fruit." I say save the low-hanging fruit for the children. We have to do some difficult adult things requiring real work, change, and good old goal-driven accomplishment for the common good now. One never won the day with the battle cry, "Relax!"

In your opinion, how serious is China's commitment to the environment?

Well, there is enormous commitment by some but simultaneously a desperate desire to create jobs and implement projects that will unavoidably degrade their ecology. I got involved with Huaibei, which has a population of 2 million and a history as an energy city. It's about 200 miles southwest of Shanghai. They're on the Grand Canal, and have been producing coal, undermining a big lake to extract the coal beneath. It's almost a surrealistic scene, fascinating really, because the land is now collapsing and all the suburbs and environs are slanting into the lake. You can actually paddle boats and look down at houses underwater.

In Huaibei, they are interested in ecological city planning, or say they are. And so I've been going there for about four years. All the solutions are now on the supply side: more solar energy, more wind energy to replace oil, more machines, more everything, faster, more. I'm on the demand side. Let's make a city that doesn't need much energy. It could be a world leader, the case study in a coal-energy city taking the lead in energy conservation by design of the city itself. Perhaps, on the supply side, it may be a little more expensive to get solar energy, but you don't need anywhere near as much energy if you design the city, either from the start or as it changes over the years, in the ecocity mode.

I made a number of designs for a model project for their city that I considered complete, with all the elements of a healthy town close together, mainly pedestrian and coordinated with adjacent natural restoration. In my design, basically the ecocity "fractal" idea I mentioned earlier, there were the three-dimensional connections, the bridges between buildings and exterior glass elevators and ecocity architectural features like rooftop uses, multistory attached solar greenhouses, street orchards, a plaza facing the waterfront to celebrate local nature.

Instead they wanted more energy from biofuels, something called phragmites, a cane, a perennial tall grass. It's very invasive, so people usually don't like it, but it's very hardy and grows quickly. How much do you have to cut down to get serious quantities of fuel? So now, Huaibei wants to become the champion of biofuels, continue being an energy-supplying city, shifting from underground production of energy to replacing farming land and lakes for fish with organic product to burn and send into the atmosphere. I think this is a disaster. I can appreciate that my friends in China want to create jobs, and want to open the subject of ecocities, but I don't think we can afford the time to make all these bad mistakes on the way to an ecological solution.

Tianjin Eco-city is making progress in the meantime, and I've learned more recently about national policies pushing in the right direction and even declaring the intent of building more ecocities using the term. (I use "ecocities" without a hyphen because I want it to look like a normal word in common usage.) If there's a serious problem around biofuels, don't go there. Instead, work as fast as you can on reshaping the city and saving energy; put in solar energy as fast you can, and reduce demand at the same time.

So what is the answer? Leadership?

Well, leadership is helpful but followship is better because most of us are followers; we don't have the disposition to take big public risks. And there are things that everybody can do.

It is about doing, but doing what? I've suggested that one of the most important things is building cities. For that, something else is indispensible and it's called vision. To make the vision into a mental picture, imagine a metropolis busily damaging its bioregion, filling its watershed air basin with smog, and busily spreading across the countryside, farmland, and

梯, 以及一些生态城市的建筑特征, 比如: 屋顶使用多层的太阳能温室、街道果园, 以及可以临水欣赏自然的购物中心。

然而, 他们想要从生物燃料中获取更多的能量, 这种生物燃料叫芦苇, 一种根茎植物, 多年生的长草。它极富侵略性, 所以人们通常并不喜欢它, 但它极为耐寒, 长得很快。那么, 要获得一定数量的燃料, 你究竟需要割掉多少芦苇呢? 所以现在, 淮北市想要成为生物燃料方面的冠军, 想要继续成为一个能源供应城市, 从地下能源产品转移到替代性的农耕用地和养鱼湖泊, 这些地方有可以燃烧的有机产品, 并排放到空气中。我认为这是一场灾难, 我会赞赏我中国的朋友, 他们想要创造工作的机会, 想要打开生态城市这个话题, 但是我不认为在找到一种生态解决方案的路上, 我们有时间来犯所有这些错误。

天津的生态城市在同时取得进展, 最近我了解了更多关于中国的一些政策, 他们想在正确的方向上进行推动, 甚至在说明建立更多生态城市的意愿时使用了这个术语 (我使用"生态城市"一词时中间并不加上连字符, 因为我想让它在日常使用中看起来像一个普通词汇)。如果生物燃料存在严重的问题, 就不要再往那个方向走了。相反, 尽可能快地工作以重塑城市, 节省能源; 尽可能快地投入到太阳能中去, 并减少能源需求。

那么答案是什么? 领导力?

嗯, 领导力很有帮助, 但追随者是更好的答案, 因为我们大多数人都是追随者; 我们并不愿意去冒巨大的公共风险, 这里有许多东西是每个人都能做的。

关键在于做, 但做什么? 我已经提出建议, 最重要的一项就是建设城市。因此, 还有一些东西是必不可少的, 它被称为愿景。让愿景成为一幅意识中的图像: 想象一座大都市正忙于毁坏它的生态区, 让它的河流水域充满烟雾, 而且漫向乡村、农田, 甚至还翻山越岭。然后想象这种"聚合体", 就像蒙特利尔人不自觉地指向大型的都市行政区域, 发现他们主要的城市中心、大区中心和街区中心。

然后想象整座大都市区域转变成一个生态大都市, 由许多生态城市组成, 这些城市过去是中央商务区; 现实中能生存的城镇我们称生态城镇, 那里过去常是区域的中心; 以及真实的乡村 (称之为生态乡村），那里现在是居民区的中心。所有这些是更多的"混合使用""均衡发展"，像小型广场之类的公共空间在那里兴旺繁荣。

水道得到了修复，农田从一些社区公园、公共花园、运动场扩展到一个互相连接的，不断增长的公开区域中，改变了原有的形式——在那里，城镇向外扩张，成为连续的、以柏油路连接在一起的一大片区域。自然和农田来到了公寓形式的边缘——而且比我们郊区散乱的、功能单一的独立家庭住房，呈现出更多样的发展模式。

改变是缓慢的，需要数十年的时间。但是在一种清晰的生态城市愿景引导下，随着年老破旧的基础设施被拆除，无论是哪种可回收的建筑材料，都有助于建立发展中的生态信息中心，自然和农业回归到距离我们生活不远的地方，而非远离我们上班的交通路线。在这种意识图景中，我们能看到一种真正健康的生活方式正在形成，并通向未来。

伴随着不均衡的高回收率，以及长期以来燃料价格的上涨，扩展延伸的遥远边际已经开始萎缩，城市中心也开始变得越来越繁华和受欢迎。大约从五年前开始，年轻人开车的次数少了，对汽车的欲求被电子化、直接私人的一对一、面对面联系所代替，这一联系发生在所有事物都是生动真实的中心里。自行车越发受到欢迎，而尽管经济下挫，停车费用则越发增高。许多标志都让人感觉很有希望，无论我们是不是领袖，我们中的一些人还是有着联合起来的愿景。

up the hills and over beyond. Now imagine this "agglomeration," as the Montrealers refer unselfconsciously to larger metro regional administrative districts, finding their main city centers, large district centers, and neighborhood centers.

Now imagine the whole metropolitan area turning into an ecotropolis made up of ecocities where Central Business Districts used to be, real viable towns we can call ecotowns where the district centers used to be, and real villages (call them ecovillages) where neighborhood centers are now located. All these are far more "mixed-use" "balanced development"; public open spaces like small plazas thrive there.

Waterways have been restored and farmland has spread from a few random community gardens and official parks and playing fields to join one another in growing fingers of open space, reversing the earlier pattern where towns spread to become one continuous asphalt-linked mass. Nature and farms come right up to the edges of apartment-style—and much more diverse—development than we see in the scattered, functionally identical single-family houses of the suburbs.

The change has been slow, taking many decades. But guided by a clear ecocity vision, with the removal of aging, disintegrating infrastructure, recycling whatever building materials can help build the growing ecologically informed centers, and the return of nature and agriculture near where we do our living instead, far from where we do our commuting. In this mental image we can see a truly healthy way of life shaping up deep into the future.

Already the distant fringes of sprawl, with disproportionately high foreclosure rates and a long-term increase in fuel prices, have begun to wither. Already city centers are getting more popular and prosperous. Starting about five years ago, young people have been driving less and are replacing desire for cars with direct personal connections electronically one-on-one, and face-to-face in the centers where things are lively. Bicycles are growing in popularity, and parks are getting more money despite the economic downturn. Many signs are hopeful. And some of us have the vision to unify, whether we are leaders or not.

Maria Sevely
FORM | PROFORMA

Maria Sevely, Project Director of FORM | Proforma, was steeped in architecture from an early age through exposure to her architect/professor father, his colleagues, and mentors, including many of the most renowned figures in the field. She entered architectural school at fifteen and went on to study architecture, architectural history, and fine art at Wellesley, Harvard, and the Rhode Island School of Design (RISD). Maria's background includes substantial work as a design associate with three of the world's leading architectural firms—Pei Cobb Freed & Partners, Richard Meier, and Philip Johnson—often working closely with the founding principals. She was Senior Project Designer for the United States Holocaust Memorial Museum in Washington, D.C., among other projects with PCF&P, and Project Architect for Philip Johnson's last major project, the Cathedral of Hope. With FORM | Proforma, she is working to expand upon this foundation, utilizing computer capabilities to develop new geometries and advance the application of an analytical design approach for current projects in New York, the northeastern United States, and China.

玛丽亚·塞弗莉
FORM | PROFORMA

玛丽亚·塞弗莉受到她那位作为建筑学教授的父亲，以及他的同事和导师——包括很多这一领域最有声望的人物的影响，玛丽亚·塞弗莉很小的时候就沉浸于建筑之中，并在15岁时进入建筑学院，然后继续在威尔斯利学院、哈佛和罗德岛设计学院研究建筑、建筑史和美术。玛丽亚的背景包括在三个世界顶级建筑公司（贝聿铭、考伯、弗里德及合伙人建筑师事务所；理查德·梅尔建筑事务所；菲利普·约翰逊建筑事务所）中从事大量的设计工作，而且，她还经常与这些公司的创始负责人进行密切的合作。她是华盛顿的美国大屠杀纪念博物馆的资深建筑设计师，同时也参与了贝聿铭事务所的其他项目，并作为项目工程师参与了菲利普·约翰逊建筑事务所最近一个大型项目——希望教堂工程。她现在正致力于在纽约、美国东北部和中国开发一种新的形式基础，应用计算机设计开发新的几何形式，提升分析设计方法在项目中的应用程度。

For complete biography, please refer to the Appendix.

完整的传记，请参阅附录。

LEED Platinum Chang'An Mixed-Use Development Proposal, Beijing, China. (FORM, 2012; studio amd rendering)
中国北京长安多功能综合开发方案，获得能源与环境设计白金认证。（形式，2012年）

What has New York City done to meet new energy standards?

NYC recently passed legislation requiring all buildings over 50,000 square feet [4,645 square meters] to benchmark their energy usage. In the next twenty years they have to verify consumption, and then bring it up to meet certain energy standards. The focus is on existing buildings, as new construction already has to certify certain performance levels.

The city has an active list of buildings and has already sent out the first round of questionnaires. In the first phase, they just want to know current usage as a baseline for going forward: what your building systems are, the amount of circulation space, the number and type of windows, the number and location of tenants, down to the number of computer terminals, whether you have a refrigerator or a sink in your space, and so on. In addition to office and commercial buildings, the mandate includes apartment buildings, co-ops, and condominiums. Owners are supposed to indicate whether they have a low-flush or regular toilet, dishwasher, washing machine, etc.

Building managers and owners then have to document their electric bills every month for a year. Usage has to be verified, based on an ASHRAE [American Society of Heating, Refrigerating and Air-Conditioning Engineers] tier-2 report, and then the buildings have to be brought up to meet certain energy standards. This requirement took effect in 2013. There are actually architects and engineers who are sending out missives saying, "You have to benchmark; we will help you meet the new energy standard."

The ordinance is being handled through the Department of Buildings in association with the Department of Environmental Protection, the same people responsible for the city's no-smoking ordinance. Building owners/managers were supposed to file by May 1, 2011, but many decided to just wait and see what happens. Others are doing it themselves and just sending in haphazard data, but they're going to get into trouble because they are going to be required to stand by the data they submit, and improve from that point. Quarterly fines will be issued for noncompliance.

The final phase of the law has to do with the efficiency of lighting. This isn't a requirement until 2020, but for whatever reason, a lot of building owners are very aware of this one requirement. They're skipping over everything before, but this one aspect has somehow gotten through, with many buildings already taking measures to adjust their means of illumination!

纽约市采取了哪些措施来适应新的能源标准？

纽约市最近通过立法，要求所有面积超过5万平方英尺（4645平方米）的建筑对它们的能源使用状况进行基础测试。在下一个20年，这个城市必须检查自身的能源消耗，然后将其提升到符合特定能源标准的程度。此次关注的焦点是现存的建筑，而新的建筑已经被要求符合特定的性能水准了。

这个城市有一个现在运营建筑的列表，而且已经发了第一轮的调查问卷。在第一个阶段，他们只是想知道当前的能源使用情况，将之作为未来发展的基线：从什么是建筑系统，交通空间的面积，窗户的数量和类型，租户的数量和位置，到计算机终端的数量，以及你的空间中放的是冷冻机还是散热器等等问题。除了办公和商业建筑，还包括住宅楼、合作公寓、共管公寓。建筑所有者被要求说明，他们拥有的洗手间、洗碗机、洗衣机等设备是节水的还是普通的。

建筑管理者和所有者必须在一年中的每个月都记录电费清单，并确认使用量，基于ASHRAE（美国供暖、制冷和空调工程师协会）的二级报告，建筑必须提升以适应特定的能源标准。这项政策已经在一些建筑上开始实行了，并将于2013年成为一项硬性要求。实际上有一些建筑师和工程师已经发出信函说："你必须进行基准测试，而我们会帮你达到新的能源标准。"

通过与环境保护部门联系在一起的建筑部门，这一条例已经被执行，环境保护部门也为城市中的禁烟条例负责。建筑的所有者/管理者会被要求在2011年5月1日前上交文件，但有许多人决定再等等，看看会发生什么。其他人则开始行动，并且仅仅是送上一些非常随意的数据，不过，他们将会有麻烦了，因为他们会被要求支持所提供的数据，并从那个点上改进，不服从者将会受到季度性罚款。

法律的最后阶段要处理照明设备的效率问题。在2020年前，这并不是一项必须的要求，但不管出于什么原因，还是有许多建筑所有者非常明确地认识到这一要求。之前，他们忽略了所有东西，但是这一方面的问题却以某种方式得到了解决——许多建筑已经采取措施，调整它们的照明方式！

在评估中，建筑的表皮有多重要呢？

他们确实会看窗户、建筑材料以及外层，但那并不是重点。我认为是工程师——工程实践手册——撰写了部分法规，因此评估十分侧重于机械系统。

实际上我发现了一个漏洞，如果你已经获得了LEED（能源与环境设计指导）认证，那么就可以不遵守条例。以这栋建筑为例（麦迪逊大道185号），它被要求在2014年之前达到规则的要求。如果它在2014年前获得了LEED认证，那么它就可以不用参加能源使用监控的整个过程了。

从理论上说，LEED space认证象征了它自身的健全性。

是的，但是就评价费用和要求之间的差异而言，ASHRAE协会以及所谓的复古型性能检验会耗费更多的时间和金钱，比LEED认证要求更为严格。在麦迪逊大道185号的例子中，要获得LEED认证，它们需要做许多重大的改进，但比起另外一个标准而言，LEED认证还是相对容易的。

新的立法有什么广泛的应用吗？

纽约市总是为自己有一套先进的建筑规范而感到骄傲，上溯到1870年，纽约市建立了这个国家第一套建筑规范，其中要求建筑要达到防火的标准。从那时起，它就一直演变着。消防部门依然深度参与其中，实际上，是他们在做一些检测工作。

纽约的建筑规范很重要，因为它是作为美国其他地区模板的——尽管现在它的性质已经全然改变了，因为在2008年，纽约市采用了一套国际规范标准（颇具讽刺意味的是，这套规范原先就是在纽约市规范的基础上发展起来的）。采用国际规范的决定，是为了让国外设计师在面对纽约时感到更舒适，不那么拘束受限。实际上，"国际"这个词使用得并不恰当，因为它主要是在美国被使用的。这就好像是我们说的"世界职业棒球大赛"，可实际上只有美国的队伍参加比赛。

你是如何成为新能源规范方面的权威的？

几年前，我们受邀参加面试，成为纽约市绿色建筑顾问的三家公司之一，这种顾问的工作是，

How important are building envelopes in the evaluation?

They do look at windows, building materials, and the exterior envelope, but that's not what's being stressed. I think it was engineers—MEPs—who wrote parts of this law, and it's heavily weighted toward the mechanical systems.

I actually found a loophole, which is that you can be exempted from compliance if you are LEED certified. For example, take this building [185 Madison Avenue], which is going to be required to have met the laws by 2014. If it were to become LEED certified prior to 2014, it would be exempted from the whole exercise of monitoring energy usage.

In theory, LEED certification comes with its own badge of integrity.

Yes, but in evaluating the difference between the cost and the requirements, the ASHRAE study and the so-called retro-commissioning takes more time and money, and is more demanding than LEED certification. In the case of 185 Madison they would have to make a number of major improvements to become LEED certified, but it would still be easier than the other route.

What are the broad implications of the new legislation?

New York City has always prided itself on having a progressive building code, ever since establishing the first building code in the country, back in 1870, with the requirement for fireproof buildings. It has just kept evolving ever since. The fire department is still heavily involved; they actually do some of the inspections.

New York's building code is important because it serves as a prototype for the rest of the United States, although now it's been turned on its head because in 2008 NYC adopted a version of the international code (which ironically had originally been based on the NYC code). The decision to adopt the international type was so that foreign architects would feel more comfortable and less intimidated by the NYC requirements. Actually, the term "international" is a misnomer, since it's used primarily in the United States. It's like calling it the World Series but with only American teams competing.

How did you become an authority on the new energy code?

A couple of years ago we were one of three firms invited to interview for the position of green architectural consultant to New York City, whose job would be to review every building

Monticello from the west lawn, Charlottesville, Virginia. (Thomas Jefferson, 1769–1808)
弗吉尼亚州夏洛茨维尔蒙蒂塞洛的西草坪。（托马斯·杰斐逊，1769年—1808年）

project in the context of sustainability. The question was how to do that while working with a wide range of projects and different architects with different design intentions and different budgets. We had to think through how we'd approach that and determine the most important things to evaluate. While ultimately the city chose to work with a single source, we were commended for our specialist approach, including a green specifications writer, green contractor as adviser, and engineers experienced in achieving sustainable solutions.

If on-site labor really understands the objective, they're going to bring more to it, so we recommended workshops on city projects where construction workers could actually get some training in how to realize the intentions. It's not just, "Here are the specs; now build." It's about transforming an idea into practice.

在可持续性的语境中评估每一个建筑项目。问题是，面对众多的项目，面对有着不同设计目的的建筑师以及不同的预算，如何评估。我们必须思考，如何做到评估以及确定评估中最重要的东西。最终，纽约市选择与其中的一方合作，我们被推荐并提出了一些专业的方法，包括：一套绿色说明书；作为顾问的绿色承包者；以及在获得可持续性方案中经验丰富的工程师。

如果现场施工的工人真的能够理解目标，他们就会为绿色建筑增加更多的东西，所以我们推荐关于城市项目的研讨会，在那里，建筑工人们能够真正在实现设计目的上得到一些训练。那不等同于"这是说明图纸，开始建设吧"。它是关于观念向实践的转化问题。

Monticello and connected South Pavilion, where Jefferson and his family lived while the main house was under construction. (Thomas Jefferson, 1770)

蒙蒂塞洛和贯通的南楼，当主楼还在施工时，杰弗逊和他的家人就住在那个地方。

（托马斯·杰弗逊，1770年）

使我们独树一帜的是：我们强调节约能源、减少费用的努力不能仅仅是一项技术活动，人们还必须关注空间的宜居性、舒适度和适用性。从历史上说，这是一条准则。早期人们住得离自然资源很近，他们知道在冬天之前砍下足够多的木头；他们建造的房子冬天可以充分燃烧木头保持温度，夏天时又十分凉爽。今天，人们已经变得非常依赖系统，依赖中央制热和空调系统，依赖电力照明。先前，由于建造系统还不是高度发达，主要的关注点放在建设平台［一栋建筑从地表的温度（寒冷与湿气）影响中隔离出来的方式］与建筑外层上。而且，这两个要素使建筑的内外季节性地保持凉爽和温暖，窗户则为日光照明提供了最重要的途径。在当前的实践中，恰当地设计平台和表皮以便最大化地回应自然位置条件，与高性能的系统结合起来，才能达到最有效地利用能源。今天的建筑师越来越忽略表皮的设计与建筑能耗之间的密切关系，因此，他们经常将设计奠基于其他的方面。

想想住在蒙蒂塞洛的托马斯·杰弗逊，蒙蒂塞洛是一个美丽的居住地，即使从今天的标准来看仍是奢侈的，但是杰弗逊直到后来才住在那里。当杰弗逊第一次决定要结婚时，他建造了一座单间小屋，他的家人在那里住了很多年。他在一个窑里烧砖，砍伐树木，修饰木材，建造房屋，房屋大约有20平方英尺。他们并不认为这是不同寻常的困难。在蒙蒂塞洛的房子完成之后，杰弗逊将原先的

What set us apart was in stressing that efforts to save energy and reduce costs shouldn't be merely a technical exercise, but that one has to keep in mind the livability, comfort, and suitability of the space. Historically, this was the norm. In the early days people lived close to natural resources. They knew to cut enough wood before winter; they built houses that burned logs efficiently and stayed cool in summer. Nowadays, people have become very systems oriented, relying on central heating and air-conditioning, and on electricity for lighting. Previously, because building systems were not highly developed, the primary focus was the building platform—the means by which a building is separated from the thermal impacts of the ground (cold and moisture)—together with the building envelope. Together, the two components kept the seasonal cold and heat in or out. And windows provided the principal means of illumination through daylighting. In current practice, appropriately designing the platform and envelope to maximally respond to natural site conditions in conjunction with high-performance systems can lead to the most efficient use of energy. Today's architects have tended to forget that envelope design has a lot to do with a building's efficiency and therefore have often based their designs on other considerations.

Think about Thomas Jefferson living in Monticello, which is a beautiful residence, luxurious even by today's standards, but he didn't live there until later. When Jefferson first decided to get married, he built a one-room house where his family lived for many years. He made the bricks in a kiln on site, felled the trees, dressed the wood, and built the house, about

Ephraim Hawley House in 2011, Trumbull, Connecticut (c. 1690)
2011年的以法莲霍利大楼, 康涅狄格州特兰伯尔 (大约1690年)

20 feet [6 meters] square. They didn't consider this unusual hardship. Once Monticello was completed, Jefferson incorporated the small original structure into the overall design and continued to use it as his law study and guest house.

In New England you see a lot of traditional salt box houses with small windows facing southwest. The house slopes down to the ground along the north to minimize impact from storms and prevailing winds, while fireplaces built into the stone wall provide a heat block. Snow tends to safely slide off the roof.

小型结构扩充为整体的设计, 继续用它作为自己的法律研究室和客房的模板。

在新英格兰地区, 你会看到许多传统的、带有西南朝向的小窗户的盐罐屋。这些房子从北面向地上倾斜以减少飓风和盛行风的影响, 而壁炉则建在石墙之中来提供加热。雪也会安全地从屋顶滑落下来。

相反, 在像新奥尔良之类的地区, 房子被抬升, 离开地面以抵御洪水。人们通过在泥土中增

Raised house in New Orleans' 9th Ward, designed by Shigeru Ban for Make It Right. (2009)
坂茂建筑事务所为"正确之举基金会"设计的新奥尔良第9街区的楼房。（2009年）

加10到20条编织的藤垫，建构起地基，这样的方式十分独特，然后房屋就建设在上面，基本上可以说是漂浮的。它们有一个前廊，内部的天花板很高并带有楼梯，以排除热空气，就像一种烟囱效应产生的结果。重点是，人们必须根据当地的条件进行建设。这是有关从一栋建筑的位置和使用进行建设的恰当性问题，而不是强行采用一种普遍的玻璃盒子结构（仅仅因为玻璃是一种简单的、容易获得的漂亮建筑方式）。

这样的态度在什么时候发生改变?

改变发生在20世纪中叶，当建筑系统变得更有效时。直到二战之后，空调变得普及，供热系统得到长足进步，这些才变成现实。

　　在创造一种室内环境的过程中，人们需要考虑建筑平台、建筑，以及使建筑正常运作的系统。正如我前面提到的，信赖通常在建筑平台和外层上占据重要地位，因为系统——比如说，用

By contrast, in places like New Orleans, houses are raised up off the ground against flooding. Foundation footings were constructed in a unique manner by pushing some 10 to 20 woven cane mats into the mud and then building on top, basically floating the house. They included a front porch and had high-ceiling interiors with big staircases to exhaust hot air, like a chimney effect. The point is that people had to build relative to local conditions. It was all about the appropriateness of a building for its location and use, and not just imposing a universal glass box because it's an easy way to achieve a handsome building.

At what point did such attitudes change?

It happened in the mid-twentieth century, when building systems became much more efficient. It wasn't really until after World War II that air-conditioning became commonplace and heating systems became much more developed.

In creating an indoor environment, one needs to consider the building platform, the building, and the systems that make the building work. As I mentioned previously, reliance used to be heavily weighted on the building platform and envelope, since the systems—for example, fireplaces for heating—were really primitive, even in great palaces. There's been tremendous development over the past century. Interestingly, around the new millennium, American scientists were asked to name the greatest scientific advance of the past 1,000 years. They concluded it was air-conditioning, as it has saved and prolonged life more than any other advance.

As Gropius said of revolution, there is huge enthusiasm and the pendulum swings way over toward new ideas, so you forget about past ideas for a while. Then people slowly start to remember and re-incorporate the past. In the same way, we've gone through a period where we've been very heavily dependent on building systems, and we forgot the value of good siting, and the value of really looking at building platforms and envelopes in a way that minimizes dependence on energy. Some people are now beginning to look at that with fresh eyes. It's mentioned in LEED standards, and in various other programs, like Enterprise Green Communities, which looks at buildings in a more total way.

How did these standards come about?

I think the people who started the United States Green Building Council anticipated the new energy concerns and realized the need for leadership. If you just do it in a grassroots way, there's no organization, no standards. Some people looked at

Energy Star, the federal government's program, which is very demanding. The Empire State Building somehow managed to become Energy Star certified!

The Energy Star list for the entire NYC area is very small, only about 20 buildings. It's not that the LEED list is all that large, but many people aspire to it. However, if you look at those who register to become LEED certified, the number that actually succeeds is much smaller, just a fraction of the buildings originally registered. Costs during construction play a significant part. The construction manager invariably says, "We don't really need that; we've got plenty of sustainability already."

What is the real premium for green building?

It's very small, typically between 3 and 8 percent of construction cost.

Have energy concerns led to a more holistic approach to building?

When we went to the municipal government, we actually used the word holistic because we wanted a total building approach, a total work environment. We wanted to make sure that buildings are good buildings and the overall space is good on every level: that people feel satisfied using it, that it's comfortable and healthful. This is a component of some of the European standards, as in the Bau Biologie program that originated in Germany, although it's not stressed here. Switzerland has a similar program, Minergie, which places human health concerns above technical considerations or cost savings.

I think those standards really led to people in the U.S. becoming more aware of the formaldehyde content in some building materials. Now you can get beautiful plywood or wallboards that don't contain formaldehyde. For years, people were doing LEED projects and getting very high ratings, using standard wallboard filled with these toxins. Think about air quality in those spaces; it might not be all that great.

Another example is fluorescent lighting, of which the U.S. is enamored. It's routinely used in schools and offices across the country, yet for decades, there have been medical studies showing how bad it is for people, not just for eyesight, but it actually interrupts brain wave patterns. For people with dyslexia and certain neurological conditions, the effects are exacerbated under fluorescent lights.

People also get sick from the glues and preservatives in building materials, carpets, wallboards, ceiling panels.

于制热的壁炉——是非常基础的，即使是在大型宫殿中也是如此。过去的一个世纪取得了重大的发展，有趣的是，在新世纪开始前后，美国科学家被要求指出过去1000年最伟大的科学进步。他们得出结论说是空调，因为它挽救和延长人们的生命超过任何一项进步。

正如格罗皮乌斯谈到革命时说的，对于新观念存在着巨大的热情和摇摆不定的态度，所以你暂时忘记了过去的想法。然后人们慢慢地开始记起并重新融入过去。以同样的方式，我们正经过一段时期，在这个时期，我们严重依赖于建筑系统，而忘记了出色选址的价值，忘记了以一种最低限度依赖能源的方式考虑建筑平台和外层的价值。现在，有些人以一种新的眼光开始考虑这些问题。需要以一种更全面的方式考虑建筑，这在leed标准，以及像企业绿色社区类的其他项目中都被提及。

这些标准是如何产生出来的？

我认为成立美国绿色建筑委员会的人，预料到了对新能源的关注，并且认识到了引导的需要。如果你只是以一种很底层的方式来做这些的话，那么就不会有组织，不会有标准。有些人着眼于能源之星，联邦政府项目，这个要求非常高，帝国大厦就尽力获得了能源之星的认证！

能源之星为整个纽约市区开出的名单非常少，大概只有20栋建筑。当然并不是LEED的名单就很长，但确实有很多人在追求这个。然而，如果你看看那些注册登记想要获得LEED认证的名单，你就会发现实际上成功的数目是很小的，只占注册登记建筑的一小部分。建设期间的花费扮演着重要的角色，建设项目的管理者总是说："我们并不是真的需要那样做，我们已经获得很多可持续性了。"

对于绿色建筑而言，哪些是真正额外的费用呢？

额外费用的数目很小，一般来说占建设费用的3—8%。

能源方面的考虑会不会促成一种对建筑更加全面的处理方法呢？

当我们去市政府的时候，确实使用了"全面"这个词，因为我们想要一个整体的建设方法，整体的工作环境。我们要确认建筑是好的，整个空间在每个层面上都是好的：人们心满意足地使用它，它是舒适且健康的。这是一些欧洲标准的构成部分，就像起源于德国的建筑物生机项目——虽然在这里它并没有受到重视。瑞士有个类似的项目，迷你能源，它将人类健康问题置于技术考虑或成本节约之上。

我认为这些标准确实引导着美国人，让他们变得越来越关心一些建筑材料中的甲醛含量。现在，你能得到漂亮的、不带有甲醛的胶合板和墙板。许多年来，很多人都在做LEED项目，他们用充满着有毒物质的标准墙板获得很高的评级。想想这些空间中的空气质量吧，可能并不是那么好。

另一个例子是美国人热衷的荧光灯照明。它例行公事般的被应用到整个国家的学校和办公室中，然而数十年来，已经有医学研究表明它对于人类是多么的有害——不仅是对人的视力，而且它实际上损毁了人的脑波模式。对那些有阅读障碍和特殊神经问题的人而言，在荧光灯下他们的情况会更加恶化。

人们也会因为建筑材料、地毯、墙板、天花板中的胶水和防腐剂而得病。各种添加剂可能会增加建筑材料的耐久度或降低成本，但它不一定是个非常健康的环境。有一些机器可以测量空气质量，但是从根本上说，每一种单独的成分都必须被评估。

可持续性的定义是什么？

从环境运动的角度看，对可持续性的看法是，由于人越来越多，所以保护自然资源不被消耗殆尽是很明智的。有一些人开始思考建筑的耐久度问题，这也是可持续性的一个方面：更长的寿命。当我看到一栋漂亮的老建筑被拆毁的时候，我真是伤心透了。我不介意如果它是被拥有相同的可持续百年的标准来建造，但是一般来说不会这样。

从根本上说，在可持续性中，临时性建筑以及它们是否能被重新改作他用，也是一个需要考虑的东西。在过去这是一个关键点，但现在，我们已经变得对那些根据特定目的而建造的建筑更加着迷。这没问题，但过些年后，人们就必须把房子用作它途。那时的意见将是，把它们拆掉重建。

那是广泛意义上不可回收概念的一部分。

对。我们将为此付出代价，因为建筑不是砸钱而是投资。在19世纪，美国人很节省。但是从二战

The various additives may increase the durability of building materials or reduce costs, but it's not necessarily a very healthy environment. There are machines that can measure air quality, but basically, every single component needs to be evaluated.

What is the definition of sustainability?

From the environmental movement, the perception is that since there are more and more people it would be wise to try to conserve natural resources so they don't run out. A few people are starting to think about the durability of buildings, which is an aspect of sustainability: longer life. It breaks my heart when I see a beautiful old building demolished. I wouldn't mind if it were replaced with something of integrity built to the same kind of 100-year standard, but often it's not.

Essentially, temporary buildings should be a consideration in sustainability and whether they can be repurposed. That used to be a concern, but we've become fascinated with purpose-built buildings. That's fine, but often after a few years they have to house different needs. The attitude is to tear them down and rebuild.

That's part of a broader mentality of disposability.

Right, but we're going to pay for it because buildings are much more of an investment than toss-aways. In the nineteenth century, Americans were savers. But ever since World War II, the country has been living high, with increasing amounts of consumer goods and credit. We got accustomed to high energy use. It's part of that whole mindset of having inexhaustible supplies, which, of course, no one has.

People build what they think they want based on aesthetics or certain spatial characteristics, not energy usage. They want big glass walls, but then they put up curtains. They want a view, but they want their privacy more.

The builder-house that predominates across the country is not sustainable. It's not an enduring, energy-efficient structure. As architects, we don't find it aesthetically pleasing either. There are millions of them, and their construction shows no sign of letting up.

Are New York's energy guidelines universally applied?

No. They try to tailor them to the spaces. There's been some criticism that LEED certification can become a point-chasing exercise. It's very easy to take a project and tick off the easiest means for compliance, technically achieving "sustainability"

434

without having the true spirit. In the most extreme cases, one can pay carbon credits, i.e., paying for the energy you use with an offset surcharge that theoretically negates your own energy usage. It's like during the Civil War when people would pay a surrogate to fight for them. It's that mentality.

There are also trade-offs, so you can, for example, exceed guidelines for allowable amounts of glass by making other parts of your building more sustainable. I was looking at a house in New England where, at fairly high cost, they were able to document zero energy use for an entire year. To maximize a scenic view they used triple-glazed, low-E, argon gas-filled glazing units. Because of that big glass wall, they had to double or triple the R-value of every other assembly in the house to reduce energy usage. There is a self-generating energy source for heating and cooling, but without all that glass, it could have been achieved more easily at lower cost.

In any project, the architect has to meet client requirements and the constraints of the particular site before ever talking about curtain wall. I would think that, given the amount of glass exposure on the typical tower, it would be very difficult to meet projections for significantly reduced energy usage no matter how you orient the building. Early curtain walls weren't engineered; it was a feat just to have them not break apart when putting so many glass units together. People love the look of the glass tower. Maybe we'll come up with some fantastic new engineered material!

Big architects typically have clients who demand LEED certification and are able to pay for it. What happens with smaller firms?

It's budget related. Among smaller clients, only those with healthy budgets want to broach the subject because they just view it as added cost. That may change because of recent legislation and steep energy prices. In the New York area, some builders are starting to offer sustainable or LEED-certified housing. But most times, it's the first thing that gets jettisoned because it's not seen as necessary. Of course it's more achievable on a smaller project, since costs are relatively smaller. On big projects any one decision can add a lot to the cost and it's magnified by sheer scale. It will be interesting to see how people react now that benchmarking is the law. With regard to housing, according to the National Association of Home Builders, the average home size in the U.S. was 1,400 square feet [130 square meters] in 1970; recently it has been reported to be 2,700 square feet [215 square meters]. It will be interesting to see the impact of new energy concerns.

开始，这个国家开始过奢侈的生活，有着不断增长的消费物品和信贷。我们已经习惯了很高的能源消耗，这是我们有享之不尽的物资这种心态的一部分，当然，实际上不会如此。

人们是基于美学或者特定的空间特性建造他们认为自己想要的东西，而不是基于能源使用。他们想要大的玻璃墙，但很快又拉上了窗帘。他们想要一个好的视野，但更想要隐私。

建筑商的房子遍布全国，但并不是可持续性的，它不是一种耐久、节能的结构。作为建筑师，我们也没有在其中找到美学上的愉悦。这些建筑成千上万，而且这种建构方式显然没有停下来的迹象。

纽约的能源指导方针是否被普遍应用了呢？

没有。他们想要修改它使之适应不同的地区。已经有一些批评指出，LEED认证会成为一种赶时髦的举措。接管一个项目，然后标示出符合规定的最简单方式，从技术上而不是在真正的精神意义上达到"可持续性"，这是很容易的。在一些最极端的例子中，人们可以花钱购买碳信用等能源，花钱购买所使用的能源，这种费用从理论上说可以抵消你自己的能源消耗。这就像在内战期间，那时人们花钱雇用一些人代替他们作战。人们就是这种心态。

也有一些权衡取舍，因此你可以通过令建筑的其他部分变得更有可持续性，而超出指导方针上关于玻璃数量的规定允许。我过去考察了新英格兰地区的一栋房子，它造价昂贵，但其全年使用的能源记录为零。为了将视野中的景致最大化，他们使用了三层低辐射、充满氩气的玻璃元件。由于大玻璃墙的缘故，设计师必须为房屋中的其他设备增加两倍到三倍的热阻，以减少能源消耗。有一种自我生成的、用于加热和制冷的能量来源，如果没有所有这些玻璃，就能以较低的费用实现。

在任何一个项目中，建筑师在谈论幕墙之前必须满足顾客的要求，对特定的位置进行限制。我会想，由于在一座普通高楼上玻璃排列的数量，无论你如何确定建筑的方位，项目也很难达到大量减少能耗的要求。早先的幕墙不是设计出来的；它只是一种特殊技术，让我们在将许多玻璃单元放在一起时不会分裂开来。人们喜欢玻璃大厦的外表。可能我们将会提出一些新奇的工程化材料！

一般来说，大牌建造师的客户要求LEED认证，而且他们也能够支付得起。那么小一点的公司呢？

这与预算有关。在小一点的客户中，只有那些有着良好预算的公司想要讨论这个问题，因为他们只是把Leed认证当作额外的花费。由于当前的立法以及不断上涨的能源价格，这种情况可能会有所改变。在纽约地区，一些建筑商开始提供可持续性或LEED认证的住房。但是大多数时候，Leed认证都是第一个被舍弃的东西，因为它看起来不是那么必要。当然，在一个小一点的项目上更有可能做到这一点，因为费用相对也较小。在大项目上，任何一个决定都会增加一大笔费用，而且会成倍地增加。现在基准测试已经是一项法律了，人们对此的回应将是非常有趣的。在全国住房建筑商协会看来，就住房而言，20世纪70年代，美国的平均家庭面积是1400平方英尺；有报告指出，这一数据现在已经上升到2700平方英尺了。看到新的能源所产生的后果将是非常有趣的。

　　整个历史都充满着各种行为和反应。许多第一代的现代派已经失宠了，但我记得我父亲（他曾和格罗皮乌斯一同学习）喜欢说，并没有所谓的现代建筑，有的只是好建筑。好建筑的本质是要负责任。

　　现在，建筑这个职业的一个问题是，尽管人们想要的是便宜、恰当的解决方案，但我们却与这种文化的规范背道而驰。像格罗皮乌斯之类的建筑师反抗的就是这种东西。他们并不一定要憎恨所有具有历史意义的东西，而只是讨厌那些浪费金钱的廉价山寨品。他们喜欢一些简单而不太昂贵的东西，因为那是人们能负担得起，并且运作良好、效率更高的解决方案。有一些建筑行业的人，喜欢看到更小的、与地点和环境相适应的小型楼房，这些楼房所使用的建筑材料可能是可再生的，用量较小，制造它们需要的能量也更少，同时能持续的时间更长。但这并不一定是公众想要的东西。

　　从能源使用的情况看，纽约市看起来还不错，这主要是因为人们不开车。纽约市民住在一个相当拥挤的区域内；他们挤在一起，如果分散于整

Throughout history, there have been actions and reactions. A lot of the first-generation modernists have fallen out of favor, but I remember my dad (who studied with Gropius) liked to say that there really isn't modern architecture; there's good architecture. Being responsible is the essence of good architecture.

One of the problems the architectural profession has right now is that we're running counter to the cultural norm whereby people want the cheap, fake expedient solution. It's the same thing that architects like Gropius were rebelling against. They didn't necessarily hate everything historical; they just hated the cheap imitations that people were wasting their money on. They preferred something simple and less expensive, which would hold up and work well and be a more honest solution. There are those in architecture who would like to see smaller buildings that are responsive to their location and to the environment, and that use materials which can be regenerated, and in smaller rather than larger amounts, requiring less energy to create them, and which would last a long time. But that's not necessarily what the public wants.

NYC looks very good in terms of energy usage largely because the people don't drive. New Yorkers live in a fairly compact area; they're stacked and have fewer buildings for the population than if we were spread out all over the place. While total energy use is high, per capita amounts are low.

It's somewhat related to paying for carbon usage. We're still using energy, maybe somewhat more efficiently, but if big buildings could be designed to have more integrity and make better use of everything, it would be even better than just saying we're getting pretty good efficiency just by virtue of high density.

What is the city's preference in selecting architectural firms?

They like the idea of one-stop shopping. They are more resistant to the notion of coordination among separate firms even if there's a history of their having worked together successfully, and even if a group of associated firms produces a more tailored response with true specialists all in one place. Some of the larger companies may actually be relying on employees in far-flung areas to staff their supposedly comprehensive in-house teams.

Will that mentality trickle down to developers to the point where it might hurt independent architectural offices?

436

Yes. Some firms have completely stopped doing Construction Documents (CDs), or working drawings. They only take projects through Design Development (DD) because it's not economically worthwhile to do otherwise. They sell themselves based upon their planning and design ability. They provide some guidance for CDs, but since most clients aren't willing to adequately compensate them for drawing the CDs, architects defer. The client takes the DDs or minimal CDs and finds a contractor to do the so-called shop drawings. Even residential clients do that: "I'll take care of it myself." And then they run into problems later. They don't understand why they need the architect's working drawings, much less how vitally important specifications are, and don't want to pay for them.

Increasingly, there is the idea that you go to one entity and they give you a product. If you read the newspapers or weekly journals, the people who cover New York real estate and architecture tend to treat buildings like consumer products. I even saw an article a year or so ago in the *Wall Street Journal* where they were advising people to buy houses the same way they would buy a car. Shop for the features you want, and then look for a reputable manufacturer or dealer, the General Motors of home building.

Apparently the public doesn't care if things are individually tailored anymore. They are perfectly happy buying a Louis Vuitton bag that everybody else likes, and they do the same with buildings. There are some high-end residential clients who still want an individually designed house, but they are few and far between. The average middle-class family can't afford it.

Maybe it will change if, depending on the world economy, there's a real demand to go back to concerns for low-cost housing. In the course of his career, my father served as a consultant for the UN. He was asked to look at 112 countries and to come up with guidelines for a basic family living unit for each: how to build an efficient, bare-bones dwelling.

Given today's global economy, we have to look at how to deal with housing at very basic income levels; maybe that will influence thinking overall. One of my good friends from school now heads Food Aid, the largest Catholic charity in the U.S. He's testified before Congress and written to the president because our current policy of using food for commercial production, like using corn to produce ethanol, is contributing significantly to raising food costs to the point where hundreds of thousands of people who five years ago could be fed, are now literally starving because of our

个区域的话，人群所需要的楼房将要多得多。尽管总体的能源消耗很大，但单个人的平均能耗是很低的。

这有点儿和支付碳使用费有关。我们依然在使用能源，可能某种程度上确实是更有效率了，但如果大型建筑可以设计得更具有整体性，能够更好地利用所有东西，这要比说我们只通过高密度就得到非常好的能源效率会好得多。

在选择建筑公司时，城市有什么偏好吗？

他们喜欢一站式服务。他们比较抵触在不同的公司之间合作的想法，即使这些公司有成功进行协同工作的历史，即使一个联合的公司集团有各方面的专家，能够做出更恰当的回应也是如此。有些大公司实际上可能依赖于遍布各地的雇员，以充实他们被认为全面的内部团队。

这种开发商的心境会不会损害到独立的建筑办公室？

是的。有些公司已经完全不再做施工文件(CDs)或施工图纸了。他们只是通过设计深化图(DD)展开项目，因为如果不这么做，在经济上就会不划算。他们基于自己的规划和设计能力销售自己。他们为施工文件提供指导，但由于大多数客户并不想因为做出施工文件就付给他们足够的补偿，所以建筑师们总在拖延。客户拿着设计开发图或者最原始的施工文件，然后找到一个承包商做所谓的施工图。即使是住宅客户也是这么做的，"我会自己照顾好自己"。然后他们很快就遇到麻烦了。他们不理解为什么需要建筑师的施工图纸，更不用说理解那些数据参数有着重要的意义，而且他们也不想为此付钱。

逐渐地，出现了这样一种想法，你去到一个地方，然后他们给你一个产品。如果你读报纸或者周刊，那些报道纽约房地产和建筑的人倾向于把建筑物当作是消费品。一年前左右，我甚至在《华尔街日报》上看到过一篇文章，上面说他们建议人们像买一辆汽车一样去买房子。购买那些你想要的东西，然后找到一家信誉好的制造商或经销商，就像是房屋建设方面的通用公司一样。

从表面上看，公众不再在乎是否事物是量身定做的。他们在购买任何其他人都可能会喜欢的LV包时感到非常高兴，他们也是这么对待建筑的。有一些高端的住宅客户依然想要一栋定制的房屋，但这些人数很少。普通的中产阶级家庭则买不起这些。

根据世界的经济情况，如果人们真的需要重新关注低价住宅，事情可能会发生改变。我的父亲作为顾问为联合国工作过，在他的职业生涯中，他被要求视察112个国家，然后为每个国家提供关于基本家庭住宅单元的指导：如何建设一座简洁有效的住宅。

根据今天的全球经济，我们必须发现如何在最基本的经济层面上处理住房问题，这一点可能会对全盘考虑产生影响。我在校时的一个朋友现在掌管食物救助组织——美国最大的天主教慈善组织。他在国会做过证并写信给总统，因为我们当前将食物当作商业生产的政策，就像在用谷物生产酒精，实际上对食物成本的提升贡献颇多——五年前，成千上万的人可能在这一成本水平上获得温饱，但是现在由于我们富足的假定和政策，他们只有忍饥挨饿了。如果你必须为贫穷的人找到解决方案，并以一种有效的方式实现它，那么，你需要一些新主意。亨利·福特说，为每个人提供汽车的方式就在于大规模生产。

我想你需要根据区域和其他限制，从头开始重新评价目标和发展思路。然后，与那些参与到建筑过程中的人（特别是开发商，投资方，银行和政府部门）协同工作并说服他们，这是极为根本的。同时，你需要与这个团体一起开发出新的发展模式，这种模式将超越可持续性——环境方面的意识：支持清洁的空气、水、土地和能源。目标是要提供一种整体的方法，不仅达到环境方面的目标，而且要达到经济和社会的目标，然后才有可能获得一些更大的收益。

affluent preferences and policies. If you have to find solutions for people who have very little, and do it in an efficient way, you can perhaps arrive at some new ideas. Henry Ford said the way to provide a car for everyone was mass production.

I think you have to start from the beginning to reevaluate objectives and development guidelines in light of zoning and other constraints. Then, it's essential to work with and convince the people who are most involved in the building process, especially developers, investors, banks, and government officials. And, together with this group, develop new models for development that will be above and beyond sustainable—environmentally conscious: supporting clean air, water, land, and energy. The goal is to provide a holistic approach that achieves not only environmental objectives but also economic and social goals. Then maybe there will be some larger benefit.

SECTION THREE / FORUM OF EXPERTS

Quality of Life
Rhonda Phillips, Arizona State University
Jorge Eduardo Rubies, Preserva São Paulo Association
Alfredo Trinidade, Secretariat of the Environment

第三篇 / 专家论坛

生活质量

郎达·菲利普斯，亚利桑那州立大学

豪尔赫·卢比斯，圣保罗乡村保护协会

阿尔弗雷德·特林达德，巴西库里提巴环境秘书处

Rhonda Phillips
ARIZONA STATE UNIVERSITY

Rhonda Phillips, Ph.D., AICP, CEcD, is Associate Dean of Barrett, The Honors College, and Professor at the School of Community Resources & Development at Arizona State University. Community investment and well-being comprise the focus of her work, including community-based education and research initiatives for enhancing quality of life. She holds dual professional certifications in economic and community development (CEcD from the International Economic Development Council) and urban and regional planning (AICP from the American Institute of Certified Planners). Previously, she served as the founding director of the Center for Building Better Communities at the University of Florida, and was the 2006 UK Ulster Policy Fellow Fulbright Scholar. Dr. Phillips is author or editor of fifteen books, including *Community Indicators Measuring Systems* and *Introduction to Community Development*, and co-editor of leading professional serials. She is president-elect for the International Society for Quality-of-Life Studies. She serves as a Senior Sustainability Scientist, Global Institute of Sustainability, at ASU and is on the affiliate faculty for the School of Geographical Sciences and Urban Planning.

郎达·菲利普斯
亚利桑那州立大学

郎达·菲利普斯，博士，美国注册规划师协会，注册经济开发者，她是巴雷特荣誉学院的副院长，也是亚利桑那州立大学社区资源与开发学院的教授。社区投资和福利是她关注的焦点，这其中包括基于社区的教育工作以及提高生活质量的研究项目。她获得了许多专业认证，包括经济和社区开发领域（国家经济发展委员会认证的注册经济开发者），以及城市和区域规划领域（美国注册规划师协会的认证）。之前，她是佛罗里达大学建立更好社区研究中心的创始负责人，也是2006年英国阿尔斯特大学政策研究方面的富布莱特学者。郎达是十部书籍的作者或编者，如《社区开发指标评估系统》、《社区开发指南》；她也是一些前沿专业丛书的主编。她是国际生活质量研究协会的候选主席，亚利桑那州立大学全球可持续性研究院的资深可持续性科学家，也是地理科学和城市规划学院的兼职人员。

For complete biography, please refer to the Appendix.

完整的传记，请参阅附录。

People walking in Hyde Park, Sydney, Australia. (2007)
澳大利亚悉尼，人们在海德公园散步。（2007年）

Tall, mixed-use buildings have been proposed to stem suburban sprawl, but how does one assess the impact of high density?

Planners have wrestled with this for quite a while. Other than some qualitative and anecdotal perspectives on the negative effects of too high a density, what comes to mind, of course, are the U.S. housing projects of the 1960s and '70s, where there was a very high density, for example in New York or Chicago. These didn't work from a social perspective and ultimately ended up being demolished; note that these were targeted at affordable low-income housing. It harkens back to some of those old psychological experiments about what happens when too many rats are put into the maze. Researchers have looked at things like crowding and density, but typically from a perspective of psychological well-being rather than the built environment, per se.

Beyond high density is the issue of height and its social consequences. What happens to people living on the 400th floor? Do they bother to come down or does there exist an entire segment of the population that remains indoors?

It brings up some really fascinating questions. As a community developer I always look for ways to connect people, groups, and organizations and how to build a sense of community. One could speculate that if there were spaces to encourage interaction and well-being, the question then becomes whether it matters if people are removed from the ground plane or in a different dimension, literally in the space dimension instead of on the ground. That would be very interesting to explore.

In a new development, which features would be most important from a community perspective?

Public space, obviously. That's a given. It is vital to create a sense of place and connection. One of the major criticisms about the way we have grown and sprawled is that we've lost our sense of connection to place and to each other, so that would be the number one priority. This can be achieved with increased density in the housing units, for example, but still having ample open space, park space, or community gathering space and optimizing those common areas instead of the individual building lot. It's not just a question of where we live, but also where we work, where we play, where we interact, where we shop. Is there a way to integrate the neighborhood commercially? We've literally lost our neighborhood stores.

人们提议利用功能混合型高层建筑，来遏止郊区的蔓延趋势，但我们如何评估高密度所带来的影响呢？

长久以来，规划者们一直在思考这个问题。与过高密度会带来负面影响的一些定性和臆测观点不同，他们想到的是美国20世纪60年代和70年代的住房项目，当时像纽约或芝加哥这样的城市，其住房密度非常高。从社会的角度来看，这些项目并没有起到什么作用，并且最终被拆毁，但应注意，这些项目当时都是以向低收入人群供应住房为目标的。这让人联想到以前的一些心理学实验：将太多大鼠放进迷宫时会发生什么？研究人员关注的是拥挤和密度之类的问题，但通常是从心理幸福感的角度来考虑，而不是从建筑环境的角度。

除了高密度之外，还有高度问题，以及它所带来的社会后果。对于居住在第400层的人们，会发生什么？他们是否愿意下楼，或者是否会出现一部分每天宁愿待在室内的人群？

这带来了一些真正引人关注的问题。作为一名社区开发商，我总是寻找方法将人、团体和组织联系起来，并考虑如何建立一种社区归属感。人们会考虑是否有足够的空间，促进彼此互动并增进幸福感，因此，问题就变成，如果人们不在地面居住或者在不同空间层面生活，即居住在空中而不是地面上，情况又会怎么样？这将会是非常有趣的探索。

对于一个新的发展项目，从社区角度来看，哪些特点最为重要？

很明显，公共空间最为重要。这是不争的事实。营造一种场所感和连结感是很关键的。对于成长和扩张的方式，最主要的批评之一就是：我们失去了场所感，以及彼此之间的连结感。因此这应该是最优先考虑的事项。比如说，要实现这点，可以增加住房单位的密度，同时保留足够的公共空间、公园空间或社区聚集空间，并优化这些公用区域，而不是独立的建筑用地。这不仅仅涉及到我们在哪里居住，还涉及到在哪里工作、娱乐、互动和交往以及在哪里购物的问题。有没有什么方法以商业的方式整合住宅区？准确地说，我们已经失去了邻里商店。

对于过去几十年间迅速发展的较大型新兴城市来说，其中大部分并不支持混合用途建筑项目。

Pruitt-Igoe, St. Louis, Missouri. (Minoru Yamasaki, 1955)
密苏里州圣路易斯普鲁伊特一戈。（山崎实，1955年）

事后证明，我们的确需要邻里商店，这样你可以步行或骑自行车去购买一些物品，而不必开车前往距离很远的大型零售综合超市。我们再次想到传统街区，这些街区可提供部分零售服务和商业服务，并拥有公共空间，而且你住在这里，你也希望你的孩子在这里上学。这些都已成为了过去，尤其是在美国。我认为，如果中国正面临着类似我们在二战之后所出现的增长和扩张压力，那么它将不得不解决这些问题，以防止出现一些与我们一样的、不具有可持续性的发展模式。这已经被多次证明过了。平均的往返时间太长。现在，一些大型房地产公司将步行便利性也当作一个服务的亮点，也就是预估你是否可以步行到商店、公园和学校。令人惊奇的是，有些点之间分隔得非常开，这意味着你需要行走如此之远。大规模的城市扩张让你除了在自己的社区内行走之外，没有其他地方可去。

Most of our larger, newer cities that have been growing rapidly over the last several decades did not support mixed-use development. In hindsight, it is clear that we really do need a neighborhood store that you can walk or bike to for a few items instead of having to drive to large retail complexes that may be many miles away. We're coming back to the idea of traditional neighborhoods, where there is a bit of retail, some commercial services, and public space, but also where you live and hopefully where your kids attend school. That's what we've lost, particularly in the U.S. I would guess that if China is facing similar pressures to grow and sprawl like we did after World War II, they'll have to address those issues to prevent some of the same development patterns, which are just not sustainable. That's been proven over and over again. Average commuting times are too long. Now, some of the large real estate listing services actually assign a walkability score to properties, literally rating whether you can walk to a store, to a park, to school. It's astounding how atrociously off mark some of those scores are, meaning how very far you'd have to walk. There are vast expanses of urban development with no possibility of walking anywhere other than within your own neighborhood.

Jogging in Central Park, New York. (2006)
在纽约中央公园慢跑。（2006年）

Historically, townspeople shared common interests, even common backgrounds, which reinforced a collective identity and strong sense of place. Can one expect the same kind of social interaction in today's more transient, fast-paced, and diversified communities, particularly bedroom communities, if adequate public spaces are provided?

In most of our modern developments since World War II we haven't built an interaction space, or a sense of place. People don't have the opportunity to interact because most everyone drives. They pull in their driveway and go into their home. There will be those who never engage. In bedroom communities particularly it's a pervasive issue because the people are not part of the community in any sense other than they have a place to sleep at night, and then they go back to the city.

On the other hand, there are places very close to large cities that have retained a definitive sense of place and a strong sense of community. I think part of it goes back to having the opportunity to engage but also having a need

从历史上看，市民具有共同的利益，甚至具有共同的背景，这增强了集体认同和强烈的地方感。在当今更加瞬变、节奏更快和多元化的社区中，尤其是近郊居住区，如果提供了足够的公共空间，那么人们可以期望类似的社会交往吗？

对于第二次世界大战以来的大部分现代开发项目，我们尚未建立互动空间，或营造一种地方感。人们没有互动的机会，因为大多数人总是驾车出行。在停放车辆之后，人们径直回家，一些人并不交往。尤其在近郊居住区，这是一个普遍的问题，在一定意义上，人们不再是社区的一部分，这些社区仅仅是他们晚上睡觉的地方，白天他们会返回城市。

另一方面，有一些非常接近大城市的地方，保留了明显的地方感和强烈的社区感。我认为部分原因是人们具有交往的机会，并且具有这样的需要或意图：无论是出于保护、改善和维持一种文化特色，还是仅仅让人们聚在一起。从较小的单独农村居民点一直到市中心小型居住区，我们都可以看到这样的现象：人们出于一种目标聚到一起。因此，这也是驱使城市规划专家们的推动力：如果具有一定的目标和场所，人们会聚集在一起。如果物理环境支持这种目标感，就会促进人们的交流，如果没有，那么就很难做到。例如，当州际公路穿越并隔断了我们许多城市街区时，将不会再有强烈的连结感或互动关系。各种类型的重大建设项目同样隔断了这种物理连结。如果没有物理连结，就很难有社会连结，所以在这个意义上来说，它是至关重要的。

在相同的地方生活和工作，对于许多新的社区来说，是一个理想。这一目标具有什么影响？

如果你在一个社区内拥有自己的房屋或居住在一个公寓里，并拥有自己的生意或者为社区里的公司工作，那么市民的参与意识就会上升。还会有一些人，虽然他们不在社区居住，但也会参与社区生活，他们可能是企业主或者有历史联系的人们，存在着各种各样的原因。但一个人获得的越多，参与程度就越高，这非常合理，并且很明显，在同一社区工作和居住，就没有过多往返的需要。

商住两用的环境是否会减少多样性，并遏制某种活力，由于其他地区人口、文化和观念的注入，这种活力会给社区带来生机。

这取决于几个因素。一个是当地的经济实力，是否能够吸引别人来投资、就业以及开拓事业等；另一个是该社区是否能成为人们有能力生活于其中的社区，至少要对广泛的年龄段、收入和背景的人群具有吸引力。

美国有少数城市的确正在努力做这些事情，但数量屈指可数，例如弗蒙特州的伯灵顿，科罗拉多州的博尔德以及俄勒冈州的波特兰市。这些城市都领会了规范的意义。这也就是说，对于可负担得起的居住房屋，他们都有采取措施。各种社会政策和公民参与性的确让这些地方成为了理想的居住地。从经济意义上来说，它们做得非常好，但有限的供房抬高了住房成本，这三个地方都出现了这种情况。这是一种权衡，是一个很难解决的问题。

or purpose to do so, whether it's for protection, improvement, maintaining a cultural feature, whatever it may be that people coalesce around. We see this in small isolated rural settlements all the way to inner city pockets: people coming together for a purpose. Again, this is a lot of what has driven the urbanists' push: if there is a purpose and a place, people will come together. If the physical environment supports that sense of purpose, it fosters the exchange. Without the physical connection it's much harder to do. For example, when the Interstate Highway System came through and divided many of our urban neighborhoods, there was never again the strong sense of connection or interplay. Major construction projects of any sort similarly sever the physical bond. It's hard to have a social connection without a physical dimension, so it is vital in that sense.

Living and working in the same place is an ideal in many new communities. What are the consequences of that goal?

If you own a home or live in an apartment in a community, own a business or work for an organization there, the sense of civic engagement goes up. There are also those who are engaged in the community who do not live there, maybe business owners, or people with historical connections; there are all kinds of reasons. But the more vested a person is, the higher the level of engagement. That makes total sense. And clearly, having the job and bedroom in the same community negates the need to commute.

Does a live/work environment lessen diversity, stemming a certain dynamic that vitalizes a community with infusions of people, cultures, and ideas from elsewhere?

That depends on a couple of factors. One is the strength of the local economy, and whether it's attracting others to come in and invest, and take jobs, open businesses, and so on. Another is the ability to be an affordable community, at least to be attractive to a wide range of ages, incomes, and backgrounds.

A small number of cities in the U.S. really try hard to do these things—you could easily count them on one or two hands—for example Burlington, Vermont, and Boulder, Colorado; also Portland, Oregon. All of these cities have a strong grasp of planning. That said, they all do have issues with affordable housing. The various social policies and civic engagement make them clearly desirable places to live. They do very well in an economic sense, but then limited supply drives up the housing costs and that's what has happened in all three places. It's a trade-off. It's a hard equation to solve.

In very large and tall mixed-use buildings housing a wide variety of incomes and residential needs, how is social integration achieved and segregation prevented?

Again, that would have to be a carefully thought through. Do you have more affordable units interspersed with the higher-end units? It's a careful balancing act. Some cities, for example, require a certain number of affordable units in every project of a certain size. Sometimes the units are in one section or just on one floor, and everyone knows it so there is instant delineation between the less fortunate and those who have more. You wouldn't want to have certain floors only for the wealthy, but at the same time you have to be able to balance that with market appeal and ask who will buy this unit that costs three times as much as the one next door? In a project of the magnitude we're discussing, I would say it would probably be large enough that you could overcome segregation with some diversity of units on various floors.

It is an interesting idea. It's part of the sustainability movement; it really is. We've gotten too disconnected from nature. That's what we're seeing in this crushing need to have a sense of place and connection, and to have everything from the local food market, to people wanting to know where their food is grown, and on and on and on. Maybe it's more of a return to something that we have lost from our past, of being connected to nature and understanding its cycles better and working with them instead of trying to conquer them. The underlying ethos of sustainability is that we have to partner with the Earth. I don't know if a really large-scale high-density building would truly have less impact on the planet versus being a little more dispersed into ecovillages or a setting where people can grow their own food, or at least part of it, or have composting with every three units or in some way being connected back to nature. I just don't know, I think we're so early into this that no one will really be able to answer definitively at this stage.

Cultural facilities are variously planned in clusters à la Lincoln Center, or dispersed throughout the community. Is there a preferred approach in modern community development?

The cultural aspect as well as the social, environmental, economic, and every other aspect needs to be integrated within the community. I'm more for the dispersement model, meaning easy access on both the neighborhood and community

对于满足各种收入人群和住房需要的大型混合用途高层建筑，如何实现与社会融合，并防止隔离？

这必须仔细思考。你要考虑是否在人们更容易负担的众多住宅单元中穿插着一些高端单元？这是一种审慎的平衡做法。例如，一些城市要求具有一定规模的项目必须提供一定数量的可负担得起的住宅单元。有时，这些住宅单元位于建筑物的某个部分或仅仅一个楼层，并且每个人都知道它在哪里，那么立即就会在普通人与富人之间形成隔阂。你不希望一些楼层专供富人使用，但同时，你必须要能够平衡与市场吸引力的关系，并面对这样的一个问题：谁愿意以相当于隔壁住宅单元三倍的价钱购买这个住宅单元？对于我们正在讨论的这种规模的项目，我会说，它应该足够大，以便你能够克服楼层住宅单元的多样性所带来的隔离问题。

这是一个有趣的想法。这是成为可持续发展运动的一部分，真的。我们已经失去了与大自然的亲密接触。在这种断裂中我们看到，我们需要一种场所感和连接感，需要从本地食品市场购买每一样东西，需要知道食物在哪里生长等等问题。这可能不仅仅是回到某个我们已经失去的过去，我们要和自然联结在一起，更好地理解其周期规律，与它亲密合作而不是想要去征服它。可持续发展的基本理念是：我们必须成为地球之友。与相对分散的生态村或居住点相比，我并不知道真正的大规模高密度建筑物是否能够减少对地球的影响，在这些比较分散的生态村或居住点，人们可以在那里种植自己所需的粮食，或至少一部分粮食，让每三个住宅单元的住户堆制堆肥，或以某种方式与大自然亲密接触。我并不确定，我想现在探讨这个问题还为时过早，在这个阶段，没有人能够明确回答这个问题。

文化设施可以以集群的方式进行规划，例如林肯中心，也可以分散于整个社区。那么在现代社区开发项目中，是否有首选的方法？

社区中需要整合文化以及社会、环境、经济和所有其他方面的内容。我更支持分散模式，这意味着邻里和社区可以更方便地接触这些设施。或许有一个大型演艺中心，这样游客可以来到这里观看节目，居民可以带他们的客人来到这，但除此之外，并没有别的用途。有证据表明，以社区为基础的设施将可以得到更好的利用，并更深地扎根于社区。人们会觉得自己成了其中的一部分，成了构成场所感的某种东西的一部分。我们知道，有活力的场所感有助于形成更强大的社区，这反过来又意味着更佳的可持续性。

社区越强大，就越具有弹性，并可以更好地应对危机。在这次经济衰退中，许多表现不佳的社区没有很强的联系和社区关系，而另外一些社区由于有着很强的联系，过渡就要好的多。对于这些问题，现在人们已经进行了大量的研究。这些社区为什么能够做得更好呢？这可以归结为社区关系的力量。对我来说，这是一种可持续性：具有弹性，能够作为一个社区应对出现的各种情况。如果你不参与并且没有场所感，那么你就无法建立这些关系。这两者密不可分，缺一不可。可持续性是一个过程，而不是一个最终结果。我们永远不可能做到完全的可持续性，但我们可以参与这个过程，这个过程充满反复和联系。它完全不是一个直线进程。

可持续性社区的标准是什么？

可持续性具有三个方面：经济、公平（包括社会、文化和政治方面）和环境（它是我们现实、自然的联系，我们在自然世界中的系统和过程），必须将这三者联系起来。如果我们充满活力的经济与环境循环系统发生脱轨，那么它将不会长久。这样的例子有很多，例如未能被保护的采矿城镇或自然资源城镇。还包括一些新兴城市，其中之一就是凤凰城，也就是我目前所在的地方。我们周围也有很多例子，比如拉斯维加斯，这里的社区与周边环境在很大程度上出现分离，并且忽视自然环境，因此它并不具有长久的可持续性。同时，还存在着公平方面的脱节，这意味着社会关系、地方感和政治体系等存在着或多或少的缺陷，因此无法应对危机。如果一个社区是可持续的，它可以在上述三个方面有效的应对危机。

如果你询问问环保人士，你会得到一个略微不同的观点，因为他们着眼于能够支持人们互动的自然系统。以我的观点来看，这涉及到我们如何长期的在任何及所有维度上系统的工作。

经济学家会告诉你一些完全不同的东西，他们将会谈论持续增长、持续创收和持续繁荣，其实，这并不是可持续的。没有一个有机体可以持续增长和生存。这是一个被证明的事实，一个科学范式，不是社区开发商可以随便臆测炒作的概念。对此，有这样的一个习语：只有傻瓜和经济学家相信持续增长。在自然科学领域，持续增长是不可能的。这同样适用于我们的社区、经济和政治系统。

levels. There may be one large performing arts center where tourists come and residents bring their guests but really isn't used otherwise. Evidence has shown that community-based facilities will be better used and more ingrained in the community. People will feel part of them, and being part of something contributes to a sense of place. And again, we know that a dynamic sense of place leads to stronger communities, which in turn means greater sustainability.

The stronger a community is, the more resilient it is, and the better it can respond to crisis. A lot of the communities that have fared poorly in this economic recession did not have strong connections and community relations; those that did, have fared far better. There's quite a bit of research into such issues right now. Why have they done so much better? It boils down to the strength of community relationships. To me, that's sustainability: being resilient, being able to respond to whatever comes your way as a community. You can't build those relationships if you don't have engagement and a sense of place. You can't have one without the other. To me sustainability is a process, not an end product. We will never be completely sustainable but we can embrace the process, which is very reiterative and connected. It's not a linear thing at all.

What are the criteria for a sustainable community?

Sustainability has three dimensions: Economic, Equity (including social, cultural, and political dimensions), and Environmental (which is our physical and natural connections, our systems and processes in the natural world). The three E's, and they have to be connected. For example if we have a dynamic economy that is disconnected from the environmental cycle, it will not last. There are so many examples of that, like mining or natural resource towns that couldn't persevere. And boom towns, one of which—Phoenix—I'm sitting in right now. You don't have to look far to find others, Las Vegas for instance, where the community is so disconnected from its surroundings and so disregards the natural environment that there is no way it could be sustainable for the long term. At the same time, there's been disconnection from the equity realm, meaning that the social connections, the sense of place, the political systems, etc., are defaulting in one way or another. They are not able to respond to a crisis. If a community is sustainable, it can weather crisis in any of the three dimensions.

If you ask environmentalists, you'll get a slightly different perspective, as they look at natural systems capable of supporting the level of human interaction. From my perspective, it's how we work as a system in any and all of these dimensions for the long term.

Economists are going to tell you something very different. They'll talk about continued growth and generating continued income and continued prosperity. That's not sustainable. No organism can have continued growth and survive. This is proven fact, a scientific paradigm, not speculation by community developers. There's an old saying to the effect that only fools and economists believe in continued growth. In the world of hard science, continuous growth is impossible. The same goes for our communities, and our economic and political systems. Yet, particularly in America, we've always fostered a mentality of continued growth. You still hear it when people here say, "when things get back to normal, when we start growing again." There's no realization that may never be, that you have to get to a point of being sustainable and that may not mean continued growth. It may instead mean development, or redevelopment, or staying in static point, but certainly rethinking what we are doing.

Capped growth has a direct parallel in population control but goes against Wall Street's premise that you've got to keep getting more, and bigger.

It's totally false, this premise of unlimited growth.

Can a community be developed by outsiders without the input of would-be residents?

That is what's called the top-down approach: when outside experts parachute in, design, and then leave. It rarely works well. The reason is that people are not necessarily vested or committed to the project. What's wanted is civic engagement so that people can say, "This is what we need as a community, or this is what we think we need." At the same time, you need the architectural experts, the designers, the engineers. You can't do it without their technical input, but there needs to be a very strong community voice in this process.

We need only look as far as mid-century urban renewal to see what happens without it. In all honesty, urban renewal did offer some benefits, but by and large, it was awful. It was a top-down approach where people would come in and say, "We're going to build a sea of high-rises right here in the middle of this African American neighborhood." It literally displaced entire communities.

That's what we see in most U.S. cities, particularly in the older high-rise areas. Large urban precincts were cleared out and communities destroyed. There was pushback from people like Jane Jacobs in the 1960s who literally saved Greenwich Village, now one of the most vital neighborhoods in New York. The key was people having a voice. I'm strongly of the mind

然而，特别是在美国，我们一直怀有一种持续增长的心态。你仍然可以听到这样的言语："当一切恢复正常，当我们开始再次增长……"。人们并没有认识到这可能不会发生，没有意识到你必须达到可持续性，而这并不意味着持续增长。相反，这可能意味着发展或恢复发展，或保持当前的发展，并反思我们所做的事情。

高速增长与人口控制直接平行，但却有悖于华尔街的看法：你需要持续获得更多，并变得更加强大。

这种无限增长的假设是完全错误的。

社区可以由外来人开发，而不采纳潜在居民的意见吗？

这就是所谓的自上而下的方法：外部专家进入、设计，然后离开。但它很少奏效，原因是人们不一定会支持或致力于该项目。真正需要的是公民参与，这样人们可以说，"作为社区，这就是我们所需要的，或者我们认为这就是我们所需要的。"与此同时，你需要建筑专家、设计师和工程师。没有他们的技术贡献，就无法完成建设项目，但在这个过程中，需要社区发出强有力的声音。

我们只需回顾一下美国上世纪中期的城市更新运动，就可以知道如果没有社区发出的声音会出现什么结果。说实话，城市更新的确带来了一些好处，但总的来说，它是可怕的。当时的城市更新采用了自上而下的方法，有人来到社区，告诉那里的人，"我们要在这个非洲裔社区，建造大量的高层建筑。"这可以说是整个翻新了社区。

这就是我们在大多数美国城市所看到的情况，特别是旧的高层建筑区。大型的城市街区被清除，社区遭到破坏。在20世纪60年代，有很多人反对这样的做法，例如简·雅各布斯(Jane Jacobs)，正是由于这些人的努力，格林威治村才得以幸存，而且这个村现在成为现纽约最重要的街区之一。关键是人们具有发言权。我认为，具有重大意义的任何项目都需要征集潜在居民和机构的意见。征集过程需要较长的时间，这是个不争的事实，因为可能会出现突发事件或者意见争执。开发商不想这样做，因为他们知道这会耽搁不少时间。但它是至关重要的，从长远来看，可以带来更好的结果，而且更容易被接受。这已经经过了反复证明。

Sunset view of Tropicana Avenue from Blue Diamond Hill, Las Vegas, Nevada. (2006)
美国内华达州拉斯维加斯从蓝钻石山上看热带大道的落日景色。（2006年）

你并不是说新规划的城市包括生态城市会失败，而是这需要一定的时间才能形成真正的社区意识。

同样，这取决于具体情况。一些城市取得了成功，而另一些则没有。当你规划一个新的城市时，会出现困难，因为那里没有常住人口，你需要将人们从其他地方吸引过来。但是，将潜在人口作为目标并努力让他们流入城市，这仍然是非常重要的。生态村的概念是更加典型的小规模社区，甚至在城市重建计划中也出现了生态村概念，并不一定在开发新项目过程中才有。新的城市无章可循，需要自己去摸索，因为可能并没有常住人口去帮助其规划，并找出可行和不可行之处。对于你正在讨论的项目，它是现有城市的一个重建项目，还是一个新区呢？

that any project with any major implication has to have input from potential residents and organizations. The process takes longer, that's a given, because there will be contingencies or maybe dissension. Developers may not want to do it because they know it's going to slow things down. But it's vital and in the long run, it makes for a better product and better receptivity. That's been proven over and over again.

You're not saying that newly planned cities, including ecocities, will fail, but that it will take time for any true sense of community to develop.

Again, that depends. Some have been successful, and others have not. It's a little harder when you're planning a new city and there is no resident population and you're going to be bringing people in from elsewhere. But it's still important to target a would-be population and try to get its input. The ecovillage concept is more typically small scale; there are even ecovillage concepts emerging within inner city redevelopments, if not necessarily new development. New cities are their own animal because there may not be a resident population to help plan and figure out what works and what doesn't. The project you're discussing, is it a redevelopment in an existing city or is it a new area?

Since a new community typically doesn't open with all of its institutions and services in place, is the first generation of residents a casualty?

It may not be a lost generation but it may take some serious time to develop—and it's just going to be that way for a while. In America, there were plans to develop a couple of hundred new towns after World War II to accommodate the very high rate of population growth and new family formation. Fewer than a dozen were built before the program was scrapped. But those communities have actually done really well and seemed to have worked really well right from the beginning. One of the reasons is that they were planned prior to sprawling suburbanization patterns. They were very walkable, so you could get to places, see neighbors, and socialize within your little area and its public spaces. That's why they keep going back and looking at these older cities like Bethesda, Maryland, and Reston, Virginia. It may have been a social air of contingency, before television and the Internet. But they still work really well. The physical design made them successful but they weren't very high density.

The basic idea goes back to Ebenezer Howard and the Garden City movement. Reston was designed loosely around the garden city concept. But basically these towns were designed for a maximum capacity of about 50,000 people, that's it. So they are not practical for a large mixed-use building. But if you apply the formula for 50,000 people and replicate it, maybe it can be stacked vertically. That would probably be the key right there, trying to emulate what works so well with a vertical application.

The anticipated population of new mixed-use towers is primarily upwardly mobile young couples with a more diverse cross-sectional mix. What is the impact of the aging population?

The aging population tends to want to cluster, and they want more density so that they can easily get to medical services, the grocery store, social activities, etc., so there is clustering in communities targeted for older populations. Again, walkability or at least accessibility becomes crucial. There's a whole dimension of planning for the aging that is emerging. The sprawling suburban neighborhoods don't work for elderly

由于新建社区在初期阶段，各种机构和服务设施一般情况下无法全部到位，那么第一代居民算是受害者吗？

他们可能不会成为失落的一代，但这的确需要一些时间来发展，事情在一段时间内确实都是这样。美国在二战结束后，有开发数百个新城镇的计划，以适应非常高速的人口增长和新家庭的组建。但仅在开发了数个城镇之后，该计划就被取消。然而这些社区做得非常好，似乎从一开始就非常好。原因之一就是，这些社区的规划早于郊区化模式。它们具有很好的步行便利性，居民可以步行前往一些地方，拜访邻居，在自己的家里和公共场所开展社交活动。这就是人们为什么喜欢回去参观旧城市的原因，例如马里兰州的贝塞斯达和弗吉尼亚州的雷斯顿。在有电视和互联网之前，它构成了一种随意轻松的社会氛围。这些城市今天运转的仍然很好，具体的设计使它们获得了成功，而且它们的密度并不高。

其基本思想可追溯到埃比尼泽·霍华德(Ebenezer Howard)和花园城市运动。雷斯顿城镇正是根据花园城市的理念而设计的。但基本上这些城镇所设计的最大容量约为5万人。因此，对于大型混合用途的建筑物来说，这种城镇是不可行的。以一种垂直应用的形式仿效如此运作良好的东西，这个可能是解决今天问题的关键。

新型混合用途塔楼的预期人口主要是社会和经济地位处于上升趋势的年轻夫妇，他们的背景更加多样化。人口老龄化会有什么影响？

老龄化人口往往希望集群，他们想要更高的密度，这样他们可以轻松地获得医疗服务、前往杂货店并参加社会活动等，因此，在社区中有专门针对老年人口的集群区。同时，步行便利性（至少是无障碍）变得至关重要。对于老年人，有一种全新的规划特征。快速增长的郊区居民区并不适合老年人，当前出现了让老年人搬回市中心或传统社区中心的趋势，这样他们可以享受更便利的交通，并方便地获得所需要的服务。如果在郊区，这些就不可能。现在，随着越来越多的人搬回城市，较年轻的一代开始搬到郊区，因为那里的花销变得更低。在某种程度上，这是一个反向趋势，我认为非常高密度的项目既可以服务于老年人，也可以服务于年轻人。

Elderly people traffic sign, Melford, UK. (2004)
英国梅尔福德老年人交通标志。（2004年）

针对年轻人的规划与针对老年人的规划有什么差异呢？

针对年轻人的规划完全是另一回事。在某些方面，它与针对老年人的规划非常不同，但在另一些方面，却很相似。年轻人想要社会互动、开放的空地，并希望能够步行、骑车或乘坐公共交通工具上班，而不是等着让人开车载他们去上班。所以，年轻人和老年人会遇到了一些相同的问题。当然，年轻人所希望的一些活动可能会影响到老年人，例如噪声、音乐和滑板等。所以有必要将一些使用空间分开，但其中秉承的理念是非常相似的:无障碍、步行便利性和社会互动都在其中。两者皆是如此。

people so there's a trend for the aged to move back to the city center, or to traditional community centers where they have transportation and easy access to the services they need. They cannot get that in the suburbs. The younger generation is now moving out into the suburbs because it's becoming more affordable as more and more people move back into the city. It's a reverse trend in a way. I would think that a very high-density project could serve an aging population, as well as a young population.

What's the difference between planning for the young and old?

Planning for youth is a whole other dimension. In some ways it's very different from planning for the aged, and in some ways it's very similar. Youth want social interaction, open space, and want to be able to walk, bike, or take public transit and not have to wait for someone to drive them. So some of the same issues come up for both youth and the elderly. Of course, some of the activities that youths want can come into conflict with an aging population, like noise, music, skate-boarding, etc. So there has to be separation in some of the uses of space, but the concepts are very similar: accessibility, walkability, social interaction, it's all there. At both ends.

Jorge Eduardo Rubies
PRESERVA SÃO PAULO ASSOCIATION

Jorge Eduardo Rubies, Esq., is the founder and president of Preserva São Paulo Association, the only historic preservation nongovernmental organization in São Paulo, Brazil's largest city. Frequently consulted by journalists and researchers, he is a respected authority on historic preservation, Brazilian architecture, and urbanism. His bilingual Portuguese-English website, www.piratininga.org, is a key source of information about São Paulo and its architecture. Mr. Rubies' dual professional degrees in business administration and law have allowed him to organize and litigate on behalf of the public realm for the protection of endangered historic buildings. He also leads the recently founded "São Paulo Citizens Front against Real Estate Speculation and for Ethics in Politics." This organization encompasses dozens of grassroots movements, united to promote in a nonpartisan, orderly, and pacific way the common goal of a sustainable, inclusive, pluralistic, and green city for every citizen.

豪尔赫·爱德华多·卢比斯
圣保罗保护协会

豪尔赫·爱德华多·卢比斯是圣保罗保护协会的创立者和会长，这个组织是巴西最大城市，圣保罗唯一关于历史文物保护的非政府组织。他是历史文物保护、巴西建筑和城市化方面受人尊重的权威，记者和研究人员经常向他咨询相关问题。他的网站www.piratininga.org 有葡萄牙语版和英文版，是关于圣保罗及其建筑方面非常重要的一个信息资源。卢比斯先生有两个专业学位，分别是商业管理和法律，这让他能够作为公共领域的代表在保护濒危历史建筑方面展开组织活动，并进行诉讼工作。他也领导着现在刚刚成立的"反对房地产炒作，争取政治伦理的圣保罗公民阵线"。这个组织发起许多草根运动，以一种无党派、有秩序、和平的方式联合起来以推动一个共同的目标，这个目标是为每个人争取一个可持续、包容、多元和绿色的城市。

For complete biography, please refer to the Appendix.

完整的传记，请参阅附录。

Downtown São Paulo, Brazil. (2004)
巴西圣保罗市中心。（2004年）

QUALITY OF LIFE Jorge Eduardo Rubies

What is Preserva SP and why was it established?

Let me begin by saying that I am absolutely passionate about my city. I love São Paulo. I wanted to do something about my city, for my city, particularly historic preservation. That was my main concern so I formed Preserva SP, or Preserve São Paulo. But I started to realize that this struggle was useless unless I began to understand politics and how things work here. It's impossible to separate one thing from the other. So I am organizing a group of grassroots movements to reclaim the city for the citizens.

Citizens must fight for their city, in a pacific way naturally, in an ordered and democratic way. They must select decent, honest politicians, people who really represent them, not just corporate interests or campaign donors.

Unfortunately, corruption is a big issue in Brazil. The government doesn't work for the citizens but for campaign donors. Real estate speculators are the biggest contributors so they make the laws. We couldn't do anything because the power of the real estate speculators is so huge here in São Paulo that it becomes a fight between David and Goliath. They have control but they are very shortsighted. The speculators just want to make money. When the city is destroyed, they move elsewhere. That's how it works. They don't think about their children or grandchildren, not at all.

How is Brazil planning architecturally for the future?

The U.S. and most of the European countries, and even China, are all trying to plan their urban growth. Brazilians typically don't plan. And when we do plan, if a new mayor or new governor comes in from a different party, they start all over again from zero. Good things are lost.

For example, take the Hearst Building in New York City. When a new tower was called for, Norman Foster built on top of the original 6-story building (1928). In Brazil, they would just demolish the base to build a new building. They wouldn't think of reusing the old part. In the U.S. and in Europe, old factories are transformed into lofts or apartments or repurposed as museums. Here they just demolish them to build horrendous new money-making housing. The new architecture is skin-deep, mostly fake classical facades, and it is usually surrounded by high walls so there's no dialogue with the city; the new apartment buildings are more like fortresses.

How do you hope to reclaim the city?

We don't want to stop development; development is good and necessary. But we want sustainable development that is good for the citizens and for the environment, not the current

什么是圣保罗保护区(Preserva SP)？为什么要建立它？

让我先申明一下，我绝对热爱我的城市，我爱圣保罗。我想要为我的城市做点什么，特别是在历史保护方面。这是我主要关注的东西，因此我建立了圣保罗保护区(Preserva SP)或者你可以称之为Preserve São Paulo。但是我开始认识到，除非我理解政治以及这里的事情是如何运作的，否则这样的努力就是徒劳。你不可能将事物分割开来。所以我正在组织一系列草根运动，以便能够为市民们改造城市。

市民们应该为他们的城市而斗争——当然，是以一种和平的方式，以一种有序和民主的方式进行。他们必须选择正值诚实的政治家，选择那些能真正代表他们的人，而不是代表企业利益或者竞选捐助者利益的人。

不幸的是，腐败在巴西是个大问题。政府不是为市民工作，而是为竞选捐助者工作。房地产投机者是最大的捐助者，所以法律是他们制定的，我们什么都做不了，他们的权力在圣保罗如此之大，所以，这成为了一场大卫与歌利亚之间的战斗。房地产投机者已经控制了局面，但他们的目光太短浅了，只想着赚钱。当一个城市被毁掉时，他们就转移到其他地方去，这就是他们的运作方式。他们不会为子孙着想，一点也不会。

巴西如何在建筑上规划未来？

美国和大多数欧洲国家，甚至中国，都在努力规划他们的城市发展。一般来说，巴西人是不怎么进行规划的，当一个来自不同政党的新市长或新地方长官执掌政权，确实要进行规划的时候，他们总是又一次从零开始，好的东西总是消失殆尽。

例如，以纽约市的赫斯特塔楼为例。当人们要求一座新的塔楼时，诺曼·福斯特(Norman Foster)在原有六层建筑的楼顶上展开建设（1928年）。而在巴西，他们只会拆掉地基，盖座新楼，不会想着重新利用旧有的建筑。在美国和欧洲，老旧的工厂被改造为阁楼、公寓或博物馆。在这里，他们就拆除这些工厂，以便建立可以赚大钱的新住宅楼。新建筑是很肤浅的，大多数都是仿古典式的外观，而且经常用高墙围起来以致与城市之间并无交流，新的公寓建筑更像是城堡一样。

你希望如何改造城市？

我们不想停止发展，发展是好的而且是必须的。但我们想要对市民和环境都有益的可持续性发展，而不是当前这种只产生更多污染、更多交通拥堵、更

Typical new apartment building isolated by high street walls, São Paulo, Brazil. (2012)
巴西圣保罗典型的被高街墙隔开的新公寓大楼。（2012年）

多城市问题的发展。

在工作日，我们都不能进入圣保罗，因为这里有超过700万辆的汽车，却没有任何可行的公共交通系统。我们的地铁系统非常安全、非常现代，充满了艺术作品，你甚至可以获得地铁站的免费导游。但这个系统对于这样规模的城市而言，实在是小得可怜。我们没有足够的地铁站点，也没有足够的地铁线路。政府应该在基础设施和公共交通上投入经费，但实际却没有这么做。圣保罗每年的预算大约是200亿美元，这对于一个新兴国家而言是一大笔钱，但我们看不出它花在了哪里。

kind of development that just creates more pollution, more traffic congestion, more problems for the city.

During the week, we can't move in São Paulo because there are over 7 million cars and we don't have any viable public transportation. Our subway system is very safe, very modern, full of art work; you can even take free guided tours of the stations. But the system is ridiculously small for a city of this size. We just don't have enough stations, not enough lines. The government should be spending money on infrastructure and public transportation but it is not doing that. The annual budget of São Paulo is about $20 billion. That is a lot of money for an emerging country, but we don't see any of it.

456

Let me make a parallel with China, which is another emerging country. In China, the government is doing some things that are not very good, in my opinion. For instance, the architectural heritage of cities like Beijing is being destroyed. But at least the Chinese government is spending money on the infrastructure, public transportation, subway networks, railways, new roads, etc. Here in Brazil the government spends money on football stadiums, not projects like that.

Right now, the government is going to build a $2 billion tunnel in São Paulo just for cars. That's completely absurd, especially because there are two avenues parallel to the tunnel that are not congested all the time. Why build this tunnel? They should put money into public transportation, not to encourage car use. Seven million cars make it impossible to move in the city. They are not doing this kind of thing anywhere else in the civilized world, not in the U.S., Europe, not in China! But the Brazilian government continues to make projects for cars rather than people.

Worse yet, the tunnel will destroy an entire neighborhood. They have razed hundreds of homes in very traditional neighborhoods in the southern area of São Paulo. That's the drama that is going on here.

What has happened to all those people?

They are taking part in this movement I am organizing. We are struggling, we are fighting, but in Brazil, money is more important than people. It's not just a problem of architecture or urbanism. It's a problem of politics, the lack of democracy, lack of transparency, lack of dialogue between the citizens and government.

There is a writer I like a lot: Jane Jacobs, the great urbanist. What they are doing in São Paulo is contrary to everything in her book. What the government is doing here is what Robert Moses did 50 or 60 years ago, but worse. At least Moses did some very useful things for New York.

How has São Paulo changed in past decades?

Until the 1950s São Paulo was gorgeous, very European, extremely French, almost Parisian, with wide, tree-lined boulevards. And that's another thing. They cut down all of the trees, which were full grown and very big, to make more lanes for the expressway. Two years ago!

São Paulo has very big problems of urban decay. Downtown is still extraordinary, in my opinion. The downtown infrastructure is not that bad, but the buildings are old and are not being maintained. Crime and vandalism are big problems.

The government has urban renewal programs that intend

让我将其与另一个新兴国家——中国做一个对比。在中国政府正在做一些不太明智的事情。例如，在像北京这样的城市，传统的建筑遗产已经被毁掉了。但至少中国政府在基础设施、公共交通系统、地铁网络、铁路以及新的道路等方面投入了资金。在巴西这里，政府不是把钱投在这些项目上，而是花在足球场上。

现今，政府打算在圣保罗建造一座耗资20亿美元供汽车使用的隧道。这完全是浪费，特别是因为有两条与此隧道平行的道路并非总是处于拥堵状态。为什么修建这条隧道？他们应该把钱用于公共交通系统，而不是去鼓励汽车的使用。700万辆车让人们在城市里寸步难行。在文明世界的其他地方，人们并不如此行事，美国、欧洲没有，中国也没有！但是巴西政府继续为汽车，而不是为人民进行项目建设。

然而，更糟糕的是，隧道将摧毁整个街区。他们已经将位于圣保罗南部的传统街区中数以百计的家庭夷为平地，这就是在那里发生的剧情。

那些人身上都发生了什么？

他们正参与到我组织的这场运动中。我们在奋斗，我们在斗争，但是在巴西，钱比人要重要得多。这并不仅仅是一个建筑或者城市化的问题，这是一个政治问题，缺乏民主，缺乏透明度，缺乏市民与政府之间的交流对话。

有一个我很喜欢的作家，简·雅各布斯(Jane Jacob)，伟大的城市规划专家。巴西在圣保罗正做的，与她书中的一切都截然不同。政府在这里所作的，正是罗伯特·摩斯(Robert Moses)在50或60年前做的事，但前者还要更糟糕一些。至少摩斯为纽约做了一些非常有益的事情。

圣保罗在过去十年间有何变化？

直到上世纪50年代，圣保罗还是极美的，非常欧洲，尤其是非常具有法国风格，大部分地区都是巴黎式的，有宽阔的林荫大道。但两年前却发生了另一幕：他们砍掉所有的树，这些树茂盛而且高大——只为了给高速公路腾出更多的地盘。这一切就发生在两年前！

圣保罗存在着城市衰败的大问题。在我看来，市中心依然显眼突出，基础设施也算过硬，但建筑老旧却没有得到维护。犯罪是一个问题，而故意破坏公物则是一个大问题。政府有一个城市重建工程，试图拆毁市中心所有的旧街区，从而为大型商业建筑腾出空间。这个项目令人沮丧，因为所有我们试图抢救的旧建筑都被摧毁了。

这种对于一个城市至关重要部分的摧毁，在文明世界的其他地方从未发生。他们不这么做因为那是不民主的。生活在这些社区的人们，并不想被赶走，他们想要待在那里。但政府只想着要铲平一切，他们通过征用权法律将人们从街区中赶走——在有些地方，此类街区是生机勃勃的，而非破败不堪。

当地居民对当前的发展有多关心呢？

这要看不同情况。那些直接受到影响的人，比如，直接受到隧道、城市改造项目影响的人是很愤怒的。但其他人，即使是受过良好教育的人，都不太在乎，因为这些对他们没有直接影响，他们不会考虑团体的利益，人们一般不考虑城市。

巴西其他地方的人如何看待这些问题？

在里约热内卢，由于奥运会和世界杯，情况差不多一样——甚至还要更糟一些。在巴西，腐败的程度是难以想象的，用于基础设施建设的资金只会被用来修缮足球场，而不是用于大众交通系统，不是用于公共卫生、医疗保健或者教育。可悲的是，大学、建筑师和城市规划专家，大多数人都不讨论这个问题。他们应该对这些问题有所意识，但不幸的是，很多人向商业利益屈服，因为他们自己就为许多新建筑工作。

市政建设部门，城市区划委员会或者保护机构没有进行控制吗？

我们拥有所有这些官方机构，它们都发挥作用，但没有执行能力。例如，我们的地标委员会是在纽约市地标保存委员会的启发下设立的，但它却没有权威。政治家被允许攻击和限制这个机构的权力，所以现在的情况很困难。

圣保罗的情况非常具有戏剧性，特别是对那些直接受到城市改造项目影响的人而言。已经住在同一个街区数十年的人，现在失去了他们的家园、邻居和朋友。而拆迁补偿金却并不公平合理，那真是一幕真实而又悲惨的场景。

由于有限的信息，错误的信息或者简单的程序差异，人们有时会批评政府的行为或夸大情况的严重性——即使实际上真实情况并非如此。

很不幸，我没有夸大。这不是一个政治家和一个政党的问题。这是全部政党中所有政治家的问题。他们的所作所为是一样的，可能在圣保罗情况更为

---

to destroy old neighborhoods in the downtown area to make space for big commercial buildings. And it's been very frustrating because all the old buildings we tried to save were being demolished.

This kind of thing, where they destroy vital parts of a city, doesn't happen in other parts of the civilized world. They don't do that anymore because it's undemocratic. The people who live in these neighborhoods, they don't want to be expelled; they want to stay. But the government just wants to bulldoze everything; they use eminent domain laws to expel the people from the neighborhoods, which in some cases are very dynamic, not run-down.

How concerned are local residents about recent developments?

Well, it depends. People who are directly affected, for instance by the tunnel, by the urban renewal projects, they are furious. But other people, even well-educated people, don't care because it doesn't affect them directly. They don't think about the community. People don't think about the city in general.

How does the rest of Brazil perceive these problems?

Well, in Rio de Janeiro it's the same situation, even worse because of the Olympics and the World Cup. The scale of corruption here in Brazil is unthinkable and the money for our infrastructure is going only to reform the football stadium, not for mass transportation, not for sanitation, not for health care or education. Sadly, the universities, the architects, the urbanists, most of them are not discussing this. They must be aware, but unfortunately most of them surrender to commercial interests, since they are working on many of the new buildings.

Doesn't the municipal building department, zoning board, or preservation agency exert control?

We have all of these official bureaus but they don't work. There's no muscle. For instance, our landmarks commission was inspired by the Landmarks Preservation Commission in New York City, but it doesn't have authority. Politicians are allowed to attack and limit the agency's power, so it's a very difficult situation.

The situation here in São Paulo is very dramatic, particularly for the people who are directly affected by the urban renewal projects. People who have lived in the same neighborhoods for decades now lose their homes, their neighbors, their friends. And the compensation is not fair. That's very,

very sad and a real drama.

Because of limited information, misinformation, or simply a different agenda, people sometimes criticize government actions or exaggerate the circumstances, when in fact the reality is quite different.

Unfortunately, I am not exaggerating. It is not a problem of one politician and one party. It's all politicians in all parties. They do the same things. Maybe in São Paulo it is harsher, more extreme, because really bad things are going on here.

What's happening in São Paulo and in Brazil is the same thing that's happening in Europe, in Spain, in Greece, in Ireland. People are finding out that their democracies are not as democratic as they thought, and they are beginning to rebel. It's a clear case of a lack of transparency and democracy where all the laws are made to benefit big business, particularly real estate speculation. They make buildings in places where they shouldn't be, far away from the center of town, where there is no infrastructure. It's irresponsible to allow the current building boom in such areas. Let's build high-rises in the downtown center, not 10 or 20 miles away. There's no logic in that.

Does the government have a public-housing program?

Yes, but these programs are not that serious. There is a lot of corruption and the results are not very satisfactory.

How has the explosive growth of favelas been handled?

In fact, it has not been handled at all. This issue is one of the biggest challenges for city authorities. In 1973, 1 percent of the population of the city of São Paulo lived in favelas; today it is around 20 percent—that is, two million people. São Paulo's population grew 0.55 percent per year between 2000 and 2007, while in the favelas it grew 4.18 percent per year, roughly eight times as fast. Public spending in social housing just can't keep up with the demand.

Favelas are what we call slums or shanty towns. They are commonplace in big cities in Brazil, and originated from the migration of people from poorer, rural areas. Favelas were originally very precarious, with shacks made of improvised materials, but nowadays most of the houses are made of bricks, and they now have a lot of infrastructure such as electricity, telephone, Internet, and even cable TV. In São Paulo, favelas began to appear in the 1960s, and in the '70s and '80s they experienced explosive growth. Today the growth rate has slowed but it is still much higher than the city's growth in general and, of course, by now the favelas are already very big.

Originally, urban planners and city authorities thought that

严酷、更为极端,因为真正糟糕的东西正在这里发生。正在圣保罗和巴西发生的,也是在欧洲、西班牙、希腊和爱尔兰发生的。

人们发现他们的民主制度并不像想象的那么民主,而且他们正开始进行反抗。当所有法律的制定都是为了让大商业集团,尤其是让房地产投机者获益时,这就很明显是缺乏透明度和民主的表现。他们在本来不该盖楼的地方盖楼,那里没有什么基础设施,而且远离城市中心。当下,能允许建筑在这些地方蓬勃发展是很不负责任的,我们应该在城市中心区建设高层建筑,而不是在10或20英里以外的地方,在那些地方建设是没有道理的。

政府有没有公共住房项目?

有,但这些项目并不那么严肃。其中有许多腐败现象,而且结果也不太令人满意。

贫民窟的爆炸性增长如何控制?

实际上,它并未完全得到控制。这个问题是对于城市当局者的最大挑战。1973年,圣保罗1%的城市人口居住在贫民窟,今天大约有20%,即200万人住在那里。从2000年到2007年,圣保罗的人口增长率是每年0.55%,而在贫民窟,增长率达到4.18%,大约是前者的8倍。在社会住房上的公共花费并不能满足人口增长的需求。

贫民窟被我们称为贫民区或棚户区,在巴西的大城市中很常见,它们由来自贫困乡村地区的移民组成。贫民区最初是非常危险的,许多建筑材料搭建而成的棚户构成了贫民窟的主体,不过在今天,大多数的房子都是由砖砌成的,也有了许多基础设施,像供电、电话、因特网,甚至还有有线电视。在圣保罗,贫民窟开始出现于上世纪60年代,而在70年代和80年代它们经历了爆炸式的增长。今天它的增长率已经慢了下来,但依旧比城市总体的增长率要高许多,当然,到目前为止,贫民窟已经很大了。

一开始的时候,城市规划者和城市官方当局认为,处理贫民窟最好的方式是通过社会住房项目,简单地将其中的居民搬迁到公寓大楼中——这些大楼一般在很远的地方。但是今天,人们已经在贫民窟中生活了两到三代了。他们已经扎根在了这些地方,他们会以令人难以置信的创造性弥补贫民窟资源的匮乏,而贫民窟的艺术和文化则是丰富

Two million people reside in favelas in São Paulo. (2012)
200万人居住在圣保罗的贫民窟。（2012年）

多彩、生机勃勃和非常令人激动的，就像帕来索波里斯，在那里我们可以发现大量的艺术表现作品。

今天，我们的目标不是要去消灭贫民窟，而是要将它们融入到城市中，给它们带去公共服务，让居民们感觉自己像是真正的市民和重要的社会角色，并对他们自己社区的命运发出积极的声音。

圣保罗最主要的能源是什么？

来自大坝的水电能源。它应该是便宜的，但电能却很贵，所有东西都很贵。巴西现在在石化燃料上是可以自足的，但汽油的价格却比大量进口石油的美国还贵很多。高税收，以及大企业控制政治，他们依照自己的利益来制定法律，所以你没有别的选择，只能掏钱。

如果有的话，圣保罗在可持续性方面关心的是什么问题？

真的没有关于这些问题的商讨和争论。有一些建筑师在第一线，但一般来说，很少有城市规划专家会为城市而斗争，很少。不幸的是，商业建筑胜过了对于环境的关注。如果说有一些争论，它们也是由市民发起的，而不是那些专家。并没有在市政层次上对于可持续性、废弃物管理、污染的关心——完全没有。

the best way to deal with the favelas was simply to move the inhabitants to apartment blocks through social housing programs, usually in distant places. But nowadays, people have been living in favelas for two or three generations. They have roots in these places. They compensate for the lack of resources with an incredible amount of creativity, and art and culture in the favelas is tremendously intense, diversified, and exuberant, such as in Paraisópolis, where we can find a myriad of art expressions.

Today, the goal is not to eradicate the favelas but to integrate them within the city, to take public services to them, and to make their inhabitants feel like real citizens and important social players with an active voice in the destiny of their communities.

What is São Paulo's main source of energy?

Hydropower, from the dam. It should be cheap, but power is very expensive; everything is expensive. Brazil is now self-sufficient in fossil fuels, but the price of gas is much higher than in the U.S., which imports a lot of oil. Taxes are high, and big business controls politics. They make the laws according to their interests. So you have no choice but to pay.

What concerns, if any, does São Paulo have for sustainability?

There is really no discussion, no debate about these issues. There are some architects on the front line, but in general there are very few urbanists who are fighting for the city, very few. Unfortunately, commercial architecture prevails over environmental concerns. To the degree that there is any debate, it is by the citizens, not by the professionals. There is no concern on the municipal level about sustainability, waste management, pollution—not at all.

What about recycling?

There is no official recycling program here in São Paulo, unlike in Curitiba. It is unbelievable, but true. Aluminum cans are recycled not by the government but by poor people from the favelas. They collect the cans in the streets and sell them for the raw materials. Twenty years ago there was a public program for recycling and separation of garbage, but the government discontinued it.

São Paulo appears to be a thriving, if overly dense, modern metropolis, but the kind of thinking you've been discussing is very backward. It doesn't seem possible.

Unfortunately, it is true. I would like for it to be different but that is not the case. It's a very sad situation. But things are beginning to change. Younger people, for instance, have more concerns about environmental issues due to the Internet, Facebook, social networks, and so on. Just like in Europe and in the Arab world, people are beginning to rise up. In São Paulo, too, they are beginning to become more conscious about all these problems. It is a little more difficult here because by nature, Brazilians are really passive. They don't have this tradition to fight, to protest, to demand things from the government. It's just not in the Brazilian psyche.

In your grassroots movements you have cultural leaders, educators, and laymen all coming together. Collectively that's a very strong front. Are architects also involved?

Yes, we have architects and urbanists together with us but it's a new movement. We started less than one year ago. So we have a long way to go.

循环回收方面呢？

与库里提巴不一样，在圣保罗没有官方的循环利用项目。这是难以想象的，但却是真实的，铝罐不是由政府来回收，而是由贫民窟的穷人进行回收，他们将街上的罐子收集起来，并将其作为原材料卖掉。20年前有一个循环回收和垃圾分类的公共项目，但政府却没有持续展开。

如果说人口非常密集的话，也意味着，圣保罗表现为一个欣欣向荣的现代大都会；然而你所讨论的那些观点与此相反。这看起来不太可能。

不幸的是，这千真万确，我希望它不一样，但事实就是如此。情况非常糟糕，不过事情正在发生改变。例如，通过因特网、Facebook（脸谱网）、社交网络等新鲜事物，年轻人开始越来越关心环境问题。就像欧洲和阿拉伯世界一样，人们开始反抗，在圣保罗也是这样，他们开始越来越关心所有这些问题，但这有点困难，因为巴西人本性是非常消极的。他们没有这种去战斗、去抗议、去向政府要求什么的传统。这些东西并不在巴西人的灵魂之中。

在你的草根运动中，你让文化领袖、教育家和非专业人士集合到了一起。综合而言，它表面上看起来非常强盛。有没有一些建筑师参与其中呢？

有的，我们有建筑师和城市规划专家加入，但这还是一个新的运动。从开始到现在，尚未满一年的时间，所以我们还有很长的路要走。

Among the celebrated folk artists in São Paulo's Paraisópolis favela are the self-taught mechanic Antônio Ednaldo da Silva, known as "Berbela" (top), and the "Brazilian Gaudi" gardener Estavão Silva da Conceição.

在圣保罗的帕来索波里斯贫民区的一些著名民间艺术家中，有自学成才被称作"伯拜拉"的机械工安东尼奥·埃德纳尔多·达席尔瓦（上图），以及"巴西的高迪"、园丁伊斯塔瓦·席尔瓦·康赛桑。

Alfredo Trinidade
SECRETARIAT OF THE ENVIRONMENT
CURITIBA, BRAZIL

Alfredo Vicente de Castro Trinidade is an award-winning environmentalist and forester in Curitiba, Brazil, one of the world's greenest cities. A specialist in urban technical management, he has worked for many years in the service of the Secretariat of the Environment, and was appointed director of the Department of Research and Conservation of Fauna in 2011. Among the other agency positions he has held are environmental planning manager, technical coordinator for biodiversity, director of the Department of Research and Monitoring, technical adviser, and chief licensing officer. In local and international venues, notably as part of the Brazilian delegation of UN Habitat and as a member of the "Smog Summit" in Toronto, Mr. Trinidade has worked to balance Curitiba's modern development needs with environmental protection and improvement. He holds an undergraduate degree from the Agrarian Science Sector of Federal University do Paraná and an advanced professional degree in urban technical management from Catholic University of Paraná.

阿尔弗雷德·特林达德
巴西库里提巴环境秘书处

阿尔弗雷德·维森特·德·卡斯特罗·特林达德是巴西库里提巴（世界上最绿色环保的城市之一）一位备受赞誉的环境学家和林务官。作为城市技术管理方面的专家，他已经在环境秘书处工作服务很多年了，并在2011年被任命为动物研究和保护部门的主管。他还在其他一些机构中任职，比如担任环境规划主管、生物多样性方面的技术协调专员、研究和检测部门主管、技术顾问、首席许可证管理官员。在当地以及国际场合，库里提巴是联合国人居署的巴西代表以及多伦多"烟雾峰会"的成员，所以特林达德先生要平衡这个城市的发展需求与环境保护、环境改善的关系。他在巴拉那联邦大学的农业科学院获得学士学位，并在巴拉那天主教大学获得城市技术管理方面更高级的一个学位。

For complete biography, please refer to the Appendix.

完整的传记，请参阅附录。

Parque Barigui, Curitiba, Brazil. (2009)
巴西库里提巴的巴里吉公园。(2009年)

Curitiba, the capital of the state of Paraná in Brazil, has been recognized as the most environmentally advanced city in the world. What role do you play in city government?

My Department of Research and Monitoring on Fauna, from the Municipal Secretariat of Environment, is the sector of environment that oversees the Zoological Garden, Museum of Natural History, Environmental Education House at the Zoo, the Network for the Defense and Protection of Domestic Animals, and the animals at Passeio Público, the first park created in the city, in 1886. Until 2010 we were just the department of Zoo, but we had some problems controlling domestic and exotic species so it was proposed that city hall expand the department. I was appointed director of the Department of Research and Monitoring on Fauna. We take care of protecting animals in the city, so we work with different agencies but we focus a lot on the process of environmental education.

We have a problem with dogs and cats, which is a worldwide problem. We used to capture and sacrifice strays, but people complained, and in 2005 the mayor halted the practice. After that we had a problem with population control—we have around 450,000 dogs in Curitiba—so we had to devise programs to help people understand that they have to clean up after their pets and that they cannot just release them and let them reproduce without control.

We have another problem with native species, some of which are imbalanced right now, like the capybara, a kind of rodent, like a large dog, a Rottweiler. It lives in the river margins and no longer has natural enemies because they have disappeared with urbanization. Capybara have reproduced in large numbers, and now they cross the streets and move around freely, and cause some incompatibilities with the population. So we have to constantly monitor them, and if they pose a threat to the city and its inhabitants, we decide if we're going to intervene. But capybara are native, and the federal law states that native species must be preserved no matter what. Okay, we agree with that, but if the capybara eat someone in the middle of the night, who's going to be responsible for that?

Are capybara disease carriers?

They could be carriers, but they are not. We continue to collect blood and tissue samples; we analyze them constantly. Right now that's not a problem, but in other cities there was a kind of fever that may have been transmitted by ticks that live on the capybara. We don't have that in Curitiba, but we constantly monitor the situation.

巴西巴拿马州首府的库里提巴，被认为是全世界环境最先进的城市。你在该市政府中扮演什么角色？

我的部门来自市政府的环境秘书处，负责研究和监测动物群落，它是一个环境部门，主要负责监督动物园、自然历史博物馆、动物园环境教育机构、家畜保护网络、市政公园里的动物以及1886年建立的第一公园（Passeio Público公园）。直到2010年，我们还只是动物园的一个部门，但是我们也涉及一些控制家禽和外来物种的问题，因此有人建议市政厅扩展这个部分。我被任命为动物群落研究和监测部门的主任，我们关注城市里的动物保护，所以会与不同的机构合作，但也非常关注环境教育的进程。

我们有一个关于狗和猫的问题，这也是个世界性的问题。过去我们常常捕捉和消灭流浪狗、流浪猫，但人们对此多有抱怨，2005年，市长叫停了这一行为。在那之后，我们又遇到了数量控制问题——在库里提巴周围，有45万只狗——所以我们必须设计出一个项目，帮助人们理解：他们必须跟在宠物后面清除粪便，而且不能任意放出这些宠物，让它们没有节制地繁殖。

关于本地物种，还有另外一个问题：有些东西的数量很不均衡，比如水豚，一种像罗特韦尔犬那样的啮齿动物。水豚居住在河边，它们的天敌已经随着都市化进程而消失了，所以水豚现在没有了自然天敌，它们大量繁殖，自由地穿越街区，四处迁徙，给人们带来很多麻烦。所以我们必须持续监视它们，假如它们表现出对城市和居民的威胁，我们会决定是否进行干预。但水豚是野生的，联邦法律明确规定，野生物种无论如何都要受到保护。是的，我们同意这一点，但如果水豚在半夜吃掉某个人，谁来为此负责呢？

水豚是不是病毒携带者？

它们有可能是，但现在还不是。我们不停地收集水豚的血液和组织样本，持续对这些进行分析。目前为止这还不成问题，但在其他城市，有些发烧可能是通过寄宿在水豚身上的虱子进行传播的。在库里提巴没有这种疾病，但我们持续关注着情况的发展。

亚热带城市要面对和关注的还有哪些其他特殊的动物种群？

我们所拥有的动物中数量最大的是鸟类，因为它们飞过这里，并且会停下来。猴子总是很饿，会

Curitiba's protected native fauna includes monkeys and capybara rodents.
库里提巴的野生动物保护群有猴子和水豚等啮齿动物。

吃掉这些鸟儿的蛋。外来的猴子作为宠物被带到库里提巴，它们后来被主人释放，开始繁殖。因为它们不是本地物种，因而没有天敌，所以我们必须规划，如何来管理这些灵长类动物而不是猎杀它们。我们一直计算它们的数量，抓住其中的一些，令其绝育。在动物园里我们有两个湖和一些岛，放养着一些大猴子。它们不会游泳，所以这还不错。它们可以快乐地生活在那里而不会流窜到城市中去。

我们现在正要通过一部法律，规定任何出售动物的人都要在动物身上植入芯片，这样就会有一个记录，记下谁对此负责。从理论上讲，五到十年内，城市里的所有动物都会拥有芯片。

包括一只宠物鼠?

是的，有这个可能。也许在鱼类中并不普遍，但是鹦鹉，其它鸟类，鼠类，爬行动物、蛇、狗、

What other special fauna concerns does the subtropical city face?

The largest numbers of animals we have are birds because they fly by and stop in. Monkeys, who are always hungry, eat their eggs. Exotic monkeys were brought as pets to Curitiba. They were later released by their owners and they started to reproduce. They don't have a natural enemy because they're not from here, so we have to devise a plan of how to manage the primates without killing them. We are always counting them, capturing and sterilizing some of them. We have two lakes in the zoo and some islands where we bring large monkeys. They don't swim so it's okay. They can live there happily without spreading into the city.

We are just now passing a law that requires anyone who sells an animal to implant a microchip so there will be a record of who is responsible. In five to ten years, theoretically, all animals in the city will have them.

Including a pet mouse?

Yes, it's possible. It's not so common in fish, but for parrots, other birds, mice, reptiles, snakes, dogs, cats, and whatever else you can think of. We don't implant newborns; we wait a little bit because they may not survive and then the chip would be lost. We're organizing fairs every two months, called Pet Buddy, advertised by radio and newspapers, and invite people to bring their pets for free microchips so they don't have to pay a veterinarian. After the law passes, there will be fines for noncompliance. A second new law is a more detailed local version of a federal ban against animal abuse.

Why is Curitiba known as a green city?

Curitiba covers 432 square kilometers [167 square miles], of which 77 million square meters [30 square miles], about 17 to 18 percent, is natural forest cover. The primary ecosystem is the Atlantic rain forest. We also have 35 parks spread throughout the city, not just concentrated in the center. Most of the vegetation is still in private ownership. If you take the population and divide it by the measure of green areas in the city, there are about 50 square meters [538 square feet] per citizen. It's a very good number.

There are a lot of environmental regulations in the city. You cannot cut down a tree without a permit from city hall, and you may have tax incentives for conservation. There is a construction incentive which, for leaving trees intact, allows you to build higher depending on which zone of the city you are in. So it's in the interest of private enterprise to keep the forest in the city. It helps maintain the equilibrium.

Is pollution an issue in Curitiba?

Yes, of course. Curitiba is a rich city with a population of 1.7 million people. Lots of people move here from other places, so we have a large number of cars, almost one per two inhabitants. If I'm not mistaken, we have the third-largest fleet in Brazil. There is a state-run network for monitoring air quality control, and we have three stations that monitor the entire region. We can therefore work to avoid situations like in São Paulo, where the air can be so bad that they advise people to stay home. That doesn't happen here.

Our forest works as a sinkhole for gas emissions. We did an inventory in 2009 and found that we have around 4 million tons of CO_2 stored in our forest, which offsets almost 2 years of cars running in the city. So there is a policy to keep that in place. We are also doing an inventory of emissions. We

猫，你能想到的所有其它东西都会有。我们不在新生动物身上植入芯片，而是会等一段时间，因为它们有可能不会存活，那样的话芯片就会丢失。每两个月，我们会组织集会，称之为"宠物之友"，并通过广播和报纸进行广告，邀请人们带来他们的宠物，为它们免费植入芯片，因而人们就不用再向兽医付费进行这项工作。在上述法律通过之后，可能会针对不配合者进行罚款。而第二种新的法律则是一个更详细的，有关反对虐待动物的联邦禁令之地方版本。

为什么库里提巴被认为是一个绿色城市？

库里提巴的面积是432平方公里（167平方英里），其中有7700万平方米（30平方英里），大约17-18%为原始森林所覆盖，最主要的生态系统是大西洋热带雨林。在城市里，散布着35个公园，这些公园并非只是集中在城市中心，大多数的植物依然为私人所有。如果你先计算人口数量，再除以城市里的绿地面积，结果显示平均每个市民大约拥有50平方米（538平方英尺）的绿地。这是一个非常好的数字。

城市中有许多环境条例。如果没有市政厅的允许，你不能砍掉一棵树；而你却可能因为保护一棵树而获得税率奖励。有一种建设奖励：只要保持森林的完整，依照你所在的城市区域，你可以建设更高的楼房。因此，在城市中，保护森林是私人企业的利益所在，这有助于保持均衡。

污染在库里提巴是个问题吗？

是的，当然。库里提巴是一个富有的城市，有170万人口。许多人从其他地方搬到这里，所以这个城市有许多汽车，几乎每两个居民就有一辆。如果我没有搞错，库里提巴有巴西第三大的车辆队伍。我们有一个国家网络系统，用来监控空气质量，还有三个站点，随时监控整个区域。因此可以避免发生圣保罗那样的情景，在那里空气质量会差到使人们被建议待在家里。这里并没有发生这些事情。

我们的森林像一个解决气体排放的污水坑一样工作。2009年，我们做了一份盘点总结，发现森林里储存着近400万吨的二氧化碳，这相当于抵消了城市里汽车两年的排放量。所以，有一项政策用来保护森林，使其保存在适当的位置上。同时，我们也对气体排放做了一个总结。我们估计，森林每年清除了整座城市3%的气体排放量。

Itaipú Dam on the Paraná River, at the border of Brazil and Paraguay (2006)
巴拉那河上的伊泰普水电站，在巴西和巴拉圭边界（2006年）

库里提巴依赖于哪种能源？

99%是水能，因为我们有伊泰普水电站(Itaipú)，这是世界上最大的水坝之一。它并非是完全清洁的能源，但它比其他能源要好，虽然并不便宜，但供应不是问题。

普通人有多关心环境问题呢？

2009年，《观察周刊》杂志进行了一项调查，发现92%的巴西人非常关心气候变化以及由此对他们生活产生的影响。在库里提巴，人们非常非常关心环境问题。例如，我们是巴西第一座拥有技术完善的垃圾填埋区的城市。填埋区1989年开业，最近刚刚停工，我们正在修复这一地区。

estimate that the forest removes 3 percent of gas emissions in our city every year.

What kind of energy sources does Curitiba rely on?

Ninety-nine percent is hydraulic because we have Itaipú, which is one of the biggest dams in the world. It's not completely clean energy but it's better than the other options. It's not so cheap but there is no problem with supply.

How concerned is the average citizen about the environment?

In 2009, *Veja* magazine conducted a survey and found that 92 percent of Brazilians were very concerned about climate change and its impact on their lives. In Curitiba, people are very, very concerned. For instance, we are the first city in Brazil to have a technically sound landfill. It opened in 1989. It has just shut down and we are in the process of restoring the area. We are now bringing all our waste to another landfill, which is a private enterprise, but we follow the same technical guidelines so we won't cause pollution. There's also a program, not implemented yet, that includes mechanical plants that will separate 90 percent of the waste, much more accurately than manual separation. Ultimately we will have efficient separation, composting, and lots of other systems, leaving just a little waste that will not be able to be managed. As it is now, two-thirds of the city's daily waste is processed.

In some places, we collect waste three times a day. We have separate collection for recyclables, vegetable waste, and dangerous waste such as batteries; hospital waste is up to private companies. Still, if you ask someone from Curitiba, they're going to say this is not good enough. They are always asking for more. Most other cities in Brazil don't even have a landfill, just a dumpsite for waste. Even Rio de Janiero has just a dumpsite near the airport; you can see the vultures flying around.

Curitiba uses multiple receptacles for aluminum, glass, plastic, etc. Do people actually put waste in the right bins?

Yes. Jaime Lerner (a former mayor of Curitiba) always tells the story that if you go to an airport or somewhere, and you see a person with a paper in his hand looking for a trash can, you can be sure he's from Curitiba. Sometimes people from other places, who are not used to this infrastructure, just throw waste in the river. But most of the population does separate the waste. They know there are different dates for the collection of different materials. Schedules are posted on the Internet. If you go to our website you can read about the collection system.

Curitiba is well known for its innovative social programs. Are there any that promote waste management?

There are two different programs. In some irregularly settled areas there isn't always a proper street, or maybe the houses are very close together so the truck cannot come in for collection. In these areas—there are 35 of them—we drop off a container where everyone can put their waste. They get a receipt for their deposit which they can redeem: 5 kilos [11 pounds] of vegetables for every bag of waste. So a lot of people without an income collect waste and wait for the weekly produce delivery; they can bring home five, ten, fifteen kilos [11–33 pounds] of vegetables. Sometimes that

我们现在将垃圾放到另外一个填埋区去，这是一个私人企业所有的，但我们依照着同样的技术方案，所以不会引起污染。还有一个尚未实施的项目，包括可分离90%垃圾的机械装置，这要比人工分离精确得多。最后，我们将会获得有效的分离、制肥和其他许多系统，只留下一点点不能被处理的垃圾。像现在这样，城市三分之二的日常垃圾都得到了处理。

在有些地方，我们一天收集三次垃圾，并将收集上来的垃圾分为可回收物、蔬菜垃圾、危险垃圾（比如电池）；医疗垃圾由私人公司负责。尽管如此，如果你问库里提巴人，他们会说这还不够好。他们总是要求更多，大多数巴西的其他城市都没有垃圾填埋区，只有垃圾场。甚至里约热内卢在机场附近有一个垃圾场，你可以看到秃鹰在上空盘旋。

库里提巴使用不同的容器来装铝、玻璃、塑料等东西。人们真的会将垃圾放在正确的容器中吗？

是的。杰米·勒纳(Jaime Lerner)总在讲一个故事：如果你去机场或者其他的什么地方，然后你看到有一个人手上拿了一张纸，正在寻找合适的垃圾桶，那么你可以确定他来自库里提巴。有时候，来自其他地方的人，并不熟悉这些基础设施，往往就将垃圾扔到了河里。但是大多数库里提巴人却会分类这些垃圾，他们知道有不同的日期收集不同的物质，日期表被发布到因特网上，如果你登陆我们的网页，你可以阅读关于收集系统的内容。

库里提巴因创新的社会项目而闻名，有没有一些是旨在推动垃圾管理的？

有两个不同的项目。在一些不规则的居住区，那里可能没有一条合适的街道，或者可能房子之间非常接近，以至于卡车不能进去展开收集。在这些地区（有35个）我们放置了一个容器，所有人都可以将垃圾放在里面。他们放入的东西可以换回一张收据，这张收据可以进行兑换：每包垃圾换回5公斤蔬菜。所以许多没有收入的人收集垃圾，等待着每周一次的产品兑换交易；他们可以带5公斤、10公斤或15公斤的蔬菜回家。有时候这可能是屋里唯一的食物。这是一个非常好的项目，开始于1991年。它的意义在社会层面上是可行的，而在经济层面则不然，这意味着交易没有利润但城市却为此买单。不过，一种社会秩序和环境上的收益则是可以被证明的。可能在将来，这一项目将会有所收益。

还有一个项目被称为"生态—市民"工程，即有人推着手推车四处走动，寻找纸张、塑料等任何可以回收的东西。我们为此提供场所，所以这些垃圾不用再被带回家里。这就像一个没有公共权力参与的合作社，基本上说只是一个地方，在那里人们可以将自己的东西归类，然后卖给任何需要这些东西的人。我们只是监视良好状况，确保没有剥削。先前存在很多问题，像毒品、对妇女的盘剥，以及强迫小孩子去收垃圾等等。刚开始时，有很多对这一项目的抵制，但是现在，我们有五个"生态—市民"区，并打算在接下来的几年建设更多这样的地方。有时，当经济形势不那么好的时候，我们甚至分发垃圾，由此保证人们不会在这个过程中放弃。尤其是在一个新区域开始运作的头几个月，人们可能不会太感兴趣，所以我们带来可回收的垃圾。通过这种方式，他们可以看到"生态-市民"工程是如何运作的，也会明白这对他们有好处。

大多数社会项目开始于上世纪90年代，莱纳先生最后一次担任市长的时候。他也成立了关于环境的公开大学，杰克·柯斯托（Jacques Cousteau）来到库里提巴参加了成立仪式。这一项目主要关注成人教育、接待游客、教授专业技术课程。最初，我与他们一起做了很多工作，但他们现在已经离开了市政部门，自己独立展开工作。我们在库里提巴非常幸运，因为我们的市长大部分都是技术专家而不仅仅是政客。尽管他们来自不同的党派，但没有一位市长仅仅因为一个项目是另一个管理者开展的，就下令停止项目。运作良好的东西都得到了保持。

库里提巴是否是其他城市的一个模板？

我都数不过来每年有多少城市派遣他们的代表团来这里参观，可能有100或120个，大多数城市是来学习我们获奖的交通系统和环境项目。有一些是来学习我们的社会项目，特别是产前项目或垃圾收集和管理系统。我们的交通系统甚至被哥伦比亚的麦德林市复制下来，而且现在已经被引入到墨西哥城这样的地方。这套系统1971年被引入这座城市，特点是有供公共汽车使用的专用车道。

may be the only food in the house. This is a very old program, begun in 1991. It's socially viable, not economically viable, meaning there's no profit in it but the city pays for it. There is a social order and environmental benefit that is justifiable. Maybe it will be profitable in the future.

There's another program called Eco-citizen where somebody pushes a cart from place to place looking for paper, plastic, or anything he can recycle. We have provided buildings so the waste no longer has to be brought home. It's like a cooperative without public power, basically just a spot where someone can sort his stuff and sell it to whoever he wants. We just monitor the health conditions and make sure there is no exploitation. Previously there were lots of problems, like drugs, and the exploitation of women, and forcing children to collect waste. In the beginning there was some resistance to this program, but now we have five Eco-citizen areas and we are going to have several more in the coming years. Sometimes, when the economy is not so good, we even deliver waste so people don't give up on this process. Especially in the first months of implementation in a new area, people may not be so interested, so we bring in recyclable waste. This way, they can see how the Eco-citizen program works and that it will be good for them.

Most of our social programs began in the 1990s, the last time Mr. Lerner was mayor. He also established the Open University for the Environment, and Mr. Jacques Cousteau came to Curitiba to inaugurate the opening. This program is devoted to adult education, receiving tourists, and technical courses. I worked with them a lot in the beginning, but now they've left municipal government and gone out on their own. We are very lucky in Curitiba because most of our mayors have been technicians, not just politicians. Although they were from different parties, no mayor cancelled a program just because it was started by a different administration. Things that worked well kept going.

Does Curitiba serve as a model for other cities?

I can't even count the number of cities that send missions to visit us each year, maybe 100 or 120, mainly to learn about our award-winning transportation system and environmental programs. Some come to study our social programs, especially our prenatal program, or our waste collection and management system. Our transportation system has even been copied by Medellín, Colombia, and is now being introduced in places like Mexico City. It was introduced here in 1971, with dedicated lanes just for buses.

Bosque de Portugal, Curitiba, with 22 pillars honoring great poets of the Portuguese language. (2007)

葡萄牙博斯克公园，库里提巴，有22根纪念伟大葡萄牙语诗人的柱子。（2007年）

What deters people from driving in the bus lanes?

There is a very large fine. People don't do that except for extreme emergencies.

Is Curitiba planning future transit improvements?

Yes. You have to plan for the future. Although we have the best transportation system in Brazil, it is already overburdened. During peak hours it's very difficult to board the bus, and not everybody travels seated. A subway system is one way to speed up this process. The proposal is to construct a subway along the main bus line so no land has to be appropriated or bought. The cut-and-cover system requires minimal digging. You just cut the depth necessary and then cover it with a linear park. There would still be two lanes for cars, but the subway would replace the bus.

It's actually an integrated system. Some stations will be three stories high and combine subway, bus, and cars. We have already ordered environmental studies and are ready to proceed, but when we are going to start building is another

是什么阻止人们行驶在公共汽车专用车道上？

罚款数额很大。除非是极端紧急的情况，否则人们不会那么做。

库里提巴是不是计划在将来对交通系统进行改进？

是的。你必须为将来做打算。尽管我们现在有巴西最好的交通系统，但它也已经不堪重负了。在高峰时段，很难挤上公共汽车，而且并不是所有人都有座位。地铁系统是加速改进过程的一种方式。建议是沿着主要的公交路线建设一条地铁线路，因而也就不需要占用或购买额外的土地。随挖随填系统只需要最低限度的挖掘，你只需要挖到必要的深度，然后覆盖上一条带状公园。依然还是会有两条机动车道，但地铁将会取代公交车。

这实际上是一个综合系统。有些车站会有三层那么高，将整合地铁、公交和机动车。我们已经布置了环境方面的研究，也做好了开展这一项目的准备，但是我们什么时候开始建设却是另外一个问题，也许不会等30年或40年那么久。当然，我们也同时在对公交系统进行改进。

Bi-articulated bus in dedicated central bus lane near shopping district, Curitiba. (2006)
库里提巴在商业区附近的中心专用公交路线上的双铰接巴士。(2006年)

例如，今年市长引进了一种非常大的公共汽车。这种公共汽车是现在世界上最大的，能够容纳300人，而且完全由生物燃料驱动。交通系统也青睐这种公共汽车，因为当它接近十字路口的时候，路灯就会改变颜色。去年，我们的公交车队有1%使用生物燃料；我们期望在2012年底这个数字可以达到10%，所以这是一个先进的系统。

我们尚未有一个数字的或"智能的"系统，来显示到达时间。另外在故意破坏公物上也存在一定的问题，因此在拥有此类设备前，市民们还需要再成熟一些。另外，公共汽车的发车频率很高，因而实际上也并不真的需要这些。

2016年奥运会将对库里提巴有何影响？

更大的影响将来自2014年的世界杯足球赛，因为库里提巴将是其中的一个比赛举办城市。那么，将会有非常多的旅游者来到这里，城市的旅馆和服务业已经在做准备了，同时，城市也在改善中转机场、公共汽车等等。奥运会将对库里提巴产

question, maybe not for thirty or forty years. In the meantime, of course, we are still improving the bus system.

For instance, this year the mayor inaugurated a very large bus, the largest in the world, which carries 300 people and runs completely on biofuel. Traffic favors the bus so street lights change when it approaches the intersection. Last year, 1 percent of our bus fleet ran on biofuel; we expect 10 percent by the end of 2012, so it's a progressive system.

We do not yet have a digital or "smart" system that posts arrival times. We still have problems with vandalism, so the population has to mature a little before we can have this kind of equipment. Besides, the buses arrive very frequently so there is no real need.

How will the 2016 Olympics affect Curitiba?

The greater impact will be from the World Cup in 2014 because Curitiba is going to be one of the cities where some games will be played. So there will be lots of tourists. The hotels and services are already preparing, and the city is improving airport connections, buses, and so on. The Olympics will have a further impact on Curitiba because not all the activities will be in Rio, and athletes may come here to prepare.

Does Curitiba envision its continued role as a world leader in sustainability?

Sure. We are always trying to come up with good ideas. For instance, the dedicated bus lanes that we introduced in the '70s: now everybody is doing it. We came up with segregation of waste, and landfills, and now everybody in Brazil has started to do that. We are always revamping. Our newest program is Biocity, which was introduced in 2006 to make urban development and environmental conservation compatible. The goal is to restore biodiversity by using indigenous plants, practically and conceptually, in the daily actions of city hall.

Curitiba won two international awards last year. In Stockholm, the Globe Forum awarded us a prize for being the most sustainable city in the world. We received a second award in Mexico City after *The Economist* [the magazine] conducted a worldwide survey and determined that Curitiba was the world's greenest city. At www.biocidade.curitiba. pr.gov.br/ you will find a summary of our Biocity program, which incentivizes us to work with native fauna and flora and to work with climate change. So we have been developing several propositions, trying again to be one step ahead.

In 2011, the Clinton Climate Initiative held a summit meeting in São Paulo for cities with more than 20 million inhabitants. Curitiba was invited to participate even though it's not a megacity. The mayor was asked to deliver a speech and he chose the subject of biofuel and how we are working with it in the city. It is possible to run on 100 percent biofuel. Biofuels diminish pollution, with 20 percent less emission than any other kind of fuel. That is why we are converting our bus fleet.

What's different about Curitiba is that we don't just stand on the philosophical point of view, we go for it. So if we want to develop a new park, we build one. We don't just talk about the project, and then merely study and survey the area. That way, you can spend ten years preparing and you never do it. Sometimes we go in and do a survey at the same. It may not be the best technical way to do it, but it does give results.

What are the consequences of a "do it now and fix it later" approach?

During the 1970s in Brazil there was lot of that, but I find it very troublesome. We talk about that a lot. For example, there are too many cars on the streets. There are people 50 to

生进一步的影响，因为不是所有的活动都在里约进行，运动员们可以来这里做准备。

库里提巴是否期待在可持续性方面，继续扮演一个世界领先者的角色？

当然。我们总是试图提出好的主意。例如，上世纪70年代我们引入的公共汽车专用道：现在所有城市都这么做了。我们提出了垃圾分离和垃圾填埋的观念，现在巴西的所有人也都开始做这些。我们总是不断进行修正，我们最新的项目是2006年引入的生态城市工程，它旨在使城市发展与环境保护相协调。通过在市政厅的日常行为，无论是实践上还是概念上，使用本地植物，来恢复这一地区的生物多样性。

库里提巴在去年获得了两个国际性奖项。在斯德哥尔摩，全球论坛授予我们全球最可持续性城市奖。《经济学人》杂志发起了一项全球性调查，在评定库里提巴为全球最绿色城市之后，我们在墨西哥城获得了第二个奖项。假如你点击 www.blocidade.curitiba.pr.gov.br/，你会发现一个关于我们生态城市工程的概要，这激励我们在本土为本地动植物群落以及气候变化而努力工作。因此，我们已经发展了许多议题，力图再次领先一步。

2011年，克林顿气候行动组织在圣保罗举行了一次峰会，这次峰会是为人口超过2000万的城市举办的。库里提巴也被邀请参加，即使它并不是一个大城市。市长被邀请发表一个演讲，他选择的主题是关于生物燃料，以及我们如何在城市中利用它们。百分之百依赖生物燃料是可能的。生物燃料会减少污染，相比其他的燃料会降低20%的气体排放量。这就是为什么我们正在改变公共汽车车队使用燃料的方式。

库里提巴的独特之处在于：我们并不是停留在所谓的哲学观点上，我们还努力去实现它。所以如果我们想要建设一个新公园，就会行动起来去建一个。绝不仅仅是口头谈论某个项目，然后仅仅研究和调查某个区域。那样的话，你可能会在准备工作上花费十年的时间，却永远不会去做。有时我们参与进去并做个调研，这也许不是最好的技术方式，但它确实产生了结果。

那种"现在就做，然后修理"方法的后果是什么？

上世纪70年代，在巴西出现过这种方法，但我发现它是很麻烦的。我们就此谈过很多。例如，有太多汽车在街道上。因为，原来有许多50到60岁的人都没有过汽车，而现在经济状况好了，他们能够买得起一辆车了。如果别人一辈子都拥有汽车，他们怎么能说："不，不允许你那么干？"我认为中国现在的情况与此相似，在那里人们能够做许多以前不被允许做的事情。中国的建筑进度，单单是推倒一切的速度，要比巴西过去快得多。如果没有细心的规划，也没有人决定应该拆除什么、保留什么，那么就会有大问题。

例如，三峡大坝就将淹没一大片区域，改变许多城市的位置，这是很疯狂的，世界上没有任何其他地方会这么做。我们现在正在巴西北部新建另一座大坝，但是那里有许多人抱怨政府，说政府曾经宣布过不会再有新大坝了。30年前，政府会直接做，不用担心什么，所以抱怨是一个进步的过程。我认为中国也将经历相同的过程，很快就会有人说："不，不要再那么干了，你必须做一个不一样的规划。"我们看到在四年前的世界杯上，人们就开始关注，是否该让运动员在受污染的空气中奔跑。

我真的希望中国能够意识到并慢下来一点儿，因为他们将面临废弃物带来的巨大问题。我能想象那里的问题有多么大，因为太过积极地进行消费了，他们如何管理生产出来的所有废弃物呢？这定然会是一个大问题。

库里提巴有319岁了。我们的第一个总体规划可以追溯到上世纪40年代，然后在1965年进行了更新，因此人们对这一进程的理解是一种文化。我们告诉每个人，并不存在唯一的城市方案。我们不能只是告诉你做这个做那个，你必须修改你自己的解决方案。例如：里约，一个非常复杂的城市，在上世纪80年代和90年代，试图引进莱纳先生以期对市政厅有所帮助，但是现在，城市已经非常固定了，很难做出一些不一样的东西。在城市发展的同时，你必须将各个系统置于恰当的位置上，修复一些损坏的东西是很难的。

在库里提巴的未来，什么是富有重要意义的提升呢？

如果我们被允许执行新的废弃物管理系统，未来没有垃圾填埋区，没有机械分离和循环装置，这个城市将成为许多其他城市的模板，因为我们将证明，存在着一种对废弃物进行燃烧和掩埋处理的替代方法。另外，我认为整合交通系统将是非常重要的，因为由此我们将置身于大都市区域，从库里提巴延伸到25个其他城市——这个都市区域

60 years old who never owned a car, and now that the economy is good, they're able to buy one. How can you say "No, you're not allowed to do that" if other people have had them their entire lives?" I think China has a similar situation where people now can do a lot of things that they weren't allowed to do before. The building process, just bulldozing over everything, is a lot faster in China than it was in Brazil. If there isn't careful planning and someone deciding what to demolish and what to preserve, it's going to be a major problem.

The Three Gorges Dam, for instance, is going to flood a huge area and dislocate entire cities; this is crazy. Nowhere else in the world would they do that. We're in the process of building another dam in northern Brazil, but so many people there complained that the government has declared this will be the last one. Thirty years ago the government would have just done it and wouldn't have cared. So it's an evolving process. I think China is going to go through the same process. Soon someone is going to say, "No, don't do that anymore, you have to plan differently." We saw that in the last World Cup where there were concerns about whether the athletes could run in the polluted air.

I really hope China can see that they have to slow down a little bit because they are going to have major problems with waste. I can imagine how big a problem it is there, because they are consuming very actively. How are they managing all the waste they are producing? That must be a major concern.

Curitiba was founded in 1693. Our first master plan dates from the 1940s and then was updated in 1965, so it's kind of the culture of the population to understand this process. We tell everyone there is no single urban recipe. We can't just tell you to do that and that. You have to adapt your own solutions. For instance, Rio, a very complicated city, tried to bring Mr. Lerner in to help in the city hall in the '80s and '90s, but the cities were already consolidated; it's very difficult to do something different now. You have to put systems in place while the city is growing. To restore after something's damaged is very difficult.

What significant improvements are in Curitiba's future?

If we are allowed to implement the new waste management system, not having landfills, mechanical segregation, and recycling in the future, it will be a model for a lot of other cities because we will prove there are alternatives to burning and burying waste. Also, I think the integration of our transportation system will be important because we are going to expand from Curitiba to 25 other cities in the metropolitan region that will benefit and use our subway and buses. Some of the system is already integrated, but it's going to be faster and more efficient in the future.

We also have many projects for restoring riverbeds and river qualities. In 10 to 15 years, we are going to have some good examples of rivers really restored. We have a project, financed by the French Development Agency, which is called Viva Barigui and is the first project I have ever seen that reconstructs the forest and the river margins. It's not a question of just planting trees. We have done studies of the architecture of the forest before it was removed for urbanization and we are trying to reimplement it. So in 15 years, if this project works, we're going to have results that might inspire other cities.

Right now we have fairly well-preserved parts of the forest at the edges of the city, then small parks further in, one here, there, and there, with a lot of urbanization in between. What we are trying to do is connect them with the forest and with the river. Some areas will be perhaps 10 meters [33 feet] wide, while others are going to have 300 meters [985 feet] of river margins. If we are able to do that, the fauna will live there naturally, water quality is going to improve, and erosion will be controlled. And the population will benefit and be able to use the reclaimed areas for leisure.

Does this program require relocation?

It's part of the process. In Brazil, we have a federal law that says in the first 30 meters [98 feet] of the river margin there cannot be any kind of construction. It's been a law since 1965, but people still settled there, especially those who had nothing. For many years, nobody did anything, but it is understood that they cannot be there because if there is a 30- or 50-year flood, they are going to be affected. So we have an official program that removes people from these areas and gives them a new home, maybe a new apartment or sometimes just property.

We are in the process of implementing two parts of this park now. The people have already been relocated. To do the whole thing would consume the entire city budget, so we are working with what is possible. The idea is to interconnect with the forest and with the river. If we can implement this, it's really going to be something.

将使用我们的地铁和公交车，并从中受益。有一些系统已经进行整合了，但在将来，它将变得更快速更有效。

我们也有许多项目，旨在修复河床和河流的水质。在未来10-15年内，我们将会有一些关于修复河流的好例子。我们有一个被称为"生活巴瑞基"（Viva Barigui）的项目，它是由法国开发署资助的，是我看到的旨在重建森林以及河岸的第一个项目。这并不仅仅是植树的问题，在森林被都市化移除前，我们已经对森林的构架进行了研究，现在试图重现它。所以在15年内，如果这个项目运作良好，我们可以获得一个令其他城市备感鼓舞的结果。

现在，在城市的边缘，我们拥有森林中一些保持得相当好的部分，然后是小型的公园零散地分布在各地；而在这些公园之间，则有许多地方已经城市化了。我们想要做的，就是将这些公园与森林、河流联系起来。有些地方可能会有10米宽，其他一些则可能是距河边有300米的区域。如果我们能够做到这些，动物的族群将自然地生活在那里，水质也会得到提高，侵蚀则会受到控制。而人们将从中收益，并能够利用重新改造、再次回收的区域进行休闲活动。

这个项目是否需要对人口进行重新安置？

这是进程的一部分。在巴西，联邦法律规定，在河边第一个30米处，不能够进行任何的开发建设。这个法律从1965年就开始生效了，但是人们依然在那里安营扎寨，特别是那些一无所有的人。许多年来，人们什么也没做，但是大家都明白，那些人不应该待在那，因为如果来一次30年一遇或50年一遇的洪水，他们将受到重创。所以我们有一个官方项目，会让人们离开这一区域，并给他们一个新家，这可能是一栋新的公寓，或者有时只是一些财物。

我们正在对公园的两部分进行实施工程，人们已经被重新安置了，完成所有这些会消耗掉整个城市的预算，所以我们只是尽可能地展开工作。我们的想法是，既与森林联系，也与河流相互联系起来。如果能够实现这一想法，那真的是非常有意义的！

Curitiba skyline. (2005)
库里提巴天际线。(2005年)

SECTION FOUR / PROPOSAL FOR A VERTICAL CITY

New Urban Form
Design Approach
Proposed Schemes for a Vertical City
Sky Lobby / Change of Lifestyle

第四篇 / 关于垂直城市的建议

新的城市形式

设计方法

垂直城市问题上已经提出的方案

空中大堂——改变生活方式

新的城市形式

New Urban Form

在之前的访谈章节中，我们已经发现，正如我们所知道的，文明正处于多年环境破坏带来的危难时刻。当我们将追求更好的生活放在第一位时，我们引发了全球变暖。不管人们信不信，其严重的后果已然迫在眉睫：我们经历的恶劣天气比我们先辈要更糟更频繁，极地冰川融化、海平面上升、干旱、湖泊干枯断流，还有很多。

作者们恰好属于关注自我的一代，这一代养成了和周边人过一种"相互攀比"的消费型生活模式，这些一步一步地持续，导致了当前的危机。我们现在需要在如下两种极端情形之间做出抉择：

一，将这些变异现象仅仅当作一种地球演化的自然循环，并采取一种观望的自得态度，直到越来越多具体的科学证据表明，人类确实对此负有责任。总之，最终的结果是无可奈何，把这些问题留给后代人去解决。

二，我们还是小心一点，接受立即展开行动的呼吁，以便扭转、减少或至少延缓无可逃脱的命运。

作为后一种方案的坚定支持者，就像在第3页中的"8-8-8"宣言所描述的那样，我们对此的回应是，人们需要一种新的城市模式，选择一种新的生活方式。

新城市形式概要——一个垂直城市

总则

在第7页概述中构思的垂直城市，清楚地说明需要创造一种崭新的、自给自足的，方便行人的城市模型，它既有人口在6万到15万的"卧室社区"，也有人口上限在15万到25万的"生活／工作社区"。一些设施有助于培育和强化对一种新生活方式的选择，而提供这些设施需要仔细地考虑如下问题：

位置选择

在选择位置时必需尽一切努力最大限度地保护：本地动植物作为家园的现存自然景观。

In the previous Overview section, we observed that civilization, as we know it, is at a crisis point brought on by years of disregard for the environment. In our pursuit of a better life for numero uno, we have triggered global warming, the effects of which (for both believers and nonbelievers) are powerfully demonstrated at our very doorsteps: severe weather that is more violent and frequent than that experienced by previous generations; melting polar caps; rising sea levels; drought; and the drying up of ponds and lakes. The list goes on.

The authors, who happen to be of the self-absorbed generation that nurtured neighbors with a "keep up with the Joneses" lifestyle of material consumption and excess, incrementally contributed to the current crisis. We are now confronted by the need to choose between two extremes:

One: Regard the unfolding phenomenon as just a natural cycle of Earth's evolution and assume a wait-and-see complacency while demanding more and more concrete scientific proof that humankind does indeed have a hand in its making. Conclude with resignation that, in any case, it is a problem for future generations to solve.

Two: Err on the side of caution and accept the call for immediate action to reverse, reduce, or at least postpone irreversible demise.

As vigorous proponents of the latter course, our response, as outlined in the 8-8-8 Manifesto on page 3, points to the need for a new urban form and the choice of a new lifestyle.

A Suggested Brief for a New Urban Form—A Vertical City

General Description

The Vertical City, as envisioned in the Overview on page 7, makes clear the need to create a new type of self-sustaining, pedestrian-friendly city that will accommodate both a "bedroom community" of 60,000 to 150,000 people and also a "live/work community" with a maximum population of 150,000 to 250,000. To provide facilities that will nurture and reinforce the choice of a new lifestyle, the following must be carefully considered:

Site

Site Selection

In selecting a site, every effort must be made to preserve to the greatest degree possible:

The existing natural landscape as home for native flora and fauna;

    Arable land for cultivation of food for local consumption;

    Open space for parks and recreation;

    Natural greenbelts, or buffer zones, between neighboring communities;

    Natural water sources, protected from contamination.

Site Size

For the larger live/work community, dimensions are determined by the maximum distance one can travel comfortably on foot in 10 to 15 minutes. The smaller bedroom community requires a smaller area, but of sufficient size to accommodate shopping and other nonresidential support facilities.

Site Location

The live/work community must be located near transportation routes for connectivity to other cities. The bedroom community requires easy-access rapid transit to employment opportunities in nearby cities (the lures for migration from rural areas).

Site Planning:

Built-on land area is to be mitigated by creating a pedestrian-friendly community. The main level as well as all sky lobbies, reserved exclusively for pedestrians, is segregated from vehicular traffic on a lower level. The main entrance to all buildings is located on the aforementioned pedestrian level. The new Vertical City is to be surrounded by open land for cultivation of produce and operation of a fishery for local food consumption. The open space will also provide a green buffer between other cities.

Horizontal Circulation

Vehicular

Vehicular circulation within the Vertical City is limited to the lower level. To preserve fast-disappearing arable land, all external vehicular traffic to and from the project ideally should be elevated.

Pedestrian

As one of the premier features of the Vertical City, the

为当地生产消费食物的本地农田。

作为公园和休闲区的空地。

邻近社区之间的自然绿化带或缓冲区。

保护自然水源免受污染。

位置面积

对更大的生活/工作社区而言，面积大小由人们可以舒适地用十分钟步行穿越的距离来确定。较小的卧室社区需要的面积也较小，但也要有足够的面积提供购物和其他非住宅保障设施。

位置场所

生活/工作社区必需位于交通路线上，以和其它城市相联。卧室社区需要有能够便捷快速换乘到附近城市工作点的路径（这是吸引乡村移民的地方）。

位置规划

通过营造一个方便行人的社区，用于建设的土地面积变小了。水平层面和空中大堂只供行人使用，和较低层中的车辆交通分离开来。通向所有建筑的主入口位于上述行人层上。新的垂直城市被空地所包围，这些空地用于耕种农作物，经营渔场以满足当地的食物消费需求。空地也提供了一条各个城市之间的绿色缓冲带。

水平面上的交通

车辆

垂直城市内的车辆交通被限制在较低层中。为了保护日益减少的耕地，所有外层的车辆交通理想上都应被抬升到水平面之上。

行人

作为垂直城市的一个首要特征，交通的主要方式是步行或自行车，最长的距离十分钟内也可以走到。通过被遮盖的道路网络，交通路线可以免受恶劣天气的影响。

自行车

自行车可以通行在专用路线上，也可以进入人行道。

延伸的空中大堂
位于接近一百层左右的垂直区间处，延伸的空中大堂允许行人在所有高楼之间流通联系（参见下文的"高楼"）。

自动人行道
必要时，可以在交通拥挤的特定路线上开通自动人行道系统。

快速轨道交通
必要时，将来可以考虑在主要人行道层下方的设备服务层，整合进快速公共轨道交通系统。这一交通系统还将把垂直城市和其他邻近社区联结起来。

垂直交通

除了楼梯、坡道、电梯，高楼还有高速无绳电梯，把不同的空中大堂联系在一起。每个楼层都提供来回往返换乘到本地电梯的服务。

基础设施

道路
由于在主裙楼层上，内部主要的交通模式是步行，传统的道路在数量上要进行限制，并限制于供服务车辆和少数私人车辆使用的较低水平。高速公路和铁路与其他社区相连接以便保护珍贵的耕地。

服务
为了安全起见，也为了便于维护和修理，电力、天然气、水、暴雨和污水处理系统等服务设施，以及相关的储油罐，装置设备位于裙楼平台中。这种布置将防止上方的步行层或下方的车辆层产生的破坏行为。

primary modes of circulation are by foot or bicycle, the maximum distance being that which can be walked in 10 minutes.

Bicycles
Bicycles are encouraged, in dedicated lanes, in conjunction with foot traffic.

Extended Sky Lobbies
Located at vertical intervals of approximately every 100 stories, extended sky lobbies allow pedestrian interconnection among all towers. (See "Towers," below.)

Moving Sidewalks
As necessary, a system of radiating moving sidewalks can be installed along selected heavily traveled routes.

Mass rapid transit
If necessary, consideration will be given in the future to the incorporation of a rapid-transit system in the utility service level below the main pedestrian level. The same transit system will connect the Vertical City to neighboring communities.

Vertical Circulation

In addition to stairs, ramps, and escalators, the towers are served by ropeless high-speed express elevators connecting one sky lobby to another. Transfers to banks of local elevators provide service to individual floors above and below.

Infrastructure

Roads
Since the primary internal circulation mode is by foot on the main podium level, conventional roads are limited in number and relegated to the lower level for service vehicles, privately owned cars and rentels. Highways and rail links to other communities are raised in order to preserve precious arable land.

Services
Service utilities such as electricity, natural gas, water, storm and waste sewers, and related storage tanks, plants, and equipment are located within the podium platform for security and easy access for maintenance and repairs. Such a disposition prevents disruption of activities on the pedestrian level above or the vehicular level below.

Buildings and Structures

Basements

It is believed that with the introduction of the raised podium platform, described below, the cost of basement excavation and construction can be avoided.

First or Grade Level

The lowest level of the podium is located on grade for deliveries and vehicular access and parking.

Podium Platform

A raised podium houses all the services and infrastructure required by the Vertical City. The roof of the podium is the main platform, or "ground" level of the Vertical City, from which the main entrances of all buildings are accessed.

Podium Buildings

Comparatively low support buildings, averaging 6 to 10 stories, house schools, shopping, theaters and cinemas, restaurants, cafés, bars, medical facilities, government agencies and other functions that are not suitably accommodated in the towers. Collectively, the support buildings constitute the Central Business District (CBD) found in all conventional communities. This primary commercial area will be supplemented by smaller shops, stores, and service outlets in tower sky lobbies and extended sky lobbies (described below).

Towers

The iconic towers of the Vertical City are primarily residential, serving both the bedroom and live/work communities. In the latter, the towers also serve, significantly, as centers of employment, housing offices, laboratories, health care, and even certain clean light industries. A limited number of mixed-use towers will serve each community according to market demand.

Tower Sky Lobbies and Extended Sky Lobbies

Key to the success of ultra-high-rise living are the multistory elevator transfer lobbies, or sky lobbies, that connect and structurally reinforce the Vertical City's towers. Effectively "stacked village squares," the extended sky lobbies and their landscaped roofs offer unlimited possible uses, as might be found in vibrant town centers: shops of all kinds, food and goods outlets, open markets for local vegetables and seafood, sidewalk cafés, newsstands, vendors, and even street hawkers. Other functions include schools, medical services, and civic and government facilities, as well as a wide range of fitness centers, pools, and jogging paths. Landscaped vest-pocket parks invite leisurely strolls or just

建筑与结构

地层或地下室

可以确信，如下所描述的，由于采用抬升的裙楼平面层、开挖和建设地基的费用可以节省下来。

地面层

裙楼的最底层位于水平面上，用于货物运输，车辆通道和停车。

裙楼区平面层

在一个抬升的裙楼中装下了所有垂直城市需要的服务和设施。裙楼的屋顶是主要的平面层，或者是垂直城市的"地面"层，所有建筑的主要入口都在那里。

裙楼建筑

相对低的配套建筑平均从6-10层不等，容纳了学校、商店、剧院、影院、餐馆、咖啡厅、酒吧、医疗机构、政府部门和其他不适宜放在高楼中的功能设施。总的来说，在所有传统的社区中，人们都会发现，配套建筑都构成了中央商务区(CBD)。在高楼的空中大堂和延伸的空中大堂中，一些更小的商店、铺面、服务网点会补充到这个基本商业区中。

高楼

标志性的高楼首先是住宅楼，用于卧室社区和生活/工作社区。在后者中，高楼的另一个重要功能是作为一个包含各种工作岗位、家庭办公室、实验室、医疗服务甚至清洁轻工业的中心。根据市场的需求，有限的多用途高楼将服务于每个社区。

高楼空中大堂和延伸的空中大堂

超高层生活成功的关键在于多层电梯换乘大堂，或者空中大堂，这些将垂直城市的各座高楼连接起来，并强化其结构。高效的"集成式村落广场"，延伸的空中大堂及屋顶景观，提供了无限的应用可能，就像我们可以在一个生机勃勃的市镇中心看到的那样：各式各样的商店、食物和商品折扣店、出售本地蔬菜和海产品的开放市场、路边咖啡馆、报亭、小店，甚至是街头小贩。其他功能还包括学校、医疗服务、城市和管理设施，以及大量的健身中心、游泳池和慢跑道路。风景秀丽的小公园引来大量悠闲的漫步者，或者是一些仅仅想要放松一下的人。另外，也有地方供人们交际，享受当地的艺术、工艺、节日和其他社交活动，这些大大提升了人们的都市感受。人们不需要到更低的、位于任何基础之上的裙楼"平面"层去，生活就有无限的可能性。

结构系统

在这本书中，高达千米或英里的高楼都被构思出来，有些已经处于设计阶段了。结构系统和施工能力是垂直城市所需要的。作者意识到，最大的挑战是设计应对高空的风，因此作者提出，最佳的支撑方案是通过交叉联结将高楼绑在一起。通过这种策略，设计师就可以跳出现存高楼往常采用的圆锥形状。

机械系统

当前的技术将让垂直城市成为节能城市。实现这一点也让人们有机会应用新兴的、之前在超高层建筑领域尚未采用的技术。

电源

推荐使用非化工燃料发电。用来防止能源在长距离运输中有所流失，发电厂应当位于离垂直城市很近的地方。另外，设计还要求整合光伏幕墙、高空风能、燃料电池、太阳能热水，最大化地发挥各种切实可行的替代性能源供应方式。

垂直交通

正如我们所讨论的，垂直城市采用了传统和创新的两种垂直交通模式。

电源，空调冷气

建筑方向，遮阳设备和建筑外墙都是设计来提升居民舒适度的。最佳的自然通风和阳光由最佳能效的机械和照明系统来提供，这些系统的运行可以根据居民个人的需求进行智能调节控制。（超高层建筑中的其他工程挑战和机遇参见"新材料与技术"）

relaxing alfresco. Additionally, there are gathering places for personal mingling and enjoyment of local arts, crafts, festivals, and other forms of social engagement that so richly enhance the urban experience. The endless possibilities for activities inside the vertical city as well as the exploration of farms and natural landscape outside enriches the lives of the Vertical City residents.

Structural System

At this writing, kilometer- or mile-high towers are being contemplated or are already on the drawing boards. The structural system and construction capabilities required for the Vertical City already exist. The authors recognize that the biggest challenge is design for high-altitude wind, and we therefore propose using the structure of the sky lobbies as bracing for optimum support. By this strategy, designers are freed from the predictable tapering shape of existing tall towers.

Mechanical Systems

The Vertical City will be as energy efficient as current technology permits. Realization also presents the opportunity to engage new and emerging technologies that may not have been employed previously in the realm of ultra-high-rise buildings.

Power Sources

Non–fossil fuel sources are recommended for the generation of electricity. To avoid energy losses from long-distance transmission, power plants should be located in close proximity to the Vertical City. Additionally, the design calls for the integration of photovoltaic curtain walls, harnessing high-altitude wind, the use of fuel cells, solar hot water, and the maximization of other practical alternative energy supplies.

Vertical Transportation

As discussed, the Vertical City uses both conventional and vanguard modes of vertical circulation.

Electrical and HVAC

Building orientation, shading devices, and building enclosures are designed to optimize the comfort of occupants. Optimized natural ventilation and sunlight are supplemented by mechanical and lighting systems of the greatest energy efficiency, with operations overseen by smart controls that respond to individual occupant requirements. For other engineering challenges and opportunities in ultra-tall buildings see "New Materials and Technologies,"

Plumbing

The Vertical City incorporates plumbing fixtures that use a minimum of water, and also drainage capture systems that avoid wasting the cold water that precedes the flow of desired hot water from a faucet. Infrastructure includes the necessary piping and apparatus for the collection, storage, treatment, recycling, and reuse of:

Rainwater, for use as potable water
Brown water, for use in flushing or irrigation
Kitchen waste, for conversion to organic fertilizer
Human waste, for conversion to fertilizer

Recycling Systems

In addition to the aforementioned water systems, the Vertical City aims for 100 percent recycling. The towers' vertical advantage is maximized by trash chute recycling systems that segregate glass, metal, plastics, paper, organic compostables, hazardous substances, and residual waste.

Construction Materials

For both interior and exterior, the Vertical City uses energy-efficient construction materials that are, to the greatest degree possible, renewable, from local sources, and nonpolluting in their manufacture. It is important that the materials contain no toxic chemicals, and that they be recyclable for subsequent uses.

Landscape

The natural landscape should be preserved for beauty and as a habitat for native plants and animals. Every apartment in the Vertical City has a landscaped private terrace, and all pedestrian paths are lined with trees and open onto numerous vest-pocket parks. Similarly, the roofs of all buildings, especially those with relatively low podiums, are landscaped, as are the roofs of the sky lobbies and extended sky lobbies to afford a sea of green comfort and color to tower occupants. Native plants are used wherever possible; the application of chemical fertilizers and insecticides is prohibited.

Cost

As architects, the authors have on too many occasions found our design imagination hamstrung by budgetary concerns. A rare exception occurred while working on the preliminary phase of the refurbishment of the United Nations Headquarters in New York City. In evaluating and prioritizing the design

水管设施

垂直城市中水管设施的固定装置使用极少量的水，排水收集系统可以杜绝浪费水龙头流出热水之前空放的冷水。相关的基础设施包括必要的管道，和用于收集、储存、处理、循环回收、重复利用的设备：

雨水用于饮用

中水用于冲洗和灌溉

厨房垃圾用于转化为有机肥料

人类粪便用于转化为肥料

回收再循环系统

除了之前提到的水处理系统，垂直城市力求实现100%的循环利用。利用废物滑槽回收系统（它可以分离玻璃、金属、塑料、纸张、有机化合物、有毒物质和废物残渣），高楼的垂直优点被最大化。

建筑材料

无论内外，垂直城市都使用节能的建筑材料，它们是最大限度来自于当地资源的可再生物质，而且在制造它们时不会产生污染。这些材料中不含有毒的化学物质，可以继续循环利用，这一点非常重要。

景观

为了美丽的景致，也为了动植物有个居所，自然景观应该得到保护。垂直城市中的每个公寓景观阳台，所有的人行道都绿树成荫，通向许多小公园。与此相似，所有建筑的屋顶，特别是相对较低的裙楼建筑屋顶，都有景观；空中大堂和延伸的空中大堂，其屋顶也是如此。这给高楼的住户提供了一片舒心的绿海。本地植物可以用在任何可能的地方；化学肥料和杀虫剂是被禁用的。

成本费用

作为建筑师，作者有太多次设计方案因为预算问题而被迫放弃的经历。不过，纽约市联合国总部大厦翻修的初步设计阶段是我们工作中为数不多的例外。在为联合国的想法评估和优化设计意见时，我们被要求仅仅专注于技术标准问题；原本非常重要的成本等非技术问题，反而不需我们考虑。这种方法的聪明之处在于，它为解决问题创造了一个机会而不是设置了障碍，它让想象力充分发挥，探索那些尚未发现的领域。

对于垂直城市而言，在详尽的内部辩论之后，我们承认在如下二者之间游移不定：找个理由忽视敏感的成本问题或者在财政的实际问题上绝不妥协。我们决定分享如下假定的成本费用比较，将传统城市与我们提出的垂直城市这一新城市模式进行权衡对照：

赞成传统郊区：
按建设上层结构每平方英尺的单位成本来算，更偏向单一的郊区家庭，不管是采用木材（像在南北美洲，非洲和西伯利亚）、石料还是轻型钢材，都要比垂直城市中耐火的超高层建筑便宜。

赞成垂直城市
垂直城市在如下方面节约出来的资金大大抵消了郊区家庭建筑在成本上的优势：

土地：垂直城市的土地面积是传统城市和郊区所占面积的一小部分。当人需要将每一英寸土地都用于种植食物维持生活目的时，当可用的耕地由于干旱或上升的海平面日益减少时，土地就成为最重要的问题了。

基础设施：由于在水平面上很紧凑，垂直方向上堆叠整合，在垂直城市中，基础设施——道路、人行道、公共设施和其他相关费用的总量大大降低了。

建筑外墙：对于每个独立单元来说，更少的暴露型外墙意味着更少的成本。

机械设备：集中式、节能、智能，这些设备减少电耗，运营成本也更低。

运营：训练有素的人操作高效、最先进的设备，将减少运营人力成本，大大节约能源。

options for the UN's consideration, we were instructed to limit the process to technical criteria; consequently the non-technical issue of relative cost, while extremely important, was not to be included. The wisdom of that approach was that it created an opportunity rather than a hindrance to the problem-solving process, allowing the imagination to pursue avenues that might otherwise have remained unexplored.

For the Vertical City, we confess, after exhaustive internal debate, to having oscillated between finding justifications to bypass the sensitive cost issue or to put forth a defense of the financial practicalities. We decided to share the following hypothetical cost comparison, weighing a conventional city against the new urban form of the proposed Vertical City:

In favor of conventional suburbs:

The unit cost of building superstructure per square foot favors the single-family suburban home, whether framed in wood (as it is in North and South America, Africa, and Siberia), concrete, or steel over the fireproof high-rise superstructure of the Vertical City.

In favor of the Vertical City:

The cost advantage of suburban home building is more than offset by the Vertical City's savings in such aspects as:

Land: The land area of a Vertical City is a small fraction of that occupied by a conventional city or suburb. When mankind reaches the point of needing every square inch of land for the life-sustaining purposes of growing of food, and when available lands are further lost to drought or rising seawater, this issue will preempt all others.

Infrastructure: Being horizontally compact and vertically stacked, the amount of infrastructure— roads, sidewalks, utilities, and other associated costs—is greatly reduced in the Vertical City.

Building Enclosure: Less exposed exterior surface translates into less cost per individual units.

Equipment and Mechanical Plants: Centralized, energy efficient, and smart, such utilities reduce power consumption and yield lower operating costs.

Operation: Trained personnel operating efficient state-of-the-art equipment results in reduced

486

operational manpower and great energy savings.

Maintenance. The coordinated practice of ongoing preventive maintenance will prolong the useful lives of the equipment and mechanical plants.

Energy Use: The good practices followed in the abovementioned areas—building enclosure, equipment, operations, and maintenance—will result in greater energy savings over the individual suburban home.

Alternate Energy Sources: There are more clean alternative energy options available to the Vertical City, and at a lower unit cost, by virtue of great height, particularly through the benefits afforded by sheer volume and contiguous exterior surfaces.

Clvic Services: Administration of a compact, consolidated community requires fewer civil servants and engenders quicker emergency response.

Automobiles: The cost of private automobile ownership, insurance, parking, and maintenance is avoided.

More Efficient Food Use: Locally available fresh produce and convenient shopping reduce spoilage and waste.

Reduction in Taxes: Efficiencies throughout the Vertical City translate into reduced taxation for residents

维护：持续展开的预防性维护协调工作将延长机械设施、设备的使用寿命。

能源使用：和单一的郊区家庭相比，上述提到的那些建筑外墙、设备、运营和维护方面的良好表现，将大大节约能耗。

替代性能源：有许多更清洁的替代性能源可供垂直城市选择，由于其高度，特别是总体积和连续整体外墙带来的好处，至少能源的单位成本会降低。

城市服务：管理一个统一紧凑的社区需要的市政服务人员更少，其应急反应也会更快。

汽车：私人车辆、保险、停车和维护上面的费用将节省下来。

更有效的食物利用：当地可以获得的新鲜产品和方便的运输减少了损耗与浪费。

减税：对于居民来说，垂直城市中的高效能会给他们带来减税的好处。

Reduced Medical Costs: The long-term health benefits of a community focused on walking, nutritious

医疗费用减少： 步行、营养完善的食物、清洁的能源带来积极健康的生活，也减少医疗方面的需求，而这将为社区带来长期的健康收益。

还有一些金钱无法衡量的无形效益，包括：

当地的就业岗位

可持续性

环境兼容

更少的污染

清洁的空气与水

高质量的生活

非常方便就可以享受各种娱乐、艺术、音乐、其他文化和休闲形式带来的快乐

社会互动

社区感

节约交通时间，有更多时间用于休闲放松

对于个体住户而言，同样的社区维护工作转变为更少的小修小补

安全环境中个人的安全感缓和了压力和焦虑感

认识到我们正在为后代人的福祉，甚至可能是他们的生存承担自身的责任，认识到这一点带来的满足感

unadulterated foods, and clean energy will result in an active, healthy lifestyle and reduced need for medical care.

Intangible benefits, to which monetary sums cannot be directly assigned, include:

Local availability of employment

Sustainability

Environmental compatibility

Less pollution

Cleaner air and water

Improved quality of life

Convenient access to the excitement of varied entertainment, art, music, and other forms of recreational and cultural engagement

Social interaction

Sense of community

More leisure time as a result of shorter commutes

Coordinated community maintenance translating into fewer maintenance chores for individual residents

Sense of personal safety in a secure environment, leading to less stress and anxiety

Satisfaction in the knowledge that responsible action is being taken for the well-being, and perhaps the survival, of future generations

487

NEW URBAN FORM

一个垂直城市的设计方法

Design Approach

最终形成这本书的整个探索历程让我们思考高层可持续生活可能采取的，最适宜的物质形式。依照根本的指导方针（参见第3页，8-8-8-宣言），在提出这些关于新城市模式概要形式的问题后，我们想要探究答案，想要看看这样一种解决方案在一个超高的垂直城市里是否确实可行。在接下来的章节中，各种方法将按部就班、合理有序地呈现出来。每个结论都自然地来自于前面的结论，而且我们确实相信，这些结论将共同导向我们的设计方案。我们从考虑基本的人类需要和居住形态开始，然后从那里前进拓展，超越提升。

The course of inquiry that resulted in this book led us to speculate about the physical form that high-rise sustainable living might optimally assume. Informed by a vision of essential guidelines (see the 8-8-8 Manifesto, page 3), and having asked the questions that led to the formation of a Brief for a New Urban Form, we wanted to explore the answers and to see whether such a solution might actually be possible in the form of a super-tall Vertical City. Several approaches appear on the following pages in a step-by-step rational progression, each conclusion growing naturally from those that came before and leading collectively—and we believe inevitably—to our design proposal. We began by considering elemental human needs and residential patterns and proceeded onward, and upward, from there.

注释

在接下来逐步展开的陈述中，作者提出一种理性的方法，来描绘想象中的垂直城市。那些规则和特定的数值只供参考，并不意味着可以普遍适用。因此如果读者有特定人口统计学数值，建议用当地特定的数据来进行替换。

Note
In the following step-by-step progression, the authors present a rationalized approach for giving form to a hypothetical Vertical City. The formulas and assigned numerical values are intended for reference only and are not meant to be universally applicable. It is therefore recommended that appropriate local data be substituted to represent the reader's specific demographic interests.

基本概念

除了食物和水，人类还需要一个安身之所。对于流浪者而言，它是人们在山边发现的一个洞穴，或者可能是覆盖于树枝上的棕榈叶，亦或是更耐用、更便携的鹿皮帐篷。随着人类定居下来，人们开始需要更坚固的、土木砖石建成的居所。在各种不同的情形中，一个未来共同体的种子萌芽于一个独立的住宅单元，并随着时间的推移在大小、功能和复杂程度上不断增长，这一点今天依然如此。在农村或郊区的场景中，整个演变过程依然是从一个独立的家庭住宅开始的——这是所谓的美国梦——而在人口稠密的城区，所谓的种子就是那些多住户的公寓建筑。当它们位于大城市、位于员工上下班枢纽附近的时候，这样一种集中化的住宅建筑就成了所谓的"卧室社区"。通过一种自然的增长序列，当一个"卧室社区"发展到为其居民提供工作机会时，它就成为了"生活／工作"社区。

目前，还没有一种通用的解决方案可以应用到每一个"生活／工作"社区中去——在一个工薪阶层有权利去其他地方寻找更好工作机会的社

The Basics

In addition to food and water, human beings require shelter for survival. For the nomad, it was a cave discovered on the side of a mountain, or perhaps palm leaves overlaid on branches, or maybe a more durable and transportable buckskin tent. With human settlement came the need for more permanent shelters of mud, logs, bricks, or stone. In each case, the seed of a future community began with a single residential unit and grew over time in size, function, and complexity. The same holds true today. In a rural or suburban setting, the process begins with a single-family dwelling—the so-called American Dream—while in densely populated urban areas, the seed is the multifamily apartment building. A concentration of such residential buildings becomes a bedroom community when located near a larger city and the workforce commutes to jobs. When, through a natural progression of growth, a bedroom community develops to the point of providing employment opportunities for its residents, it becomes a "live/work" community.

There are no one-size-fits-all solutions that are equally applicable to every live/work community, especially in a society where resident breadwinners have the right to seek better job opportunities elsewhere. Likewise, within the original live/work community, various employment enterprises could fail, thereby compelling laid-off employees to seek work in neighboring communities.

Beyond the "bedroom" and "live-work" paradigms are larger, more complex and diverse cities that provide, in addition to employment, other services and amenities. In the case of world-class cities such as New York, Chicago, Shanghai, Beijing, Paris, Frankfurt, and London, it is a question of accommodating not just the local population but also the varied needs of international businesses, visitors, and tourists.

Bedroom Communities

Typical of "bedroom" or "dormitory" communities are the suburban towns and villages that dot large and small cities the world over. Radiating out from the urban hubs are transportation arteries—be they highways, railways, or waterways—that connect neighboring settlements and provide service routes for the conveyance of needed food and goods. Lands bordering the access corridors become prime sites for new bedroom communities.

As the title implies, the bedroom community's primary function is to provide shelter for its inhabitants. In addition to such residential ("R") functions are the corollary nonresidential ("non-R") functions of providing food, energy, security, health care, education, entertainment, houses of worship, and other support systems.

Typically, most of the nonresidential support facilities, such as stores, restaurants, and municipal offices, are located on a single Main Street that serves surrounding streets with their row after row of single-family homes. Introduced here is what we call the "R/Non-R ratio" as a numeric expression of residential buildings in relation to buildings for nonresidential use. This ratio varies greatly from one community to another. In a well-served town, for example, it might be 15:1 (15 housing units to one store or other nonresidential support facility), while in more extreme cases there might be several hundred residential units per store, in which case that one store would be very important.

Conclusion 1: The vertical cities in our case study will include solutions for a bedroom community with an R/Non-R ratio of 20:1 (20 residential units for each nonresidential unit).

Aside from the work required for the operation and upkeep of a traditional bedroom community, employment for

会中尤其如此。类似的，在一个原初的"生活/工作"社区中，有很多提供工作的企业都会失败破产，因而会迫使失业人员在其他临近社区中寻找工作。

在"卧室"和"生活/工作"的社区范式外是一些更大、更复杂、更多样的城区，它们提供工作之外的其他服务和设施。在诸如纽约、芝加哥、上海、北京、巴黎、法兰克福、伦敦等这样的世界性都市中，供应是一个大问题，这不仅仅是因为当地众多的人口，还因为国际商务、参观者和游客提出的诸般需求。

卧室社区

典型的"卧室"或"住宿"社区是一些郊县和村落，它们坐落在世界上大大小小的城市周边。从城市中心辐射出来的是一些交通动脉（高速公路、铁路或水路），连接起临近的定居点，并为定居点所需的食物和货物运输提供服务路线，往返于城市之间的走廊通道，使得这些走廊通道边缘成为新"卧室社区"的最佳区域。

正如标题所指出的，卧室社区的主要功能是为其居民提供庇护所。除了这一住宅功能（"R"），还有其他一些非住宅功能（"非R"），比如提供食物、能源、安全、医疗保健、教育、娱乐、教堂，以及其他一些保障系统。

一般来说，大部分非住宅的保障设施——比如商店、餐厅和市政府办公室——都坐落在唯一的一条城镇大街上，它为周边街区成排成排的独栋楼房提供服务。在这里要介绍的是我们所谓的"R/非R比率"，这是一个反映住宅建筑与非住宅用建筑关系的数值。在不同的社区，这个比率变化很大。比如说，在一个完善的小镇上，这个比率可能是15：1（15个住宅单元对应一家商店或其他非住宅保障设施），而在更极端的案例中，这个比率可能会达到数百个住宅单元对应一家商店，此时，这样的一家商店就将变得非常重要。

结论1：我们对垂直城市的研究将包括为一个卧室社区提供解决方案，我们决定将其"R/非R比率"确定为20：1（20个住宅单元对应1个非住宅单元）。

除了一个传统卧室社区需要的维持和运营工

作外，大部分工作者的工作岗位通常都是在起初产生这些卧室社区的母体城市中找到的。朝来夕往，连同日复一日固定不变的交通车流（流动可以获得当地没有的货物和服务），是与交通相关污染的罪魁祸首。

在纽约这样的大都会，可以确定居住着有1万到4万人的卧室社区已经拓展进入了三州地区，这些地区包括纽约州、康乃提克州和新泽西州，以及数量稍少的宾夕法尼亚州。纽约州的塔里敦（1万人口），康乃提克州的特兰波尔（3.5万人口），新泽西州的蒙特克莱尔（3.9万人口）都是典型的范例。

拥有超过6万人口的更大城市，比如康乃提克州的格林威治（6.2万人口），纽约州的白原市（5.7万人口），都是不太符合一般标准的卧室社区，它们可能还处于转变为自给自足的生活/工作城市的早期阶段。

结论2：就我们的研究而言，我们将垂直城市卧室社区的人口上限定为6万到15万（平均10万）人。

生活/工作社区

一个卧室社区转变为一个自给自足的生活/工作社区，需要的不只是为大多数工薪阶层居民提供工作机会，也包括相应的在非住宅保障措施上的提升，在充足的商业办公建筑、教育、社会和文化机构、文化设施、洁净的工业消耗，及所有必要的公共设施上实质性物质能力的提升。在这里，各个社区间R/非R的比率依然是变化的，一般来说是3：1或4：1。

结论3：就我们研究的生活/工作垂直城市而言，我们确定了一个R/非R比率为4：1的社区（4个住宅单元对应1个非住宅保障设施）。

曼哈顿的常住人口接近160万，而每天的流动人口则有150万，我们可能需要注意这里的R/非R比率是颠倒性的1：4。在这里，住宅单元的平衡被城市无数的办公楼、零售和商业区、影院、餐厅、博物馆、医疗中心、教育机构及其他非住宅用途的保障设施严重打破了（这些为组成纽约市五个行政区，也为来自更广阔三州地区，甚至三州之外地区的通勤上班族提供了就业机会）。吸引远近游客的旅游景点也在很大程度上为非住

the majority of breadwinners is usually found in the parent city that spawned such bedroom communities in the first place. Morning and evening commutation, together with a steady flow of chore traffic throughout the day—running errands and securing goods and services not available locally—are major contributors to transportation-related pollution.

In metropolitan New York, clearly defined bedroom communities of 10,000 to 40,000 people extend far into the tristate area, comprising New York State, Connecticut, and New Jersey. (Pennsylvania is included to a lesser degree.) Tarrytown, NY (population 10,000), Trumbull, CT (35,000), and Montclair, NJ (39,000) are representative examples.

Larger cities with populations of +60,000, such as Greenwich, CT (62,000), and White Plains, NY (57,000), are borderline bedroom communities that may be in the early stages of becoming self-sufficient live/work cities.

Conclusion 2: For our case study, we targeted a maximum population of 60,000 to 150,000 (100,000 average) for the bedroom community of the Vertical City.

Live/Work Communities

Transformation of a bedroom community into a self-sufficient live/work community requires not only employment opportunities for most wage-earning residents, but a corresponding increase in nonresidential support facilities and the physical capability to adequately house commercial offices; educational, social, and cultural institutions; cultural facilities, and clean light-industrial usages, plus all the necessary infrastructure. Here again, the R/non-R ratio varies from community to community, typically 3:1 or 4:1.

Conclusion 3: For our case study live/work community in the Vertical City, we determined an R/Non-R ratio of 4:1 (4 residential units for every nonresidential support facility).

It may be worth noting that the Borough of Manhattan, with a resident population of approximately 1.6 million and a daily transient population of another 1.5 million, has an inverted R/Non-R ratio of 1:4. Here, the balance of residential units is far outweighed by the city's numerous office towers, retail and commercial spaces, theaters, restaurants, museums, medical centers, educational institutions, and other nonresidential facilities, all of which provide employment for the five boroughs that make up New York City, as well as for commuters from a large part of the tristate region, and beyond. Attractions for tourists from near and far also contribute to the hefty nonresidential uses.

To establish the size of our case study Vertical City, we started with a list of 30,000 American cities, extracting those with a population above 100,000 for the top 300 American cities.

Among these, 22 percent have populations of 300,000 or more, including New York, Los Angeles, Chicago, Houston, San Francisco, and Boston. Each of these major metropolitan centers has been created by complex forces, histories, and other dynamics that make the cities atypical and which defy replication. They are cited here for informational purposes only.

We determined that 15 percent of the top cities with populations between 200,000 and 300,000 are less relevant to our study, and therefore we purposely do not include any examples.

Among the cities that are most germane to our study are those with populations between 100,000 and 200,000 (representing 63 percent of the top 300 U.S. cities). Included are Yonkers, Providence, Syracuse, and New Haven, among others. We concluded that the 100,000–200,000 range is a reasonable population to use in our brief.

Conclusion 4: For our case study we targeted a maximum population of 150,000 to 250,000 (200,000 average) for the live/work community of the Vertical City.

Residential area per person

The following are approximate residential building floor areas per person (derived by dividing published total residential building areas by the population of the particular city):

Manhattan: 700 square feet [70 square meters] per person

Providence, RI: 2,000 square feet [185 square meters] per person

Oklahoma City: 2,500 square feet [230 square meters] per person

From published data concerning the size of average residential units and also the average number of persons per household, we arrived at a rough guide for the average number of residential square feet per person in the following countries:

United States:	884 sq. ft. [82 m²] per person
Australia:	646 sq. ft. [60 m²] per person

宅设施的利用做出了贡献。

为了确定我们现在所研究的垂直城市的规模，我们研究了3万座美国城市，并抽样了前300座人口超过10万的美国城市。

在这些城市中，有22%人口达到或超过30万，包括纽约、洛杉矶、芝加哥、休斯顿、旧金山和波士顿。这些大都市的中心都由一些复杂的力量、历史和动因创造而成，这些东西使得各个城市与众不同，难以复制。在这里引述它们只是为了提供一些信息。

我们确定大城市中那些15%人口在20万到30万之间的与我们的研究并不相干，因此我们打算不将这类例子囊括在我们的研究中。

与我们的研究最贴近的城市，是人口在10万到20万之间（意味着美国大城市前300名中的63%）的那些。其中包括扬克斯、普罗维登斯、锡拉丘兹、纽黑文和其他一些城市。我们得出结论认为，就研究结果来说，人口在10万到20万之间是一个适当的范围。

结论4：就我们所研究的垂直城市中生活/工作社区而言，我们将15万—25万（平均20万）人当作我们的人口上限。

人均居住面积

下面是大致的人均住宅建筑面积（用特定城市已经公布的总居住面积除以城市人口可以得出这些数据）：

曼哈顿：700平方英尺（70平方米）每人

普罗维登斯，罗德岛州：2000平方英尺（185平方米）每人

俄克拉荷马式，俄克拉荷马州：2500平方英尺（230平方米）每人

从已经公布的关于平均住宅单元大小以及每户家庭平均人口的数据中，我们对下列国家的人均住宅面积有了大概的了解：

美国：　884平方英尺（82平方米）每人

澳大利亚：　646平方英尺（60平方米）每人

英国：　510平方英尺（47平方米）每人

法国：　480平方英尺（45平方米）每人

日本： 340平方英尺（32平方米）每人
中国： 180平方英尺（17平方米）每人
韩国： 175平方英尺（16平方米）每人

平均 500平方英尺（46平方米）每人

结论5：面对如此大的差异，我们决定给读者提供如下两种选择：

　　1. 采用平均500平方英尺（46平方米）的方案

　　2. 采用美国和澳大利亚土地空间分配额较高的方案（综合平均下来为750平方英尺〈70平方米〉每人），我们相信提出要求越高的空间挑战，我们越能解决小的空间要求。正在讨论中的区域只是住宅区，并不包括下面会讨论的非住宅区域。

两种研究中的建筑面积

1a.
具有10万人口的卧室社区其R/非R比率为20：1（20个住宅单元对应1个非住宅保障设施；见上述的结论1和结论2）

住宅建筑面积：
10万人口×750平方英尺（70平方米）每人=总计：7500万平方英尺（700万平方米）

非住宅建筑面积：
0.05（1/20）×7500万平方英尺（700万平方米）=总计：375万平方英尺（34.8万平方米）

总计为：7870万平方英尺（735万平方米）

1b.
人口为10万，面积为每人500平方英尺（46平方米）的卧室社区：

总计为：5250万平方英尺（483万平方米）

人口在20万，R/非R比率在4：1（4个住宅单元对应1个非住宅保障设施；见上述的结论3和结论4）

住宅建筑面积：

England:	510 sq. ft. [47 m²] per person
France:	480 sq. ft. [45 m²] per person
Japan:	340 sq. ft. [32 m²] per person
China:	180 sq. ft. [17 m²] per person
Korea:	175 sq. ft. [16 m²] per person
Average	**500 sq. ft. [46 m²] per person**

Conclusion 5: Confronted with such a broad differential, we decided to offer the reader the following two choices:

1. Use the average of 500 square feet [46 m² per person], or

2. Use the higher space allotments characteristic of the United States and Australia (a combined average of 750 square feet [70 square meters] per person) in the belief that by addressing the higher and therefore more demanding spatial challenges, we would also resolve smaller spatial requirements. The area in question is exclusively residential and does not include the nonresidential areas discussed below.

Floor area for the two case studies

1a.
Bedroom community with an average population of 100,000 and an R/Non-R ratio of 20:1 (20 residential units for each non-residential support facility; see Conclusions 1 and 2 above)

Residential building area:
100,000 population x 750 square feet [70 square meters] per person = TOTAL: 75 million square feet [7 million square meters]

Nonresidential building area:
0.05 (1/20) x 75 million square feet [7 million square meters] = TOTAL: 3.75 million square feet [348,000 square meters]

COMBINED TOTAL: 78.75 million square feet [7.35 million square meters]

1b.
Bedroom community with a population of 100,000 @ 500 square feet [46 square meters] per person:

COMBINED TOTAL: 52.5 million square feet [4.83 million square meters]

2. Live/Work community with a population of 200,000 and an R/Non-R ratio of 4:1 (4 residential units for each nonresidential support facility; see Conclusions 3 and 4 above)

493

Residential building area:

2a.

200,000 population x 750 square feet [70 square meters] per person = TOTAL: 150 million square feet [13.9 million square meters]

Nonresidential building area:

0.25 (1/4) x 150 million square feet [13.9 million square meters] = 37.5 million square feet [3.5 million square meters]

COMBINED TOTAL: 187.5 million square feet [17.4 million square meters]

2b.

The identical population of 200,000 but at 500 square feet per person:

COMBINED TOTAL: 125 million square feet [11.6 million square meters]

Sustainable Planning and Design Principles

As stated in the introduction to this book, the authors wish to test whether a Vertical City construct offers a viable solution to curtailing haphazard and unsustainable urban expansion and suburbanization, in both developed countries and developing countries around the world. It is a truism that to find solutions one must first define the problems. What better means than to consult recognized authorities in the field? We found two sources particularly valuable. *City Building: Nine Planning Principles for the Twenty-First Century,* by John Lund Kriken with Philip Enquist and Richard Rapaport—this is a superb book that addresses city planning from the perspectives of sustainability, accessibility, diversity, open space, compatibility, incentive, adaptability, density, and identity. *The HOK Guidebook to Sustainable Design* by Sandra Mendler, AIA, William Odell, AIA, and Mary Ann Lazarus, AIA also covers planning issues but more specifically addresses building design, and includes a very accessible guide to "Ten Simple Things You Can Do."

Drawing from these two sources with great respect and appreciation, but with no attempt to apply their guidelines in full, we endeavored to test our Vertical City hypotheses. Our objectives and the steps in this feasibility study are outlined below.

Objectives of the Sustainable Vertical City

The principal objectives of sustainable Vertical Cities are to

2a.

20万人口×750平方英尺（70平方米）每人=总计：15000万平方英尺（1390万平方米）

非住宅建筑面积：

0.25（1/4）×15000万平方英尺（1390万平方米）=总计：3750平方英尺（350万平方米）

总计为：18750万平方英尺（1740万平方米）

2b.

同样是20万人，但人均面积为500平方英尺：

总计为：12500万平方英尺（1160万平方米）

可持续性规划设计原则

正如这本书的前言所说的，作者想要检验是否存在一种垂直城市构想，能为减少随意、不可持续的城市扩张及市郊化提供一个可靠的解决方案——这种情况不仅发生在发达国家，也包括遍布全球的发展中国家。大家都知道，要找到解决方案，首先必须确定问题。还有什么比向这一领域公认的权威咨询更好的呢？我们发现有两种解决方案特别有价值：约翰·隆德·希里更和菲利普·恩奎斯特以及理查德·拉帕波特撰写了《城市建设：21世纪城市规划的九条原则》。这部非凡的作品从可持续性、亲近性、多元性、开放空间、兼容性、刺激性、适应性、密度和身份认同的角度阐述城市规划问题。桑德拉·F·门德勒（美国建筑师协会）和威廉·奥德尔（美国建筑师协会）撰写的《HOK可持续设计指南》也涉及了一些规划问题，但却更明确地阐述了建筑设计问题，而且还包括一个非常容易上手的、关于"十件你能够做的简单事情"的指南。

这两种解决方案非常令人满意、备受赞赏，但我们没有打算完全采纳他们提出的指导方针，我们努力想要检验我们的垂直城市假设。关于这种可行性研究的目标和步骤将在下面的章节中得以勾勒。

可持续性垂直城市的目标

可持续性垂直城市的主要目标是，最大限度地降低对不可再生资源的使用，并创造出一个健康、富有成效的生活工作场所。可持续性规划开始于一种坚定的信念，要尽可能地保护自然，包括珍贵的水资源，因而最大限度降低人类居住产生的消极影响。在我们的能力范围内，我们最迫切需要做的是保护在持续不断城镇化进程中被蚕食的耕地。根据当前的速度可以预测，所有可耕种土地将在约300年内消失殆尽！保护耕地因而必然成为我们当前最迫切的事。

规划进程延伸到通过提高内部环境的质量，为居民的安宁福祉而设计建筑的性能。通过采用对环境负责的产品和措施，这一规划在建设期间依然持续着。在一栋建筑的生命中，合格的、训练有素的工作人员需要操作机械厂房和设备，采用预防维护的方式优化建筑的长期性能。一栋建筑的整个生命周期都必须被考虑在内，将建设材料和建造方法结合起来，使之易于报废、拆除，易于对其上层架构、机械装置、设备和组合材料进行再利用。尽管卧室社区和生活/工作社区之间有很多共同之处，但先找出二者之间一些关键的差异将是有益的。

对所谓垂直城市主要特征的一般描述

为了我们对卧室或生活/工作垂直城市的研究能够达到上述的主要目标，我们设定了高度不同的一系列高密度高层楼房（主要是为了居住的目的），这些楼房从一个基座上升起，在6-10层的建筑中安置了一批非住宅设施。整栋建筑位于一个没有汽车、适于步行的裙楼区之上，其最大半径为1/4英里，而最大的步行直径为1/2英里。这样的平台适合于全世界各种地形和场所，不管是在空旷的草原、沼泽、林地、沙漠、水下，还是在一个现成的都市环境中。

为了方便参考，我们已经选择了著名的曼哈顿城市网络作为衡量单位，它具有如下特点：

一个街区的衡量是从一条街道的中心线到另一条街道的中心线=260英尺（79米）

minimize the use of nonrenewable resources, reduce waste in the built environment, and create a healthy, productive place to live and work. Sustainable planning starts with the firm resolution to preserve as much of Nature as possible, including precious water resources, and thereby minimizing the negative impact of human habitation. We urgently need to do all in our power to preserve arable land, which is being lost to ever-increasing urbanization. There are those who project that at the current rate, all cultivatable land will be gone in approximately 300 years! Preservation must therefore become our highest priority.

The planning process extends into the design of building performance for the protection and welfare of occupants by enhancing the quality of interior environments. It continues during construction by means of environmentally responsible products and practices. During the life of a building, qualified trained personnel are needed to operate mechanical plants and equipment, optimizing long-term performance with preventive maintenance. The full life cycle of a building must be taken into account, incorporating construction materials and erection methods that allow for ease of decommissioning, repurposing, demolishing, and recycling of the superstructure, its mechanical plant, equipment, and component materials. While there are many common aspects shared by bedroom communities and live/work communities, perhaps it will be instructive to first identify critical differences.

General Description of the Major Features of Proposed Vertical Cities

In order for our case study bedroom or live/work Vertical Cities to achieve the abovementioned primary objectives, we propose a series of high-density high-rise towers (principally for residential purposes) of varying heights. The towers rise from a podium that houses an array of nonresidential facilities in 6- to 10-story buildings. The whole is positioned on a car-free pedestrian-friendly platform of maximum quarter-mile [400-meter] radius and maximum walkable diameter of a half mile [800 meters]. The platform is suitable for a variety of terrains and locations around the world, whether on open grasslands, marshes, woodlands, desert, in water, or in an existing urban context.

For ease of reference, we have selected the well-known Manhattan city grid as a unit of measure, with the following characteristics:

One block measured from centerline of street to centerline of street = 260 feet [79 meters]

Street width = 60 feet [18 meters]

One-block buildable plot = 200 square feet [18.5 m²]

Gross area of one square block (260 ft. x 260 ft.) = 67,600 square feet [6,280 m²]

Net area of one square-block buildable plot (200 ft. x 200 ft.) = 40,000 square feet [3,716 m²]

Maximum walkable site dimensions: quarter-mile [400-meters] radius, or approximately 5 blocks, or

half-mile [800 meters] diameter (approximately 10 blocks)

Gross area of maximum walkable site = 10 x 10 blocks, or 100 square blocks

Commonalities of Bedroom and Live/Work Communities

Following are the many common aspects of bedroom and live/work communities. Also noted are differences regarding site location and site size limitations.

Site Size Determinants

Bedroom Community

The proposed Vertical City bedroom community with a population of 100,000 and an R/Non-R ratio of 20:1 would consist primarily of eight to ten high-rise residential towers. The towers would occupy approximately 24–30 city blocks; podium buildings supporting nonresidential functions would occupy approximately 12–16 square blocks, for a total build-out of 36–46 blocks, well shy of the 100 square blocks given above as the generally accepted maximum limit for comfortable traveling by foot.

Live/Work Community

The live/work community, with its larger population and an R/Non-R ratio of 4:1, requires a larger site than the bedroom community. In our case study, the maximum population is determined by the amount of nonresidential building area that can be housed in the 6 to 10 story podium buildings, all within a maximum walkable site of 10 blocks by 10 blocks, or 100 square blocks. Once the podium building area is fixed, that result is multiplied by 4 (the R/Non-R ratio of 4:1) to determine the maximum residential population that the podium buildings can support. In our case study, the result is a population of approximately 200,000, requiring from 16 to 20 high-rise towers.

街道宽=60英尺（18米）

一个街区可供建设的土地=200平方英尺（18.5平方米）

一个方形街区的总面积（260英尺×260英尺）=67600平方英尺（6280平方米）

一个方形街区内可供建设的净面积（200英尺×200英尺）=40000平方英尺（3716平方米）

最大程度上适合步行的区域长度=半径1/4英里400米的区域或接近5个街区，或直径1/2英里800米的区域（接近10个街区）

最大程度上适合区域的总面积=10×10个街区或100个方形街区

卧室和生活/工作社区的共同点

下面是卧室和生活/工作社区的许多共同之处。同样记录了与区域位置，区域大小限制相关的差异。

区域面积决定了

卧室社区

我们提出的人口10万，R/非R比率为20:1的垂直城市卧室社区首先要有4—6座高层住宅楼。这些住宅楼将占据约24—30个城市街区，提供非住宅功能的裙楼将占据增建的总计12—16个街区中约36—46个方形街区。远未达到上述普遍接受的、可舒适步行的最大限度，即100个方形街区。

生活/工作社区

生活/工作社区有更多的人口，其R/非R比率为4:1，相比于卧室社区，这需要更大的区域。在我们的研究中，最大限度的人口数将由非住宅建筑区域的大小来决定，这些区域可以被置于一群6—10层的裙楼中，在最大为10个街区乘以10个街区或100个方形街区的步行区域内。一旦确定了裙楼区域，将其乘以4得出的结果（R/非R比率为4:1）就可以确定裙楼能够保障的最大居民数量。在我们的研究中得到的结果是接近2万左右的人口数，这需要16—20座高层楼房。

区域位置

卧室社区

新卧室社区存在的首要理由（尤其是在发展中国家）在于要安置民工潮，这些人移居到城市，想要寻找就业机会，同时提高生活标准；相应的，这些卧室社区的位置要非常靠近一座现有的城市，并且可以便捷地进入已有的公共运输系统。

生活/工作社区

相反，一座新的生活/工作社区显然更加独立，因为它有能力雇佣其工薪阶层。另一方面，如果它是从原先没有汽车的卧室社区发展起来的，那么它将非常依赖于步行道网络，而不是依赖公共运输系统以换乘到其他社区进行工作。公共运输系统的运输潜能，乃是为依附于生活/工作社区的临近社区中那些寻找工作的居民提供服务。

位置选择

关键的是，在预期的各种位置中选择一个特定的位置，我们需要了解现存的水域、集水区和野生动物栖息地。首先，对可耕地的入侵必须尽量减少或得到制止。在具体情形中，如果这样一个社区位于高产农田区旁边，那就必须留出足够的面积用于耕种新鲜的农产品——这些农产品是供给新社区居民的。

食物的需求从小麦、大米、玉米这类主食到水果和蔬菜，设想新鲜农产品的完整供应都可以在本地种植获得，这是不现实的。然而，本地耕种应该尽可能达到最大限度以减少浪费。研究已经表明，在美国和欧洲，超过一半以上的食品变质腐坏，有些在从农场到餐桌的过程中就已经有损失了。专家给出的报告称，一般购物车中超过15%的食物被浪费，其中足足有三分之二是在长期储存的过程中浪费的。每天直接获取当地种植的农场品，将有助于减少这些无谓的浪费。

由于本地种植的水果是季节性的，农田首先用于蔬菜耕种，蔬菜的生长期较短，即使是在最

Site Location

Bedroom Community

The primary reason for the existence of new bedroom communities, particularly in developing countries, is to house the influx of rural workers who have migrated to the city in search of employment and improved standards of living. Consequently, such bedroom communities need to be located in relatively close proximity to an existing city and have ready access to public transit.

Live/Work Community

By contrast, a new live/work community is, by definition, more independent, since it has the capacity to employ its wage earners. If started from scratch, it can be located some distance from existing cities. If, on the other hand, it evolves out of a former car-free bedroom community, there will already be greater reliance on pedestrian networks and less dependence on public transportation to commute outside the community for work. There is still the possibility for job-seeking residents to commute to neighboring communities.

Site Selection

Critically, in selecting specific sites within would-be locations, there needs to be an awareness of existing watersheds, catchment areas, and wildlife habitats. Above all, encroachment on arable farmland is to be minimized or avoided. In instances where such a community is located adjacent to productive farmland, sufficient acreage needs to be set aside for the cultivation of fresh produce for the occupants of the new community.

Since food requirements include such staples as wheat, rice, and corn, as well as fruits and vegetables, it is unrealistic to assume that a complete supply of fresh produce can be grown locally. Nevertheless, local cultivation should be pursued to the maximum degree possible in order to reduce waste. Studies have shown that in the United States and Europe, up to one-half of all food products are spoiled, damaged, or otherwise lost en route from the farm to the dinner table. Experts also report that up to 15 percent of food in the typical shopping cart is wasted, fully two-thirds from prolonged storage. Direct access to locally grown crops would offer fresh produce daily to residents and help to reduce such needless waste.

Since locally grown fruit is seasonal, cropland is devoted primarily to vegetables, which have short growing seasons and, with even basic greenhousing, can be cultivated year-round. The challenge is to arrive at a realistic estimate for the amount of dedicated cropland required.

In the United States and Europe, the minimum amount of arable land needed to feed one person (including land for livestock grazing) is 1.25 acres [.5 hectares]. The world average is .6 acres [.25 hectares]; the bare minimum is .17 acres [.07 hectares]. For demonstration purposes we used the world average of .6 acres [.25 hectares] as a starting point, and then picked a somewhat arbitrary portion, say one-third of the world average, or .2 acres [.08 hectares] per person for the year-round growing of vegetables. With favorable climatic conditions, improved farming technology, and elimination of the waste typically associated with shipping and storage, it may be possible to further reduce the land required to .01 hectares, [.025 acres] per person. Therefore, for a bedroom community of 100,000 people, 2,500 acres [1,000 hectares] would be needed.

Similarly, for the live/work community with a population of 200,000, approximately 5,000 acres [2,000 hectares] would be required. The existence of such green cushions of land would also serve as the boundaries for adjacent future communities.

Site Planning

In our case study, all buildings rise from a car-free podium that is either on grade level or raised, depending on subsoil conditions. Utilities are located in the overhead space of the grade level for ease of secure access and maintenance. Automobiles and service vehicles are relegated to a dedicated lower level. To house our smaller bedroom community (population from 60,000 to 100,000) or the more complex live/work communities (population of up to 250,000), the high-density community consists of ultra-tall residential towers that range in height from 80 to 400 stories, or a mile high [1.6 kilometers]. They are complemented by additional towers for mixed-use or office occupancy. The nucleus of the scheme is a concentrated tower cluster surrounding a central outdoor plaza that can be partially or fully roofed.

A tree-lined street system is fully accessible to pedestrians and bicycles but off-limits to vehicular traffic. The site is organized in short, easily walkable blocks that accommodate a wide variety of uses. Small parks shorten and enliven pedestrian routes, as they enhance the environment and inhabitants' overall quality of life. Colonnades, overhangs, and ground-floor setbacks provide protection from the weather. For the larger live/work community, selected streets are protected from inclement weather by mechanically operated skylights. Parks and the natural Great Outdoors are never more than 5 blocks away. If there is sufficient need for a local

普通的温室中，它们也可以全年种植。我们需要的是一个切实的评估，评估总共需要多少专属的农田。

在美国和欧洲，养活一个人所需的最低限度的耕地面积（包括放牧需要的土地）是1.25英亩（0.5公顷）。世界的平均值是0.6英亩（0.25公顷），最小值为0.17英亩（0.7公顷）。为了证明，我们将世界的平均值0.6英亩（0.25公顷）当作起点，然后任意选择一个区域，把每人应有世界平均值的三分之一或0.2英亩（0.08公顷）用于一年的蔬菜种植。上佳的气候条件，先进的农业技术，通过运输和储藏杜绝浪费现象。所有这些都有可能进一步将所需土地减少到0.01公顷或0.025英亩每人。因此，对于一个有10万人口的卧室社区而言，需要2500英亩（1000公顷）的土地。

类似的，对于人口在20万的生活/工作社区而言，需要的土地将接近5000英亩（2000公顷）。这样一些土地绿色缓冲区的存在也将作为划分未来临近社区的边界。

位置规划

在我们的研究中，所有建筑都从一个没有汽车的裙楼区上拔地而起，要么在地平面上，要么在升高的平面上，这取决于泥土的条件。为了便于取用和维护，各种设备都位于水平面上的空间中。汽车和服务车辆被置于一个专属的、更低的水平面上。为了安置我们较小的卧室社区（人口从6万到10万）或更复杂庞大的生活/工作社区（人口会增加到25万），高密度的社区会有超高的住宅楼，高度从80到400层或者说有1英里（1.6千米）高。它们配备有其他混合用途或办公用的附属楼房。规划的核心是一个集中的楼层群落，它环绕着一个户外中心广场，这个广场可以部分或全部装上顶盖。

绿树成荫的街道系统完全对行人和自行车开放，但却对车流有所限制。简而言之，我们要选择的位置都是便于步行、能够提供一系列广泛用途的街区。用小公园缩短和点缀人行道，同时也可以改善环境和居民的生活质量。柱廊、挑檐和地面上的缩退型阶梯帮助人们躲避恶劣天气。对于更大的生活/工作社区而言，特定的街道可以通过机械控制的天窗帮助人们躲避恶劣天气。公园和户外自然区之间的距离不会超过五个街区。如果对于当地的运输系统有充分的需求，那么它可以被容纳进裙楼区下方的汽车服务层，或者上方的人行道层，沿着周边一条特定的街道排列。

水资源及其保持

接近三分之一的世界人口忍受着缺乏清洁饮用水之苦。即使是现在，美国西南部和东南部的很大一片区域依然处于干旱状况下，这让人想起上世纪30年代的沙尘暴毁坏了美国"产粮区"中的农作物。水资源缺乏的情况也进一步对航运造成了广泛的影响，即使是在汹涌的密西西比河上也是如此。

在一个可持续性世界中，水的重要性从未如此明显。相应地，屋顶和平整地面上暴雨降水的径流也应该被收集起来，并被当作生物处理系统中的一个自然结果。这包括了一系列含有绿色植物的过滤池、池塘湿地，他们可以对迳流进行处理和提纯，并沉淀排出物使之变成适合灌溉的中水，用于建筑项目景观绿化中的草地、灌木、树丛和其它植被。类似地，中水也可用于给自然地下蓄水层补充水分，可用于垂直城市中的抽水马桶，以及附近的鱼塘和农场。整个垂直城市中都在使用各种节水装置、器具、装备、调节器和系统。

建筑设计

可持续性建筑的设计实践或多或少已经成为全球范围内负责任建筑师和工程师的规范。英国的建筑研究所环境评估法(BREEAM)，德国的可持续建筑协会(DGNB)，日本的建筑物环境效率综合评估系统(CASPEE)，澳大利亚的绿星标准，以及美国的绿色建筑认证体系(LEED)，都为评估、分类和可持续认证设定了目标。所有这些认证标准，尤其是LEED和BREEAM标准已经超出了国界，影响到亚洲和俄罗斯等其他地方。而且，发达国家和发展中国家一样，都在不断通过建筑法规和准则强制推行这种规范。然而，至少在这里，勾勒出可持续性建筑设计的确切目标是很重要的。

transit system, it can be incorporated into the vehicular service level below the podium, or located above, on the pedestrian level, along one of the dedicated peripheral streets.

Water Resources and Conservation

Approximately one-third of the world's population suffers from the lack of a reliable source of clean drinking water. Even at this writing, a large part of the southwest and southeast United States is undergoing drought conditions reminiscent of the 1930s Dust Bowl. Low water conditions further pose widespread threats to shipping navigation, even on the mighty Mississippi River.

The importance of water in a sustainable world has never been so apparent. Consequently, storm water runoff from landscaped roofs and paved surfaces should be collected and processed as a matter of course in bio-treatment systems. This would involve a series of filtering pools containing green plants and wetlands to process and purify the runoff and to receive the discharge until it becomes suitable gray water for irrigation of the project's landscaping. Similarly, gray water can be used for recharging the natural groundwater aquifer, for toilet flushing, and for use in the nearby fish ponds and farms. Throughout the Vertical City, water-saving fixtures, appliances, equipment, controls, and systems are employed.

Building Design

The practice of sustainable building design has become, to one degree or another, the norm for responsible architects and engineers the world over. Goals are set for the assessment, rating, and sustainability certification by the British BREEAM (Building Research Establishment Environmental Assessment Method); the German DGNB (Deutsche Gesellschaft für Nachhaltiges Bauen/German Society for Sustainable Building), CASBEE (Comprehensive Assessment System for Built Environment Efficiency) in Japan; Green Star in Australia, and LEED (Leadership in Energy and Environmental Design) in the United States. Such certification standards, particularly LEED and BREEAM, have begun to spread into Asia, Russia, and beyond. Certification is increasingly mandated in building regulations and codes in developed and developing nations alike. Nevertheless, it is important that the salient goals for sustainable building design at least be outlined here.

We begin by designing the building skin, or exterior enclosure, to be efficiently insulated against the elements, to admit optimum natural light and ventilation, and to harness solar or wind-powered energy. Internally, we strive to ensure a healthy, comfortable, and productive environment for the occupants via efficient use of energy by the lighting, heating, ventilation, and air-conditioning systems. We seek out other clean-energy sources to reduce reliance on fossil fuels, and avoid specifying products, fixtures, or equipment whose manufacture follows the theory of planned obsolescence.

We also adamantly support the idea that malfunctioning motors, "black box" and other faulty components of otherwise functioning equipment, should be easily replaced by simply unplugging the defective part and plugging in the replacement. We wholeheartedly embrace the revolutionary idea, as advocated by William McDonough and Michael Braungart in their must-read book *Cradle to Cradle: Remaking the Way We Make Things*, that we seek out products whose components, "after their useful life, provide nourishment for something new—either as 'biological nutrients' that safely re-enter the environment or as 'technical nutrients' that circulate within closed-loop industrial cycles without being 'downcycled' into low-grade uses (as is the case of most current 'recyclables')". In other words, waste in all of its many manifestations is to be avoided.

Building Construction

As for the construction of the buildings themselves, we advocate using construction methods that are the most energy efficient and pollutant free, whose composition and components are nontoxic and are from renewable sources, and whose manufacturing process does the least harm to the environment. Also, we encourage the selection of locally available materials whenever possible.

Smart Building Occupancy

We strive to improve efficient energy use by allowing occupants individual control over their environment. In addition, our goal is to optimize the energy efficiency of building systems and services (including elevator use, delivery of goods, transportation needs, and systems operation) in direct response to individual needs through real-time response to information gathered and stored in portable personal communications systems.

A building's wastewater is to be recycled as gray water; in severe conditions, advanced technologies can be employed to convert gray water into potable water. Because individual apartment units are stacked vertically in high-rise towers, we

我们通过设计建筑表层或外层开始，使之能够有效地抵御恶劣天气，获得最佳的自然光照和空气流通，并利用太阳能或风能。在建筑内部，通过照明、加热、通风和空调系统，对能源实现有效利用，为住户营造一种健康、舒适、高效的环境。我们寻找其他清洁的能源以减少对石油的依赖，并拒绝其加工制造依照一种"计划报废"理论而进行的特定产品、装置和设备。

我们也坚定地支持下面一种观点：不能正常工作的发动机、"黑盒子"和其他工作装置的故障组件应当很容易通过拔去故障部分，插入替代部分而更替。我们由衷地支持威廉·麦克唐纳和迈克尔·布朗嘉特在他们的必读书《从摇篮到摇篮：重塑我们的生产方式》中提出的革命性观点，即：我们想要找到这样一些产品，它们的组件"在有效期之后，能为一些新的东西提供营养——要么是非常安全地作为重新进入环境的'生物养分'，要么是在封闭的工业循环中继续流转而不是'衰退'到低层次应用的'技术养分'"（就像现在大多数"可回收物"一样）。换句话说，浪费在所有的地方都销声匿迹了。

建筑构造

对于建筑自身的构造而言，我们主张使用最节能最无污染的方法，建筑的成分和组件都是无毒的，来自可再生能源，其制造的过程也不会破坏环境。而且，只要有可能，我们就鼓励选择当地可用的材料。

聪明的建筑居住者

我们想允许居住者个人掌控他们的环境，来提高能源的使用率。另外，我们的目标是，通过对便携式个人交流系统中收集和储存的信息做出实时的反馈，以直接回应个人需要的方式，优化建筑系统和服务设施的能源效率（包括电梯的使用、货物传递、运输需求和系统操作）。

我们期望回收建筑的废水以获得中水，而且在情况严峻时，可以利用先进的技术将中水转化为饮用水。由于个人的公寓单元在高层建筑中垂直堆叠起来，所以我们建议包含一个个分离的系统将被浪费的食物收集起来，作为当地农庄中的肥料。相似的，现在是时候探究一些方法，以清除人类粪便中的毒素，使之也能成为当地农民的肥料，就像数个世纪以来在很多亚洲国家所发生的那样。再说一次，必不可少的潜在前提是杜绝浪费。

recommend incorporating a separate system to collect food waste for composting in local farmlands. Similarly, the time has come to explore ways to remove toxins from night soil so that it also becomes usable as fertilizer by local farmers, as it has in many Asian countries for centuries. Again, the essential underlying premise is doing away with waste.

1/4 mile = 5 blocks @260' ea. [80 meters]
1/4英里（1.6公里）= 5块@260'（80米）

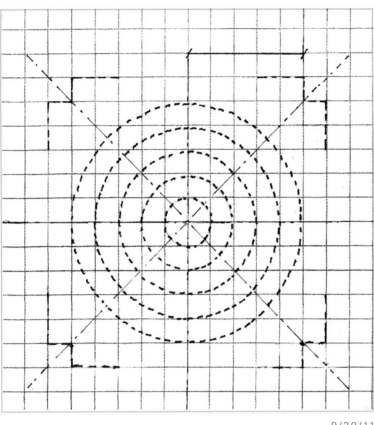

9/20/11

Illus. 1: Maximum size (as applied to a new self-sustaining live/work community)
图1：最大尺寸（适用于一个新的自我维持生活/工作的新社区）

关于垂直城市的推荐方案

一般描述

许多社区功能，比如住房（包括宾馆和酒店式公寓）、教室、实验室和办公室，很大部分都是重复的，因而理论上说适合安置在高层楼房中（高楼的功能）。不过，还有一些在操作、空间或形式上有特殊要求的功能，比如影院、博物馆、礼堂、展览和会议厅、体育场甚至是清洁产业，适合放在裙楼区宽阔的水平面上（裙楼功能），在高楼自身的基底之外。正是在这些各种各样的裙楼功能中，会发现多样性、意外惊喜和同步性。另外，作者们相信需要将自然地平面连接起来——为了锻炼、片刻远离人群的休憩以及重新与自然联系在一起。

区域的规模

方便行人的区域最大半径为1/4英里（0.4公里），或者说接近5个纽约的社区，每个社区从边界到中心线有260英尺（79米），边界是宽60英尺（18米）的街道。这一区域的直径是1/2英里（0.8公里），或接近10个城市街区。假定舒适的步行速度是4.5英尺（1.35米）每秒，一个社区可以用近1分钟穿过，而半径1/4英里的5个社区需要5分钟。人们可以在10分钟内走过1/2英里（0.8公里），穿过城市的10个街区。每个方形街区可用于建设的土地面积是长200英尺（60米）的正方形，或者40000平方英尺（3720平方米）。

Proposed Schemes for a Vertical City

General Description

Many community functions, such as residential (including hotels and service apartments), classrooms, laboratories, and offices, are largely repetitive and therefore ideally suited for housing in high-rise towers (tower functions). Yet there are other functions with special operational, spatial, or form requirements—for example, theaters, museums, auditoria, exhibition and convention halls, sports arenas, or even clean industries—that are better suited for positioning in the broad-based horizontal spaces of the podium (podium functions), beyond the footprints of the towers themselves. It is also within the cornucopia of such podium functions that diversity, serendipity, and synchronicity are to be found. Additionally, the authors believe that there needs to be a reconnection with the natural ground plane—for recreation, moments of quietude away from the masses, and a revitalizing bond with nature.

Site Dimensions

The pedestrian-friendly site has a maximum radius of a quarter mile [.4 kilometers] or approximately five New York City blocks, each block measuring 260 feet [79 meters] from the centerlines of the bounding streets (60 feet [18 meters] wide). The diameter of the site is one half mile [.8 kilometers] or approximately 10 city blocks. Assuming a comfortable walking speed of 4.5 feet [1.35 meters] per second, a block can be traversed in approximately one minute, and the quarter-mile (5-block) radius in 5 minutes. One could walk the half-mile [.8-kilometer] site, covering 10 city blocks edge to edge, in 10 minutes. The buildable lot area is 200 feet [60 meters] square or 40,000 square feet [3,720 square meters] per square block.

Podium: Building Platform

In order to create a car-free pedestrian-friendly city, we advocate the creation of a dedicated raised platform whose upper level is reserved exclusively for pedestrian and bicycle use; a lower level accommodates vehicular use. Additionally, depending on the terrain, ground, and subsoil conditions, the podium platform can be variously positioned, as indicated in Illus. 2, below:

A. Above grade
B. On grade (most suitable for insertion in existing under developed urban fabric)
C. On uneven or hilly terrain
D. Offshore or over flooded land, whether the result of global warming, megastorms, or tidal surges. In cases where flooded conditions become permanent, the lower (water) level would be for water taxi or service barge use.

裙楼区或建筑平台

为了创造一个没有汽车，对行人友好的城市，我们建议建造一个专用的凸起的平台，其较高层保留供人行道或自行车使用；较低层则供汽车使用。另外，根据地形、地面和地下的具体情况，裙楼区平面如图2中所示可以被置于不同的地方：

A. 水平面上方
B. 在水平面上（最适合嵌入到现存落后的城市建筑物中）
C. 在不平坦或丘陵地带
D. 离岸的或在河滩地之上，不管是因为全球变暖的热浪还是因为潮水上涨带来的结果。在洪水的环境已经变得无法改变的地方，较低的（水）层面将会用于水上出租车和驳船服务。

Podium: Building Platform Maximum Size
裙楼：建筑平台的最大尺寸

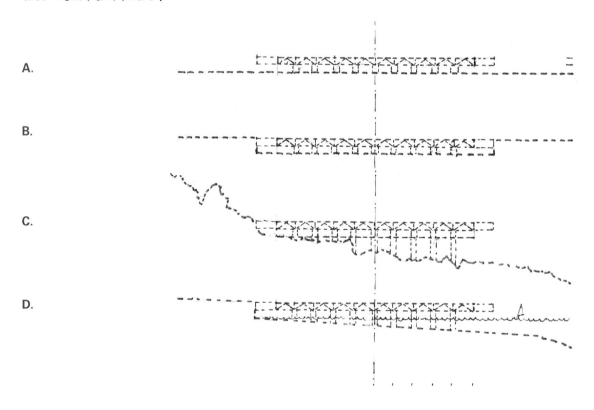

A.

B.

C.

D.

Illus. 2: Maximum size (as applied to a new self-sustaining live/work community)
图2：最大尺寸（适用于一个新的自我维持生活/工作的新社区）

裙楼的较低层

停车与服务的道路

在一个自给自足、对行人友好的社区，汽车并不是必须的。不过，现代的推动力——无论是消费者的心理、独立性、便利性、以拥有汽车为荣、自我等等——使得人们不可能完全摒弃汽车。不过，希望政府的政策，比如提高油价，或人们最终认识到与粗心大意、不合格的司机共用道路必定有风险，这些可以抑制私人使用汽车。刚刚出现的分享汽车的观念将进一步帮助减少私人汽车拥有量。对一辆汽车没有迫切的需要；趋向更小汽车、混合动力或电力汽车，机械式停车；以及中央区域停车库作为快速交通枢纽的主意，这些将进一步减少停车的需要（位于所建议垂直城市的地下较低层）。运转递送和服务也被限于垂直城市的地下较低层，以便使地平面没有障碍物。

裙楼的中间层

服务和设备

挖开街道和人行道往往是为了修理或修复掩埋的管道电缆，为了抑制这种需要，人们建议将这些设备放在地下汽车层与上方主要人行道层之间的中间层。当需要的时候，一个轻轨交通系统也可以被置于这一中间层中。

裙楼的顶/主层

人行道、自行车和高楼大堂的入口，以及独立的裙楼

垂直城市裙楼的顶层或主层位于水平面上或在水平面之上，这要取决于土壤条件，它只能严格用于人行道和自行车。唯一的例外是公共交通系统，依照需要，它可以坐落于垂直城市的外部边缘。

Podium Lower Level

Parking and Service Roads

In a self-sustaining pedestrian-friendly community, cars are not necessary. Nevertheless, modern impulses—consumer psyche, independence, convenience, pride of ownership, or ego, for example—make it unlikely that cars can be completely eliminated. It is hoped, however, that governmental policy, escalating fuel costs, or the eventual recognition of the inherent danger of sharing the road with unqualified or careless drivers would discourage private use. The budding idea of car sharing will help to further reduce individual ownership. The infrequent need for an automobile; the tendency toward smaller cars, hybrid or electric; mechanized parking; and the wisdom of a central regional parking garage to serve rapid-transit hubs will further reduce the need for parking (located in the lower level of the proposed Vertical City). Deliveries and services are also limited to the lower level in order to keep the ground level unobstructed.

Podium Intermediate Level

Services and Utilities

To eliminate the need for digging up the streets and sidewalks in order to make repairs or modifications to conventionally buried service pipes, cables, or ducts, it is proposed that such utilities be located in the intermediate level between the vehicular level below and the main pedestrian level above. When required, a light-rail transportation system can also be located on the intermediate level.

Podium Top/Main Level

Pedestrians, bicycles and entrance to tower lobbies and individual podium buildings

The top or main level of the Vertical City platform is located on or above grade, depending on soil conditions, and is strictly for use by pedestrians and bicycles. The only exception is a public transportation system, which, depending on need, might be located along the Vertical City's outer edge.

Tower Schemes

High-rise and ultra-high-rise buildings (exceeding 1,000 feet [305 meters] in height) are the predominant building type in our case study Vertical City. Their primary use is residential, but they also encompass offices, hotels, and mixed use. Because the requirement for views, natural light, and ventilation are more critical in apartments, their floor configurations are relatively more shallow than in offices. As illustrated below, our towers are configured with three or four such arrays, or "wings," around a central ropeless express-elevator core. The typical office tower also consists of a cluster of wings around a central express-elevator core, but the wings are deeper.

Irrespective of their usage, tall buildings confront three major obstacles: wind, elevators, and budget.

Wind

In 1956, when Frank Lloyd Wright envisioned a mile-high [1.6-kilometer] building, construction technology was not yct up to the challenge. Occupants on the upper floors would have experienced excessive sway, inherent in skyscrapers constructed with structural steel frames, the predominant material available for skyscrapers of that era.

More recently, with new building technologies and procedures, expert structural engineers have indicated that there are no longer any engineering limitations imposed by construction materials or methodologies on super-tall towers, a mile [1.6 kilometers] or higher. Freed from such constraints, we have set one mile (5,280 feet) [1.6 kilometers] at 13 feet [3.9 meters] floor-to-floor, or 400 stories as the hypothetical height of our case study towers. Admittedly, the choice of a mile-high [1.6-kilometer] limit is an arbitrary one, yet it seems to poetically realize the foresight (some would say the audacity) of the great American architect who first suggested it some 56 years ago, intriguing architects with the notion ever since.

We do not take lightly that the slenderness of the tower's wings and their large surface areas will add to the already significant challenge of finding sustainable structural solutions for the ultra-tall buildings. We maintain that any structural problems will be more than offset by the energy generated by the building's high-performance photovoltaic façade panels. Additional energy is generated by wind turbines located at the junctures of the wings and the central core, toward which the sail-like wings will channel air flow.

The challenges are exacerbated by the fact that winds are smoother and more constant at high altitudes, whereas

高楼规划

高层和超高层建筑（高度超过1000英尺）是我们研究垂直城市时主要的建筑类型。它们的主要用途是居住，但它们也可用作办公室、宾馆和多种混合用途的需求。因为对风景、自然光和通风的要求在公寓中更为重要，因此它们的楼层布局相对来说比办公室要稀松些。正如后面的图例所示，我们的高楼排成三或四列或"侧翼"，环绕着中心一个无绳快速电梯核心。典型的办公楼也有一组环绕一个中心运输电梯核心的"侧翼"，但这些侧翼要更宽。

不考虑它们的用途，高层建筑面临三个主要障碍：风、电梯和预算。

风

1956年，当弗兰克·劳埃德·赖特预期一英里高（1.6千米）的建筑时候，建设技术尚未做好准备接受挑战。高层的居民将体验到剧烈的摇晃，这是由钢结构建成的摩天大楼的固有特征之一，也是那个时代的摩天大楼能够获得的主要材料。

最近，伴随新的建筑技术和方法，行内的结构工程师已经指出，建设的材料或方法不再是"为什么摩天大楼不能达到1英里（1.6千米）或更高"这类工程上的原因或限制。从这种局限中挣脱出来，我们已经将1英里（5280英尺或1.6千米）分为每层13英尺（3.9米）或总计406层，并将其当作我们所研究高楼的设定高度。需要承认的是，选择对1英里（1.6千米）高建筑的限制，这是较为随意的，不过这似乎很好地实现了美国伟大建筑师的远见（有些人可能会说这是很大胆鲁莽的），他在56年前就第一次提出了这个看法，令当时和现在的建筑师都颇感兴趣。

高楼细长的侧翼及其巨大的表面积，会增加现有的、为超高建筑寻找可持续性结构方案的重大挑战，我们并未对此掉以轻心。我们坚持，任何结构问题都会被建筑高性能光伏外层板产生的能量大大抵消。位于侧翼和中央核心连接处的风力涡轮机将产生出更多的能量，而像帆一样的侧翼则会引导气流的运动。

挑战被如下事实所加重：在高海拔地区，风更平稳更恒定，而在低海拔地区，风的流动会被不规则的地形、山水景致、人造物体所搅乱。较低位置上不规则、走走停停的风能让建筑结构恢复原样，但恒定的高海拔风将使高层建筑处于持续的压力下。

为了克服剧烈摇晃，我们建议通过"连接物"将相邻的高楼连起来，由此可以将临近的建筑连结、固定成一个牢靠的整体。主要的连接物位于空中大堂的同一层，以创造出一系列抬高架空的社区广场（将在下文描述）。根据马来西亚吉隆坡双子塔连结第41层和第42层双层空中天桥的工程和设计细节判断，这样一些连接物在工程上的困难挑战是真实的，但也是可以完成的。

电梯

我们在这里介绍超高层楼房设计上的一个重大突破，即由无绳快速电梯提供的独立中央循环核心。随着预期的建筑越来越高，垂直运输的科学一直努力试图跟上步伐。今天，电梯从理论上说可以以60英尺（18米）每秒的速度运行到最高2000英尺（600米）或151层上的办公楼。更加普通的电梯都能在60秒内到达1855英尺（556米）的高度或142层每层13英尺（3.9米）的地方。很明显的是，电梯的最大提升高度受到其井道电机的限制，这个电机要拉动轿厢和货物的混合重量，以及沉重的起重钢索（"绳子"）。楼层越高，需要的钢索就越多，重量就越增加。这些实际的限制要求乘客在140层的地方走出电梯，在"空中大堂"换乘另一趟电梯，从而到达更高的楼层。

垂直运输研究表明，一般用户不会忍受多于一次的这种换乘。就像在一座典型高楼的首层大厅时，人们被要求乘坐低速、中速、高速等不同类型的电梯到达他们的目的地，类似的过程也发生在到达一个空中大堂时。幸运的是，无绳电梯的不断研究和进步使我们越来越接近打破这些技术上的制约。就像其名字所表明的那样，这种先进的系统没有沉重的钢索，因而理论上说没有高度限制。无绳电梯的另一个重大益处在于，多个轿厢可以在同一个竖井中往返运行，也节省了大量的空间。我们的目的是设计一个适合无绳快速电梯的中央核心，这种电梯能够上升到大楼的最高处，为到达间距接近100层左右的不同空中大

wind flow is disrupted by irregular terrain, landscaping, or man-made objects at lower altitudes. The irregular, stop-and-go effect of lower-level winds allows building structures to recover, but constant high-altitude winds subject ultra-tall buildings to continuous stress.

To help overcome excessive sway, we propose to link adjacent towers via "connectors," thereby interlocking and bracing neighboring buildings into a rigid whole. The main connectors are positioned at the same levels as the sky lobbies to create together a series of elevated village squares (described below). We have been informed that the engineering challenges of such connectors would be substantial but achievable.

Elevators

We introduce here a major breakthrough in ultra-high-rise tower design with an independent central circulation core served by ropeless express elevators. As taller and taller buildings are envisioned, the science of vertical transportation has striven to keep pace. Today, elevators exist that can theoretically travel at speeds of 60 feet [18 meters] per second to a maximum height of 2,000 feet [600 meters], or 151 office floors. More common are elevators that can ascend 1,855 feet [556 meters], or approximately 140 floors at 13 feet [3.9 meters] floor-to-floor in 60 seconds. The problem, simply put, is that an elevator's maximum rise is limited by the capacity of its hoistway motor to carry the combined weight of the cab, its cargo, and the heavy steel hoisting cables ("ropes"). The higher the floor, the more cable required and the greater the weight. Such practical limits require any passengers going beyond 140 floors to transfer at a "sky lobby" to a second bank of elevators serving the upper levels.

Vertical transportation studies show that regular users will tolerate no more than one such transfer. Just as people in the ground-floor lobby of a typical high-rise are required to take the low, mid-range, or high bank of elevators to reach their destination, a similar process takes place on reaching a sky lobby. Fortunately, ongoing research and development of ropeless elevators brings us closer to breaking the technical shackles. As the name implies, this advanced system is free of heavy steel cables and thus it theoretically has no height limitations. Another significant benefit of ropeless elevators is that multiple cabs can travel in the same shaft, freeing up considerable amounts of space. Furthermore, ropeless elevators can generate electricity in the down mode, resulting in substantial energy savings. Our intention is to have a central core dedicated to ropeless express elevators that rise the full height of the towers, providing access to sky lobbies located approximately every 100 floors. On arrival in the sky lobby of choice, passengers go up or down to their destination floors by transferring to local low, middle, or high banks of traditional elevators located within "wings" of the various towers.

The central service core will also contain the following:

Bunker-like construction for added protection against a potential repeat 9/11-type attack

Redundant means of emergency egress

Dedicated fire-fighting staging areas at vertical intervals, as may be required by the relevant authorities

Receiving and distribution stations for deliveries of goods to individual apartments

Budget

In order to encourage unencumbered discussions of theoretical possibility, it is often necessary to put aside the practicalities of cost. We are confronted here with just such a situation. It is beyond the scope of our theoretical case study to arrive at construction costs, particularly when so many factors of detail remain unknown. Rather, we pose our case study as an argument to invite discussion and perhaps to inspire readers with the requisite knowledge and resources to work up a budget exercise to their own satisfaction and maybe even to share with others.

We believe the total construction cost (including site preparation, and installation of infrastructure such as roads, sidewalks, street furniture, utilities and services, but excluding land cost) for a Vertical City compares favorably with a conventional new suburban subdivision consisting of single-family homes on individual building lots. It is self-evident that land cost weighs heavily in favor of the Vertical City approach, as it occupies only a fraction—less than 1 percent—of the land required for a typical single-family subdivision with the same population.

Some maintain that the operational and maintenance costs between the vertical and horizontal paradigms provide another endorsement for the Vertical City. Add to this the differences in energy use for daily transportation (and its impact on the environment of a typical car-oriented suburb compared to that of a car-free community), and the outcome, once again, will appear to favor the Vertical City.

堂提供入口。在到达所选择的空中大堂后，乘客可以换乘那里位于不同楼层"侧翼"中的传统低速、中速或高速电梯，然后或上或下到达他们的目的楼层。

中央服务中心将包括如下要素：
类似堡垒的结构，为防止911事件的再次发生提供额外的保护
多种方式的紧急出口
垂直间隔上专用的消防工作区，可能是相关权威部门所需要的
接受和分发传送来的物品给特定的私人公寓

预算

为了鼓励对理论可能性不受限制的讨论，经常要暂不考虑实际的费用。我们在这里遇到的就是这样一种情形。工程造价问题超出了我们研究的理论范围，特别是当如此多的细节因素依然不可知的时候。相反地，我们将研究当作一种引起争议的讨论，这些讨论既有正面的也有反面的，而且可能用一些必要的知识和资源激励读者，让他们制定满足自己需求的预算编制，甚至还可能与其他人分享这种预算编制。

我们相信一个垂直城市所需的全部建设费用（包括场地准备像道路、人行道、街道设施、公共设备公共服务之类基础设施的建设，但排除了土地费用），与传统意义上新郊区中分散的住宅小区（在一个独立的建筑单元中包含独栋家庭住宅）相比要划算很多。很显然，在支持垂直城市方案的考虑中，土地费用占据了很大的比重，因为相比于安置同样数量人口的典型独栋家庭住宅小区，前者的土地需求量只是后者的一小部分，少于1%。

有些人认为，垂直与水平建筑样式之间的使用和维护成本，也为垂直城市增加了另一个筹码。另外，每天交通（一个典型的以汽车为主导的郊区，同一个没有汽车的社区对环境的不同影响）耗费的能源差异以及获得的结果，也再一次有利于垂直城市。

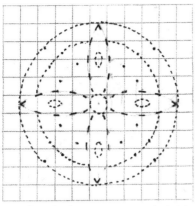

3A Quad-wing apartment skylobby
四翼公寓空中大堂层

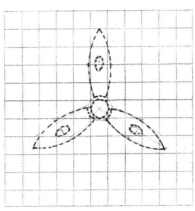

3D Tri-wing apartment typical level
三翼公寓标准层

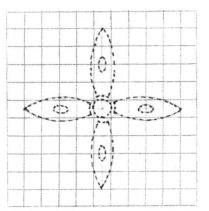

3B Quad-wing apartment typical level
四翼公寓标准层

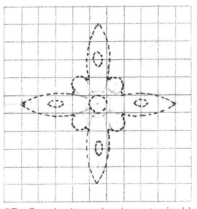

3E Quad-wing mixed-use typical level
四翼多功能标准层

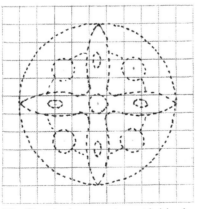

3C Quad-wing apartment lobby level
四翼公寓大厅层

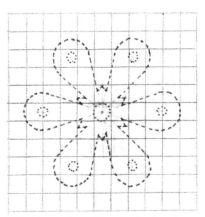

3F Hex-wing office lobby level
六翼办公标准层

Illus. 3: Tower plan configurations
图3：高楼的计划配置

TOWER PLAN CONFIGURATIONS

No attempt has been made to design the typical apartments, although drawing inspiration from Le Corbusier, it is envisioned that each apartment will occupy two floors, with public circulation corridors on alternate floors leading to the apartment's front entrance and main level (including a living room, dining room, and kitchen). The upper level of each duplex will have floor-through bedrooms, affording cross ventilation from front to back. Each apartment will have a private garden terrace, with access to natural light and ventilation for the back rooms.

Depth of the Apartment Wings
There has been no attempt to establish the depth (exterior wall to public corridor) in the apartment wings. Depending on market conditions, it can vary from 25 feet [7.5 meters] to 35 feet [10.5 meters]. The same applies to hotels.

The diagrams at left serve primarily for comparison between the shallower apartment wings and the deeper office wings, where the distance from exterior wall to public corridor typically measures 25 feet [7.5 meters] in Europe and 50 feet [15 meters] in New York City *(Illustration 3F)*.

For the purpose of our case study, each residential wing is assigned a length of 200 feet [60 meters] and a total width of 65 feet [19.5 meters]. Together with the core, each is assigned the following approximate gross area per floor:

Tri-wing plan: 43,200 square feet [4,015 square meters] per floor

Quad-wingplan:56,200squarefeet[5,223squaremeters] per floor

Juxtaposition of the Tower Wings
While one- or two-story homes in many U.S. suburban subdivisions are separated by driveways or planted beds as narrow as 10 to 15 feet [3 to 4.5 meters], typical 3- to 5-floor brownstone apartment buildings in Manhattan abut each other to form rowhouses that are separated from facing rowhouses by 60-foot [18-meter] -wide streets. Later, as taller apartment buildings were constructed, that same 60-foot [18-meter] separation was maintained.

Conclusion
60 feet [18 meters] is set as the absolute minimum separation between apartment towers.

As for the positioning of the tower wings relative to each other, every effort should be made to protect the panoramic

塔楼平面配置

尽管人们从勒·柯布西耶获得灵感，认为每个公寓都将有两层，每层都有流通走廊通向公寓的前门入口和主层（包括一个起居室，餐厅和厨房），但并没有人想要设计出这种典型的标准公寓。每个双层公寓的上层都会有占据整个楼层面积的卧室，使得前后的空气流通起来。从侧翼包住公寓的将是一个私人的花园露台，可以为后面的每个房间带来自然光和流通空气。

公寓侧翼的深度
人们并不能确定公寓侧翼的深度（从外墙到公共走廊）。根据市场条件，它可以从25英尺（7.5米）到35英尺（10.5米）不等（宾馆中也是如此）。

左侧的图表主要是为了在较浅的公寓侧翼和较深的办公楼侧翼之间进行比较，就后者而言，在欧洲，从外墙到公共走廊之间的距离一般是25英尺（7.5米），而在纽约市这一数值是50英尺（15米）（图3F）。

为了我们研究的目的，每个住宅侧翼都被设定为长200英尺（60米）宽65英尺（19.5米）。与核心区加起来，每层的总面积都被设定为接近如下的数值：

三翼规划：43200平方英尺（4015平方米）每层

四翼规划：56200平方英尺（5223平方米）每层

高楼侧翼的并置
在美国郊区的住宅社区中，一层或两层的家庭住户被车道或植被分开，它们一般较为狭窄，从10到15英尺（3到4.5米）不等；而曼哈顿一般3-5层的褐色公寓楼相互毗邻，形成行列式房屋，通过60英尺（18米）宽的街道，这种行列式房屋同面对面的组屋区分开来。后来，随着越来越高的公寓建筑建设起来，这60英尺（18米）的间距被保留了下来。

结论
60英尺（18米）被当作公寓楼之间绝对的最小间距。

对于在位置上相互关联影响的高楼侧翼来说，所有的努力都应该致力于保护每个公寓的全景视域。基于这种考虑，三翼结构毫无疑问是第一选择，紧随其后的就是四翼结构。

当市场行情发挥作用时，一栋公寓楼的较低层可以被转为办公之用。当办公室被嵌入四个侧翼的连接点时，使用面积会得到增加，因为在办公建筑中，自然光和流通空气乃是首要的关注点，而非景致风光。

不同办公楼的侧翼或者说"花瓣"可以置于相互之间较为接近的位置，就好像它们都是从中心的"主干"中抽离出来的一样。由于它们从窗户到核心的距离较长，办公楼的花瓣比它对应的住宅部分要宽。

实用性

下面是一些可能的关于高楼/裙楼构造的建议。一个卧室社区的构造相当的简单直接，所有的住宅功能都在高楼中，而非住宅的保障功能则位于裙楼及高楼延伸的空中大堂中。

生活/工作社区的构造更为复杂。根据市场行情，尽管所有的住宅功能也都包含在高楼中，高楼还是有其他一些功能，例如，宾馆或办公室，因此需要创造出许多混合用途的大楼，另外还有其他一些单一功能的办公楼或公寓楼。同样地，生活/工作社区的基座也更为复杂，它不只有社区住房部分需要的非住宅保障设施，也有很多能够提供工作机会的商业、工业、教育、市政、娱乐、轻工业和其他需要的功能。

裙楼的建筑用地在一个260英尺（79米）街区组成的网格内规划设计，在扣除道路之后，只剩下一块4万平方英尺（3700平方米）的建筑用地。

view from each apartment. The tri-wing configuration is unquestionably the first choice in this regard, followed closely by the quad-wing.

When dictated by market conditions, the lower portions of an apartment building can be adapted for offices. Usable area can be increased by the insertion of office pods at the points of intersection of the four wings, since natural light and ventilation, not views, are the primary concern in office buildings.

The wings, or "petals," of the different office towers can be positioned relatively close to each other as they branch out from the central "stem." In consequence of their deeper window-to-core dimension, office-tower petals are broader than their residential counterparts.

Applications

Following are suggestions for possible tower/podium compositions. The composition of a bedroom community is rather straightforward in that all of the residential functions are in the towers, and the nonresidential support functions are located in the podium and in the extended sky lobbies of the towers.

The composition of the live/work community is more complex. Here too all residential functions are contained in the towers, but as market conditions dictate, there may be other tower functions as well, for example, hotels or offices. Likewise, the podium of the live/work community is more complex in that it not only contains the nonresidential support facilities required by the bedroom portion of the community, but also a substantial amount of job-producing commercial, institutional, educational, civic, entertainment, light industrial, and other functions as required.

The podium's buildable lots are organized within a grid of 260-foot [24-meter] square blocks, which, after deducting for roads, yields a buildable area of 40,000 square feet [3,700 square meters].

BEDROOM COMMUNITY

Specific concerns governing bedroom communities and podiums have been discussed elsewhere in this book. For simplicity, we now limit ourselves to representative examples with the following briefs:

Basic examples:

Population:	100,000
Area per person:	750 s/f [70 m^2]
Residential building area:	75 million s/f [7 million m^2]
Residential/Nonresidential ratio:	20:1
Nonresidential support area:	3.75 million s/f [348,000 m^2]

Alternate examples:

Population:	100,000
Area per person:	500 s/f [46.5 m^2]
Residential building area:	50 million s/f [4.6 million m^2]
Residential/Nonresidential ratio:	20:1
Nonresidential support area:	2.5 million s/f [232,250 m^2]

BEDROOM COMMUNITY PODIUM

Buildable lot area of square block:	40,000 s/f [3,700 m^2]
Podium, average building height:	6 stories
Total buildable area per block:	240,000 s/f [22,300 m^2]

Nonresidential support area:

Basic example:	3.75 million s/f [348,000 m^2]
Alternate example:	2.5 million s/f [232,250 m^2]

Square blocks required:

Basic example:	15
Alternate example:	10.4

卧室社区

关于卧室社区的一些具体问题已经在本书其他地方讨论过了。为了简单易懂，在这里我们仅限于有代表性的范例，它具有如下一些要求：

基本范例：

人口：	10万
人均面积：	750平方英尺（70平方米）
住宅建筑面积：	7500万平方英尺（690万平方米）
住宅/非住宅比率：	20：1
非住宅配套区面积：	375万平方英尺（34.8万平方米）

替代范例：

人口：	10万
人均面积：	500平方英尺（46.5平方米）
住宅建筑面积：	5000万平方英尺（460万平方米）
住宅/非住宅比率：	20：1
非住宅配套区面积：	250万平方英尺（23.225万平方米）

卧室社区裙楼

广场街区的可用面积：	4万平方英尺（3700平方米）
裙楼，平均楼层建筑高度：	6层
每个街区的可用面积总量：	24万平方英尺（2.23万平方米）

非住宅配套区：

基本范例：	375万平方英尺（34.8万平方米）
替代范例：	250万平方英尺（23.225万平方米）

需要的广场街区：

基本范例：	15
替代范例：	10.4

TRI-WING SCHEME/BEDROOM COMMUNITY
三翼方案/卧室社区

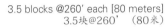

3.5 blocks @260' each [80 meters]
3.5块@260'（80米）

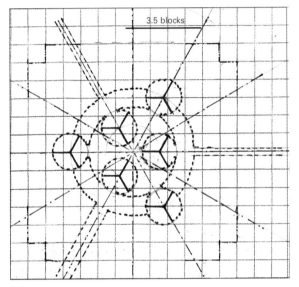

Illus. 4: Site plan
图4: 总图

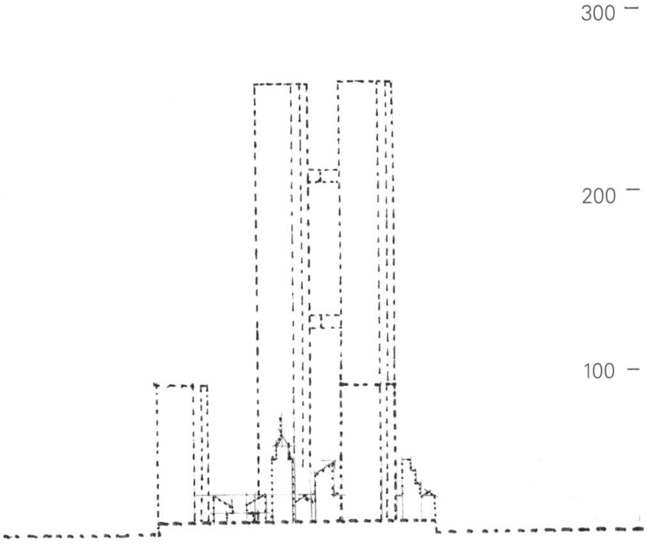

Illus. 5: Elevation
图5: 立面图

基本范例		
人均面积：	750平方英尺（70平方米）	
住宅建筑面积：	7500万平方英尺（690万平方米）	
每层的楼面面积：	4.32万平方英尺（4000平方米）	
需要的层数：	1736层	
需要的大楼数量：	6座	
大楼的高度：	240—360层	

替代范例		
人均面积：	500平方英尺（46.5平方米）	
住宅建筑面积：	5000万平方英尺（460万平方米）	
每层的楼面面积：	4.32万平方英尺（4000平方米）	
需要的层数：	1157层	
需要的大楼数量：	6座	
大楼的高度：	180—240层	

Basic example

Area per person:	750 s/f [70 m2]
Residential building area:	75 million s/f [6.9 million m2]
Tower area per floor:	43,200 s/f [4,000 m2]
Number of floors required:	1,736
Number of towers required:	6
Heights of towers:	240 to 360 stories

Alternate example

Area per person:	500 s/f [46.5 m2]
Residential building area:	50 million s/f [4.6 million m2]
Tower area per floor:	43,200 s/f [4,000 m2]
Number of floors required:	1,157
Number of towers required:	6
Heights of towers:	180 to 240 stories

QUAD-WING SCHEME/BEDROOM COMMUNITY
四翼方案/卧室社区

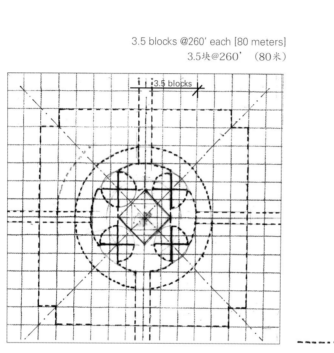

3.5 blocks @260' each [80 meters]
3.5块@260'（80米）

Illus. 6: Site plan
图6: 总图

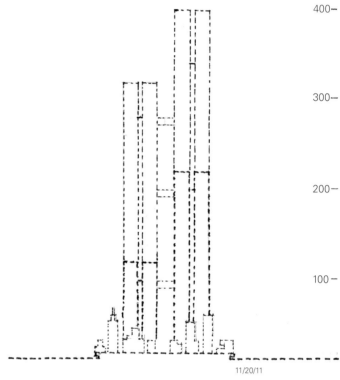

11/20/11

Illus. 7: Elevation
图7: 立面图

Basic example

Area per person:	750 s/f [70 m²]
Residential building area:	75 million s/f [6.9 million m²]
Tower area per floor:	56,200 s/f [5,220 m²]
Number of floors required:	1,334
Number of towers required:	4
Heights of towers:	320 to 380 stories

Alternate example

Area per person:	500 s/f [46.5 m²]
Residential building area:	50 million s/f [4.6 million m²]
Tower area per floor:	56,200 s/f [5,220 m²]
Number of floors required:	890
Number of towers required:	4
Heights of towers:	180 to 320 stories

基本范例

人均面积:	750平方英尺（70平方米）
住宅建筑面积:	7500万平方英尺（690万平方米）
每层的楼面面积:	5.62万平方英尺（5220平方米）
需要的层数:	1334层
需要的大楼数量:	4座
大楼的高度:	320—380层

替代范例

人均面积:	500平方英尺（46.5平方米）
住宅建筑面积:	5000万平方英尺（460万平方米）
每层的楼面面积:	5.62万平方英尺（5220平方米）
需要的层数:	890层
需要的大楼数量:	4座
大楼的高度:	180—320层

生活/工作社区

之前关于生活/工作社区的讨论涉及不同的人口数量和人均面积，最终得出不同的建筑面积总量。在这里我们仅限于有代表性的范例，这些范例有20万人口，还有如下一些要求：

基本范例：

人均面积：	750平方英尺（70平方米）
住宅建筑面积：	1.5亿平方英尺（1400万平方米）
住宅/非住宅比率：	4:1
非住宅配套区面积：	3750万平方英尺（350万平方米）

替代范例：

人均面积：	500平方英尺（46.5平方米）
住宅建筑面积：	1亿平方英尺（930万平方米）
住宅/非住宅比率：	4:1
非住宅配套区面积：	2500万平方英尺（230万平方米）

生活/工作社区裙楼

广场街区的可用面积：	4万平方英尺（3720平方米）
裙楼的平均楼层建筑高度：	10层
每个街区的建筑面积：	40万平方英尺（3.716万平方米）

非住宅配套区：

基本范例：	3750万平方英尺（350万平方米）
替代范例：	2500万平方英尺（230万平方米）

需要的广场街区：

基本范例：	95
替代范例：	63

LIVE/WORK COMMUNITY

Previous discussions of live/work communities involved different population totals and areas per person, resulting in different building areas. Here we limit ourselves to represen- tative examples, each with a population of 200,000 and the following briefs:

Basic examples:

Area per person:	750 s/f [70 m2]
Residential building area:	150 million s/f [14 million m2]
Residential/Nonresidential ratio:	4:1
Nonresidential support area:	37.5millions/f[3.5millionm2]

Alternate examples:

Area per person:	500 s/f [46.5 m2]
Residential building area:	100 million s/f [9.3 million m2]
Residential/Nonresidential ratio:	4:1
Nonresidential support area:	25 million s/f [2.3 million m2]

LIVE/WORK COMMUNITY PODIUM

Buildable lot area of square block:	40,000 s/f [3,720 m2]
Podium, average building height:	10 stories
Total buildable area per block:	400,000 s/f [37,160 m2]

Nonresidential support area:

Basic example:	37.5millions/f[3.5millionm2]
Alternate example:	25 million s/f [2.3 million m2]

Square blocks required:

Basic example:	95
Alternate example:	63

TRI-WING SCHEME/LIVE-WORK COMMUNITY

三翼方案/生活工作社区

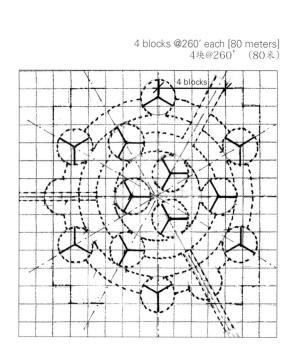

4 blocks @260' each [80 meters]
4块@260' （80米）

4 blocks

400 –
300 –
200 –
100 –

Illus. 8: Site plan
图8: 总图

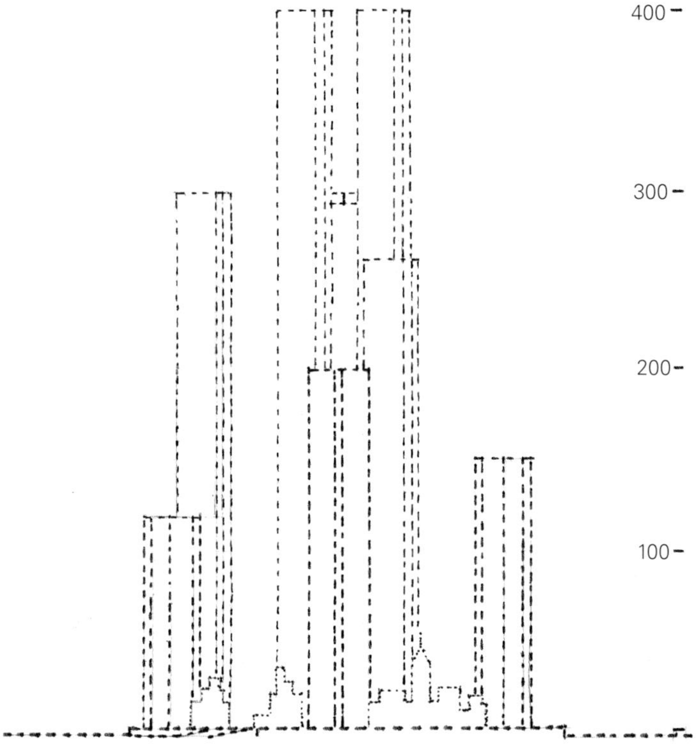

Illus. 9: Elevation
图9: 立面图

Basic example

Area per person:	750 s/f [70 m²]
Residential building area:	150 million s/f [14 million m²]
Tower area per floor:	43,200 s/f [4,000 m²]
Number of floors required:	3,500
Number of towers required:	12
Heights of towers:	200 to 400 stories

基本范例

人均面积：	750平方英尺（70平方米）
住宅建筑面积：	1.5亿平方英尺（1400万平方米）
每层的楼面面积：	4.32万平方英尺（4000平方米）
需要的层数：	3500层
需要的大楼数量：	12座
大楼的高度：	200－400层

Alternate example

Area per person:	500 s/f [46.5 m²]
Residential building area:	100 million s/f [9.3 million m²]
Tower area per floor:	43,200 s/f [4,015 m²]
Number of floors required:	2,300
Number of towers required:	12
Heights of towers:	180 to 300 stories

替代范例

人均面积：	500平方英尺（46.5平方米）
住宅建筑面积：	1亿平方英尺（930万平方米）
每层的楼面面积：	4.32万平方英尺（4015平方米）
需要的层数：	2300层
需要的大楼数量：	12座
大楼的高度：	180－300层

QUAD-WING SCHEME/LIVE-WORK COMMUNITY
四翼方案/生活工作社区

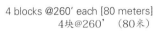

4 blocks @260' each [80 meters]
4块@260' （80米）

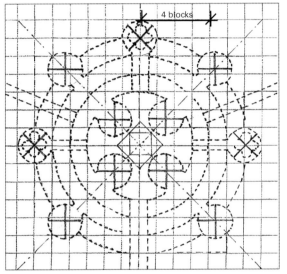

4 blocks

Illus. 10: Site plan
图10: 总图

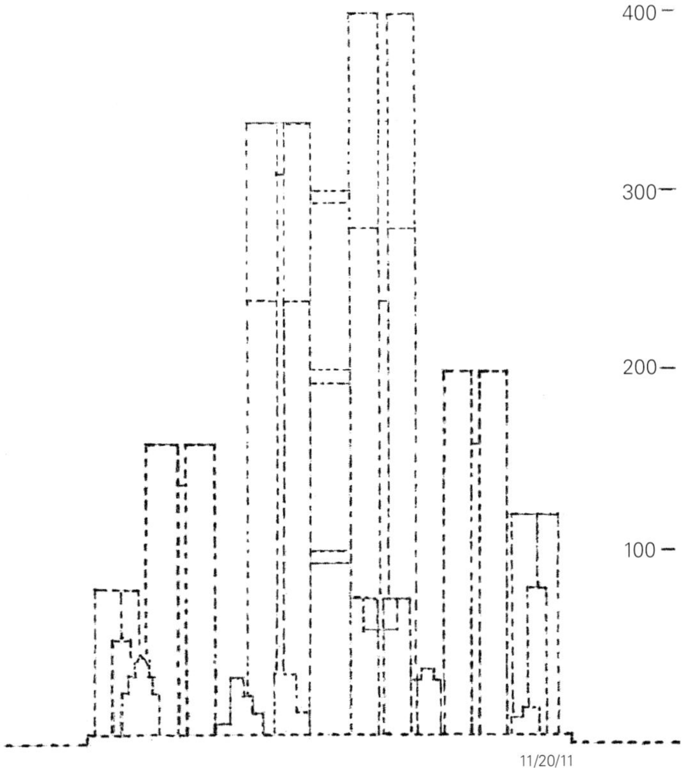

400 —

300 —

200 —

100 —

11/20/11

Illus. 11: Elevation
图11: 立面图

基本范例

人均面积：	750平方英尺（70平方米）
住宅建筑面积：	1.5亿平方英尺（420万平方米）
每层的楼面面积：	5.62万平方英尺（5220平方米）
需要的层数：	2700层
需要的大楼数量：	11座
大楼的高度：	100－400层

替代范例

人均面积：	500平方英尺（46.5平方米）
住宅建筑面积：	1亿平方英尺（420万平方米）
每层的楼面面积：	5.62万平方英尺（5220平方米）
需要的层数：	1800层
需要的大楼数量：	11座
大楼的高度：	180－300层

Basic example

Area per person:	750 s/f [70 m^2]
Residential building area:	150 million s/f [14 million m^2]
Tower area per floor:	56,200 s/f [5,220 m^2]
Number of floors required:	2,700
Number of towers required:	11
Heights of towers:	100 to 400 stories

Alternate example

Area per person:	500 s/f [46.5 m^2]
Residential building area:	100 million s/f [9.3 million m^2]
Tower area per floor:	56,200 s/f [5,220 m^2]
Number of floors required:	1,800
Number of towers required:	11
Heights of towers:	180 to 300 stories

DEVELOPED LIVE/WORK COMMUNITY

Large detailed site plan *(Illus.12 at right)*

This large-scale site plan of the live/work community shows the quad-wing residential towers described above.

Of note are:

- The distance from the center of the raised-podium site to the entrances of buildings located on the add-on perimeter 6th block is a quarter mile [.4 kilometers], or approximately 5 city blocks.

- The tree-lined street system dotted here and there with vest-pocket parks is intended for pedestrians and bicycles only.

- The short-block street pattern results in shorter pedestrian origin-to-destination routes.

- The maximum walk is about a half mile [.8 kilometers], or 10 city blocks, which can be comfortably negotiated in approximately 10 minutes.

- The intimate street pattern encourages building diversity and variety, in welcome relief from the more ordered lifestyle within the towers.

- The four central towers share a roofed landscaped plaza.

- Other towers are located on the perimeter in their own settings for identity and breathing spaces.

- The second ring road, radiating out from the center, is covered with a mechanically operated skylight system to offer protection from the weather. Radiating branches of the covered road connect all the towers.

- All occupants are within a short walk of the continuous perimeter park and thence to the open countryside.

- Under the third ring road, on the intermediate level of the podium, is a public transit system that circumnavigates the site with stops at each outer-ring tower and provides connections to the nearest rapid-transit system.

- Alternatively, if dictated by budget constraints, in lieu of the below-podium transit system it is possible to incorporate a bus system (with elevated prepay boarding stations) on the podium level of the third ring road.

成熟的生活/工作社区

大区域总图（图12）

生活/工作社区的大比例尺总图规划包含上面描述过的四翼住宅楼。需要注意：

- 从凸起的裙楼区中心到建筑入口（位于附加的周边第六个街区处）的距离是1/4英里（0.4公里），或者接近5个城市街区。

- 点缀在四处的绿荫街道系统，以及小型公园仅供行人与自行车使用。

- 街区较小的街道模式形成从出发点到目的地更短的人行道。

- 最大步行距离大约为1/2英里（0.8公里）或者10个城市街区，可以在10分钟左右较为舒适地走完。

- 紧密的街道模式促进建筑多样性和复杂性，而且欢迎人们从高楼中井然有序的生活方式中解放出来。

- 四座中心高楼分享一个屋顶景观广场。

- 其他的高楼位于周边自己独特的环境中，以获得各自的特征和空间。

- 从中心发散出去的第二条环路，覆盖了一个机械控制的天棚系统，以抵御恶劣天气。所遮盖道路的辐射分支将所有高楼连结起来。

- 所有住户都很快可以走到环形公园之间，并可以从那里到郊外去。

- 在第三条环线下，在裙楼的中间层之上，是一个公共交通系统，它环绕着这一区域，并在每个环线外的高楼都有站点停靠，而且与最近的快速运输系统连结起来。

- 另外，如果受到预算的制约，为了替代裙楼下方的交通系统，有可能将一个公交系统（架高的车站可预先支付车费）整合到第三条环线的裙楼层中。

SITE PLAN FOR LIVE/WORK COMMUNITY
生活/工作社区位置计划

Illus. 12: Raised-podium site plan

图12：抬高的平台总图

Site Cross Section

All the features described in the large-scale site plan *(Illus. 14)* can be seen clearly in this cross section of one-half of the site. The features are the elevated pedestrian podium; the relation of the residential towers to the podium; the enclosed central plaza (seen at left); the variety of relatively low nonresidential buildings on the podium; tree-lined streets; the covered second ring road; the perimeter park; the below-podium vehicular and service level; the terraced connection from the podium to the surrounding ground plane; and the preserved natural landscape, parks, playing fields, and sports facilities surrounding the Vertical City. Beyond the limits of the podium and outside the boundaries of the drawing are the fields, orchards, and fishponds of sufficient size to provide fresh produce and seasonal fruits to the community.

站点截面

如图14所示的在大比例总图所述的所有功能，可以清楚地看到，在这个截面的1/2的基地：行人登上平台；到平台上的住宅大楼之间的关系；封闭中央广场（见左）在平台上的品种相对较低的非住宅建筑，绿树成荫的街道，覆盖二环路周边公园；平台下的车辆和服务水平，从平台上的梯田连接周围地平面和保存完好的自然景观、公园、运动场，和周围垂直城市的体育设施。超越边界的平台和绘图边界之外，田间地头、果园和鱼塘大小足以向社会提供新鲜的农产品和时令水果。

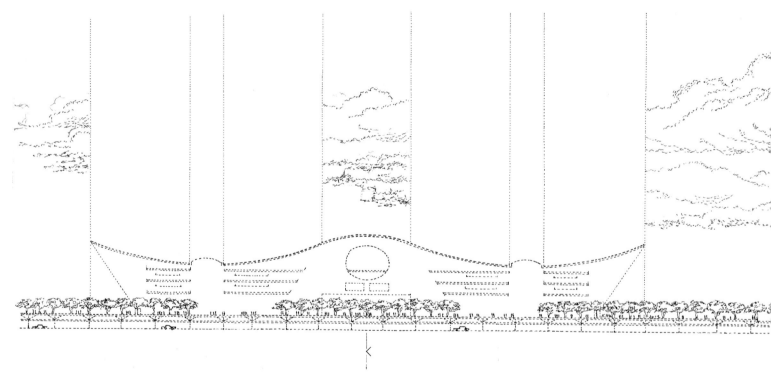

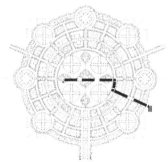

Illus. 14: Site cross section and key plan
图14：基地剖面和主要平面

空中大堂
——改变生活方式

Sky Lobby /
Change of Lifestyle

从某种意义上说，垂直城市和普通城市只有一个差别：它没有街道。垂直城市的设计能容纳25万居民，而维持这么多人生活的有各种商业、工业和政府部门。

空中大堂是垂直城市的一个重要因素。从结构上说，它把所有的垂直高楼连在一起，将其变成一个内部相互支持的稳定结构形式。这是一个天然的聚集场所，为居民的社交和商业活动提供了一个平台，是有助于于改变生活方式为一种健康的居住条件，它在节能环保的环境中转变人们原有的生活形式。

空中大堂位于每个高楼的电梯换乘层，通过无绳高速电梯，乘客可以从地面到达那里，然后通过换乘区间电梯或上或下去往自己想去的目标楼层。根据规模和需要，一个垂直城市可以有2至3或4至5个多层空中大堂。空中大堂每隔80至100层有一个，连接所有的垂直结构，这样居民就可以经过最短的距离到达垂直城市的任何一个地方。

空中大堂的封闭部分可以提供各种各样的商业功能，类似于一个郊区城市的主街道，有种类繁多的商店、服装店、洗衣店、修鞋店、美容店、杂货店和便利店，也有小型律师事务所、会计师事务所、医生和牙医办公室、药店和诊所等，还有满足居民需要的特殊政府部门。

空中大堂的屋顶是室外开放式的，距离下一个空中大堂的高度在80到100层之间。这是一个天然聚集场所，居民们可以在这里接触到其他人；大堂中有一个小城市应有的各种设施，例如公园、慢跑和自行车道、游泳池、篮球和足球场、网球场，甚至还点缀着各种餐厅、酒吧、咖啡馆、居民可以在这里和家人共享晚餐或者在酒吧和咖啡馆和朋友喝上一杯。

在一个科学和技术占主导地位的世界中，我们往往忘记了民间工艺的存在。过去的50年中，我们看到了许多民间工艺已经消失了，这其中有使用纸张、竹、电线、树叶和秸秆做成画作和玩具的手艺；我们也见证了许多流行多年的地方美

Vertical City is in every sense a city with only one exception, i.e., no streets. It is designed to house up to 250,000 residents and is supported by varies commercial, institutional facilities and governmental departments.

Sky Lobby is an important element in the Vertical City. Structurally, it ties all the vertical towers together and turn it into an inter-supporting stable structural form. It is a natural gathering place and provides its residents a forum for social activities and businesses and it facilitates a Change of Lifestyle into a healthy living condition, as well as an energy efficient and ecologically friendly environment.

Sky lobbies are located at the elevator transfer floors of each tower and it is reached from the ground floor via ropeless high-speed express elevators, from which passengers transfer to local elevators going up or down to reach their destination floors. Dependent upon the size and need, a Vertical City can have 2 to 3 or 4 to 5 a multi-story Sky Lobbies. Sky Lobbies are located approximately 80 to 100 stories apart, and provide links to all vertical structures so residents can reach any part of the VC via shortest distances.

The enclosed part of the Sky Lobby accommodates a variety of commercial functions and is similar to the main street of a suburban city, providing a wide variety of shops, clothing stores, laundries, shoe repair, beauty parlors, groceries and convenient stores, it also has small law offices, accounting offices, doctor and dental offices, pharmacies and clinics etc... as well as certain governmental branches serving the needs of its residents.

The roof of the Sky Lobby is open to weather, with a clear height of 80 to 100 stories before reaching the next Sky Lobby. It is a natural gathering place for its residents to make social contact with fellow residents; it has all the amenities of a small city such as parks, jogging and biking trails, swimming pools, basketball and football fields and tennis courts, it is also dotted with various restaurants, bars and coffee houses where residents can enjoy a quite dinner with family or have a drink with friends at bars or coffee houses.

In a world where science and technology dominates, we tend to forget the existence of local folk arts and crafts. In the past 50 years, we have already seen many local arts and crafts such as using paper, bamboo, wires, leaves and straws turning them into images and toys have disappeared,

Illus. 15: Artist's visualization of a Vertical City
图15：垂直城市的虚拟表现图

Artist Vision of Sky Lobby
艺术家幻想的空中大堂

We have also witnessed the disappearances of various traditional local foods that were popular many years ago and have now been replaced by McDonald, and Kentucky Fried Chicken. An area dedicated to promote local arts and crafts, local and regional foods will add color and interest to the life of its residents, it will also help artisans to maintain their craft and keep its tradition alive for generations to come.

The impact of the environment on an individual in a Vertical City is indisputable; the same is true for the ability of individuals to influence their environments. It can therefore be argued that the characteristic of a desired lifestyle should be clearly stated in the architect's or urban designer's brief, since architecture and its supporting spaces have great power to inspire change.

With this firmly in mind, we offer key goals and aspirations that people, moving beyond primordial needs for shelter, safety and comfort, will give weight to, determining their impact on the environment and on society as a whole.

First and foremost is the requirement to live in harmony with nature. This can be achieve by:

Meeting one's needs without further depleting limited natural resources beyond nature's ability to replace them.

Striving for energy efficiency in all facets of life, including:

Seeking alternative energy means

Localizing power and services

Shifting away from cheap fossil fuel mentality, the main cause of urban pollution, and finding alternatives to automobile transportation. Ideally, creating car-free pedestrian- friendly communities

Eliminating cost of shipping foods from distance locations through local harvesting

Eradicating the cost and pollution of long daily commutes by promoting local employment and clean mass transit

Ensuring that building design and construction specifications for materials, methodologies, systems, equipment, and fixtures meet or exceed energy-efficiency standards

Eliminating waste through conservation and recycling

Selecting building sites that will have the least impact

食的消失，它们已经被麦当劳和肯德基所取代。空中大堂这样一个致力于弘扬当地民间工艺和地方美食的场所，将会为居民的生活增光添彩，而且也有助于手工艺人保存他们的手艺，让传统能世代相承。

在垂直城市中，环境对个人的影响是显而易见的；同样的，个人能力也会影响他们所生活的环境。因此可以说，一个理想生活方式的特征应该在建筑师或城市设计师的概要陈述中清晰地表达出来，因为建筑及其配套空间有引起改变的巨大力量。

由于对此深信不疑，我们给出了一些关键的目标和愿景，即希望超越了原始遮蔽风雨、安全、舒适之需求的人类，能够作为一个整体，开始重视并决定影响环境和社会。

首要的是与自然的和谐共处，这可以通过如下方式实现：

满足人的需求，但别不顾自然是否有修复自身的能力，过分消耗有限的自然资源。

争取在生活的各个方面进行节能，包括：

寻找可替代的能源工具

使电力和服务本地化

摆脱对廉价化工燃料的依赖心理，它们是城市污染的主要原因，找到汽替代汽车交通运输的方法。在理想的情况下，创造一个没有汽车，对行人友好的社区

通过收获本地作物来消除从较远地方运输食品产生的成本

通过增加本地就业和清洁的公共交通工具来根除每天通勤产生的费用和污染

确保建筑设计和建设规划在施工材料、方法、系统、设备、装置上达到或超过节能降耗标准

通过节约和回收利用消除浪费

选择对现有的自然生态系统、水资源、耕地影响最小的建设位置

充分发挥积极的生活方式，确保社区的所有点与公共空间或公园之间在步行可及的距离之内

在可能的地方引入园林绿化——公共空间、街道两旁、人行道、屋顶等；利用各种可以发挥植物净化空气的机会

开辟足够的土地，以便在当地为居民的营养和喜好生产花卉、蔬菜、海鲜、家禽和其他各种食物

为了保持健康的土壤，采用并改进有机农业的作业方法，如轮作、免耕、绿色肥料和其它不使用石油的操作。减少温室气体排放，在总体上收获生态效益。

不再使用化肥和杀虫剂

通过培养人们对邻居的愿望、需求和权利有意识，鼓励个人参与和社会责任感，包括：

保证各行各业的人享有平等的居住条件和机会

和其他居民相互熟络，彼此做好邻居

追求正义、平等的行为和一种公平意识

通过相互联系和互动提升人们的幸福感

在致力于促进共善的同时，尊重个体差异、个人的价值观和理念

鼓励一种超出个人具体需求的生活目的

鼓励一种个人为社会大众福祉做贡献的态度

通过共同的经历、活动和目标创造一种社区意识

on the existing natural ecosystem, water resources, and arable land

Maximizing active lifestyle and assuring that all points of the community are within walking distance of open spaces or parks

Introducing landscaping wherever possible–in open spaces, along streets and walkways, on rooftops, and taking advantage of other opportunities for air cleansing plantings

Creating sufficient land preserves for the local production of flowers, vegetables, seafood, poultry, and other foodstuffs for the nourishment and enjoyment of its residents

Adopting and improving methods of organic farming such as crop rotation, non-tilling, green manure, and other non-petroleum based practice for healthy soils, reducing greenhouse emission and ecological benefits in general

Eliminating the use of chemical fertilizer and insecticides

Encouraging personal engagement and society responsibility by cultivating an awareness of the wishes, needs, and rights of one's neighbor, including:

Ensuring that equal accommodations and opportunities are available to people in all walks of life

Becoming familiar with, serving as good neighbors to, other residents

Pursuing the practice of justice, equality, and a sense of fair play

Promoting a sense of well-being through human connection and interaction

Respecting individual differences and personal values and ideas while promoting commitment to the common good

Aspiring to a purpose in life beyond that of one's personal needs

Encouraging an attitude of personal contribution for the benefit of society at large

Creating a sense of community through shared experience, activities and goals

The goal of an improved lifestyle embraces the desire for a society that is safe and crime-free, and that nurtures healthy and sustainable daily practice. The physical environment:

Is conceived securely without dark alleys, permitting easy monitoring and policing

Is pollution free, offering ample clean air and water

Practices segregated recycling of paper, plastic, glass, metal, toxic materials, organic compostable, and residual waste. Also recycled are the rainwater, brown water, and waste water, as well as human waste

Allows individual to control their own environments through advance technologies for increased energy efficiency, conservation, and reduced cost. Due to the inherent compactness of the Vertical City, its operation, management, maintenance, and governance should result in raised standards of living for residents while lowering carrying costs and taxation

In addition to the foregoing, the search for a community that offers opportunities for employment, self-improvement, enrichment, enjoyment, and fulfillment prescribes:

The local availability of diverse jobs without the need to commute elsewhere

The richness and excitement inherent in a thriving city. By virtue of the concentration of people from all walks of life in the dense environment of the Vertical City, unprecedented opportunities exist for commercial, social, cultural, entertainment, sports, recreation, and leisure activities

The celebration, exhibition, and conservation of local art, craft, music, dance, cultural traditions and ceremonies, many of which are disappearing in today's fast-paced world. While the Vertical City looks boldly forward, it proudly retains important cultural traditions from the past

530

一种改进的生活方式, 其目标包括: 渴望一个安全、无犯罪、鼓励健康、可持续的社会。其自然环境:

被认为是安全的, 没有黑暗的街巷, 很容易进行监控和管治

是无污染的, 提供充足、清洁的空气和水分离可回收纸张、塑料、玻璃、金属、有毒物质、有机堆肥和残余废物。同时回收的还有雨水、污水、废水和人类排泄物

允许个人通过先进的技术控制自己的环境, 从而提高能源效率, 节约能源, 降低成本。由于垂直城市固有的密度, 其运作、管理、维护和治理会在降低运输成本和税收的同时提升居民的生活标准

一个社区要为居民提供就业机会, 让他们可以自我提升、充实、享受、满足, 它就必须:

有在本地可以获得的各种工作机会, 居民不需要到其它地方去上班

一个蒸蒸日上的城市内在的繁荣和兴奋点是: 通过将各行各业的人集中到垂直城市密集的环境中, 使商业、社会、文化、娱乐、体育、各种休闲活动都充满前所未有的机遇

庆典、展览、保护当地的艺术、手艺、音乐、舞蹈、文化传统和仪式——其中许多在当今快节奏的世界里正在消失。在垂直城市看起来大步向前的同时, 它也充满自豪地保留了一些过去重要的文化传统

垂直城市是一个真实的无车的城市；它具有任何一个传统城市所能提供的功能。大型的公共场所，如体育馆，会议中心，娱乐场所能被建立在距离垂直城市不远之处。可以预见当居民从一个传统城市移居到垂直城市的时候，他们会经历一些心理的焦虑。本书的作者们的意图是考虑新环境的所有方面，并当新的居民搬到垂直城市时，他们不会感觉到任何影响，从而实现一种新的生活方式的平稳过渡。

Vertical City is truly a city without cars; it can accommodate everything that a conventional city accommodates. Large functions such as sports stadiums, convention and entertainment centers can be housed a short distance away from the Vertical City. It is anticipated that residents will have some psychological anxieties when relocating from a conventional city to a Vertical City. It is the intent of the authors to consider all aspects of this new environment and when new residents moving into Vertical City, they will have little or no impact for a smooth transition to this new lifestyle.

SECTION FIVE / THEORETICAL APPLICATIONS

Applied Principles of the Vertical City
Conclusion

第五篇 / 理论应用

垂直城市的应用原则

结论

垂直城市理论的可行性

Applied Principles of the Vertical City

垂直城市的原则针对的是现实城市中存在的诸多问题，为了测试这些原则的有效性，我们首先要选择一些城市来展开实验。在这一过程中，我们敏感地意识到这样一个事实，城市居民厌恶那些外来人员，他们总是肆意地挑毛病而不是热爱这座城市。这种情绪是可以理解的，这就好像，虽然我们纽约人相互抱怨纽约没有多少大家能够负担的便宜房屋，抱怨下雨时没有出租车，抱怨暴风雪后肮脏的污水，但当一个外来者发出同样的抱怨声时，我们也会感到厌烦。因此，我们希望一开始就指出，我们不是故意挑毛病，而是想要在特定城市中，准确地认识那些显而易见的情形，而且相信在这些城市中，通过应用垂直城市理论，我们能够找到解决问题的方法。

在世界范围内，政府投入了大量人力物力，想要认清、评估、解决他们各自的问题，看不到这一点是很不公平的。我们的资源和能力让我们只限于指出在美国，像波士顿、芝加哥、纽约、宾夕法尼亚和旧金山这样的城市，政府确实在做这类领导工作。

这本书强调的重点并非是要阻止在发达国家和发展中国家普遍存在的城市扩张问题。而且，尤其是在发达国家，如美国，问题在最古老的郊区日益显现：它们建立伊始是令人非常满意的社区，现在在建筑、服务和道路却衰败退化，而因为远离城市中心，新家庭也不愿意待在这里。规划者、管理者和城市理论家将来都要谈到这些问题，而我们相信垂直城市的原则是值得考虑的，因为它是拥有强烈的环境意识、高密度、设备齐全、无车辆、方便行人的社区。

我们意识到，自满是人类本性的一部分。在不断上升的海平面淹到我们家门口，或者在我们撤离到巴克明斯特·富勒穹顶的保护罩之前，人们都会按照旧有的习惯生活。这些习惯是由陈旧的信念、原则、规范所主导的。人们总是希望，土地消失的时候，科学可以让我们依赖丰富的海洋资源而继续生存——我们可以采取某种方式，开采肥沃的海底层来制造氧气，而生产食物的水生植物可以弥补消失的雨林。其他人则坚持认为，我们过于专注地球变暖的负面意义。为什么

To test the validity of the principles of the Vertical City against the problems of actual cities, we had first to choose the cities. In doing so, we were sensitive to the fact that residents often resent outsiders who, instead of giving a balanced view of their beloved city, have had the audacity to find fault. Such feelings are understandable and are no different from when we New Yorkers complain to each other about the scarcity of affordable housing, the dearth of taxis when it rains, and the dirty slush after a snowstorm, but resent it when outsiders voice the same complaints. Therefore, we wish to state at the outset that it is not our intention to find fault but to accurately identify certain well-documented situations in selected cities that we believe can find relief in the application of the theories of the Vertical City.

It would be a great injustice not to acknowledge that municipalities throughout the world channel tremendous talent and resources into the identification, evaluation, and resolution of their respective problems. It is beyond our resources to do more than state that in the United States, certainly in cities like Boston, Chicago, New York, Philadelphia, and San Francisco, such leadership exists.

The emphasis of this book thus far has been on halting the universal problem of urban sprawl, in both developed and developing nations. In addition, particularly within developed nations and especially in the United States, there is the growing problem posed by the oldest suburbs. They were highly desirable communities when first established, but with their deteriorating buildings, services, and roads they are now being abandoned for newer homes even farther away from the city center. In going forward, as planners, administrators, and urban theorists address these issues, the principles of the Vertical City for an environmentally conscious, high-density, self-contained, car-free, pedestrian-friendly community are worthy of consideration.

We are aware that complacency is part of human nature, and until the rising sea level is at our very doorsteps, or we have to retreat within the protective shelter of a Buckminster Fuller geodesic dome, mankind will live out its existence following old habits governed by antiquated beliefs, principles, and regulations. There is always hope that, as land disappears, science will enable our survival by tapping into the riches of the sea—that we can somehow mine the layers of fertile ocean floor to cultivate oxygen- and food-producing water plants to compensate for the loss of the rainforests.

Others maintain that we have concentrated too much on the negative side of global warming. Why not celebrate the shorter winters and longer growing seasons? Soon we will all be living in one big tropical paradise!

About one thing we are sure: We need to talk!

It is thus that we—brazenly, yes—superimpose images of Vertical Cities on existing cityscapes as a means to generate dialogue and test the air, as it were. We are fully aware that only through a miracle would any of the illustrations come to fruition. On the other hand, if, after making the required adjustments, major or minor, in a location most suitable for the specific conditions, the collective wisdom of just one urban center saw fit to test the validity of the Vertical City's principles, there would be cause for celebration.

With few exceptions, most of the selected cities are experiencing unprecedented population increases, resulting in sprawling suburban expansion. Additionally, many are suffering from the deterioration of older suburbs caught in limbo between the urban center and new development along the city's ever-spreading perimeter. Most of the urban poor are to be found in these older suburbs.

For each city, we address one or two problems on the following pages and then offer as a potential solution the application of selected principles of the Vertical City.

不为冬天变短，植物的生长季节变长而欢呼呢？很快我们都可以生活在一个大型的热带天堂之中！

有一件事我们十分确定，我们得好好谈谈这个问题！

因此，我们十分大胆地将垂直城市的景象叠加到现存的城市风光上，将它当作一种手段，用以产生对话，做做实验。我们完全明白，这些图景要成为现实，需要奇迹的出现。另一方面，只要有一个城市中心认为可以测试一下垂直城市原则的有效性（在一个最切合具体情况的地方，可以做必要的、或大或小的调整），就已经值得庆祝了。

除了少数例外，大多数被选中的城市都在都市化扩张浪潮中经历了前所未有的人口增长。另外，大多数城市都忍受着陈旧郊区衰退败坏、前途未卜的困境；这些郊区位于城市中心与新兴的城市周边区域之间。大多数的城市贫民都位于这些陈旧的郊区中。

对于每座城市，我们都指出一个或者两个迫切的问题，并应用特定的垂直城市原则给出可能的解决方案。

Moscow, the capital of Russia, was ravaged during the war with Napoleon in 1812. It was gradually rebuilt over the years, but the greatest change came after the breakup of the Soviet Union, in 1991. Since then, rapid development has made Moscow the largest city in Europe, with a population of more than 11 million and some 4 million cars on the road. Infrastructure is stretched to the limit, yet people continue to pour into the city by the thousands. Moscow's population is expected to double in the next 20 years.

To accommodate the expected new arrivals, Moscow planned to expand its physical boundaries toward the southwest in an area about twice the size of the existing capital city, doubling its horizontal footprint. Historically, such lateral expansion has been the norm in most major cities, but it has proved to be unsustainable.

Applying the principles of the Vertical City to Moscow

A series of bedroom communities (each accommodating upward of 150,000 residents) plus live/work communities (each accommodating up to 250,000 residents) within a square footprint of one half mile per side, or roughly .25 square miles [.64 square kilometers], measured against Moscow's current population density of 9,772 persons per square kilometer, would occupy only 2.5 percent of the land, leaving 97.5 percent for agriculture, recreation, and other productive uses. Since the pedestrian-friendly communities are car-free, there is no requirement to build new roads. Every Vertical City constructed would eliminate the use of 90,000 cars.

Such an approach would stem the tide of destructive urban sprawl. It would allow Moscow a chance to update its mass transit system and other aging infrastructure while completing its urban renewal efforts in a less hectic fashion than is currently the case.

莫斯科，俄罗斯首都，在1812年与拿破仑的战争中被破坏。数年之间，莫斯科逐渐重建，不过最大的改变发生于1991年解体之后。从那时开始，快速的发展让莫斯科成为欧洲最大的城市，拥有超过1100万人口和400万辆汽车。基础设施的承载能力已经到了极限，但还有成千上万的人继续涌入这里。莫斯科的人口在未来20年间有望增加一倍。

为了给预期中的新到人口提供住所，莫斯科计划向西南方向一处面积有首都两倍大的区域拓展边界，使其水平面积翻番。从历史上说，这种向侧翼的拓展在多数大城市都发生过，但它也已经被证明是不可持续的。

应用垂直城市的原则

在一块边长0.5英里的正方形平地或大约0.25平方英里（0.64平方公里）的平地上，可以有一连串的的卧室社区（每个社区容纳人口可达15万）加上生活/工作社区（每个社区容纳人口可达25万）。莫斯科现在的人口密度为9772人每平方公里，与此对照，上述垂直城市的方案只需占据当前2.5%的土地，而其他97.5%的土地则可用于农业、休闲娱乐和其他生产活动。由于方便行人的社区是没有车辆的，因此也不需要建设新道路。每建设一个垂直城市，就可以减少9万辆车。

这种方法将抑制城市扩张带来的破坏浪潮。它也将让莫斯科以一种与当前不一样的、不那么狂热的方式完成城市翻修工作，并让莫斯科有机会升级其公共交通系统和老迈的基础设施。

Moscow / 莫斯科

Bordeaux
波尔多

Bordeaux is a city in southwestern France famous for its agreeable climate, fine wines, and good quality of living. But these very characteristics have recently drawn a tremendous influx of people to the region, requiring some 50,000 new dwellings, which, in turn, imposes a tremendous strain on the environment. Because arable land for Bordeaux's prized vineyards is scarce and ecologically fragile, none should be surrendered to urban development.

Applying the principles of a Vertical City to Bordeaux

A Vertical City housing a single bedroom community would occupy only 36 square blocks, or .09 square miles [23 hectares] and would easily accommodate Bordeaux's total new population requirement for 50,000 dwellings.

　　波尔多，位于法国西南部的一个城市，以其舒适气候、美酒佳酿、品位生活闻名于世。但这些优秀的品质现在吸引众多人涌入这一区域。这些人需要约5万套新住房，也给环境带来巨大的压力。因为用于波尔多珍贵葡萄园的耕地不仅在生态上很脆弱，而且在面积上很稀少，所以波尔多不能向城市发展低头。

应用垂直城市的原则

　　一个垂直城市中，单单一个卧室社区只占据36个街区，或0.09平方英里的面积，却很容易就可以容纳波尔多所有的新人口，满足5万套新住房的需求。

Tokyo, Japan's largest city, is experiencing a population decline as the populace ages and young couples give birth to fewer children. While the city's economy remains robust, it is predicted that by 2100, Tokyo's population will plunge to less than half of its current 13.23 million. Meanwhile, the urban planning objective of metropolitan Tokyo is improvement of the urban infrastructure. The focus is especially on transforming the regional transportation system to create a low-carbon, healthy, and appealing environment that will continue to attract investors and increase Tokyo's international competitiveness. Less predictable are the environmental and economic consequences of the government's reconsideration of a zero-nuclear policy.

Applying the principles of a Vertical City to Tokyo

Replacing sprawling low-density neighborhoods with carefully planned Vertical Cities can help Tokyo solve the problems stemming from its antiquated infrastructure, reduce its energy demands (irrespective of energy sources), and create a healthier, more efficient, and competitive international city of the future.

东京，日本最大的城市，由于老龄化，以及年轻夫妇生的孩子越来越少，正经历着人口的下降。城市的经济依然健康，因此预计到2100年，东京的人口将比现在1323万的一半还少。同时，作为大都市的东京，其城市规划的目标是进一步改进城市基础设施。现在关注的焦点特别放在：转变区域交通系统，以营造出一个低碳、健康、有魅力的环境，而这样的环境将继续吸引投资者，提升东京的国际竞争力。不过，政府重新考虑其核能政策将会带来什么样的环境和经济结果，尚未可知。

应用垂直城市的原则

用细致规划的垂直城市替换庞杂的低密度社区，这将帮助东京解决陈旧基础设施的问题，在不考虑能源资源的同时减少能源需求，并且在将来创造出一个更健康、更高效、更具有国际竞争力的城市。

APPLIED PRINCIPLES OF THE VERTICAL CITY

Tokyo / 东京

Chicago
芝加哥

Chicago, home to the world's first skyscraper and to many of the tall buildings that followed, has urban problems similar to those of other major cities. That it has its own philosophy—the Chicago School—to study and to help find solutions to those problems is to be admired and respected. Any attempt to contribute to that effort without the requisite local expertise would be presumptuous. Instead, we have selected a relatively mundane topic as a basis for supporting principles of the Vertical City, namely the Urban Farming Ordinance, passed in 2011 by the Chicago City Council to allow cultivation previously prohibited within the city limits. The legislation stipulated that products of the new farms can be sold commercially.

Applying the principles of a Vertical City to Chicago

As the need for urban expansion arises, principles of a Vertical City could offer a possible solution in a new self-sustaining live/work community that is car-free and pedestrian-friendly. Housing up to 250,000 people, such a community would occupy a site of no more than one-half mile by one-half mile [0.8km x 0.8km], roughly equivalent to only 1 percent of the land consumed by a conventional city. The land thus saved would support urban farms and the production of fresh organic produce or seafood for local consumption.

芝加哥，世界第一座摩天大楼的故乡，有许多高层建筑。芝加哥有类似于其他大城市的城市问题。它甚至有自己的思想学派——芝加哥学派——来研究和帮助找到解决这些问题的方法，这一点令人钦佩和尊重。任何想要在这方面有所助益却不请教当地专家的做法，都是冒失的。我们选择了一个相对通俗日常的主题，"城市农业条例"，作为支撑垂直城市原则的基础。条例在2011年被芝加哥城市委员会通过，进而允许在城市特定地区展开之前禁止的农业活动。立法规定，新农庄出产的产品可以进行商业销售。

应用垂直城市的原则

随着城市扩张需求的出现，垂直城市原则能够通过新的、自我维持的生活/工作社区为此提供一种可能的解决方案，这一社区是没有汽车且方便行人的。它可以容纳25万人，占据的位置将不会超过1.5英里×1.5英里（0.8公里×0.8公里），大约只相当于一个传统城市土地面积的1%。由此节约的土地可以用来支持城市农场，生产供当地消费的新鲜有机农产品和海产品。

San Francisco, in northern California, is unquestionably one of the most beautiful cities in the world. Through dedicated local efforts, San Francisco was recently removed from the list of the 25 Most Polluted Cities in America, yet its traffic congestion continues to grow. Studies are underway to encourage ridership of the underutilized rapid-transit system.

More than 200 municipalities make up the San Francisco Bay Area, with a combined population of approximately seven million. In 2007, the San Francisco Bay Area State of the Urban Forest Study was undertaken to assess the ecological and economic benefits of the region's trees with regard to land-use planning, water and energy conservation, air quality, public health and safety, land value, tax base, employment, and city services. The study concluded with an urgent need to address the accelerating air pollution and the increased water and energy demands that have resulted from the region's fast-paced urban growth, particularly in outlying areas.

Applying the principles of a Vertical City to San Francisco

In selecting suitable sites, there needs to be a full awareness of existing watersheds, catchment areas, and wildlife habitats. Above all, encroachment on arable farmland is to be minimized or avoided.

旧金山，加州北部的旧金山毫无疑问是世界上最美丽的城市之一。通过当地的不懈努力，旧金山最近从美国污染最严重的25个城市榜中被划去，不过，旧金山的交通拥堵情况还在恶化。正在进行的研究要求增加未被充分利用的快速公交系统的客流量。

超过200个城市共同构成了旧金山湾区，其人口总计近700万。2007年，"旧金山湾区城市森林协会"着手从土地使用规划、水资源与节能、空气质量、公共健康与安全、土地价值、课税基础、工作岗位、城市服务等方面，评估区域树木的生态和经济效益。协会得出结论，认为非常需要立即指出，由于这一地区（尤其是郊区）城市的快速扩张，空气污染正在加重，需要的水和能源越来越多。

应用垂直城市的原则

在选择合适地点时，需要意识到现存的水流域、集水区和野生动物栖息地。最重要的是，应当尽可能少或避免占用农田耕地。

San Francisco
旧金山

São Paolo, one of the fastest-growing cities in Brazil, has a population of 20 million. One-third of the people live in favelas, or slums, crammed into a mere 10 percent of the city's land area. In the 1980s, the government of São Paulo overturned its policy of corrective eradication and instead adopted an easier, more economical and humane approach of improving living conditions within the favelas.

Applying the principles of a Vertical City to São Paolo

There are no slums in the Vertical City; affordable housing is provided for people of all economic and social strata.

　　圣保罗，巴西发展最快的城市之一，人口达2000万。有三分之一的人口生活在贫民区，挤在只占城市面积10%的地方。在20世纪80年代，圣保罗市政府不再使用纠正、消除的政策，而是采用一种更便捷、更经济和更人道的方法来改善贫民区人民的生活条件。

应用垂直城市的原则

　　在垂直城市中没有贫民区，各个经济和社会阶层的人都负担得起住房。

São Paolo
圣保罗

Mumbai/孟买

Mumbai (formerly known as Bombay). In the decades since 1947, the year of India's independence, squatters have settled on virtually every scrap of empty land in Mumbai, India's wealthiest and most populous city, with more than 20.5 million people in the metropolitan region. The population has almost doubled in the last 20 years and continues to grow, with an estimated 60 percent of the people living in makeshift slums. Mumbai encompasses 603 square kilometers [233 square miles] of land, resulting in a density of more than 31,000 persons per square kilometer [more than 80,000 people per square mile].

Despite abject poverty, there is a strong sense of community in Mumbai's slums, which have evolved into the city's recycling hub, responsible for processing some 7,000 tons of garbage daily, separating paper, plastic, metals, and electronic equipment. In and of themselves, the slum dwellings are of no economic value, yet they occupy precious land worth billions of dollars. In the last ten years, Mumbai's Slum Rehabilitation Authority has implemented reforms that allow developers to build much-needed market-rate projects on land previously occupied by slums, with the stipulation that housing for the displaced squatters be provided onsite.

Applying the principles of a Vertical City to Mumbai

Mandatory recycling includes paper, plastics, glass, metals, rainwater, brown water, kitchen and human waste, and, ultimately, the building equipment and materials at the end of their useful lives.

It is estimated that 48 Vertical Cities would be needed to accommodate Mumbai's Slum Rehabilitation, freeing up over 300 square kilometers [116 square miles] of land, which can be turned into fields, ponds, orchards, and recreation areas. Of course, we cannot expect Mumbai to build 48 Vertical Cities all at once; however, even a modest three or four Vertical Cities as a start would make a big difference to the Mumbai landscape and vastly improve the quality of life, in a dignified setting.

孟买，1947年印度独立后的几十年，非法占用公地者实际上已经占据了孟买每一寸空地：这是印度最富有、人口最稠密的城市，在都市区中，有超过2050万人。在过去20年间，孟买的人口几乎翻了一倍而且还在增长，而大约60%的人住在临时贫民窟。孟买有603平方公里（233平方英里）的土地，而人口密度超过了3.1万人每平方公里（8万人每平方英里）。

尽管一贫如洗，孟买的贫民区还是有一种强烈的社区感，实际上它已经变成城市的循环回收中心，负责处理7000吨的生活垃圾，分离纸张、塑料、金属、电子设备等物品。就其自身而言，贫民区的住房没有经济价值，但却占据了价值数十亿美元的珍贵土地。在过去的十年间，孟买的贫民区改造局已经开始着手改革，允许开发商在之前被贫民区所占据的土地上，修建更合乎市场行情的项目，并且规定占用公地的无家可归者可以就地获得新建住房。

应用垂直城市的原则

强制回收的东西包括：纸张、塑料、玻璃、金属、雨水、棕水、厨房垃圾和人类废弃物，最后是不再有使用价值的建筑设备和材料。据估计，要重新安置孟买的贫民区，需要48个垂直城市，而这将腾出超过300平方公里（116平方英里）的土地，这些土地可以用作田地、池塘、果园和休闲区。当然，我们不能期望孟买立刻建成48座垂直城市，不过，即使一开始的时候有三四座垂直城市，也会让孟买的景观大为不同，而且可以堂堂正正地大幅提升人们的生活质量。

Mongolia is certainly not going to run out of land any time soon. With an area of 1,560,000 square kilometers [606,000 square miles], it is more than four times the size of Japan, yet it contains only three million people.

Ulaanbaatar, The population of its capital, has been expanding at an unprecedented rate, having risen some 70 percent over the last 20 years. It now stands at 1.5 million, fully half of the country's total population. This increase is due to migration by rural natives, as well as foreign nationals pouring into the city, seeking a share of its newfound mineral wealth. Government authorities, with the help of international agencies, are desperately working to find solutions to previously unforeseen urbanization problems.

Applying the principles of the Vertical City to Ulaanbaatar

Blessed with land to spare, it is likely that Ulaanbaatar will undergo several generations of less dramatic urban planning experimentation than that advocated by the Vertical City, notably the use of ultra-high-rise towers. However, the basic principles of the Vertical City—responsible environmental stewardship, car-free and pedestrian-friendly communities, energy efficiency, achieving higher-density living, preservation of land, and self-sustainability—should be kept in mind during that process.

蒙古当然不会很快耗尽土地。它的面积达到156万平方公里（60.6万平方英里），比日本的四倍还多，而它养育的人口只有300万。

蒙古首都乌兰巴托的人口以一种前所未有的速度在增长，过去20年增加了约70%。现在乌兰巴托的人口是150万，占了全国人口总数的一半。农村居民和外籍人士涌入城市导致了人口的剧增，他们都想从新发现的矿产财富中分一杯羹。在国际组织的帮助下，政府正在竭力寻找对策，以解决这些前所未有的城市化问题。

应用垂直城市的原则

因为拥有得天独厚的广阔土地，看起来乌兰巴托后世好几代人都不怎么需要去关注激动人心的城市规划实验，这些实验是垂直城市提出来的，其中显而易见的是要新建超高层建筑。不过，垂直城市的基本原则还是应当在乌兰巴托的城市化进程中被牢记：负责的环境管理工作，没有汽车和方便行人的社区，节能、多元化的生活，保护土地，以及自身的可持续性。

Ulaanbaatar
乌兰巴托

Hong Kong. By 1945, at the end of World War II, the population of Hong Kong had shrunk to 600,000, less than half of its pre-war total, due to food scarcity and mass emigration. The Chinese Revolution of 1949 reversed the tide as millions of people fled the Communist regime on the mainland. The trend continued during the Great Leap Forward. Along with capital and skills from China there came an influx of cheap labor, swelling the population. The resulting squatter camps were dangerous firetraps and, when built on Hong Kong's craggy hills, highly susceptible to fatalities from monsoon rains and landslides.

Housing was badly needed in land-scarce Hong Kong, a 31-square-mile [80.4-square-kilometer] island off southern China's coast. As most land in Hong Kong is "queen's" or government-owned, authorities resolved the issue by building low-income housing away from the central business district (CBD) and then "selling" prime land to developers via long-term leases in order to build housing for the more prosperous. Two problems were resolved in one stroke. Currently, Hong Kong's population is over 7 million and still rising.

Hong Kong was returned to China in 1997, China promised to allow Hong Kong to be self-goverd for 50 years. For a short while before 1997, many Hong Kong residents emigrated overseas, mostly settled in Vancouver and real estate plummeted. However, this period was short lived, with the support of China, foreign investments continued to pour in and Hong Kong prospered.

With surging investments from all over the world, real estate prices have risen to the highest in the world, placing the housing market far beyond the reach of most of the middle class.

Applying the principles of the Vertical City to Hong Kong

Located only a few miles from Hong Kong's CBD, the abandoned Kai Tak International Airport (decommissioned in 1998), with its 3.4-kilometer [2.1-mile-long] runway, would be an ideal site for the construction of a new Vertical City. All of the principles already discussed would serve crowded Hong Kong well.

香港，二战末期的1945年，由于食物匮乏和大规模移民，香港的人口缩减至60万，比战前的一半还少。1949年中国的解放战争扭转了这一趋势，数百万中国人离开大陆来到香港。在大跃进时期，这一浪潮还在持续。伴随着大陆资本和技术而来的，还有大量的便宜劳动力，这让香港的人口开始膨胀。由此产生的贫民聚集区也是火灾高发区，而且它们建在香港崎岖陡峭的山上，很容易受季风雨、山体滑坡等自然灾祸的影响。

在地少人多的香港，一个中国南海岸31平方英里（80.4平方公里）的小岛，对住房的需求极为迫切。由于当时香港的大多数土地都是"女王的"，或者为政府所有，当局于是在远离中央商务区的地方修建廉价房，以长期租赁形式将优质土地"卖给"开发商，让开发商去修建住房以促进繁荣，从而解决了问题，可谓是一举两得。现在，香港的人口超过700万，而且还在增加。

香港在1997年回归中国，中国承诺香港自主管理50年。在1997年前夕，很多香港居民移民海外，大部分人定居在温哥华，房地产暴跌。但是这个时期很短，在中国的支持下，外国投资继续涌入香港，香港又变得繁华。随着世界各地投资的增长，香港的房地产上升为世界最高的价格，使得地产市场远高于大多数中产阶级所能承受的范围之内。

应用垂直城市的原则

废弃的启德机场（1998年停止使用）距离香港中心区仅有数英里，跑道长3.4公里（2.1英里），是建设一个新垂直城市的理想场所。所有上述讨论过的原则都可以在拥挤的香港发挥作用。

Chongming Island, the third-largest in China after Taiwan and Hainan, covers 1,040 square kilometers [400 square miles] of land. A totally rural island with a population of 740,000, it is situated at the mouth of the Yangtze River, barely a stone's throw from downtown Shanghai. Chongming is a perfect place for humans to coexist with nature.

As it is located so close to Shanghai, and recently connected by both bridge and tunnel, there is considerable pressure to develop the island, even as there is an equally strong impulse to preserve its natural qualities. Many Shanghai residents already talk about future visits to the Chongming countryside for fresh air—to "wash one's lungs" (洗肺).

Applying the principles of the Vertical City to Chongming Island

Six to eight Vertical Cities could add the capacity for transient populations of a million or more while occupying a footprint of less than two percent of the island. Local farmers could continue organic farming and fishing, or find better-paying jobs by training for work in newly developed leisure industries. Income collected by the government could help implement the master plan prepared by Skidmore Owings & Merrill (SOM), and help turn the island into a paradise retreat away from the stresses of one of the world's largest and busiest urban centers.

崇明岛，台湾和海南之外中国的第三大岛，面积达1040平方公里（400平方英里）。这个乡村岛屿有74万人，位于长江口处，与上海市中心只有一箭之遥。崇明岛是人和自然和谐共处的绝佳地方。

由于离上海如此之近，而且最近又通过大桥和隧道与上海联结起来，于是，有一股强大的推动力想要开发崇明岛——即使也有同样的压力要求保护其自然品质。许多上海居民已经谈到，将来要去崇明乡下呼吸新鲜空气，去"洗肺"。

应用垂直城市的原则

6点到8点，垂直城市可以为上百万短暂停留的人提供歇息的地方，他们占据的面积不会超过岛屿面积的2%。当地的农民工可以继续有机耕种和养鱼，或者通过培训，在新开发的休闲产业区中找到报酬更好的工作。政府获得的收益可以帮助完成SOM公司提出的总体规划，把崇明岛变成一个天堂岛，人们到这里来释放城市带来的压力（上海是世界上最大、最忙碌的城市中心区之一）。

Chongming Island
崇明岛

Conclusion

结论

The world needs no more urgent a wake-up call than the realization that the current global population of 7 billion has the potential to reach 9 or even 10 billion by 2050—less than four decades from now! Such alarming growth forces the question about the point at which Earth will be no longer able to feed such a mass of humanity. Putting political correctness aside, there is an obvious solution: population control/birth control/family planning. Whatever the name, there is a grave need to stabilize growth. Research has shown a direct correlation between education and population control; therefore, we implore experts in related fields—teachers, medical providers, sociologists, religious leadership, politicians, everyone!—to urgently redouble their efforts to find a solution and prevent the unchecked, ever-increasing growth of humanity from dragging us all into the abyss.

We often refer to the delicate balance of nature—exemplified by the need for the perfect mix of temperature, atmosphere, and nutrients in the ooze of the tidal pool—for life as we know it to begin. So too was the delicate mutation of our forbears' DNA to enable us to walk upright, carrying proportionally the largest brain of all Earth's creatures. On the other hand, there is no delicacy in the brutal forces that the universe is capable of when it hurls an asteroid—for instance, Little Ida—a mere 36 miles [58 kilometers] in diameter (about the size of the island of Singapore), for a direct hit on our planet; or when the melting ice caps and ice sheets of the Arctic and Antarctic raise the sea level by some 195 feet [58.5 meters], drowning most of Earth's coastal cities; or when the entire planet is blanketed in mile-thick ice [1.6 kilometers], thereby wiping out mankind—the mankind who unthinkingly continued to tip the balance until it reached a catastrophic slide into oblivion. But in time, perhaps millennia later, spring will return, and nature, being the nurturing mother that she is, will give a wiser mankind another chance.

As architects, we can help point out to men and women across the world that we need desperately to pause and take stock of our actions before steering the course irreversibly in the wrong direction. We need to stop releasing so much carbon dioxide into the atmosphere that we cannot ever recover; killing our streams, rivers, and oceans by the criminal dumping of pollutants; and mindlessly continuing to waste natural resources or otherwise consume more than our fair share of Earth's rapidly diminishing bounty. Some would argue that it's already too late. Before we are forced to live

从现在开始30多年后，到2050年时，全球人口有可能从当下的70亿猛增到90亿甚至100亿。对于整个世界来说，没有比这更响亮更急迫的警钟了。如此惊恐的增长不禁令人质疑，地球是否还能再养活这么多人。抛开政治正确性不谈，很直接的解决方案就是控制人口/控制出生/家庭计划。不管以什么样的名义，稳定增长率都是迫切的要求。研究已经表明，教育和人口控制之间有直接的关联。因此，我们恳求相关领域的专家——教师、医疗人士、社会学家、教会人士、政治家以及每个人！——进一步努力，找到一个解决方案，不要让不受控制、不断增长的人口数量把我们所有人拽入深渊。

我们经常提到自然的微妙平衡，例如潮汐池中的软泥，其温度、空气和养分要达到恰当融合才有利于生命的诞生。同样地，正是因为我们祖先DNA的突变，才让我们拥有地球生物中比例最大的大脑，能够直立行走。另一方面，宇宙的可怕力量并不优雅美妙，它能够掷出一颗小行星直接击中我们的星球〔比如，小伊达长只有36英里（58公里），大约有新加坡那么大〕。或者冰山，南北极冰层融化，把海平面提升195英尺（58.5米），淹没地球上大多数沿海城市。又或者整个星球被冰冻在一个1英里（1.6千米）厚的冰层中，从而彻底毁灭人类——人类正轻率地持续打破均衡，直到有一天灾难将这一切埋葬。但最后，可能是数千年后，春天回归，自然又重新成为万物之母，它会再给一种更聪明的人类一个机会。

作为建筑师，我们可以向全世界的人指出，在走上向空气中排放大量二氧化碳这条不归路之前，在犯罪般地向我们的溪流、江河、海洋倾倒废弃物之前，在愚蠢地无度浪费自然资源，消耗大大超过应当得到的地球恩赐之前，我们迫切需要停下来，评估我们的行为。有些人会说已然太晚了。在我们被迫生活在海上漂流的浮筒中，被迫生活在地面上电缆悬吊的豆荚中之前，我们需要用心探究垂直建设的可能性，不一定要受1英里高的限制，足够的高度将让我们不会再因为扩张失去更多的耕地、树林和珍贵的自然资源。我们需要重新评估原先认为不适合种植的地方，以前干燥现在被海水淹没的地方，以及城市中某些人口稀少、无法自我修复的区域。无论如何，让我们先开始探究那些我们最熟悉的问题，然后延伸到我们的安乐窝之外，一小步一小步，探究所有有助于保护地球（目前所知是唯一适合我们生活的星球）的方法。这本书正是以此为目标的。

on the ocean atop floating pontoons or aboveground in pods suspended from cables, we need to diligently explore the possibilities of building vertically, not necessarily at mile-high limits, but at sufficient heights that will ensure against losing to urban sprawl yet more arable land, yet another clump of trees, yet another bit of precious nature. We need to reevaluate terrain that is not suitable for farming, previously dry ground now inundated by rising sea levels, and parts of existing cities that lack the critical population mass to rejuvenate itself. By all means, let us start by exploring issues with which we are most familiar and then, reaching beyond our comfort zone, explore all avenues that may—small step by small step—help us find ways to sustain the only habitable planet we know. This book has been one such attempt.

APPENDIX

Kenneth S.H. King
Kellogg H. Wong
Experts' Biographie
Photo Credits

附录

Kenneth S. H. King

Kenneth S. H. King, M.Arch., NCARB, was born in 1933 to a family of modest means, the youngest of four children. They lived in the Yu Yuan section of Shanghai, a tourist area in the eastern part of the old city. When Ken was just a toddler, his father, an accountant, was posted to Singapore for a year, and then to Hong Kong for an additional three years. By the time the family returned to Shanghai and settled in the French Concession in 1941, the city was under Japanese occupation.

Toys were expensive and scarce, so Mr. King spent his childhood playing chess, making origami, and raising silk-worms. In the process, he learned to think strategically, to be dexterous and see geometrically in three dimensions, and to appreciate Nature and the delicate balance of its living systems. He says of his early years, "In retrospect, I realize that these are the very qualities that predisposed me to become an ecologically driven architect."

When Japan surrendered in 1945 and brought World War II to a close, Mr. King remembers vividly how, after years of air raids and huddling under the dining room table, Shanghai exploded with life, flooded by all manner of American war surplus, "including everything from C-rations and chocolates to thousands of U.S. sailors." The excitement lasted about two years, until civil war broke out between the Kuomintang (the Nationalist Party), headed by Chiang Kai-shek, and the Communist Party under Mao Zedong.

The Kuomintang was riddled with corruption and started printing currency without the necessary reserves. When inflation soared, Chiang Kai-shek sent his son Chiang Ching-kuo to Shanghai to restore control. But the goods on store shelves suddenly disappeared and the black market became the only source of provisions. People started losing faith in the Nationalists. Despite hefty support from the United States, they could not prevail, and in 1949, the Kuomintang retreated to Taiwan.

金世海，建筑学硕士，美国国家建筑注册委员会会员

1933年，金世海出生于一个小康家庭，他是四个孩子里面最小的。金家当时住在上海豫园区，现在，那里已经是城市南部的一个观光区了。当金先生还是一个小孩子时，他的父亲，一位会计师，被派去新加坡工作一年，然后又到香港停留了三年。当1941年金家回到上海并在法租界安顿下来时，上海已经被日军占领了。

在那个时代，玩具昂贵且稀少，因此，金先生在下棋、折纸、养蚕中度过了自己的童年。玩耍过程中，他学会了策略性思考、敏捷的技巧以及从几何学角度三维地观察事物，他甚至还学会了欣赏自然及生态系统中那微妙的平衡。在谈及自己的童年时，金先生说："回想起来，我发现，正是这些气质让我成为了一个生态型的建筑师。"

1945年日本投降，二战宣告结束。金先生记得很清楚，经历了多年空袭和在餐桌下避难之后，上海重新焕发了生机。大量美军战时供应源源不断地流入上海，"包括各种各样的东西：从干粮和巧克力到美国水兵"。美好时光大约持续了两年，直到蒋介石领导的国民党和毛泽东领导的共产党之间爆发内战。

国民党被贪腐击垮了，在没有物资储备的情况下，他们就开始滥印货币。当通货膨胀无法控制时，蒋介石派他的儿子蒋经国到上海来稳定经济。但是，商店货架上的货物忽然被一扫而光，黑市成了物资供应的唯一来源。人们开始对国民党失去信心。尽管有美国的大力支持，国民党还是未能取胜，并在1949年退守台湾。

之后不久，金老先生在一家英国公司得到一个职位，并决定搬回那时依然在英国统治下的香港。金先生在香港完成高中学业，1953年他到了

伦敦，并在伦敦北岛理工学院（伦敦城市大学的前身）学习建筑。

在所有老师中，爱德华·柯蒂斯教授对金先生影响最大，他年轻、富有远见，鼓励学生进行创造性思考，在20世纪50年代早期就向他们介绍鲍豪斯，而那时，许多建筑学院依然深受学院派美术风格的影响。这一进步运动将美术与工艺联系起来，通过在简单的逻辑形式中整合排列多种功能，创造了一种全新的建筑风格。

柯蒂斯教授在许多方面都领先于他的时代。1956年，他邀请课上的学生到他刚刚设计并建造的伦敦以外的房子去。这是一个大小合宜适中的结构，坐落在半英亩（2000平方米）的土地上，但它的突出之处在于，它是一项早期的生态设计项目。所有居住空间都朝南，房屋外面有反光装置，将更多的自然光线反射到屋内。在晴天时，无需供暖系统，房屋也温暖舒适。

房屋外面，在草坪下，水管被安置在6英尺（2米）深的位置，那里土壤的温度大约是60华氏度（15.5摄氏度），因而水管中的水可以一直保持着这个温度。所以，在冬天为房屋加温时，夏天为房屋降温时，可以少消耗一些能源。柯蒂斯教授在那些最具环境意识的建筑师产生可持续性观念之前，就设计了这个房屋。

在柯蒂斯教授的鼓励下，金世海先生申请了芝加哥的伊利诺伊理工学院（IIT）继续自己的学业，密斯·范·德·罗那时是学院建筑系主任。他的申请被接受了，但由于经费问题他不得不放弃IIT转到位于纽约特洛伊的伦斯勒理工学院，那里为他提供全额奖学金。一年之后，他获得了他的建筑学硕士学位。在麻省一家小型建筑公司待了一段时间后，金先生定居纽约并于1965年获得了美国公民身份。

在其职业生涯早期，金先生专注于医院建筑的设计。这一专门领域对综合功能的要求，特别是对建筑关键效能和分层循环系统的要求，让金先生自然而然地在规划上学到了很多东西。之后，金先生又参与了位于埃及亚历山大蒙塔扎宫的规划。蒙塔扎宫占地300英亩（120公顷），位于可俯瞰大海的悬崖之上。他也参与莫卡塔姆城的总体规划工作，这个项目在一座能俯视开罗的山上，占地1万英亩（4000公顷）。今天它以"垃圾城"闻名于世。正是在这里，札巴林社区（Zabaleen），（在阿拉伯语中是拾荒者的意思）发展了一种有效、环保的废物处理系统，它领先于大多数现代绿色环保方案，能分类和循环

Shortly afterward, the elder Mr. King was offered a position in an English firm and decided to move back to Hong Kong, which was then still under British sovereignty. Ken King finished high school there and, in 1953, went to London, where he studied architecture at Northern Polytechnic Institute (now London Metropolitan University).

Among all of the school's teachers it was Prof. Edward Curtis who inspired him most. Young and visionary, he encouraged students to think creatively, introducing them to the Bauhaus in the early 1950s, when many architecture schools were still strongly under the influence of Beaux-Arts. Linking arts and crafts, the progressive movement created a new architectural style by organizing diverse functions in simple logical forms.

Professor Curtis was ahead of his time in many ways. In 1956, he invited his class to the house that he had designed and recently built just outside London. It was a modest structure sitting on about a half acre [2,000 square meters] of land, but its significance loomed large as an early ecological design. All the living spaces faced south, with exterior reflectors to direct more natural light inside. On sunny days, the house was comfortably warm even without the use of the heating system.

Outside, under the lawn, water pipes were installed to a depth of 6 feet [2 meters]. Earth's temperature at that level is about 60 degrees Fahrenheit [15.5 degrees Celsius], hence water in the pipes was kept at a constant temperature so the house could be heated in the winter and cooled in the summer, using much less energy. Professor Curtis designed his house before the notion of sustainability existed, even among the most environmentally conscious architects.

With the encouragement of Professor Curtis, Ken King applied to continue his studies at the Illinois Institute of Technology (IIT) in Chicago, where Ludwig Mies van der Rohe headed the architecture department. He was accepted but had to decline due to financial concerns, and instead attended Rensselaer Polytechnic Institute in Troy, New York, on a full scholarship. He received his Master of Architecture degree a year later. After a brief stint in a small Massachusetts architectural firm, Mr. King settled in New York City. He became a naturalized U.S. citizen in 1965.

Ken King's early career focused on hospital design. The complex functions of this specialized field, particularly with respect to critical efficiencies and layered circulation systems, led him naturally to learn more about planning. He became involved in the planning of Montazah, Egypt, a 300-acre [120-hectare] recreational development on the rocky bluffs of

Alexandria overlooking the Mediterranean Sea. He also participated in the master-planning of Mokkattam, a 10,000-acre [4,000-hectare] development in the hills above Cairo, best known today as the home of "Garbage City." It was here that the Zabaleen (Arabic for garbage collectors) developed an effective and environmentally friendly waste disposal system far ahead of most modern green initiatives, sorting and recycling some 80 percent of the city's trash. (By contrast, Western garbage collectors recycle only 20–25 percent.)

Mr. King subsequently became involved in many other projects, including the design of the first Bank of China building in New York City, yet it was his early planning experience together with Professor Curtis's house that had the most enduring impact on his career, as they fueled his interest in sustainable design. "It is the planner's responsibility to think ahead and to envision possibilities," he explained. Today, as the world confronts massive traffic congestion, widespread pollution, the irreplaceable loss of arable land, and the destruction of the rainforests and other natural resources, this greater vision has compelled Kenneth King to join with the millions of environmentally conscious people around the world in finding solutions for a more sustainable existence.

利用80%的城市垃圾。(相比而言，西方的垃圾收集站只能循环利用20%—25%的垃圾。)

金先生后来还参与了许多其他项目，包括纽约第一家中国银行的规划工作，然而，早期他与柯蒂斯教授那所房屋相关的规划经验，对他整个职业生涯产生了最持久的影响，让他对可持续性设计极感兴趣。他解释说："规划者的职责就是预见、构思各种可能性。"今天，世界面临着大规模的交通拥堵，大范围的环境污染，可耕种土地无可挽回的流失，雨林和其他自然资源的破坏，这种规划者的超前视野迫使金世海先生加入到全球成千上万有环保意识的人群中，并努力为一种更可持续的生活寻找解决方案。

Kellogg H. Wong

Kellogg H. Wong was born in Rosedale, Mississippi, in 1928, a year before the Great Depression. At the age of five, he and his two siblings were sent to Nanjing for a "better" life, but returned to the United States when Japan invaded Shanghai in 1937.

He attended the bilingual Chinese Mission School in Cleveland, Mississippi, established by Chinese parents in the segregated Mississippi Delta, before rising through the much superior high school system in Memphis, Tennessee. He earned a Bachelor of Architecture degree from the Georgia Institute of Technology in 1952 and was awarded a Fontaine-bleau Traveling Fellowship, which he had to decline because his attendance at a land-grant college required him to serve two years in the military. His tour of duty began as a Guided Missile Briefing Officer with a top secret clearance at the U. S. Army Ordnance Corps facilities at White Sands Proving Grounds in New Mexico, followed by a stint as commander of one of the largest U.S. Army ammunition depots in Korea.

After military service, Mr. Wong practiced architecture in Houston, Texas, and then enrolled at the acclaimed Cranbrook Academy of Art, founded by Eliel Saarinen, in Bloomfield Hills, Michigan. He earned his M.Arch. degree and, at the urging of his brother, Pershing Wong, also an architect, moved to New York City in order to gain the invaluable experience of working for a year or two in the office of the increasingly prominent Chinese American architect I. M. Pei. At the time, Pei was employed by the legendary developer William Zeckendorf as head of the architectural department of Webb & Knapp, one of the largest real estate companies in New York.

Except for a two-year sabbatical as Assistant Professor at Rice University School of Architecture, Mr. Wong remained with the Pei office for some forty years—nearly his entire professional life. During that time he worked almost exclusively with I. M. Pei on such award-winning projects as the School of Journalism at Syracuse University, National Airlines Terminal at Idlewild (now JFK International) airport, and most important, Everson

黄慧生1928年出生于密西西比州的罗斯代尔，一年之后股票市场的崩溃导致了大萧条。五岁时，为了过上"更好"的生活，他和两个兄妹被送到了中国南京。1937年日本入侵上海时，他们回到了美国。

在进入田纳西州孟菲斯更好的高中之前，黄先生在密西西比州的克利夫兰上了双语的中国教会学校。这所学校是华裔父母们在密西西比三角洲隔离区建立起来的。1952年，他在乔治亚理工学院获得了建筑学学士学位。毕业时，黄先生获得了枫丹白露旅行奖学金，但为了服两年的兵役，他只能无奈地放弃了，因为奖学金要求他到提供资助的大学去。作为一名导弹项目简报官，他开始了自己的服役期。服役地点在新墨西哥州白沙导弹靶场的美国军械部基地，这是涉及美国军队军火设施的绝密工作。后来有一段时间，他曾短暂担任驻韩国的美军最大军火库之一的指挥官。

服完兵役之后，黄先生在德克萨斯的休斯顿开始从事建筑行业的工作。之后，他进入广受赞誉的克兰布鲁克艺术学院深造，这所学校由伊利尔·萨里南建立，位于密歇根州的布隆菲尔德山。在那里，他获得了建筑学硕士学位，而且，在同为建筑师的哥哥黄培生(Pershing Wong)的鼓励敦促下，黄慧生先生搬到了纽约，以便能在声望日隆的美籍华裔建筑师贝聿铭的办公室工作一两年，进而收获宝贵的经验。那时，贝聿铭先生受雇于传奇开发商威廉·齐肯多夫(William Zeckendorf)领导的齐氏威奈公司建筑部，这家公司是当时纽约最大的房地产公司之一。

除了在莱斯大学建筑学院担任了两年助理教授外，黄先生在贝聿铭工作室工作了近40年——这几乎相当于他的整个职业生涯。在此期间，他单独与贝聿铭先生合作完成了多个获奖项目，像雪城大学新闻学院和位于爱德怀德机场（现在的肯尼迪国际机场）的国家航空公司航站楼等项目，还有他们最出名的项目——位于纽约锡拉丘茨的艾弗森艺术博物馆，这是贝聿铭先生的第一个文化项目，也是后来诸多博物馆项目的先驱。另外，黄先生也参与了纽约市四季酒店的规划设计工作。

在新加坡独立和国家建设的早期阶段，黄先生负责拉夫尔斯国际中心和后来拉夫尔斯城一系列雄心勃勃、多功能的规划设计，这是一个占据400万英尺（合37万平方米）的综合项目，包括酒店、会议中心、办公室、餐厅和商业空间。另外，他也在高52层的OCBC（华侨银行集团总部）项目中扮演重要角色。1976年完工时，这曾是亚洲最高的建筑。他也是港汇广场一个双子办公设施的负责人，并参与了新加坡河流总体规划和南码头总体规划项目的工作，这两个项目共同构成了新加坡中心商业区最终拓展方案的指导方针。

在中国，黄先生致力于北京外围香山饭店的规划和选址工作。他还参与了香港中银大厦的工作——1989年完工时，中银大厦是世界第六高的建筑。另外，他参与了香港新宁大厦的规划设计，这座大厦具有办公和居住的双重功用。

黄先生在贝聿铭工作室最后参与的国际项目分别位于马来西亚、中国台湾和印尼，也包括位于雅加达金融银行的一系列规划项目。这些项目包括主题设计、景观优化、酒店、办公楼、住宅楼、公共空间及配套设施的设计规划。

作为贝·考伯·弗里德及合伙人建筑师事务所（之前的贝聿铭建筑设计事务所）的名誉合伙人，黄慧生先生于2000年退休。在接下来的几年间，他被米兰的雷纳托塞尔诺集团聘请，参与了纽约联合国总部大楼主体翻新计划前期预备方案的工作。这一方案规划得到了联合国大会全会的一致认可，并成功实施。

之后，黄先生完成了位于加利福尼亚州莱克县的弥勒佛香巴拉修道院的设计工作，并开始担任纽约市FORM公司形式上的负责人至今。他同时也是位于北京的龙安建筑规划设计顾问有限公司纽约设计中心的高层人员。

多年来，黄先生在很多著名大学、组织及会议上发表演说，并为各个重要的国际设计委员会服务，同时，他还积极参与到很多主管部门、专业机构委员会、社区和教堂规划咨询委员会的工作中去。

过去十年间，黄先生对于中国的兴趣大大增加，这是一个正承受持续发展压力的国家。他坚定地相信，耐心的理解和专业的指导将使这里的社会和建筑实现长足的发展。黄先生决定参与到这本书的工作中来，这本身就深刻地阐述了他对公共服务的态度：为了人类更大的福祉，为了环境而努力奋斗，这是他一向视为荣耀和责任的事。

Museum, Pei's first cultural project and the sculptural precursor of the many museums that followed. Mr. Wong also worked on the Four Seasons Hotel in New York.

In the early days of Singapore's independence and nation-building, Wong worked on an ambitious mixed-use scheme for Raffles International Center and its subsequent realization at Raffles City, a 4-million-square-foot complex encompassing hotels, convention spaces, offices, restaurants, and commercial space. In addition, he played an important role in OCBC, the 52-story Overseas Chinese Banking Corporation headquarters, which was the tallest building in Asia upon completion in 1976; the twin-tower Gateway office complex; and both the Singapore River Master Plan and Marina South Master Plan, which together formed the guidelines for the eventual development of Singapore's expanded central business district.

In China, Mr. Wong worked on the Fragrant Hill Hotel outside Beijing (planning and site selection phase), as well as on the Sunning Plaza office and apartment project, and the Bank of China Tower in Hong Kong, the world's sixth-tallest building upon completion in 1989.

Mr. Wong's last international works in the Pei office were located in Malaysia, Taiwan, and Indonesia, including a series of projects at Anggana Danamon in Jakarta that encompassed master planning, landscape optimization, hotels, office buildings, residential towers, public spaces, and supporting facilities. Mr. Wong retired from Pei Cobb Freed & Partners in 2000; he remains an Associate Partner Emeritus. In the following year he was enlisted by the Renato Sarno Group of Milan, Italy, to perform the preliminary phase of the Capital Master Refurbishment Plan of the United Nations Headquarters, New York. The project was successfully completed upon unanimous approval of the full session of the United Nations General Assembly.

Subsequently Mr. Wong completed the design for the Buddha Maitreya Shambhala Monastery in Lake County, California, and assumed his current role as a principal of FORM | Proforma of New York City, also heading up the New York Design Center of J.A.O. Design International, Beijing.

Over the years Mr. Wong has lectured at numerous leading universities, organizations and conferences, served on major international design juries, and actively participated on the governing bodies and committees of professional agencies and community and church planning advisory boards.

During the past decade his interests have focused increasingly on China, where development pressures continue to grow and where, he firmly believes, with patient understanding and professional guidance, great social and architectural advances can be realized. Mr. Wong's decision to participate in this book speaks deeply of his commitment to public service, which he sees both as an honor and as a duty for the greater good of humanity and our environment.

Experts' Biographies

Mark S. T. Anderson
Peter C. O. Anderson

Anderson Anderson Architecture, San Francisco

Mark Anderson, FAIA, and Peter Anderson, FAIA, have broad experience in building design and construction as architects, builders, and educators. In partnership in the firm Anderson Anderson Architecture, the Anderson brothers have designed and constructed numerous building projects in the United States and Asia and have also directed construction technology research projects, exhibitions, and public art installations in the United States, Europe, and Asia. Their work has received many competition prizes and design awards, including three Progressive Architecture Honor Awards; national honors from the American Institute of Architects, the Boston Society of Architects, the American Wood Council; the Emerging Architects honor from the New York Society of Architects; and particular recognition for design work in service to society and the environment, from such groups as the Danish Royal Index Awards, Copenhagen; the Zumtobel Awards, Zurich; and the Holcim Awards, Zurich.

Their drawings, design models, and industry prototypes have been widely exhibited, including at the Venice Biennale of Architecture, the Museum of Modern Art in New York, the Hamburg Museum für Kunst und Gewerbe, the Los Angeles Museum of Art and Design, the San Francisco Museum of Modern Art, the Panama City Museum of Art, and the Danish Design Museum. Their work has appeared frequently in books and professional publications in the United States, Asia, and Europe—including in the journals *Architecture*, and *Architectural Record* (New York), *PLAN* (Milan), Deutsche BauZeitung (Berlin), *Architecture Review* (London), and *L'Industrie delle Costruzioni* (Rome). A monograph on their work, *Anderson Anderson: Architecture and Construction*, was published by Princeton Architectural Press in 2001. A book on their design and construction technology work, *Prefab Prototypes; Site-Specific Design for Off-Site Fabrication*, was published by Princeton Architectural Press in 2007, and nominated for a Royal Institute of British Architects Book Award. A fifty-page retrospective and critical review on their design work was published by *Taiwan Architect*, May 2009. Mark Anderson is a member of the architecture faculty at the University of California, Berkeley, and an appointed Peer member of the United States Federal Commission for Excellence in Art and Architecture. Peter Anderson is a member of the architecture faculty at the California College of the Arts in San Francisco. Mark and Peter are both Fellows of the American Institute of Architects.

Ajmal Aqtash

form-ula

core.form-ula, New York

In late 2009 Ajmal Aqtash co-founded form-ula, an award-winning multidisciplinary design practice that seeks to understand the intersection of design and engineering and its collaborative possibilities to produce culturally rich and high-performance architecture for large- and small-scale projects. The website that core.form-ula launched in July of 2007, is now the research and development wing of form-ula in its goal to encourage innovation in design, architecture, engineering, and art.

Ajmal has an extensive background in morphology and is an assistant director at the Center for Experimental Structures (CES) at Pratt.

He worked with Dr. Haresh Lalvani on exploring folded structures, particularly using AlgoRhythms Technologies, a trademark of Milgo-Bufkin.

Before founding form-ula, Ajmal worked on several projects for Skidmore, Owings & Merrill LLP (SOM), ranging from super towers to large complex proposals for airports and universities. During his years at SOM, he worked within the Design Group, Technical Group, and Digital Design Group and Performative Design Group, pushing advanced design methods, computational techniques, and sustainable design. Toward the end of his tenure at SOM, he founded the Advanced Design Group and the Performative Design Group; both groups were new additions to the 80-year history of SOM.

Ajmal is currently an adjunct professor at Pratt Institute School of Architecture, has taught at Stevens Institute School of Mechanical Engineering, and has been an invited juror at Columbia, Penn, Pratt, SCI-Arc, and Stevens. Ajmal currently lives and works in New York City.

Erich Arcement

Sam Schwartz Engineering, New York

Erich Arcement is a senior vice president and the general manager of Sam Schwartz Engineering's New York Office. He is directly responsible for all Sam Schwartz Engineering (SSE) projects based in the New York region. Prior to his current role, Mr. Arcement was the Director of SSE's Traffic and Transportation Engineering group. He has extensive expertise in engineering ranging from geometric design and analysis of roadway, bicycle and pedestrian facilities to traffic impact studies, environmental assessments, pavement marking and signage design. He has been with the firm since it opened in July 1995 and has managed hundreds of projects ranging from small impact studies to multi million-dollar on-call contracts. Mr. Arcement's responsibilities include delegating work to SSE staff, quality control of projects, and oversight of project costs and budgets. He is also responsible for client contact and project budget management.

Erich is a Professional Engineer and Professional Traffic Operations Engineer and holds a Bachelor of Engineering in Civil Engineering (Cooper Union for the Advancement of Science and Art) and a Master of Science in Transportation Planning and Engineering (Polytechnic University).

Rick Barker

Barker Mohandas, Bristol, Connecticut

Rick Barker is involved in most of the work of Barker Mohandas. Prior to co-founding the firm, he was Director of Technical Services at Otis World Headquarters, where his responsibilities included preconstruction services for major projects. He also chaired the Otis product strategy group for dispatching control products and simulation tools, and co-led strategy for the company's tall-building elevator product called Skyway™. He led a major study to improve elevator product energy efficiency, co-led the Otis Odyssey™ system development that integrated elevators and horizontal automated people movers (APMs), and was a key liaison between Otis World HQ and United Technologies Research Center.

Prior to joining Otis, Rick led vertical transportation at Jaros Baum & Bolles Consulting Engineers (JB&B) in New York, and was involved in the company's largest projects at the time. At Westinghouse Elevator (now Schindler) he was involved in the company's largest projects in western New York, and at Delta Elevator in Boston (now Otis) in the company's largest modernization contracts.

Rick has authored papers on super-speed elevators, on an integrated horizontal-vertical transportation system, and on elevators and fire safety for firefighters and disabled persons. He is a member of the National Interest Review Group of the ASME-A17 Safety Code for Elevators and Escalators, and a former member of the A17.1 Emergency Operations Committee and similar NYC and FDNY committees on fire and elevators. He has co-chaired the Vertical and Short-Distance Horizontal Transportation Committee for the Council on Tall Buildings and Urban Habitat, guest lectured at MIT School of Architecture, and was a board member of the Building Owners and Managers Association of Buffalo. Rick Barker is named, singly or jointly, in 24 patents held by Otis.

Thomas J. Campanella

Department of City and Regional Planning
University of North Carolina, Chapel Hill

Thomas J. Campanella is Associate Professor of Urban Planning and Design at the University of North Carolina, Chapel Hill, where he is also a Faculty Fellow at the Institute for the Arts and Humanities. He taught previously at the Massachusetts Institute of Technology, and was a visiting professor at the Harvard Graduate School of Design in 2008. He has also lectured at Nanjing University in China, and was a Fulbright Fellow at the Chinese University of Hong Kong from 1999–2000.

A native of Brooklyn, Mr. Campanella is an urbanist and historian whose work focuses primarily on the evolution of the urban civic landscape of the United States. He has also studied and written about the wholesale transformation of Chinese cities in the post-Mao era. His most recent book is *The Concrete Dragon: China's Urban Revolution and What It Means for the World* (Princeton Architectural Press, 2008), which the *Washington Post* has called "a powerful overview of China's huge building boom and its social and environmental consequences."

Professor Campanella is also the author of *Republic of Shade: New England and the American Elm* (Yale University Press, 2003), the first study of the origins and significance of "Elm Street" in America. The book received the Spiro Kostof Award from the Society of Architectural Historians, and was featured in the *New Yorker, Los Angeles Times*, and *Boston Globe*, which named it one of its "ten best non-fiction books" of 2003. His other works include *Cities from the Sky: An Aerial Portrait of America* (Princeton Architectural Press, 2001) and *The Resilient City: How Modern Cities Recover from Disaster* (Oxford University Press, 2003), edited with Lawrence J. Vale. Mr. Campanella also writes frequently for the popular and professional press. His essays have appeared in the *Wall Street Journal, Metropolis, Salon, Wired, Orbit*, and *Architectural Record.*

Thomas Campanella holds a Ph.D. in urban planning from MIT and a master's degree in landscape architecture from Cornell University. He published an award-wining article in *Landscape Journal* during his second year at MIT, and his doctoral dissertation received the John Reps Prize from the Society for American City and Regional Planning History. At Cornell, he was awarded the Michael Rapuano Memorial Medal for design excellence in the College of Architecture, Art, and Planning.

Mr. Campanella is currently working on two books: *The Last Utopia* will chronicle the rise and fall of Soul City, North Carolina, an intentional "new town" planned and partially built in the 1970s by the civil rights lawyer Floyd B. McKissick; *Designing the American Century* will examine the careers of two of the most important American landscape architects of the twentieth century—Gilmore D. Clarke and Michael Rapuano—creators of many of the parks and public works associated with Robert Moses in New York.

Scott Ceasar, P.E.

Cosentini Associates, New York

Scott Ceasar joined Cosentini Associates in 1988 and is now a senior vice president with over 20 years of project experience that includes major hotels and convention centers, courthouses, corporate headquarters, and high-rise office buildings. Mr. Ceasar was one of the first engineers in the country to receive LEED accreditation from the U.S. Green Building Council and is responsible for assisting Cosentini's engineering teams with the development of sustainable engineering solutions for their projects.

Mr. Ceasar holds a Bachelor of Science in mechanical engineering from The Johns Hopkins University in Baltimore and is a licensed professional engineer in New York. He is an active member of the U.S. Green Building Council and the American Society of Heating, Refrigerating, and Air-Conditioning Engineers. ity.

Dickson Despommier

Columbia University, New York

Dickson Despommier was born in New Orleans in 1940, and grew up in California before moving to the New York area, where he now lives and works. He has a Ph.D. in microbiology from the University of Notre Dame. For 27 years, he conducted laboratory-based biomedical research at Columbia University with NIH-sponsored support. He is now an emeritus professor.

Professor Despommier has always been interested in the environment and the damage we have caused by the simple act of encroachment. At present, he is engaged in a project whose mission is to produce significant amounts of food crops in tall buildings situated in densely populated urban centers as described in his book *The Vertical Farm: Feeding the World in the 21st Century* (2011). This initiative has grown in acceptance over the last few years to the point of stimulating planners and developers around the world to incorporate vertical farms into their vision for the future city. To date, there are seven up and running in Japan, Korea, Seattle, Chicago, and Vancouver, with many more in the planning stage. The hope is that vertical farming will become commonplace throughout the built environment on a global scale.

Dr. Despommier has received numerous teaching awards (eight Best Teacher of the Year awards at his medical school), and the national American Medical Student Golden Apple Award for Teaching Excellence in 2003. He has published four books, over 80 peer-reviewed scientific papers, and numerous review articles on a wide variety of subjects. He has lectured on the subject of vertical farming at MIT, Harvard, NYU, Cornell, Rutgers, Brigham Young University, and Singularity University, to city agencies in Chicago; New York City; Seattle; Newark, New Jersey; Los Angeles; Seoul, Korea; Amman, Jordan; Beijing, China; Bangalore and Coimbatore, India, and to federal government agencies, including the United Nations, IMF, USDA, and USAID. He has presented at TED, TEDx Chicago, TEDx Bermuda, TEDx Washington, DC, TEDx Engineering Columbia University, and TEDx Youth, PopTech, Ars Electronica, IdeaCity, Monterey Design Conference, Seoul Digital Forum (2008, 2011), the Indian Institute for Architecture annual meeting (2009), and The Colbert Report. Dickson Despommier lives with his wife in Fort Lee, New Jersey.

Dr. Chengri Ding

Urban Studies and Planning Program
University of Maryland, College Park

Dr. Chengri Ding is an associate professor of the Urban Studies and Planning Program, at the University of Maryland, specializing in urban economics, urban and land policies, urban planning, and China studies. He has published articles in the Journal of Urban Economics, Journal of Regional Science, Urban Studies, Environment and Planning B, Housing Policy Debates, and Land Use Policy. He has edited three books on China in land and housing policies, urbanization, and smart growth. Dr. Ding also has numerous publications in Chinese (manuscripts and journal articles). He has been P.I. for many international policy projects on China, including urban master planning, farmland protection, property taxation reform and local public financing. Reports and publications have been widely received among both high level officials and scholars. Dr. Ding has given more than 50 invited or keynote speeches and presentations.

He has been a consultant to the World Bank, Global Business Network, FAO, and many leading Chinese agencies. He serves on the advisory board for the International Institute of Property Taxation. Dr. Ding is the founding director for the Lincoln Institute of Land Policy's China Program.

Philip Enquist, FAIA

Skidmore, Owings & Merrill LLP, Chicago

Philip Enquist is leader of the global city design practice of Skidmore, Owings & Merrill LLP (SOM), the world's most highly awarded urban planning group. Phil and his studios have improved the quality and efficiency of city living on five continents by creating location-unique strategic designs that integrate nature and urban density within a framework of future-focused public infrastructure.

The scale of Phil's design perspective continues to expand from innovating sustainable urban forms that enhance city living with walkable, transit-enabled districts humanized by their natural amenities to rapidly changing urban clusters within regional ecosystems like North America's Great Lakes basin and China's Bohai Rim.

Phil is committed to the profession through one-on-one mentorships, his recent teaching of a studio for architecture and urban design students at Harvard University's Graduate School of Design, and as the Charles Moore Visiting Professor at the University of Michigan's Taubman College of Architecture and Urban Planning.

He was honored with the 2010 Distinguished Alumnus Award from the Architectural Guild of the University of Southern California (USC) School of Architecture for his dedication to strengthening the physical, social, and intellectual infrastructure of cities. The previous year, the *Chicago Tribune* named him and his studio Chicagoans of the Year in Architecture, citing "the city-friendly designs of Phil Enquist."

Philip Enquist passionately believes that the world's explosive growth in cities and population must be managed by humanely bold and holistically sustainable thinking at the national, regional, and metropolitan scale, and that human habitat design will become the alpha design science of the twenty-first century.

Carl Galioto

HOK, New York

Carl Galioto, FAIA is a senior principal at HOK, managing principal of the New York office, and a member of HOK's board of directors. With more than 30 years of management and project delivery experience, Carl possesses industry-leading expertise in the design and implementation of large, complex projects in the New York metropolitan area and throughout the world. His experience includes such high-rise and super-high-rise projects such as One World Trade Center; 7 World Trade Cente; 7 Times Square; Lotte Tower in Seoul; Evergrande Tower in Zhengzhou, China; and the Greenland Tower in Dalian, China. His experience extends to numerous projects in aviation and the health sciences.

Throughout his career, Carl has used building science to develop better building designs—better aesthetically, functionally, and from the standpoints of sustainability and life safety. He has been involved in the development of building codes and improvement of high-rise life safety in particular. After 9/11 he was influential in initiating changing the New York City Building Code to a customization of the International Building Code and was involved in the analysis of the NIST investigation of the WTC collapse and the translation of those findings into model building codes. Those provisions and more appear as innovations in his projects.

In collaboration with Rensselaer Polytechnic Institute, Mr. Galioto co-founded the Center for Architecture, Science and Ecology, a program of graduate-level research to create and implement innovate sustainable solutions. As a member of HOK's board and chair of its Project Delivery Board, Carl also holds firm-wide responsibilities for project delivery, including HOK's virtual design and construction initiative: buildingSMART.

Gordon Gill

Adrian Smith + Gordon Gill, Chicago

The architect Gordon Gill is one of the world's preeminent exponents of performance-based design. His work, which ranges from the world's largest buildings to elements of a single home, is driven by his philosophy that there is a language of performance: a purposeful relationship between design and the performance criteria placed upon the subject.

A founding partner of the award-winning firm of Adrian Smith + Gordon Gill Architecture, Gordon's work includes the design of the world's first net-zero-energy skyscraper, Pearl River Tower (designed at SOM Chicago), and the world's first large-scale positive-energy building, Masdar Headquarters. These landmark projects achieve energy independence by harnessing the power of natural forces on site, striking a balance with their environmental contexts. Gordon has also designed performing arts centers, museums, schools, civic spaces, and urban master plans across the globe.

Gordon's work has been published and exhibited widely in the U.S. and internationally. His designs have repeatedly been recognized by the American Institute of Architects. In 2009, he was selected as Chicago's Best Emerging Architect by the Chicago Reader. Prior to founding AS+GG in 2006, Gordon was an associate partner at Skidmore, Owings & Merrill and a director of design for VOA Associates. Most recently, he co-founded PositivEnergy Practice, a consulting firm that designs and implements energy- and carbon-reduction strategies for clients around the world.

T.J. Gottesdiener

Skidmore, Owings & Merrill LLP, New York

T.J. Gottesdiener, FAIA, is an architect and managing partner of the New York office of Skidmore, Owings & Merrill (SOM). A graduate of Cooper Union's Chanin School of Architecture, Gottesdiener joined SOM in 1980 and was made partner in 1994.

The diversity of Mr. Gottesdiener's work is demonstrated by the number of projects he has done across the United States and around the world. Although he has a particular expertise in high-rise and super-tall towers, he has also been involved in some of SOM's most visible master plans, renovation, corporate, and transportation projects.

Internationally, his experience is exemplified by the completed Tokyo Midtown project, one of the largest private developments in Japan's history, which includes office, hotel, commercial, retail, residential, and museum components. Other examples include the landside terminal at Ben Gurion International Airport in Tel Aviv; his extensive work in Brazil and The Philippines, and master plans and buildings for Digital Media City Landmark Tower in Seoul, Korea, among other projects in Asia.

Mr. Gottesdiener is committed to enhancing the built environment of New York City and has been responsible for some of SOM's most complex and challenging projects in Manhattan. Closely involved in the revitalization of Lower Manhattan and the redevelopment of the World Trade Center site, Mr. Gottesdiener has overseen the design and construction of two of the site's most recognizable projects—7 World Trade Center, completed in 2006, and One World Trade Center, which is scheduled to be finished in 2015 and will be North America's tallest building.

Other important New York City projects include the mixed-use Time Warner Center; the former Bear Stearns Headquarters Building; Times Square Site One; and the renovation of the landmark Lever House building. Along with his project responsibilities, Mr. Gottesdiener supervises the management and operations of SOM's New York office.

Mr. Gottesdiener is also involved in some of New York City's most important real estate and design organizations. He is the vice chairman of the Metropolitan Museum of Art Real Estate Council and serves on the Urban Development/Mixed-Use council at the Urban Land Institute, the Board of the International Center of Photography, and the President's Advisory Board at the Cooper Union. He is a member of the Council on Tall Buildings and Urban Habitat, the Real Estate Board of New York, and the Skyscraper Museum. He is also a Fellow of the American Institute of Architects and an Honorary Fellow of the Philippine Institute of Architects. In 2007, The Cooper Union honored Mr. Gottesdiener with the President's Citation in Architecture in recognition of his substantial contribution to this field.

Robert Heintges

Heintges & Associates, New York

Robert Heintges, FAIA, is the founding principal of Heintges & Associates, and, with over 30 years of experience, is an internationally recognized authority on the design and implementation of the curtain wall.

Mr. Heintges is also an adjunct professor of architecture at Columbia University's GSAPP. He has contributed his extensive knowledge on curtain wall theory and practice as well as his expertise in environmentally responsible design to numerous publications and organizations. Mr. Heintges is personally involved in virtually every project the firm undertakes and has developed long-standing relationships with many of the world's greatest architects.

James C. Jao

J.A.O. Design International, Beijing

James Jao was born in Taiwan in 1957; he became a U. S. citizen in 1980, and is currently president and CEO of the Long On Group, headquartered in Beijing, China. He has served as Expert for the State Administration of Foreign Experts Affairs of the P.R.C.; Strategic Consulting Member of China International Urbanization Developing Community; Second CEO of China Real Estate Design League; Member of "Nation Elites" and planning and economic development consultant for more than 30 provinces. In 2012, Mr. Jao was appointed Vice President and Standing Member of the Overseas Promotions Association of the Inner Mongolia Chinese People's Political Consultative Committee (CPPCC). He also serves as Special Overseas Representative to the CPPCC of Lishui, Zhejiang Province. Since 2007, he has been supervisor for the Beijing Jing Rui Housing Technology Foundation and is guest professor at the National Mayor's Training Center, Tsinghua University, Nanchang University Adjunct Professor; and Suzhou University Graduate Student Advisor. He is also a member of the UN Habitat World Urban Campaign Steering Committee.

Mr. Jao received his B.Arch. degree from Pratt Institute and an MBA from Pace University in New York. He is registered in New York, California, New Jersey, Connecticut, Pennsylvania, Georgia, and Massachusetts. He is also certified by the Chinese National Administrative Board of Registered Architects (NABAR) as a Class A architect.

In 1987, he was elected to the Board of Directors of the New York Society of Architects, and honored as an "Outstanding Asian American Architect." In 1989 he was recognized as a top 10 young architect. From 1990 to 1994, he served on the New York City Planning Commission, its first Asian American and youngest commissioner, before resigning and moving to Asia.

Mr. Jao is a special columnist for *China Daily* and consultant for mainstream media. He has lectured to thousands of Chinese officials and leaders, providing them with the latest planning theories. Among his publications are *A Decade of Planning Experience in China* (2005), and *Searching the Souls of China City* (2009). In 2012, Mr. Jao published the Chinese book *Your City and Mine* and also, for Western readers, *Straight Talk about Urbanization in China*, co-authored with Dr. Janet Adams Strong.

James Jao is president of J.A.O Design International Ltd., Associated Design Institute, Design & Technical Services, Hannuo City (Beijing) Investment Consulting Co., Ltd.; and Design Director of J.A.O. Design International Architects & Planners Ltd. The Long On Group, founded by Mr. Jao by 1984, consists of urban planning, architectural design, capital and real estate investment and charity. J.A.O.D. has been recognized repeatedly as one of the "Ten Most Influential Design Firms in China," and its projects have received over 117 distinguished awards in China and abroad.

Gregory Kiss

Kiss + Cathcart Architects, Brooklyn, New York

Gregory Kiss has been working to advance the art and technology of environmentally responsible architecture for over 25 years. After receiving a Bachelor of Arts from Yale University and a Master of Architecture from Columbia University, he founded Kiss + Cathcart Architects in 1983. He was a founder in 1994, and President until 2004 of the not-for-profit Native American Photovoltaics, which brings solar power to remote homes and schools on the Navajo Reservation.

Mr. Kiss has designed many ground-breaking high-performance building projects in the Americas, Europe, and Asia. His research into the functional and aesthetic improvement of photovoltaics for buildings has led to several new products and systems. He has authored technical manuals for the Department of Energy and lectures frequently on recent advances in solar technologies and their potential for practical integration into architectural design. Recently he has been developing integrated biological systems for buildings, for energy, habitat, and aesthetic benefits, and has extended this work to architecturally integrated food production.

Current projects include Bushwick Inlet Park, in Williamsburg, Brooklyn, and the Bronx River Greenway River House, both of which have won awards from the New York City Public Design Commission. Other projects include the photovoltaic glass train shed for New York City Transit's Stillwell Avenue Terminal in Coney Island, solar and sustainable housing in the Netherlands, the Bocas del Toro Station for the Smithsonian Tropical Research Institute in Panama, the PV system at 4 Times Square, and a photovoltaic manufacturing facility for Heliodomi in Greece.

Jaime Lerner

Jaime Lerner Arquitetos Associados, Curitiba, Brazil

Jaime Lerner is an architect and urban planner, founder of the Instituto Jaime Lerner, and chairman of Jaime Lerner Arquitetos Associados (JLAA). President of the International Union of Architects 2002–2005, and three-time mayor of Curitiba, Brazil, he led the urban revolution that made the city renowned for urban planning, public transportation, environmental social programs, and urban projects. Mr. Lerner served as governor of Paraná State twice and conducted an economic and social transformation in both the urban and rural areas.

Jaime Lerner's international awards include the highest United Nations Environmental Award (1990), Child and Peace Award from UNICEF (1996), the 2001 World Technology Award for Transportation, and the 2002 Sir Robert Mathew Prize for the Improvement of Quality of Human Settlements. In 2010 Lerner was named one of the 25 most influential thinkers in the world by *Time* magazine. In 2011, in recognition of his leadership, vision, and contribution in the field or sustainable urban mobility, he received the Leadership in Transport Award, granted by the International Transport Forum at the Organization for Economic Co-operation and Development.

Jaime Lerner Arquitetos Associados develops projects for the public and private sectors for cities in Brazil and abroad, such as Porto Alegre, São Paulo, Rio de Janeiro, Brasília, Florianópolis, Recife, Luanda (Angola), David (Panama), Durango, Oaxaca, Mazatlán (Mexico) and Santiago de Los Caballeros (Dominican Republic).

Liu Thai Ker

RSP , Singapore

Dr. Liu Thai Ker is an architect-planner. Since 1992, he has been director of RSP Architects Planners & Engineers Pte Ltd., a consulting firm of over 1,000 people, with 11 overseas offices and projects in 17 countries. Dr. Liu is also the chairman of Centre for Liveable Cities, which he founded in 2008.

Dr. Liu has served as adjunct Professor in the School of Design and Environment and the Lee Kuan Yew School of Public Policy at the National University of Singapore. He is also adjunct professor in the College of Humanities, Arts & Social Sciences at Nanyang Technological University. He is a member of several governmental bodies in Singapore, and planning adviser to over 20 cities in China.

As architect-planner and CEO of the Housing & Development Board, 1969–1989, he oversaw the completion of over half a million dwelling units. As CEO and chief planner of Urban Redevelopment Authority, 1989–1992, he spearheaded the major revision of the Singapore Concept Plan and key direction for heritage conservation.

In the cultural arena, Dr. Liu served as the Chairman of the National Arts Council from 1996 to June 2005 and the Singapore Tyler Print Institute from 2000 to 2009. He has served as the chairperson of the External Review Panel, Arts Quality Framework appointed by the Ministry of Education, and a founding member of the board of trustees, Arts & Culture Development Fund, Ministry of Information, Communications and the Arts.

Dr. Liu obtained his Bachelor of Architecture with First Class Honors and University Medal from the University of New South Wales in 1962 and Master in City Planning with Parson's Memorial Medal from Yale University in 1965. He attended INSEAD Advanced Management Program in Paris in 1980. In 1995, he was conferred Doctor of Science honoris causa by the University of New South Wales.

Among his awards are the Public Administration Medal (Gold) 1976, the Meritorious Service Medal 1985, Singapore Institute of Architects Gold Medal, and the Medal of the City of Paris, France in 2001. In 1993, he was given the Second Asian Achievement Award for Outstanding Contributions to Architecture.

Marvin Mass, P.E.

Cosentini Associates, New York

Marvin Mass began his career at Cosentini in 1952 and since then has been responsible for the ongoing development and enhancement of the firm's standards of design, construction, and supervision.

Mr. Mass's distinguished career includes such pioneering achievements as the design of New York City's first total-energy and chilled water storage plants; the application of waterside economized air-conditioning systems for office buildings; eutectic salts and vacuum tube solar collector systems; as well as the systems within the renowned Solar Energy House at the University of Delaware. He is a long-standing faculty member at the Harvard Graduate School of Design and Yale University, and has lectured at the University of Pennsylvania, Pratt Institute, and Cooper Union.

Among the awards bestowed on Mr. Mass are the Brown Medal of the Franklin Institute and the 1990 AIA Institute Honors, as well as a lifetime achievement award from the American Society of Heating, Refrigerating, and Air Conditioning Engineers.

Mr. Mass holds a Mechanical Engineering degree from New York University and is a licensed mechanical engineer in 23 states. He is a member of the National Society of Professional Engineers; the New York State Association of the Professions; the American Society of Heating, Refrigerating, and Air-Conditioning Engineers; the Building and Architecture Technology Institute; and the Institute for the Study of the High- Rise Habitat. He is also an Honorary Member of the American Institute of Architects.

Mr. Mass's recently published book *The Invisible Architect*, is an account of the many prominent architects and important buildings that have filled his distinguished career.

Bruno Roberto Padovano

FAUUSP , Paulo, Brazil

Bruno Roberto Padovano was born in Milan, Italy, in 1951. He is married to Suzana Mara Sacchi Padovano, an industrial designer, and they have two sons.

He settled in São Paulo, Brazil, in 1969 after an eight-year stay in South Africa, where he studied at the School of Architecture of Witwatersrand University, in Johannesburg.

He received a professional degree in Architecture and Urbanism from Faculdade de Arquitetura e Urbanismo da Universidade de São Paulo, Brazil, two master´s degrees from the Harvard Graduate School of Design, and a Ph.D. from FAUUSP, where he has taught at the undergraduate and graduate levels since 1978. He is currently vice head of the Design Department and co-editor of the magazine "Projetos".

Mr. Padovano has practiced planning and design in Brazil and other countries, especially China with the firm Architecture of Metropolitan Post. His works include a university campus (UNIFIEO), industrial buildings (APC in Barueri, Brazil, and Nature Factory, Ningbo, China with AMP), a sport and leisure center (SESC/Nova Iguaçu), several large public and private housing complexes, and the urban regeneration of the Alphaville Commercial Center and the Alameda/Monteiro da Silva Gabriel in São Paulo. During his professional career, he has won eleven public competitions in Brazil in these areas, and several other prizes and honorable mentions.

Bruno Padovano has published books and articles on related subjects, and was the curator of the 8th International Biannual of Architecture of São Paulo, in 2009. He currently holds the position of Scientific Coordinator of Nucleo de Pesquisa em Tecnologia da Arquitetura e Urbanismo da Universidade de São Paulo, and works as consultant to public agencies and private groups.

William Pedersen

KPF/New York

William Pedersen is the founding design partner of Kohn Pedersen Fox Associates (KPF), which he established in 1976 with A. Eugene Kohn and Sheldon Fox. For his achievements and contributions to the built environment, Mr. Pedersen has personally received seven AIA National Honor Awards (333 Wacker Drive in Chicago, Procter & Gamble General Offices Complex in Cincinnati, DZ Bank Headquarters in Frankfurt, World Bank Headquarters in Washington, DC, New Academic Complex at Baruch College in New York, Gannett/USA Today Corporate Headquarters in McLean, VA, and One Jackson Square in New York). Also of note, the Council on Tall Buildings and Urban Habitat (CTBUH) honored the Shanghai World Financial Center as the Best Tall Building Worldwide in 2008, and MIPIM Asia honored the International Commerce Centre as the Best Business Centre (Gold Award) in 2011.

In addition, Mr. Pedersen earned the Lynn S. Beedle Lifetime Achievement Award from the CTBUH in 2010. He was awarded the Gold Medal for lifetime achievement in architecture from Tau Sigma Delta (the National Honor Society for Architecture and the Allied Arts), and the Arnold W. Brunner Memorial Prize in Architecture for Contributions in Architecture as an Art—awarded by the American Academy of Arts and Letters. Additional honors include the Rome Prize in Architecture, awarded by the American Academy in Rome; the Prix d'Excellence from L'Ordre des

Architectes de Québec; and the U.S. General Services Administration's Building Design Excellence and Honor Awards.

Mr. Pedersen lectures internationally and has served on academic and professional juries and symposia throughout the world. He is on the Board of the University of Minnesota Foundation and has been a visiting professor at the Rhode Island School of Design, Columbia University, and Harvard University. He has held the Eero Saarinen Chair at Yale University and has also been the Otis Lecturer in Japan. Honored with the Herbert S. Greenward Distinguished Professor in Architecture by the University of Illinois at Chicago, he is currently a member of the University of Minnesota Foundation's Board of Trustees and a recipient of the University's Alumni Achievement Award.

Rhonda Phillips

Arizona State University, Phoenix

Rhonda Phillips, *Ph.D., AICP, CEcD,* is Associate Dean, Barrett, The Honors College, and professor at the School of Community Resources & Development at Arizona State University.

Community investment and well-being comprise the focus of her research and outreach activities, including community-based education and research initiatives for enhancing quality of life. Rhonda works with faculty, staff, students, and organizations to expand the reach of community-based education and research initiatives. Her focus is using community indicator and evaluation systems for monitoring progress toward community and economic development revitalization goals.

Dr. Phillips holds dual professional certifications in economic and community development (CEcD from the International Economic Development Council) and urban and regional planning (AICP from the American Institute of Certified Planners). Previously, she served as the founding director of the Center for Building Better Communities at the University of Florida, and was a 2006 UK Ulster Policy Fellow Fulbright Scholar.

Rhonda is author or editor of ten books, including *Community Development Indicators Measuring Systems*, and *Introduction to Community Development*. She is co-editor of the *Community Quality-of-Life Indicators: Best Cases* series (Springer) and editor of *Community Development*, the journal of the Community Development Society (Taylor & Francis), as well as the book series *Community Development Research* and Practice (Routledge). Rhonda is the Vice President, Programs, for the International Society for Quality-of-Life Studies. She serves as a Senior Sustainability Scientist, Global Institute of Sustainability for ASU's School of Sustainability, and is on the affiliate faculty for the School of Geographical Sciences and Urban Planning.

Dennis Poon

Thornton Tomasetti, Inc., New York

Dennis Poon is a vice chairman at Thornton Tomasetti, Inc., a 550+ person, privately held structural engineering firm. Throughout his 30+ years' experience he has been responsible for the design and construction of high-rise offices, hotels, airports, arenas, and residential buildings throughout the U. S. and abroad. He has expertise in the application of state-of-the-art engineering technologies for building analysis, design, construction, and project management, including project delivery strategies.

High on the list of his accomplishments is leading the structural engineering team for the design of the Taipei 101 in Taiwan, a 101-story office tower with a unique profile, a five-story-deep basement and a surrounding six-story retail structure. Mr. Poon is the Principal-in-Charge for the structural design of the Shanghai Tower, Ping An Tower, Wuhan Greenland Tower in China, and the Signature Tower in Jakarta, Indonesia.

Other notable projects include the World Financial Center in New York; the 66-story high-rise mixed-use Plaza 66 in Shanghai; O'Hare International Airport in Chicago; MGM CityCenter Block A in Las Vegas; Kamal Mixed Use Development West Bay Diplomatic District in Doha, Qatar; Shangri-La Hotel, Futian in Shenzhen, China; Doha Convention Center + Tower in Doha, Qatar; and Qatar Education City Convention Center, Qatar.

In addition to his responsibilities as Vice Chairman of the firm, Mr. Poon is in charge of the firm's international operation with branch offices in London, Hong Kong, Shanghai, Dubai, Abu Dhabi, Saudi Arabia and Vietnam. He has lectured in Universities worldwide, including MIT, Columbia, Delft University in Holland, NJ Institute of Technology and Hong Kong City University on such topics as high-rise building design and steel and concrete structure design.

Mr. Poon holds a bachelor's degree in civil engineering from the University of Texas at El Paso and a master's degree in civil engineering from Columbia University. He serves on the board of trustees and the editorial board of the Council on Tall Buildings and Urban Habitat. Mr. Poon was named one of *Engineering News Record*'s Top 25 Newsmakers of 2010 for working with Expert Review Panel members in China on the design criteria and codes that lead to more efficient high-rise structural design, part of his effort to align a China code with international standards.

He has received the ACEC Diamond award for MGM City Center Block A, the ACEC National Grand Award for the design of Taipei 101, the Diamond Award from NYACE, the National Recognition Award from ACEC as well as a US CIB Award of Merit for the Plaza 66 Structural System, two James F. Lincoln Arc Welding Foundation Awards for Bally's Park Place Casino Hotel in Atlantic City, NJ, and the Pedestrian Bridges at Fifth Avenue Place in Pittsburgh, a Structural Engineers Association of Illinois Merit Award for the Chicago Board of Trade Building, and the Outstanding 50 Asian Americans in Business Award.

Mr. Poon has been a speaker at many engineering conferences worldwide, such as the Third and Fourth Annual Ultra High-Rise Building Summits, Council on Tall Buildings and Urban Habitat conferences in Melbourne, Australia, Chicago, and Seoul, and at the Structures Congress in Florida.

Ashok Raiji

Arup, New York

Ashok Raiji, P. E., LEED AP, is a principal in the New York office of Arup and has led the planning of city-scale developments as well as the design of many high-performance buildings all over the world. He is Arup's commercial, residential, and retail business leader in the Americas, and leads Arup's global residential business.

He has led teams for various recent high-performance projects, including the Songdo Convention Center and the 320-meter Northeast Asia Trade Tower in Korea, Boston Seaport Sustainability Master Plan, and the Kresge Foundation Headquarters (LEED Platinum rated), as well as master planning of new eco cities in Asia.

Mr. Raiji lectures frequently on sustainable design and has served as technical adviser to the National Building Museum and the Museum of the City of New York on matters involving sustainability.

He is a LEED Accredited Professional and a licensed professional engineer in 25 states. He is also a Visiting Professor of Architecture at the Irwin S. Chanin School of Architecture at The Cooper Union in New York.

Richard Register

Ecocity Builders, San Francisco

Richard Register, founder and president of Ecocity Builders, is one of the world's great theorists and authors in ecological city design and planning, having advocated the building of better cities for more than four decades. Among his many firsts was convening the inaugural Ecocity International Conference in Berkeley, California, in 1990, followed by others in Adelaide, Australia, 1992; Yoff, Senegal, 1996; Curitiba, Brazil, 2000; Shenzhen, China, 2002; Bangalore, India, 2006; the 7th International Ecocity Conference, San Francisco, California, 2008; Istanbul, 2009, Montreal, 2011; Third International China Eco-city Conference, Tianjin, China, 2012; and the 10th International Ecocity Conference in Nantes, France, in September 2013.

Mr. Register is founding president of the not-for-profit organization Urban Ecology (1975) and founder and current president of Ecocity Builders (1992), an educational and research not-for-profit corporation engaged in ecological community design and planning.

Mr. Register has lectured frequently in some 29 countries and 29 U.S. states. He was recently appointed to the International Scientific Advisory Committee on Active Ecological Urban Development to the Scientific Committee on Problems in the Environment (SCOPE), an international association of nation-states and two dozen major scientific associations.

Mr. Register is also editor of *Village Wisdom: Future Cities* (1997). His acclaimed self-illustrated books include *EcoCities: Building Cities in Balance with Nature* (2002; Chinese edition, 2010); *Ecocity Berkeley: Building Cities for a Healthy Future* (1987), and *Another Beginning* (1978). In progress is *World Rescue: An Economics Built on What We Build*, a book on the role of ecocities in economics..

Leslie E. Robertson

LER, New York

Leslie E. Robertson, P.E., C.E., S.E., D.Sc., D.Eng., Dist.M.ASCE, NAE, AIJ, JSCA, AGIR, Chartered Structural Engineer (UK and Ireland), First Class Architect and Engineer, Japan. Dr. Robertson established his own firm, Leslie Earl Robertson, Structural Engineer, LLC, in January 2013 after more than a half century leading the New York practice that still bears his name (Leslie E. Robertson Associates/LERA.

Since beginning his engineering career in 1952, his practice has embraced the guiding principle of providing an imaginative and responsible approach to engineering problems. Robertson's groundbreaking structural designs that have influenced the design and construction of tall buildings include:

• The first high-rise building to use a composite megastructure space frame to resist all loads imposed by typhoon winds and the weight of the building (Bank of China Tower, Hong Kong, 1989);

• The creation of mechanical damping units to reduce wind-induced swaying (World Trade Center, New York, circa 1968);

• The first use of prefabricated multiple-column and spandrel-wall panels to resist the lateral force from hurricane winds and to allow column-free interior space (World Trade Center, New York, circa 1972);

- The first use of a space-frame megastructure and outrigger or hat system for a high-rise building (United States Steel Headquarters, Pittsburgh, Pennsylvania. - now USX, 1970); and

- The creation of the shaftwall system now almost universally used for fire-resistive walls in high-rise buildings.

Dr. Robertson was elected to the National Academy of Engineering in 1975. *Engineering News Record* named him Construction's Man of the Year in 1989 and, in 1999, listed him among its Top 125 People of the Past 125 Years. He has earned numerous awards and honors and has served on the boards of several cultural and engineering organizations, including the Skyscraper Museum and Council on Tall Buildings and Urban Habitat. Lehigh University in Bethlehem, Pennsylvania, and Rensselaer Polytechnic Institute in Troy, New York, have awarded him honorary doctorates in engineering, and the University of Western Ontario in Canada presented him with an honorary doctorate in science.

Dr. Robertson developed the structural designs for the World Trade Center (New York), the Bank of China Tower (Hong Kong), Miho Museum Bridge (Shigaraki, Japan), and Shanghai World Financial Center (Shanghai). He currently directs the structural design for the Jamsil Lotte Tower (Seoul) and the KL 100 Tower (Kuala Lumpur). A member of the National Academy of Engineering and Distinguished Member of the American Society of Civil Engineers, he has served in leadership positions on these and many other professional organizations. The IABSE International Award of Merit in Structural Engineering, IStructE Gold Medal of the UK, and Fazlur Rahman Khan Medal are but a few of the honors he has received for his professional contributions. A Distinguished Alumnus of UC Berkeley (B.S., 1952), Les has been academically acknowledged as well through honorary doctoral degrees from Notre Dame, Lehigh, Western Ontario universities, and Rensselaer Polytechnic Institute.

Jorge Eduardo Rubies

Preserva São Paulo Associação, São Paulo, Brazil

Jorge Eduardo Rubies, Esq. is the founder and president of Preserva São Paulo Association, the only historic preservation NGO in São Paulo, the largest city in Brazil. His passion and knowledge about his city made him a respected authority in the fields of historic preservation, Brazilian architecture, and urbanism, frequently consulted by journalists and researchers about these issues. His website www.piratininga.org, in Portuguese and English, is one of the main sources of information about São Paulo and its architecture. Preserva São Paulo Associação has also developed a unique approach in the challenge to preserve the rich architectural heritage of one of the fastest-changing cities in the world, adopting original ways to awaken the need for historic preservation among the citizens, through actions such as urban explorations of abandoned buildings.

Preserva São Paulo is currently leading a campaign to save a 4-acre public area of great architectural and environmental importance, which the São Paulo municipal government intends to sell for real estate development.

Mr. Rubies' dual professional degrees in business administration and law have allowed him to organize and litigate on behalf of the public realm for the protection of endangered historic buildings. He also leads the recently founded São Paulo Citizens Front against Real Estate Speculation and for Ethics in Politics. This organization encompasses dozens of grassroots movements, united to promote in a nonpartisan, orderly, and pacific way the common goal of a sustainable, inclusive, pluralistic, and green city for every citizen.

Maria Sevely

FORM | Proforma, New York

Steeped in architecture from an early age through exposure to her architect/ professor father, his colleagues and mentors, including many of the most renowned figures in the field, Maria Sevely entered architectural school at fifteen and went on to study architecture, architectural history, and fine art at Wellesley, Harvard, and RISD. Maria's background includes substantial work as a design associate with three of the world's leading architectural firms—Pei Cobb Freed & Partners, Richard Meier, and Philip Johnson—often working closely with the founding principals. She was a senior project designer for the United States Holocaust Memorial Museum in Washington, D.C., among other projects with the Pei firm, and project architect for Philip Johnson's last major project, the Cathedral of Hope. With FORM | Proforma, she is working on this foundation, utilizing computer capabilities to develop new geometries and advance the application of an analytical design approach for projects—to date— in New York, the northeastern United States, and China.

Adrian Smith

Adrian Smith + Gordon Gill, Chicago

Adrian Smith, FAIA, RIBA, has been a practicing architect for 45 years. His philosophy is to engage the history, art, landscape, climate, and indigenous materials of the places where he is designing. His goals are to interpret and honor the societies that his buildings serve; to forge a unique dialogue between culture and place; and to foster a strong connection to the people who see and use his buildings. His contextual approach has produced buildings recognized throughout the world for their beauty, elegance, and subtlety of their cultural references. His work has won more than 110 local, national, and international awards.

Adrian's extraordinary body of work as a designer includes four of the world's current 11 tallest buildings, including Burj Khalifa in Dubai, the world's tallest building; Jin Mao Tower in Shanghai; the Trump International Hotel & Tower in Chicago; and Zifeng Tower in Nanjing, China. Other landmark structures in his portfolio include the Broadgate Tower in London, recently named the Best Tall Building in Europe; and Rowes Wharf, the iconic structure that revitalized Boston's waterfront district.

Most recently, Adrian's portfolio has expanded to include Kingdom Tower, to be the world's tallest building when completed in 2016 in Jeddah, Saudi Arabia; and Wuhan Greenland Center, to be the world's fourth-tallest building when completed the same year in Wuhan, China. Prior to founding Adrian Smith + Gordon Gill Architecture in 2006, Adrian was a design partner at Skidmore, Owings & Merrill and its chairman, 1993–1995. He is the author of two books on his work and co-author (with Gordon Gill) of the newly published *Toward Zero Carbon: The Chicago Central Area DeCarbonization Plan.*

Alfredo Vicente de Castro Trinidade

Secretariat of the Environment, Curitiba, Brazil

Alfredo Vicente de Castro Trinidade is an award-winning environmentalist and forester in Curitiba, Brazil, one of the world's greenest cities. A specialist in urban technical management, he has worked for many years in the service of the Secretariat of the Environment, and was appointed director of the Department of Research and Conservation of Fauna in 2011. Among the other agency positions he has held are environmental planning manager, technical coordinator for biodiversity, director of the Department of Research and Monitoring, technical adviser, and chief licensing officer.

In local and international venues, notably as part of the Brazilian delegation of UN Habitat and as a member of the "Smog Summit" in Toronto, Mr. Trinidade has worked to balance Curitiba's modern development needs with environmental protection and improvement. He holds an undergraduate degree from the Agrarian Science Sector of Federal University do Paraná and an advanced professional degree in urban technical management from Catholic University of Paraná.

Antony Wood

Council on Tall Buildings and Urban Habitat, Chicago

Antony Wood, B. Arch., Ph.D., has been Executive Director of the Council on Tall Buildings and Urban Habitat since 2006, responsible for the day-to-day running of the Council and steering in conjunction with the Board of Trustees, of which he is an ex-officio member. Prior to this, he was CTBUH Vice Chairman for Europe and Head of Research. His tenure has seen a revitalization of the CTBUH and an increase in output across all areas.

Based at the Illinois Institute of Technology, Antony is also an associate professor in the College of Architecture at IIT, where he convenes various tall building design studios. A UK architect by training, his field of specialization is the design, and in particular the sustainable

design, of tall buildings. Prior to joining the Council and IIT, Antony was an associate professor and lecturer in architecture from 2001 to 2006 at the University of Nottingham, where he ran the third- and fifth-year programs, respectively, and was an active member of various research teams. While at Nottingham, he founded the Tall Buildings Teaching and Research Group.

Before becoming an academic, Antony worked as an architect in practice in Hong Kong, Bangkok, Kuala Lumpur, Jakarta, and London, between 1991 and 2001. It was during this time that he developed his passion for, and professional background in, tall buildings. His projects include the mixed-use tower project of SV City, Bangkok (completed 1995), the 44-story condominium project of Kuningan Persada, Jakarta (1997), and the prestigious Kuala Lumpur Central International Railway Terminal, Malaysia (completed 2001).

Antony is the author of numerous books and papers in the field of tall buildings, sustainability, and related areas, including *Tall & Green: Typology for a Sustainable Urban Future*. His Ph.D. thesis explored the multidisciplinary aspects of skybridge connections between tall buildings. He is associate editor of both the *CTBUH Journal* and *The Structural Design of Tall and Special Buildings* journal, and co-chair of the CTBUH Tall Buildings and Sustainability working group. He is currently engaged in numerous books, including: *The CTBUH Guide to Sustainability for Tall Buildings in Urban Environments*; *New Paradigms in High-Rise Design*; and *The Tall Buildings Reference Book*.

Photo Credits

Huang Shulin: 283.

NASA:

NASA identifier: China_OSE2003010; www.dvidshub.net/image/ 753124/pollution-over-china-natural-hazards#US42M1rwLjs%23i xzz2M7ObzjLz: 44 top.

http://earthobservatory.nasa.gov/IOTD/view.php?id=7949: 51 top & bottom.

William L. Stefanov, NASA-JSC833233: Tianjin_and_Beijing_ ISS026-E-010155_lrg.jpg: 88.

Portman Archive: 197 left & right

Shutterstock:

Copyright: Boris Diakovsky: 20.

Copyright: trekandshoot: 30 top.

Copyright: Lee Prince: 30 bottom left.

Copyright: Tim Roberts Photography: 30 bottom right.

Janet Adams Strong:

12, 26, 52, 206 left, 221, 262, 264 left & right,

331, 467 left, 469 top & bottom, 586

Visualizations:

FORM | Proforma: 56, 63. FORM | Proforma. Background: Wade Shepard, The Vagabond Network: cover & 534.

FORM | Proforma. Background: Широкоформатные обои длярабочего стола: 546. FORM | Proforma. Background: Roy Philippe/Corbis: 550.

FORM | Proforma. Background: Mike Sheetal: 554.

FORM | Proforma. Background: J. Crocker: 558.

FORM | Proforma. Background: Robert Campbell (U.S. Army Corps of Engineers Digital Visual Library): 562.

FORM | Proforma. Background: Felipe Borges: 566.

FORM | Proforma. Background: Panoramic Images: 568.

FORM | Proforma. Background: Herbert Spichtinger/Corbis: 571.

FORM | Proforma. Background: Bruno Morandi/Grand Tour/Corbis: 574.

FORM | Proforma: 577.

Wikimedia Commons:

Andysonic777 at the English language Wikipedia: http://commons. wikimedia.org/wiki/File:Garibalbi_Lake_BC.jpg: 9.

BriYYZ from Toronto, Canada (Hello New York City!): http:// commons.wikimedia.org/wiki/File%3AHello_New_York_City!_ (8020113376).jpg: 10.

LeRoy Woodson, Environmental Protection Agency, NARA record: 2368875: 23.

Basykes: http://commons.wikimedia.org/wiki/File:Beijing_traffic_ jam.jpg: 24.

Curtis Palmer: http://commons.wikimedia.org/wiki/File:Electronic_ waste.jpg: 27.

IDuke: http://commons.wikimedia.org/wiki/File:Markham-suburbs_ aerial-edit.jpg: 28.

Caiguanhao: http://commons.wikimedia.org/wiki/File:Deng_Xiaop ing_billboard_at_Shunde.jpg: 34.

JialiangGao: http://commons.wikimedia.org/wiki/File:Subsistent_ Farming_Southern_China.jpg: 36.

Wilson Dias/Agência Brasil: http://commons.wikimedia.org/wiki/Fil e%3AMadeiraDesmatamento2WilsonDiasAgenciaBrasil.jpg: 41.

Diorit: http://commons.wikimedia.org/wiki/File%3ASlash_and_ burn_Ambalapaiso.JPG: 42.

Brocken Inaglory: http://commons.wikimedia.org/wiki/File%3A Vog_from_Sulfur_dioxide_emissions_.jpg: 44 bottom left.

Ludovic Hirlimann: http://upload.wikimedia.org/wikipedia/com mons/5/50/Electric_car_charging_Amsterdam_%282%29.jpg: 44 bottom right.

Wilfredor: http://commons.wikimedia.org/wiki/File%3A Contaminaci%C3%B3n_en_el_Lago_de_Maracaibo%2C_ Estado_Zulia.jpg: 46.

Doug Wilson, Environmental Protection Agency, (NARA record: 8464433): http://commons.wikimedia.org/wiki/File:Pollution_ of_the_Snohomish_river,_Everett,_Washington_State._-_NARA_- _552248.jpg: 48.

Lydur Skulason from Iceland: http://commons.wikimedia.org/wiki/ File:Geothermal_borehole_house_2333875782.jpg: 58.

http://www.new-territories.com/blog/architecturedeshumeurs/ ?p=108: 66 top.

Rynnolohmus: http://commons.wikimedia.org/wiki/File%3AKatal %C3%BCsaatorosakesele_kasvanud_s%C3%BCsiniknanotoru mets.jpg: 66 bottom.

Jeffrey G. Katz: http://commons.wikimedia.org/wiki/File:Wind_Tur bines_in_Washington_State_IMG_7956WMC.tif: 69.

OhWeh: http://commons.wikimedia.org/wiki/File:Solarpark Thüngen-020.jpg: 73.

Patrick Nagel: http://commons.wikimedia.org/wiki/File%3A Shanghai_from_the_SWFC.jpg: 74.

http://cnreviews.com/business/research-insights/stefano-negri- china-urbanization_20090616.html; http://creativecommons.org/ licenses/by-nc-sa/3.0/us/: 78 top.

Hawyih (self-made (自己的作品): http://commons.wikimedia.org/ wiki/File%3AShenzhen_huizhanzhongxin.jpg: 78 bottom.

calflier001: http://commons.wikimedia.org/wiki/File%3ACHINA_ RAILWAYS_HIGH_SPEED_(CRH)HARMONY_TRAIN_SET_AT_ BEIJING_SOUTH_STATION_ABOUT_TO_DEPART_FOR_ SHANGHAI_CHINA_OCT_2012_(8154122913).jpg: 81.

U.S. Naval Historical Center Photograph: http://commons.wikime dia.org/wiki/File:Shanghai_1939.tif: 81 top.

Nicor: http://commons.wikimedia.org/wiki/File%3APudong%2C_ Shanghai_Panorama_02.jpg: 81 bottom.

Luke Watson (Lukeaw); http://commons.wikimedia.org/wiki/ File%3APetronas_Towers_by_Day.jpg: 161 top.

Captain: http://commons.wikimedia.org/wiki/File%3A BurjSnowOld.jpg: 165.

MacRudi (talk): World_Trade_Center_Building_Design_with_ Floor_and_Elevator_Arrangment.svg: MesserWoland; http://commons.wikimedia.org/wiki/File%3AWorld_ Trade_Center_Building_Design_with_Floor_and_Elevator_ Arrangment_m.svg: 167.

Chris73: http://commons.wikimedia.org/wiki/File:Double_elevator_ at_Midland_Square_Nagoya.JPG: 170.

Stankn: http://commons.wikimedia.org/wiki/File%3AReflections_ at_Keppel_Bay_(dawn).jpg: 200.

Jakub Hałun; Florian Fuchs: http://commons.wikimedia.org/wiki/ File%3AThe_Bow_in_Calgary.jpg: 207.

http://commons.wikimedia.org/wiki/File%3A20091003_Macau_ Grand_Lisboa_6700.jpg: 220.

Beyond My Ken: http://commons.wikimedia.org/wiki/File%3A Portal_to_Park_Avenue.jpg: 256.

Raidarmax: http://commons.wikimedia.org/wiki/File%3ATraffic_in_ New_York_City.JPG: 259.

Jean-Christophe BENOIST: http://commons.wikimedia.org/wiki/ File%3ANYC_-_Time_Square_-_From_upperstairs.jpg: 267.

WiNG: http://commons.wikimedia.org/wiki/File%3AHKWoCheEst overview_20070828.jpg: 277.

McKay Savage from London, UK: http://commons.wikimedia.org/ wiki/File%3AChina_-_Yangshuo_29_-_Rice_Paddy_Terraces_ (140905203).jpg: 286.

Drolexandre: http://commons.wikimedia.org/wiki/File%3APaddy_ field_Longsheng.JPG: 288.

Anna Frodesiak: http://commons.wikimedia.org/wiki/File%3A Traffic_jam_in_Haikou%2C_Hainan%2C_China_01.jpg: 295.

Adam Jones Adam63: http://commons.wikimedia.org/wiki/ File%3ABotanical_Gardens_-_Curitiba_-_Brazil.jpg: 322.

Júlio Boaro: http://commons.wikimedia.org/wiki/File%3AS %C3%A3o_Paulo_City.jpg: 349.

Milton Jung: http://commons.wikimedia.org/wiki/File%3AFavela_ do_Moinho_Brazil_Slums.jpg: 351.

JMGM/Jurema Oliveira: http://commons.wikimedia.org/wiki/ File%3AComunidade_bairro_do_Lim%C3%A3o3.jpg: 352.

Fulvio souza: http://commons.wikimedia.org/wiki/ File%3AImagem_Santo_Amaro_Station.jpg: 357.

Rodrigo Soldon: http://commons.wikimedia.org/wiki/File%3AS %C3%A3o_Paulo_skyline_city.jpg: 359.

http://commons.wikimedia.org/wiki/File%3AMatogrosso_ I5_1992219_lrg.jpg: 381 left.

http://commons.wikimedia.org/wiki/File%3AMatogrosso_ ast_2006209_lrg.jpg: 381 right.

YF12s: http://commons.wikimedia.org/wiki/File%3AMonticello_ 2010-10-29.jpg: 436.

Tomticker5: http://commons.wikimedia.org/wiki/File%3AEphraim_ Hawley_House_04_2011.JPG: 438.

Dcoetzee: http://commons.wikimedia.org/wiki/File%3APeople_ walking_in_Hyde_Park%2C_Sydney.jpg: 449.

United States Geological Survey: http://commons.wikimedia.org/ wiki/File%3APruitt-igoeUSGS02.jpg: 451.

Gruban / Patrick Gruban from Munich, Germany: http://commons. wikimedia.org/wiki/File%3ACentral_Park_jogging.jpg: 452.

Stan Shebs: http://commons.wikimedia.org/wiki/File%3ALas_ Vegas_Tropicana_Avenue_1.jpg: 457.

Ziko van Dijk: http://commons.wikimedia.org/wiki/File%3A2004_ melford_elderly_people.JPG: 459.

By User:Jurema Oliveira: http://commons.wikimedia.org/wiki/ File%3ASaopaulo_copan.jpg ; http://www.fotosedm.hpg.ig.com. br/: 461.

Marcos Guerra: http://commons.wikimedia.org/wiki/ File%3ABarigui.JPG: 471.

Herr stahlhoefer: http://commons.wikimedia.org/wiki/ File%3AItaipu_171.jpg: 475.

Leonardo Stabile: http://commons.wikimedia.org/wiki/ File%3ABosque_de_Portugal%2C_Curitiba.JPG: 478.

Mario Roberto Duran Ortiz Mariordo: http://commons.wikimedia. org/wiki/File%3ACuritiba_04_2006_03_RIT.jpg: 479.

henribergius: http://commons.wikimedia.org/wiki/ File%3ACuritiba_skyline_(Nov_28%2C_2005).jpg: 483.

Other images courtesy of:

RSP Architects Planners & Engineers: 2. / Anderson Anderson Architecture: 94-107. / Skidmore, Owings & Merrill: 108-117. / SOM /@Crystal CG: 109. / Kohn Pedersen Fox Associates: 118-129 (credit John Chu: 118; credit Tim Griffith: 122, 123 right; credit Nacasa & Partners: 123 left; credit Chang Kim: 129 left & right). / Adrian Smith + Gordon Gill: 130-145 (credit James Steinkamp; © AS+GG: 131, 136, 137, 140-41). / Antony Wood, Council on Tall Buildings and Urban Habitat: 146-161. / ©Zbigniew Piech, Linear Synchronous Motors – Transportation and Automation Systems, 2nd ed. (2011), Fig. 7.7, Chapter 7, p. 230: 173. / HOK: 178-191. R.A. Heintges & Associates: 192. Pei Partnership Architects: 193. Cosentini Associates: 206 right, 213. Gensler: 211. Thornton Tomasetti: 222-239. Leslie Earl Robertson, Structural Engineer: 240-253. Pei Cobb Freed & Partners: 246. Sam Schwartz Engineering: 256. Thomas J. Campanella: 270, 271, 276, 279. Chengri Ding: 282. SOM: 296-305 (©SOM | Crystal CG: 297; 303 row 1: SOM, rows 2 & 3: SOM | by Khalid Al-muharraqi | © Muharraqi Studios, row 4: SOM |© Panto-Ulema). J.A.O. Design International: 308-321 (Beijing American Club: 308). Instituto Jaime Lerner: 322-327 (credit Adam Jones adamjones.freeservers.com: 322). IPPUC, Instituto de Pesquisa e Planejamento Urbano de Curitiba: 328. RSP Architects Planners & Engineers: 334-347. Bruno Padovano: 348. Ajmal Aqtash, Form-ula, Core.form-ula: 362-375. Dickson Despommier: 376, 383. Wonwoo Park: 377, 379 left. Plant Chicago, NFP/Rachel Swenie: 379 right. Blake Kurasek: 385 (all). Kiss + Cathcart Architects: 388-401. ARUP: 402-412. Richard Register: 416-427. Maria Sevely, F O R M | Pr o forma: 432, 433, 437 right. James A. Marshall: 437 left. Kevin Scott, Make It Right: 439. Rhonda Phillips: 448. Jorge Rubies: 460, 463, 467 right. Alfredo Vicente de Castro Trinidade, Secretariat of the Environment Curitiba , Brazil: 470, 473 left & right. Kenneth King: 574.